Pearson New International Edition

Urbanization:
An Introduction to Urban Geography
Paul L. Knox Linda M. McCarthy
Third Edition

Pearson Education Limited
Edinburgh Gate
Harlow
Essex CM20 2JE
England and Associated Companies throughout the world

Visit us on the World Wide Web at: www.pearsoned.co.uk

ISBN 10: 1-292-03916-7
ISBN 13: 978-1-292-03916-9

British Library Cataloguing-in-Publication Data
A catalogue record for this book is available from the British Library

ARP impression 98

Printed in Great Britain by Ashford Colour Press Ltd.

Table of Contents

GLOSSARY

The glossary words and terms defined below are highlighted in bold text throughout the book as well as in the index. All words in italics below are defined.

A

acid rain Precipitation that has mixed with airborne pollutants (particularly sulfur dioxide and nitrogen oxides from burning fossil fuels) to produce rain that contains levels of acidity—often in the form of sulfuric acid—that are harmful not only to vegetation and aquatic life but also to some building surfaces and statues in cities.

advanced capitalism The term for the most recent or "disorganized" phase of capitalism in which the relationships between business, labor, and government are more flexible, largely because a great deal of corporate activity has escaped the framework of national states and their institutions that still constrain organized labor and most government functions. See also *flexible production systems.*

agglomeration diseconomies The negative economic effects of urbanization and the local concentration of industry, including higher rents, traffic congestion, and air pollution.

agglomeration economies Economic advantages that individual companies enjoy because of their location clustered together among functionally related activities, thereby sharing the use of specialized servicing industries, financial and other services, and public utilities. See also *localization economies* and *urbanization economies.*

American Renaissance The literary culture of the American Renaissance—rooted in the works of Ralph Waldo Emerson and Henry David Thoreau—that strongly influenced the American design professionals of the mid-nineteenth century. The American Renaissance took nature as a fundamental spiritual wellspring, defining the ideal as a setting in which mankind and nature had achieved a state of balance. This attitude led to a vision of ideal urban landscapes that combined the morality attributed to nature with the enriching and refining influences of cultural, political, and social institutions.

annexation The addition of unincorporated land into a municipality, especially during the nineteenth century in the United States. Pioneer suburbanites gained better services from being annexed by the municipality; the municipality achieved important *economies of scale* and prevented the loss of their middle income groups from the tax rolls and political life.

anomie Literally "normlessness"—often associated with the isolation of urban life, but it is more accurate to think of those social conditions in which the norms of personal and social behavior are so weak or muddled that some people become socially isolated, confused, or uncertain about how to behave, while others are easily able to challenge or ignore social conventions. One of the consequences of such circumstances is an increase in *deviant behavior.* See also *Gesellschaft.*

Art Nouveau A movement in the 1890s and early 1900s in France (called *Jugendstil* in Central Europe) that emphasized sensuous shapes and organic themes as an alternative to Victorian extravagance. The Art Nouveau style became widely used in graphic art, but in architecture it was limited to the embellishment of buildings rather than being reflected in their form.

Arts and Crafts Movement Led by William Morris in England, this movement sought to reestablish the importance of craftsmanship during the increasing mechanization and mass production of the second half of the nineteenth century. Morris wanted to bring artists and craftspeople together to synthesize honest, simple, and popular forms. In architecture, this meant drawing heavily on vernacular themes, something that proved convenient and attractive to the planners of the first Garden Cities.

B

balloon-frame construction Balloon framing uses standardized, machine-cut 2″ × 4″ studs nailed together by inexpensive, machine-cut nails. By spreading the stress over a large number of light boards, the balloon frame has strength and stability far beyond its insubstantial appearance. Using standardized components and requiring only semi-skilled labor, balloon framing can be around 40 percent less expensive than traditional methods that used heavy beams and corner posts held together by mortise and tenon joints.

Bauhaus School Founded by Walter Gropius in 1919, the Bauhaus School did more than any other Modernist movement to promote the ideal of architecture and design as agents of social redemption. Through industrialized production, modern materials, and functional design, architecture could be produced inexpensively, become available to all, and so improve the physical, social, moral, and aesthetic condition of cities. The unifying themes were simplicity of line, plain surfaces, and suitability for mass production.

behavioral assimilation The process in which members of a minority group adopt the culture—language, norms, and values—of the wider society (or charter group), and so become acculturated to the mainstream life of the city. Contrast with *structural assimilation.*

bid-rent theory William Alonso's neoclassical economic model of urban land use based on the idea that companies and households compete for spaces in a way that maximizes their utility. The bid-rent curve on a graph plots the relationship between how much rent companies and households are prepared to pay for land relative to accessibility to a specific point—usually the center of a city. Commercial establishments are expected to be able and willing to outbid households for central sites because the extra income accruing to a central location through increased trade is likely to outweigh the savings in commuting costs obtained at the same site by a household. The bid-rent curve slopes downward with decreasing accessibility, reflecting decreasing rents with distance from the center of a monocentric (single-centered) city. See also *concentric zone model.*

Bilbao Effect After *starchitect,* Frank Gehry's, Guggenheim museum in Bilbao, Spain led the economic and cultural revival and re-branding of this old industrial port town, many city planners from around the world have tried to replicate the "Bilbao Effect" by securing an iconic new building designed by a starchitect in an effort to spur their own urban revitalization.

block busting The practice by some real estate agents in which prices are deliberately driven down, temporarily, allowing them or their associates to buy up as many properties as possible before restoring equanimity to the market and then selling to a new group of purchasers. In U.S. cities, for example, some real estate agents have introduced black purchasers into predominantly white areas in the hope that the latter will move out and sell to the realtors or their agents at deflated prices, and have then resold the properties to new black families at higher prices. See also *steering.*

boomburb The fastest-growing suburban jurisdictions during 1990 and 2000, typically located along the interstate beltways that ring large metropolitan areas primarily in the western United States. Defined using 1990 and 2000 U.S. Census data by Robert Lang of Virginia Tech's Metropolitan Institute as: places with more than 100,000 residents; not the largest city in their metropolitan areas; maintaining double-digit rates of population growth during 1990 to 2000. Although possessing elements found in most cities (housing, retailing, entertainment, and offices), they are not typically patterned in a traditional urban form (e.g., they almost always lack a dense business core). Distinct from traditional cities not so much in function but in their low density and loosely configured spatial structure. Contrast with *edge cities.*

boosterism Attempts by local governments to develop their local economies by attracting companies and investment into the city from elsewhere and through partnerships with private-sector sources of capital. See also *civic entrepreneurialism, growth machines, progrowth coalitions, public-private partnerships.*

Broadacre City Frank Lloyd Wright's response to the challenge of machine age, automobile-based urbanization. Based on the idea that, through planned metropolitan decentralization, cities could be spared from the sprawl and congestion threatened by the automobile, allowing people to live instead in affordable "Prairie Style" homes designed to be in harmony with their low-density semirural surroundings. His ideal city—Broadacre City—was to be built on the basis of two new technologies: the automobile and mass-production building technology using high-pressure concrete, plywood, and plastic.

brownfields Abandoned or underutilized industrial and commercial facilities where expansion or redevelopment is complicated by real or perceived contamination. Many brownfields are found in central cities and industrial suburbs with a history of traditional manufacturing. These sites can be abandoned industrial and railroad facilities or manufacturing plants that are operating but show signs of pollution. Brownfields can also be small commercial or even residential lots with only suspected contamination. Contrast with *greenfields* and *greyfields.*

business services See *producer services.*

C

CBD The central business district of an urban area. This central nucleus of commercial land uses has a concentration of office, government, retail, and cultural activities. As the most accessible part of a monocentric (single centered) urban area, land values and building densities are high.

central business district See *CBD.*

central places Urban centers—hamlets, villages, towns, and cities—that provide goods and services to their surrounding *hinterlands.* A central place system comprises a hierarchy of central places—ranging from a small number of very large central places (cities) offering higher-order goods (expensive and infrequently purchased items, such as designer furniture and jewelry) to a large number of small central places (hamlets) offering low-order goods (inexpensive, frequently purchased, everyday necessities, such as newspapers and milk). Walter Christaller sought to explain this size and spacing of towns and cities in his central place theory.

centrality The functional dominance of cities within an *urban system.* Cities that account for a disproportionately high share of economic, political, and cultural activity have a high degree of centrality.

centrifugal movement Outward movement within an urban area as a result of forces—high land costs, congestion, and so on—that displace economic and other activities toward the periphery. Contrast with *centripetal movement.*

centripetal movement Inward movement within an urban area as a result of forces—accessibility to suppliers and customers, specialized services, workers—that attract economic and other activities toward the center. Contrast with *centrifugal movement.*

chain migration A migration process in which migrants who were encouraged and assisted in their move by friends and relatives from their place of origin encourage other friends or relatives to join them, helping them to find accommodation and jobs when they arrive.

circuits of capital investment Geographer David Harvey's notion of circuits of capital investment within economies between which capital is switched depending on market and other economic conditions—for example, from the *primary* circuit of industrial production to the *secondary* circuit of fixed capital assets (machinery and equipment, buildings) or to the *tertiary* circuit of education or technology research.

city-state An independent political unit comprising a city that controls a surrounding *hinterland.* Early examples included the Sumerian and Greek city-states and medieval city-states such as Florence, Genoa, and Venice in Italy and Bremen, Hamburg, and Lübeck in Germany. A contemporary example of a city-state is Vatican City.

civic entrepreneurialism This term captures how local governments have become more proactive in promoting economic development since the early 1970s. They now negotiate with large corporate investors and attempt to stimulate and attract private enterprise with fiscal and other incentives and by attempting to create attractive conditions for profitable investment. See also *boosterism, growth machines, progrowth coalitions, public-private partnerships.*

class factions Narrower categories ("the professions," for example) within the overall concept of *class structure.* See also *class formation, class structuration, class structure.*

class formation A process that results in conscious collectivities of people who experience class as it is built into their lives in particular ways. They come to realize the force of class in and through the immediate circumstances that they can experience and understand directly. See also *class factions, class structuration, class structure.*

class structuration Dimensions, in addition to *class structure* and *class factions,* that David Harvey has identified as contributing to people's experiences that constitute the process of *class formation.* These include the division of labor that determines the formal class structure, institutional barriers to social mobility, the system of authority, and the dominant consumption patterns of a particular time and place. See also *class formation, class factions, class structure.*

class structure The formal categorization of class positions in a society at any given time. It is based on the positions people hold within the *division of labor* and the framework of economic organization. The broad categories of class structure are cast in terms of fairly heterogeneous groups: the "middle class," for example, consisting of a great variety of occupations. See also *class formation, class factions, class structuration.*

cognitive distance A measure of the perceived (rather than just the physical) distance that is generated from the brain's perception of the distance between visible objects in an urban area, a perception that is influenced by land use patterns, the distinctiveness of objects, and the impact of symbolic representations of the environment such as maps and road signs. An important concept in understanding mental maps.

collective consumption Usually refers to goods and services provided by the public sector. Less often refers to services that, literally, have been consumed by a group of people in a collective manner (such as a lecture). The term originated in a neo-Marxist (or Marxian) theory formulated by Manual Castells that argues that there are certain services (schools, hospitals) that are crucial for the maintenance of capitalism but that can be too expensive for individual capitalist enterprises to provide and so require provision through nonmarket means by the public sector.

colonial cities Cities that were deliberately established or developed as administrative or commercial centers by colonial or imperial powers.

command and control centers Cities that have key corporate management, government, and financial and business services functions. For example, these cities have a high proportion of company headquarters where a large amount of corporate decision-making is done and from where national and transnational business operations are controlled. See also *world cities.*

comparative advantage A principle to explain patterns of trade and specialization in which places and regions specialize in activities for which they have the greatest advantage in production relative to other regions—or for which they have the least disadvantage. Contrast with *competitive advantage.*

competitive advantage The advantage acquired in economic competition by some locations because of the benefits that accrue from an early start in production of a particular good and the continuing defense of that historic base through superior organization and adaptability. Other locations can suffer from a competitive disadvantage due to the absence of an *initial advantage* that would have allowed them to develop and maintain an ongoing competitive advantage. See also *initial advantage.* Contrast with *comparative advantage.*

competitive capitalism The earliest phase of industrial capitalism—from the late eighteenth to the end of the nineteenth century—the heyday of free enterprise, with markets characterized by comparatively high levels of free-market competition between small-scale businesses, consumers, and workers who acted almost completely independently, and with few constraints or controls imposed by government. See also *organized capitalism.*

concentric zone model The idealized model of city structure devised by Ernest Burgess based on Chicago in the 1920s in which socioeconomic status increases in a series of concentric zones moving outward from the *CBD.* Compare with the *sector model* and *multiple-nuclei model.*

congregation The residential clustering of an ethnic minority through choice, rather than the involuntary segregation brought about by structural constraints and discrimination.

consumer services Personal services, including retailing, medical care, personal grooming, leisure and recreation, eating out, and entertainment. Contrast with *producer services.*

contagion effect The negative contagious process that contributes to the *spiral of decay* in neighborhoods in which abandonment and long-term vacancy have a depressing effect on the value and desirability of adjacent properties. Nearby property owners seek to sell or move out and remaining property owners are discouraged from investing in the maintenance and upkeep of their depreciating properties—that can lead to further abandonment. See also *spiral of decay.*

containerization The process in which automated cranes efficiently load and unload massive standardized containers filled with large amounts of cargo between the ships and the flatbeds of nearby trains and trucks—improving the operation of cargo shipping and handling by reducing the turnaround time at ports. Previously, ships, trains, and trucks sat in port while major items of cargo were loaded and unloaded individually by a sizable workforce.

control centers See *command and control centers.*

corporatist model of urban governance A model of local power structure that rests on a symbiotic relationship between the various arms of local government and private organizations (labor organizations, community groups, etc.). Key organizations become incorporated into the formal decision-making process and city governments delegate a certain amount of authority in return for cooperation and support. This model sees society in terms of segmented socioeconomic organizations that are taken under the wing of professional politicians and technocrats who are then able to expand and consolidate the scope of their power and authority. Contrast with *elitist model, managerialist model, neoelitist model, pluralist model, hyperpluralist model.*

cultural transmission The process in which distinctive values and norms of behavior, sometimes ones that would be seen as different by society at large, are passed from one generation to another within local cultures within local environments. See also *culture of poverty, cycle of poverty, neighborhood effects.*

culture of poverty The argument that emphasizes the role of values and attitudes in perpetuating the *cycle of poverty* so that a culture of poverty eventually emerges. A vicious cycle links lack of opportunity and lack of aspiration. The poor see the improbability of achieving any kind of material or social success and so adapt their expectations and behavior, becoming so accustomed to deprivation, so defeated, and so withdrawn that they become unable and unwilling to seize opportunities for educational or occupational advancement. Indolence, distrust, suspicion, and introversion become a way of life, defining a distinctive culture of poverty. See also *cultural transmission, cycle of poverty, neighborhood effects.*

cumulative causation In Gunnar Myrdal's model, the self-propelling spiral of growth that occurs in specific settings like cities as a result of the build-up of advantages from the development of *economies of scale, agglomeration economies,* and *localization economies.* Cumulative causation results in some other cities and regions experiencing negative consequences called backwash effects. See also *spread effects.*

cycle of poverty The transmission of poverty from one generation to another through a cycle that begins with unhealthy conditions: low incomes, poor housing, and overcrowding. Overcrowding contributes to susceptibility to physical ill-health, which is compounded by poor diets that are also a result of low incomes. Ill-health causes absenteeism from work, which results in decreased income. Absenteeism from school through illness may also contribute to the cycle of poverty by constraining educational achievement and limiting occupational skills, leading to low wages. See also *contagion effect, neighborhood effects, spiral of decay.*

D

Dalits Also a reference to outcastes, this term is used for those people who were traditionally regarded as "Untouchables," those at the bottom of all caste systems whose members dealt with tasks related to human waste and dead animals. Mohandas Gandhi, the inspirational leader of India before independence, crusaded for this group of people. He called them Harijans, meaning "children of God," but today most of these people prefer to be referred to as Dalits, meaning "the oppressed"; the Indian government refers to them as "Scheduled Castes." Traditionally, the Dalits were forced to live outside the main community because they were deemed capable of contaminating food and water by their touch. They were denied access to water wells used by other groups, refused education, banned from temples, and subject to violence and abuse. Although these practices were outlawed by India's constitution in 1950, discrimination and violence against Dalits still occurs in some rural areas.

Dark Ages The period of several centuries of stagnation and decline in economic and city life in Western Europe following the collapse of the Roman Empire in the fifth century.

debt financing The issuing of bonds by municipalities to finance new infrastructure and improved services to support more urban growth. Bond repayments are met from property taxes, the revenues from which are expected to increase as a result of the growth stimulated by the spending financed by the bond issues.

defensible space Physical spaces and settings that people can identify with or exert control over. Related to architect Oscar Newman's attack on Modern architecture in which he argued that much of the vandalism, burglary, and criminal violence of central city neighborhoods was related to the "designing out" of territorial definition and delineation in Modernist apartment blocks: Once the space immediately outside a dwelling becomes public, nobody feels obliged to supervise it or defend it against intruders.

deindustrialization A relative decline in industrial employment associated with a shift to services in a country or region where industry had traditionally been a significant component of the economy. It may result from secular shifts in an economy that are related to technological change and/or the globalization of the economy in which some mass production and assembly line manufacturing are moved to lower cost locations in less developed countries. Such trends may involve not only a relative decline but also an absolute one and may involve declining industrial output as well as employment. See also *Fordism, flexible production systems.*

deinstitutionalization A public policy, partly to reduce spending, of moving the mentally ill and other people from large public institutions such as psychiatric hospitals and orphanages to grassroots care that typically involves a combination of independent living and local service delivery points.

Demographic Transition A trend in birth rates and death rates—from high to low levels—over time. This model suggests that improved diets, public health, and scientific medicine cause a steady decline in death rates with increasing levels of economic development. Birth rates decline later, and more slowly, as sociocultural practices take time to adjust to these new circumstances. The result is a sharp increase in population growth until birth rates fall to relatively low levels. Based on the experience of developed countries like the United Kingdom, not all less developed countries should be expected to follow this demographic path.

depreciation curve A decline in the value of housing as it ages over time within each new subdivision or tract. Such curves are averages that mask local unevenness in physical deterioration. Housing of the same age and initial quality wears unevenly because of variations in maintenance and improvements and the localized effects of road works, abandonment, and so on. See also *spiral of decay.*

Deutscher Werkbund A future-oriented school of thought, led by Peter Behrens (architect and chief designer for the German electrical manufacturer AEG) founded in 1906 in response to the challenge of the Machine Age. This loose coalition of architects, artists, and craft firms wanted to reform the relationship between artists and industry, their guiding principle being that quantity and quality should complement one another. Behrens saw industrialization as the manifest destiny of the German nation, and his factory designs were muscular temples to technological power. Other members of the Werkbund, such as Walter Gropius, emphasized the need to overcome the alienating aspects of traditional society by leaving behind all the architectural refinements and symbolism of the old order and replacing them with a no-nonsense style. See also *Secessionists.*

deviant behavior Social behavior, such as criminal activity, that deviates from social norms.

diagonal economic integration A form of business organization in which a company diversifies its interests through corporate mergers or acquisitions of companies that are engaged in separate and distinct enterprises, producing different goods or services for different markets. An automobile manufacturer, for example, may buy into energy, advertising, or entertainment companies.

Digital Divide The gap in opportunities, among different segments of the population and different places, to access advanced information and communication technologies, particularly the Internet, for a variety of activities.

disorganized capitalism See *advanced capitalism.*

distance-decay effect The rate at which a particular process or activity diminishes with increasing distance, such as the tendency for those who live furthest away from the sources of goods and services to consume them less often because of the increased travel costs or the increased time involved in visiting the source of supply.

divisions of labor The specialization of different workers, countries, or regions in particular kinds of tasks and economic activities. Social divisions of labor reflect the social characteristics of the people who undertake different types of work (e.g., related to age, ethnicity, gender). Spatial divisions of labor involve regional economic specialization, based on the distribution of resources and markets and on the exploitation of *economies of scale, agglomeration economies,* and *localization economies.*

Dual City A large metropolitan area characterized by disparities in wealth and status or a trend toward increasing social inequality, or both. Manuel Castells argued that the economic, social, and metropolitan restructuring since the 1960s has resulted in both qualitative and quantitative changes in American slum and poverty-area settings that have produced new patterns of vulnerability, fragmentation, and disadvantage while intensifying the occupational and spatial polarization of cities. See also *world cities.*

dust dome See *urban dust dome.*

E

economies of scale Cost advantages for companies from large-scale production; equivalent to increasing returns to scale in which an increase in inputs (raw materials, labor, etc.) by x percent results in an increase in output of more than x per cent. A crucial part of mass production in *Fordism.*

economies of scope Cost advantages from large-scale flexible organization of a range of commodities; created by the capacity of companies to provide entirely new products or services, or both, through the flexible use of the same production or service network. A crucial part of *flexible production systems.*

edge city A term coined by journalist Joel Garreau to describe recent decentralized urban development—that sometimes overshadows the old downtown—comprising nodal concentrations of shopping and office space situated on the outer fringes of large metropolitan areas, typically near major highway intersections. Contrast with *boomburbs.*

egalitarian liberalism See *liberalism, egalitarian.*

elitist model of urban governance A model that sees most important decisions made by a handful of powerful business people with elected officials as the "understructure" of power, dependent on the patronage and tolerance of these powerful individuals. The legitimacy of this leadership is ceded from below rather than imposed from above; the passivity of most people reflects their voluntary acceptance of the leadership's domination in return for its pursuit of the public interest, with the ballot box providing a register for any serious abuse of power. Compare with *neoelitist model* and contrast with *corporatist model, managerialist model, pluralist model, hyperpluralist model.*

eminent domain The right of government to take private property under due process of law for legitimate public purpose with just compensation. See also *police powers.*

EMR See *extended metropolitan region.*

enabling technologies New technologies that are critically important in facilitating urban and economic restructuring. Since the 1980s, for example, three kinds of enabling technologies have been significant: production process technologies (e.g., robotics); transaction technologies (e.g., computer-based just-in-time inventory control systems); and circulation technologies (e.g., e-mail).

enterprise zone (EZ) An officially designated economically distressed area where public fiscal and other incentives—tax exemptions or reduced planning regulations—are available to promote economic development by inducing companies to locate and create jobs there.

entrepôt A port that specializes in the trade of goods for reexport. They operate primarily as intermediary trading centers—receiving goods from foreign countries for transshipment to other countries. Hong Kong, Singapore, and Rotterdam are the world's top three entrepôts.

environmental determinism The theory that human behavior is determined by physical environment (especially climate). In an urban context, an approach that draws on behaviorist notions to argue that city living affects behavior. See also *Gesellschaft.*

EPZ See *export processing zone.*

ethnoburb Suburban ethnic clusters of residential areas and business districts characterized by vibrant ethnic economies that depend on large numbers of local ethnic minority consumers; they also have strong ties to the global economy that reflects their role as outposts in the international economic system through business transactions, capital circulation, and flows of entrepreneurs and other workers. They are multiethnic communities in which one ethnic minority group has a significant concentration but does not necessarily comprise a majority. The clear ethnic imprint on the suburban landscape differentiates the ethnoburb from the typical U.S. suburb in which members of ethnic minorities are more dispersed among the white charter group.

exchange value The value of a commodity, such as housing, as measured by the amount that it can command on the market when it is sold or exchanged. Compare with *use value.*

exclusionary zoning Zoning tactics invented in San Francisco to discriminate against the Chinese and refined in New York in 1916 to discriminate against undesirable factories that were soon deployed by suburban communities to exclude undesirable social groups and unwanted land uses. By carefully framing their *land use zoning* ordinances, suburban jurisdictions can restrict certain types of activity and people from moving into a locality. See also *restrictive covenant.*

export processing zone (EPZ) A small, closely defined area in which government creates especially favorable investment and trading conditions to attract export-oriented, usually foreign-owned, industries. These conditions include the availability of factory space and warehousing at subsidized rents, low tax rates, and exemption from export duties. See also *free trade zone.*

Expressionism A style of art and literature that, instead of depicting objective reality, depicts the subjective emotions aroused by objects or events. Accomplished by distorting and transforming nature, rather than imitating it.

extended metropolitan region (EMR) The recent extension of *megacities* in parts of Asia and Latin America beyond their city and metropolitan boundaries to form extended metropolitan regions. This results in an urban form without set boundaries that can stretch for 50 miles from the urban core. These have emerged as single, economically integrated regions comprising the central city, developments within transportation corridors, satellite towns, and other peripheral developments, but no single government body is responsible for overall land use planning.

external economies Benefits that translate into cost savings that accrue to producers from associating with similar producers in places that offer the services that they need, such as specialist suppliers.

exurban development Residential and other development that occurs outside a city and beyond its suburbs. See also *centrifugal movement, edge cities.*

EZ See *enterprise zone.*

F

FDI See *foreign direct investment.*

Federal Home Loan Bank Act Legislation passed in 1932 by which the Hoover administration created a reserve for mortgage lenders in an effort to respond to declining residential construction activity with the onset of the Great Depression and the large number of households losing their properties through foreclosure.

Federal Housing Administration (FHA) An agency created by the Roosevelt administration in 1934 in an effort to reduce unemployment resulting from the Great Depression by stimulating construction jobs in the private sector. It stabilized the mortgage market and facilitated sound home financing on reasonable terms. Instead of lending money, it insured mortgage home loans made by private institutions. The insurance gave banks and savings and loan associations the confidence to disburse more mortgages and to charge one or two percentage points less than before, stimulating demand for home ownership. FHA guarantees also stimulated demand as a result of terms that required smaller down payments and extended repayment periods.

feminization of poverty The process in which women are more likely to be poor, malnourished, and otherwise disadvantaged due to gender inequalities within households, communities, and countries.

Fertile Crescent A region of early agriculture and early urbanization in a crescent-shaped area of fertile land stretching from Mesopotamia (the land between the Tigris and Euphrates rivers) in modern Iraq to the lower Nile valley in Egypt.

FHA See *Federal Housing Administration.*

filtering Homer Hoyt's thesis that the primary motor behind residential mobility is the chain of moves initiated by the construction of new dwellings for the wealthy and involving the filtering of households up the housing scale and the consequent filtering of houses down the social scale. The construction of new housing for the affluent leads to their out-migration from older properties and the subsequent occupation of this housing by persons of lower socioeconomic background. See also *sector model* and *vacancy chain.* Contrast with *gentrification.*

first modernity See *modernity, first.*

fiscal mercantilism Efforts by local governments to increase local revenues by attracting—through the use of fiscal and other incentives—companies that create lucrative taxable land uses. Similar to *boosterism* and *civic entrepreneurialism*.

fiscal retrenchment Cutting back by governments on expenditures for public welfare and services as a result of trends such as economic downturns or voter aversion to higher taxes. See also *new conservatism* and *privatization*.

fiscal squeeze The unavoidable difficult fiscal situation that many central cities in the United States find themselves as revenue capacity has fallen (e.g., due to the suburbanization of businesses and middle-income people), while expenditure demands have increased (e.g., due to the growing need to supply services to aging low-income people).

flexible labor strategies A set of policies designed to increase the capacity of companies to adjust their outputs to variations in market demand and/or to reduce the costs of production through industrial organization involving labor practices involving part-time temporary workers, flexible working hours, and so on. See also *flexible production systems*.

flexible production systems Practices that allow flexibility in manufacturing in terms of what, when, how, and where production occurs. These practices include exploiting various enabling technologies, subcontracting arrangements, different labor markets, different market niches for products, and the development of new labor processes using flexible working hours, part-time workers, etc. See also *flexible labor strategies*.

Fordism A method of organizing economic production, income distribution, consumption, and public goods and services that centers on the mutual reinforcement of mass production and mass consumption. Named after Henry Ford because of his innovations and philosophy concerning automobile manufacture, it features a highly specialized and differentiated division of labor based on scientific management and assembly-line production geared to the provision of standardized, affordable goods for mass markets. Associated with *Keynesianism* and *Taylorism*.

foreign direct investment (FDI) The total overseas investment made by private companies involving direct investment in a company or companies in one or a number of foreign countries (e.g., takeovers, new subsidiaries, etc. rather than portfolio investment) to achieve managerial and production control. See also *transnational corporations*.

free trade zone An area delimited by a government unit in an effort to promote jobs and investment in which goods may be manufactured or traded without customs duties. See also *export processing zone*.

Futurists A subset of the Modern Movement in architecture that developed in the 1920s that wanted nothing to do with the past. Led by Filippo Marinetti and represented in the field of architecture and urban design by Antonio Sant'Elia, the Futurists sought to provoke social and institutional change using cities as the stages for permanent revolt, with huge and spectacular edifices that were monuments to both the masses and technology.

G

galactic metropolis The term, coined by geographer Pierce Lewis, to capture the disjointed and decentralized urban landscapes that have resulted from the splintering urbanism of globalized capitalism. The galactic metropolis is fragmented and multinodal, with mixed densities and unexpected juxtapositions of urban form and function. It is characterized by edge cities—suburban hubs of shops and offices that sometimes overshadow the old downtown.

gateway cities Cities that because of their location serve as links between one country or region and others.

GDP See *Gross Domestic Product*.

Gemeinschaft Roughly translated as "community," tight-knit social relationships that are based around family and that are of the kind that Tönnies argued were evident in traditional agrarian environments. Contrast with *Gesellschaft*.

generalized housing market The housing market—identified by historical geographer James Vance, Jr.—fueled by the ownership and control of rental property by landlords comprising entrepreneurial shopkeepers, small merchants, and people in professional occupations.

gentrification A process involving the renovation of housing in older, centrally located lower-income neighborhoods through an influx of more affluent people (e.g., middle-income professionals) seeking the character and convenience of less expensive and well-located residences. It displaces poorer people through eviction, rising real estate values and rents, and increasing property taxes. Contrast with *filtering*.

gerrymandering The practice in redistricting of manipulating the boundaries of electoral subdivisions to gain political advantage for a particular party or candidate.

Gesellschaft Roughly translated as "society," loose-knit social relationships that are of the kind that Tönnies argued were evident in urbanized environments. Contrast with *Gemeinschaft*.

GI Bill See *Servicemen's Readjustment Act*.

globalization The increasing interconnectedness of different people and places around the world through common processes of economic, political, and cultural change. Also, the tendency for the emergence of a global culture (that is, a universal trend that is sweeping all countries). See also *world cities*.

green urbanism A process that involves planning for cities in a way that facilitates major lifestyle changes (such as walking, bicycling, and reductions in material consumption), the preservation and restoration of the natural environment, and the application of new or collective technologies (such as public transit, district heating, and green building and design).

greenbelt cities Cities containing government-sponsored low-cost housing that were intended to allow the U.S. government to plan suburban development by drawing people from central cities to make room for slum clearance and redevelopment. Three were built in the United States in the 1930s and sold off to nonprofit corporations after World War II, to be swallowed up in the sprawl of automobile suburbs. See also *Resettlement Administration*.

greenfield site Building sites—former farms, golf courses, and so on—at the edge of a city's built-up area that are ripe for development. Contrast with *brownfields and greyfields*.

greenhouse effect The process in which the earth's atmosphere is warmed due to the trapping of solar radiation by water vapor and so-called greenhouse gases, such as carbon dioxide. The resulting "global warming" may have dire consequences due to the melting of the polar ice caps, rising sea levels, and climatic changes.

greyfield An abandoned or underutilized site such as an old shopping mall with huge parking and paved surfaces but without the environmental issues of brownfields. Contrast with *brownfields and greenfields*.

gridiron street pattern A street pattern in which the streets are laid out at right angles to one another. Contrast with *organic growth*.

gross domestic product (GDP) An estimate of the total value of all materials, foodstuffs, goods, and services produced by a country in a particular year. It does not include the value of profits from overseas investments and profits accruing to foreign investors.

ground rent Also known as economic rent or location rent, the surplus paid (to land owners, for example) above the minimum amount that would be necessary to use the land at all. Often defined in practice as the total revenue that can be generated by a particular activity on a particular parcel of land, less the total production and transportation costs associated with that same parcel of land.

growth machine Partnerships of private-sector and public-sector interests that implement strategies to enhance the economic development of cities and regions, largely through attracting inward investment, mostly from the private sector but also from public funds. See also *boosterism, civic entrepreneurialism, progrowth coalitions, public-private partnerships.*

growth pole Introduced as a concept by the French regional economist François Perroux in the 1950s, a growth pole is usually planned around one or more highly integrated high-growth industries that are organized around a propulsive leading sector, benefits from *agglomeration economies*, and can spread prosperity to nearby regions through *spread effects*. For example, eight *metropoles d'équilibre* (balancing metropolises) were planned in France (including Lyon, Marseille, and Bordeaux) to stimulate regional development while redirecting some economic activity away from Paris to reduce its *primacy.*

H

habitus A term coined by French sociologist Pierre Bourdieu to describe the culture associated with people's life-world that involves both material and discursive elements. A social group has a habitus if it has a distinctive set of values, ideas, and practices: a collective perceptual and evaluative schema that derives from its members' everyday experience that operates at a subconscious level, through commonplace daily practices, dress codes, use of language, and patterns of consumption.

heat island The microclimate of a city typically is slightly warmer than the temperature of the surrounding countryside. The built environment has a lower albedo than natural earth surfaces, is less able to reflect incoming solar radiation, and so absorbs more heat. The atmosphere in cities is also warmed by the release of heat from fossil fuels and, in summer, from air conditioning units. The aggregate effect is for cities to become heat islands, with average temperatures one or two degrees Fahrenheit above those of the surrounding countryside. See also *urban dust dome.*

hinterland A market area—the sphere of economic influence of an urban area. The urban area serves its hinterland with goods and services and its hinterland in turns supplies the urban area with products for processing or for export.

horizontal economic integration A form of business organization in which a company tries to capture the market for a single stage of production, a single good or service or an entire industry, and achieve *economies of scale*, by using corporate mergers or acquisitions of companies that formerly competed in the same market(s) with similar goods or services. A successful automobile manufacturer, for example, might buy out other automobile manufacturers.

horsecar suburb Suburbs that sprang up in the nineteenth century at the edge of the walking city around the terminals of radial horsecar and omnibus routes.

household life cycle A life cycle that reflects the relationship between household types and residential segregation. For middle-income households, for example, there are recognizable stages, each with a distinctive household composition that is associated with particular space needs, specific preferences in terms of accessibility, type of housing tenure, and locational setting, and different propensities to move.

housing submarkets Distinctive types of housing in localized areas of cities that, through various institutional mechanisms, tend to be inhabited by people of a particular type (in terms of socioeconomic background, age, ethnicity, etc.). See also *culture of poverty.*

hyperpluralist model of urban governance The sociocultural fragmentation and metropolitan spatial restructuring since the 1980s can turn the pluralism of *progrowth coalitions* into a hyperpluralistic situation in which unstable power relations are reflected in unstructured and multilateral conflict. Power is exercised over narrow areas and for limited periods by a variety of special interest groups who go their own way in seeking narrow gains and who are less restrained in their conduct and less likely to accommodate one another than under the *pluralist model.* Coalitions are essential to the hyperpluralist model, but they are short-lived and ad hoc. Compare with the *pluralist model* and contrast with the *corporatist model, elitist model, managerialist model, neoelitist model.*

I

IDL See *international division of labor.*

IMF See *International Monetary Fund.*

incorporation The right of a group of people to petition their state for articles of incorporation as a village, town, or city in the United States. This confers limited liability and, therefore, the opportunity for *debt financing* through issuing bonds to pay for new urban infrastructure and services to support further urban growth. See also *suburban incorporation* and *annexation.*

industrial capitalism The phase of capitalism, beginning with the Industrial Revolution and lasting until the early 1970s, that comprised (1) an early phase, *competitive capitalism*, characterized by comparatively high levels of free-market competition, with many small-scale producers, consumers, and workers who acted almost completely independently with relatively little government intervention, and (2) a later phase, *organized capitalism*, characterized by comparatively highly structured relationships between labor, government, and corporate enterprise.

informal sector The sector of an economy that involves a wide variety of activities that are not subject to formalized systems of regulation or remuneration. In addition to domestic labor, these activities include strictly illegal activities such as drug peddling and prostitution as well as a wide variety of legal activities such as casual labor in construction crews, domestic piece work, street trading, and providing personal services such as shoe-shining.

initial advantage The critical advantage acquired in economic competition by some locations because of the benefits that accrue from an early start in economic development.

international division of labor (IDL) The idea of the organization of spatial *divisions of labor*, principally at the national scale until the late twentieth century, in which each country specialized in certain sectors of the economy, such as industry in Western Europe or agriculture and raw materials in many African countries.

International Monetary Fund (IMF) A United Nations affiliate established in 1945 to help encourage international monetary cooperation, ensure international currency exchange stability, promote economic and employment growth, and, while not a development bank, provide temporary economic assistance to countries experiencing balance of payment problems. In 2004 the IMF had about US$107 billion in credit and loans to 87 of its 184 member countries. See also *World Bank.*

inverse concentric zone pattern The land use pattern found in some less developed countries, such as in Latin America, that contrasts with the concentric zone pattern identified by Burgess for cities in the United States in the 1920s. In the inverse pattern residential quality decreases with distance from the *CBD*. Contrast with the *concentric zone model*.

K

Keynesian suburb Suburbs that were developed by the U.S. government in the period of Keynesian macroeconomic management (during and after the Great Depression) in an effort to stimulate the economy by promoting housing construction by the private sector and to provide affordable housing to U.S. households. See also *Keynesianism*.

Keynesianism A doctrine of macroeconomic management closely associated with the British economist John Maynard Keynes, who advocated for managed capitalism: the use of fiscal policy (e.g., budget deficits) and economic *multiplier effects* to achieve and maintain full employment. This set of policies underpinned the welfare state in developed countries such as the United States in the 1950s and 1960s. The objective was to manage economies by countering the lack of demand during recessions through government spending—hence the term "demand management." This approach was undermined by inflation and high unemployment in the 1970s. A key element of *Fordism*.

L

labeling A situation in which all residents of a neighborhood are characterized based on the poor image of their surroundings. This can have negative consequences, such as restricting their employment opportunities. See also *cycle of poverty*.

land banks Land bought and held by companies mainly to ensure a supply of developable land. Many of the parking lots on the edge of downtown areas, for example, are held primarily for their speculative value rather than for their earning capacity as parking lots.

land use zoning The public regulation of land and building use to control the character of a place.

liberalism Liberalism, from the Latin *liberalis*, of freedom, is a belief in the importance of freedom. Liberalism includes a number of different variants, a dominant one being classical liberalism, that became popular in the eighteenth century. In economic terms, liberalism puts faith in the operation of the capitalist system and free trade.

liberalism, egalitarian The market failures that had triggered the Great Depression in the early 20th century undermined the legitimacy of classical liberalism and led to its eclipse by an egalitarian liberalism that relied upon government to manage economic development and soften the unwanted side effects of free-market capitalism for those people most adversely impacted.

localization economies Cost savings that accrue to companies as the output of their particular industries increase as a result of clustering together at a specific location. See also *agglomeration economies* and *urbanization economies*.

M

machine politics A form of paternalistic urban governance that emerged with incipient industrialization and rapid urbanization in the late 1800s in the United States that was characterized by charismatic leaders controlling hierarchical political organizations that drew on working class support. See also *municipal socialism*.

malapportionment The creation of electoral subdivisions of unequal size. See also *gerrymandering*.

managed capitalism See *Keynesianism*.

managerialist model of urban governance A model that sees civil servants effectively as social gatekeepers and "street-level" bureaucrats who mediate policy implementation. It emphasizes the autonomy of the technocrats and bureaucrats in the local civil service and the fact that urbanization had become so complex that elected officials had to rely increasingly on the expertise of civil servants; as a result, the local state is controlled by civil servants whose professional ideology and departmental allegiances are crucial in determining a good deal of the shape of local government activity. Contrast with *corporatist model, elitist model, neoelitist model, pluralist model, hyperpluralist model*.

maquiladora Literally, "mill" in Spanish. These factories take advantage of Mexico's low-cost labor and lax environmental regulations and locate in *export processing zones* (*EPZs*) in Mexico, mainly along the U.S. border. They are often owned or built with foreign capital, and assemble U.S. components for reexport to the United States as finished products free from customs duties.

mechanical solidarity One of the two forms of social "solidarity" identified by Emile Durkheim, it is based on similarities between people. Contrast with *organic solidarity*.

megacity A very large city—with a population of 10 million or more—characterized by both primacy and centrality within its national economy. Manuel Castells used this term for large cities in which some people are connected to global information flows while others are disconnected and "information poor." See also *Dual City* and *world city*.

megalopolis French geographer Jean Gottman's term for multicity, multicentered, urban regions. The name was first given to the dominant urban corridor in the United States that extends along the eastern seaboard from Boston to Washington, D.C.

merchant capitalism The initial phase of capitalism. As the feudal system disintegrated, it was replaced by an economy that was dominated by market exchange, in which communities came to specialize in the production of the goods and commodities that they could produce most efficiently in comparison with other communities. The key actors were the merchants who supplied the capital required to initiate the flow of trade—hence the label merchant capitalism. See also *comparative advantage*.

metropolitan consolidation The process in which key economic functions (e.g., headquarter offices and research and development [R&D] laboratories) tend to become increasingly localized in larger metropolitan areas in response to corporate reorganization following mergers and acquisitions and the differential growth and changing patterns of economic specialization within urban systems. Contrast with *regional decentralization*.

modernity, first The first modernity coincided with the two major phases of capitalism: competitive capitalism and managed (Keynesian) capitalism (comprising organized and advanced capitalism). Urban development during the first modernity was framed by competitiveness within closed geographic systems (national states) that were competing with one another. So, before the current globalized phase of capitalism, the first modernity was associated with social institutions such as family, politics, science, and religion that were more able to offer protection from social and environmental risks (e.g., unemployment and pandemics). See also *second modernity*.

modernity, second The second modernity coincides with the current globalized phase of capitalism. Urban development is now subject to competitiveness at the global scale. German sociologist, Ulrich Beck, who coined the term, argues that in the face of globalization forces, social institutions such as family, politics, science, and religion can no longer offer protection from social and environmental risks (e.g., unemployment and pandemics) as in the past (first modernity). See also *first modernity.*

multiple-nuclei model Harris and Ullman's model of urban city structure characterized by decentralization and nodes of different economic and residential areas. Contrast with the *concentric zone model* and the *sector model.*

multiplier effects The extra industries, companies, incomes, and employment in various sectors of the economy generated by new economic activity in one sector of the economy can be said to result from that activity's multiplier effects. Primary multiplier effects comprise a number of *localization economies* including the pool of labor in the existing interrelated industries that has skills and experience attractive to additional firms, and the likelihood of new innovations that provide further stimulus to growth as a result of the linkages between existing industries that promote interactions among professional and technical personnel and sustains R&D facilities, research institutes, and so on. Secondary multiplier effects are the result of the spiral of growth resulting from the initial spiral of growth caused by the primary multiplier effects.

municipal socialism A form of urban governance that emerged with incipient industrialization and rapid urbanization in the late 1800s in the United States and Europe that was characterized by local government intervention in the marketplace in order to impose standards and to ensure the provision of key services and basic amenities. See also *collective consumption* and *machine politics.*

N

natural increase The surplus of births relative to deaths in a country or place (calculated as the difference between the crude birth rate and crude death rate). It does not take into account migration.

neighborhood effects The concept that residential environments both influence and reflect local cultures within cities: People tend to conform to what they perceive as local norms in order to gain or maintain the respect of their local peer group. See also *cultural transmission.*

neighborhood unit American sociologist-planner Clarence Perry's concept that became one of the most widely adopted city planning building units for planned new towns throughout the world. The neighborhood unit was centered on a local elementary school, shops, and community institutions within walking distance of homes and bounded by roads carrying heavier flows of automobile traffic in order to provide a safer, traffic-free environment at its center.

neoclassical economics A conceptualization of how economic activity operates in capitalist society. The economy comprises many small producers and consumers who act rationally, although none is large enough to affect significantly the operation of the market. Companies are seen as atomistic agents with full information in a world of pure markets (with no entry barriers) and all have exactly the same resources, technological capability, and market power with deviations regarded as market "imperfections." Companies utilize factors of production (land, labor, and capital) to maximize their profits; consumers sell theirs (especially labor) to purchase goods and services that maximize their individual preferences. The market is an abstract space in which companies and consumers set the prices and the forces of supply and demand cause economic resources to be used in the most efficient way possible. Involves normative model-building—constructing simplified versions of how the real world ought to operate.

neoelitist model of urban governance This model accepts the premise of the *elitist model* that in some cities nearly all important decisions are made by a handful of powerful business people but questions the assumption that a consensus on the public interest is consciously and freely agreed to by the bulk of the population or whether rather it is the product of the systematic suffocation of opposition. Compare with the *elitist model* and contrast with *corporatist model, managerialist model, pluralist model, hyperpluralist model.*

Neo-Fordism See *flexible production systems.*

neoliberalism A market-driven approach to economic and social policy that stresses so-called free trade, cuts in government spending on social programs, privatized public services, and reduced regulation of private companies.

neoliberal ideologies A particular conceptualization that economies can be made more competitive through the implementation of various types of New Right strategies based on the desirability of free markets as the ideal condition not only for economic organization but also for political and social life.

neoliberal policies Policies designed to make economies more competitive through the implementation of various types of New Right strategies based on the desirability of free markets as the ideal condition not only for economic organization but also for political and social life. These policies involve reducing the role and budget of government, including removing subsidies, deregulation, and the *privatization* of formerly publicly owned and operated operations, such as utilities.

new conservatism More authoritarian than neoliberalism (New Right) ideologies, new conservatism involves a commitment to high military spending and the global assertion of national values (especially in the United States by, for example, the Reagan and both Bush presidencies). See also *neoliberal ideologies* and *neoliberal policies.*

new international division of labor (NIDL) The idea of the reorganization of the *international division of labor,* formerly principally at the national scale, to a global scale based on international production and marketing systems. Involves the decentralization of traditional manufacturing production from regions such as Europe and North America to some less developed countries such as in Asia and Latin America. Contrast with *international division of labor.*

newly industrializing economies (NIEs) Less developed countries (also known as newly industrializing countries (NICs)) that have acquired a significant industrial sector, usually through *foreign direct investment (FDI)* by *transnational corporations.* Examples are Mexico, Brazil, and Taiwan.

NIEs See *newly industrializing economies.*

NIDL See *new international division of labor.*

NIMBYism Not In My Backyard-ism. Characteristic of middle-income suburbs that are averse to the location of locally unwanted land uses (LULUs) such as refuse burners within their jurisdictions. The more extreme version of NIMBYism is BANANAism (Build Absolutely Nothing Anywhere Near Anything).

nodal center Urban centers that are important nodes—in terms of their concentrations of businesses and economic activities—for a region or urban system.

O

organic growth Urban growth that evolves in an unplanned manner—as when homes are built along preexisting winding pedestrian paths—rather than in a predetermined way based on some planned approach. Contrast with *gridiron street pattern.*

organic solidarity One of the two forms of social "solidarity" identified by Emile Durkheim, it is based on the existence of differences stemming from specialized economic roles—the *division of labor.* This term was based on a biotic analogy, the idea that in a complex organism (modern, urban society), every single "organ" (social group) is mutually dependent on the rest. Contrast with *mechanical solidarity*.

organized capitalism The later phase of industrial capitalism that began in the late nineteenth century and was characterized by comparatively highly structured relationships between labor, government, and corporate enterprise. These relationships were mediated through legal and legislative instruments, formal agreements, and public institutions. See also *competitive capitalism*.

overaccumulation crises Distinctive crisis phases in the long-term dynamics of capitalist economies, characterized by unused or underutilized capital and labor. An inevitable outcome of the difficulty of matching supply to demand under changing conditions, these crises represent critical moments for the political economy of capitalism. They can be recognized by the appearance of idle productive capacity, excess inventories, gluts of commodities, surplus money capital, and high levels of unemployment.

overurbanization A condition experienced in many contemporary less developed countries in which cities grow more rapidly than the jobs and housing they can sustain. See also *megacities*.

P

parkway A limited-access recreational highway developed by master planner Robert Moses in the United States.

place A specific geographic location with distinctive physical and human characteristics, such as a city. A place can also be associated with a sense of place that people associate with that location such that it influences their decisions about where to live, for example.

pluralist model of urban governance This model suggests that power is dispersed, with different kinds of interests dominant at different times over different issues. The structure of power is essentially competitive, drawing on a wide range of participants and ensuring a fundamental element of democracy through the need for elites to acquire mass loyalty. Society is kept in rough equilibrium by this competition, with labor acting as a counterbalance to business, consumers offsetting the power of retailers, tenants constraining the power of landlords, and so on. These groups also have overlapping membership, which promotes intergroup contact and makes for tolerance and moderation in local politics. Compare with the *hyperpluralist model* and contrast with the *corporatist model, elitist model, managerialist model, neoelitist model.*

police powers The powers of government to take actions (including exerting control over the development of privately owned property and land by local governments) to protect the health, safety, and public welfare. See also *eminent domain*.

primacy A condition in which the population of the largest city in an urban system is disproportionately large in relation to the second-largest and third-largest cities in that system. See also *primate city*.

primary products Products derived from natural resources, as in agriculture, mining, forestry, and fishing. These products are often important as raw materials in manufacturing.

primate city A country's leading city as evidenced by measures of its *primacy* including its significantly larger population compared to other cities and by other characteristics reflecting the primate city's national importance and influence, such as economic and political activity and power. See also *primacy* and *megacities*.

privatization A diverse set of policies designed to introduce private ownership and/or private market allocation mechanisms to goods and services previously allocated and owned by the public sector. Involves the sale of government assets (such as key industries) to private owners and the contracting out of services formerly provided by government to private companies in an effort to increase government efficiency and save public money.

producer services Also known as business services, these services enhance the productivity or efficiency of other companies' activities or enable them to maintain their specialized roles. Examples include advertising, personnel training, recruitment, finance, insurance, and marketing. Contrast with *consumer services*.

progrowth coalition Another name for *growth machine*. See also *boosterism, civic entrepreneurialism, public-private partnerships*.

public-private partnership Coalitions of private-sector businesses and/or business leaders and public-sector officials and agencies, with others such as unions and chambers of commerce, that seek to promote economic growth and the prosperity of their urban area. See also *boosterism, civic entrepreneurialism, growth machines, progrowth coalitions*.

pull factors In migration studies, forces of attraction that induce migrants to move *to* a particular location. Usually operate in combination with *push factors*.

push factors In migration studies, negative events and conditions that impel an individual to move *from* a location. Usually operate in combination with *pull factors*.

R

rank-size rule A statistical regularity in city-size distribution, set out by Zipf in 1949, such that the population of a particular city is equal to the population of the largest city divided by the rank of the particular city. So if the largest city in an urban system has a population of 1 million, the fifth-largest city should have a population of 200,000, the hundredth-ranked city should have a population of 10,000, and so on.

redlining The practice by commercial lenders (e.g., banks) of withholding loans for properties in areas of cities that are perceived as high risk. The practice involves literally drawing a red line around high-risk neighborhoods on a city map and then using the map as the basis for determining loans. This results in a bias against minorities, female-headed households, and other vulnerable groups because they tend to be localized in high-risk neighborhoods. Redlining can also become a self-fulfilling prophecy, as neighborhoods starved of mortgage funds become progressively more run-down and increasingly high-risk for loans. Compare with *steering*.

regional decentralization The process in which improved transportation and communications infrastructures, coupled with the attraction of cheaper land, lower taxes, lower energy costs, *boosterism*, and cheaper and less-militant labor, allow cities located in traditionally underdeveloped regions of a country to grow rapidly as a result of their growing relative attractiveness to private investment compared to the cities in established industrial regions. Contrast with *metropolitan consolidation*.

relative location The location of an urban area, although absolute (as measured, for example, by latitude and longitude), can also be relative, fixed in terms of its "situation." Situation refers to the location of a city relative to other places; its accessibility to routeways, for example, or to other cities.

rent gap The disparity between the rents actually charged for run-down central city areas and their potential market rents (as predicted by *bid-rent theory*) following renovation. If large, the rent gap can lead to urban redevelopment and *gentrification*.

Resettlement Administration An agency created in 1935 by the Roosevelt administration to help plan suburban development and draw people from central cities in order to make room for slum clearance and redevelopment. See also *greenbelt cities.*

restrictive covenant A contractual provision in the sale of a property that limits the nature of subsequent development. Such covenants have usually been used to discriminate against low-income groups and locally unwanted land uses (LULUs). See also *exclusionary zoning.*

revanchist city Based on the French word *revanche* meaning revenge, a revanchist city reflects a new form of revenge by the powerful in society for the moral and economic decline of city life following the social reforms of the 1960s and involves deregulation, *privatization*, and the other *neoliberal policies* since the 1970s, as well as *gentrification*.

rural-to-urban migration The movement of rural residents to larger towns and cities in search of a better life, as during the Industrial Revolution in England. In many less developed countries today, rural residents migrate to urban areas because of the desire for employment and the prospect of access to public facilities and services that are often unavailable in rural regions. See also *overurbanization.*

S

Secessionists A future-oriented school of thought that emerged in the early 1900s in response to the challenge of the machine age. Headed by Austrian architect Adolf Loos, the principal motivation of the Secessionists was the elimination of all "useless" ornamentation from architecture. See also *Deutscher Werkbund.*

second modernity See *modernity, second.*

sector model The model of urban residential land use structure advocated by Homer Hoyt that suggests that class differences in residential areas are arranged in wedges (sectors) based on transportation routes. See also *filtering*. Contrast with the *concentric zone model* and *multiple-nuclei model.*

service-dependent neighborhoods A concentration of dependent groups such as ex-psychiatric patients in central city close to community-based services. Also known as "asylum-without-walls."

Servicemen's Readjustment Act (GI Bill) Legislation passed in 1944 to facilitate the readjustment of U.S. veterans returning from World War II. Created the Veterans Administration, one of the major goals of which was to facilitate homeownership for returning veterans.

setbacks *Land use zoning* ordinances, beginning with New York in 1916, that required the upper stories of skyscrapers to be stepped back beyond a certain height in order to allow light and air to reach the street. This restriction resulted in the characteristic wedding-cake profile of interwar skyscrapers. See also *land use zoning.*

shock city A city that is seen as the embodiment of surprising and disturbing changes in economic, social, and cultural life—such as Manchester, England, during the Industrial Revolution.

single-room occupancy (SRO) hotel Hotels in the run-down parts of cities that rent out units for occupancy by one person. These units may contain food preparation or bathroom facilities, or both.

slum UN-HABITAT defines a slum as a run-down area of a city that is characterized by substandard housing and squalor and a lack of tenure security.

social construction The idea that most differences between people are the result not of their inherent characteristics but of the way they are treated by others in society. Race, for example, is a social construction that artificially categorizes people into separate groups based on characteristics that include physical appearance (particularly skin color), ancestral heritage, and cultural history.

social disorganization A situation in which some communities find themselves weakly equipped to maintain social order because of high levels of poverty or inequality, high rates of population turnover, or high levels of diversity in social, demographic, cultural, and ethnic attributes that can inhibit the formation of local ties and weaken the effectiveness of local social institutions.

social distance The difference between people based on factors such as socioeconomic background and power leading to separation in social life. This may be the result of mutual desire or predominantly the wishes of the powerful. It is often expressed in terms of physical distance and residential differentiation.

social ecology The social and demographic composition of neighborhoods.

social reproduction The various elements that are necessary to reproduce the workforce and the consumers needed to keep a capitalist society functioning (e.g., family, schools, health services, welfare state).

sociospatial dialectic Ed Soja's term for the mutually interacting process in which people shape the structure of cities and at the same time are themselves affected by the structure of those cities.

space A term that in general usage refers to location (as in absolute location [measured by latitude and longitude] or *relative location*). But for urban geographers, space is not simply a medium in which economic, social, political, and historical processes are expressed, it is socially constructed. As such, it influences the nature of the relationships between different social groups within cities. See also *territoriality, defensible space.*

spatial mismatch The mismatch in the location of low income lower-skilled people and jobs that has developed as many of the low-skilled jobs traditionally found in central city areas where many low income people are concentrated have been relocated to suburban areas, to be replaced mainly by jobs requiring higher skills.

special district A single-purpose jurisdiction, such as a special sanitary district, established as an additional layer of government for a specific purpose. Special districts are not subject to the statutory limitations on financial or legal powers that apply to municipalities. They also have the potential advantage of being customized to correspond closely to specific functional areas.

spiral of decay The process in which deferred or makeshift maintenance and repair, in conjunction with an accumulation of garbage and graffiti, can set a low-income neighborhood containing substandard housing on a downward spiral of decay that results in it becoming a *slum.* See also *contagion effect.*

splintering urbanism The term devised by geographers Stephen Graham and Simon Marvin to describe the fragmentation of the economic, social, and material fabric of cities as a result of the selective impact of new technologies and networked information and communications infrastructures.

spread effects In Gunnar Myrdal's model, the positive spillover effects on an urban area or region—including capital investment and the inmigration of skilled workers—from the economic growth of some other urban area or region. Also known as *trickle-down effects.* See also *cumulative causation.*

squatter settlement Residential development on land that is neither owned nor rented by its occupants. Often found in the cities of less developed countries, squatting also occurs in developed countries, including in some European cities.

SRO hotel See *single-room occupancy hotel.*

stagflation An episode of economic recession accompanied by comparatively high rates of price inflation (a decline in the value of money because prices keep rising).

starchitects Star architects who have gained worldwide celebrity for their iconic building designs, some of which have helped spur economic and cultural revival (as in Bilbao, Spain). Starchitects can also become public intellectuals, involved in discourse on a wide range of topics, contributing to influential cultural ideas and trends, and fuelling and reinforcing economic and cultural globalization. See also *Bilbao Effect*.

stealth cities Urban developments at the edge of metropolitan areas, including some *edge cities*, that have grown so quickly that they have not been *incorporated* as separate jurisdictions and do not have official names that appear on street maps or government statistical tables. They are called stealth cities because they are invisible administratively and politically, without their own chambers of commerce, libraries, town halls, or courthouses. But they are very real, showing up clearly both on the skyline and on plots of land values.

steering The practice by realtors of deterring households from moving into neighborhoods occupied by households of a different socioeconomic background or ethnicity in order not to jeopardize local prices. Until 1968, when it was made illegal in the United States, it resulted in "slammed door" discrimination, mainly against African Americans. Compare with *redlining*.

streetcar suburb A suburb that sprang up at the edge of the city around the terminal of a radial streetcar line. By making it feasible to travel up to 10 miles from the *CBD* in 30 minutes or so, the streetcar greatly increased the territory available for residential development and opened the way for the growth of streetcar suburbs.

structural assimilation A process involving the diffusion of members of a minority ethnic group through the social and occupational structure of the wider society (or charter group). Contrast with *behavioral assimilation*.

subsistence activities Activities that comprise an economic system, usually farming, in which the producers (farmers), and their families, consume most of what is produced, leaving little surplus for trade or sale.

suburban incorporation The petitioning for articles of incorporation by suburban communities to protect their reputation, status, and independence—often a preemptive strategy to avoid annexation of their community by a central city. As the pace of suburbanization quickened from the 1920s in the United States, middle-income suburbanites sought to escape not only proximity to central city *slums* but also central city tax burdens and to establish a distinctive setting for governance and politics in which middle-income values and preferences might flourish. See also *incorporation* and *annexation*.

sustainable urban development A vision of urban development and resource use that seeks a balance among economic growth, environmental impacts, and social equity in a manner that can be sustained in the long run for future generations.

T

tax increment financing (TIF) A mechanism used by cities in the United States to finance redevelopment efforts that is directly tied to the success of those efforts. If an area within a city can be made more attractive to private developers and new development occurs, the tax revenue collected from that area would be expected to rise. Tax increment financing taps into any increase in tax revenues by using the tax "increment" (the difference between the taxes after redevelopment and the expected taxes without redevelopment) to finance the improvements and other activities that stimulated the redevelopment in the first place.

Taylorism The name (after analyst F. W. Taylor) for a form of labor organization in manufacturing industries in which the planning and control of work are given over entirely to management, leaving production workers to be allocated specialized tasks that are subject to careful analysis—"scientific management" using techniques such as time-and-motion studies. Adopted by Henry Ford in the early twentieth century to mass produce automobiles in Detroit. See also *Fordism*.

technology systems Distinctive "packages" of technologies, energy sources, and political-economic structures that represent the most efficient means of organizing production at any given phase of economic development. Based on key sets of interdependent technologies, they represent the underpinnings of successive modes of organizing such things as economic production, income distribution, consumption, and public goods and services over several decades at a time.

territoriality The tendency for particular groups within a society to attempt to establish some form of control, dominance, or exclusivity within a localized area. Group territoriality depends primarily on the logic of using space as a focus and symbol for group membership and identity as a means of regulating social interaction.

TIF See *tax increment financing*.

TNC See *transnational corporation*.

transnational corporation (TNC) Similar to a multinational corporation (MNC), a company that engages in production, distribution, and marketing that span international boundaries, with subsidiary companies, factories, offices, or facilities in a number of countries. TNCs have the power to coordinate and control operations in several countries; control may be exercised without the necessity for legal ownership, such as through subcontracting arrangements. Production is carried out on a global scale in such a way as to maximize profits. For example, as part of a global assembly system, a low-skill labor-intensive stage of the production process may be located in a less developed country where wages and unionization levels are low. Many headquarters of the largest TNCs are concentrated in the *world cities* of London, New York, and Tokyo. See also *globalization*.

trickle-down effects The economic growth that spreads—trickles down—to more remotely located cities and regions that is induced by high levels of demand in more centrally located economically vibrant cities and regions. Also known as *spread effects*.

U

underemployment A situation in which people work less than full time even though they would prefer to work more.

uneven development The spatial outcome, within and between countries, of the continuous seesawing of capital from one set of opportunities to another on the basis of particular local mixes of skills and resources. Capital is invested unevenly over time and across space because, whenever possible, development will occur wherever businesses judge that their investment will yield the highest return. When businesses try to exploit differences between places, they create a continuously variable geometry of labor, capital, production, markets, and management.

"Untouchables" See *Dalits*.

urban dust dome The *heat island* effect leads to a distinctive pattern of local air circulation on days when there is no regional air movement. Light surface winds are drawn toward the city center, then rise and descend slowly at the edge of the built-up area. This pattern is associated with the development of an urban dust dome in which fumes, soot, and chemicals are trapped in the air over the city. This phenomenon becomes particularly pronounced when a temperature inversion develops and a lid of relatively warm air effectively flattens the dust dome, keeping pollution low over the city. When exposed to strong sunshine, such concentrations of pollution can be transformed through photochemical reactions into *smog*. See also *heat island*.

urban heat island See *heat island*.

urban realms The term coined by geographer James Vance, Jr., to capture the fact that many of the world's largest metropolitan regions consist of a loose coalition of urban realms or economic subregions bound together through urban freeways that tend to function semi-independently, and contain a broad mix of land uses (retail, commercial, and residential). As a result, the central urban core loses its economic dominance and becomes just another realm.

urban regime The term coined by Clarence Stone to capture the fact that local *public-private partnerships* involve power structures that rest on slowly changing coalitions of dominant groups and interests that are represented by city officials who sustain both their own power and that of the coalition by ensuring a variety of benefits and policy outcomes to the key groups involved.

urban renewal The revitalization of run-down sections of central cities. In the United States the federal Urban Renewal Program was a slum clearance and public housing program initiated under the 1937 Housing Act. The 1949 Housing Act subsequently provided assistance to central cities toward the costs of clearing "blighted" areas and assembling land for redevelopment. Due to problems including the negative reactions of residents of neighborhoods that were officially designated as blighted, it was terminated by Congress in 1973. In 1974 the Housing and Community Development Act introduced the block grant as the principal form of federal aid for local community development.

urban social movements Pressure groups and organizations with varying degrees of public support that petition for change, often outside conventional political channels. Sometimes termed new social movements. These form an important part of the theory of *collective consumption*.

urban system The complete interdependent set of urban settlements of different sizes that exists within a particular territory, such as a region or a country.

urbanism The forms of social interaction, patterns of behavior, attitudes, values, and ways of life that develop in urban settings.

urbanization economies The economic benefits that companies enjoy because of the package of infrastructure, ancillary activities, labor, and markets typically associated with urban settings. See also *agglomeration economies* and *localization economies*.

use value The value of a commodity, such as housing, as measured by its owners when they consume it. Compare with *exchange value*.

Usonia Architect Frank Lloyd Wright's physical framework for a new American way of life in which the working class would be emancipated from the trap of congested but expensive cities, living instead in a semi-rural setting spread out at low densities in differentiated and individualized houses designed to be in harmony with their natural surroundings.

V

vacancy chain The chain of movement resulting from properties becoming available through factors such as new building, the subdivision of properties, and the death or out-migration of existing occupants. See also *filtering*.

vertical economic integration A form of business organization in which a single large company tries to control all aspects of the same industry and capture a greater proportion of the final selling price by using corporate mergers or acquisitions of companies that formerly were engaged in different stages of the same industry (from research and design through production to sale). An automobile manufacturer, for example, might take over companies that make specialized components like engines or car navigation systems or that distribute or sell automobiles. See also *Fordism*.

W

White City The temporary plaster city built for the World's Columbian Exposition in Chicago in 1893, designed by a team lead by architect Daniel Burnham, that showcased the Beaux Arts style of architecture and later City Beautiful movement and belief in the role of the built environment as an uplifting and civilizing influence. The White City had uniform building heights and imposing avenues with dramatic perspectives.

World Bank Along with its main component, the International Bank for Reconstruction and Development, a United Nations affiliate established in 1948 to finance productive projects that further the economic development of its 184 member countries from Afghanistan to Australia and from the United States and the United Kingdom to Ukraine, the United Arab Emirates, Uganda, and Uruguay. In 2003, for example, the World Bank loaned about US$11 billion to less developed countries. See also *International Monetary Fund*.

world cities A term coined by Patrick Geddes to describe those cities in which a disproportionate share of the world's most important business—economic, political, and cultural—is conducted and that serve as headquarters to *transnational corporations*. Typically, London, New York, and Tokyo are identified as the leading tier of world cities, although other cities such as Paris, Frankfurt, Chicago, Los Angeles, and Zurich also have important global roles. Characterized by social polarization. See also *command and control centers*.

world-system A term coined by historian Immanuel Wallerstein to describe any spatially extensive interdependent economic system comprising a number of countries that have a single *division of labor* but multiple cultural systems.

Z

zone in transition The name given by Ernest Burgess and the Chicago School of Human Ecology to the concentric ring of land uses lying between the *CBD* and the inner ring of working class residential areas. Characterized by a mixture of industry, low-status commercial activity, and high-density poor-quality rental accommodation, often inhabited by recent immigrants. The *concentric zone model* identifies it as a zone of disinvestment despite its central location, and consequently the likely location of a *rent gap*.

zoning See *land use zoning*.

Urbanization and Urban Geography

John Elk III/Alamy

Urbanization and Urban Geography

This text introduces urban geography by focusing on the processes and outcomes of urbanization that are so important for the people who live in cities. Cities are products of many forces. They are engines of economic development and centers of cultural innovation, social transformation, and political change. At the same time, urban areas vary in everything from employment opportunities for job seekers to patterns of land use in neighborhoods, racial composition in metropolitan regions, and social behavior in urban society. Given the considerable range of issues that urban geography encompasses, we need to try to develop a consistency in our approach. Understanding theories about cities and the way they change—rather than simply listing their various attributes—will help to ensure that we maintain a consistency and will give us greater insight into the way cities work. Although we will sometimes need to look closely at specific people and events, our goal is to focus on understanding how to read the economic, social, and political "blueprints" that give shape and character to various kinds of cities and urban life. By "generalizing" in this way, we will have a more immediate and richer understanding of each new aspect of urbanization that we encounter.

LEARNING OUTCOMES

After reading this chapter, you should be able to:

- Assess how and why the concepts of space, territoriality, distance, and place influence everyone in cities.
- Describe why several approaches to urban geography have arisen, and how it is possible to gain urban insights from each.
- Explain how urban geographies result from long-term changes involving the interrelated dynamics of economic, demographic, political, cultural, technological, environmental, and social change.
- Compare and contrast the broad phases in the nature of capitalism and their importance for cities in the United States since the 1700s.
- Understand the ways in which the rapidly increasing interdependence of the world-system has impacted cities and the people living in them.

CHAPTER PREVIEW

The fundamental task of the student of urban geography is to make sense of the ways that towns and cities have changed and are changing, with particular reference to the differences both between urban places and within them. As a student, you will understandably find yourself asking some fundamental questions. What exactly is urban geography, and how does it relate to other aspects of geography and to other social science subjects? What is involved in urbanization, and what outcomes of urbanization that affect people are important in studying urban geography? This chapter answers those questions.

First, we address the question of urban geography as a subject for academic study, emphasizing the importance of concepts of space, territoriality, distance, and place that influence everyone in cities. We then review the various approaches that have been taken to the study of cities from a geographical perspective. We will see that

Cities are hives of human activity and crucibles of social, cultural, and political change, where there is always something happening. Urban geography can help us to understand, analyze, and interpret urban landscapes such as in this photograph of Woodruff Park and the people enjoying a wonderful view of the downtown skyline of Atlanta, Georgia.

they are all encompassed by the overall framework for study that is adopted in this text: a framework that allows us to deal with various aspects of urban geography in terms of their relationship to urbanization as a process.

The second half of the chapter addresses this question of urbanization as a process. Here we see how urban geographies and urban change associated with every aspect of people's lives are the result (and sometimes the cause) of long-term changes in economic development and in the interrelated dynamics of technological, demographic, political, social, cultural, and environmental change. Because these changes are being framed increasingly at the global scale, our framework for studying urban geography also views urbanization from a global perspective involving people and cities from all over the world.

THE STUDY OF URBAN GEOGRAPHY

Like other aspects of human geography, urban geography is concerned with "local variability within a general context."[1] This means that it is concerned with understanding both the *distinctiveness* of individual places (at the scale of towns and cities or particular neighborhoods) and the *regularities* within and between urban areas in terms of the spatial relationships between people and their environment (see Urban View 1 entitled "The Art of Taking Back a Neighborhood: The Heidelberg Project"). We should immediately note that environment here includes not only the natural physical environment but also the built environment (everything from homes, factories, offices, and schools to roads and bridges), the economic environment (economic institutions, the structure

URBAN VIEW 1

The Art of Taking Back a Neighborhood: The Heidelberg Project[2]

Multicolored polka dots decorating one house, a boat filled with stuffed toy animals nearby, and a vacant lot with rows of car hoods painted with faces (Figure 1). The Dotty Wotty House, Noah's Ark, and Faces in the Hood are just some of the guerrilla art installations on Heidelberg Street in Detroit by Tyree Guyton, a trained artist, whose philosophy is:

> I believe that my job as an artist is to help people to see! I wanted to use my talents to bring about positive change in my community . . . I use art as a catalyst for social change. I chose to start right here in my own neighborhood and yet I realized that the first change had to start with me. Changing my mind and seeing with my eye of understanding helped to eradicate my fears and limitations. Social change must start with self and then you can change the entire world around you.

And you cannot help notice the change immediately as you arrive at Heidelberg Street, having passed city block after city block with as many vacant lots as houses, uncut grass and invading weeds, and trash littering the crumbling sidewalks. For the urban geographer, this is an unforgettable example of "local variability within a general context." The distinctiveness of the Heidelberg Project reflects the kind of local variability that is possible in a particular city despite the general context of urban decline due to regularly repeated stories of deindustrialization, suburbanization, and poverty in Rustbelt cities like Detroit.

What started with decorating Tyree Guyton's grandfather's house with polka dots 25 years ago has become the nonprofit Heidelberg Project, which attracts 275,000 visitors of all ages each year. But it has not only changed former crack houses and abandoned homes into art that explodes with bright paint and discarded items, it has also helped keep the street free of trash and crime, create a strong sense of community, and provide educational opportunities for neighborhood children to experience the transformative potential of art.

The Heidelberg Project's social importance was recognized with a 2004 Places Award by the Environmental Design Research Association (EDRA):

> Clearly this work is not only about what you see. It's about the dialogue it engenders . . . The Heidelberg Project offers an alternative vision to young children in one of America's most blighted urban areas; it broadens community awareness of the power of art; and it brings a new sense of important social realities to the consciousness of visitors.

Spencer Grant/Alamy

FIGURE 1 An abandoned house is covered with discarded toy stuffed animals to become art as part of the Heidelberg Project in Detroit that has also helped to keep this street free of trash and crime, create a strong sense of community, and provide educational opportunities for neighborhood children to experience the transformative potential of art.

and organization of economic life, and so on), and the social environment (including norms of behavior, social attitudes, and cultural and political values that shape interpersonal relations among people).

For urban geographers some of the most important questions therefore include the following: What attributes make cities and neighborhoods distinctive? How did these distinctive identities evolve? Are there significant regularities in the spatial arrangement of towns and cities across a country or region of the world? Are there significant regularities in the spatial organization of land use within cities and in the spatial patterns of people in neighborhoods by social status, household type, or race? The urban geographer will also need to know about the causes of any regularities that do exist. How, for example, do people choose where to live and what are the constraints on their choices? How do people's areas of residence affect their behavior? What groups, if any, can manipulate the spatial organization of towns and cities? And who profits from such manipulation?

In posing these questions, urban geographers have learned that the answers are ultimately to be found in the wider context of economic, social, and political life. Cities must be viewed as part of the economies and societies that maintain them. The study of cities cannot be undertaken in isolation from the study of history, economic development, sociocultural change, or the increasing interdependence of places within the world economy. A proper understanding of cities requires an interdisciplinary approach. The traditional focus of geography—the interrelationships between people and their physical and social environments—requires geographers to draw on the work of researchers in those related disciplines.

Space, Territoriality, Distance, and Place

Urban geography is a coherent and distinctive framework of study through the central themes of *space, territoriality, distance,* and *place*. For the geographer, *space* is not simply a medium in which economic, social, political, and historical processes are expressed. It is also a factor that influences patterns of urban development and the nature of the relationships between different social groups within cities. From this perspective, cities are simultaneously the products and the shapers of economic, social, and political change. Partitioning space through the establishment of legal boundaries is also important because it affects the dynamics of cities in several ways. For example, the establishment of municipal boundaries restricts a city's capacity to raise revenue to its own territory, while electoral boundaries affect where people vote and the outcome of local elections and politics.

Territoriality is the tendency for particular groups within society—ethnic groups, gangs, gated communities—to attempt to establish some form of control, dominance, or exclusivity within a localized area. Group territoriality depends primarily on the logic of using space as a focus and symbol of group membership and identity as a means of regulating social interaction. It is important for the geographer because it is often the basis for individual and group behavior that creates distinctive spatial settings within cities. These spatial settings in turn mold the attitudes and behavior of the people living in cities.

Distance is important for several reasons. It affects the behavior of both producers and consumers of all goods and services. It influences patterns of social interaction and the shape and extent of social networks. Variations in people's physical accessibility to opportunities and to amenities such as jobs, schools, stores, parks, and hospitals are important in determining the local quality of life.

Finally, *place* is important because of the geographer's traditional and fundamental concern with areal differentiation and the distinctiveness of regions and localities. The distinctiveness of particular metropolitan regions, cities, districts, and neighborhoods is central to the analytical heart of urban geography: mapping variability, identifying regularities in spatial patterns, and establishing the linkages that constitute functional regions and subareas. In addition, the sense of place (Figure 2) that people associate with certain

Jonathan Burkham

FIGURE 2 A wall mural. A distinctive mural in a Hispanic neighborhood in Milwaukee, WI, near Cesar Chavez Drive and across from the Sixteenth Street Community Health Center that provides health care, health education, and social services to low-income residents in the neighborhood. Community art such as this provides a remarkably clear expression of the sense of place and territoriality that are fundamental to the spatial organization of people in cities.

cities and localities is important, because it can influence their decisions: where to live, where to locate an office or a factory, whether to hire someone from a particular place, or whether to walk alone through a certain part of town, for example.

Approaches to Urban Geography

Urban geography has evolved to encompass a variety of approaches to its subject matter. This is the result of a more general intellectual evolution of ideas in the social sciences. For example, a widespread "quantitative revolution" has occurred both in urban geography and in the social sciences as a whole. Two developments spurred this revolution. Large quantities of reliable socioeconomic data about cities and city neighborhoods became available from sources such as the censuses of population and housing in many countries. At the same time, tools to analyze and shape this information were becoming widely available in the wake of new digital technologies and geographic information systems (GIS). Modern analytical and modeling techniques have made a decisive contribution to the social sciences. They have allowed the urban geographer to see farther, with more clarity, and have provided the means by which to judge theories about urbanization.

Like other fields, urban geography has also been influenced by changing social values. As each society's comprehension of urban problems has grown, attitudes toward research in urban geography have become more flexible. Much of the research undertaken today in urban geography has relevance far beyond the ivory towers of academia and involves a more active engagement with the needs and concerns of local communities, private companies, and government. Urban geographers are likely to be consulted on issues that range, for example, from optimal political redistricting (redrawing the boundaries of local voting districts) to the evaluation of government policies aimed at enhancing local economic development in distressed central city neighborhoods.

Finally, changes in cities themselves and in the nature of urbanization have also contributed to the evolution of approaches to urban geography. As we have become aware of changes in cities and have looked more closely at those changes, new topics for study have emerged. A good example is the increased interest that geographers have taken in community well-being after the cuts to many public services in richer countries like the United States that have hit some poorer and older people very hard over the past couple of decades. An equally striking, more recent example is the interest by geographers in the link between local housing markets and international finance after the financial meltdown of 2008–09 and the personal misfortunes of families losing their homes due to foreclosure.

The details of the evolution of urban geography as an academic subject are beyond the scope of this text. It is important, however, to establish a few central points about that evolution. Several decades ago, work in urban geography (or "settlement geography," as it was more commonly known) saw towns and cities as adaptations to natural physical circumstances. Attributes of urban settlements were interpreted as responses to local sites, regional resources, and the opportunities and constraints surrounding them. So, for example, the growth of Pittsburgh as a steel town can be interpreted spatially, from this perspective, in terms of the availability of local sources of coal, iron ore, limestone, and water, along with proximity to large markets for iron and steel products that benefited local businesspeople and workers.

Another body of work within this *spatial description approach* that dominated the early development of the subject was focused on the "morphology" of towns and cities—their physical form, their plan, and their various townscapes and "functional areas" (that is, districts with a distinctive mixture of interrelated land uses). Keeping with the example of Pittsburgh,

Kablonk/SuperStock

FIGURE 3 An urban geographer looking at this neighborhood of young families with children would want to discover what it has in common with similar neighborhoods in other cities and what makes it distinctive from them.

these studies would likely have emphasized the influence of the city's hilly topography on the layout of its streets and neighborhoods, and perhaps shown how urban growth and land use were "fixed" by the limited amounts of flat land (along the Monongahela River) that the captains of industry at the time saw as suitable for large ironworks.

Gradually, scientific principles influenced attitudes toward knowledge, and both settlement and morphology studies fell out of favor. In their place there emerged a *spatial analysis approach* based on the philosophy and methodology of *positivism* that had been developed in the natural sciences. This philosophy was founded on the principle of verification of facts and relationships through accepted scientific methods. The rise of positivism affected all of geography and most of the social sciences, and the "quantitative revolution" reinforced it. Its practitioners came to redefine urban geography as the *science* of urban spatial organization and spatial relationships, and, as a science, it focused on the construction of testable models and hypotheses. One example of the many possible positivistic approaches to aspects of Pittsburgh's urban geography would be an attempt to quantify the relationship between neighborhood social status and proximity to steel mills. The hypothesis might be that neighborhood social status tends to increase steadily with distance from a steel mill. If this were demonstrated with verifiable evidence, the next step might be to begin to develop a theory by replicating the study in other industrial cities.

This more abstract approach has contributed a great deal to our understanding of cities, and it continues to be a mainstay of urban studies. But sometimes its abstractions and overdependence on statistical data can seem flat and lifeless in the face of the urban realities of people's lives to which they are applied. The abstractions tend to leave unanswered many of the important questions concerning underlying *processes* and *meanings*. So, although it might be useful to be able to use census tract data to establish and quantify a tendency for neighborhood social status to increase with distance from an industrial plant, we are left without any sense of how the relationship came about: Who made what decisions, and why, in the process of establishing the relationship? How might any exceptions to the overall relationship be explained?

In response to such questions, a *behavioral approach* emerged. This approach focuses on the study of individual people's activities and decision making in urban environments. Although the behavioral approach continued to use a positivist methodology, explanatory concepts and analytical techniques were also derived from social psychology and some key ideas came from social philosophy, with its insights into human needs and impulses.

The behavioral approach's relative neglect of the importance of cultural context for understanding people's actions and the meanings attached to those actions led to the emergence of the *humanistic approach*. The behavioral approach's positivist methodology was replaced by methods, such as ethnography and participant observation, that attempted to answer questions that capture people's subjective experiences. For Pittsburgh, what kind of meaning does the presence of a steel mill carry for different individuals within the city? How do people's feelings about steelworkers' neighborhoods affect their decisions about where to live in Pittsburgh? Like the behavioral approach, though, the humanistic approach has been criticized for not paying enough attention to the constraints on people's decision making and behavior.

As a result, another approach, generally referred to as the *structuralist approach*, gained momentum within urban geography. This approach is cast, in contrast to the behavioral and humanistic approaches, at the scale of macroeconomic, macrosocial, and macropolitical changes. It focuses on the implications of such changes for urbanization and on the opportunities and constraints they present for the behavior and decision making of different groups of people. At its broadest level, this approach draws on a combination of macroeconomic theory, social theory, and the theories and concepts of political science, and includes the *political economy approach*. For example, a political economy approach to Pittsburgh's urban geography would certainly want to relate the patterns of growth and decline of both steel mills and their associated blue-collar neighborhoods to the broader structure of, among other things, access to capital by businesses for investment in industry, the availability of skilled labor, the framework of government policies affecting industrial and residential development, and **deindustrialization** associated with corporate restructuring within the global economy.

The structuralist analyses of the structure of social inequality paved the way for explicitly incorporating the experience of women into urban geography. The *feminist approach* deals with the inequalities between men and women, and the way in which unequal gender relations are reflected in the spatial structure of cities. For Pittsburgh, the feminist approach might be interested in examining changing gender roles as the urban labor force has restructured following the closing of the steel mills, such as the trend for some women in two-earner households to be overrepresented in certain fast-growing part-time occupations.

The *structure-agency approach* was an attempt to unite the structuralist approach's concern with macrolevel social, economic, and political structures with the humanistic approach's emphasis on human agency. Structuration theory sees society's social structures as created and recreated by the social practices of human agents, whose actions are themselves constrained by these social structures. As a result, it is impossible to predict the exact outcome of the interactions between social structure and human agency. Despite the elegance of this theorization, empirical investigation has proved difficult because it is not easy to analyze the continuous and complex interrelationships between structure and agency. A study of Pittsburgh might involve questions about how the **gentrification** of some central city neighborhoods is the result of an intricate set of interactions between human agents, including landowners, mortgage lenders, planners, and realtors; institutions such as the city government; and social structures involving planning regulations such as **land use zoning** and building codes.

Finally, the influence of literary theory has led to the emergence of *poststructuralist approaches*, including the *postmodern approach*. The postmodern approach strongly opposes the idea that any general theories can explain cities and the people who live in them. Instead, it accepts the shifting and unstable nature of the world and concentrates on questions of who defines meaning, how this meaning is defined, and to what end. It is concerned with understanding the power of symbolism, images, and representation as expressed in language, communication, and the urban landscape. Again, using the example of Pittsburgh, a postmodern approach would likely examine the city government's attempts to "reinvent" the city within the global economy as it restructures away from steel and toward jobs in the high-tech and services industries. This approach would draw attention to the city government's use of language and communication to deliberately construct certain images of the city (that may not reflect reality) that, in turn, are designed to influence the views and decisions of potential investors and residents.

For the most part, all these approaches can be regarded as potentially complementary. Although it is neither possible nor desirable to merge them into some kind of all-encompassing model or theory of urbanization, it is possible to gain insights from each. To that extent, all these approaches are represented (with classic examples pointed out). This does not mean, however, that the reader will be faced with sudden or unexpected shifts. The material is organized to emphasize urban geography as the outcome of urbanization as a process.

URBAN VIEW 2
Census Definitions

Just what is meant by an "urban" settlement varies a good deal from one country to another. For the Bureau of the Census in the United States, the term urban applies to the territory, people, and housing units located within urbanized areas and urban clusters.[3]

- An *urbanized area* (UA) is a densely settled area (whether or not the territory is legally incorporated as a city) with at least 50,000 people and a density of at least 1,000 people per square mile at the urban core and at least 500 people per square mile in the surrounding territory.
- An *urban cluster* (UC) consists of an urban core with a population density of at least 1,000 people per square mile and at least 500 people per square mile in the surrounding territory that together encompass a population of at least 2,500 people, but fewer than 50,000 people.

Because such definitions of *urbanized* depend on administrative boundaries, they do not capture the concentrations of people that live in a number of contiguous jurisdictions that form one continuous metropolitan sprawl. The U.S. Bureau of the Census began to use standardized definitions of metropolitan areas in the 1950 census under the designation of standard metropolitan area (SMA). The term was changed to standard metropolitan statistical area (SMSA) by the 1960 census and to metropolitan statistical area (MSA) in 1983.

Within urban areas, detailed census information is available by *census tracts, block groups,* and *census blocks*:

- *Census tracts.* These geographical subareas have boundaries that were drawn with the objective of delineating small populations that are relatively uniform in terms of their demographic and socioeconomic characteristics. They vary a good deal in territorial extent and population size, although tracts within metropolitan areas contain between 1,000 and 8,000 people. As metropolitan America has grown, so has the number of census tracts recognized by the Census Bureau. The boundaries of many longer-established census tracts have remained the same over several decades, making it possible to use census data to analyze neighborhood change in certain areas. Elsewhere, however, modifications to tract boundaries and the addition of new tracts make intercensal comparison difficult.
- *Block groups.* Every census tract is divided into as many as 9 block groups, each of which contains an average of 10 census blocks. A block group consists of all blocks whose numbers begin with the same digit in a census tract. Block groups generally contain between 300 and 3,000 people. Block groups are important because they are the smallest geographical area for which detailed data are tabulated.
- *Census blocks.* Census blocks generally correspond to the physical configuration of city blocks and are bounded by streets or other prominent physical features. They are the smallest statistical unit for which census data are available, although the range of data is not detailed, being limited to basic population and housing characteristics.

In preparing the 1990 census, the U.S. Bureau of the Census developed an electronic database called the Topologically Integrated Geographic Encoding and Referencing (TIGER) system. TIGER files contain address ranges, latitude and longitude coordinates, and the location of roads, railways, rivers, and other physical features. They can be linked with small-area census data to form the basis of sophisticated Geographic Information System (GIS) applications that offer a great deal of potential for research in urban geography. GIS—organized collections of computer hardware, software, and geographic data that are designed to capture, store, update, manipulate, and map geographically referenced information—has grown rapidly to become an important method of urban geographic analysis. GIS technology allows an enormous range of urban problems to be analyzed. For instance, it can be used to identify the most efficient evacuation routes for people from all or part of a city in the event of a terrorist attack, to monitor the spread of infectious diseases within and between cities, to analyze the impact of proposed changes in the boundaries of legislative districts, to identify potential customers in the vicinity of a new business, and to provide a basis for urban and regional planning.

The changing realities of urbanization present a continual challenge to the census in delineating geographic units that reflect actual patterns of urban and metropolitan change. So census tabulations include data not only for MSAs but also metropolitan areas (MAs), for primary metropolitan statistical areas (PMSAs), consolidated metropolitan statistical areas (CMSAs) and core based statistical areas (CBSAs), each designed to provide a standardized framework for comparisons and analysis at different spatial scales.

URBANIZATION: PROCESSES AND OUTCOMES

Figure 4 provides a useful outline of urbanization as a process. It is clear from the diagram that urbanization involves much more than a mere increase in the number of people living and working in cities and metropolitan regions. It is driven by a series of interrelated processes of change—economic, demographic, political, cultural, technological, environmental, and social. It is also modified by locally and historically contingent factors such as topography and natural resources or an "accident" of birth that resulted in Henry Ford being born in Dearborn, Michigan, that ultimately resulted in the Ford Motor Company being headquartered in Detroit and not somewhere else in the United States. It is true that the overall result of urbanization has been a tendency for more and more people to live and work in ever-larger cities and metropolitan regions (although, as we will see, this is not a necessary condition of urbanization). At the same time, urbanization results in some important changes in the character and dynamics of the **urban system** (the complete set of urban areas regionally, nationally, or even internationally), and within cities and metropolitan regions, it causes changes in patterns of land use, in *social ecology* (the social and demographic composition of neighborhoods), in the *built environment*, and in the nature of **urbanism** (the forms of social interaction and ways of life that develop in urban settings). Certain groups of people might view some of these outcomes as problems. Government policies, legal changes, city planning, and urban management might eventually address those problems, often resulting in changes (sometimes unanticipated) that in turn affect the dynamics that drive the overall urbanization process.

As suggested by Figure 4, the relationships among these processes are complex. Urbanization is not only influenced by the direct effects of these dynamics, but it also experiences "feedback" effects. Meanwhile, almost every aspect of urban change is itself to some extent interdependent with some of the others. This complexity can be confusing! We will eliminate much potential confusion with concepts and examples introduced in this chapter. Bear in mind, though, that Figure 4 presents a very broad framework that covers the entire subject matter of urban geography. The economic, demographic, political, cultural, social, technological, and environmental processes and relationships implied in the diagram will be elaborated, and the urban outcomes that affect people will be detailed. It is recommended, therefore, that you refer to Figure 4 from time to time, putting each new set of material into overall perspective.

Economic Change

At the heart of the dynamics that drive and shape urbanization are economic changes. The sequence and rhythm of economic change will be a recurring theme as we trace and retrace the imprint of urbanization. It is the evolution of capitalism itself that has structured this imprint that affects everyone living in cities. Figure 5 summarizes the main features of this evolution. In the United States, there have been several broad phases in the nature of capitalism, and we are now in the early stages of a significantly different phase, framed in the context of **globalization.**

The earliest phase, lasting from the late eighteenth century until the end of the nineteenth, was a phase of **competitive capitalism**, the heyday of free enterprise and laissez-faire economic development, with the political economy of the country characterized by classical **liberalism**: competition between small family businesses and with few constraints or controls imposed by governments or public authorities. In the earlier years of this phase, the dynamism of the entire system rested on the profitability of agriculture and, increasingly, manufacture and

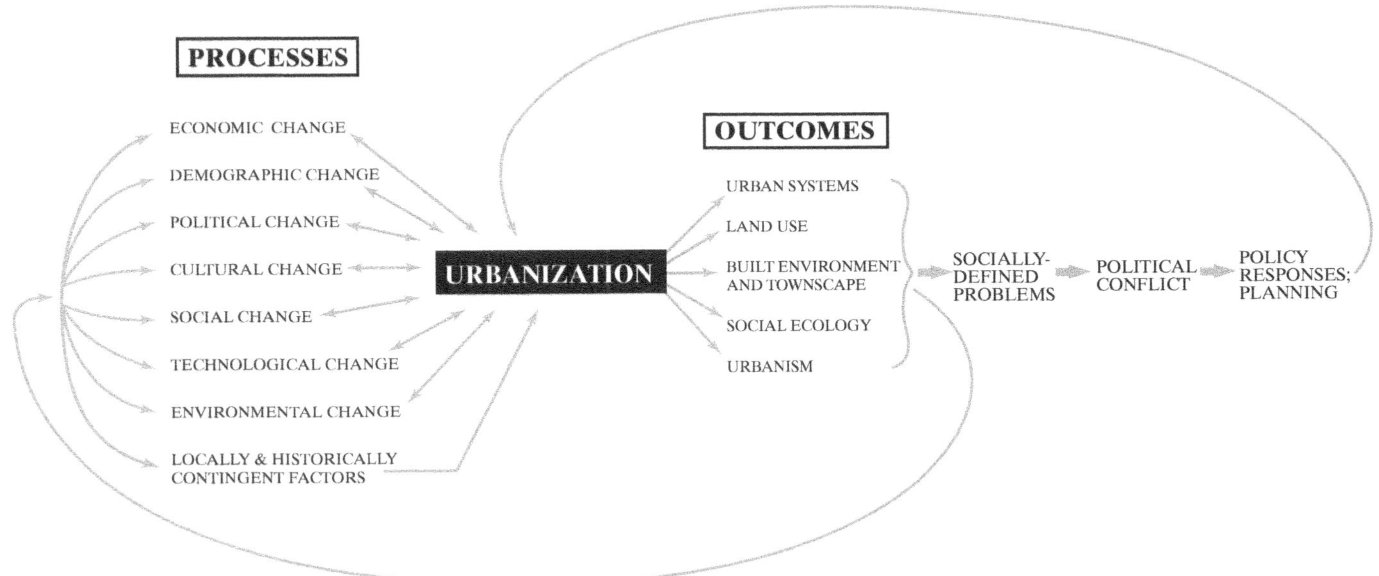

FIGURE 4 A framework for the study of urban geography: urbanization as a process.

	← First Modernity →			Second Modernity →
Major phases of capitalism	**Competitive**	**Managed (Keynesianism)** *Organized*	*Advanced*	**Globalized**
Technology systems	Water power Steam engines Cotton textiles Iron works	Coal-powered steam engine Steel Railways Machine tools World shipping	Internal combustion engine Oil and plastics Electrical engineering Aerospace Radio Telecommunications	Microelectronics Digital telecommunications Biotechnology Informatics
Infrastructure development	Canal building First railway boom	Second railway boom Streetcar boom	First automobile boom Motorways and airports	Broadband and satellite systems
Urban governance and politics	Laissez-faire Economic Liberalism Municipal Socialism and Machine Politics	Boosterism and the Politics of Reform	Cities as Growth Machines Metropolitan Fragmentation and Progrowth Coalitions	Re-modernization of Institutions, Entrepreneurialism, Neoliberalism
Role of central government	Negligible (liberalism) Increasing: regulator	Strong (direct): manager/regulator (egalitarian liberalism)	Strong (indirect): partner/facilitator	Decreasing: broker (neoliberalism)
Urban systems	Mercantile urban systems (regional)	Organization of industry (national frameworks)	Metropolitan spatial decentralization	World Cities and international networks
Metropolitan form	Pedestrian City Transitional City	Industrial City Cities of sectors and zones	Suburban infill Metropolises of central cities surrounded by urban realms	Megapolitan regions

FIGURE 5 The long-term view. A summary of some of the major features of economic change and urban development in the United States.

"machinofacture" (industrial production that was based less on handicraft and direct labor power than on mechanization, automation, and intensively used skilled labor).

After the Civil War, the extensive railroad system and a strengthened federal system helped to create an economy that was truly national in scale. The most successful family businesses grew bigger and began to take over their competitors. Business became more organized as corporations set out to serve regional or national consumer markets rather than local ones. Labor markets became more organized as wage norms spread, and government began to be more organized as the need for regulation in public affairs became increasingly apparent.

By the turn of the twentieth century, these trends had reached the point where the nature of capitalist enterprise had changed significantly. It could now be characterized as **organized capitalism**—a label that came to be increasingly appropriate with the evolution of the economy over the next 75 years or so. In the early decades of the twentieth century, the system's dynamism (that is, the basis of profitability) shifted away from industrial manufacture and machinofacture as a new labor process took hold. This process was **Fordism**, named after Henry Ford, the car manufacturer who was a pioneer of the principle of mass production, based on assembly line techniques and "scientific" management (known as **Taylorism**), together with mass consumption, based on higher wages for workers and sophisticated advertising techniques targeted at consumers.

The success of Fordism was associated with the development of a rather tense but nevertheless workable relationship between business interests and the labor unions, whose new strength was in itself another important element of "organization." Meanwhile, the role of government had also expanded—partly to regulate the unwanted side effects of free-enterprise capitalism and partly to mediate the relationship between organized business and organized labor. After the Great Depression of 1929–1934, government's role expanded dramatically to include responsibility for full employment, the management of the national economy, and the organization of various dimensions of the social well-being of people. The market failures that had triggered the Depression undermined the legitimacy of classical, laissez-faire liberalism and led to its eclipse by an **egalitarian liberalism** that relied on government to manage economic development and soften the unwanted side effects of free-market capitalism for those people most adversely impacted. This top-down approach to economic management, with a commitment to low unemployment, is often referred to as **Keynesianism**, after the British economist John Maynard Keynes.

After World War II, another important transformation in the nature of capitalist economies became evident. Developed countries like the United States and Canada experienced a shift away from industrial production and toward services, particularly sophisticated business and financial services, as the basis for profitability. It is denoted in Figure 5 as an evolution from **organized capitalism** to **advanced capitalism**. This shift began to transform occupational structures for workers, sparking deindustrialization—a decline in manufacturing jobs but *not* in manufacturing production.

Broadly speaking, the key dynamic of urbanization during the two centuries spanning the 1770s and the 1970s was interurban competition for jobs and investment, with the unwanted side effects of uneven development being cushioned increasingly by local, regional, and national government intervention. This long evolution of the capitalist economy saw the emergence of institutions that mutually confirmed and supported one another in an overall process of modernization: the nation-state, the Fordist company, the nuclear family, the system of industrial relations, the welfare state, and the formal institutions of science and technology.

But from the mid-1970s on, cities in the United States, along with others around the world, were caught up in a very different dynamic. The increasing globalization of the economy allowed huge **transnational corporations** to outmaneuver the national scope of both governments and labor unions by moving routine production and assembly operations to lower-cost, less developed parts of the world as part of a **new international division of labor (NIDL)**. This contributed to a profound destabilization of the relationship between business, labor, and government in developed countries like the United States. Meanwhile, Fordism began to be a victim of its own success, with mass markets for many products becoming saturated. As it became increasingly difficult to extract profits from mass production and mass consumption, many enterprises sought profitability through serving specialized market niches. Instead of standardization in production, specialization required variability and, above all, **flexible production systems**. There was a rapid decline of the old base of manufacturing industries and the onset of a "new economy" based on digital technologies and featuring advanced business services, cultural products industries, and knowledge-based industries, all framed within the context of a new international division of labor and international finance. The internationalization of economic geography weakened the leverage of both big government and big labor, destabilizing the organized capitalism of the mid-twentieth century and allowing a fundamental intensification of the economic and spatial logic of capital—especially big capital. Much of this was directly at odds with the top-down **managed capitalism** and planned modernization of the previous two centuries, tipping the U.S. space-economy into a significant new phase dominated by **neoliberalism**: a selective return to the free-market ideas of classical liberalism. The Reagan administration of the 1980s dismantled much of the Keynesian welfare state, deregulated industry, ushered in an era of public-private co-operation in place-making and economic development, and rekindled libertarian ideas about the primacy of private property rights. Urban planning has morphed into public-private cooperation, while state and local governments increasingly behave like businesses in their attempts to attract economic development and balance the books. In attempts to recapture some control over the global scale of the new economic logic and its social, cultural, and environmental implications, national governments have become increasingly collaborative, supranational entities have emerged, many institutions have extended their focus from a national to an international frame of reference, and many local and regional organizations have become involved in cross-border collaborative networks of one sort or another.

For some observers, this points to nothing less than the onset of a second major phase of modernization in which the structures and institutions of nineteenth- and twentieth-century modernization are both deconstructed and reconstructed (Figure 5). Whereas urban development during the **first modernity** was framed by competitiveness within closed geographic systems (national states) that were competing with one another, urban development at the onset of this **second modernity** is subject to competitiveness at the global scale.

The significance of this historical evolution for urbanization, urban geography, and cities and the people living in them is fundamental. Each new phase of capitalism saw changes in what was produced, how it was produced, and where it was produced. These changes called for new kinds of cities, while existing cities had to be modified. At the same time, of course, cities themselves played important roles in the transformation of capitalist enterprise. As centers of innovation, cities and towns have traditionally functioned as engines of economic growth that provide opportunities for new forms of livelihood and improved prosperity. But despite producing the bulk of national wealth, cities are also locations of exploitation and unemployment. Within a global economy in particular, the costs and benefits of globalization are unevenly distributed among the people that live in cities. The challenge for cities in the global economy of the twenty-first century is to function not only as engines of economic growth but also as agents of change for greater social justice and urban sustainability.

Demographic Change

One of the most important subsets of interdependence suggested in Figure 4 is that between demographic change and urbanization. Cities are, in a fundamental way, the product of their people. Put another way, the size, composition, and rate of change of urban populations significantly shape the character of urbanization. Yet the condition of cities themselves can in turn influence those characteristics. Crowded and degraded slums, for example, can lead to higher death rates; cities with good amenities tend to attract particularly large numbers of migrants; and border towns and big cities with international ports and airports tend to attract a disproportionate share of

URBAN VIEW 3
Globalization and Cities

The urbanization processes that produce the urban outcomes shown in Figure 4 operate at different spatial scales. The rapidly increasing interdependence of the **world-system** means that the economic and social well-being of cities and the people living in them depends increasingly on complex interactions that are framed at a global scale. Globalization has had profound effects on cities and systems of cities because of the close interaction between global and local forces—a process that has been called *glocalization* or the *global-local nexus*. The process and its outcomes involve **uneven development** both within and among cities. Globalization has led, for example, to the emergence of so-called **world cities**—command centers such as New York, London, and Tokyo that are key players in the new concentrated financial system.

Although globalization is a complicated and controversial topic, we can identify a number of interrelated dimensions—economic, cultural, and political—associated with its processes and urban outcomes. Economic globalization reflects the fact that, although urban, regional, and national circumstances remain very important, what happens in any given city and how it affects, for example, workers or shoppers, is broadly determined by its role in systems of production, trade, and consumption that have become global in scope. The term globalization is usually associated with the growing importance of transnational corporations operating across a number of countries. The activities of these companies, in the spheres of both production and marketing, are increasingly integrated at a global scale in a new international division of labor. Products are made by workers in multiple locations from components manufactured by workers in other places to take advantage of the full range of geographical variations in costs. With a global assembly system, labor-intensive work can be done where labor is cheap, raw materials can be processed near their source of supply, and final assembly can be done close to major urban markets.

Cultural globalization is associated with the development of a broader global culture. This is a controversial idea, but it essentially involves the widespread diffusion of Western values of materialism. Globalization can be seen in the popularity of Hollywood films or the spread of the hamburger. Indeed, Ritzer[4] coined the term "McDonaldization" to denote the ways in which processes of mass consumption are eroding cultural differences among people around the world. It is argued that globalization involves the homogenization of culture—the development of cultural interrelatedness of people throughout the world. New telecommunications systems—especially within urban areas—that allow rapid transmission of information and images to people's homes and electronic devices have facilitated this process. But there has been resistance by some people to these global forces through the assertion of local cultural identities, including various popular social movements seeking greater autonomy at the level of regions and cities.

Political globalization has been associated with the reduced power of national governments to shape their own destinies. In large measure this has been bound up with the globalization of financial markets so that money can now flow rapidly across national boundaries. In addition, because transnational corporations can switch investment from one country to another, public bodies, including city governments, have little choice but to adopt incentive strategies to compete in attracting and keeping internationally mobile companies and their investment in an effort to promote jobs and economic prosperity for workers and residents.

Altogether, these various dimensions of globalization present an enormous challenge to the effectiveness and appropriateness of the economic, political, social, and cultural structures and institutions that had evolved over the two centuries beginning with the Industrial Revolution in the late 1700s. Globalization has brought hybridity and cosmopolitanism to the national cultures and social systems of people in many parts of the world. As a result, many social and cultural structures and norms have been subject to change or obsolescence, while new, transnational, social and cultural phenomena have emerged. Similarly, globalization has intensified transnational economic and political interdependence, prompting the reform of many economic and political institutions and the emergence of new supranational institutions and networks. For many observers, it all amounts to the onset of a second modernity, a distinctive break from the national frameworks of urban and economic development to a globalizing framework with an entirely new set of possibilities and competitive opportunities; and, of course, the attendant conflicts and contradictions of a new world order.

immigrants. Meanwhile, urban economic well-being often mediates the relationships between demographic change and urbanization. So, for example, both birth rates and migration rates depend a great deal on people's perceptions and expectations of economic opportunities.

Political Change

The broad ideological swings and shifts that occur from time to time are an important aspect of the influence of political change on urbanization. One well-known example is the reform movement that emerged in the United States in the 1870s and 1880s in response to a variety of social problems. The reform movement had an important and lasting influence on urban affairs. A very different and more recent example is that of the political shift at the national and international level before the beginning of the "War on Terrorism": the end of the Cold War, which had a marked effect on the economies and workers in some Sunbelt cities that had been heavily dependent on defense-related industries.

In this last example we see again the mediating role of economic change. Indeed, politics has become intimately related to economic development. Economic issues are important to people and almost always appear in local elections, while both the need for local services and the ability to pay for them by people living in cities are functions of local

Myrleen Ferguson Cate/Photo Edit

FIGURE 6 People disembarking the ferry at Coronado Island, with the skyline of San Diego, California, in the background. The economic and demographic profile of San Diego has been influenced by its military bases, climate, scenery, recreational opportunities, and proximity to the Mexican border.

economic prosperity. As suggested by the direction of the arrows in Figure 4, urbanization also directly affects political change in some ways. One example is the way that coalitions of urban voters shaped the basis of modern party politics at the national level during the 1930s and 1940s in the United States. Another is the electoral significance of suburban voters who have more recently formed an important foundation of support for the Republican Party. Urbanization also affects political change indirectly through people's perceptions of the problems associated with various dimensions of urban change, because their perceptions inform and frame many of the issues that are contested in the political arena.

Cultural Change

We can find parallel examples of the interdependence of urbanization and cultural change. The broad cultural shift of the "postmodernity" of the 1980s, for example, brought, among other things, a renewed interest in the past that has engaged many people and found expression in urban form through historic preservation and the recycling of architectural styles. Meanwhile, urbanization has contributed to cultural dynamics through the youth subcultures—distinguished by features such as clothing, slang, music, and vehicles, including skateboards—that have flourished in certain urban settings. Other aspects of the interdependence between urbanization and culture involve still further processes of change. The materialism of mainstream American culture, for example, has affected urbanization through home ownership trends and patterns of residential development, but it has depended on processes of economic change that have been associated with workers' wages not keeping pace and rising personal debt. It is not only (or always) economic change that is important in mediating such interrelationships. Demographic change is also important, the "baby boom" generation having been the innovators and "carriers" of successive aspects of cultural change from the counterculture of the 1960s to the yuppie materialism of the 1980s and the "suburban bling" of the private master-planned developments of the 2000s.

Technological Change

At the broadest level, and in parallel with the overall evolution of capitalism, we should recognize that the economy has been carried along by a succession of **technology systems** that have been fundamental to the changing conditions that producers have had to confront. These are indicated in Figure 5 in terms of the clusters of energy sources, transportation technologies, and key industries that characterized each system:

- Early mechanization based on waterpower and steam engines; the development of cotton textiles, pottery, and iron working; and the development of river systems, canals, and turnpike roads for the assembly of raw materials and the distribution of finished products for sale to consumers.
- The development of coal-powered steam engines, steel products, railroads, world shipping, and machine tools that affected workers everywhere.
- The development of the internal combustion engine, oil and plastics, electrical and heavy engineering, cars, aircraft, radio, and telecommunications whose legacy continues for people in cities today.
- The exploitation of nuclear power, the development of limited access highways, durable consumer goods industries, aerospace industries, electronics, and petrochemicals were a further breakthrough affecting people in cities.

- The most recent (and still incomplete) technology system, which is based on microelectronics, digital telecommunications, robotics, biotechnology, fine chemicals, and information systems, that is affecting every aspect of urban life.

These technology systems gave shape and direction not only to the evolving national economy but also to the pace and character of urbanization and urban life. They have been imprinted, layer by layer, on American cities and the people living in them.

There are many other, more specific examples of the interdependence of urbanization and technological change, although it is often difficult to disentangle cause and effect. Many technological changes, although not strictly causing—or being caused by—changes in urbanization, have been important preconditions for change: the streetcars that facilitated the first widespread suburbanization of urban residents in the United States, for example. To take a very different example, the broader web of interdependence among technology, culture, economics, demographics, and so on implied by Figure 4 is well illustrated by the impact of new birth control technology (the contraceptive pill) in the mid-1960s. Not only did this technology help put an end to the baby boom, it also helped change attitudes toward sex, marriage, and female participation in the labor force—simultaneously affecting several dimensions of urban life.

Environmental Change

The complexity of the interactions between urbanization and environmental change creates problems of local through global proportions. The area of Earth's surface needed to absorb the waste products of the people who live in a large city is likely to exceed that city's boundaries—the ecological footprint—although this can depend on the fuels used for heating, energy generation, and manufacturing; the amount of motorized traffic; the technologies used for disposing of solid and liquid wastes; and local climatic conditions. Most large cities cannot assimilate their waste products, and by burning them they contribute to pollution. The role of people in cities in producing greenhouse gas emissions from increased car use and coal-fired power plants has global climate change implications. At the local scale, cities involve changes in land use and land cover that can produce a diversity of environmental problems. In the United States, Europe, and Russia, for example, countless **brownfields**—abandoned or underused traditional manufacturing facilities with actual or potential contamination—are a legacy of a weakly regulated early industrialization process that now complicates redevelopment efforts in many central cities.

Social Change

Still following Figure 4, we move now to some brief illustrations of the interdependence between urbanization and social change. Here we can cite the changes that have occurred over the past 30 years in terms of people's behavior toward racial minorities—changes that have carried over to affect educational achievement, occupational composition, and, ultimately, urban residential patterns. Black and Hispanic suburbanization, for example, is largely attributable to such changes. Urbanization can also induce social change. The physical and socioeconomic attributes of urban settings, for example, foster certain behavioral changes, such as the social isolation and withdrawal that seems to be generated among the "lonely crowd" of central city districts. The most important changes in American society, for example, have been changes in social status that have been driven by occupational changes resulting from the structural transformation of the economy: the growth of the middle-income workers that accompanied the emergence of organized capitalism and their subsequent decline with the shift toward advanced capitalism (Figure 5).

Key Terms

advanced capitalism and flexible production systems
competitive capitalism
deindustrialization
egalitarian liberalism
first modernity
globalization and globalized capitalism
managed capitalism (Keynesianism)
neoliberalism
new international division of labor
organized capitalism and Fordism
second modernity
territoriality
uneven development
urban system
urbanism

Review Activities

1. Consider the statement that "cities are simultaneously the products and the shapers of economic, social, and political change" and review the discussion of such changes. Can you identify examples that illustrate how cities are both the products and the shapers of change? Can you think of any additional examples that affect young people in particular?

2. Begin an *electronic portfolio* of your own work that builds on your reading. An e-portfolio can be compiled in many ways; yours should reflect your own reactions to the material in the text and your own urban interests, experiences of cities, and impressions of city life.

   It is a good idea to compile the e-portfolio in such a way that you can easily add new material. If you do not have e-portfolio software, word-processing programs will let you integrate images, sound, online links to YouTube and other Web sites, and other multimedia in a fairly seamless manner. Another possibility would be to build your e-portfolio using some sort of Web publishing tool, with the ultimate goal of displaying your work on a Web site. You could also consider creating your e-portfolio using Microsoft PowerPoint or some other presentation software.

The content of the e-portfolio should consist of three elements (although they do not necessarily have to be kept separate):

a) A summary of the most important aspects of the material, together with a record of the questions that the material raises for you. Note how the material is related to the overall framework provided by Figure 4 and how it is related to topics and ideas covered elsewhere. Note any issues that you think need further clarification and record people's ideas and opinions that you believe are particularly provocative or challenging.

b) Online, library, and other material that illustrates or explains the issues you have identified in item (a) above. This material might include maps, photos, mp3 music, or videos related to the key sources and suggested readings that are available from your library or on the Web, such as newspaper or magazine articles, reports from government or research institution Web sites, documentaries and mockumentaries on YouTube, and so on.

c) Additional material that reflects your own interests and reactions to the themes and ideas that you have encountered. This material might take the form of prose commentary, short essays, poetry (your own or others'), quotations or short extracts, drawings, photographs, brief video animations, short sound recordings, or data in the form of maps, charts, or graphs that you can digitize and include in your e-portfolio.

Subsequent follow-up sections will contain specific suggestions on what you can do to make your e-portfolio an interesting project.

3. Get in the mood—watch a movie! There are many movies, documentaries, and mockumentaries with "urban" themes that you can download, rent, or borrow. A great documentary is Julien Temple's 2010 BBC production called "A Requiem to Detroit?" that uses contemporary music and images to tell the story of urban decline in the Motor City—as described by the BBC: "a vivid evocation of an apocalyptic vision: a slow-motion Katrina that has had many more victims."

Log in to **www.mygeoscienceplace.com** for self-study quizzes, *MapMaster* layered thematic and place name interactive maps, *Urban View* Google Earth™ tours, key resources and suggested readings, related websites, "In the News" RSS feeds, and additional references and resources to enhance your study of urbanization and urban geography.

NOTES

1. R. J. Johnston, "The World Is Our Oyster," *Transactions of the Institute of British Geographers* 9 (1984): 444.
2. See the Heidelberg Project's Web site at *http://www.heidelberg.org/* and Bruner Foundation at *http://www.brunerfoundation.org/rba/pdfs/2005/5_Heidelberg.pdf.*
3. See the U.S. Department of Commerce census web pages (*http://2010.census.gov/2010census/index.php*).
4. G. Ritzer, *The McDonaldization of Society: Revised New Century Edition* (Thousand Oaks, Calif.: Pine Forge Press, 2004).

The Origins and Growth of Cities and Urban Life

From Chapter 2 of *Urbanization: An Introduction to Urban Geography*. Third Edition. Paul L. Knox, Linda McCarthy.

Linda McCarthy

The Origins and Growth of Cities and Urban Life

A fascinating aspect of urban geography involves trying to understand the processes that led to the development and growth of individual cities and systems of cities. Because urban areas are the result of a long evolution, we need to develop a historical perspective. We can see history's powerful legacy in the surviving fragments that document the sequence associated with the people and events that helped produce today's towns and cities. As Ildefons Cerdà, a nineteenth-century Spanish town planner, put it: "Our cities are like historical monuments to which every generation, every century, every civilization has contributed a stone." Since the evolution of the first cities about 5,500 years ago, changes in social, cultural, economic, political, technological, and environmental processes—including long-distance trade, overseas colonization, and industrialization—have helped fuel urban growth and change. These changes are visible in the internal structure and in the land uses of people in cities and in the development of regional, continental, and, later, global urban systems, as well as world cities.

LEARNING OUTCOMES

After reading this chapter, you should be able to:

- Explain what makes a place "urban."
- Describe the possible explanations for the emergence of urban economies and societies.
- Describe how and why Europe developed a city-based economy after the Dark Ages.
- Outline and explain the impact of the Industrial Revolution on towns and cities in Europe and North America.

CHAPTER PREVIEW

This chapter follows the evolution of cities from their earliest origins about 5,500 years ago through the *Industrial Revolution* that began in the English Midlands in the mid-1700s. During this long span of urban development and redevelopment, changes in social, cultural, economic, political, technological, and environmental processes—including merchant capitalism, overseas colonization, and industrialization—helped drive urban growth and change. The impacts of these changes can be seen in the internal structure and patterns of land use of people in cities, and in the development of regional, continental, and later global urban systems, as well as world cities.

The term **urban system** refers to the complete set of urban settlements of different sizes that exists within a given territory. Territorial limits set the bounds of an urban system, though only a global scale can be justified as defining a true system in the sense that it captures all the functional relationships among cities. Conventionally, however, urban systems are studied at the scale of regions or countries.

With cities and urban life being such relatively recent features in the long span of human existence, we need to consider first the environmental, demographic, and other preconditions needed before cities could even begin to emerge. We then review the various theories of urban origins that together offer the reasons for why cities actually

Because cities are the result of a long evolution, urban geographers need to develop a historical perspective. We can see history's powerful legacy in the surviving fragments that document the sequence associated with the people and events that helped produce any city, such as Rome, Italy. The white marble Arch of Septimus Severus in the Roman Forum was built in 203 C.E. to commemorate this emperor's victories in Parthia (present day Iran).

originated. The earliest towns and cities were developed independently in regions of the world where people were transitioning to agricultural food production. Five regions provide the earliest evidence for urbanization and urban civilization: Mesopotamia; Egypt; the Indus Valley; northern China; and the Andes and Mesoamerica. A look at the internal structure of these cities—street patterns, religious precincts, different neighborhoods, and so on—reveals a great deal about their evolution and the political, economic, and social changes that went on in them.

Urbanization spread out from the five regions of urban origin so that by about 1000 C.E. successive generations of city-based empires—including those of Greece, Rome, and Byzantium—had emerged in Southwest Asia, China, and parts of Europe. But urban expansion was a precarious and uneven process. For example, although urbanization continued in other parts of the world, the **Dark Ages** that followed the collapse of the Roman Empire in Western Europe was a time of stagnation and decline in economic and city life. Not until the eleventh century did the regional specializations and long-distance trading patterns emerge that provided the foundations for a new phase of urbanization based on merchant capitalism. Colonization and the expansion of trade around the world eventually allowed Europeans to shape the world's economies and urban societies. The Industrial Revolution later generated new kinds of cities—and many of them. Together, European colonization and the Industrial Revolution created unprecedented concentrations of people in cities that were connected in networks and hierarchies of interdependence around the world.

THE DEFINITION OF A CITY

Although most people recognize a city when they see one, no single definition can apply to all cities across space or even to the same city through time. Certainly a description of Mexico City today by a local resident would be very different to Bernal Diaz del Castillo's description of Tenochtitlán during the Spanish conquest of the Aztecs (see Urban View 1 entitled "Bernal Diaz del Castillo's Description of Tenochtitlán in 1519").

Wheatley's definition of **urbanism** captures the remarkable social and political changes surrounding the emergence of cities that resulted in a

> particular set of functionally integrated institutions which were first devised . . . to mediate the transformation of relatively egalitarian, ascriptive, kin-structured groups into socially stratified, politically organized, territorially based societies.[1]

Sjoberg's definition highlights important physical and economic attributes that define a city:

> It is a community of substantial size and population density that shelters a variety of nonagricultural specialists, including a literate elite.[2]

V. Gordon Childe[3] attempted to characterize the distinctive features of cities with a list of distinguishing features of urban civilization:

- *Size.* Settlements were significantly larger in population size than anything that had existed previously.
- *Structure of the population.* Occupational specialization—with the transition from the old agricultural order—meant that the employment of full-time administrators and craftspersons was possible. Consequently, residence rather than kinship became the qualification for citizenship. Inevitably, the rule of the priest-kings, who guaranteed peace and order, involved social stratification.
- *Public capital.* The emergence of public capital allowed monumental public buildings to be erected and full-time artists to be supported.
- *Records and the exact sciences.* The need to keep records promoted the beginnings of a written script and mathematics, both of which became intimately bound up with urban civilization.
- *Trade.* By no means an urban innovation, a network of trade routes has become a hallmark of urbanization.

URBAN VIEW 1

Bernal Diaz Del Castillo's Description of Tenochtitlán in 1519

In the middle of Lake Texcoco, Tenochtitlán was the island capital of the Aztecs (just north of modern-day Mexico City). At an estimated 200,000 people, it was one of the largest cities in the world at the time, and larger than most European cities except Paris and Constantinople.

Bernal Diaz del Castillo was a conquistador who wrote an eyewitness account of the conquest of the Aztecs in Mexico by the Spanish.[4] The reason Montezuma initially welcomed the Spanish into Tenochtitlán may have been that he thought that Hernán Cortés was the fair-skinned god Quetzalcoatl whose return was predicted by Aztec prophecy. Diaz served as a swordsman under Cortés and described their entry into Tenochtitlán on November 8, 1519:

> As we neared the vicinity of Tenochtitlán, we saw many towns and villages built on artificial islands in the lakes. We were amazed and said that it was like the enchantments they tell of in Spanish legend, on account of the great towers and temples and buildings rising from the water, and all built of masonry. And some of our soldiers even asked whether the things that we saw were a dream.

URBAN VIEW 1
Bernal Diaz Del Castillo's Description of Tenochtitlán in 1519 (*continued*)

Library of Congress

FIGURE 1 The Spanish conquest in 1519 of the Aztec city of Tenochtitlán (just north of modern-day Mexico City) showing the three causeways into the city and the great temple.

We entered Tenochtitlán along a wide causeway that was crowded with people who came out of the city to see us; and the towers and temples were full of people, as well as the canoes from all parts of the lake (Figure 1). It was not surprising, because they had never before seen horses or men like us.

When Cortés was told that the Great Montezuma was approaching, he dismounted from his horse, and they paid great reverence to one another. We were offered accommodation in some large houses that had belonged to Montezuma's father and given a sumptuous dinner. That night our orders were to be much on the alert, both the cavalry and all of us soldiers. So this was our lucky and daring entry into the great city of Tenochtitlán!

The next day we climbed to the top of the great temple. From there we saw the three causeways that led into the city and the aquaduct that supplied the city with fresh water (Fig. 1). On the great lake we saw a multitude of canoes, some coming with supplies of food, and others returning loaded with cargos of merchandise from the marketplace.

There were crowds of people in the marketplace, some buying and others selling, so that the murmur and hum of their voices and the words that they used could be heard from a great distance. Some of the soldiers among us who had been in many parts of the world, in Constantinople, and all over Italy, including Rome, said that they had never before seen so large a marketplace that was so full of people, and so well organized and regulated.

Each kind of merchandise was sold in a different part of the market. There were dealers in gold, silver, and precious stones, feathers, and embroidered goods. There were slaves for sale, both men and women, tied to long poles, with collars around their necks so that they could not escape. There were traders who sold great pieces of cloth and twisted thread, and others selling ropes or sandals or pottery or cacao. Yet others sold sweet cooked root vegetables. In another part of the market there were the skins of lions and tigers, otters and jackals, deer and mountain cats, some tanned and others not. One part of the market was full of people selling vegetables, fruits, and herbs, and meat, fowl, and fish, as well as honey and nut paste. But why do I waste so many words in recounting what they sell in that great market?—for I will never finish if I tell it all in detail.

Eyewitness accounts like this and the archaeological evidence are all that remain of Tenochtitlán. Cortés returned two years later and laid siege to the city. He cut off the causeways and destroyed the aqueduct to prevent food and water from getting to the inhabitants. By the end of the eight month siege, the city had been almost completely destroyed by cannon shot and fires, and many of the Aztec people had died from smallpox, a new disease brought from Europe.

Childe used the term *urban civilization* because civilization and cities historically have gone hand in hand—the Latin word *civitas* (cities) is the word from which *civilization* is derived. From the beginning, cities have been crucibles of innovation that have produced some of the most incredible breakthroughs in human achievement. Ancient Sumer (in southern Iraq) is called the Cradle of Civilization because of the countless inventions of the people in the earliest cities there. The legacies of these early urban innovators continued throughout subsequent civilizations and are seen in contemporary cities—in the use of writing, mathematics, the wheel, and recording time in multiples of 60.

PRECONDITIONS FOR URBANIZATION

The preconditions necessary for cities to emerge came about with the transition of people from mobile food collection—hunting, gathering, and fishing—to sedentary food production based on agriculture. The increased volume and reliability of an agricultural food supply allowed higher numbers and densities of people to live permanently in one place. This population increase promoted the proliferation of agricultural villages.

The villages had to be located in regions where the environmental conditions—climate, water supply, topography, natural resources, and soil conditions—were favorable for agriculture. Early breakthroughs in technology and in farming practices—innovations in river and water management, crop and animal strains, and food transportation and storage techniques—were needed to support improvements in food production. Increasingly complex social organizational structures were required to handle the growing population and exchange of agricultural and other products among the village communities.

THEORIES OF URBAN ORIGINS

Although it would have been difficult for cities and urban life to emerge in the absence of certain basic environmental, demographic, social, and other preconditions, there are various explanations for why cities originated. Although no one theory provides a full account, each allows insights into the role of different factors in promoting early urbanization.

Agricultural Surplus

Archaeologists, including V. Gordon Childe[5] and Sir Leonard Woolley,[6] argued for the importance of an agricultural surplus. Once early farmers could each produce more food than was needed to feed their own families, they could support a growing sedentary population. The need to administer the agricultural surplus called for the more centralized structures of social organization found in cities. New, stratified social structures and institutions were needed to assign rights over resources, exact tributes and impose taxes, deal with the ownership of property, and administer the formal exchange of goods. Elite groups stimulated urban development because they used their wealth to build palaces, arenas, and monuments as displays of their power and status. This building construction demanded a greater degree of occupational specialization in nonagricultural activities, such as people working in crafts, engineering, and administration, that could be organized effectively only in an urban setting.

This interpretation has been criticized as simplistic—an agricultural surplus alone may not have been enough to trigger all the societal and other changes necessary to produce cities. Some experts disagree on the cause-and-effect relationship and believe that fundamental changes in social organization would have been required before an agricultural surplus could be produced.

Hydrological Factors

Karl August Wittfogel[7] pointed out that many early cities emerged in areas of agriculture that often depended on irrigation and the control of the regular spring floods. He contended that elaborate irrigation projects demanded that people adopt new **divisions of labor**, cooperation on a large scale, and the intensification of cultivation. These changes would stimulate urban development by promoting occupational specialization, a centralized social organization, and population growth based on the production of an agricultural surplus.

Again, this interpretation has been criticized by those who believe that major changes in social organization would have been necessary, if not before, then at least in tandem with, the development of major irrigation projects. Other experts have questioned whether a complex social organizational structure was even necessary in order for people to undertake large-scale irrigation. Still others have pointed out that not all early cities, including some in Mesoamerica, depended on massive irrigation schemes.

Population Pressures

Ester Boserup[8] believed that increasing population densities and/or a growing scarcity of wild food sources from hunting and gathering—that previously had provided adequate levels of subsistence for people from a relatively low workload—brought on the transition to agricultural food production and urban life. Again, the relationship is unclear in terms of whether food production and urban life caused—or were the result of—increased population densities. In certain cases demographic growth pressures may have disturbed the balance between population and resources and forced some people to move to areas with more marginal environmental conditions for agriculture. This scenario could have promoted early breakthroughs in agricultural technologies and practices or the establishment of nonagricultural activities, such as trade, defense, or religion, that would have supported the establishment of further urban settlements.

Trading Requirements

Some experts, observing how countless urban centers had evolved around marketplaces, have interpreted the emergence of cities primarily as a function of long-distance trade.[9] Participation in large-scale trading networks would call for a system to administer the formal exchange of goods that in turn would promote the development of centralized structures of social organization. Increasing occupational specialization by people and growing economic competitiveness by cities would encourage more urban development. What remains unclear is the extent to which trade was the cause or the consequence of urban development.

Defense Needs

Some theorists, including Max Weber,[10] have contended that cities originated because of the need for people to gather together for protection inside the safety of military defenses. Wittfogel[11] pointed out that a comprehensive system of defense was needed to protect valuable irrigation systems from attack. But despite widespread evidence of walls and other fortifications, not all early cities had defenses. As Wheatley acknowledged, although not necessarily being the initial reason for the evolution of cities, "warfare may often have made a significant contribution to the *intensification* of urban development by inducing a concentration of settlement for purposes of defense and by stimulating craft specialization."[12]

Religious Causes

The presence of temples and other religious structures reflects the importance of religion in the lives of the people living in the earliest cities. Sjoberg[13] suggested that the control of altar offerings by the religious elite conferred economic and political power that allowed this group of people to influence the social changes that helped initiate urban development. Wheatley[14] maintained that a pervasive institutional structure like religion would have been needed to reinforce for people the changes in social organization that were associated with the economic, technological, and military transformations involved in early urban growth.

A More Comprehensive Explanation

More recently, the consensus has been that our understanding of the origin of cities and urban life should be based on a combination of these separate yet interrelated explanatory factors. As Wheatley put it:

> It is doubtful if a single autonomous, causative factor will ever be identified in the nexus of social, economic, and political transformations which resulted in the emergence of urban forms.[15]

Understanding both the complexity of the many changing processes and the interactions among them are more important than identifying the cause-and-effect relationship for any one explanatory factor. The desire for this kind of comprehensive understanding reflects a growing conceptualization that the origin of cities represented a gradual transformation involving incremental change affecting people over time rather than an abrupt urban revolution.

URBAN ORIGINS

The earliest towns and cities developed independently in regions of the world where people had transitioned to agricultural food production. Five regions provide the earliest evidence for urbanization and urban civilization (Figure 2). Over time, the regions of urban origin produced successive generations of urbanized world empires.

Mesopotamia

Mesopotamia (the land between the Tigris and Euphrates rivers), in the area of modern Iraq, provides the earliest evidence for urbanization—from about 3500 B.C.E. This was the eastern

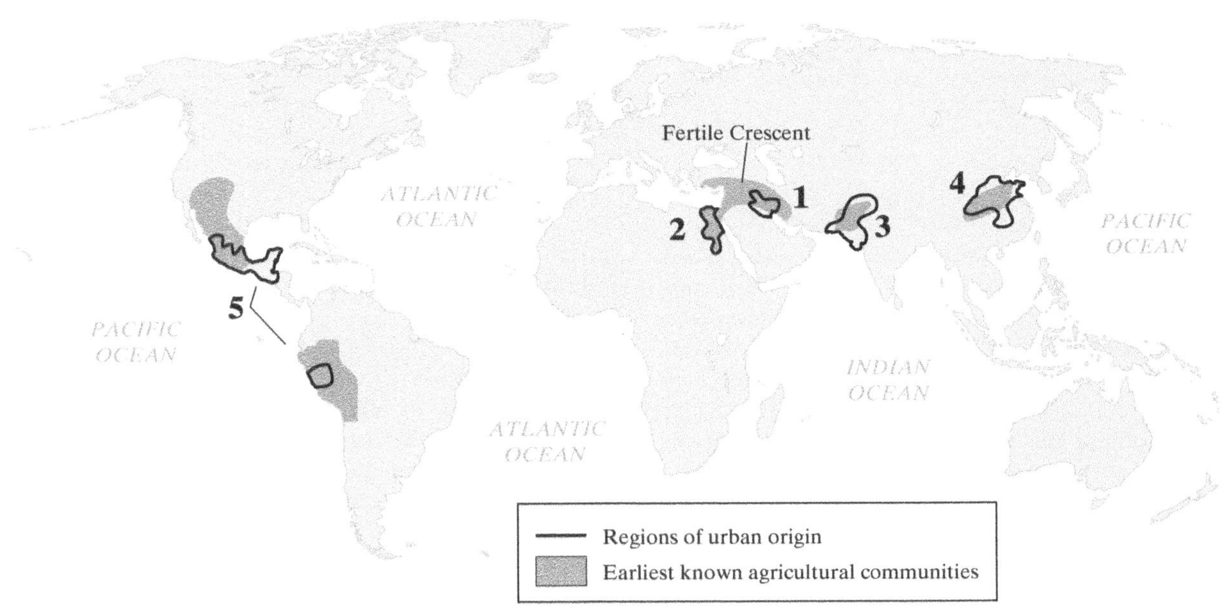

FIGURE 2 The regions of urban origin (in heavy outline) and the earliest known agricultural communities (shaded areas): (1) Mesopotamia, (2) Egypt, (3) Indus Valley, (4) Northern China, and (5) Andes and Mesoamerica.

part of the so-called **Fertile Crescent** (Figure 2). The significant growth in size of some of the agricultural villages on the rich alluvial soils of the river floodplains formed the basis for the large, relatively autonomous, and often-rival **city-states** of the Sumerian Empire from about 3000 B.C.E. They included Ur, in southern Iraq, the capital from about 2300 to 2180 B.C.E., as well as Eridu, Uruk, and Erbil (ancient Arbela). These fortified city-states contained tens of thousands of people, social stratification, with religious, political, and military classes, innovative technologies, including massive irrigation projects, and extensive trade connections. By 1885 B.C.E., the Sumerian city-states had been taken over by the Babylonians and then the Neo-Babylonians, who governed the region from their capital city, Babylon.

FIGURE 3 Erbil (ancient Arbela) in northeast Iraq is located atop a *tell*, a mound, visible as a hill rising high above the surrounding plain—representing the remains of generations of sun-dried mud-brick buildings—that reflects millennia of human occupation in one place involving new structures being built upon the collapsed rubble of demolished ones. The 100-foot high Erbil tell is believed to represent perhaps 6,000 years of continuous occupation.

Egypt

The Fertile Crescent stretched in an arc as far west as Egypt, which became a unified state from about 3100 B.C.E. (Figure 2). Large irrigation projects were built to control the Nile's waters for agricultural and other uses. Despite the importance of early urbanization in the Nile Valley, only limited archaeological evidence of the first towns and the people who lived in them has survived.[16] Internal peace in Egypt meant that there was no need for people to occupy the same site continuously to justify massive investments in a city's defensive fortifications. Given this potential for urban mobility, the lifespan of the largest city, the capital, was relatively short. Each pharaoh was free to locate a new capital at any site he selected for his tomb. After his death the city was usually abandoned to the priests. The main surviving structures are the stone tombs and temples that were the primary focus of construction. Few of the other buildings—public, commercial, and residential—survive, having been constructed of more perishable materials, such as sun-dried brick and timber. Without a tradition of long-term building and rebuilding, Egyptian cities did not generate the rich evidence of urban development and redevelopment characteristic of the tells of Mesopotamia (see Figure 3). Nevertheless, between 2000 and 1400 B.C.E., urbanization continued with the founding of capital cities like Thebes, Akhetaten (Tell el-Amarna), and Tanis.

The Indus Valley

As in Mesopotamia, but later, by about 2500 B.C.E., the Indus Valley in modern Pakistan contained relatively large urban communities that were supported on the fertile alluvial soils and extensive irrigation systems of the river plains. This region had a single ruler and two capital cities: Harappa in the north, after which the civilization was named, and Mohenjo-Daro in the south (Figure 2). A network of trade extended as far as the Sumerian Empire in Mesopotamia. Much about Harappan civilization, including its origin, is unknown, partly because the Indus Valley script that the people used has not been deciphered.

Northern China

The Shang dynasty developed in the fertile plains of the Huang He (Yellow River) by about 1800 B.C.E. (Figure 2). As in the Fertile Crescent and Indus Valley, Shang cities, such as Chengchow (Zhengzhou) and An'yang, the capital by 1384 B.C.E., were supported by irrigated agriculture. There is evidence of social stratification and occupational specialization, including a hereditary leader as well as a warrior elite with absolute control over the agricultural peasants.

The Andes and Mesoamerica

The oldest known center of urban civilization in the Americas is a group of 18 settlements in the Central Andes situated on a dry desert terrace overlooking the green valley of the Supe river in present-day Peru. The sacred city of Caral-Supe was the principal settlement, and it has been dated to 3000 B.C.E. In contrast, the earliest urban settlements in Mesoamerica date to only about 500 B.C.E. (Figure 2). The Zapotec civilization was based on small-scale irrigated farming and was centered on Monte Albán in Mexico. This city was surrounded by a wall and contained pyramids and temples. The later city of Teotihuacán, near modern Mexico City, was larger and at its height—between about 300 and 700 C.E.—contained about 200,000 people.

Mayan cities such as Tikal and Uaxactun date from about 100 B.C.E. The Mayan cities were located in lowland areas of modern Mexico, including the Yucatan peninsula, as well as in Guatemala and Belize. Some reached populations as large as 50,000. These settlements were the centers of small states that united occasionally into loose confederations.[17] A supreme ruler and an urban elite, including religious and military groups, administered the settlements. Some of the breakthroughs in technology and farming practices in the other regions of urban origin are not found in Mesoamerica. The agriculture here was based on maize (corn) cultivation that did not require metal agricultural implements, domesticated animals to pull plows, or extensive irrigation systems. Mayan civilization reached its peak between about 300 and 900 C.E.; by the time of the Spanish conquest in the sixteenth century, it had been in decline for several centuries because of droughts, warfare, and population pressures.

INTERNAL STRUCTURE OF THE EARLIEST CITIES

A common approach to examining the internal structure of cities is to identify whether the layout of an urban area was largely unplanned or deliberately planned by one or a number of people. This categorization distinguishes between cities that evolved in an unplanned—**organic growth**—process and those that were laid out in a predetermined way based on some planned approach, such as a **gridiron street pattern**. The planned layout of streets and transportation arteries may indicate the presence of significant central control from an early stage. But an unplanned street layout does not mean the absence of a central authority. In Mesopotamia, for example, the street patterns were not planned but reflected the organic street layouts that survived as the early agricultural villages grew into towns; the archaeological evidence of massive walls and irrigation systems, however, indicates central planning of defense and water management (see Urban View 2 entitled "Internal Structure of the Earliest Cities").

A city's internal structure is never static, being the product of development and redevelopment by people over time. Cities that were founded with a strong hand of planning can contain later sections of organic growth, as in some cities of Roman origin, such as London, where later unplanned urban

URBAN VIEW 2
Internal Structure of the Earliest Cities

Mesopotamia

Sir Leonard Woolley's excavations at Ur in the 1920s and early 1930s revealed the organic growth pattern characteristic of the Mesopotamian city-states (Figure 4). Woolley described Ur in about 1700 B.C.E.:

- *The walled city* contained about 35,000 people. The mud-brick wall was about 25-feet high and at least 77-feet thick. It was oval shaped and about three-quarters of a mile long by half-a-mile wide. The Euphrates River ran along the wall in the west, a navigable canal along the north and east, and harbors were in the north and west. Near the center were the palaces and residences of the royal officials. Social differentiation was less orderly in other sections of the city. Woolley's description of an excavated middle-income neighborhood illustrates how income and climate were reflected in the size and design of the housing, while the unplanned nature of the streets provided a measure of privacy and defense:

  > The unpaved streets are narrow and winding, sometimes mere blind alleys leading to houses hidden away in the middle of a great block of haphazard buildings; large houses and small are jumbled together, a few of them flat-roofed tenements one storey high, most of them of two storeys . . . The basic plan was of a house built round a central courtyard . . . that gave light and air to the house.[18]

- *A religious area* measuring about 270 by 190 yards, surrounded by a huge mud-brick wall, was located in the northwest of the walled city. A 68-foot high terraced tower (ziggurat) containing the shrine of Nannar, the Moon god, the owner of the city-state, would have been visible for miles. This religious and administrative core—reserved for the priests and royal household—had a great courtyard surrounded by temples, a court of law, tax offices, and storage buildings for religious offerings.
- *The outer city* or *suburbs* comprised the remainder of the city-state. The houses and farms there contained an estimated 200,000 people.

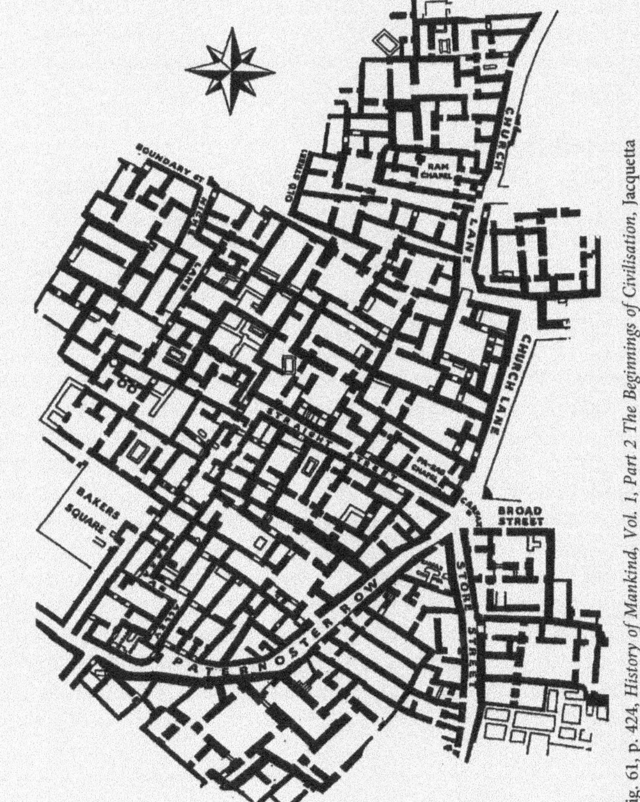

FIGURE 4 A residential section of Ur to the southeast of the religious area in about 1700 B.C.E. Note how the narrow and winding streets and the irregular size and shape of the lots reflect an unplanned—organic growth—process of urban development in Mesopotamia.

Fig. 61, p. 424, *History of Mankind, Vol. 1, Part 2 The Beginnings of Civilisation*, Jacquetta Hawkes & Sir Leonard Wooley. Paris: UNESCO, 1963.

Egypt

Towns, and even capital cities, in Egypt remained small because most people were agricultural workers who lived on the land. The cities contained the markets and the residents who worked

(continued)

URBAN VIEW 2
Internal Structure of the Earliest Cities (*continued*)

in retail, craft, government, and religious activities. With civil war and armed invasions rare, urban settlements in Egypt had little need for defensive walls or fortifications. The capital was the home of the pharaoh and the royal court. Akhetaten (modern Tell el-Amarna), built about 1400 B.C.E., is representative of capital cities in Egypt. This city ran for about 5 miles along the east bank of the Nile and was about a half mile to a mile wide. Social stratification is reflected in the city's layout:

- A walled temple and palace at the center, with other temples, government offices, military barracks, and storage buildings nearby.
- Areas of slum housing throughout the city.
- Northern and southern suburbs.
- A workmen's village based on a gridiron plan to the east of the city.

The roughly rectangular city blocks were laid out when the city was founded, with infilling left to the residents. The wealthiest residents took the best lots that fronted onto the two main streets that ran parallel to the river. The typical middle-income house was built in the center of a walled enclosure and had a porch and central living room.

The Indus Valley

Although not the first to use a gridiron plan, the Harappan cities likely were the first planned towns because they are believed to be the first system of cities that used the same town planning approach (Figure 5). Despite being located hundreds of miles apart, cities like Harappa and Mohenjo-Daro shared similarities in their basic internal structure. Each covered at least a square mile in area and contained about 35,000 people. An imposing walled citadel built on a mud-brick platform was located to the west of each city. In contrast to the religious area of Mesopotamian cities, this citadel did not contain a dominant religious building like a ziggurat. The citadel contained some structures that might have had a ritual use (such as the bath at Mohenjo-Daro), as well as buildings used for administrative purposes (including offices and grain storage buildings). The citadel's western location would have allowed urban residents to see the rooftop civic and ceremonial gatherings silhouetted dramatically against the backdrop of the setting sun.[19] The main east-west streets led to the citadel. The houses varied from small one-story one-roomed buildings to larger two-story houses with central courtyards. The workmen's quarters contained rows of identical two-roomed houses.

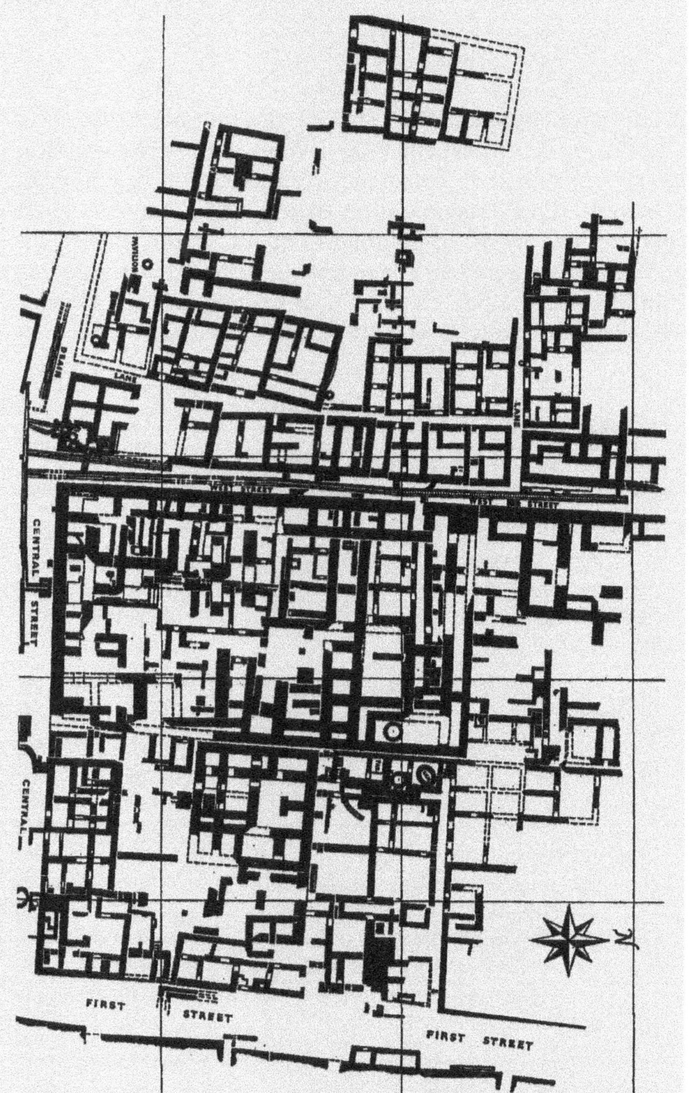

FIGURE 5 Part of the town plan of Mohenjo-Daro. Note how the use of a gridiron plan in Mohenjo-Daro (as in Harappa and other Indus Valley cities) reflects a planned approach in which fairly wide and straight main streets intercept cross streets at right angles to form large city blocks that contained a number of houses.

Fig. 73, p. 456, *History of Mankind, Vol. 1, Part 2 The Beginnings of Civilisation*, Jacquetta Hawkes & Sir Leonard Wooley. Paris: UNESCO, 1963.

Northern China

The archaeological evidence is relatively limited, but excavations at the capital city of An'yang have revealed thick walls of beaten earth surrounding the city. At the center was a walled palace. The nearby houses of the wealthy were wooden and built on raised platforms of beaten earth. Poorer residents lived in pit-dwellings. The city's layout was probably planned because all the buildings that were excavated were oriented toward the north.

The Andes and Mesoamerica

The most imposing buildings in Mayan cities were the religious and other ceremonial structures, such as the tall pyramids that were planned around broad plazas. The temples and the palaces and residences of the ruler and the military and religious elite were laid out at the center of the city. Nearby were the homes of the wealthy citizens. Farther out were low-density areas of simple thatched wooden huts that housed the majority of the people who worked in agriculture or crafts.

growth during the medieval period obliterated the gridiron street pattern of the former Roman core. Similarly, cities that evolved with an organic growth pattern can have later planned sections, as in Vienna where the planned redevelopment of the area of the former wall in the late nineteenth century contrasts with the organic growth of the earlier medieval street pattern.

URBAN EXPANSION FROM THE REGIONS OF URBAN ORIGIN

The spread of urbanization outward from the regions of urban origin involved **uneven development**—over time and within and between different parts of the world (Figure 6).[20] There was a progression of growth, expansion, and succession of early urban empires that was gradual and incremental. The spread of urbanization was a precarious process in which many civilizations lapsed into ruralism before being revived or recolonized.

The Persians helped spread cities from Mesopotamia to Central Asia. To the north, the Assyrians established a system of cities that extended west from their capital city, Assur, to Syria and Asia Minor (the Asian part of Turkey, also called Anatolia). By about 1700 B.C.E., other groups, including the Hittites, had displaced the Assyrians and established their own cities. Farther west, by 1600 B.C.E., Mycenaeans had founded urban settlements, including the legendary cities of Mycenae and Thebes in modern-day Greece.

The small "Canaanite" city-states that grew up in what is now Israel and Syria by 2000 B.C.E., such as Tyre, Beirut, Jericho, Gaza, and Damascus, were taken over by the Egyptians and Hittites. After 1200 B.C.E., with the collapse of Egyptian and Hittite control, the Israelites founded small urban centers that grew into large cities such as Jerusalem. To the west, the Phoenicians helped spread urbanization by sea as far as Spain.

Farther east in what became modern India and Pakistan, it took a thousand years for urban life—dislodged by the Aryan invaders in about 1750 B.C.E.—to reestablish itself. Hindu cities were founded, and from the end of the fourth century B.C.E., the Maurya Empire built cities across India and laid the foundations for urban life throughout southeast Asia. The Arab invasions of the eighth century C.E. began the period of Muslim rule in India; cities like Lahore were established, and Delhi became an important administrative and cultural center.

In China the Chou dynasty succeeded the Shang dynasty in 1122 B.C.E. and over the next nine centuries spread urbanization from the Huang He region to the east and south of Chang Jiang (Yangtze River). Between the third century B.C.E. and the third century C.E., the Ch'in and later Han dynasties helped spread urbanization throughout East Asia, including along the Silk Road (see the Urban View 3 entitled "The Silk Road: Long-Distance Trade and Urban Expansion"). The next major period of urbanization in China came at the time of the Mongol invasions with the establishment of cities as part of a new Mongol empire. Urbanization came later to Korea and Japan through Chinese influence.

In Japan urbanization began with the founding of Osaka in 400 C.E., followed by a succession of royal capitals in the seventh and eighth centuries. This culminated by the ninth century with Kyoto, whose status as capital for almost a millennium led to its growth to an unprecedented size for cities of the time. Following a period of decline, there was a resurgence of urbanization in the late fourteenth century with the establishment of "castle towns," including some, like Edo (Tokyo), that would become huge urban centers.

In Mesoamerica, the Maya and neighboring groups, including the Zapotecs and, later, Aztecs to the north and west, as well as the Incas farther south, continued to build cities. As we will see later in this chapter aggressive European colonization, beginning with the Spanish conquest in the sixteenth century, brought drastic changes to these urban civilizations.

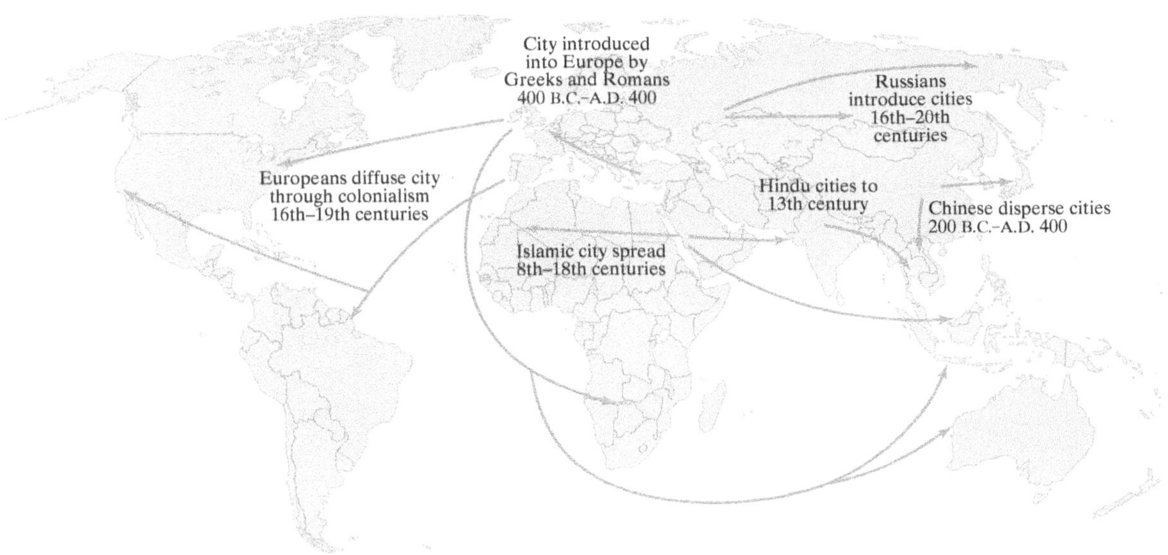

FIGURE 6 Urban expansion in conjunction with the expansion of selected empires.

URBAN VIEW 3
The Silk Road: Long-Distance Trade and Urban Expansion

The *Silk Road* evolved out of trade routes that developed initially along the Fertile Crescent in Mesopotamia in the west and later in the area of the Ch'in dynasty in the east. The Silk Road is an excellent example of how long-distance trade fueled urban expansion and produced an extensive system of cities along a route that eventually connected China with Europe between about 500 B.C.E. and 1500 C.E. (Figure 7). The overland portion started in the east from the ancient Chinese capital of Changan (modern Xi'an), a bustling center of trade and industry. This collection and distribution node served the surrounding region, from which goods were assembled and some processed before being exported westward along the Silk Road. Leaving Changan, the route followed the Great Wall of China until it forked around the Takla Makan Desert before crossing the Pamir Mountains and continuing on through Afghanistan and Iran to reach the Mediterranean, where the goods continued on to Europe by ship.

The caravans could be enormous, with as many as 1,000 camels carrying up to 500 pounds of goods each. In addition to silk, which initially appeared so extraordinary to Europeans, green and white jade, blue lapis lazuli, ceramics, gunpowder, iron, spices, fruit, and flowers went west toward Europe, while gold, silver, amber, ivory, cotton, and wool headed east. Perhaps as important as trade was the exchange of ideas about new technologies, scientific skills, language, art, and religion—not least, writing, printing, and papermaking—among the people along these routes.

The need for safe overnight campsites for the caravans and their valuable cargos led to the growth of cities that offered security and trade opportunities for people traveling along the route. Caravan cities that grew to considerable size, like Samarkand, now in eastern Uzbekistan, and Kashgar (Kashi) at the foot of the Pamir Mountains in extreme western China, were heavily fortified. People came to the regular markets that were held in these cities, especially near the city gates serving the Silk Road.

Flourishing trade along the route also generated rich profits that helped support entire empires and urban civilizations. By about 100 B.C.E., the Roman Empire in the west, the Han dynasty in the east, and the Parthian Empire in Persia (modern Iran) in between were all benefiting from the commercial activity of people that crisscrossed the Silk Road.

With Europe's emergence from the Dark Ages, steady population growth and limited amounts of usable land helped trigger the transition from feudalism to merchant capitalism. With time, the Europeans' growing technological superiority on the seas allowed them to increasingly dominate world trade. European naval discoveries opened up new trade opportunities, including a sea route to India around Africa that bypassed southwest Asia and ultimately contributed to the demise of ancient trade networks like the Silk Road.

The political and military situation in countries like Iraq, Iran, and Afghanistan would make any attempt at crossing Central Asia along the 5,000 miles of the Silk Road an incredibly difficult, if not impossible, journey today. Besides, most of the Silk Road is gone. Only some surviving remnants attest to the importance of this ancient trade network of caravan routes that were studded with towns and cities like strings of pearls extending across the inhospitable deserts and mountain ranges of Central Asia.

Vera Sweeney

FIGURE 7 Children looking at a map of the Silk Road during a visit to the American Museum of Natural History in New York. This ancient trade network of caravan routes that extended across Central Asia from as early as 500 B.C.E. until about 1500 C.E., is an excellent example of how long-distance trade fueled urban expansion and produced a extensive system of cities that connected the four major regions containing the great empires at the time: Europe, Southwest Asia, India, and China.

Urban growth and expansion outward from the regions of urban origin were fueled by critical innovations, especially in technology and economic organization, allied with changes in social organization. Demographic changes—including the deaths of many people in epidemics and wars—were a factor. Adequate numbers of workers were needed to maintain the social and economic infrastructures that supported urbanization. This proved critical in the Indus Valley in about 1750 B.C.E., where population decline allowed the Aryan invaders to bring the Harappan urban civilization to an abrupt end.

Changes in the balance between people and their resource base could help fuel urban growth or promote decline. Ultimately, self-propelled urban growth is limited by the size of a society's resource base. The maintenance of irrigation systems, for example, combined with the necessity for increasing productivity to sustain a growing population, could put incredible pressures on the agricultural workforce. After a while, investments could be neglected, armies reduced in size, and an empire's strength and cohesion fatally undermined. This kind of sequence may have been behind the eventual collapse of the Sumerian Empire and may have contributed to the abandonment of many Mayan cities hundreds of years before the Spanish conquest.

One response to a limited resource base—territorial expansion and colonization—often reinforced and extended the urbanization process. The need for increasing numbers of control centers and improved transportation networks to support colonization and long-distance trade tended to produce a hierarchical urban system. In this way the expansion of the Greek and Roman empires laid the foundations for an urban system in Western Europe.

THE ROOTS OF EUROPEAN URBAN EXPANSION

Greek Cities

The Greeks originally came into the Aegean Sea region from the north. They built upon the city-building ideas that had spread into the Mediterranean from the Fertile Crescent. By 800 B.C.E., the Greeks had founded cities such as Athens, Sparta, and Corinth.

The importance for the Greeks of religion, commerce, administration, and defense were reflected in the layout of their city-states. At the center was the high city—the "acropolis"—the defensive stronghold that contained temples, government offices, and storehouses. Below the high city ("sub-urbs" in Latin), were the "agora" for the markets and political gatherings, more government and religious buildings, military quarters, and residential neighborhoods, all surrounded by a defensive wall. Athens and the older mainland cities had an organic growth pattern characteristic of unplanned Mesopotamian urban development. The street systems of later Greek city-states were based on a gridiron pattern, usually on a north-south axis, regardless of site conditions (Figure 8).

The Greeks located many of the earliest cities along coastlines, reflecting the importance of long-distance sea trade for this urban civilization. Population growth combined with limited cultivable land on the mainland drove overseas colonization and the establishment of a Greek system of cities. Bands of colonists and their families established new independent city-states that stretched from the Aegean Sea to the Black Sea, around the Adriatic Sea, and along the Mediterranean as far west as modern Spain (Figure 9).

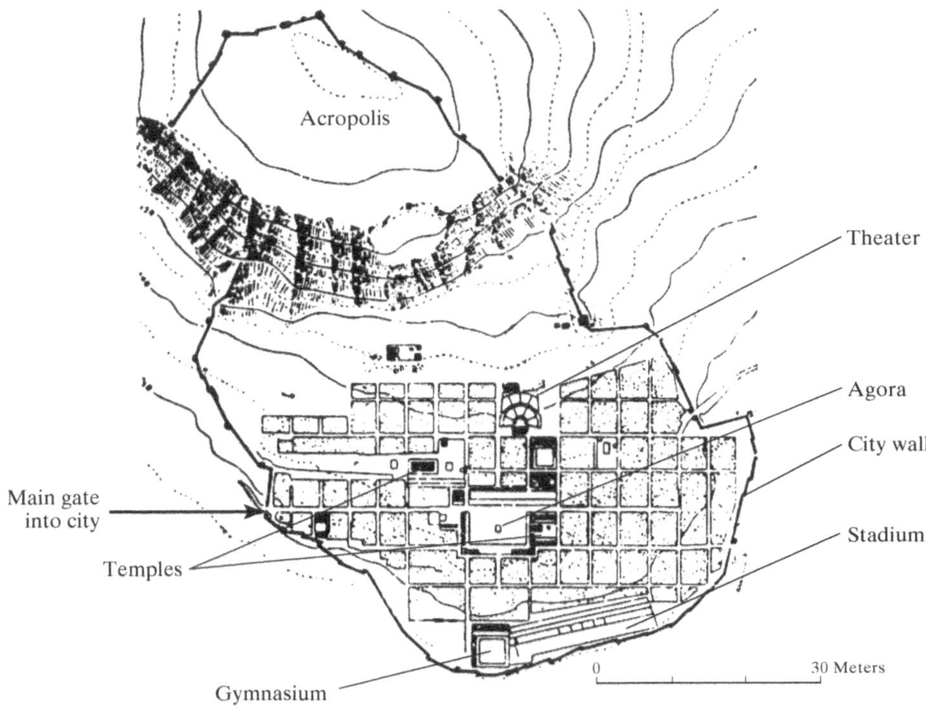

FIGURE 8 General plan of a typical Greek city-state—Priene (in modern-day Turkey). Note how the use of the gridiron plan reflects a planned approach.

History of Urban Form: Before the Industrial Revolution 3rd ed., A E J Morris, Pearson Education, 1994.

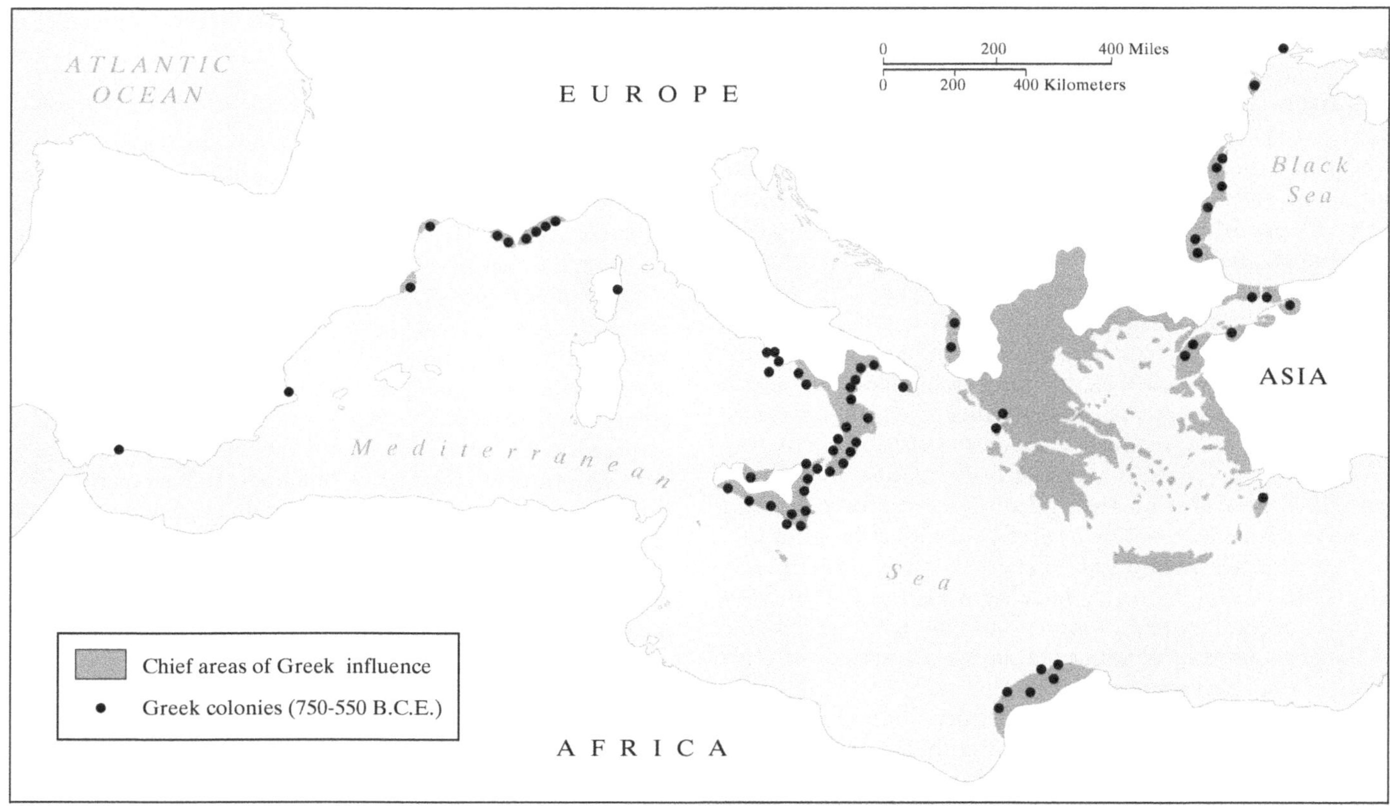

FIGURE 9 Greek city-states in the Mediterranean. Note how the earliest Greek cities were located along coastlines, reflecting the importance of long-distance sea trade for this urban civilization.

The Greeks developed new forms of government whose influence is reflected in subsequent democratic and participatory modes of urban governance throughout the world. Partly as a product of enlightened Greek culture, political authority came to reside in an assembly of—albeit male—citizens who elected a city leadership. Although Greek civic life continued to be conducted within a religious context, the laws and political decisions were no longer presented as unchallengeable divine commands, as they had been in Mesopotamia and Egypt.[21]

After the rest of Greece lost its independence to the Macedonians in 338 B.C.E., overseas colonization became more centrally organized and based on a "mass production of cities" that extended eastward toward Central Asia:

> Alexander the Great and his Successors founded on geopolitically important sites and built up as strategic strongholds, a whole network of towns and cities, which evolved into centres spreading Greek culture and civilization, over the greatest part of the then known world. It was one of the greatest 'colonial' town planning and building actions which was ever accomplished in history.[22]

But Greek cities remained quite small by today's standards. Although Athens probably reached a population of about 150,000, many of the larger cities ranged from 10,000 to 15,000 inhabitants, while most urban settlements had only a few thousand people.

Roman Cities

The expanding Roman Empire displaced Greek civilization during the second and first centuries B.C.E. By the second century C.E., the Romans had established towns across southern Europe and laid the foundation for the Western European urban system (Figure 10).

Roman cities were similar in certain respects to their Greek predecessors (Figure 11). They were based on the grid system, contained a central "forum" for markets and political gatherings, were encircled by a defensive wall, were deliberately established in newly colonized territories, were part of an extensive system of long-distance trade, and remained fairly small. Although the population of Rome likely reached the one million mark by 100 C.E., large Roman towns usually contained only about 15,000 to 30,000 inhabitants, while most places had no more than 2,000 to 5,000 people.

An important difference between Greek and Roman cities was that Roman cities were not independent. They functioned within a well-organized empire centered on Rome and were designed along hierarchical lines, reflecting the Roman rigid class system. Another difference was the greater concentration of Roman cities in inland locations, reflecting

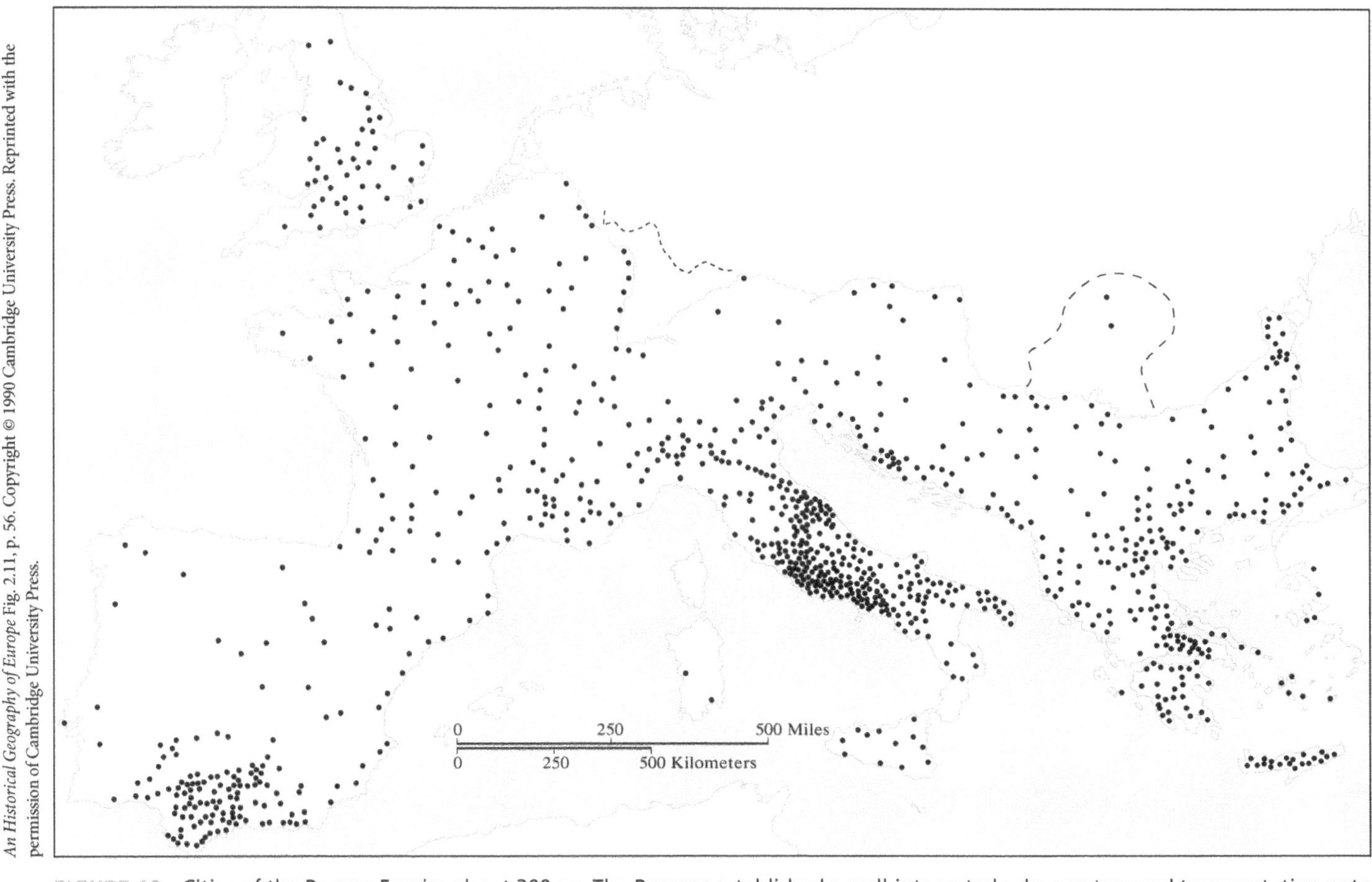

FIGURE 10 Cities of the Roman Empire about 200 C.E. The Romans established a well-integrated urban system and transportation network that laid the foundation for the Western European urban system.

their predominant function as control centers. Many modern European cities can trace their origins to the Roman period, including London, Brussels, Paris, Cologne, Vienna, Sofia, and Belgrade.

The Romans achieved impressive feats of civil engineering. The most important towns were directly connected to one another and to Rome by a magnificent system of roads that facilitated strategic military and trading communications. The underground sewer and surface water supply infrastructures in Roman cities contributed to impressive health improvements that set the standard for later cities. In Rome public latrines served the majority of people. The system of elevated aqueducts and fresh-water reservoirs that brought water for drinking and bathing into the city ran for over 300 miles and carried 60 million cubic feet of water per day.[23]

The Romans used cities as a mechanism to impose their authority and legal system throughout their vast empire. They understood that any attempts to hold newly acquired territories by military force could invite guerrilla warfare that would distract the army from its task of maintaining and extending the empire's borders; it could also hurt the development of commerce. Native tribes, therefore, had to be brought into the empire on advantageous terms—by equating Romanization with urbanization. Tribal centers were redeveloped as Roman towns of varying political status. Other towns were also established for economic and political reasons and populated by ex-soldiers and settlers from Rome and other older towns.

At the time of the fall of Rome in the fifth century C.E., the Romans had established a well-integrated urban system and transportation network that stretched from England in the northwest to Babylon in the east. By as early as the second century C.E., however, the empire's population had already begun to decline, causing labor shortages, abandoned fields, and depopulated towns and allowing the incursions of "barbarian" tribes from the Germanic lands of east-central Europe that helped topple the empire.

Dark Ages

The Dark Ages in Western Europe was a period of stagnation and decline in city life after the collapse of the Roman Empire until about 1000 C.E. Of course, urban life continued to flourish in other parts of the world, including the city building associated with the "explosion of Islam"[24] beginning in the latter half of

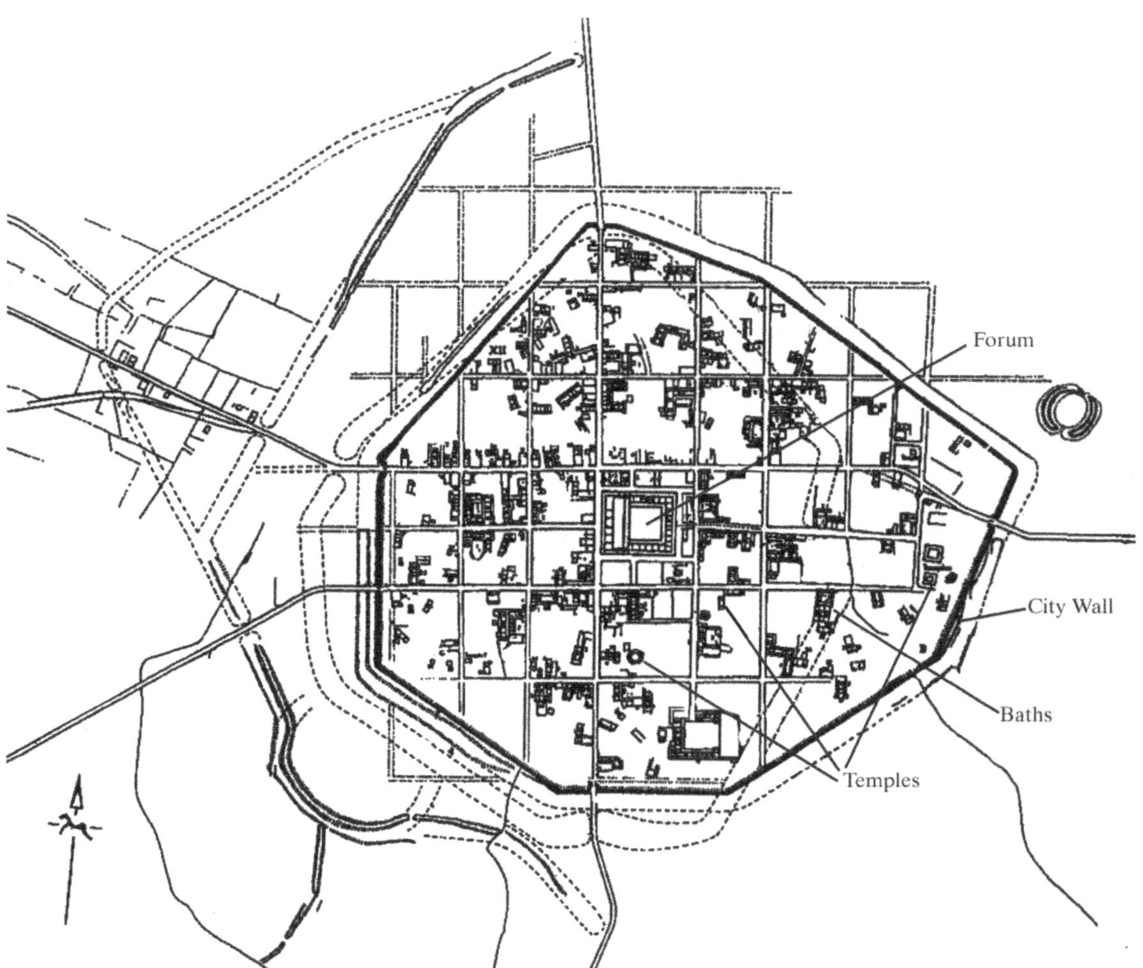

FIGURE 11 General plan of a typical Roman city—Calleva (Silchester in England). Note how the use of the gridiron plan reflects a planned approach.

the 7th century C.E. In the following centuries, existing towns like Mecca, Medina, Baghdad, and Damascus experienced a dramatic rebirth, while new cities were founded, including Teheran in Iran; Basra, Mosul, and Karbala in Iraq; and Cairo and Tangiers in North Africa; and farther south in sub-Saharan Africa, cities in the west, such as Timbuctoo (Timbuktu) in modern Mali and Kano in Nigeria; as well as cities in the east, such as Mombasa in Kenya.

Urban life flourished in parts of Europe where long-distance trade continued, as in those cities under Muslim influence, including Cordova, Granada, and Seville in Spain, or in cities under Byzantine control, most notably Byzantium (Constantinople, later Istanbul) in modern Turkey. By the end of the fourth century, as Rome was falling into decline, Constantine had moved the capital of the Roman Empire to Constantinople. As the capital of the Byzantine Empire between about 360–650 C.E., with its strategic location for trade between Europe and Asia, this city grew to become the largest in the world at the time, with about half a million people.

In the rest of Europe the Germanic invaders and, later, Viking raiders from the north took advantage of the vacuum created by the Roman Empire's collapse. This politically unstable situation made long-distance trade virtually impossible, cutting off the lifeblood of cities, and creating isolated, crumbling, and depopulated urban centers. Most of the urban places that survived were ecclesiastical or university centers, defensive strongholds, or administrative hubs.

- *Ecclesiastical or university centers:* Some Roman towns continued to be occupied because of the church practice of designating certain urban centers as bishoprics—the seats of bishops—and building a cathedral within the old Roman walls. Other towns survived because of their importance as educational and subsequently university centers. Examples include St. Andrews in Scotland; Canterbury and Cambridge in England; Rheims and Chartres in France; Liège in Belgium; Bremen in Germany; Trondheim in Norway; and Lund in Sweden (Figure 12).
- *Defensive strongholds:* The constant threat of attack spurred the construction of castles and other fortifications in cities in general and in some parts of Eastern Europe that had previously been underrepresented by urban development. Examples include the hilltop towns of

Paul Knox

FIGURE 12 St. Andrews, Scotland, an important ecclesiastic center. The cathedral was built in the twelfth century, the castle (an episcopal residence), about 1200. The university was founded in 1410.

central Italy such as Foligno, San Gimignano, and Urbino (Figure 13).

- *Administrative hubs:* Administrative centers for the upper tiers of the feudal hierarchy included Cologne, Mainz, and Magdeburg in Germany; Winchester in England; and Toulouse in France (Figure 14).

Feudalism curtailed the development of European cities because its highly structured and self-contained nature favored the self-sufficient country manor as the basic building block of settlement. Feudalism was a rigid, mostly rural form of economic and social organization based on the communal chiefdoms of the Germanic tribes who had invaded the disintegrating Roman Empire. Each feudal estate was more or less self-sufficient in the production of food, and each kingdom or principality was more or less self-sufficient in the provision of raw materials needed to craft simple products. Yet in spite of this unlikely beginning, an elaborate urban system evolved whose largest centers eventually grew into what would become the **nodal centers** of a global world-system.

Urban Revival in Europe during the Medieval Period

Beginning in the eleventh century, the feudal system weakened and began to collapse in the face of successive demographic, economic, and political crises. The fundamental cause of these crises was steady population growth in conjunction with only modest technological improvements and limited amounts of cultivable land. To bolster their incomes and raise armies against each other, the feudal nobility began to levy increasingly higher taxes. As a result, the peasants, most of whom were serfs (descended from slaves and not free) or tenants (whose freedom to move, marry, leave property to their heirs, buy goods, or sell their labor was very much constrained by public law), were forced to sell more of their produce for cash on the market. This fostered a more extensive money economy

LianeM/Shutterstock

FIGURE 13 San Gimignano, Italy, on a classic hilltop defense site, was originally settled by the Etruscans in the third century B.C.E. The town was named for Gimignano, the bishop of Modena and future saint, who saved the town from barbarian pillaging in the tenth century. It grew prosperous because of the nearby Francigena Way, a busy trade and pilgrim route, and became a city-republic in 1199. Rival families built the towers during the eleventh and thirteenth centuries as symbols of their wealth and status—the higher and more imposing the better. Fifteen of the original 72 medieval towers survive.

FIGURE 14 Cologne, Germany, a medieval administrative hub. When this woodcut was made in the late 1400s, Cologne had a population of less than 25,000 but was already an important administrative, commercial, and manufacturing center, with an important cathedral and a university that was already more than 100 years old.

and the beginnings of a pattern of trade in basic agricultural produce and craft manufactures. Some long-distance trade even began in luxury goods, such as spices, furs, silks, fruit, and wine. This trade caused towns to begin to grow in size and vitality.

Medieval towns can be classified into five categories on the basis of their origin.[25]

- *Towns of Roman origin* survived during the Dark Ages or were reestablished after being deserted. Examples include London and York in England and Regensburg in Germany (Figure 15).
- *Burgs* were fortified military bases that evolved into towns as they acquired commercial functions. Examples include Oxford and Nottingham in England and Magdeburg in Germany (Figure 16).
- *Towns that evolved from village settlements through organic growth.* Examples in England include Wycombe in Buckinghamshire and Wickham in Hampshire (Figure 17).

Bing Maps

FIGURE 15 This aerial view of Chichester in West Sussex, England, shows the legacy of its Roman origin: the main north-south and east-west streets and a later defensive wall that replaced the original Roman one.

Bing Maps

FIGURE 16 An aerial view of Wallingford in Oxfordshire, England, a Medieval burg whose origin was a *fortified military base* that evolved into a town as it acquired commercial functions. The market place was located just west of the intersection of the two main streets (at the center in the photo). The castle (in the lower right of the photo) is located beside the Thames where William the Conqueror crossed this river.

- *Bastides* were planned new towns in France, England, and Wales. The initial motive for establishing these towns was usually strategic, but their sponsors also saw them as an investment that could yield income. Bastide towns were typically laid out around a castle and protected by heavily fortified walls. The towns provided essential services for their military garrisons and helped stabilize the surrounding countryside, but they also provided a source of income for their sponsors through market tolls, rents, and court fines. In this regard, bastide towns were part of a deliberate and fairly widespread policy of town plantation that coincided with an unprecedented boom in urban growth in the twelfth and thirteenth centuries as increased trade across northern Europe marked the transition from feudalism to merchant capitalism. As planned towns, most were laid out with a gridiron street plan. The main inducement to attract settlers was the grant of a house plot within the town and farming land nearby. Examples include Ludlow and Kingston upon Hull in

Skyscan Photolibrary/Alamy

FIGURE 17 This aerial photo of Warkworth in Northumberland, England, gives an idea of how a medieval town that was evolving organically from a village settlement might have looked. The linear plan reflects how this village grew up along a winding roadway within the defensive protection of a meander along the Cloquet river. The legacy of the importance of the church and castle in the lives of the people in this village during medieval times is reflected today in the fact that they are still the most imposing structures, towering over all nearby buildings.

ImageGap/Alamy

FIGURE 18 Aigues-Mortes, France, a medieval bastide. This aerial view shows the gridiron street pattern inside the rectilinear walls of the town and the surrounding fields and vineyards.

England; Caernarvon in Wales; and Aigues-Mortes and Carcassone in France (Figure 18).

- *Planted towns* included other planned new towns throughout Europe, with or without a predetermined layout. Most were founded on a roadside or riverside location for commercial purposes to take advantage of the general reestablishment of long-distance trade. Examples include Offenburg and Freiburg im Breisgau in Germany and Berne in Switzerland (Figure 19).

All medieval towns shared common features of internal structure. At the center was an open square for markets, surrounded in larger cities by the cathedral or church, the town hall, the guildhalls, the palaces, and the houses of prominent citizens. Close to the center were streets or districts that specialized

Switzerland Tourism

FIGURE 19 Berne, Switzerland, a medieval planted town. This aerial view shows the gridiron street pattern of the original town that formed the nucleus of the contemporary city and how the layout reflects the influence on urban form of its location on a riverbend in the Aare.

in particular functions, such as banking or the production and sale of items like furniture or metalwork. The organic growth towns could have streets and alleys that were unplanned and quite narrow. The city defenses became probably the most important determinant of urban form. Urban development had to take place in stages, each of which was normally preceded by the construction of a new wall.[26]

With urban growth concentrated inside the city wall, population density was high, and the constrained space caused different socioeconomic groups to be stratified *vertically* within the same building. For example, a merchant's or craftsperson's shop would be on the main floor, and because there were no elevators, the family's living quarters would be above on the next floor, the apprentice quarters would be above, with the servants up the stairs in the attic (Figure 20). To maximize street frontage, residential and commercial buildings were often aligned with their narrow or gable end fronting onto the main streets.

The emerging regional specializations and trading patterns provided the foundations for a new phase of urbanization based on **merchant capitalism**. The key people in this system were the merchants who supplied the capital required to reestablish a vibrant system of long-distance trade—hence the label *merchant capitalism*.

> The merchants made the towns. They needed walls and wall builders, warehouses and guards, artisans to manufacture their trade goods, caskmakers, cart builders, smiths, shipwrights and sailors, soldiers and muleteers. They needed farmers and herdsmen outside the walls to feed them; and bakers, brewers and butchers within. They bought the privilege of self-government, substituting a money economy for one based on land . . . . Towns recruited manpower by offering freedom to any serf who would live within their walls for a year and a day.[27]

Beginning with commercial networks established by the merchants of Venice, Pisa, Genoa, and Florence (in northern Italy) and the trading partners of the *Hanseatic League*, a federation of city-states around the North Sea and Baltic coasts (see the Urban View 4 entitled "Hanseatic League Cities"), a trading system of immense complexity soon spanned Europe from Bergen to Athens and from Lisbon to Vienna. Long-distance trade became firmly reestablished, but it was based more on bulky staples such as grain, wine, salt, wool, cloth, and metals rather than on the luxury goods of the pioneer merchants. The increased volume of trade fostered a great deal of urban development as the merchants began to settle at locations to take advantage of the major trade routes that crisscrossed Europe and as local economies everywhere came to focus on market exchange.

At the close of the medieval period in the late thirteenth century, Europe had about 3,000 cities containing some 4.2 million people, representing between about 15 to 20 percent of the total population. Most of these urban centers were small—with fewer than 2,000 people. Paris was the dominant European city, with a population of about 275,000. Besides Constantinople and Cordoba, only Milan, Genoa, Venice, Florence, and Bruges had more than 50,000 people. This then was the Europe that stood poised to extend its grasp to a global scale.

Linda McCarthy

FIGURE 20 A view of a narrow, now pedestrianized, street in Canterbury in England. Note how the "vertical" social stratification within the buildings is reflected in the size and ornateness of the windows on each floor.

URBAN VIEW 4
Hanseatic League Cities

With its roots dating back to the twelfth century, the Hanseatic League was a trading association of independent German city-states around the North Sea and Baltic coasts. A trading arrangement initially between Hamburg and Lübeck provided a cooperative model for the merchants in other German cities. It involved an alliance between these two northern German towns that were located on either side of the Danish peninsula, Hamburg in the west and Lübeck in the east. Lübeck had access to the Baltic herring spawning grounds. Fish, which could be eaten on Fridays when meat could not, made up a high percentage of the Christian European diet at the time. Without refrigeration or canning, shipping this perishable product needed salt for salting, which was easily accessible to Hamburg from the nearby salt mines at Kiel. The merchants in Lübeck and Hamburg opened a trade route along the canal that was constructed between them and named for the source of the salt, Kiel.

Although not approaching the level and extent of economic and political cooperation among the countries of today's European Union, the Hanseatic League became the first great cooperative effort that united cities across a region of Europe into an economic association. These city-states entered into commercial agreements to promote trade through special privileges for members and to protect themselves against pirates and robber barons. Over time, more cities joined in pursuit of the trade security and increased opportunities that membership provided. Conversely, the League engaged in negotiation, bribes, blockades, embargoes, and even war against port cities that were hostile to the organization and wanted to break its monopoly.

At its height, as many as 200 towns participated in this association that extended from Amsterdam to Reval (Tallinn) in Estonia and from Stockholm on Sweden's Baltic Sea coast to important inland port cities along rivers, such as Lübeck, Hamburg, Köln (Cologne), and Magdeburg. Their foreign trading outposts, or counting houses, the forerunners of today's stock exchanges, extended the influence of this powerful trading association to Bergen on Norway's North Sea coast, Visby on the island of Gotland, London, Bruges in Flanders (Belgium), and Novgorod in Russia (Figure 21).

Merchant capitalism and long-distance trade fueled urban expansion and the development of this system of trading cities, some of which, like London, later grew to global dominance. The trading association controlled the shipping of fish, salt, grain, timber, amber, fur, flax, and honey from Russia and the Baltic coast to the west, and cloth and manufactured goods from Flanders and England to the east. The Hanseatic League had its own financial and legal systems, as well as strong traditions of civic and individual rights.

By the early sixteenth century, the Hanseatic League had begun to disintegrate—a casualty of factors that weakened its power, such as rivalry and internal struggles among League members, the new trade opportunities opened up by the discoveries of Columbus and da Gama, declining Baltic fish stocks, the social and political insecurities of the Reformation, the growing strength of trade competitors like the Dutch and later the English, and the encroachment into its trade routes by the Ottoman Empire. The difficulties of long-distance trade during the Thirty Years War (1618–48) further weakened the League and led to its demise.

Interestingly, although the Hanseatic League met for the last time in 1669, it was never officially dissolved. It lives on in the names of German cities like "Hansestadt Lübeck" and "Freie und Hansestadt Hamburg." Even today, Hamburg

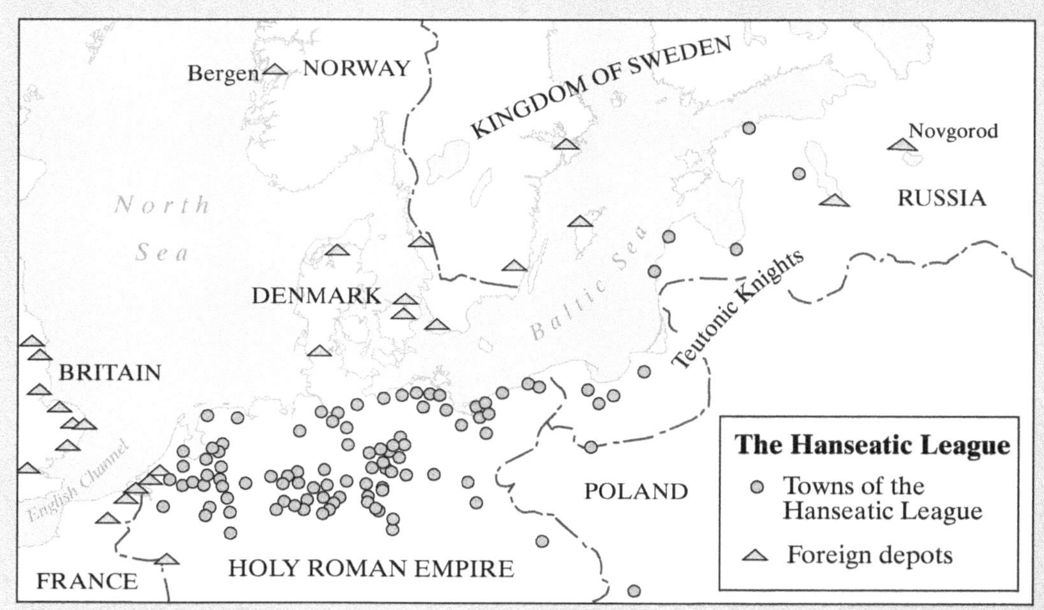

FIGURE 21 Hanseatic League cities. The towns of the Hanseatic League and their foreign trading depots.

(continued)

URBAN VIEW 4
Hanseatic League Cities (*continued*)

and Bremen are individual city-states within Germany. The League's legacy can been seen in the widespread use of German merchant and administrative terminology and in standardized sea travel and trade regulations, a first for the League. The League's strong architectural legacy is also visible in the "stepped gable" design in the Hanseatic League towns as a way to make the buildings appear taller and show off their wealth (Figure 22).

Styve Reineck/Shutterstock

FIGURE 22 The facades of buildings in Brugge (Bruges), Belgium, with the characteristic stepped gables of Hanseatic League cities. Note how, in order to maximize street frontage, residential and commercial buildings were aligned with their narrow or gable end fronting onto the main streets.

Urban Expansion and Consolidation during the Renaissance and Baroque Periods

Between the fourteenth and eighteenth centuries, fundamental changes transformed not just the cities and urban systems of Europe but also the entire world economy. The Protestant Reformation and the scientific revolution of the Renaissance (from the fourteenth to sixteenth centuries) stimulated economic and social reorganization. While the Church and religious doctrine had been dominant in people's lives during medieval times, the glorification of human reason and achievement was foremost for people during the Renaissance. The scale and sophistication of merchant capitalism increased. Aggressive overseas colonization allowed Europeans to shape the world's economies and societies.

Spanish and Portuguese colonists were the first to connect other parts of the world into the European urban system. Beginning in 1520, it took just 60 years for these colonial powers to establish the basis of a Latin American urban system. The Spanish colonists founded their cities in the western parts of Latin America by rebuilding at the sites of conquered indigenous centers like Oaxaca and Mexico City in Mexico, Cuzco in Peru, and Quito in Ecuador or in regions of dense indigenous population, as in Puebla and Guadalajara in Mexico and Lima in Peru. These **colonial cities** were established primarily as administrative and military centers from which the Spanish Crown could occupy and exploit the New World. In contrast, the cities of the Portuguese colonists farther east, such as São Paulo and Rio de Janeiro, were more commercial in nature. Although also motivated by exploitation, the Portuguese colonists located their colonial towns in the best places commercially for collecting and exporting the products of their mines and plantations.

A major aspect of this urbanization and expansion of trade was the establishment of **gateway cities** around the world to act as links between one country or region and others. These **control centers** commanded entrance to, and exit from, their particular country or region. The Europeans founded or enlarged thousands of towns in other parts of the world as they extended their trading networks and established their colonies. The great majority were ports protected by fortifications and European naval power. Beginning as colonial trading posts and administrative centers, some grew rapidly as gateways for colonial expansion into the continental interiors. European settlers came in through these cities, and produce from the continental interiors went out. Rio de Janeiro grew on the basis of gold mining; São Paulo on coffee; Buenos Aires on mutton, wool, and cereals; Kolkata (Calcutta) on jute, cotton, and textiles; Accra (in Ghana) on cocoa; and so on.

These products in turn fueled urban growth in Europe. In terms of location, the exploitation of the New World gave a decisive advantage to the port cities along the North Sea and the Atlantic Coast. By 1700 London had grown to 500,000 people, while Lisbon and Amsterdam had each reached about 175,000. The cities of continental and Mediterranean Europe grew at a more modest rate. Between 1400 and 1700 Venice grew only from 110,000 to 140,000 people; Milan's population did not grow at all.

More integrated national urban systems evolved along with the centralization of political power and the formation of national states that characterized the Renaissance period. Best exemplified by the capital cities of Paris and Madrid, their central location nationally facilitated the process of political consolidation, which, in turn, gave both cities a further impetus for growth as they acquired important administrative functions. Regional capitals and seats of county government emerged to fill out the evolving national urban systems.

The overall appearance and internal structure of cities in Europe changed during the Renaissance with the introduction of new forms of art, architecture, and urban planning. Especially in the capital cities, the flourishing of artistic and architectural expression brought about greater use of sculpture in public places, other forms of urban beautification, such as fountains, and the embellishment of monumental buildings that reached a peak with the Baroque period that began in the late sixteenth century. Merchant capitalism generated great wealth for the nobility of the various monarchies and countries, and their burgeoning spending power was used to build opulent palaces in many cities. City walls were built with more complex and costly designs during the Renaissance, such as star shaped, allowing the use of greater firepower against an attacking army (Figure 23).

URBANIZATION AND THE INDUSTRIAL REVOLUTION

Large-scale manufacturing began in the English Midlands in the mid-1700s. The Industrial Revolution was a powerful impetus for urban growth because it brought fundamental changes in how and where goods—from textiles to machine tools—were produced. Previously, individual rural workers, like spinners and weavers, had carried out the different stages of the production process by hand in their own cottages. Now for the first time, all the stages of production were mechanized and combined under one roof—in factory buildings. Initially, access to a water source that could produce power for the machinery dictated the factories' locations. In previously rural areas, urban settlements grew up around the new factories as the early industrialists provided housing to attract workers whose long working hours forced them to live nearby.

Industrialization and cities grew hand in hand (Figure 24). The Industrial Revolution generated new kinds of cities—and many of them. Industrial economies needed what cities had to offer: the physical infrastructure of factories, warehouses, stores, and offices; the transportation networks; the large labor pools; and the consumer markets. In turn, industrialization

Paul L. Knox

FIGURE 23 Sabbioneta, Italy, was built by Vespasiano Gonzaga in the mid-sixteenth century as an ideal town, with a central piazza, a ducal palace, churches, garden palace, theater, and residences all encompassed within a star-shaped plan, bounded by thick walls bearing the Gonzaga family crest.

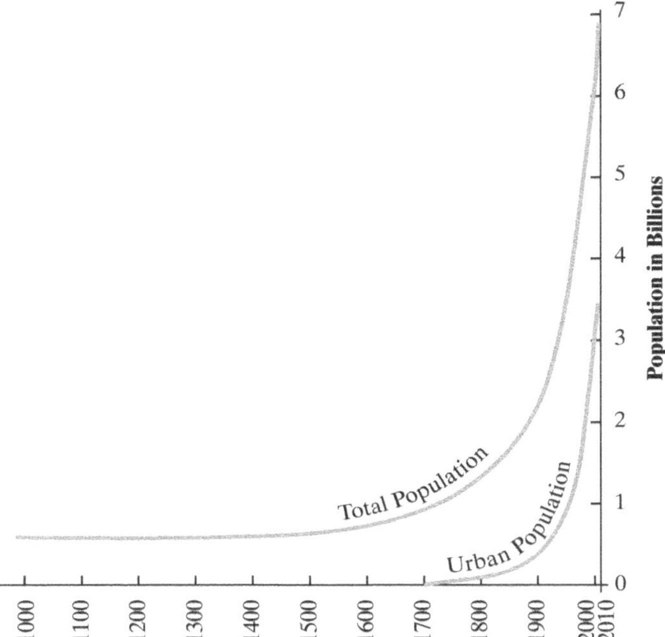

FIGURE 24 World urban population growth over time. Note the tremendous explosion in the number of people living in towns and cities that was triggered by the Industrial Revolution.

changed the appearance, internal structure, and functioning of cities. Whole districts of factory buildings grew up with their grimy smokestacks, deafening machinery, and general hustle-and-bustle of industrial activity (see the Urban View 5 entitled "Manchester: Shock City of European Industrialization"). With industrial inputs and products transported mainly by trains, new tracks, stations, and rail traffic began to play a significant role in these cities, as did the new public transportation systems of trolleys and subways. The industrial period also heralded the development of the **central business district (CBD)** with its office buildings and corporate headquarters for the new companies. Large tracts of worker housing were built. This row housing, typical of English cities, was often cramped and poorly constructed.

Rural development and urban growth were intimately connected in the industrialized regions of Europe and North America (Figure 26). Agricultural productivity benefited from the mechanization and the innovative techniques that had been developed in cities. The higher productivity released rural dwellers to work in the growing manufacturing sector in the towns and cities, at the same time providing the additional food to support a growing urban population. This process, reinforced by the agricultural tools, machinery, fertilizer, and other products made in the cities, allowed even greater increases in

URBAN VIEW 5
Manchester: Shock City of European Industrialization

Manchester was the **shock city** of European industrialization in the nineteenth century. It grew from a small town of 15,000 people in 1750 to a city of 70,000 in 1801, a metropolis of 500,000 in 1861, and a world city of 3 million by 1911. A shock city is seen at the time as the embodiment of surprising and disturbing changes in economic, social, and cultural life. Manchester was an archetypal form of an entirely new kind of city—the *industrial city*—whose fundamental reason for existence was not its military, political, ecclesiastical, or trading functions, as in earlier generations of cities. Instead, Manchester existed to assemble raw materials and to fabricate, assemble, and distribute manufactured goods. The city had to cope with record rates of growth and associated unprecedented economic, social, and political problems for many of its residents (Figure 25). Manchester was also a **world**

Charles Roberts/Mary Evans Picture Library

FIGURE 25 Manchester, the shock city of European industrialization.

(continued)

URBAN VIEW 5
Manchester: Shock City of European Industrialization (*continued*)

city, in which a disproportionate share of the world's most important business—economic, political, and cultural—is conducted. At the top of a global urban system, such cities experience growth largely as a result of their role as key nodes in the world economy.

Friedrich Engels surveyed the dreadful living conditions of the working poor in the shock city of Manchester in 1844:

> One walks along a very rough path on the river bank, in between clothes-posts and washing lines to reach a chaotic group of little, one-storied, one-roomed cabins. Most of them have earth floors, and working, living and sleeping all take place in one room. In such a hole, barely six feet long and five feet wide, I saw two beds—and what beds and bedding!—which filled the room, except for the fireplace and the doorstep. Several of these huts, as far as I could see, were completely empty, although the door was open and the inhabitants were leaning against the door posts. In front of the doors filth and garbage abounded. I could not see the pavement, but from time to time, I felt it was there because my feet scraped it. This whole collection of cattle sheds for human beings was surrounded on two sides by houses and a factory and on a third side by a river . . . All along the Irk [river] slums of this type abound.[28]

Johann Kohl, who traveled through the English Midlands in the 1840s, captured how the profound changes associated with industrialization played out in the working lives of the countless factory workers in Manchester:

> It was a cold, damp, foggy morning in December, that I took my leave of Manchester. I rose earlier than usual; it was just at the hour when, from all quarters of the busy town, the manufacturing labourers crowded the streets as they hurried to their work. I opened the window and looked out. The numberless lamps burning in the streets, sent a dull, sickly, melancholy light through the thick yellow mist. At a distance I saw huge factories, which, at first wrapt in total darkness, were brilliantly illuminated from top to bottom in a few minutes, when the hour of work began. As neither cart nor van yet traversed the streets, and there was little other noise abroad, the clapping of wooden shoes upon the crowded pavement, resounded strangely in the empty streets. In long rows on every side, and in every direction, hurried forward thousands of men, women, and children. They spoke not a word, but huddling up their frozen hands in their cotton clothes, they hastened on, clap, clap, along the pavement, to their dreary and monotonous occupation. Gradually the crowd grew thinner and thinner, and the clapping died away. When hundreds of clocks struck out the hour of six, the streets were again silent and deserted, and the giant factories had swallowed the busy population. All at once, almost in a moment, arose on every side a low, rushing, and surging sound, like the sighing of wind among trees. It was the chorus raised by hundreds of thousands of wheels and shuttles, large and small, and by the panting and rushing from hundreds of thousands of steam-engines.[29]

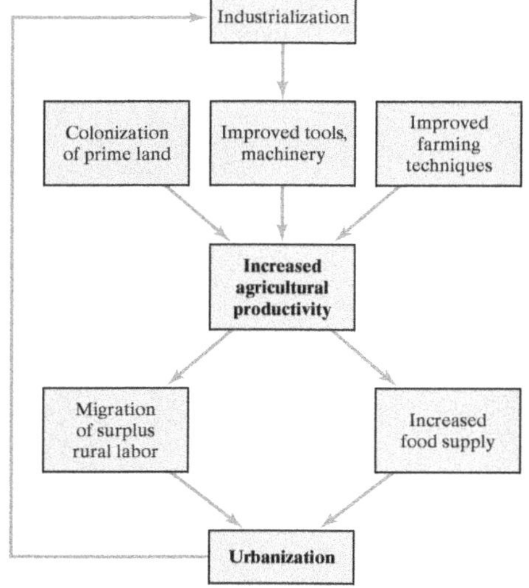

FIGURE 26 The urbanization-industrialization process.

agricultural productivity. This kind of urbanization is a special case of **cumulative causation**, where particular places enjoy a spiral buildup of advantages due to the development of **external economies, agglomeration economies,** and **localization economies.**

As industrialization spread across Europe, the pace of urbanization increased (Figure 27). The higher wages and greater opportunities for people in the cities attracted a massive influx of rural workers. Declining death rates associated with the **Demographic Transition** in Europe contributed to the rapid urban population growth. This growth in population in turn provided a huge increase in the labor supply during the nineteenth century, further boosting the rate of urbanization not only in Europe but also in the United States, Canada, Australia, New Zealand, and South Africa, as emigration spread industrialization and urbanization to the frontiers of the **world-system.**

With time, however, the breakthrough technology of coal-fired steam power freed industry to locate in existing population centers or close to resources, like coal mines. Industrialists no longer had to attract workers to their factories. There was also an overabundance of workers due to fewer farm workers being needed because of increased agricultural productivity and the large number of small landholders left landless by the consolidation of smaller farms into larger, more efficient ones. Improvements in technology caused overproduction,

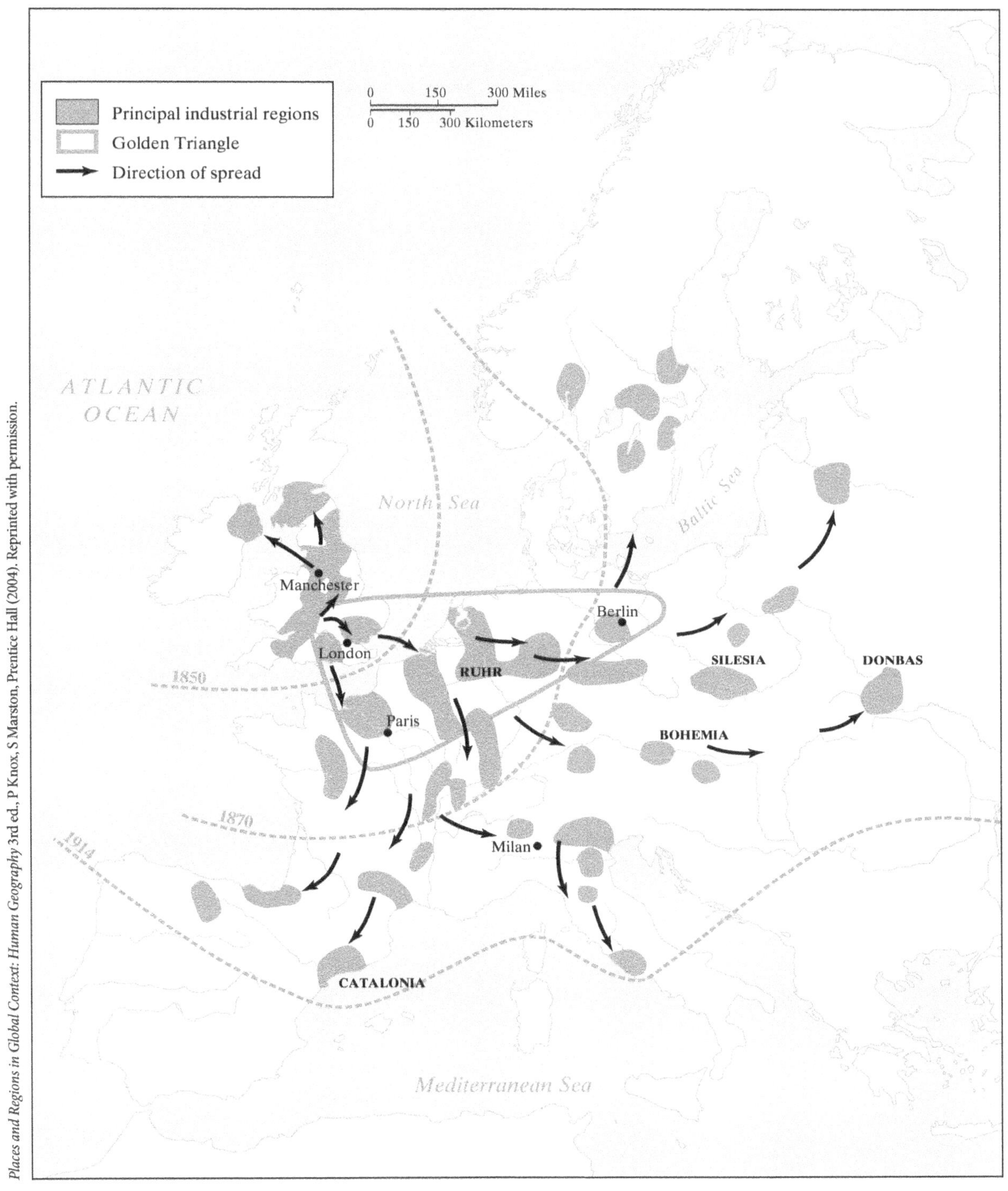

FIGURE 27 The spread of industrialization and industrial cities in Europe. European industrialization began with the emergence of small industrial regions in several parts of Britain, where early industrialization drew on local mineral resources, water power, and early industrial technologies. As new rounds of industrial and transportation technologies emerged, industrialization spread to regions with favorable locational attributes (access to raw materials and energy sources, good communications, and large labor markets). The "Golden Triangle" is Europe's economic core region that centers on the area between London, Paris, and Berlin, and includes the early industrial regions of southeastern England, northeastern France, and the Ruhr district of Germany.

while international competition increased as industrialization spread from England to the European mainland and beyond. Prices dropped, producers cut costs, and wages fell. Long hours for little pay made it a struggle for workers to pay for housing, which led to overcrowding. Living conditions of people in the slums were appalling (see the Urban View 6 entitled "Residential Segregation in Mid-Nineteenth Century Glasgow, Scotland"). Public sanitation and water systems were poor or nonexistent. Outbreaks of water-borne diseases such as cholera and typhoid were common.

URBAN VIEW 6

Residential Segregation in Mid-Nineteenth Century Glasgow, Scotland

As the Industrial Revolution progressed, so too did residential segregation. Michael Pacione provides a stark description of living conditions in the tenement slum areas of central Glasgow by the middle of the nineteenth century:

> The westward migration of the wealthy from the old town not only introduced socio-spatial polarisation into what had been in the eighteenth century a heterogeneous urban structure, but freed land and housing for other uses. Those parts of the old city abandoned by the elite were colonised by a working-class population that was burgeoning in response to the new industries' demand for labour. Residences previously occupied by a single wealthy family were 'made down' to accommodate large numbers of the poor in grossly crowded conditions . . . . In 1861 two-thirds of Glasgow's population of 394,864 lived in houses with only one or two rooms, and of these 60 per cent shared a room with at least four others. Overcrowding was intensified by property speculators building on any available open space to produce 'backjams' and 'backlands' tenements, which were either added to existing buildings or erected in the erstwhile gardens of the formerly wealthy burgher residences. By the mid-nineteenth century the old city was a congeries of poor-quality housing. One shelter visited [in 1858] in a dark ravine of a close housed a husband, wife and two children and comprised a 'sort of hole in the wall' measuring six shoe lengths in breadth, between eight and nine in length from the bed to the fireplace, and of a height which made it difficult to stand upright. Densities of 1,000 persons per acre were commonplace.[30]

By the nineteenth century, urbanization had become an important dimension of the world-system in its own right. In 1800 less than 5 percent of the world's 980 million people lived in towns and cities. By 1850, 16 percent of the world's population was urban, and there were more than 900 cities of 100,000 or more around the world. The Industrial Revolution and European colonization had created unprecedented concentrations of people in cities that were connected in networks and hierarchies of interdependence. As was the case for merchant capitalism, the industrialization of Europe and North America depended on the exploitation of other regions. The **international division of labor** that inevitably resulted from this relationship had a significant impact on the patterns and processes of urbanization that affected people outside Europe and North America.

FOLLOW UP

Key Terms

central business district (CBD)
city-state
colonial cities
Dark Ages
Demographic Transition
Fertile Crescent
gateway cities
gridiron street pattern
Hanseatic League
Industrial Revolution
merchant capitalism
Mesopotamia
organic growth
shock city
Silk Road

Review Activities

1. Do a search of YouTube to find the first episode in the six-part *Legacy: The Origins of Civilization* series that was written and hosted by historian Michael Wood and produced by Maryland Public Television and Central Independent Television, UK (1991). This episode, *Iraq: The Cradle of Civilization*, provides an excellent account of the beginning of urban civilization in southern Iraq more than 5,000 years ago. It illustrates some of the important changes associated with the shift to urban life such as occupational specialization, organized religion, bureaucratic government, and international trade.
2. "A common approach to examining the internal structure of cities is to identify whether the layout of an urban area was largely unplanned or planned." Go online and search for a map of an ancient or contemporary city from a region of the world that particularly interests you. While being aware that all cities are subject to change over time, study the layout of the streets, transportation routes, and other features of your city's internal

layout to try to determine if the city was largely planned or unplanned, or whether over time different parts appeared to have been planned while others were unplanned. Think about what kinds of political, economic, social, technological, and environmental processes may have influenced the layout of your city. Then search some more to find out about the history of urban planning for your city that can shed light on what you have concluded from your analysis of the map.

3. Chicago was the shock city of North American industrialization. Do some research online or in the library to find out why. How important was this city's location? Then find some information about the tremendous changes that were associated with Chicago's industrial growth—in population growth, modes of transportation, and the kinds of raw materials that were brought in for the city's milling, meatpacking, tanning, and woodworking industries.

4. Work on your *e-portfolio*. Look for additional materials that can help you flesh out what you have learned in this chapter. Go online and research some examples of Mesopotamian, Greek, Roman, or medieval towns to find out how and why they were established and whether they have survived and why or why not.

   Think about how you can make the aspects of urbanization described in this chapter meaningful to you personally. Find some contemporary accounts of life in cities at different points during the extensive span of time covered in this chapter. Think about how very different life was for different urban residents in earlier centuries compared to your life today.

Log in to **www.mygeoscienceplace.com** for self-study quizzes, *MapMaster* layered thematic and place name interactive maps, *Urban View* Google Earth™ tours, key resources and suggested readings, related websites, "In the News" RSS feeds, and additional references and resources to enhance your study of the origins and growth of cities and urban life.

NOTES

1. P. Wheatley, *The Pivot of the Four Quarters* (Edinburgh: Edinburgh University Press, 1971), xviii.
2. W. Sjoberg, "The Origin and Evolution of Cities," *Scientific American*, September (1965): 55–56.
3. See V. G. Childe, "The Urban Revolution," *Town Planning Review* 21 (1950): 3–17; H. Carter, *An Introduction to Urban Historical Geography* (London: Edward Arnold, 1983), 2.
4. The account in Urban View 2.1 is paraphrased from B. D. del Castillo, *The Discovery and Conquest of Mexico, 1517–1521* (London: RoutledgeCurzon, 1928).
5. Childe, "The Urban Revolution."
6. L. Woolley, "The Urbanisation of Society," in *History of Mankind*, eds. J. Hawkes and L. Woolley, *The Beginnings of Civilisation*, Vol. 1, Part 2 (Paris: UNESCO, 1963), 109–67.
7. K. A. Wittfogel, *Oriental Despotism: A Comparative Study of Total Power* (New Haven, Conn.: Yale University Press, 1957).
8. E. Boserup, *Population and Technology* (Oxford, UK: Blackwell, 1981).
9. J. Jacobs, *The Economy of Cities* (New York: Random House, 1969).
10. M. Weber, *The City*. Translated and edited by D. Martindale and G. Neuwirth (New York: The Free Press, 1958).
11. Wittfogel, *Oriental Despotism*.
12. Wheatley, *The Pivot of the Four Quarters*, 298–99.
13. W. Sjoberg, *The Pre-Industrial City, Past and Present* (Glencoe, Ill.: Free Press, 1960).
14. Wheatley, *The Pivot of the Four Quarters*, 298–99.
15. Ibid., 318.
16. A. E. J. Morris, *History of Urban Form: Before the Industrial Revolution*, 3rd ed. (London: Pearson Education, 1994).
17. Sjoberg, *The Pre-Industrial City, Past and Present*.
18. L. Woolley, *Ur of the Chaldees: A Record of Seven Years of Excavation* (New York: W. W. Norton & Company, 1965), 423–25.
19. Morris, *History of Urban Form*.
20. See Sjoberg's excellent account of urban expansion in *The Pre-Industrial City, Past and Present*.
21. C. Chant and D. Goodman, eds., *Pre-Industrial Cities & Technology* (London: Routledge, 1999), 56–58.
22. A. Kriesis, *Greek Town Building* (Athens: F. Constantinidis & C. Michalas, 1965), 126.
23. Morris, *History of Urban Form*, 56–61.
24. Sjoberg, *The Pre-Industrial City, Past and Present*, 55.
25. Morris, *History of Urban Form*, 92.
26. Ibid., 98.
27. M. Bishop, *The Penguin Book of the Middle Ages* (Harmondsworth, UK: Penguin Books, 1971), 209.
28. F. Engels, *The Condition of the Working Class in England* (translated and edited by W. O. Henderson and W. H. Chaloner, Stanford: Stanford University Press, 1968).
29. J. G. Kohl, *England and Wales* (London: Frank Cass, 1844. Reprinted in 1968 by Augustus M. Kelley, Publishers, New York), 146.
30. M. Pacione, *Glasgow: The Socio-Spatial Development of the City* (Chichester, UK: John Wiley & Sons, 1995), 84–86.

Foundations: The U.S. Urban System and Its Cities

From Chapter 3 of *Urbanization: An Introduction to Urban Geography*. Third Edition. Paul L. Knox, Linda McCarthy.

ClassicStock/Alamy

Foundations: The U.S. Urban System and Its Cities

In the United States urbanization has evolved in step with economic and societal changes. Towns and cities have played a central role in the economy since the seventeenth century, when Europeans established their first outposts of settlement on North American soil. Urban areas have been crucial in a rapid transition from a dependent, preindustrial society and economy to a globalized form of capitalism where cities function as nodes in a *world* economic system. During each major phase of this evolution, people developed new resources, technologies, and business organizations and cities were changed to accommodate the new economic order. In some cities these changes occurred sequentially and the outcomes were superimposed one on the other. In others the new economic systems were not always profitable or appropriate. As a result, the impact of each phase of change was felt in different ways and to different degrees by people in different cities. Nevertheless, it is possible to identify several distinctive periods in the evolution of the urban system, each reflecting changes in the rhythm of the economic, social, governance, technology and infrastructure systems.

LEARNING OUTCOMES

After reading this chapter, you should be able to:

- Recognize the existence of broad phases in the historical development of the U.S. urban system, and how these phases relate to ongoing process of economic, technological, demographic, political, and social change.
- Appraise how **gateway cities, entrepôts,** and **hinterlands** played complementary roles in the frontier urbanization phase of U.S. urban development.
- Understand the basic principles of urban growth and change that have shaped the U.S. urban system.
- Assess the role of streetcar and railroad development in changing the geography of land use within the city.
- Describe the impacts of economic reorganization and demographic change on the urban system in the early 20th century.

CHAPTER PREVIEW

This chapter follows the evolution of the U.S. urban system through five distinctive periods that together established the foundations for the contemporary **urban system.** As we will see, this set of urban settlements is both the *product of,* and a continuing *framework for,* processes of economic, technological, demographic, political, and social change. Each successive phase of urbanization brought new patterns of settlement, new kinds of towns and cities, and new patterns of trade and migration of people between towns and cities. Each change also brought new challenges for understanding the underlying processes of change. As we follow the development of the urban system, therefore, we will also follow the emergence of key ideas, concepts, and theories about urbanization.

We begin with the earliest period of urban development, the frontier urbanization around which the U.S. economy was organized until independence. The second distinctive period (1790–1840) was one of merchant trading, or mercantilism, during which there emerged a more extensive system of **central places,** or local

---

In the 1920s the commercial development of the internal combustion engine unleashed social, economic, and political forces that changed the physical shape of urban areas. Just when downtown areas had established their unrivaled dominance and internal functional specialization, a new logic of transportation and location associated with cars and trucks led to the decentralization of many of their retailing, wholesaling, manufacturing, and office functions. This left CBDs like the one in Philadelphia, Pennsylvania, shown here in 1935, more specialized and less dominant.

URBAN VIEW 1
Frontier Urbanization and Some Problems of Daily Life

Life for people during the period of frontier urbanization came with some social problems that we might recognize today.[1] As in urban centers elsewhere, a good water source was vital for the colonists, but not necessarily for drinking! Breweries were some of the earliest buildings to be constructed. The colonists usually drank water as a last resort; they preferred beer (and so drunkenness was a serious problem). Water was largely reserved for domestic uses and for putting out fires. In fact, the Massachusetts laws of 1638 and 1646 that forbid smoking "out of dores" or near Boston's Town House were dictated less by Puritan intolerance and more by the fact that "fires have bene often occasioned by taking tobacco."[2]

Sanitary regulations in 1652 in Boston required that all garbage be buried and included a fine for anyone who threw "any intralls of beast or fowles or stinkeing thing, in any hie way or dich or Common."[3] In Newport, the problem of privies and sewage disposal made the streets very unpleasant, "as several Privy houses sett against ye Streets" emptied into the streets where passersby were always in danger of "Spoiling & Damnifying" their clothes.[4]

A police force was soon needed to protect the community against disorderly conduct, public drunkenness, and criminal activities by some inhabitants. The titles and functions of the earliest police officers were, not surprisingly, European in origin, and included the (English) constable in Boston. His range of duties could be seen from a typical Massachusetts law of 1646:

> Evry cunstable . . . hath, by virtue of his office, full powr to make, signe, & put forth pursuits, or hues & cries, after murthrers, manslayrs, peace breakrs, theeves, robbers, burglarers, where no ma[gis]strate is at hand; also to apphend without warrant such as are ovr taken with drinke, swearing, breaking ye Saboth, lying, vagrant psons, night walkers, or any other yt shall break o[u]r laws; . . . also to make search for all such psons . . . in all houses licensed to sell either beare or wine, or in any othr suspected or disordered places, & those to apphend, & keepe in safe custody . . .[5]

In the performance of their duties, some people ignored the authority of the constables. In 1643, when Job Tyler was summoned to court by the constable, "he sd he car'd not a fart [or] turd for all their warrants."[6] Both men and women broke the law. In 1672 the Court convicted Gillian Knight of "Enticing Danll Herring to her house & there Embracing him pick't his pocket and Stole Seven Shillings from him."[7] That same year, the Widow Alice Thomas was convicted of running a brothel by giving "frequent secret and unseasonable Entertainment in her house to Lewd Lascivious & notorious persons of both Sexes, giving them opportunity to commit carnall wickedness."[8]

marketing and service centers. The third period (1840–1875) was characterized by an expansion and realignment of the urban system in response to early industrialization, the mechanization of agriculture, and immigration. With these changes in mind, we will discuss some of the key principles underpinning urban growth, urban systems, and the spatial pattern of central places.

With the fourth distinctive period, industrialization (1875–1920), we will see the effects of principles of industrial location on the development and adaptation of the urban system. We will also notice how and why urban-industrial development is inherently uneven and unstable—part of a constantly changing landscape of investment, disinvestment, and reinvestment by people.

The fifth period corresponds with the emergence of Fordism and mass-produced cars, trucks, and airplanes (1920–1945), which significantly changed the spatial organization of the urban system. This was also a period that saw some important changes in corporate organization by businesspeople and a severe economic depression. We will see how these developments affected the fundamental organization of the U.S. political economy and, therefore, the very foundations of urban development and urban life.

FRONTIER URBANIZATION

Although there were many small urban settlements in North America before the sixteenth century (Figure 1), the first large towns and cities were those that Europeans established

Buddy Mays/Corbis Images

FIGURE 1 Tyuonyl Anasazi Pueblo in Frijoles Canyon, Bandelier National Monument, New Mexico. Tree-ring analyses date construction of this settlement of community houses and central plaza within a stone oval-shaped enclosure to between 1383 and 1466 C.E.

as outposts of their economies. Spanish colonialism was the first to leave its imprint. Guided by the Laws of the Indies (dating from 1583), Spanish settlers in Florida and the Southwest planted towns that were laid out with a rectangular grid of streets surrounding a large central plaza. The earliest of these planned communities was La Villa Real de Santa Fe (Santa Fe, New Mexico), founded in 1610 as the administrative center for New Spain's northernmost frontier. Over the next 150 years or more, Spanish settlers founded a series of "pueblos" (centers of commerce and administration), "missions" (centers for religious conversion), and "presidios" (military outposts), all of which, as they grew, acquired a mixture of commercial, administrative, religious, and military functions. Among the settlements founded by Spanish colonists were Saint Augustine, San Antonio, Santa Barbara, Los Angeles, San Diego, and San Francisco. Not long after the founding of Santa Fe, Dutch settlers sailed into the Hudson estuary and established a fur trading post they called New Amsterdam. The French were less interested in colonization than in opportunities for trade, but as they foraged through the Great Lakes and along the Mississippi River system they established trading posts that slowly grew into small towns. These included Québec, Montréal, Detroit, St. Louis, and New Orleans.

It was English colonization, however, that established the most vigorous roots of the U.S. urban system. In Virginia, the Jamestown colony, founded in 1607, began the cultivation of tobacco for export to Europe and established the first representative government on the continent. Nearby Williamsburg, established in 1663 as a stockaded refuge from Native Americans, became the capital of the Virginia colony in 1699 after fire destroyed much of Jamestown. Although profitable, tobacco cultivation was labor intensive. In 1619 the first Africans were brought to Virginia to work on the tobacco plantations. By 1807, when Britain banned its subjects from participating in the slave trade, 600,000 to 650,000 Africans had been forcibly transported to North America.

In New England, Boston, founded in 1630 as a "City upon a Hill" in a celebration of spirituality, soon became a prosperous center of trade and commerce. Newport, on Rhode Island, was founded by a group of religious dissenters in 1639 as a haven for the persecuted, but, with the best natural harbor in southern New England, it too developed into a trading port. In 1664 the British took New Amsterdam from the Dutch, renamed it New York, established it as the capital of the New York colony, and set about developing the best natural harbor on the entire Atlantic seaboard into a major port. Charles Town (later Charleston) was established as the capital of the Carolinas in 1680, and William Penn established Philadelphia as the capital of Pennsylvania in 1682 (Figure 2).

The embryonic urban system operated as a string of **gateway cities**: control points for the (1) assembly of staple commodities for export, (2) distribution of imported manufactured goods, and (3) civil administration of the new territories. For some time each gateway port operated quite independently, having more linkages with European cities than with each other. As colonization extended, a hierarchy of settlements began to develop. Places with better resources and accessibility became larger and acquired a broader range of services. Meanwhile, the need to rationalize transatlantic shipping schedules prompted the development of a network of coastal shipping between the largest ports. As a result, a few cities—Boston (Figure 3), Charleston, Newport, New York, and Philadelphia—were able to establish themselves as major **entrepôts**: intermediary centers of trade and transshipment. As these entrepôts grew, they came to dominate larger and larger **hinterlands** (market areas) in which smaller settlements emerged as local market towns. In time these market towns became *inland gateways* that acted as bulking points and provided an array of services for the frontier farmers. By the time of the Revolution in 1775, the most notable were Hartford, Middletown, and Norwich (Connecticut), Albany (New York), Lancaster (Pennsylvania), and Richmond (Virginia). As the 13 colonies merged into an American union, the largest city was New York, with about 25,000 inhabitants. Philadelphia was almost as large, with about 24,000; Boston was next biggest, with about 16,000; and Baltimore, Charleston, and Newport all had between 10,000 and 12,000 inhabitants. The inland gateway cities were all relatively small, none exceeding 10,000 people.

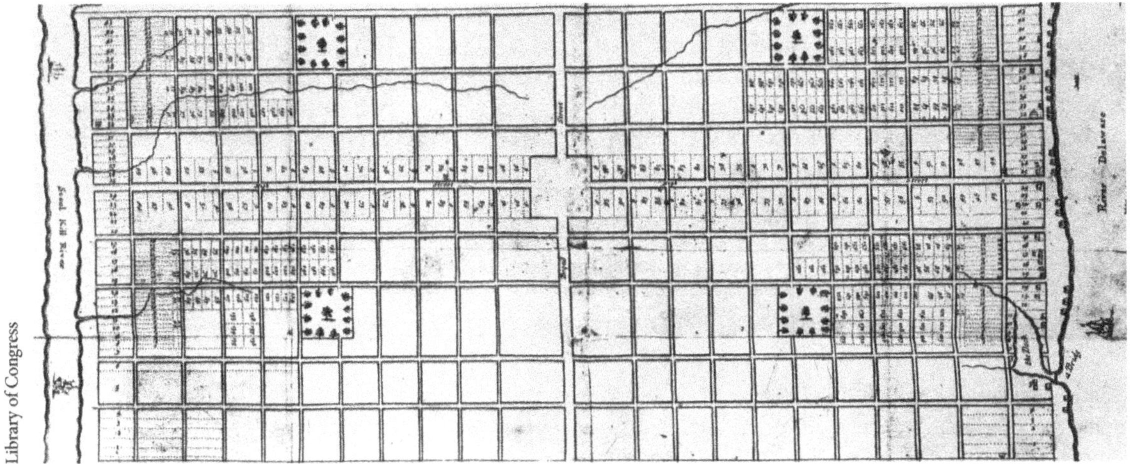

Library of Congress

FIGURE 2 Plan of Philadelphia, 1683, laid out with primary north-south and east-west streets intersecting at a central town square containing public structures such as government buildings and churches, with grid blocks delineated by secondary streets and four minor squares. The preplanned layout facilitated settlement by allowing colonists to select a parcel for their new home before leaving Europe.

Francois Havermann/Corbis Images

FIGURE 3 Boston in 1770. Boston was important during the early development of the U.S. urban system because of its role as an entrepôt, an intermediary center of trade and transshipment between colonial America and cities in northwestern Europe.

THE MERCANTILE PERIOD (1790–1840)

Although only 1 in 20 people lived in towns or cities, city-based newspapers, lawyers, and merchants greatly influenced the Constitution that was framed in Philadelphia in May 1787. The result was a Constitution that favored city-based manufacture and trade—*and therefore stimulated city growth*—by preventing the fragmentation of the economy were states allowed to set their own import taxes, coin their own money, or issue their own bills of credit. The political independence gained by the colonies as a result of the Constitution stimulated the development of the urban system in other ways:

- It became practical—and necessary—for economic links to be established between towns and cities whose main linkages under the colonial system had been with European ports.
- It meant that a much greater proportion of investment was financed by U.S. capital, with the result that fewer profits "leaked" back to the European urban system.
- It required a proliferation of government functions, from county courthouses and town halls to state capitals and, of course, the development of the District of Columbia, chosen in 1790 as the permanent seat of federal government.
- It was associated with westward expansion (following the Northwest Ordinances of 1785 and 1787, the Louisiana Purchase of 1803, and the addition of Texas and the territories acquired from Mexico in 1846) which required frontier towns like Santa Fe that subsequently matured into local service centers.

The most striking growth occurred in gateway cities—New Orleans (Figure 4), St. Louis, and other river ports—that were located at strategic points along rivers that linked the new western territories with the larger cities of the Atlantic seaboard. East Coast merchants, however, were not content to miss out on the lucrative trade. Their response was to tap the waterways of the Great Lakes and the Hudson, Ohio, and Mississippi river systems with networks of canals. As a result, two east-west corridors of trade merged. The first stretched from New York up the Hudson and the Erie Canal (opened in 1825) to the eastern Great Lakes, where Buffalo, Cleveland, Detroit, Chicago, and Milwaukee emerged as important wholesaling centers.

National Archives and Records Administration

FIGURE 4 The most striking urban growth during the Mercantile period took place in river ports. New Orleans grew very rapidly because of its situation as a gateway to the entire Mississippi system. This view is from about 1862.

The second stretched across the mountains from Philadelphia and Baltimore to Pittsburgh and the Ohio Valley, where Cincinnati and Louisville became important inland gateways.

As the urban system expanded and trade between cities increased, particular cities and regions were able to specialize according to their **comparative advantage** (the economic activity, given local conditions, that could be undertaken most efficiently, compared to other places). Cincinnati, for example, specialized in hog processing, earning the nickname "Porkopolis." Manufacturing began to contribute to the growth of the leading eastern ports, while some towns in the more heavily populated and intensively developed northeast—Albany, Lowell, Newark, Poughkeepsie, Providence, Springfield, and Wilmington—became the cradle of the Industrial Revolution in the United States.

Immigration provided an important source for peopling both the frontier and the growing ports and inland gateways. By the American Revolution (1775–1783), the colonial population had reached about 2.5 million, of which more than 500,000 were African-American slaves, 250,000 were Scots-Irish, and 200,000 were German. Meanwhile, an increasingly important impetus to urban growth was improved agricultural productivity. Urbanization was fueled by advances in farm productivity that (1) provided the food to support the increased numbers of townspeople and (2) released agricultural workers who moved to towns and cities, swelling the numbers of producers and consumers.

By 1840 the U.S. urban system had become independent and was on the way to becoming integrated. New York had grown to 391,114 and had increased its lead over Philadelphia (93,665) by a significant margin. Meanwhile, Baltimore and New Orleans topped 100,000 and Boston had grown to more than 93,000. A few regional centers such as Cincinnati and Albany had populations between 25,000 and 50,000, while larger centers such as Louisville and Richmond stood at about 20,000 (roughly the size of Washington, D.C.). Most towns had populations of 15,000 or less; Chicago had fewer than 5,000 people.

Inside the Mercantile City

Just as we can consider the urban system as both the result of and the framework for processes of economic, demographic, social and other changes, so too can we view the spatial form and organization of cities themselves. Changing patterns of urban form and land use must, like the changing urban system, be understood in relation to the rhythms of the economy and society. The evolution of urban form stems not only from the history of economic development, migration, and immigration but also from the interaction of economic and demographic growth phases with changes in social structure and lifestyles, innovations in building materials and construction techniques, advances in urban transportation, and changes in the legal framework of land ownership, land use law, and land use policy.

URBAN VIEW 2
Vance's Mercantile Model

One model of the development of the North American urban system is the *mercantile model* by urban geographer James Vance, Jr. According to Vance, external (European) influences and long-distance trade were particularly important in establishing the geometry of the North American urban hierarchy through five distinctive stages (Figure 5).

1. *Exploration.* Voyages of exploration to North America by entrepreneurs from Europe in search of economic opportunities.
2. *Harvesting natural resources.* Colonists establish settlements in order to exploit natural resources such as the codfish of the Grand Banks off Newfoundland and timber and beaver pelts from New England.
3. *Emergence of agricultural production and coastal gateway cities.* The colonists' permanent settlements and farms become associated with the development of an embryonic urban system geared to exporting agricultural staples such as grain, salted meat, indigo (blue dye), tobacco, and cotton to Europe, and importing manufactured goods and luxury items from Europe. Gateway cities along the coast (Vance's "points of attachment") become the focus of the emergent system (Figure 5).
4. *Establishment of inland gateway cities.* Settlement spreads farther inland due to continued demand for agricultural exports and increased colonization. This requires the development of long-distance routes and inland gateway cities that serve as "depots of staple collection" at strategic locations along these routes (Figure 5). This stage corresponds to the "frontier urbanization," with entrepôts and inland gateway cities functioning as wholesale collection centers for agricultural products intended for export to Europe.
5. *Domestic market and urban system infilling.* The domestic market grows large and affluent enough to sustain the growth of domestic manufacturing. The gateway ports, entrepôts, and inland gateways attract much of this economic activity because of their established populations and good accessibility. Meanwhile, colonists spread out from the long-distance routes and establish agricultural settlements that begin to support subregional systems of market towns. But the established ports, entrepôts, and inland gateways, with much larger markets and better distribution networks, are the places that can offer the most exclusive and expensive goods and services. So long-distance trade, rather than local markets, establish the spatial pattern of the cities, like New York, Boston, and Chicago, that come to function as the leading centers of the maturing urban system.

(continued)

URBAN VIEW 2

Vance's Mercantile Model (*continued*)

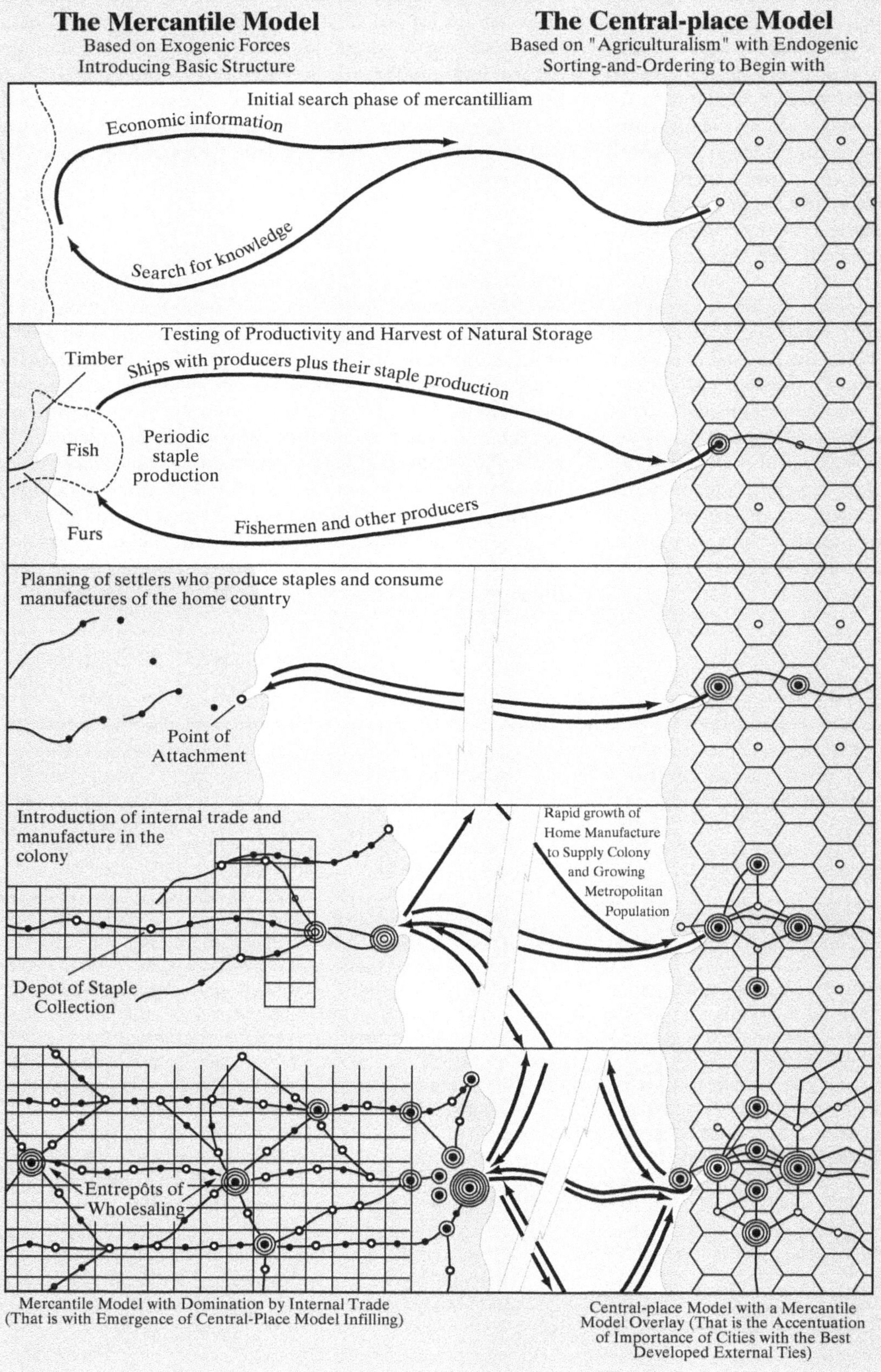

FIGURE 5 Vance's model of mercantile settlement.

The Merchant's World: The Geography of Wholesaling, James Vance Jr, Prentice Hall (1970). Reprinted with permission.

Within each phase of urban development, many consider innovations in transport systems to have been the single most significant determinant of urban form and land use. Transport systems controlled the density and areal extent of urban development and gave expression to the pent-up energies of successive phases of economic and social change.

The Pedestrian City In the mercantile city, the character of economic development and the lack of fast inexpensive forms of transportation gave rise to very compact cities with distinctive patterns of land use. People walked, and most goods were moved by hand cart or horse cart. The result was a loose intermingling of activities. Because most towns and cities of any size were seaports or river ports, the hub of economic activity was the waterfront, dominated by merchants' offices, workshops, warehouses, and wharves. Clustered nearby were hotels, churches, retail stores, and public buildings, together with the homes of prominent families. Housing for artisans, storekeepers, and laborers edged into vacant spaces and extended to the edge of town, where commercial activities that needed extra space (such as textile mills), large quantities of stream water (breweries for example), or were particularly noxious (like slaughterhouses and tanneries) located.

There was little separation between home and workplace. Factory owners often built their homes next to their factories; artisans and storekeepers lived above or behind their workshop or store; laborers and service workers lived off alleyways and in lofts; servants lived in the upper floors of their masters' houses; and, in Southern cities, slaves lived in compounds behind the main house. As cities increased in population and density, enclaves of specialization began to emerge around clusters of workshops and factories and around concentrations of ethnic groups. Boston's North End, for example, became a distinctive Irish quarter in the early nineteenth century. Nevertheless, the mercantile city was so compact that short distances separated poor from rich and artisan from laborer, and no enclave was large enough to be thought of as a separate specialized district.

Models of the Mercantile City Bearing some resemblance to the *pedestrian city*, the best-known and most influential model is Sjoberg's *preindustrial city*[9] in which the social pyramid—a large group of outcasts and laborers, a smaller group of artisans, and an even smaller elite group—is reflected in the city's spatial pattern (Figure 6).

This generalization has been questioned by James Vance, Jr.,[10] who gives greater emphasis to the *mosaic of occupational subdistricts* and downplays the extent of a fringe of low-income laborers, placing them instead as lodgers scattered within these various subdistricts. What is important, however, is not so much these details as the overall structure. Both Sjoberg and Vance describe preindustrial cities as having

- a central core dominated socially by the residences of an elite group.

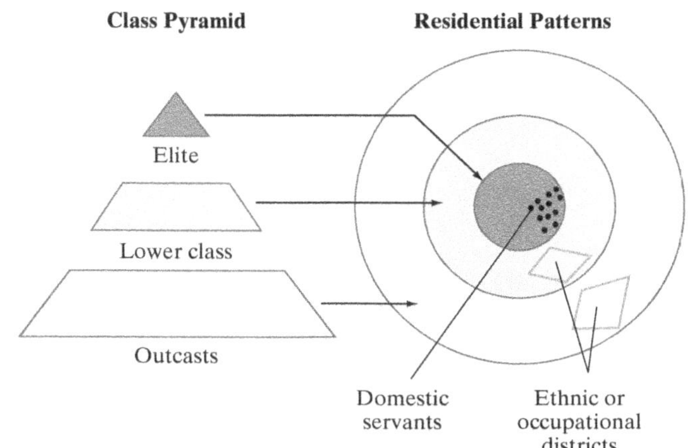

FIGURE 6 The relationship between class structure and place of residence in the preindustrial city (after Sjoberg).

"Testing the model of the pre-industrial city," JP Radford, *Transactions of the Institute of British Geographers* 4. Reprinted with permission of Blackwell Publishing Ltd.

- a number of occupationally distinctive but socially mixed districts.
- a residual population of the very poor living in the back alleys and on the fringes of the city.
- everything at a human scale: a walking city in which home and work were tightly connected by the organization of work into patriarchal and familial groupings.

These features, common to preindustrial cities around much of the world, could be found in cities of the early nineteenth century: Charleston, S. C., for example (the sixth-largest city in the United States until 1830).[11] Beyond these broad parallels, however, the truth is that U.S. cities exhibited all kinds of exceptions to any set of generalizations. They were young, still shaped by local, idiosyncratic, and contingent factors. They were growing rapidly and had no time to "shake down" into the classic preindustrial patterns of the cities of medieval Europe and feudal Asia. And before they could, they were overtaken by the revolutionary forces of industrialization.

EARLY INDUSTRIAL EXPANSION AND REGIONAL REALIGNMENT (1840–1875)

The transition from a trading economy to a mature agricultural and embryonic industrial one took place during the 1840s for several reasons. One reason was the arrival of industrial technology and methods of industrial and commercial organization from the hearth of the Industrial Revolution in Europe. Another was accelerating improvements in agricultural productivity and the ability to colonize formerly marginal land as a result of mechanization that had begun in the mercantile period. With less labor needed on farms, **rural-to-urban migration** fueled the urbanization process. At the same time, higher levels of agricultural production were able to feed the growing numbers of immigrants who were also fueling urban growth and providing the labor needed in new factories (Figure 7).

FIGURE 7 New arrivals. Crowding the deck on arrival in New York harbor, these immigrants were processed on Ellis Island before going on to find jobs in the fast-growing cities.

Beginning in 1840, a great influx of immigrants began to arrive from Europe (Figure 8). Severe famines in Ireland in the 1830s and 1840s, combined with a British "laissez faire" policy and the rationalization of agriculture from small farms to larger estates, fueled mass migration. By 1850, the economic and social changes associated with rapid industrialization and the mechanization of agriculture had prompted an even greater exodus of people from the German states, France, and Belgium, extending to Scandinavia by the late 1870s and sparking waves of Danish, Norwegian, and Swedish immigrants, and spreading to southern and eastern Europe—involving Italians and Jews, as well as Slavs from the Austro-Hungarian Empire—during the 1880s and 1890s.

Because many of the new industrial technologies had specific locational requirements (proximity to large quantities of raw materials, for example, or accessibility to a variety of different suppliers or markets), this phase of urbanization brought the emergence of some new towns and cities and the rapid growth of previously very small settlements. There were four categories of newcomers:

1. *Power sites* that attracted industries that consumed significant quantities of energy. Before the widespread use of coal-fired steam technology and, later, electricity, the power of falling or running water was important. This led to the appearance of a series of industrial towns along the Fall Line in New England and the eastern margins of the Appalachians, such as Allentown and Harrisburg in Pennsylvania.
2. *Mining towns* that sprang up to provide the coal and ores needed by the industrial economy, such as Appalachian coalfield towns like Norton, Virginia.
3. *Transportation centers* that emerged at strategic locations made accessible by the canal and railroad network, such as Roanoke, Virginia, a typical example of a railroad town.
4. *Heavy manufacturing towns* whose dependence on large volumes of raw materials tied them to the source areas of those inputs. Pittsburgh, already an important river port and wholesaling center, would become *the* steeltown thanks to its location near coalfields and deposits of iron ore.

The development of steam-powered riverboats and the railroad network were central to the evolution of the new industrial economy and urban system. Ports and lakeside cities such as Chicago, Cincinnati, Memphis, and Nashville prospered because they were able to operate as interfaces between established trading routes and the budding railroad system.

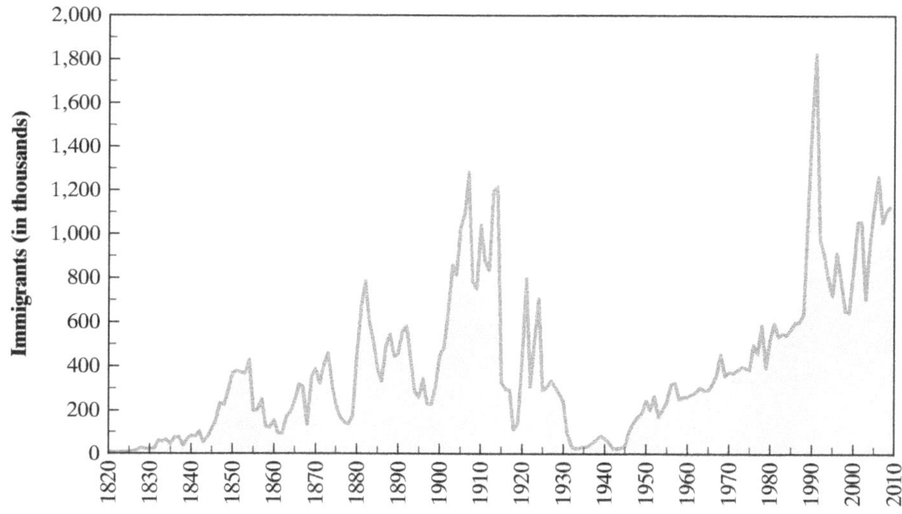

FIGURE 8 Immigration to the United States.

Within a few years of work on the first railroad (the Baltimore and Ohio, in 1828), rail networks were developed around all the established ports on the inland waterways that reached into the agricultural lands of the interior plains. Initially, therefore, the railroads were complementary to the waterways as long-haul carriers of general freight.

In 1869 the railroad network reached the Pacific, when, at Promontory, Utah, the Union Pacific railroad (built west from Omaha) met the Central Pacific railroad (built east from Sacramento). By 1875 intense competition between railroad companies began to open up the western prairies as far as Minneapolis-St. Paul and Kansas City and had extended the urban system to a continental scale. The significance for economic development—and therefore for urbanization—was enormous. The railroad allowed a loose-knit collection of regional economies to develop into a national one in which American enterprise could fully exploit the commercial advantages and **economies of scale** of a huge market and apparently unlimited resource base.

Meanwhile, the development of the railroads realigned the spatial organization of the urban system (Figure 9). The westward penetration of railroad lines allowed large quantities of corn and wheat to be moved directly eastward, rather than being shipped by water via St. Louis and New Orleans. As a result, these cities, along with smaller, intermediate ports, experienced slowed rates of growth and relied increasingly on regional trade and service functions. In contrast, cities along the two major east–west wholesaling alignments (New York–Buffalo–Detroit–Chicago–Milwaukee and Philadelphia–Pittsburgh–Cincinnati–Louisville) benefited from both the extra trade and reduced freight prices that resulted from fierce competition between the railroads and water transportation. Their growth helped lay the foundations of what would become the Manufacturing Belt.

This chronology points us to another important generalization about the development of urban systems: *Most early industrial growth, and therefore most urban growth during early industrialization, occurred in the largest existing towns and cities.*

By 1875 the urban system had expanded to the point where more than 15 cities, with their adjacent counties, each accounted for more than 100,000 people. New York, with 1.3 million, was at the top of the urban hierarchy, followed by five cities with populations in the range of 350,000 to 450,000: Baltimore, Chicago, Philadelphia, Pittsburgh, and St. Louis. Cities with between 100,000 and 150,000 included Athens (Georgia), Boston, Buffalo, Cincinnati, Cleveland, Detroit, Manchester (New Hampshire), New Orleans, Providence, Rochester, and Syracuse.

Some Principles of Urban Growth

The reasons for the concentration of growth in the largest existing towns and cities stem from the principle of **initial advantage** based on:

- The craft industries and wholesaling and transportation activities that provided the basis for industrial enterprises as their owners used their capital and accumulated profits to invest in factories and machinery.
- Traditions and skills of entrepreneurship, investment, and lending that adapted easily to industrial development.
- The largest pools of labor.
- The largest and most affluent markets.

These observations are related to fundamental principles inherent to the process of urban-industrial development. The principle of initial advantage is itself a special case of the operation of **external economies**: cost advantages that accrue to individual companies because of their locational setting. External economies can be derived from the availability of a pool of labor with appropriate skills, accessibility to good specialized business services, or the quality of the physical infrastructure of roads, harbors, and utilities—any advantage, in fact, that stems from the collective rather than exclusive use of any of the factors necessary for profitable activity.

Urban settings then provide companies with opportunities for external economies. In this context, external economies are often referred to as **agglomeration economies** or **urbanization economies**. Where external economies are limited to companies in a particular industry, they are known as **localization economies**. Examples include sharing a pool of labor with special skills (such as shipwrights, precision instrument makers), supporting specialized technical schools, collaborating to develop a research institute or marketing organization, and drawing on specialized subcontractors, suppliers, distribution agents, and legal counsel. These localization economies go a long way in explaining the tendency for cities to maintain economic specializations: the attractiveness of Pittsburgh for the iron and steel industry, of Akron to manufacturers of rubber products, and of Dayton to makers of fabricated metal and machinery, for example.

Interpreting and Analyzing the Urban Hierarchy and the Central Place System

What was emerging was an urban system that had much in common, in terms of the size and spatial patterning of towns and cities, with urban systems in many other countries. Of special importance to social scientists is that certain regularities emerged that cried out for explanation. Among the regularities in the mercantile and early industrial expansion periods were consistent relationships between (1) cities' rank in the urban system and their actual population—the so-called rank-size rule, and (2) the size and spacing of towns and cities—the basis of central place theory.

The Rank-Size Rule The relationship between cities' size and their rank within an urban system is expressed in a simple formula, known as the **rank-size rule**:[12]

$$P_i = P_1 \div R_i$$

where P_i is the population of city i, R_i is the rank of city i by population size, and P_1 is the population of the largest city in

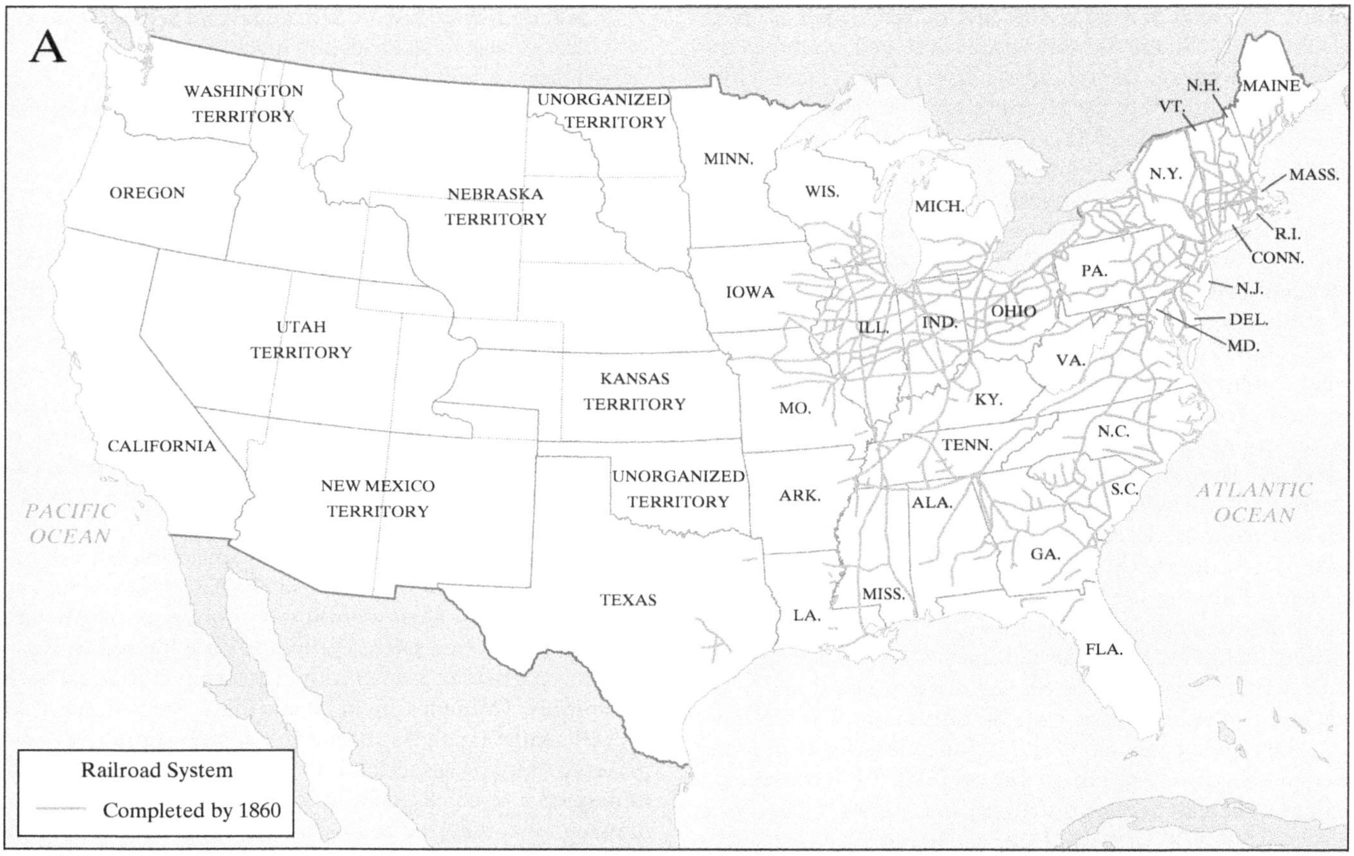

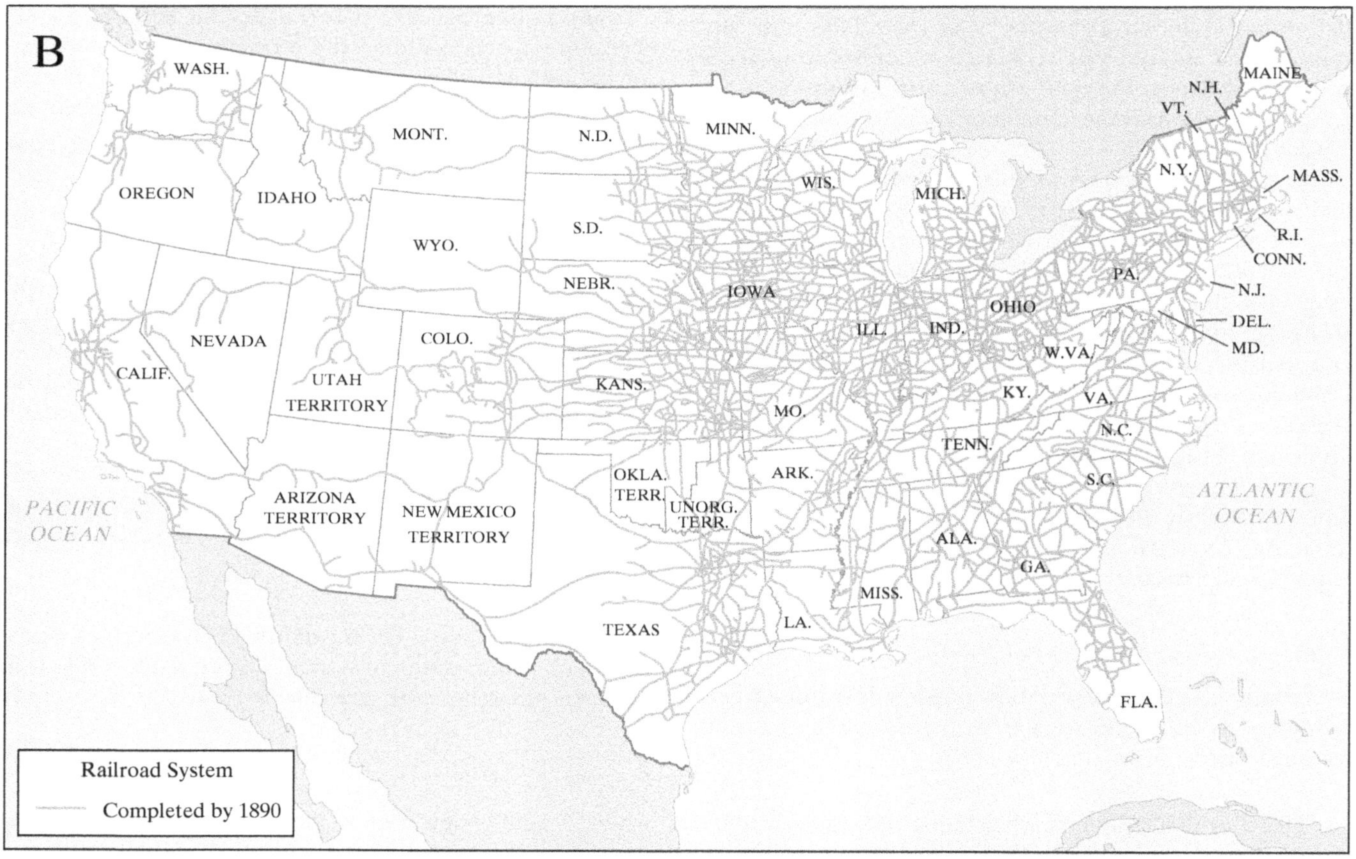

Association of American Railroads.

FIGURE 9 (A) The U.S. railroad system in 1860 with over 30,000 miles of track, much of it short lines. (B) The U.S. railroad system in 1890. The railroads had expanded rapidly in the 1880s, when more than 70,000 miles of track were added. By 1890, the total network was 163,597 miles.

the urban system. So if the largest city has a population of 1 million, the fifth-largest city should have a population of 200,000, the hundredth-ranked city should have a population of 10,000. Plotting this relationship on a graph with a logarithmic scale for population sizes would produce a straight line, reflecting a *log-linear* relationship. The rank-size distribution of the U.S. urban system plotted for several points in time in Figure 10 shows that the system as a whole has conformed fairly consistently to the rank-size rule. This conformity to the rule, however, does not necessarily mean that individual cities have maintained their relative position along each curve. The growth of some cities (such as San Diego) has sent them from the lower end of the hierarchy to the top, while other cities (like Savannah) have declined, at least in relative terms, so that they have fallen down the hierarchy. *It is the overall relationship between cities' size and their rank within the urban system that stays fairly constant.*

In some urban systems (though not in the United States), the disproportionate size of the largest city distorts the log-linear character of the rank-size distribution. In the United Kingdom, London is about seven times the size of Birmingham, the second-largest city, while in Ireland, Dublin is more than four times the size of the second-largest city, Cork. Such cases led to the "law" of the **primate city**: "A country's leading city is always disproportionately large and exceptionally expressive of national capacity and feeling."[13] But for every case of **primacy** there is an urban system where the largest city approximates to the rank-size rule. The search for explanations of primacy has focused on levels of economic development and the size of a country's inhabited area, but neither factor stands up to empirical testing. More likely, primacy is a good example of the *contingency*, with each case arising from a particular combination of circumstances.

Central Place Theory The significance of trade and marketing also introduced certain regularities in the size and spatial patterning of towns and cities within the urban system. In part, this regularity was a result of transportation technology: Settlements evolved to match the distances that could be covered by river, canal, or turnpike (toll road) in a day's travel. In addition, the role of towns and cities as local service centers—central places—brought a new logic that helped shape the spacing of towns and cities by size. Smaller settlements, offering a limited range of services for their residents and the residents of nearby areas, were quite numerous and located at relatively short, fairly consistent distances from one another. Large cities were fewer and farther between but offered a greater variety of services, catering to residents from many surrounding areas. In between were intermediate-sized central places serving intermediate-sized markets with middling packages of services.

It was not until much later that these patterns were considered in terms of underlying principles and idealized outcomes. The seminal work by Walter Christaller, a German geographer, identified striking regularities in the size and spacing of settlements in southern Germany.[14] For the sake of clarity and simplicity of argument, Christaller first assumed a uniform landscape, uninterrupted by rivers, roads, or canals: Accessibility is assumed to be a direct function of distance, in any direction. From this starting point, the foundations of central place theory depend on elementary principles concerning the "range" and "threshold" of goods and services.

The *range* of a particular product or service is the maximum distance that consumers will generally travel to obtain it. "High-order" goods and services such as specialized equipment, professional sports events, and specialist medical care (relatively costly and required infrequently) might have a range of a hundred miles or more, whereas "low-order" goods and

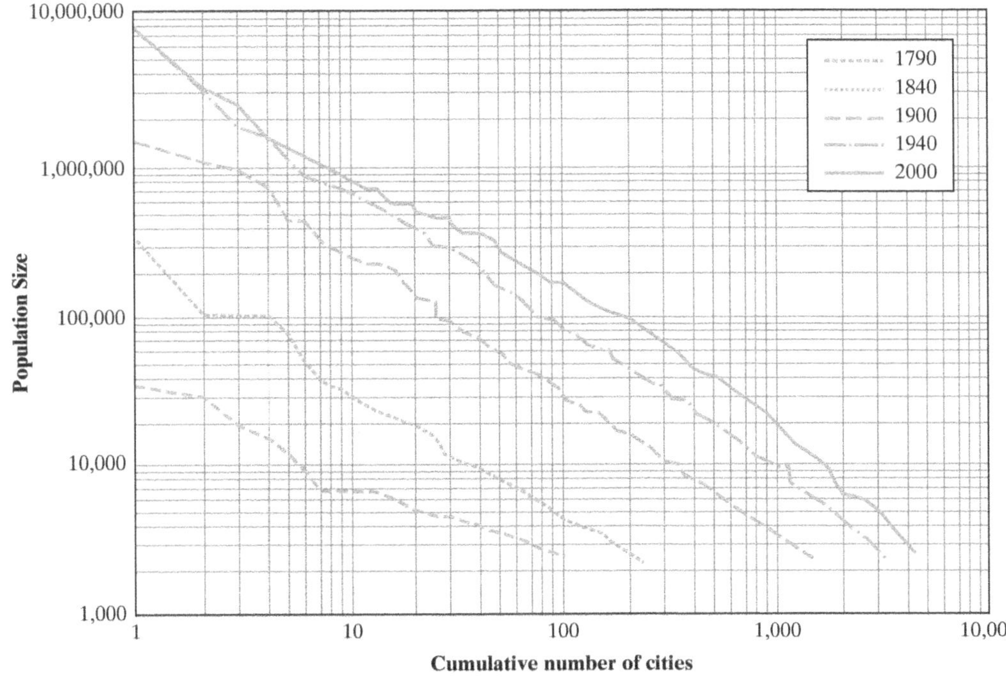

FIGURE 10 The rank-size distribution of U.S. towns and cities.

services such as grocery stores and gas stations (perishable or required in relatively large amounts at frequent intervals) will have a much shorter range. It follows that the maximum area served by a particular service from a particular central place will be a market area, or hinterland, whose radius is equal to the range of the service.

The *threshold* of demand of a product or service is the minimum market size of potential customers in a hinterland required to make it profitable to offer that product or service. High-order services such as hospitals have thresholds in the tens of thousands; low-order services such as grocery stores have thresholds of just 200–300 people.

From these principles we can hypothesize an urban system in which the smallest settlements, all offering similar packages of low-order goods and services—such as a pizza joint or coffee shop—are arrayed evenly across the landscape. Higher-order central places will, logically, offer all of the lower-order goods and services, and it is assumed that consumers will use the nearest central place that offers the goods or services they require. As a result, a *nested hierarchy* of central places can be constructed, shown schematically in Figure 11, with marked discontinuities in the population size and packages of services offered at each level of the hierarchy.

Beyond Consumer Hinterlands One of the main shortcomings of central place theory is that, by focusing on central place functions, it fails to address other important influences on the size and spacing of cities—manufacturing, for example, and the long-distance trading functions (such as wholesaling) that are essential to the development of most urban functions, whether they stem from the agricultural, industrial, or service sector of the economy. Many central place functions, even by the mid-nineteenth century, depended on a long chain of production and exchange that involved numerous steps, from harvesting or mining through processing, packaging, shipment, and storage to retail distribution. The products offered for sale in Albany, Buffalo, and Cleveland, for example, would have originated from all corners of the United States, from many parts of Europe, and from a few other parts of the world.

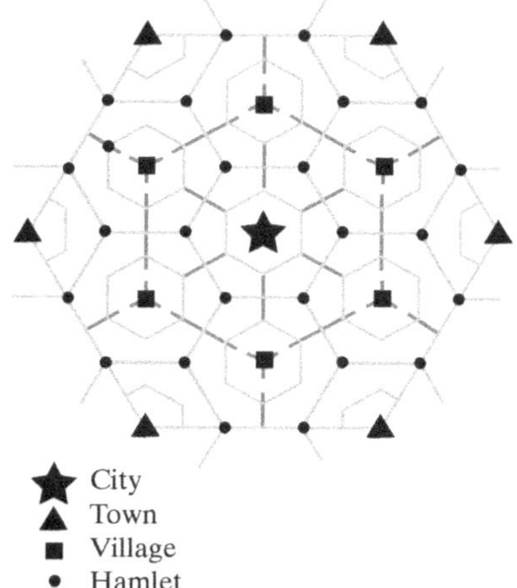

FIGURE 11 Schematic diagram of a central place system.

Another shortcoming of central place theory is that it provides only a *static* model of settlement patterns—unresponsive to changes in population densities, consumer spending power, or transportation technologies—and paints a picture of an urban system "that somehow springs into existence fully formed, with no historical antecedents."[15]

Inside the Early Industrial City

Probably the most fundamental change to emerge to the form and spatial organization of cities was the intense competition that developed for the best and most accessible sites for the factories and the warehouses, shops, and offices that depended on them. Meanwhile, industrialization generated three new social groups: an industrial business elite, white-collar managers and office workers to run the businesses, and blue-collar factory workers to operate the machinery. It was not long before these groups were in fierce competition for the best and most accessible sites for housing. Social status, newly ascribed in terms of money, became synonymous with ability to pay rent, so that a person's address came to acquire unprecedented importance.

Ultimately, these changes would turn the city inside out, with specialized industrial and commercial uses claiming most of the central area, the rich exchanging their congested and polluted central locations for the peripheral location of the poor, and occupational clustering giving way to residential differentiation by income.[16] But it took businesses, households, and landowners time to sort themselves out into these new patterns. What is more, they had to do so in an unregulated environment and within a legal framework in which land ownership was treated as a civil liberty rather than as an economic or social resource. Most of all, growth had to be accommodated without any major technological advances in transportation. For a whole generation cities were brimming with new arrivals and jammed with new factories, warehouses, and institutions, with no adequate means of transportation that might have helped to rationalize the organization of space (Figure 12). Meanwhile, crucial precedents were being set in terms of land law and the land development process.

Urbanization and the Public Interest

The early colonists from England had a strong desire to avoid the rigidity and inequality of feudal land laws. They found a continent with almost unlimited possibilities for land ownership (Native Americans notwithstanding), and established a simple approach to land law that was based on individuals' civil liberties. Any person would be able to own land and in return would be obliged to pay taxes. There were to be no constraints on who might buy the land, inherit it, or for what purpose it might be used. The Northwest Ordinances of 1784–1787 and the federal Constitution ratified and codified these principles, adding (through the Fifth and Fourteenth Amendments) the principle that no government agency should seize any private property except under due process of law (eminent domain) and then only for legitimate public purpose and with just compensation.

John L. Simbert/Corbis Images

FIGURE 12 The early industrial city: Boston. This photograph shows Quincy Market and Faneuil Hall in the heart of the city, where industrial, retail, wholesale, commercial, and residential land uses were crammed together.

URBAN VIEW 3
Immigrant Housing Conditions[17]

By the second half of the nineteenth century, many immigrants who had flooded into larger cities were living in horrific conditions. The term *slum* conveys the interrelationships among the environmental limitations, spatial isolation, and social problems of the largely immigrant populations (Figure 13). A survey of housing in Manhattan in 1867 by the Metropolitan Board of Health inspected over 18,000 buildings that housed more than three families. Many were tenements—the former homes of individual wealthy families that had been partitioned into cheap multifamily units for immigrants. More than half were in bad condition, and in one-third of these tenements neglect was the main reason.

Overcrowded rooms, subdivided buildings, congested lots, and inadequate facilities were all common strategies used by landlords to keep rents low without reducing the return on a property. Reports on the condition of housing for the poor regularly mentioned how buildings were sublet to an agent for a fixed return. The agent would then obtain the maximum return by neglecting repairs and services and demanding the highest possible rent. For their part, this arrangement allowed absentee owners to shirk direct responsibility for their property or tenants.

In 1870 the newly formed Massachusetts Bureau of Statistics of Labor concluded in its first annual report that:

> low-paid laborers are not earning sufficient wages to justify their living in [improved] tenement houses . . . and, as there are no intermediate houses between these and the lowest class of houses . . . the laborers of this class are practically compelled to crowd into the miserable refuges in which they now congregate. In fact, so far as we have been able to ascertain, there are no places within the settled portions of the city of Boston where the low-paid toiler can find a house of decency and comfort.

Jacob Ris/Corbis Images

FIGURE 13 Back alley in New York City in the late 1800s. By the second half of the nineteenth century, many of the immigrants who had flooded into larger cities were living in horrific housing conditions. Many buildings were tenements—the former homes of wealthy single families that had been partitioned into cheap multifamily units for recent immigrants.

Even so, treating land ownership as a civil liberty has had some profound consequences for the nature of urban development. When public lands came to be disposed of in the nineteenth century, much of the country was delivered up to the control of private speculators:

> Those who had private capital, or who were bankers or agents for Eastern money, had an enormous advantage. Such men could purchase a whole valley, a promising townsite, whatever they wished. They could then sell part of it cheaply to settlers and retain large tracts and await the substantial price rise that would follow in the wake of the development of adjacent land.[18]

The influence of speculators was reinforced by their role as moneylenders to pioneers, who were often able to pay cash for their land, but did not have enough for capital improvements like buildings. As a result, the lending of money gave speculators strong local control over land owned by others. A further consequence was the manipulation by speculators of the locations of key investments, both public and private. The concentration of land ownership and influence over land use decisions in the hands of speculators "settled a vast class of conservative interest across the nation: It installed men who saw the duty of government to be the defense of private property in every town in North America."[19]

The onset of industrialization began to reveal some of the disadvantages of enshrining property rights as a civil liberty. With no controls over land use and little regulation of the sale of land, urban development became a free-for-all. Although individual freedoms were protected, cities provided less stable settings for businesses, less convenient settings for their residents, and less healthy settings for everyone.[20] These considerations eventually led to the acceptance of land use zoning laws and city planning.

Instruments of Change: Horsecars and Railroads

The appearance of all kinds of industrial land uses, together with the arrival of thousands of immigrants, and the emergence of what they saw as a massive and threatening proletariat, soon gave the upper socioeconomic groups a strong motivation for moving permanently as far away from them as possible. (The wealthy had already been sending their families to the countryside for the uncomfortable and epidemic-prone summer months.) Similarly, the intensifying stresses and conflicts of city life repelled many of the new class of affluent white-collar workers.

There was, therefore, strong demand for the first forms of public transport to provide an opportunity to escape. A few families, wealthy enough to afford private carriages or hire hackney carriages, had moved to exclusive exurban settings at the first signs of industrial squalor. But it was not possible for others to join them until the development of horse-drawn omnibus and horsecar systems, and short-haul passenger railroad routes.

Horsecars The first horse-drawn omnibus service was established by Abraham Brower in New York in 1829, picking up and setting down passengers along the length of Broadway for a flat fare of 12 cents. By the 1840s there were several hundred omnibuses in New York, and transportation companies had also appeared in Baltimore, Boston, Brooklyn, New Orleans, Philadelphia, Pittsburgh, St. Louis, and Washington, D.C. Meanwhile, John Mason had established the first light rail system drawn by horses in lower Manhattan in 1832. In 1854 Paris became the first European city with a horsecar system, by which time they had become widespread in the United States (Figure 14), allowing people to commute up to three or four miles to downtown workplaces within a reasonable time

Corbis

FIGURE 14 A mule-drawn "horsecar" photographed in Washington, Georgia, in 1905.

(30 to 45 minutes). The cost of travel, however, meant that only limited numbers of better-off families could afford to move to the **horsecar suburbs** that developers built at the edge of the pedestrian city at the terminus stops of the radial horsecar and omnibus routes. Ordinary wage earners received little more than a dollar a day; most round-trip fares were between 15 and 25 cents.

Railroads Although the railroads were principally developed for transport between cities, and had a major influence on the development of the urban system, they also had some striking effects on the structure and organization of land use within cities.

- They provided the catalyst for the reorganization of land uses in city centers. The railroad companies needed large, flat tracts of land at or near the city center to accommodate passenger stations and the necessary infrastructure: locomotive sheds, shunting yards, and, often, goods-handling facilities. They also had the economic (and political) power to secure these sites.
- Once developed, the railroad facilities had the effect of realigning the pattern of city center land uses, laying the foundation for modern **CBDs** (**central business districts**). Because of the funneling of large numbers of people through the railway stations, nearby locations became prime sites for hotels, restaurants, and larger stores. And, because of the arrival and departure of bulk goods (usually from a different part of the terminus), nearby locations became prime sites for warehouses and the offices and storehouses of wholesale distributors.
- At about the same time, a critical new building technology was introduced: iron girders with curtain walls. Together with passenger elevators (another innovation of the 1850s) this new technology facilitated the beginnings of high-rise construction. The result was the location within the embryonic CBD of those few land users who could afford the costs of such technology: hotels and office buildings for insurance companies and publishing houses, for example.
- Within the central parts of cities the railroad infrastructure brought a radical change to the physical environment as odd-shaped fragments of land and inconsequential streets were left between the rail lines and freight yards. The railroad companies defined the character and projected the limits of cities. "Every mistake in urban design that could be made was made by the new railroad engineers, for whom the movement of trains was more important than the human objects they achieved by that movement."[21]The consequences of treating land narrowly in terms of the rights of individual owners (as opposed to broader notions of collective good or environmental quality) were being inscribed into the structure of cities.
- The mainline tracks ran in swathes that radiated from the city center. Where they passed through stretches of broad flat land, they attracted stockyards, factories, and warehouses and repelled all but the lowest grade of housing. In this way, pronounced linear industrial belts were established. Because of the amount of fixed capital tied up in them, these industrial belts served as fixed elements around which the future development of cities had to take place.
- Elsewhere, the shadow of the railroad condemned smaller tracts of land to permanent dereliction or blight. Between junctions and crisscrossing rail lines were junkyards and abandoned lots, while under the viaducts that were often needed to carry the tracks into the heart of cities were an assortment of makeshift and unsavory activities. Even where the tracks passed through residential neighborhoods, the railroad brought its polarizing influence, regrouping the social geography of cities around the "right" and "wrong" sides of the tracks.
- The railroad companies' greatest influence on land use, however, was in their promotion of commuter traffic. Railroad company owners recognized the potential demand among more affluent middle-income households for a means of escaping the increasingly degraded central city environment that the railroads themselves had helped create. The speed of steam trains meant that commuter stations and residential subdivisions could be established in the countryside beyond the city, creating beads of **exurban development** along the approaches to the bigger cities.

Boston was the first U.S. city around which such a pattern developed, but New York, Philadelphia, and Chicago soon followed (in the 1850s) and, by the 1860s, the pattern had spread to most other large cities. Considerable numbers of people were involved. About 20 percent of Boston's business classes traveled to work by train in 1848 from exurbs located between 12 and 15 miles from the city center. Ten years later 40 trains were running each day between downtown Philadelphia and the exurb of Germantown, and a similar number of trains were carrying the 5,000 or so commuters into Chicago from Evanston.[22] But the critical feature of the railroad exurbs was not their size so much as their exclusivity. Their cost—10 to 15 cents for a one-way ticket—made them accessible at first only to successful merchants, industrialists, doctors, lawyers, and other newly rich people.

Like the omnibus and the horsecar, therefore, the railroads siphoned off from the city the affluent and educated, depositing them in narrow concentrations of development that extended only a comfortable stroll from the station or stop. Because commuters had to walk to the station, lot sizes in railroad suburbs tended to be fairly modest (8,000–10,000 square feet), allowing for a sufficient number of dwellings to make the station or stop profitable.

All this, of course, had a profound effect on the rhythm and pattern of urban life. The railroads allowed for the rapid distribution of highly perishable foodstuffs—milk, vegetables, fish, and meat—and so facilitated marked improvements in the diet of city dwellers. This also led to the beginning of the standardization of diet and patterns of consumption generally. In the same way, the bulk transport of building materials began to introduce a certain sameness to city buildings across broad regions. The railways' need for organization and punctuality led to the standardization of time. In Britain, Manchester's municipal council decided in 1847 to adjust its clocks to London time; by 1852, every city had done the same, and clocks with standard time became prominent in every railway station.

THE ORGANIZATION OF INDUSTRY (1875–1920)

As **organized capitalism** gathered momentum and transport and communications networks became more efficient, the urban system became increasingly consolidated, with more pronounced patterns of economic specialization and greater integration between different parts. On the railroads the innovation of steel rails to replace iron ones made it possible to carry heavier loads at higher speeds (which encouraged the concentration of industry in larger cities) and provided the opportunity to standardize the railroad gauge throughout the country (which encouraged trade and specialization among cities). At the same time, the railroad system extended dramatically. The total length of railroad tracks increased from 30,000 miles in 1860 to 160,000 miles in 1890, allowing towns such as Birmingham, Jacksonville, Memphis, and Houston to emerge as central places with regional status and the revival of colonial towns such as Savannah and Charleston. Across the Great Plains, the railroad company executives, in collaboration with grain elevator companies and lumber dealers, literally dictated the pattern and design of lower-order central places. Desperate for traffic, they deliberately encouraged the urbanization of the West through propaganda that extolled the virtues of territory that earlier had been marked on maps as "The Great American Desert."

This period also saw some important demographic changes. Falling death rates (from better diets, improvements in sanitation and public health, and the introduction of scientific medicine) generated a steady **natural increase** in population. Meanwhile, from 1890 to 1910 over 12 million immigrants arrived, accounting for about one-third of the country's growth. It was not long, however, before pressure from labor unions (concerned about the effects of immigration on wages and unemployment) and from nationalistic groups such as the American Protective Association and the Daughters of the American Revolution (concerned about the impact of immigration on American values and society) led to restrictive legislation. The 1921 Immigration Act set a ceiling of 250,000 immigrants a year, with quotas for each country of origin, based on the existing composition of the white population. In 1924 another Immigration Act reduced the ceiling to 150,000 immigrants a year.

The overall effect of these demographic changes was the fueling of the growth of existing cities, the infilling of settled areas, and the colonization of the few remaining frontier regions. The emergence of small towns that accompanied the commercialization of agriculture in the West, the exploitation of major mineral deposits (copper in Montana, lead and zinc in Missouri, and iron ore around Lake Superior), and the escalating demand for coal helped to more than double the number of urban places between 1870 and 1920. At the top of the hierarchy, New York and its adjacent counties in 1920 had 4.75 million residents. Boston, Chicago, Philadelphia, and Pittsburgh each had about 1.5 million, while the next tier of cities, with populations around 500,000, now included several cities from the West (Los Angeles, San Francisco, Seattle), Southwest (Dallas), and Midwest (Kansas City, Milwaukee, Minneapolis-St. Paul, St. Louis) as well as from the Northeast (Baltimore, Cincinnati, Providence).

The most important feature, however, was the way in which the logic of *industrial location* influenced patterns of urban growth and development. Primed by the initial advantage of investment in manufacturing, urban-industrial development became a self-propelling growth process, a spiral of growth known as **cumulative causation** (Figure 15). By 1920 the Manufacturing Belt had become a national economic heartland (Figure 16), its larger cities blossoming into great metropolitan centers and its smaller cities providing the settings for highly specialized and very profitable manufacturing industries.

In overall terms, *the consolidation of the Manufacturing Belt was the result of the operation of the principles of initial advantage and cumulative causation at a regional scale.* With its large markets, well-developed transport networks, and access to excellent coal reserves, it was ideally placed to take advantage of the upsurge in demand for consumer goods, the increased efficiency made possible by the telegraph system, the postal services, and the rationalization of the banking system, and the reduced energy costs brought about by the advent of coal-burning thermoelectric generating stations.

The effects were:

1. *Local specialization became geared to national rather than local markets.* Brewers in Milwaukee and St. Louis, for example, were able to take advantage of the integrated railroad system and innovations such as refrigerated rail cars and mechanized production techniques.
2. *This specialization provided the basis for increased commodity flows between the towns and cities of the Manufacturing Belt, binding the region more tightly together.* The region's financial and commercial linkages and technical expertise made it particularly attractive to new industrial activities with national markets; this in turn stifled the chances of cities in other regions to achieve comparable levels of industrialization.

The Industrial City

It is no exaggeration to describe the changes of the industrial period as revolutionary. Not only did cities begin to grow dramatically in population and territorial extent (Figure 17),

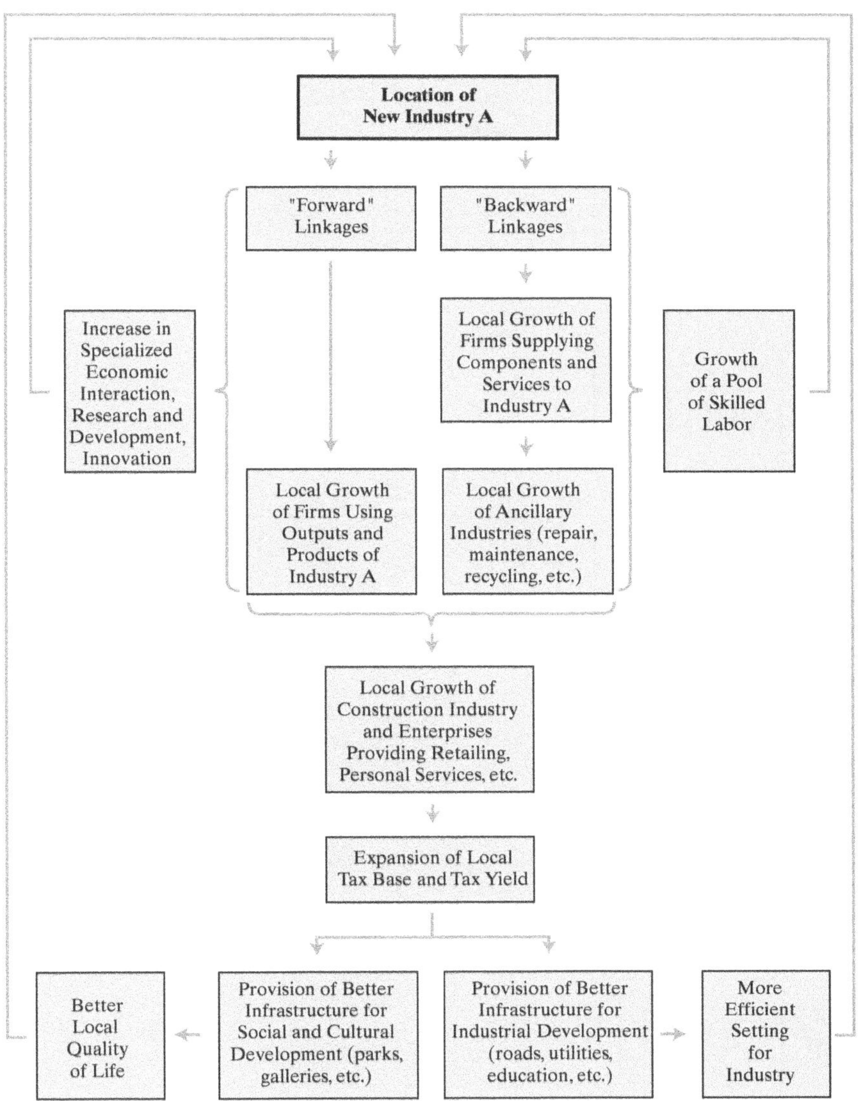

FIGURE 15 The self-propelling process of urban-industrial growth.

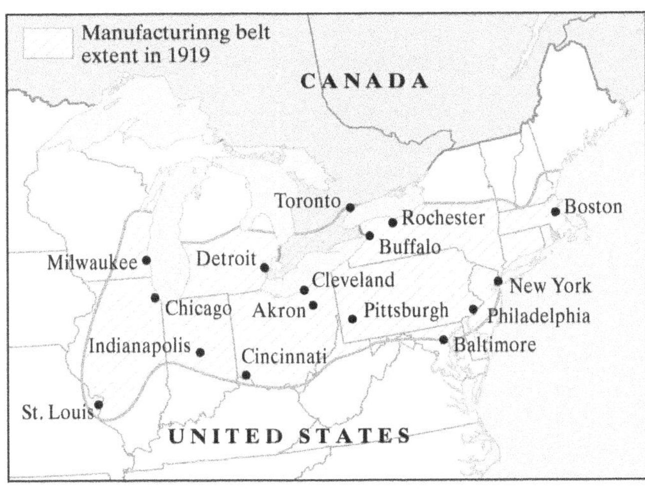

FIGURE 16 The Manufacturing Belt, 1919.

they experienced unprecedented specialization and differentiation among land uses. More sophisticated utility, transportation, and communications technologies brought the logic of economies of scale, agglomeration economies, and the **division of labor**. As a result, the organization of the economy, of society, and of urban space was radically transformed.

Economic Specialization and the Reorganization of Urban Space The pressures of increased population densities, economic specialization, competition for space, and the increasing scale of economic activities meant that urban land use became highly specialized. Users of land became spatially segregated by their ability to pay for the most attractive locations–whether for industry, business, or homes. As economic growth brought in more industry, businesses, and people, both the extent and intensity of different land uses increased, bringing acute competition and conflict over land.

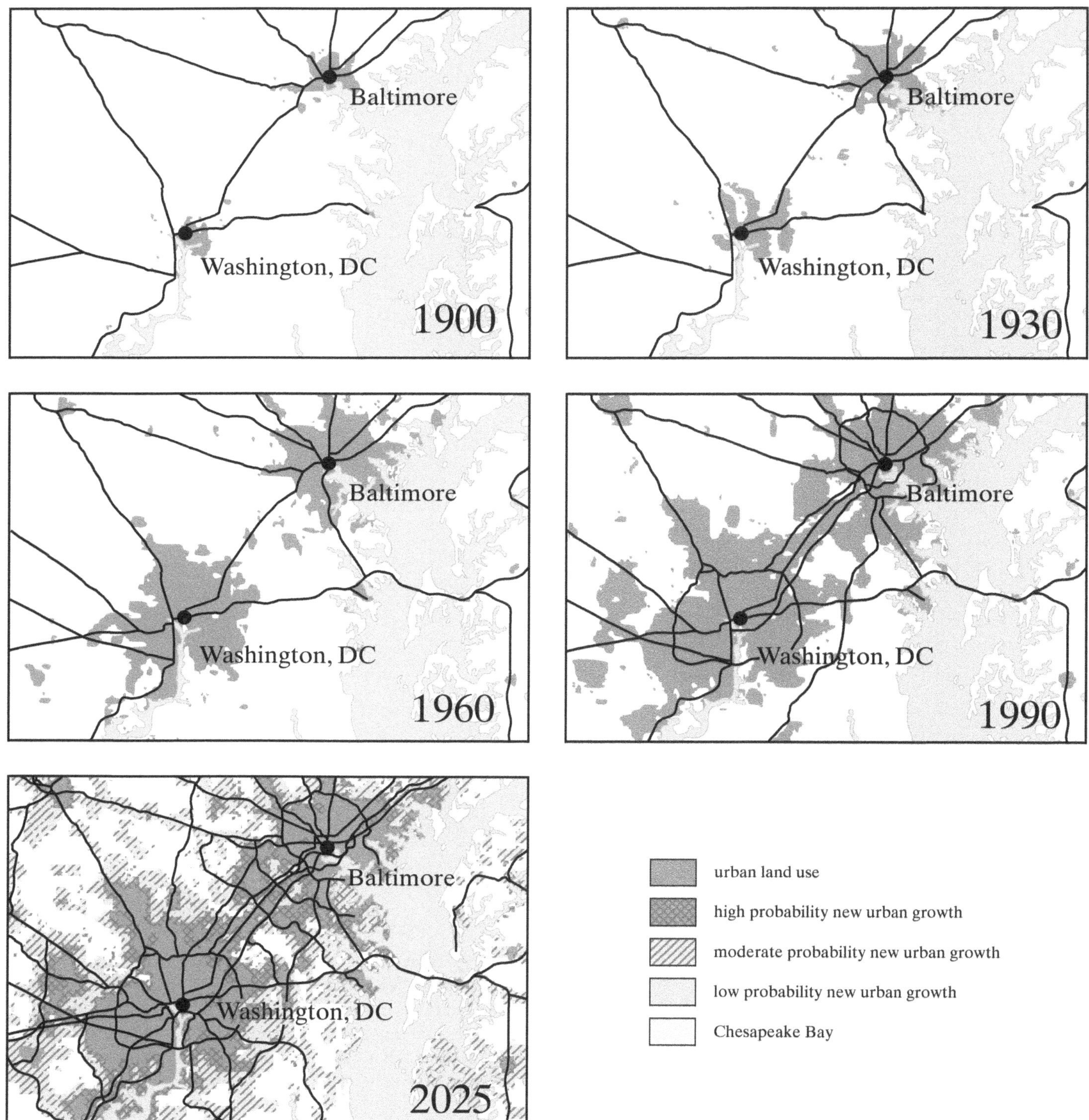

FIGURE 17 The growth of the built-up area of the Washington to Baltimore region.

Factories were the linchpin of the new land use system, taking first pick of available sites and dominating every aspect of spatial organization. For many of the pillars of industrialization—textiles, chemicals, iron, and steel—the best locations were near the waterfront, where they could secure the water to supply steam boilers, to cool hot surfaces, make chemical solutions and dyes, and, above all, dump waste.

Around the factories, speculators built housing for factory workers. This was a departure from early industrialization, when factory owners built a good deal of housing for their workers. With large-scale industrialization in larger urban settings, workers were expected to find market-rate housing. The result was a **generalized housing market**:[23] the creation of a mass market in rental housing, disconnected from any direct linkage between employment and tenancy.

The quality of housing was dictated by the wages of the workers, who found themselves the least competitive of all land users. Because most workers were also unable to afford transportation on a regular basis, the developers of workers' housing were left with little choice: the construction of cramped and spartan dwellings in the leftover spaces between the factories, dumps, and railroads. In a competitive market, builders and developers had to minimize costs and prices. This they achieved by using pattern-book standardization linked to economies of scale and by keeping lot sizes to a minimum. This resulted in the proliferation of pinched land divisions, narrow alleyways, solid blocks without parks or playgrounds, and narrow, inflexible street patterns (Figures 18, 19).

Corbis Images

FIGURE 19 Wash day in New York's tenement district.

Jacob Ris/Corbis Images

FIGURE 18 A New York alleyway ("Bandit's Roost"), photographed by Jacob Riis in 1888.

Between the first-pick of the factories and the last-pick of workers' housing, the reorganization of urban land uses involved the simultaneous operation of two tendencies: outward (**centrifugal**) and inward (**centripetal**) movement. Affluent professional and white-collar households, together with high-end retail services, were drawn to suburban and exurban locations, creating outward forces in the reorganization of land use. Most factories, warehouses, offices, hotels, and specialty retail activities, in contrast, were drawn to specialized, centrally located sites, creating inward forces.

Framing the City: Networked Infrastructures The opportunities for urban development were dramatically expanded with the emergence of new technologies: gaslight, electrification, steel-frame skyscraper construction, and elevators. These changes, in turn, brought unprecedented possibilities for entrepreneurs to accumulate wealth, introduced the consumption of all kinds of new products, and financed the creation of a modern infrastructure of roads, tunnels, bridges, sewers, and street lighting.

So the networks of water, sewage, gas, electricity, telegraph, telephone, and streetcar and subway lines were fundamental to processes of urban change. As cities struggled to grow and to adapt to new industries, new institutions, and new ways of life, they also continuously made and remade their infrastructure and built environment, applying the logic of modernization and industrialization to urbanization itself. How this played out was in part a function of new technologies, the financial capability to provide them, and highly uneven power struggles and social and political biases.

In the process, the whole culture of cities was transformed. One aspect of this was the sense of progress and optimism. Cities themselves became "electropolises," and new streets, bridges, telephone networks, gasworks, and streetcar systems became emblematic of the sense of progressive modernization. Lighting was just one of a series of radical changes to urban life. At the beginning of the nineteenth century, gas lighting had created a new urban nightlife as well as a means for securing greater law and order. By the end of the nineteenth century, electrically lit shopping arcades had become the "dream spaces" of the modern city, icons of progress and cultural transformation.

A second aspect of this transformation was that cities themselves came to be regarded as complex "machines" that could be rationally organized as a single system. Engineers and reformers drew on this notion in campaigning for improved networks of infrastructure to bring order, rationality, and sanitation to the troubled and apparently chaotic industrial metropolis.

One of the most influential early precedents was set in Paris by Napoleon III, who presided over a comprehensive program of urban redevelopment and monumental urban design. The work was carried out by Baron Georges Haussmann between 1853 and 1870 who demolished large sections of medieval Paris

Gordon Gahan/Photo Researchers, Inc.

FIGURE 20 Haussmannization in Paris, 1853–70. The comprehensive program of urban redevelopment and monumental urban design involved demolishing large sections of medieval Paris to make way for numerous public open spaces and monuments, with broad, tree-lined avenues, such as the boulevards leading to the Arc de Triomphe.

to make way for broad tree-lined avenues (Figure 20), with numerous public open spaces and monuments. He made the city not only more efficient (wide boulevards meant better traffic flow) and a better place to live (parks and gardens allowed more fresh air and sunlight in a crowded city and were held to be a "civilizing" influence), but also safer from revolutionary politics (wide boulevards were hard to barricade and facilitated troop movement and cannon fire; monuments and statues helped to instill a sense of pride and identity).

Haussmann's approach heralded a technocratically minded, comprehensive approach to urban governance and planning that quickly spread through Europe, notably to Berlin, London, and Vienna. In the newer cities of the U.S. Midwest and West, rectilinear grids of streets became the norm for organizing networks of streets and utilities, making for an unprecedented openness of urban form. Streets in industrial metropolises everywhere, previously used only for local access, informal economic activities, and social life, were now seen as part of the circulatory system of a functioning metropolitan "machine." Their size and layout were standardized; they were paved, regulated, and managed. Their construction also provided for the introduction of a subterranean system of water and sewage pipes that helped to "deodorize" and sanitize the city (Figure 21).

The Emergence of Land Use Zoning Laws Not surprisingly, this dramatic reorganization of urban space created some serious conflicts and tensions—problems that were all the more acute in the United States because of the civil liberties enshrined in private land ownership. It did not take long before some groups mobilized against the land users they perceived as threats, hazards, or nuisances. Their objective was the legal exclusion of particular land uses from certain parts of the city. The outcome was **land use zoning**, a regulatory mechanism that would become a key instrument in consolidating the specialization and segregation characteristic of land use in U.S. cities.

Land use zoning had its roots in discrimination. In San Francisco discrimination against the Chinese was focused on the laundries that by the 1880s had spread well beyond Chinatown into the other ethnic neighborhoods, where they operated as social centers for the many Chinese domestic servants who were scattered around town. The white population, regarding the laundries has havens for "undesirables," sought to close more than 300 of them in 1886 by declaring them nuisances and fire hazards. But in the federal courts the case of *Yick Wo v. Hopkins* struck down the statute because of the way it gave arbitrary power of racial discrimination to a board of supervisors. After this decision, the city of Modesto promptly came up with a simple ploy to satisfy the requirements of the Fourteenth Amendment: the division of the city into two zones, one excluding laundries and another permitting them.

Similar nuisance-zone statutes, directed variously against laundries, brothels, pool halls, dance halls, livery stables, and slaughterhouses, were quickly adopted in other cities. Los Angeles was zoned into three districts: residence-only, industry-only, and residential uses with a limited range of industries.

With the principle established, the prototype land use zoning ordinance was written in New York. The catalyst this time

FIGURE 21 Workmen repairing a sewer under Fleet Street in London. The sewers carried 87 million gallons of water daily in 1854.

was discrimination against Jewish garment manufacturers, who had begun to encroach on the territory of Fifth Avenue luxury stores from their base on the Lower East Side. Fearing that increased traffic and the presence of thousands of low-wage workers would deter wealthy customers, the Fifth Avenue Association mobilized a coalition of regulatory interest groups that sponsored zoning ordinances. These ordinances were designed to stabilize the broad pattern of land uses by encouraging specialization and uniformity at the micro scale. The New York zoning ordinance, drafted by lawyer Edward Bassett and passed in 1916, was based on the premise that *restrictions on land use are constitutional because they enable city governments to carry out their duties of protecting the health, safety, morals, and general welfare of their citizens.*

This philosophy, clearly, was a significant modification of the sanctity of land ownership as a civil right, though it did not by any means signal a retreat. The law itself was spelled out in a map where all private land was assigned a particular functional category for future land uses (it did not attempt, however, to weed out existing nonconforming uses). These categories were then listed in detail, including any restrictions such as the height, number of floors, and general characteristics of buildings, the overall density of population, and the minimum amount of public open space.

Existing land users and speculators alike appreciated the advantages of such an approach. For existing land users, the law brought security from the threat of intrusive new activities. For speculators, the law brought a welcome degree of stability and predictability to the business of land development. No longer would the developer of middle-income housing have to worry about tumbling prices resulting from the establishment of a noisy saloon on an adjacent lot. So compelling were these advantages that within 10 years of the enactment of the New York law more than 500 other cities had adopted similar ordinances. The main effect (which did not fully take hold until after 1920) was to stabilize patterns of land use and to defuse much of the tension surrounding urban growth and spatial reorganization. Later, the ancestry of zoning began to tell, and it became a major means of discrimination against social groups perceived for one reason or another as undesirable.

From the start, however, land use zoning was unable to cope with the fundamental tension generated by the growth of industrial cities: the unease of the increasingly numerous and more affluent middle-income groups cooped up with the factories, railroads, warehouses, and "huddled masses" of "ordinary" people. Suddenly, in 1888, this tension was relieved by the introduction of a simple but radically effective innovation in urban transportation: the electric streetcar, or trolley.

The Suburban Explosion: Streetcar Suburbs Drawing on the experience of numerous attempts to solve the urban transportation system—including electrified trolley systems in Cleveland and Baltimore, and, in particular, the Siemens system developed in Germany—Frank Sprague perfected an electrically driven version of the horsecar, powered by overhead cables, that opened for business in Richmond, Virginia, in the spring of 1888.

The innovation was adopted the following year by Henry Whitney, who had established in Boston an integrated horsecar system that covered the entire city. By electrifying the system, Whitney could extend the radial lines, because streetcars could travel much faster than horsecars. In Boston the total length of track in 1887 had been just over 200 miles; by 1904 it was almost 450 miles. Within five years of Sprague's success at Richmond, more than 200 cities had adopted his system. By 1902 there were 22,000 miles of streetcar tracks in U.S. cities.

By making it feasible to travel up to 10 miles from the downtown area in 30 minutes or so, the streetcar greatly increased the territory available for residential development and opened the way for the growth of **streetcar suburbs**. So much land became accessible at once that the price of land was kept down, ensuring inexpensive suburban lots. Affordable land, combined with the cheaper operating costs per passenger-mile of the streetcar (because of their large carrying capacity and efficiency of electric power), ensured that developers of streetcar suburbs instantly found a market in the solid middle-income groups, and there was a rush of speculative suburban sprawl as developers and streetcar operators worked together (Figure 22). Around the fringes of Chicago, for example, some 80,000 new residential lots were recorded between 1890 and 1920 (though many of these lots were not built on for some time).

As well as enabling the growth of suburbanization, the streetcar influenced urban form in other ways:

1. The sudden departure of thousands of middle-income households gave much-needed scope for the reassignment of space to nonresidential uses in the downtown.
2. The ease and cheapness of travel on streetcars led suburban housewives to use them on shopping trips outside rush hours, helping to sustain a specialized, high-order shopping area in the downtown.
3. The development of cross-town and circumferential streetcar lines created a series of junctions and interchanges around which grew small nuclei of commercial development. This signaled the beginning of the end of the monocentric city structure with a single focal point.
4. Larger nodes of development emerged around the terminus stops of the radial streetcar routes. To counter the lack of commuters on weekends and holidays, streetcar companies promoted the idea of outings to amusement parks, beaches, picnic grounds, and even cemeteries. These sites were often located near streetcar terminus stops (partly as a result of sponsorship by the business interests involved with streetcar development), and they soon attracted a variety of ancillary services such as restaurants and convenience stores.
5. The streetcar facilitated the growth of suburban settlements and satellite townships that were too small to sustain an economic base of their own until they acquired a streetcar system. The growing pool of local workers and customers made it possible to develop an industrial base and attract substantial clusters of shops and offices. In this way the streetcar sowed the seeds for the subsequent evolution of a *metropolitan* form of urbanization.

FIGURE 22 A streetcar in the Richmond district of San Francisco in 1909. In this new "streetcar suburb" the intersection of streetcar lines shown here had already attracted a large commercial building containing offices and stores.

Rapid Transit In a few larger cities the congestion created by streetcars converging on the downtown prompted the development of electrified rail systems that were separated from streets—on elevated tracks or in tunnels (Figure 23). These *rapid transit* systems could travel faster than streetcars while carrying more passengers, and their effect was to stretch the fingers of

Alexander Alland/Corbis Images

FIGURE 23 New York's elevated railway at Coenties Slip in the early 1800s.

urban development farther into the countryside, reinforcing and extending the radial form of the city.

Underground lines were pioneered in London, where the Metropolitan subway line was opened in 1863 and the Inner Circle line was completed in 1884, linking up the main above-ground railway stations (Figure 24). In the United States, Henry Whitney once again led the way, opening 1.6 miles of subway underneath Tremont Street in Boston in 1897. His success (over 50 million passengers in the first year) prompted the development of the New York subway system, which opened in 1904. The start-up costs of subway systems were extraordinary, however. Whitney's line cost nearly $150,000 for every 100 yards of track. Apart from a combined elevated and subway system in Philadelphia that opened in 1908, there were no more subway projects in the United States until the 1940s, when one opened in Chicago. Instead, elevated electric railways—the "els"—were the preferred solution in Boston, Brooklyn, Chicago, Kansas City, New York, and Philadelphia, where they were installed around the turn of the twentieth century.

Mass Transport and Real Estate Development There was a very close relationship between the evolution of mass transportation systems and suburban real estate development. In some cities the same company undertook transit lines and real estate development, as in the streetcar suburb of Shaker Heights in Cleveland. Transit lines often preceded land development, and in many cases the transit service was expected to lose money initially. But substantial increases in land value, because of access to transit lines, would more than make up for the early investment.

The link between transit and real estate development was particularly strong in West Coast suburban developments. In northern California, F. M. Smith bought and consolidated the trolley lines in San Francisco's East Bay and purchased 13,000 acres of land for development in the Oakland and Berkeley areas in the 1900s. In southern California, Henry Huntington, a founder of the Southern Pacific railroad, developed the Pacific Electric Interurban Transit Company in the Los Angeles area. He bought the land along his routes and laid out suburban developments, at the same time avoiding competitors' land holdings unless he was made a partner. Among Huntington's partners was Harry Chandler, the largest developer in Los Angeles. Chandler bought 47,500 acres in the San Fernando Valley, an area about the size of the city of Baltimore, and the Pacific Electric extended their lines into the valley. A $25-million water project, paid for by the city of Los Angeles, supplied the development with water after a vigorous campaign, led by Chandler's father-in-law, Harrington Grey Otis, publisher of the *Los Angeles Times*. Later Chandler bought the 270,000-acre (almost as large as the City of Los Angeles) Tejon Ranch in Los Angeles and Kern counties, land that is still controlled by Chandler interests.

For half a century, until about 1920, public transportation inexorably influenced the form and growth of cities. Transit lines consisted of the arteries along which people moved and real estate development occurred. Transit stops became the nodes of commerce and entertainment, and the radial lines converged at a hub of the metropolitan area. Residential neighborhoods grew up along the transit lines and small commercial districts were developed wherever lines intersected or branched.

Inside the Industrial City

To people at the time, the changes brought about by industrialization were shocking. Never before had there been such visible growth, such intensive building and rebuilding, such new arrangements and rearrangements of urban form and function. As we look back, we can recognize that these changes established the physical template for the development of the modern city. Three dimensions are particularly important: the growth of CBDs, the assignment of locations according to principles of rent, and the emergence of distinctive patterns of residential segregation.

Corbis Images

FIGURE 24 The world's first underground railway, in London: Baker Street Station, Metropolitan Railway, 1863.

Central Business Districts By now it was the CBD that gave visual expression to the growth and dynamics of the city. The CBD became a symbol of progress, modernity, and affluence, especially after technological advances in steel-frame construction, elevators and telephones made skyscrapers possible. The status of a city could be read from the size and degree of differentiation of its CBD. The CBD became the hub of economic, social, political, and cultural life in cities, and the organization of functions within the CBD set the framework for the next several decades of downtown development.

Department Stores and Shopping Districts The location of railroad stations and the converging of transit lines had shaped the early development of CBDs, and retailing activities dominated the land uses within CBDs. More specialized, higher-order, and exclusive retailers could generate sufficient profits to pay more for property or rent than any other potential user of downtown space. Because most shoppers were reluctant (as they are today) to walk far, the location decisions of downtown retailers were limited. The result was a distinct and compact shopping area: a zone perhaps 300 yards in radius, in which retailing was practically the only activity at street level.

Anchoring this zone, at the very center, were grand, multistory *department stores*: Macy's, Bloomingdale's, and Lord and Taylor's in New York; Jordan March and Filene's in Boston; Woodward and Lothrop's in Washington, D.C.; Wanamaker's and Strawbridge and Clothier's in Philadelphia; Carson Pirie Scott and Marshall Field's in Chicago; I. Magnin's in San Francisco; Bullock's in Los Angeles (Figure 25). These stores were landmarks, and they needed to be: Only in prominent locations at the center of things was the concentration of potential customers great enough to ensure enough daily business to carry the enormous overheads required to stock and display in proper style and quantity the best that industrialization had to offer. The classic location was a corner site at a busy intersection, but the basic requirement was being near railroad stations and, crucially, the downtown stops and transfer points of streetcar lines.

FIGURE 25 Strawbridge and Clothier's department store, Philadelphia, in the early 1900s.

Other stores were by no means located at random within the shopping zone of the CBD. Stores that were so specialized that they were one-of-a-kind could afford to let customers come to them. They could locate almost anywhere within the shopping area, or even just beyond it, finding a cheaper site and paying little attention to their neighbors.

Most stores, though, had to consider the microdetails of location. Then, as now, a significant part of consumer behavior was comparison shopping: Some people, "born to shop," seem to enjoy it for its own sake. But an equally important part of consumer behavior is based on minimizing effort. It makes sense, therefore, for particular types of specialized stores and services to cluster together: men's outfitters, for example (Figure 26). In addition, shopkeepers had to be sensitive to the functional relationships between their businesses and others. It made sense for stores selling shoes or women's fashions to locate near department stores to catch the "overspill" of shoppers. Exclusive, high-fashion stores, in contrast, need to cluster together but, equally, be away from the throng of shoppers near department stores. New York's Fifth Avenue, without a subway or elevated train in the 1920s, is a good example of how relative inaccessibility allowed exclusivity.

Downtown Office Districts In the largest cities the limited distance that shoppers would walk created a series of "pedestrian zones" that constituted specialized retail subareas. Similarly, large cities could support specialized clusters of high-order services such as banks, law offices, or medical facilities. The office district was usually beside the shopping district because it, too, had a functional dependence on the focus of transportation at routes that brought office workers on commuter trains and streetcars, and out-of-town business visitors by long-distance trains. The office workers—then still predominantly male—were described by Richard Yates in his novel *Revolutionary Road*:

> How small and neat and comically serious the other men looked, with their grey-flecked crew cuts and their button-down collars and their brisk little hurrying feet! There were endless desperate swarms of them, hurrying through the stations and the streets, and an hour from now they would all be still. The waiting midtown office buildings would swallow them up and contain them, so that to stand in one tower looking out across the canyon to another would be to inspect a great silent insectarium displaying hundreds of tiny pink men in white shirts, forever shifting papers and frowning into telephones, acting out their passionate little dumb show under the supreme indifference of the rolling spring clouds.[24]

It was the office district that gave the CBD its most prominent landmarks, with taller and taller buildings. The earliest, biggest office buildings were purpose-built for insurance companies and publishing houses. In addition to their need

FIGURE 26 The early CBD. This stretch of Kearny Street (just off Market Street, the main commercial thoroughfare) in San Francisco shows the early expression (in 1894) of clustering and functional interdependence. Along these two blocks were several men's tailors and outfitters, together with a haberdasher and shirtmaker, a store selling men's furnishings and shirts, a hatter, a watch store, a cigar store, and a men's club. It would be nice to think that the florist at the corner of Kearny and Bush was part of the functional grouping, providing the finishing touch for a gentleman's shopping trip.

for communication among large numbers of employees, these organizations found skyscrapers a valuable way to advertise. But with the emergence of the CBD as the organizational hub of a larger and increasingly complex array of urban functions, it did not take speculators long to realize the value of central office space to small companies. Speculatively built office blocks provided an "address" for smaller office functions, which could also share in the opulence of marbled lobbies and concierge services that these settings provided.

Warehouse Zones In the early industrial city wholesaling and retailing had gone hand in hand within the CBD—to the extent, in some instances, of occupying the same building. The *scale* of operations in the industrial city meant that this could no longer continue. Wholesalers increasingly had to supply not only downtown merchants but also merchants in outlying suburbs and satellite towns. At the same time, they needed to be near the railroad freight yards for the carloads of goods being shipped in, and to the handling yards of express shipping companies like American Express and Wells Fargo that had sprung up to put together carloads of smaller quantities of outgoing goods. The result was a distinctive warehouse zone, located on the edge of the CBD next to the freight yards. Two very different kinds of wholesaling tended to remain in the CBD itself, however: (1) low-volume, high-value goods such as jewelry and (2) high-volume, perishable produce such as vegetables and meat, each sold separately from central locations accessible to restaurant chefs and the owners of small food stores.

City Halls and Civic Pride Another feature in the growth of CBDs was the *city hall* and its associated functions. It was convenient for the offices of city officials to be centrally located: convenient both for them and their constituents, given that the CBD was the single most accessible place in the city. But the main reason for a CBD location was not accessibility or efficiency but symbolism. City governments, with new powers and responsibilities, needed to be identified with the hub of things. Their image was the city's image. A similar logic dictated that city halls and associated civic buildings be the equals or betters of the department stores and office buildings that were the landmarks of the commercial districts (Figure 27).

In addition, the civic pride invested in city halls dictated that they be more imposing and more splendid than those of rival cities, with landscaped malls, parks, and statues to show them off to the best advantage. The result was another distinctive pedestrian zone, dominated by the city hall itself and containing the main library, central post office, and courthouse, museum, assembly hall, opera house, and art gallery that made up the distinctive grouping of civic amenities. In between were the offices of lawyers and vocational schools and colleges, together with the premises of others who traded on the stream of workers and visitors frequenting the area: lunch counters, bars, tobacco and newsstands.

The Spatial Organization of CBDs From these observations we can put together a number of generalizations about the organization of land use within the CBD. The overall spatial structure tends to be dominated by a high-density *core* that contains the retail, office, entertainment, and civic zones and a lower-density *frame* that contains zones of warehousing, educational facilities, hotels, medical services, and

Corbis Images

FIGURE 27 Philadelphia City Hall: civic pride symbolized in bricks and mortar.

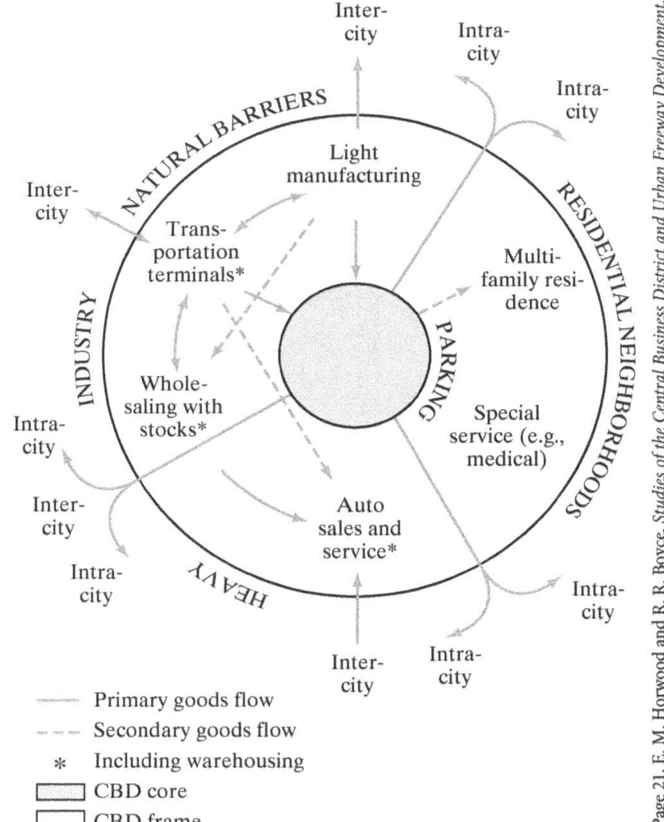

Page 21, E. M. Horwood and R. R. Boyce, *Studies of the Central Business District and Urban Freeway Development*, Seattle, Wash.: University of Washington Press, 1959.

FIGURE 28 The core–frame concept of the CBD.

a mixture of specialized shops (antiquarian bookstores, for example) and services (such as acupuncturists and picture framers) that have neither the functional linkages nor the potential profitability to justify locations within the core (Figure 28).

Nevertheless, constant changes in the relative fortunes of different kinds of activities, changes in the transportation infrastructure, and the inevitable aging and obsolescence of buildings, make the CBD very dynamic. The core tends to expand toward the higher-status, newer elements in the frame, creating a *zone of assimilation*, and to move away from lower-status, older elements, creating a *zone of discard*, which often forms a niche for another kind of economic specialization: vice and X-rated businesses. The frame itself overlaps with older structures and residential districts in a **zone in transition** ripe for conversion to CBD uses.

Land Values and Urban Land Use The spatial organization of the CBD points to the importance of accessibility and relative location in understanding patterns of land values and land use. In New York in the mid-1920s the peak land values were $22,000 per foot of street frontage on Broadway; only half a mile away land values were less than $3,000 per foot.[25] Comparable gradients could be found in every large city. In 1903 the economist Robert Hurd made the point in a much-quoted statement:

> Since value depends on economic rent, and rent on location, and location on convenience, and convenience on nearness, we may eliminate the intermediate steps and say that value depends on nearness.[26]

Hurd's reference to economic rent (sometimes referred to as **ground rent** or "location rent") requires elaboration, because the concept has become fundamental to attempts to theorize patterns of land use.

To begin with, ground rent does not correspond to the popular usage of "rent" (more properly called "contract rent"). Ground rent is the surplus paid (to land owners, in this instance) above the minimum necessary to use the land at all. It is often defined, in practice, as the total revenue generated by a particular activity on a particular parcel of land, less the total production and transportation costs associated with that same parcel of land. (Not all activities are run on business lines, however. Households, along with public agencies, do not evaluate their success in terms of profits but in terms of satisfaction, disposable income, efficiency, or some other basis. The term *utility* is, therefore, often used in place of *revenue*.) It follows that ground rent reflects the utility of a particular site for a particular activity. So it is likely that for each site there will be one activity that generates the highest possible economic rent: the *highest and best use*. Land use should, therefore, respond to land values in a predictable fashion. Activities that can derive the greatest utility from a given site should outbid all other activities for that site. Each type of land user can be thought of as having a distinctive **bid-rent** curve that reflects the prices that each is prepared to pay for sites at different distances from the CBD (Figure 29).[27]

Sectors and zones In practice, land use in industrial cities did conform to certain general patterns. The classic study was done by Homer Hoyt,[28] who made a comparative analysis of patterns of rental values in 142 U.S. cities from 1878 to 1928 on behalf of the **Federal Housing Administration (FHA)**. Hoyt developed a **sector model** of urban land use that was based on generalizations derived from his study:

- Rent, and therefore socioeconomic status, varied within cities primarily by radial sectors.
- The highest rents were found in a single sector that often extended out continuously from the CBD.
- Intermediate rents, associated with middle-income neighborhoods, were commonly found in sectors on either side of the high-rent, high-status sector.
- Low rents, associated with working class and low-income housing, were usually found on the side of the city opposite the high-rent sector.
- Over time, the high-rent sectors tended to:
  - grow outward along major transportation routes;
  - extend along ridges of high ground, free from the risk of flooding and with panoramic views; and
  - be drawn toward the homes of community leaders.

Putting these generalizations together, Hoyt came up with a generalized model of urban land use (Figure 30). The main point about this model is the **relative location** of the different sectors. Hoyt argued that corridors of industry and warehousing will tend to be surrounded on both sides by sectors of working-class housing, while middle-income housing will tend to

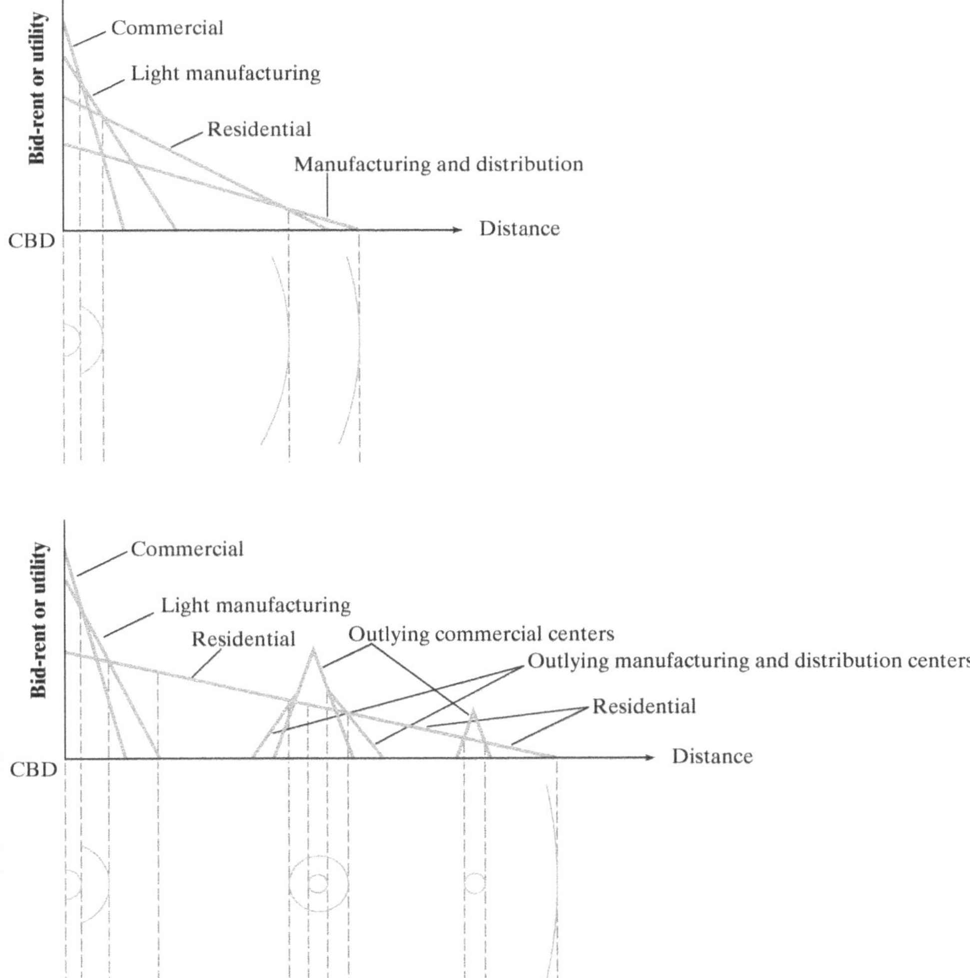

FIGURE 29 Bid-rent curves and the zonal urban structure suggested by the basic trade-off model of urban land use in a (a) single-centered city and (b) multi-centered metropolitan area.

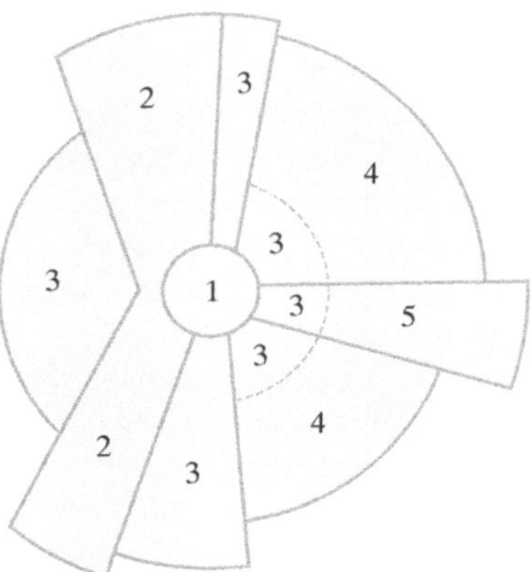

FIGURE 30 Hoyt's model of urban structure. 1–CBD; 2–wholesaling and light manufacturing; 3–lower-income residential; 4–middle-income residential; 5–upper-income residential.

act as a buffer between the industrial/working-class half of the city and the city's main sector of elite neighborhoods.

The key to the dynamics that produced these patterns was, for Hoyt, the behavior of affluent households. Once the CBD had been established and corridors of industrial development laid out, affluent households will always have first pick of the most desirable sites: away from industry and the congestion of the CBD and on high ground free from flooding risk. With urban growth, the high-status area expands along the best transportation lines, in response to, first, a desire among the most affluent to combine accessibility with suburban living and, second, a desire among the almost-as-affluent to acquire the social cachet of living in the same neighborhood as the rich and famous. The result was a *generational shift* in the residence of upper-middle-income households. Every new generation of affluent households would build or buy at the edge of the city: as far as possible from less-fortunate households and within sight of the countryside. Middle-income housing, meanwhile, would be established in surrounding sectors by speculative developers, who recognize the desirability of "good" addresses among prospective customers.

Filtering and Vacancy Chains The most critical part of this overall dynamic is a mechanism of neighborhood development and change implied by Hoyt's model. The basis of this mechanism is the chain of residential moves initiated by the construction of new homes for the affluent: the **filtering** of households up the housing scale and the consequent filtering of houses down the social scale. As a new home is custom built for a family at the top of the socioeconomic ladder, the home they leave might be purchased by a family of slightly lower socioeconomic standing, whose old home, in turn, might be purchased by a third family of still lesser means, and so on, creating a **vacancy chain** that may go on until the last vacancy is a broken-down old place that nobody wants.

Hoyt recognized that, for the filtering mechanism of growth and change to work, there have to be some special reasons to lead the most affluent to new construction: Why, in other words, should they leave well-built, well-situated homes? Hoyt's answer was *housing obsolescence*. For the affluent, several kinds of obsolescence can trigger a desire for new housing. *Functional obsolescence* might result from advances in kitchen technology or new luxury features such as home theaters. *Obsolescence of form* might result from trends in family size and organization, such as the trend away from large families that make large, free-standing mansions outmoded. *Locational obsolescence* might result from land use changes in the immediate neighborhood that are seen as intrusive: increased levels of residential development, the arrival of mass transit, or the appearance of stores. Finally, *style obsolescence* might result from shifts in architectural conventions, something to which many of the most affluent (especially the newly affluent) are especially sensitive.

Hoyt's interpretation of urban land use has been very influential in pointing to the importance of the social dimensions of spatial organization. In addition, the concept of filtering has provided a useful starting point for the analysis of neighborhood change. Urban housing markets are much more complex than Hoyt allowed. In any case, the scale, form, and dynamics of cities were about to be revised yet again by a new round of economic and technological change.

FORDISM, THE AUTOMOBILE, SUBURBAN INFILL, AND THE GREAT DEPRESSION (1920–1945)

The mass production of the internal combustion engine with the introduction of **Fordism** brought decisive changes to the balance of forces that had governed economic and urban development before World War I (Figure 31). Cars began to compete with transit systems, initiating a phase of suburban infill. The availability of tractors enabled farmers to work larger acreages with fewer farm hands, initiating a revolution in farm size and stimulating the urbanization process by releasing rural workers. Particularly significant in the 1920s was the migration of blacks from the rural South to northern industrial cities. Following the "Great Migration" of 1916–1918 that saw blacks induced to northern industrial cities because of labor shortages during World War I, this new wave of migration was in response to a number of factors: the mechanization of southern agriculture, discrimination and organized violence against blacks in the South, and labor shortages in northern cities caused by reduced levels of immigration after the immigration acts of 1921 and 1924.

The result was a decisive restructuring of the country's social geography that included the creation of first-generation black ghettos in northern cities. Cars and trucks, meanwhile, enabled rural residents to travel farther to market or to shop (or to rely on mail-order deliveries), leading to the stagnation or decline of many small towns. Another consequence of the transition from steam to internal combustion was the relative decline of coal and railroad centers and, conversely, the rapid growth of oilfield cities like Oklahoma City and Bakersfield. At the top of the urban hierarchy, air passenger transport reinforced the centralization of business management functions in a few key cities, while the development of highway networks induced the development of hundreds of satellite cities and dormitory towns,

Corbis Images

FIGURE 31 The Model T assembly line in Ford's Highland Park plant. Ford's revolutionary approach to industrial production, together with the revolutionary effects of the car itself, were to have far-reaching implications for urbanization.

reinforcing the process of **regional decentralization** in which the entire structure of U.S. cities would be reorganized.

The increased levels of economic integration and more sophisticated corporate organization also had implications for patterns of urban development. As competition to maintain profit levels intensified, strong companies began to buy out their competitors—the process of **horizontal economic integration**. The chances of new companies being successful were then reduced. Meanwhile, larger companies captured further profits through the processes of **vertical economic integration** (taking over companies that provide their inputs or purchase their outputs, or both) and **diagonal economic integration** (buying highly profitable companies whose activities are unrelated). Because of a flurry of mergers in the 1920s, just 1 percent of all companies accounted for nearly half of the country's productive capacity. *The long-term significance of this would be that the fortunes of many cities would no longer rest on the decisions of owners with local ties.*

A Critical Turn for Urbanization: The Depression and Macroeconomic Management

Before the effects of this shift could be registered, the economy had moved into a downward slide and in October 1929 the stock market collapsed. Problems of imbalance between demand and supply were unprecedented in their intensity, and every sector of the economy was involved in the recession. Economic historians differ in their explanation, but they generally agree that the basic reason was an abnormal sensitivity within the economy.[29] Agriculture had become a victim of its own success, mechanization having boosted productivity so much that the selling price of a bushel of wheat fell from more than $1.80 in 1920 to less than 50 cents in 1930. The Smoot-Hawley Tariff Act of 1930 sought to protect U.S. agriculture from cheap imports but succeeded only in contributing to a recession in world trade. Industrial and financial markets, meanwhile, had become unstable as a result of stock market speculation and complex corporate takeovers. Fearful of inflation, the Hoover administration placed tight restrictions on credit, which further dampened demand for industrial products.

The result was that the unemployment rate rose from around 3 percent in 1929 to more than 25 percent by 1933 (Figure 32). The unevenness of urban-industrial growth in previous years was mirrored in the pattern of decline, so that the intensity of the Depression was greatest in cities that had come to specialize in manufactured goods. The rate of unemployment in some specialized manufacturing towns, for example, exceeded 60 percent; in large, heavily industrialized cities like Detroit and Pittsburgh the unemployment rate in the spring of 1933 was about 50 percent; while in large but more economically diversified cities such as Philadelphia and Seattle it was between 20 and 30 percent. Meanwhile, rural-to-urban migration came to a temporary halt, slowing urban population growth dramatically.

The fundamental economic problem of the 1930s was the downswing, in which deflationary depression saw a collapse of both foreign and domestic markets. The subsequent situation, where surplus labor, productive capacity, and capital existed side by side, was an acute **overaccumulation crisis**. Too much capital had been generated, in aggregate, relative to the opportunities for its profitable investment. Viewed from a slightly different angle, the same circumstances can

Hulton Archive Photos/Getty Images

FIGURE 32 Unemployed workers lining up for jobs at a dockyard in the 1930s. The unprecedented severity of the Great Depression led to a much greater involvement by the federal government in macroeconomic management and in urban and regional policy.

be interpreted as *underconsumption*: insufficient demand by consumers in relation to the capacity of the economic system. Overaccumulation-underconsumption had occurred before, but never on such a scale as during the Depression. In previous episodes the problem had been resolved by "switching" investment into fixed capital (for production, such as machinery, or for consumption, such as housing and refrigerators) within what is sometimes called the *secondary circuit* of the economy's **circuits of capital investment**. In periods of overaccumulation, investment in some of these fixed assets—particularly those involving real estate—becomes relatively attractive. As this investment takes place, the problem of overaccumulation-underconsumption is alleviated.

In the early 1930s, however, the secondary circuit could not absorb the surplus capital. The crisis led to the emergence of an *increased level of government intervention in the economy*, using taxes to (1) channel capital into the *tertiary circuit* (investment in science and technology and in "social overhead" such as education and welfare programs; the former to maintain the economy's productivity and the latter to maintain a productive workforce) and (2) manipulate consumer demand in order to manage the economy. Public expenditures were increasingly used to stimulate demand or, through public employment, to absorb surplus labor. This approach to macroeconomic management is known as **Keynesianism** (after the British economist, John Maynard Keynes).

One of the most important long-term consequences of the Great Depression, therefore, was that the ideology of free enterprise, which had been the unchallenged basis for economic and urban development up to the 1930s, was severely shaken. Although President Hoover had felt morally obliged not to intervene in the "natural" workings of the business system, many voters favored government intervention to provide relief, reform, and recovery, electing Franklin Roosevelt to the presidency in 1932 on a platform that was a precursor of Keynesian economic management. The progressive interventionism embodied in his New Deal programs not only brought the first real experiments in city and regional planning but also introduced a general expansion of government involvement in urban development. At the same time, involving itself in macroeconomic management led to the government mediating a social pact between organized labor and big business: a more structured approach to the labor process and wage bargaining that was geared to productivity, with workers' welfare guaranteed by Social Security. This approach heralded the zenith of organized capitalism.

The government has come to play a major role in influencing the dynamics of urban development.

The Rise of Suburbia

In the 1920s the commercial development of the internal combustion engine unleashed social, economic, and political forces that have given physical shape to the evolving contemporary metropolis. Cars, trucks, and airplanes—along with electricity and telephones—helped to recast the imprint of urbanization, producing sprawling, multinodal metropolitan settings with complex patterns of land use. Just when downtown areas had established their unrivaled dominance and internal functional specialization, a new logic of transportation and location led to the decentralization of many of their retailing, wholesaling, manufacturing, and office functions, leaving the CBDs more specialized and less dominant. And just as the suburbs radiating out along transit lines had redrawn the social map, a new form of suburbanization materialized, bringing much lower densities, greater social segregation, and a greater variety of suburban land uses. Meanwhile, the cars that triggered these new developments also undermined the old mass transit and pedestrian circulation systems on which the land use patterns and functional organization of the industrial city had been based. All these changes, in turn, brought new challenges to understanding the development of urban form and land use: new concepts and theories, modified ideas, and alternative models.

U.S. cities were the pioneers and exemplars of automobile-based suburbanization. According to the records, there were 4 motor vehicles in use in the United States in 1894, 16 in 1896, 8,000 in 1900, almost 470,000 in 1910, over 9 million in 1920, and nearly 27 million in 1930 (Figure 33). This first spurt of growth in car production and ownership corresponds to the period when cars were being substituted for horse-drawn and electric-powered transport. The rate of growth decreased in the 1930s and 1940s as the process of substitution was completed, reinforced by the dampening effects of the Great Depression and World War II.

Fordism

During this time, Henry Ford's vision of mass production, coupled with mass consumption, was achieved in the auto industry—a precursor of Fordism that characterized entire economies later. Ford's vision was made possible by a combination of lower prices and higher wages and was paid for

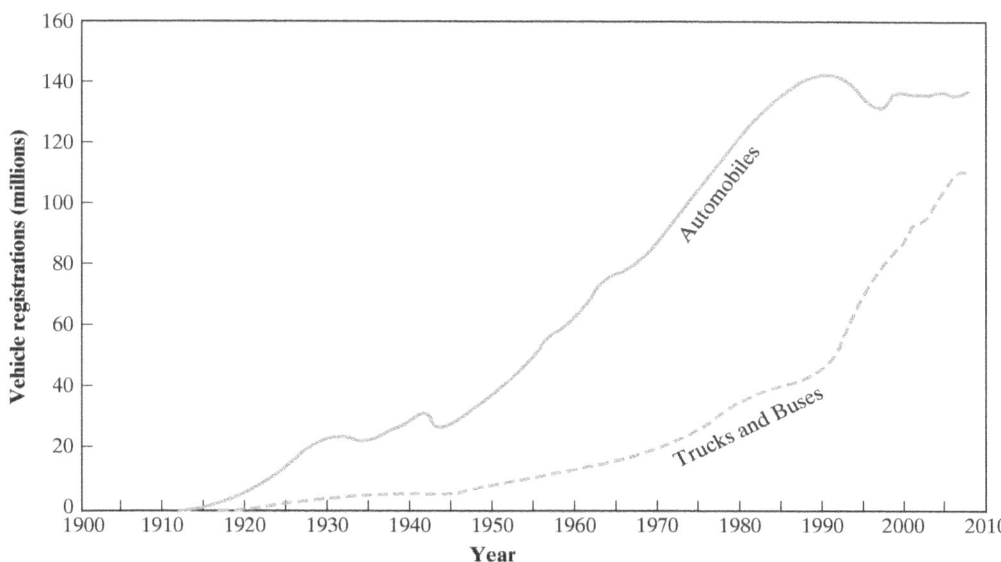

FIGURE 33 Motor vehicle registrations.

by higher productivity squeezed from economies of scale, assembly-line production, and "scientific" management (**Taylorism**). One result was that *the price of cars fell significantly*, making car ownership possible for a widening segment of society and drawing more people into the infill suburbs between the streetcar corridors and railroad exurbs. The Model T, which in 1909 sold for $950 (equivalent to 22 months pay for the average worker), cost less than $300 (less than three months pay) in 1925, when Ford's factories were producing 9,000 cars each working day: one every 10 seconds.

Paving the Way for Suburbanization

Cars, however, are of little use without good roads, and coalitions of special interests emerged to support the building and improvement of roads. The first push for good roads came in the 1890s, from bicyclists, the Post Office, and farmers' Granges—associations that wanted to promote better access to markets. This was consolidated a decade later when the car, oil, rubber, and construction industries were joined by the Good Roads Association, a confederation of urban merchants and industrialists who believed that better highways would improve business.[30] Their campaigns resulted in the 1916 Federal Aid Roads Act, which required every state, as a condition of federal aid, to establish a state highway department to plan, build, and maintain highways between cities. A second act in 1921 provided additional funds to integrate the long-distance road network.

By the 1930s federal funds were being designated for road construction and improvement *within* cities. For some time before that, though, local coalitions had persuaded local governments to step up their road-building and road-improvement programs. State governments began to levy a gasoline tax to pay for road-building—the first was Oregon's gasoline tax of one cent per gallon in 1919. Street improvement and highway construction were already the second largest item (after education) in local government budgets. Chicago, for example, improved more than 100 miles of streets between 1915 and 1930 at a cost of over $1 million per mile.[31] By 1940 about half of the country's 3 million miles of roads had acquired smooth hardtop surfaces. Parking space, however, remained a problem (Figure 34).

Parkways Some of the most imaginative and progressive initiatives were taken around New York City, guided by master planner Robert Moses. Moses developed the idea of the **parkway** as a limited-access recreational highway. The germ of the idea had been in Frederick Law Olmsted's design for New York's Central Park in 1858, and the first full-scale parkway was William K. Vanderbilt's Long Island Motor Parkway, begun in 1906. Moses saw that parkways had great appeal in the age of the automobile and could improve environmental quality in existing built-up areas by replacing blighted areas with sinuous, airy green corridors.

Associated Screen News/Getty Images

FIGURE 34 With the emergence of mass ownership of cars, many cities faced a new pressure on land use in industrial and commercial districts: space for parking lots. This example is from Montreal in the late 1930s.

Beginning with the Bronx River Parkway (1906–1923), Moses supervised the construction of over 100 miles of four-lane, limited-access, and lavishly landscaped parkways that ran through Westchester County to the north and through Long Island to the east, giving New York car owners access to beaches that were much less crowded than those served by subway and streetcar. During the 1920s and 1930s many other cities added stretches of parkway and built bridges and tunnels that brought cars to previously inaccessible and undeveloped tracts, including Chicago's Wacker Drive north of the Loop (the CBD) and the Bay Bridge and Golden Gate Bridge in San Francisco.

The Decline of Mass Transit The success of what the boosters called "automobility" was directly responsible for a downward spiral in the fortunes of transit systems. New automobile suburbs lured commuters away from the high-density corridors near transit stops. They also captured the attention of developers, ending the marriage of convenience in which development companies had subsidized transit companies. As shops, offices, wholesaling, and industry moved out to the automobile suburbs to take advantage of growing markets and cheaper land, the transit systems, still rigidly focused on the downtown, were unable to serve the consequent crosstown and intersuburban commuting flows and lost further ground.

Meanwhile, increasing numbers of car owners substituted Sunday drives for weekend streetcar outings. The transit companies' loss of revenue forced reduced maintenance, abandoning of less-profitable routes, cutbacks in weekday services, and increased fares—all of which pushed many more commuters to turn to the car. More vehicular traffic in the streets, in turn, slowed streetcars, increasing their operating costs and prompting still larger numbers of passengers to abandon the trolleys in favor of commuting by car. By the 1920s about 30 percent of the people entering the CBDs of older, more congested cities did so by car; in newer cities west of the Mississippi, the proportion was nearer 60 percent.

It has also been suggested that auto–oil–rubber corporate interests conspired to hasten the demise of transit systems:

> The way it worked was that General Motors, Firestone Tire and Standard Oil of California and some other big companies, depending on the location of the target, would arrange financing for an outfit called National City Lines, which cozied up to city councils and city commissioners and bought up transit systems like LA's. Then they would junk or sell the electric cars and pry up the rails for scrap and beautiful modern buses would be substituted, buses made by General Motors and running on Firestone Tires and burning Standard's gas.[32]

During the 1930s General Motors created a holding company through which it and other auto-related companies channeled funds to buy up streetcar systems in 45 U.S. cities.[33] In the late 1960s GM was convicted in a Chicago federal court of conspiring to destroy electric transit and to convert trolley systems to diesel buses, whose production GM monopolized. The corporate executives each received a $1 fine.

But a major factor in the demise of streetcar systems was their fixed routes could not cope with changing metropolitan form and patterns of land use. Buses were able to go around congestion that stopped the trolley in its tracks, and they were able to reach new residential subdivisions and peripheral workplaces on hard-surfaced streets whose costs were borne not by the transit companies but by everyone (through federal, state, and local taxes).

Patterns of Suburban Growth

The early investment in cars, roads, and suburban infill is widely credited with rescuing the U.S. economy from a period of stagnation, and igniting the boom of the Roaring Twenties. The place where the boom took off was the suburbs. In the 1920s, for the first time, suburbs grew faster than central cities—much faster. While central cities grew by 19 percent, adding 5 million new residents, the suburbs grew by 39 percent, adding over 4 million people.

Automobile Suburbs At first, new automobile suburbs appeared as simple additions to existing, transit-dependent suburban corridors, and they functioned that way, depending on the central city for employment and shopping. But because developers no longer had to consider the constraints imposed by people having to walk to transit stops, the form of suburban development was very different from the start. By the standards of the time, building lots were much larger (even though the houses were no bigger and often less substantial), resulting in much lower densities. Many developers got rid of sidewalks altogether, partly to save money and partly to emphasize the exclusivity of the neighborhoods. In some cases grid layouts gave way to curvilinear street patterns that minimized the number of junctions and were thought to lend a distinctive and upscale appearance to the neighborhood. In every city these infill suburbs survive as a distinctive element of urban form: what are now considered to be relatively high-density suburbs with relatively small houses and few neighborhood amenities (Figure 35). Most have now filtered down the social scale to become working-class suburbs.

But just as these new suburbs were beginning to expand, their form was influenced by the U.S. Supreme Court in its landmark case on land use zoning law: *Village of Euclid, Ohio v. Ambler Realty Co.* (1926). It ruled in favor of the municipality's right to prevent a property owner from using land for purposes other than for which it had been zoned. The Court established the power of local governments to "abate a nuisance," defined very broadly to include anything affecting the general welfare of a residential area. As a result, zoning quickly was used to exclude not only undesirable land uses from residential areas but also (by establishing large minimum lot or dwelling sizes) undesirable *people*.

The consequences for suburban development were far-reaching. Growing suburban areas rushed to become **incorporated** as municipalities so that they could control the pace and nature of growth. Speculative developers, reassured by the stability that these zoning maps gave the

Corbis Bridge/Alamy

FIGURE 35 Infill suburbs survive as a distinctive element of urban form in every U.S. city. These relatively high-density suburbs have relatively small houses and few neighborhood amenities. Most, like this one in New Jersey, have filtered down the social scale to become working-class suburbs.

land market, were emboldened to lay out larger and larger subdivisions, copying, as far as possible, Fordist techniques of mass production for mass consumption.

The overall pattern of urban development, no longer tied to the star-shaped corridors of transit lines, reverted to a more symmetrical shape (Figure 36). In these new suburbs lot sizes were larger and densities were lower. The average size of a building lot rose from about 3,000 square feet in streetcar suburbs to about 5,000 square feet in automobile suburbs; residential densities fell from about 20,000 people per square mile in streetcar suburbs to about 10,000 per square mile in automobile suburbs.

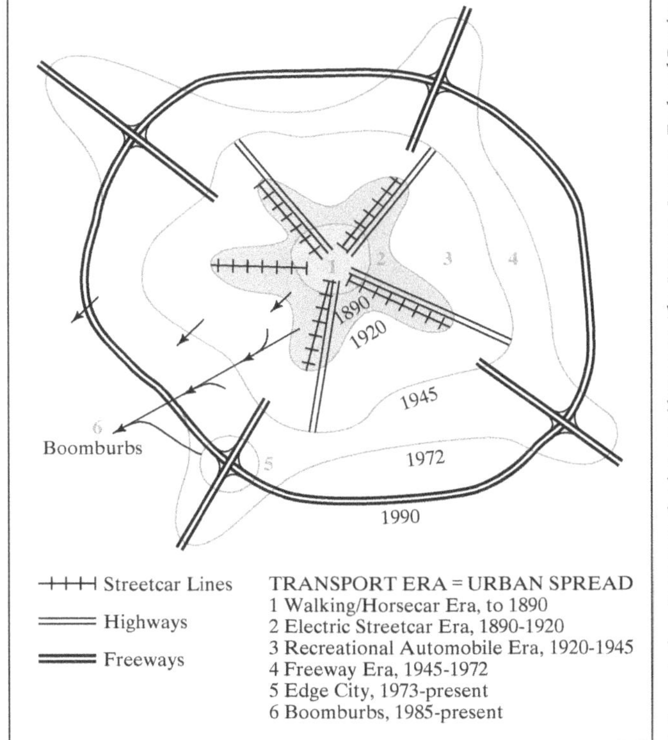

De Souza, Anthony R.; Stutz, Frederick P., *World Economy, The: Resources, Location, Trade and Development*, 3rd, © 1998. Printed and Electronically reproduced by permission of Pearson Education, Inc., Upper Saddle River, New Jersey.

FIGURE 36 Transportation innovations and changing urban and metropolitan form.

Planned Suburbs Decentralization on this scale brought unprecedented opportunities and challenges. The chief *opportunities* were to establish new patterns of development that incorporated retail, commercial, industrial, and recreational land uses in addition to housing. The chief *challenges* were to avoid unnecessary traffic congestion and preserve environmental quality. Although most suburban infill was totally unplanned and entirely without any conscious recognition of these challenges and opportunities beyond the immediate implications for profitability, there were several critical experiments and innovations that were to become models for later phases of urbanization.

In fact, the first experiments in creating planned suburban settings were those that had been designed for the upper-middle-income families that developers had earlier sought to bring to outlying railroad stations and streetcar termini before the onset of the automobile suburbs. Among the best known are Llewellyn Park in West Orange, New Jersey (1853); Chestnut Hill in Philadelphia (1854); Lake Forest, Illinois (1856); and Forest Hills Gardens, New York (1909). They were precursors for the idea of secluded, affluent neighborhoods that were self-contained at least in terms of lower-order services; but they were compact, and their layout and design were unsuited to the car.

In the mid-1920s there were several attempts to plan communities that were based on "automobility." The best-known of these are Palos Verdes Estates (Los Angeles); Shaker Heights (Cleveland); the Country Club District (Kansas City); and two "master suburbs" in Florida: Coral Gables and Boca Raton. Although each had its own innovations, they were all built by private developers for profit, with an upper-middle-class market in mind. They were characterized by very low densities (for the time) of about three dwellings per acre, high-quality landscaping, recreational facilities (golf courses, in particular), public gardens, and plazas, in addition to shopping amenities, and deed covenants aimed at preserving the character and appearance of the entire community.

The Country Club District was perhaps the most influential of these. It was the creation of developer Jesse Clyde Nichols, who later founded the Urban Land Institute, an independent research organization concerned with urban land use and development from the developers' point of view. Nichols had been impressed by the Garden City movement in Europe and the City Beautiful movement in the United States, and was determined to put together a project large enough to sustain a self-contained community. Nichols was concerned with profitability, however, so from the start set out to create a setting that would appeal to the most lucrative section of the residential market: the top end. It took him 14 years to acquire the land, before embarking, in 1922, on constructing 6,000 homes and 160 apartment buildings that eventually housed over 35,000 residents.

The centerpiece was Country Club Plaza (Figure 37), the world's first automobile-based shopping center. It featured waterfalls, fountains, and expensive landscaping, with extensive parking lots behind ornamental brick walls. Nichols carefully controlled the composition of businesses through leasing policies that brought upscale retail stores to the first floor of the development, and lawyers, physicians, and accountants to the offices on the second floor. Similarly, the residential sections of the Country Club District were carefully landscaped and controlled. Densities were kept low, streets were curvilinear, trees were preserved wherever possible, and houses were set back from the street, with driveways and garages. All sales were subject to racially restrictive deed covenants, and all purchasers were required to join the homeowners' association, the purpose of which was to ensure lawn care and to supervise the general upkeep and tidiness of streets and open spaces. It was a commercial success from the start and, despite the obvious problematic elitism of the venture, it attracted critical acclaim from builders, developers, and planners who came from across the United States to view the shape of the future.

In 1923 a small group of intellectuals founded the *Regional Planning Association of America (RPAA)*. Among them were Lewis Mumford and several architects and planners—including Clarence Stein and Henry Wright. Although RPAA members were mostly interested in big-picture, long-term scenarios for urbanization and regional development, they did manage, through Alexander Bing—a New York real estate developer and founding member—to create two planned communities. The first, Sunnyside Gardens (built between 1924 and 1928), was an undeveloped central city site in Queens only five miles from Manhattan. Here, Clarence Stein and Henry Wright designed big traffic-free superblocks that allowed vast interior garden spaces.

The second was Radburn (started in 1928), 15 miles from Manhattan in Fair Lawn, New Jersey, where Stein and Wright were able to release the Sunnyside superblock principle from the rigid grid of the central city. Traffic was channeled through a hierarchy of roads, so that most residential areas could be kept virtually free of traffic. Pedestrians and cyclists were given their own paths that crossed traffic arteries under rustic bridges; housing was clustered cozily around irregular-shaped open spaces (Figure 38). The RPAA had hoped to

Library of Congress

FIGURE 37 Country Club Plaza, Kansas City.

FIGURE 38 Part of the proposed plan for Radburn, N.J. In an effort to promote "community" and minimize the intrusive effects of cars, houses face onto common walkways; streets give access to the rear of the houses.

draw in a socially mixed group of residents, but Radburn's attractiveness quickly ensured that it became a commuter suburb for upper-middle-class families; realtors then took it upon themselves to keep out Jews and blacks to maintain property prices.[34] Nevertheless, Radburn, like the Country Club District, became an influential landmark in the history of urban design.

Suburbanization and Federal Policy Suburbanization, like other forms of economic and urban development, came to a shuddering halt with the Great Depression. Between 1928 and 1933 residential construction activity fell by 95 percent. In the same period a million households lost their homes through foreclosure. By the spring of 1933 half of all home mortgages were technically in default, and every day saw a thousand new foreclosures.[35] The Hoover administration responded by creating a credit reserve for mortgage lenders (the **Federal Home Loan Bank Act** of 1932) and a fund for making loans to nonprofit corporations formed to build or upgrade housing for low-income families (the Emergency Relief and Construction Act of 1932). Neither initiative was framed effectively, however. If they had any value at all, it was as precedents that acknowledged the need for the federal government to play a role in protecting home ownership and housing quality.

The Roosevelt administration built on these precedents in ways that not only helped revive the expansion of automobile suburbs but also had lasting impacts on the nature of urbanization in the United States. The Home Owners Loan Corporation (HOLC), created in 1933, helped stop the slide by refinancing tens of thousands of mortgages that were in danger of foreclosure and by establishing low-interest mortgages that allowed former owners to recover homes lost through foreclosure. The following year, in an attempt to reduce unemployment by stimulating construction (a labor-intensive industry), Roosevelt created the Federal Housing Administration (FHA).

This initiative was a critical part of the shift to Keynesian macroeconomic management and led to the development of so-called **Keynesian suburbs**. The FHA was given responsibility for stimulating construction by the private sector, and it chose to do so by stabilizing the mortgage market and facilitating sound home financing on reasonable terms. The FHA achieved this goal not by lending money but by *insuring* mortgage loans made by private institutions for home construction or purchase. The insurance gave banks and savings and loan associations the confidence to disburse more mortgages and to charge one or two percent less in interest than before, which stimulated demand for home ownership. FHA guarantees also stimulated demand with terms that required smaller down payments and extended repayment periods (which meant lower monthly repayments). The FHA established and enforced minimum standards for housing financed by its guaranteed loans, which helped eliminate shoddy suburban construction.

At the time, many saw federal support for home ownership as a means of defending the property system. By giving as many people as possible a stake in the system and financing their stake through long-term loans, conservatives argued, greater social and political stability would be achieved. The widespread debt represented by mortgage repayments would commit home "owners" to oppose any changes to the social and economic structure of society that might endanger the value of their property or make it more difficult for them to pay off their loan.

The immediate effect was to reignite suburban growth. Whereas housing starts had fallen to fewer than 100,000 in 1933, the number of new homes started in 1937 was 332,000, and in 1941, 619,000 (Figure 39). New Deal administrators saw a further opportunity: to plan suburban development,

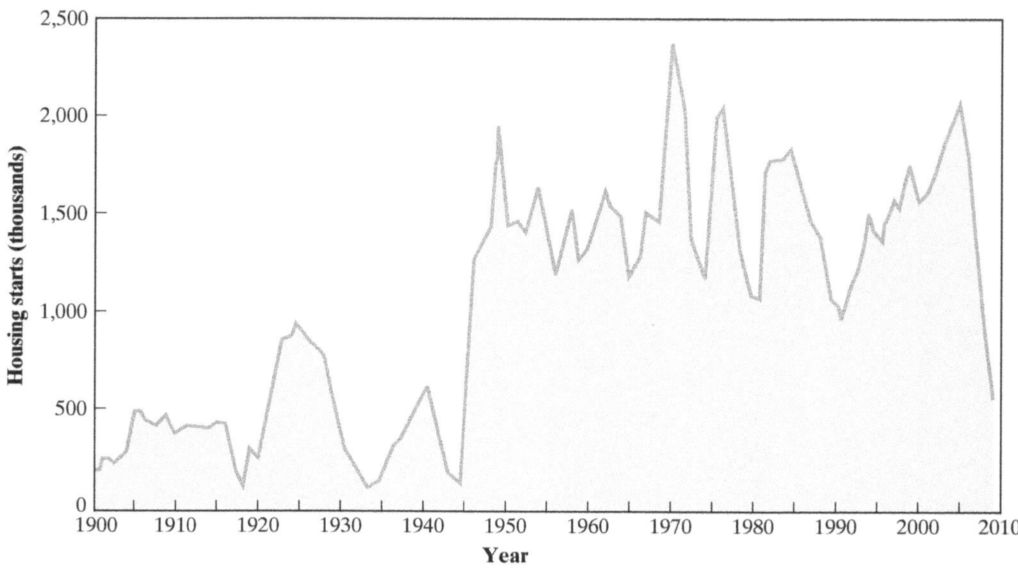

FIGURE 39 New private housing starts.

drawing people from central cities to make room there for slum clearance and redevelopment. In 1935 Roosevelt created the **Resettlement Administration** with these objectives, appointing Rexford Guy Tugwell as its head. Drawing on the design of Radburn, Tugwell envisaged some 3,000 **greenbelt cities** that would contain government-sponsored low-cost housing. Funds were allocated for only eight, however, and Congress, under strong pressure from the private development industry, whittled this number down to five. Two of the five were blocked by local legal action. The remaining three were built: Greendale, southwest of Milwaukee; Greenhills, near Cincinnati; and Greenbelt, just north of Washington, D.C. But in 1938 the Resettlement Administration was abolished and after World War II all three were sold off to nonprofit corporations, to be swallowed up in the continuing sprawl of automobile suburbs.

Suburbanization of Commerce and Industry The decentralization of commerce and industry was slower than the decentralization of households initially. The number of trucks and buses did not really begin to increase significantly until after World War II (Figure 33). But once a substantial and highly mobile population was in the suburbs, the logic of commercial and industrial location changed decisively. Suburbanites represented a large and very affluent market for everything from convenience items to expensive clothes and furniture. At the same time, they were a highly qualified workforce.

Retailing was one of the first commercial activities to decentralize to the new automobile suburbs. There were few attempts to replicate the success of Country Club Plaza, partly because of the huge investment required. Instead, automobile-related retailing took the form of a hierarchy of retail settings, each with a distinctive range of services and each with a different market area. The car, in short, enabled a sort of central place system to develop within central places.

Some *office* development decentralized to outlying business centers and along principal business thoroughfares. Car travel made these locations attractive to all sorts of business services and smaller corporate offices. Meanwhile, smaller offices for professionals such as physicians, dentists, and attorneys serving local populations, decentralized to neighborhood business streets.

During the 1920s and 1930s there were also shifts in *wholesaling*. Distributors whose market was dominated by in-town buyers were increasingly tempted to move to locations with cheaper rents. These locations had to be reasonably close to the traditional railroad freight yard warehouse district because of their reliance on rail cars for incoming goods. For some (such as department store warehouses), however, trucking made it feasible to relocate to the cheapest sites at the edge of the city.

It was the decentralization of *industry*, however, that was the most significant change brought about by trucking. By the 1920s there were strong but pent-up pressures for industrial decentralization. One pressure was the increasing cost of doing business in congested industrial districts and corridors, where rents were high. In addition, the advent of electric power made multistory factories (designed to maximize the efficiency of steam-powered machinery) obsolete. Using new assembly-line techniques made possible by electric power demanded a horizontal layout of industrial space, and this required unprecedented amounts of land. Trucks enabled factories to move out to cheaper, peripheral land, while cars and buses enabled their workers to get to them.

By the late 1930s almost no new factories were opening in traditional central factory districts, while many of the older factories were closed as their owners invested in newer, more efficient plants in the suburbs. So began the first round of *central city deindustrialization* and restructuring. Factory districts, starved of investment and with some buildings 50 years old or older, became shabby and semiderelict. Nearby working-class housing of similar vintage degenerated into slums, their problems suddenly intensified by the Great Depression. And on the edge of the central city the first-tier streetcar suburbs, having been made both socially and functionally obsolete by the car, began to tumble down the social ladder, blurring the established order of the city.

Within the CBD, the civic and entertainment zones were the least affected (though decentralization did attract away some "fringe" activities such as speakeasies, gambling dens, and brothels to roadhouses along the urban fringe). The decentralization of offices and shops caused CBD office and retail zones to become more specialized, catering increasingly to the upper end of the market. Department stores, in contrast, sought to maintain their magnetism by diversifying:

> The department store became a zoo (Bloomingdale's and Wanamaker's in New York had enormous pet stores), botanical garden (floral shops), restaurant (lavish restaurants bigger than any other in the cities), barber shop, butcher shop, museum (gift and art shops), post office, and beauty parlor.[36]

Hedging their bets, some department store companies opened new branches in the CBD frame, where they could offer free customer parking. Sears and Montgomery Ward, in Chicago and New York, were the first to adopt this strategy, and other stores in virtually every city soon followed. Because driving was seen as a male preserve, however, these new stores carried a limited range of items, dominated by household goods, automotive accessories, and building supplies. Downtown stores, in turn, came to emphasize women's fashions, cosmetics, haberdashery, and millinery.

New Patterns of Land Use It was in response to this spatial reorganization that Chauncy Harris and Edward Ullman devised a model that attempted to describe the outcomes in terms of the spatial organization of land uses.[37] This **multiple-nuclei model** is a schematic representation of the relative locations of major categories of land use (Figure 40), based on the decentralization of commercial and industrial nodes beyond the CBD. Harris and Ullman argued that new, automobile-based suburban nodes of commercial and industrial activity were not arranged in any predictable

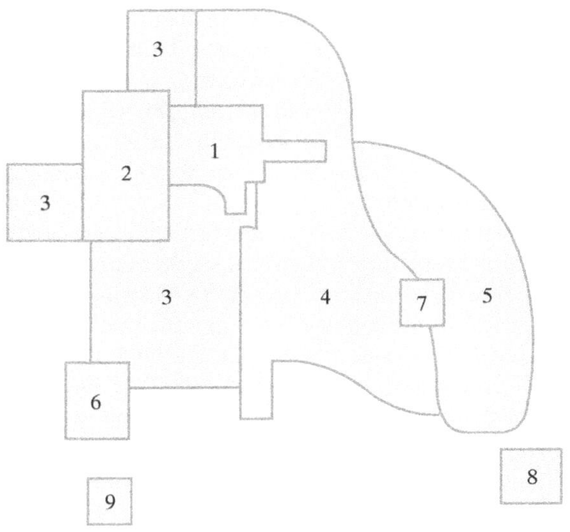

FIGURE 40 The multiple-nuclei model of urban form.

fashion, except in relation to surrounding land uses. They might develop around a transit stop or highway intersection; but if they were office and retailing centers they would attract middle-income residential development, whereas if they were industrial centers they would attract working-class residential development.

Underpinning the overall pattern was a fundamental set of functional relationships, given expression as a result of the mobility allowed by cars and trucks and by the locational flexibility allowed by telephones and electricity. Fundamental to these relationships were the mutual attractiveness of certain groups of activities and the tendency for some land uses to be repelled by others. The result is an irregular-shaped patchwork of land uses across which there is a loose functional order. Looking back, we can see that Harris and Ullman were remarkably prescient: The multiple-nuclei city was the embryo form of uniquely U.S. manifestations of contemporary urbanization—metropolitan sprawl with new "production space" and "edge cities."

FOLLOW UP

Key Terms

1926 *Euclid v. Ambler*
central places
comparative advantage
incorporated
initial advantage
Keynesian suburbs
land use zoning
Manufacturing Belt
mercantile model (Vance)
multiple-nuclei model (Harris and Ullman)
pedestrian city
primate city and primacy
rank-size rule
sector model (Hoyt)
streetcar suburbs

Review Activities

1. Think about how you can make the aspects of urbanization described in this chapter meaningful to you personally. If you live in the United States, perhaps your great-grandparents or great-great-grandparents were among the earlier waves of immigrants who contributed to urban growth. Where did they arrive, what did they do, and what cities did they and their descendants live in? Do you have family members you can ask? You might even be able to add sound recordings of people or city life to your e-portfolio.
2. Go online to find the zoning ordinances and land use zoning maps for where you live or for a city that interests you. What conclusions can you draw about the planners' objectives? Which of the models and concepts discussed in this chapter are most useful in understanding the patterns of land use shown in the maps?
3. When you have time, read a classic novel that is evocative of urban life during part of the period covered in this chapter. There are hundreds to choose from, including *McTeague*, by

Frank Norris (New York: Fawcett, 1960) and *A Tree Grows in Brooklyn*, by Betty Smith (Philadelphia: Blakiston, 1943).

4. Work on your *e-portfolio*. Research an example of a power-site town, mining town, transportation town, or heavy manufacturing town. Find out how and why it became established and what happened to it during the twentieth century. Because the topics covered in this chapter are focused on local-scale, historical events, you might wish to look for supplementary materials in old newspaper accounts. Copies of nineteenth-century and early twentieth-century newspapers might not be available online but are often available on microfilm in libraries. You can include that material in your e-portfolio with your own observations about how the story is related to themes in the text (or to your own perspective).

Log in to **www.mygeoscienceplace.com** for self-study quizzes, *MapMaster* layered thematic and place name interactive maps, *Urban View* Google Earth™ tours, key resources and suggested readings, related websites, "In the News" RSS feeds, and additional references and resources to enhance your study of the foundations of the U.S. urban system and its cities.

NOTES

1. Based on C. Bridenbaugh, *Cities in the Wilderness: The First Century of Urban Life in America 1625–1742* (Oxford: Oxford University Press, 1938).
2. Ibid., 60.
3. Ibid., 85.
4. Ibid., 239.
5. Ibid., 63–64.
6. Ibid., 64.
7. Ibid., 70.
8. Ibid., 72.
9. W. Sjoberg, *The Pre-Industrial City, Past and Present* (Glencoe, Ill.: Free Press, 1960).
10. J. E. Vance, Jr., "Land Assignment in Pre-Capitalist, Capitalist and Post-Capitalist Cities," *Economic Geography* 47 (1971): 101–120.
11. J. P. Radford, "Testing the Model of the Pre-Industrial City: The Case of Ante-Bellum Charleston, South Carolina," *Transactions of the Institute of British Geographers* 4 (1979): 392–410.
12. G. K. Zipf, *Human Behavior and the Principle of Least Effort* (Reading, Mass.: Addison-Wesley, 1949).
13. M. Jefferson, "The Law of the Primate City," *Geographical Review* 29 (1939): 231.
14. W. Christaller, *Central Places in Southern Germany*, trans. C. W. Baskin (Englewood Cliffs, N.J.: Prentice Hall, 1966). Originally published by Gustav Fischer, Jena, 1933.
15. J. U. Marshall, *The Structure of Urban Systems* (Toronto: University of Toronto Press, 1989), 278.
16. J. S. Adams, "Residential Structure of Midwestern Cities," *Annals of the Association of American Geographers* 60 (1970): 37–62.
17. D. Ward, *Poverty, Ethnicity, and the American City, 1840–1925: Changing Conceptions of the Slum and the Ghetto* (Cambridge, UK: Cambridge University Press, 1989).
18. S. B. Warner, Jr., *The Urban Wilderness* (New York: Harper & Row, 1972), 19.
19. Ibid.
20. See, for example, the analysis of locational conflict within Worcester, Mass., during the nineteenth century in W. B. Meyer and M. Brown, "Locational Conflict in a Nineteenth Century City," *Political Geography Quarterly* 8 (1989): 107–22.
21. L. Mumford, *The City in History* (London: Secker and Warburg, 1961), 461.
22. C. G. Kennedy, "Commuter Services in the Boston Area 1835–1860," *Business History Review* 26 (1962): 277–87.
23. J. E. Vance, Jr., "Housing the Worker: The Employment Linkage as a Force in Urban Structure," *Economic Geography* 42 (1966): 294–325.
24. R. Yates, *Revolutionary Road* (New York: Vintage Books, 1961), pp. 125–26.
25. R. McKenzie, *The Metropolitan Community* (New York: McGraw-Hill, 1933).
26. R. Hurd, *Principles of City Land Values* (New York: The Record and Guide, 1903), 13.
27. W. Alonso, "A Theory of the Urban Land Market," *Papers and Proceedings of the Regional Science Association* 6 (1960): 149–58.
28. H. Hoyt, *The Structure and Growth of Residential Neighborhoods in American Cities* (Washington, D.C.: Federal Housing Administration, 1939).
29. See B. J. L. Berry, *Long-Wave Rhythms in Economic Development and Political Behavior* (Baltimore, Md.: Johns Hopkins University Press, 1991), 32–33.
30. P. J. Hugill, "Good Roads and the Automobile in the United States, 1880–1929," *Geographical Review* 72 (1982): 327–49.
31. H. P. Chudacoff and J. E. Smith, *The Evolution of American Urban Society*, 5th ed. (Upper Saddle River, N.J.: Prentice Hall, 2000), 219.
32. H. Reasoner, CBS News, *60 Minutes*, cited in *Building American Cities*, eds. J. Feagin and R. Parker, 2nd ed. (Englewood Cliffs, N.J.: Prentice Hall, 1991), 156.
33. Ibid., 157.
34. D. Schaffer, *Garden Cities for America: The Radburn Experience* (Philadelphia: Temple University Press, 1982), 173–74, 177.
35. Jackson, *Crabgrass Frontier*, 193.
36. W. Leach, "Transformations in a Culture of Consumption," *Journal of American History* 71 (1984): 326, cited in Chudacoff and Smith, *The Evolution of American Urban Society*, 221.
37. C. D. Harris and E. L. Ullman, "The Nature of Cities," *Annals, American Academy of Political and Social Science* CCXLII (1945): 7–17.

Urban Systems and Cities In Transition

John G. Wilbanks/Alamy

Urban Systems and Cities In Transition

By the end of World War II, the economies and workers in industrialized countries like the United States had begun to enter a substantially different phase in terms of what they produced and how and where they produced it. This phase is sometimes referred to as advanced capitalism. Fundamental to this change was a decrease in the proportion of the workforce involved in manufacturing and a marked increase in service-related jobs. The prosperity associated with this phase of economic development, meanwhile, allowed increasing numbers of people to spend a larger proportion of their incomes on consumer services such as shopping and eating out. From the mid-1970s, however, the internationalization of economic activities had destabilized the "organized" basis of national economies and allowed an intensification of the economic and spatial logic of capital—especially big capital—and with it a significant reorganization of urban systems and a recasting of the political economy of cities within a global context.

LEARNING OUTCOMES

After reading this chapter, you should be able to:

- Explain the reasons for the sprawling form of metropolitan areas.
- Describe the key characteristics of post-World War II suburbia.
- Evaluate the impact of demographic and socio-cultural changes on patterns of urbanization.
- Discuss the mutually reinforcing problems related to the **deindustrialization** that characterized the U.S. from the 1970s onward.
- Identify and explain the main ways in which the global reach of information and communications technologies are influencing patterns of urbanization.

CHAPTER PREVIEW

This chapter deals with the changes that have occurred to the United States and other industrialized urban systems in Canada, Western Europe, Australia, and Japan as a result of the transition to an advanced form of capitalism that has now become increasingly globalized in its scope. We trace urban-system change in the United States in terms of periods of urbanization that have been distinctive in their patterns of urban growth and their patterns of interaction between cities.

The first period, spanning 1945–1973, corresponded with postwar economic recovery and growth. The principles and processes of urban change resulted in new outcomes as they operated through the new industries and technologies of advanced capitalism. We will also see how the overall prosperity of the period

Cities such as Seattle, Washington, whose corporate headquarters include Amazon.com, have been recast as the result of the needs of "new economy" industries and the capabilities of new technologies. As the birthplace and headquarters of Starbucks, Seattle's shift to advanced capitalism has been associated with increasing numbers of people spending a larger proportion of their incomes on consumer services such as eating out and shopping.

was directly reflected in the form of cities through extensive suburbanization; and how suburbanization, combined with the decline of old manufacturing enterprises, affected central city land use.

The second period, between 1973 and the present, began with a phase of economic crisis and reorganization. This was reflected within the urban system in terms of differential economic stress as city governments, companies, and workers tried to adapt to both global competition and structural economic change. A rapid decline of the old base of manufacturing industries was followed by the onset of a "new economy" based on digital technologies and featuring advanced business services, cultural products industries, and knowledge-based industries, all framed within the context of an international division of labor, international finance, and the ascendance of neoliberal politics and policy. We will see how this has resulted in a functional reorganization of the hierarchy of U.S. cities, how it has come to be reflected in variations in the quality of life, and how it has influenced—and been influenced by—recent demographic and social changes. Meanwhile, cities themselves have been recast as the result of the needs of "new economy" industries, the capabilities of new technologies, and an increasing socioeconomic polarization of metropolitan populations. We will see how these trends have resulted in the end of traditional "suburbia" and the emergence of privatized, fragmented, and disjointed settings in polycentric metropolitan regions.

REGIONAL DECENTRALIZATION AND METROPOLITAN SPRAWL (1945–1973)

As in previous periods of development in the United States, between 1945 and 1973, changes in transport and communications technologies strongly influenced the outcomes of the structural economic change to **advanced capitalism**. In this first phase the most significant developments in this respect at the scale of the U.S. urban system were the construction of the interstate highway system (Figure 1) and the growth of a network of regional and subregional airports capable of handling large passenger jets (Figure 2). The overall effects of changes in employment and transportation on the structure of the urban system can be summarized in terms of two apparently contradictory (but in fact interrelated) outcomes: **regional decentralization** and **metropolitan consolidation**. Increased accessibility, combined with the attractions of cheaper land, lower taxes, lower energy costs, local boosterism, and cheaper and less-militant labor, allowed cities in the South and West to grow rapidly.

The metropolitan areas of the "Sunbelt" were particularly attractive to labor-intensive manufacturing and to the high-growth, high-tech sectors of electronics, aerospace, and petrochemicals. These industries could establish themselves in Sunbelt cities and have the cities and their infrastructure

URBAN VIEW 1
Fast Food and Religion in the Exhaust of a Drive-in Culture[1]

In the unprecedented prosperity in the United States in the decades following World War II, color TVs, electric blenders, and automatic garbage disposals became basic household commodities in every new suburban home. But the ultimate symbol of success was the family car. It did not take long before answering the question, "What do you drive?" came to define how people saw themselves, and how others saw them, based on the cost, make, model, and age of their car. Business owners took note of ballooning car ownership and offered all kinds of drive-in opportunities for customers, including drive-in motels, drive-in banks, and drive-in movie theaters.

The first drive-in restaurant was The Pig Stand in Dallas, opened in 1921 and named after the Pig Sandwich, the popular Tennessee barbeque pork sandwich served there. Texan Jesse G. Kirby came up with the idea, having observed that: "People with cars are so lazy they don't want to get out of them to eat!" The Pig Stand offered curb service from young boys who soon were nicknamed "carhops" because they ran up to approaching cars and hopped onto the running board and started taking orders before the cars even came to a complete stop. The tips they made were their entire wages, so they wasted no time taking orders and hurrying off to collect and return with the food. Before the "golden arches" there was the silhouette of a husky pig at the more than 100 Pig Stands that soon opened, which led to many imitators right up to the present day. A legacy of the Pig Stand is that 30 percent of people in the United States now eat at least one meal in their car each week.

The drive-in church, in contrast, did not spread as widely or become a staple of U.S. consumer society like fast food. But drive-in religion was popular for a time:

> In early 1955 in suburban Garden Grove, California, the Reverend Robert Schuller, a member of the Reformed Church in America, began his ministry on a shoestring. With no sanctuary and virtually no money, he rented the Orange Drive-In movie theater on Sunday mornings and delivered his sermons while standing on top of the concession stand. The parishioners listened through speakers available at each parking space. What began as a necessity became a virtue when Schuller began attracting communicants who were more comfortable and receptive in their vehicles than in a pew. Word of the experiment—'Worship as you are ... In the family car'—spread, the congregation grew, and in 1956 Schuller constructed a modest edifice for indoor services and administrative needs. But the Drive-in Church, as it was then called, continued to offer religious inspiration for automobile-bound parishioners, and in succeeding sanctuaries facilities were always included for those who did not want a 'walk-in' church. By 1969, he had six thousand members in his church, and architect Richard Neutra had designed a huge, star-shaped 'Tower of Power,' situated appropriately on twenty-two acres just past Disneyland on the Santa Ana Freeway. It looked like and was called 'a shopping center for Jesus Christ ... By 1984, Schuller's Garden Grove Community Church claimed to be the largest walk-in, drive-in church in the world.'[2]

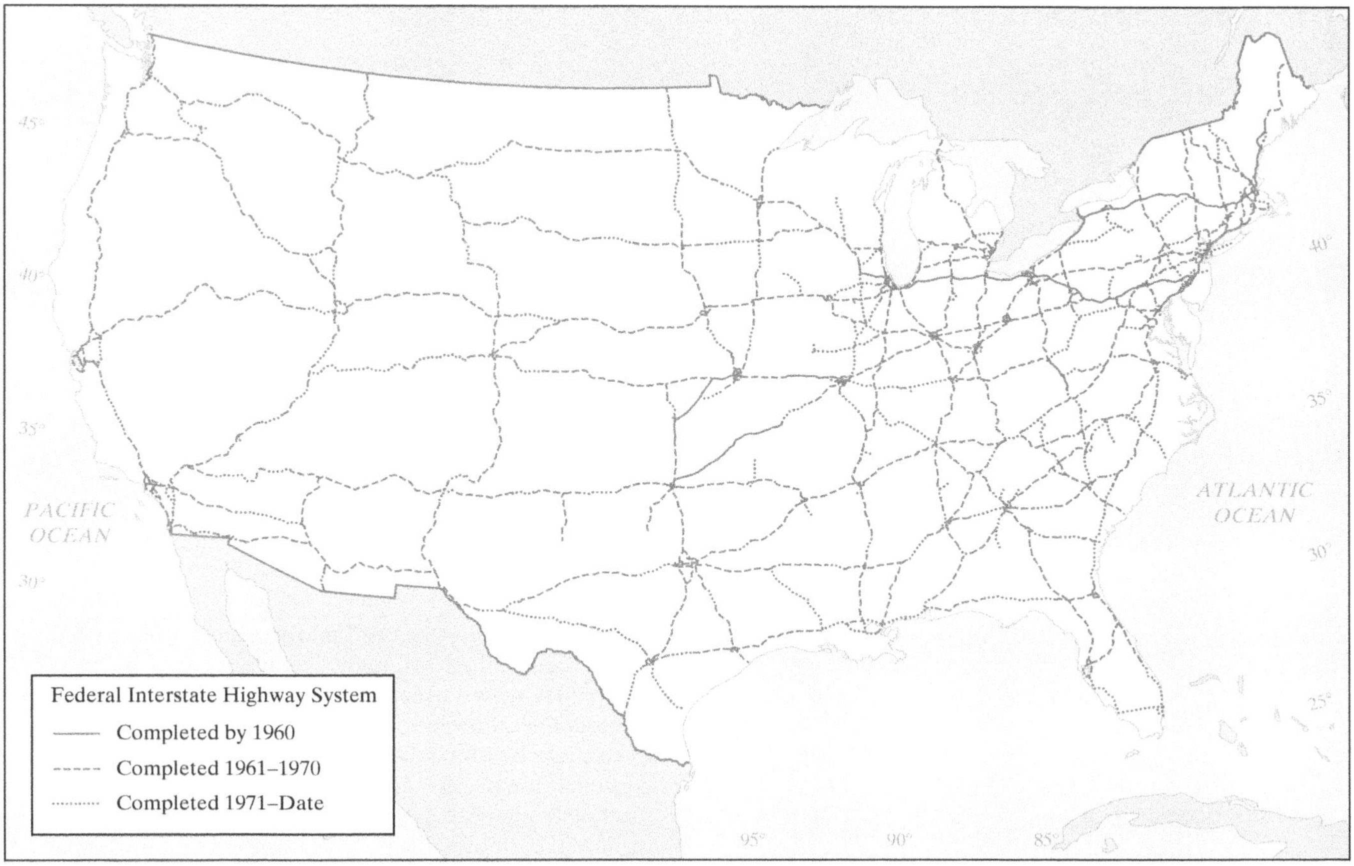

FIGURE 1 The growth of the interstate highway system.

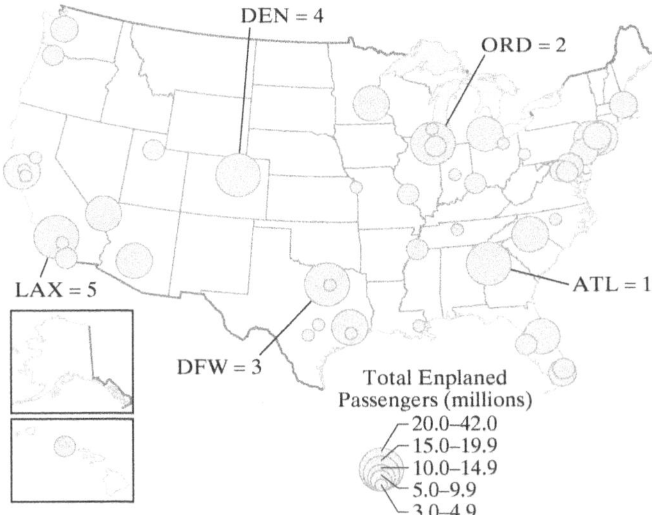

FIGURE 2 Air passenger traffic in 2009.

conform to them, rather than the other way around. Meanwhile, it seemed that the benefits of **initial advantage** and **cumulative causation** in the Manufacturing Belt were now outweighed by the effects of **agglomeration diseconomies** (the increasing costs of now-inefficient infrastructure, and the relatively low productivity of dated machinery). The outcome was a decentralization of jobs and people within the urban system, away from the Manufacturing Belt "core" toward metropolitan centers in the South and West "periphery."

But the reorganization of U.S. business that led to the regional decentralization of the economy also made for a certain amount of metropolitan consolidation. Two key economic functions—headquarters offices and research and development (R&D) laboratories—tended to become increasingly localized in larger metropolitan areas. This localization was mainly in response to (1) corporate reorganization following mergers and acquisitions and (2) the differential growth and changing pattern of economic specialization within the urban system. The pattern of **control centers**—cities with a high proportion of corporate headquarters—changed to reflect the growth of cities like Atlanta, Houston (Figure 3), and Los Angeles as regional business centers and the relative decline of Manufacturing Belt cities like Cincinnati, Detroit, and Pittsburgh.

At the same time, there was a consolidation of corporate headquarters locations in New York and Chicago. By 1970 about 25 percent of the 500 largest industrial corporations had their headquarters in New York, as did the same proportion of the 300 largest corporations in banking, insurance, retailing, transportation, and utilities. The map of Fortune 500 corporate headquarters reflects the continuing dominance of New York and Chicago as corporate control centers (Figure 4). Such dominance reflects "a process of cumulative and mutual reinforcement between relatively accessible locations and relatively effective entrepreneurship."[3] In other words, New York and Chicago were

Derek Potts/Alamy

FIGURE 3 The reorganization of U.S. business following World War II that led to the regional decentralization of the economy also made for a certain amount of metropolitan consolidation that promoted the growth of regional business centers like Houston.

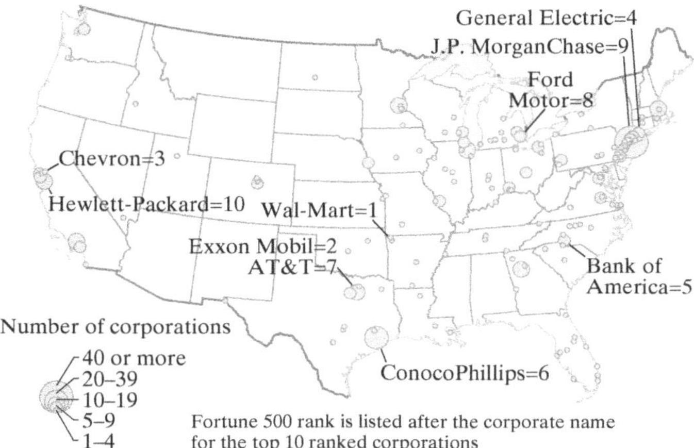

FIGURE 4 The location of Fortune 500 corporate headquarters in 2010. The size of the dot is proportional to the number of corporate headquarters in each city.

able to consolidate their importance as economic control points because of their reserves of people's entrepreneurial talent, the array of support services that they could offer, and their national and regional accessibility through the interstate highway and commercial airline networks. Meanwhile, localization is evident in the growing number and percentage of corporate headquarters in larger growth centers outside New York and Chicago—most notably the Sunbelt cities of Houston and Atlanta—that offered strong locational and entrepreneurial assets (Figure 4).

R&D facilities, another high-income, high-growth economic activity, had traditionally been dispersed, their locational pattern tied to that of parent industries. With the emergence of huge conglomerate companies under advanced capitalism, R&D activities became much more critical to corporate success. The increasingly high stakes involved made it vital to avoid market saturation, and companies could do that only by developing new products and improving old ones through R&D. Much long-range corporate R&D gravitated to central laboratories in headquarters complexes, where interaction among their brightest employees could be fostered. In companies where independent divisions produced different product lines, however, R&D laboratories tended to be located in separate divisional laboratories. Both arrangements led to a consolidation of R&D activity in larger metropolitan areas. The most striking development (though not as important in terms of numbers of jobs) was the localization of some R&D activity in a few "innovation centers"—places with strong university research facilities, a strong federal scientific presence, and a range of cultural and recreational amenities of the kind that made them attractive to highly qualified scientific workers. The best-known examples are Silicon Valley between Palo Alto and Santa Clara, California; the Route 128 area around Boston; and the Research Triangle area of North Carolina.

Metropolitan Sprawl

After 1945 a second surge of growth in car ownership occurred in the United States. From just under 26 million in 1945, the number of cars on the roads jumped to more than 52 million in 1955 and just over 97 million by 1972. During the same time, the rate of surfacing of unsurfaced roads matched this growth almost exactly, while the number of people per car fell from five to two. Advances in automotive engineering in the 1930s had brought about a dramatic improvement in power-to-weight ratios, and the average top speed of low-priced cars had also increased—from less than 50 miles an hour in the 1920s to more than 80 miles an hour in the 1950s.

The result was a dramatic spurt in suburban growth, so that the 1950s became the decade of the greatest-ever growth in suburban population. While central cities in the United States grew by 6 million people (11.6 percent), suburban counties added 19 million people (45.9 percent). In almost every metropolitan area the ring of suburban counties grew much faster than the central city (or cities). Population statistics did not always reflect this growth, however, because of the tremendous amount of **annexation** by central cities in the South and West. Corpus Christi, Dallas, Houston, Oklahoma City, Phoenix, San Diego, and San Jose each annexed over 100 square miles of territory from surrounding counties during the 1950s and 1960s, and captured the demographic growth of the suburbs.

Following Peter Hall,[4] we can recognize four preconditions for the emergence of this unprecedented suburban sprawl.

1. The principle of land use zoning, established in the *Euclid v. Ambler* case in 1926. This allowed uniform residential tracts with stable property values.
2. The backlog of demand for housing from the Depression and war years, combined with the postwar baby boom. During the war there had been a moratorium on new construction, so that by 1945 there was a backlog of between three and four million dwellings. After the war, the 16 million men and women who had been in the armed services resumed domestic life, which led to a sudden increase in the rate of household formation and in birth rates.
3. The cheap, long-term home financing that the Home Owners Loan Corporation (HOLC) and the Federal Housing Administration (FHA) had initiated during the New Deal. In 1944 the **Servicemen's Readjustment Act** (the "**GI Bill**") created the Veterans Administration, one of the major goals of which was to facilitate home ownership for returning veterans. It did this through a program of mortgage insurance along the lines of the FHA, whose own lending powers were massively increased by the 1949 Housing Act.
4. New and improved roads. The Federal Aid Highway Act (1956) authorized 41,000 miles of limited-access highway to be built, at more than $1 million per mile, with 90 percent of the funding coming from a Highway Trust Fund, established through excise taxes on vehicles, gasoline, and tires. Every major city was to be linked into the system, while the city's internal traffic was to be restructured—often according to a hub-and-spoke model, the outer rim of the "wheel" being a ring road, or interstate "beltway" (Figures 5, 6), that made outlying locations more accessible, particularly where there were intersections with major radial "spokes."

These preconditions for sprawl encouraged developers to leapfrog ahead of the suburban edge, and build large shopping malls at highly accessible sites along new freeway corridors, and cleverly use these shopping meccas as marketing tools for vast new residential communities. Industrial parks were attracted to the new highway and freeway corridors, catering to light industry and distribution/warehouse operations, while offices were attracted to prestigious, highly visible sites in landscaped, campuslike settings near freeway intersections.

The Fordist Suburb

The period immediately following World War II was characterized by a spectacular increase in housing construction. From a low point of just over 90,000 housing starts in 1933 and a prewar average of about 350,000 starts per year, construction

Tyrone Turner/Corbis Images

FIGURE 5 Circumferential beltways like this one around Washington, D.C. (I-495) were important in establishing the framework for metropolitan sprawl in the 1960s and 1970s.

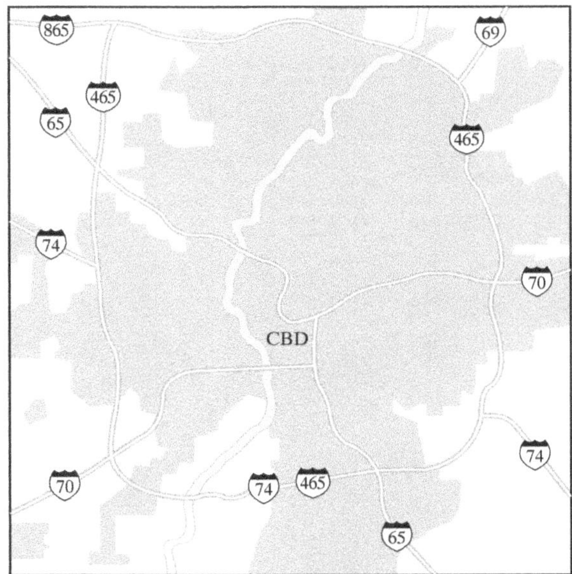

FIGURE 6 The "hub-and-spoke" layout of interstate highways in Indianapolis.

jumped to almost two million by 1950 and averaged more than 1.5 million until the recent global financial meltdown and housing crisis. Between 1945 and 1973 the FHA helped nearly 11 million households to own houses, and the overall level of home ownership increased from around 45 percent to nearly 65 percent. Nearly all of this growth was accomplished without any kind of overall planning or metropolitan management, which sowed the seeds of a number of problems and conflicts that we will examine in later chapters.

The only consistent element of planning came from developers and their chief goal, of course—profitability. The strategy of most developers was based on Fordism: mass production for mass consumption. This strategy required pursuing economies of scale, standardizing products, and perfecting prefabrication technology. The developers' equivalent to Henry Ford's assembly line was **balloon-frame construction** (Figure 7). Instead of heavy beams and corner posts held together by mortise and tenon joints, balloon framing used standardized, machine-cut 2" × 4" studs nailed together by inexpensive, machine-cut nails. By spreading the stress over a large number of light boards, the balloon frame had strength and stability far beyond its insubstantial appearance. Using standardized components and requiring only semiskilled workers, balloon framing was around 40 percent less expensive than traditional methods.

Landslides Aerial Photograph

FIGURE 7 Balloon framing. This inexpensive and efficient system of construction helped to revolutionize the residential construction industry.

The most popular form of house for the new suburbs was a single-story structure with a low-slung roof, large windows, and a carport or garage. The influential *Ladies Home Journal* had popularized this type of bungalow dwelling from as early as the turn of the century. Standardized by developers in the United States in the 1920s, the bungalow had been designed originally by British engineers for the tropical climate of colonial Bengal. It was low and well-ventilated and opened out to the garden. It could be grand or modest; using balloon framing, it was the design that made it possible for working class families to have a "dream" home (Figure 8).

Taking advantage of pattern-book bungalow designs and balloon-framing techniques, developers began to encircle every city of any significance with huge, sprawling subdivisions. The first and one of the largest was Lakewood, built by the Lakewood Park Company to accommodate more than 100,000 people on 16 square miles of former sugar beet fields south of Los Angeles (Figure 9).

Without doubt the most famous was the original *Levittown* on Long Island, begun in 1948 by Abraham Levitt and his sons William and Alfred. They were the first large-scale developers to deploy the assembly-line approach in residential development. Combined with innovative materials, new tools, standardized designs, low prices (the original Levittown Cape Cod

H. Armstrong Roberts/Corbis Images

FIGURE 8 Developers heavily marketed suburbia as the setting for the "American dream" of private home ownership.

Corbis Images

FIGURE 9 Lakewood, California, an enormous project of 17,150 homes. Work began in February 1950; by the end of the year, families were moving in at the rate of 25 each day.

design sold for $100 down and $57 per month) and, finally, slick marketing, this approach unrolled more than 17,000 homes onto the suburban fringe by 1951 (Figure 10). This was morphogenesis on a heroic scale. It marked the end of urban development as a process dominated by fine-grained accretion and infill and the beginning of a process dominated by the mass production of coarse chunks of suburban and exurban development.

This urban development coincided with a dramatic postwar increase in prosperity and the consequent rise of consumerism. Between 1948 and 1973 the economy grew at unprecedented rates. The gross national income (GNI) increased fivefold, median income more than doubled (in constant dollars), and home ownership rose by 50 percent. Historian Lizabeth Cohen has traced the development of a "consumers' republic" in the United States in this era: a society based on mass consumption of cars, houses, and manufactured household goods, all celebrated by the new medium of television. This was the era of the "Sitcom Suburb," a democratic utopia of ranch and split-level homes. Thanks to *Euclid v. Ambler*, Sitcom Suburbs were founded on local government zoning regulations that prohibited apartments, duplexes, small houses, or small lots as well as stores and offices.

Federal intervention also contributed significantly to the creation of standardized suburban settings dominated by detached single-family homes occupied by white families. The Federal Housing Administration was clearly biased from its inception toward single-family detached and owner-occupied housing. To assist local governments with planning for single-family detached homes, the FHA recommended standardized subdivision design practices that became a template for suburban subdivisions nationwide when Congress passed the landmark Section 701 planning grant program as part of the 1954 Federal Housing Act.

Within a decade or two of the end of World War II, many people in the United States had developed a distinctive way of life and a new social and spatial order known as suburbia. As industrial cities declined, Sunbelt cities and suburbia displaced them as the cradles of national personality. The new centrality of the suburbs to people's identity was reinforced during the Cold War as the United States showcased its democratic utopia of suburban lifestyles and consumer culture by way of contrast with the Soviet Union's regimented lifestyles and modest levels of living. As Robert Beauregard has observed:

> Awash in consumer goods, enjoying nearly full employment, and blessed with high wages, the daily life of the "average" American became a model for people around the globe. Suburban life anchored a standard of living commensurate with the nation's status as the leader of the "free world" and established the country's economy and form of government as the best hope for affluence, democracy, and world peace.[5]

Suburbia not only became the place where the American Dream could be realized but also the crucible for what

Corbis Images

FIGURE 10 Levittown, Long Island. Formerly potato fields, the site was developed in 1947 to house 45,000 people.

David Butow/Corbis Images

FIGURE 11 "Hamburger Row." This example is along U.S. Highway 412 in Springdale, a town north of Fayetteville in northeast Arkansas. The classic strip in Norman, Oklahoma, was described by Grady Clay in his book *Close Up: How to Read the American City* (New York: Praeger, 1974).

commentator David Brooks calls a "Paradise Spell" of relentless individual aspiration and restless consumption.

Designed for car travel, suburbia was characterized by all kinds of drive-in and drive-through services: drive-through banks and fast-food outlets, drive-up windows to drop off and pick up laundry, drive-in theaters, motels, gas stations, and minimarts. The suburban shops and shopping centers of the 1950s and 1960s not only represented distinctive new elements of urban form in their own right, but also brought with them acres and acres of parking space that imposed an unrelieved sterility to the suburban landscape. To accommodate this parking space, buildings had to be set back so far from the road that signs became larger and more outlandish in an attempt to catch the motorist's eye. The result is the classic strip (Figure 11).

There were a few exceptions to Fordist suburban development, and their successes in designing automobile-oriented environments without unwanted, unaesthetic side effects were to become influential examples for a generation of developers. In the Chicago area, communities such as Park Forest, Elk Grove, and Oak Brook were distinctive for having extremely generous green space within an otherwise standard 1950s landscape of curvilinear streets, cul-de-sacs, and shopping centers. In the Washington, D.C., area, developers returned to the **greenbelt cities** model, refining it to suit upper-middle-class markets of the 1960s in their plans for Columbia (Maryland) and Reston (Virginia) (Figure 12). Both were organized into "villages" of 10,000 to 15,000 people, and each had a town center, recreational facilities, and schools. Meadows and woods separated the villages from one another and from a system of open space corridors containing pedestrian and bicycle paths.

Paul L. Knox

FIGURE 12 Part of one of the residential "villages" in Reston, a privately developed new town in the Virginia suburbs of Washington, D.C., that became a model for many of the exclusive "master-planned" communities built around the fringes of every large metropolitan area during the 1980s and 1990s.

Meanwhile, in California there emerged a rather different recipe, featuring recreational amenities such as country clubs, golf courses, parks, ballfields, boating lakes, swimming pools, and equestrian areas scattered among vast developments. Each phase of these developments was planned as a "neighborhood" with its own school and commercial center, along with a package of amenities. Together, the neighborhoods constituted a complex of subdivisions with a distinctive, recreation-oriented environment. Examples include Irvine (carved out of the Irvine Ranch), Valencia (on the site of the Newhall Ranch), and Mission Viejo (on the O'Neill Ranch).

Suburban Production and Consumption Spaces

Continuing the logic of decentralization that began during the prewar infill era, freeways, bigger and faster trucks, and larger factories and warehouses needed the larger and less expensive parcels that only suburban land could accommodate. Although heavy industry was usually tied to older outlying industrial satellites with access to rail sidings, most industry had become "footloose," able to set up wherever local governments could be persuaded to zone land for industry. Corridors of land along major highways quickly became one common setting for footloose industries; the area around the junction of major highways became another; industrial parks built by developers a third (Figure 13); and the area around airports a fourth.

Geographer Allen Scott has described the production space of the metropolis as a mosaic of these settings, each element in the mosaic composed of particular *kinds* and *mixtures* of industrial land use: factories of different sorts, warehousing, and related offices, with retailing and services catering to the industrial workers.[6] The largest clusters of industry are further characterized by a network of linkages among plants that bind many of the land users together in an *industrial complex,* such as the aerospace and electronics complex in Orange County, outside Los Angeles.

Along with decentralized population and industry came retailing and office functions. Freeways encouraged the development of a catalytic new element: integrated shopping centers. Freeways provided the accessibility necessary to support large new department stores that could anchor shopping malls with a variety of specialized, higher-order retail stores. By 1957 some 2,000 shopping centers were scattered around U.S. metropolitan areas. This figure grew from 8,240 by 1965 to 12,170 by 1970 (and over 22,000 by 1980). Branches of major department stores anchored the largest of these, filled out with branches of chain stores catering to the middle classes and rounded off with small independent stores or franchises catering to the needs and tastes of residents in the surrounding area. By the mid-1980s about 55 percent of all retail sales (excluding motor vehicles and gasoline) were accounted for by shopping centers.[7]

Susan Van Van Etten/PhotoEdit

FIGURE 13 An industrial park in a suburb of Paris, France.

Central City Land Use

The other side of all this suburbanization was the relative decline of central cities and the further reorganization of central business districts (CBDs). Not only was manufacturing being decentralized, the whole structure of the economy was shifting away from the older industries that had traditionally been located in central areas. Between 1953 and 1970 New York City lost 206,000 manufacturing jobs, Philadelphia lost 102,000, St. Louis lost 61,000, Boston lost 30,000, and Baltimore lost 25,000. Similar losses were recorded in central cities around the country, along with losses in retailing and wholesaling (New York and Philadelphia each lost 26,000 retailing and wholesaling jobs, for example, while Boston lost 21,000, St. Louis lost 14,000, and Baltimore lost 5,000). These losses were often more than balanced by gains in white-collar service jobs, but because white-collar workers tended to prefer living in the suburbs and commuting to work rather than living in one of the aging central city neighborhoods, the population of central cities began to thin out.

This shift in the employment base and population density of central cities resulted in some radical changes in land use and urban form. Old factories and warehouses were abandoned or demolished; railroad tracks and freight yards were torn up; and slum housing was cleared. Meanwhile, new office blocks were built. But the construction often did not take place on the site of demolished or abandoned areas. The locational requirements of white-collar services were different from those of traditional industries. Office blocks crowded into the CBD in skyscrapers, leaving industrial districts in a sorry state of deterioration. Slum clearance and abandoned housing (Figure 14) added to the empty and desolate appearance of large sections of central cities.

Two other factors added to the sense of devastation in central cities. Freeway systems, as they penetrated central cities in fulfillment of the hub-and-spoke designs of highway engineers, required 200- to 300-foot rights-of-way that took the line of least (commercial) resistance, carving through low-income neighborhoods and blighting open spaces, parks, and river margins (Figure 15). Then, at the end of the 1960s, riots and civil disorder literally laid waste to large sections of some central cities. The result was that it became common for people to talk and write about U.S. cities as "doughnut cities"—empty holes surrounded by a ring of suburbanization.

Like all such characterizations, the doughnut city idea was a gross exaggeration, capturing only the most dramatic and symptomatic element of change. A great deal of the fabric of central cities was relatively unaltered during the 1950s and 1960s, while CBDs, though suffering a decline in their relative importance, continued as major hubs of retail and commercial activity. There were, nevertheless, some significant adjustments. In the CBD frame nearest to higher-status suburbs, upscale apartment houses and hotels began to replace large old mansions, drawing with them the *zone of assimilation* of the CBD core. As the CBD crept slowly in this direction, so in the *zone of discard* older hotels degenerated into rooming houses, and stores changed hands. Military surplus stores, pawnshops, and wig stores replaced small tailors and clothiers;

Visions of America/Superstock

FIGURE 14 Central city dereliction, as evidenced by this abandoned apartment building in Philadelphia, Pennsylvania, can introduce a contagious "blight" to surrounding areas, making investors reluctant to redevelop, and even leading to disinvestment by both the private and the public sector.

Paul L. Knox

FIGURE 15 The intrusive effects of interstate highways on downtown areas are clearly evident in this photograph of Miami, Florida.

stores selling cheap novelties, fire-damaged goods, and pop records replaced housewares and dry goods stores.

In more general terms, the form of the city that evolved after World War II can be seen as an extension and outgrowth of the industrial city, with a CBD that has been penetrated by industrial as well as international elements; central city districts that are subject to urban renewal and **gentrification**; and an extended metropolitan fringe containing new production space and emerging "outer city" nodes of development.

Demographic and Social Change in Cities

Parallel with all these economic, land use, and other changes, some important demographic and social changes were under way, with some important consequences for the pattern of urbanization. One of the most striking demographic events has been the maturing of the so-called *Baby Boom* generation. The boom had its origins in markedly increased birth rates after 1945, when Americans and Europeans were—literally—catching up with one another during the relative stability of the immediate postwar period. Later, the Baby Boom was sustained by the affluence of the 1950s and early 1960s, which encouraged people to marry younger, have children earlier, and have larger families. The boom ended rather abruptly in 1964, as the availability of the contraceptive pill made birth control and family planning much more effective. The economic recession of the 1970s, followed by the increasingly materialistic culture of the 1980s and 1990s, resulted in delayed marriages, deferred childbearing, and smaller family sizes. The Baby Boom was over; those born later became the "Baby Bust" generation. Many

URBAN VIEW 2
The Canadian Urban System

In case the Olympic Games had not attracted enough attention, *The Economist* Intelligence Unit's list of the most livable cities in the world in 2010 ranked Vancouver number 1 out of 140 cities (Figure 16). This ranking scored each city based on five categories: stability, health care, culture and environment, education, and infrastructure. The top ten ranked cities included three Canadian, three Australian, and four European cities, in ranked order: Vancouver, Vienna, Melbourne, Toronto, Perth, Calgary, Helsinki, Geneva, Sydney, and Zurich. Pittsburgh, in 29th place, was ranked the most livable U.S. city, which says a lot about how far it has come since its heavy manufacturing days. The highest ranked cities tended to be mid-sized, with low population densities, in developed countries. These kinds of cities allow the people living in them to enjoy cultural and recreational attractions without the higher crime levels and greater infrastructure problems of larger cities like London or New York, which were ranked 51st and 56th respectively.

The Canadian urban system was initially shaped by external demand for staple commodities like fur and lumber for export to France and Great Britain, as described in Vance's mercantile model. Subsequently, the urban system—with major cities, such as Toronto and Montréal along its southerly industrial corridor, Canada's so-called Main Street—grew based on industrialization and the country's rich resource base.[8]

Today, most Canadians still live within 200 miles of the U.S. border. The 1988 Canada-U.S. Free Trade Agreement reinforced the southern concentration of cities despite the Canadian government's efforts to achieve a more balanced distribution of industry throughout its territory. For example, national government efforts failed to expand car assembly beyond the southern Ontario car production cluster that includes Windsor and Waterloo.[9] More recently, deindustrialization in general, and the 1994 North American Free Trade Agreement (NAFTA) in particular, have involved significant outsourcing of industrial activities to Mexico and, with increasing globalization, farther afield to countries like China. In many cities, the decline in traditional manufacturing has been offset partly by the growth in services, especially business services like management and marketing.[10] This may be expected to concentrate economic growth in Canada's largest cities, augmented by population growth through immigrants, the major, and soon, only, factor driving population growth in Canada. In fact, large cities like Toronto are known for their multiculturalism and well-established ethnic communities and neighborhoods.[11]

FIGURE 16 Vancouver, Canada, ranked number 1 out of 140 cities in *The Economist* Intelligence Unit's list of the most livable cities in the world in 2010. The residents of highly-ranked cities like Vancouver enjoy wonderful cultural and recreational amenities without the higher crime levels and greater infrastructure problems of larger cities like New York or London.

cities, particularly in northwestern Europe, began to register almost zero rates of natural population increase (see Urban View 4 entitled "Contemporary European Urbanization"). Meanwhile, the availability of cheap and efficient contraceptives played an important role in the radical changes that took place in social attitudes in the mid to late 1960s. As attitudes toward sexual behavior changed, so too did people's attitudes to divorce and the family, and with these changes in attitudes there soon followed changes in patterns of household formation that were to have direct effects on patterns of demand for housing space and all kinds of urban services, from schools to hospices.

The Baby Boomers and Urban Culture At the center of these social changes were the early cohorts of the Baby Boom generation, people born in the late 1940s who had reached college age in the mid-1960s. Their formative experiences had taken place in the context of the postwar economic boom, and now, as young adults, they were presented with the new freedoms of sex, drugs, and rock 'n' roll. This freedom, in turn, helped to set in motion a widespread rebellion against the apparent complacency and rigidity of what J. K. Galbraith called "The Affluent Society"[12]—a rebellion that was given added focus by civil rights issues and dissent over U.S. involvement in Vietnam in the late 1960s. The result was an urban *counterculture* movement with radical political agendas and a strong collectivist approach to the pursuit of freedom and self-realization.

In the 1970s in the United States, these cohorts found themselves on the labor market and in housing markets, where their large numbers caused unprecedented competition. Wages stood still while house prices soared. Meanwhile, new patterns of household formation associated with changed social attitudes—more single-adult households, smaller families, and more female-headed families—produced *more* households and so even more competition for housing space. At the same time, these new patterns of household formation, coupled with greater numbers of women entering the labor force (another product of the revolutionary social changes of the 1960s) and an intensifying economic recession, led to even greater competition in labor markets.

Between 1973 and 1983 alone, the real median income of all U.S. households headed by someone under 35 fell by nearly 11 percent. One of the first casualties was the progressive, collective radicalism of the late 1960s. Liberalism gave way to neoliberalism; a survival mentality eclipsed exploration and liberation; and philosophies of self-realization came to be directed inward, fostering a more selfish, narcissistic, and materialistic culture: the "Paradise Spell" of hedonistic and competitive consumption described by David Brooks. What was most important in terms of the complexion of the overall urban system was that Baby Boomers, like previous generations who had lived through periods of economic restructuring, began voting with their feet, and moving out of cities where the economic prospects were least attractive. The strongest "magnets" for adult Baby Boomers in the United States were Sunbelt cities with expansive high-tech and defense-oriented economies, while all ten of the chief "losers" were cities in the old Manufacturing Belt.

Aging Populations Now, at the beginning of the twenty-first century, the 76 million Baby Boomers who were born in the United States have reached their late 40s to mid-60s—the generally more stable, lower-migration, middle-aged, prime earning years—and are fast approaching retirement. Thanks mainly to significant improvements in both private and public pension programs (improvements achieved during the prosperity of the 1960s) the retirement years of elderly populations in Western countries are relatively affluent compared with those of their predecessors. Until the recent global financial meltdown and housing crisis, the financial security enjoyed by many (but by no means all) of the elderly, combined with windfall gains for homeowners resulting from house price inflation, enabled large numbers of them to relocate to "places of reward and repose," where an unprecedented kind of urbanization, fueled by elderly immigration, produced some cities (Flagstaff, Arizona, Naples, Florida, and Myrtle Beach, South Carolina, for example) where a large part of the population is of retirement age, with all kinds of ramifications for local economic development, politics, and quality of life.

The Burden of Youth The Baby Bust generation, born between 1965 and 1980, has also been a significant influence on patterns of urbanization and urban culture. Also called Generation X, the cultural label placed on them is "slackers" because some of the most famous, such as Kurt Cobain of the band Nirvana, were associated with grunge music and the anti-establishment attitude it stood for. Those people born between 1981 and 1995 called Generation Y (Millennials, Echo Boomers, i-Generation, or NetGeneration) may be even less hard-working and more money-hungry than Generation X or the baby boomers.[13] The *Urban Dictionary* Web site recently added the term "Slackoisie" (pronounced slack-wah-zee), defined as "narcissistic young professionals who often complain about work, are critical of long hours, and have an exaggerated sense of self-importance and entitlement." Yet Generation X and Generation Y will be the ones that have to "mop up" after the flood of Baby Boomers moves up the career ladder toward retirement. The sheer size of the baby boom cohort (along with the relatively high salaries and benefits in line with their seniority) will affect the burden of taxes, health care, and other benefit costs that fall to Generations X and Y, as well as affecting their career prospects and job mobility.

The New Immigrants Another demographic trend with important implications for patterns of U.S. urbanization has been the increasing number of immigrants (the global financial meltdown notwithstanding). Initially, one of the main reasons for the increase was the abolition in 1965 of the ceiling and quota systems that had been introduced by the immigration acts of the 1920s. Even without taking into account illegal immigrants (estimated at about 11 million by the Department of Homeland Security), immigration accounted for nearly one-third of the population growth within the urban system for the past few decades. In marked contrast to earlier waves of immigration, the dominant nationalities among the arrivals since the 1970s have been from Latin America and Asia rather than Europe. In 2009, 40 percent of the 1,130,818 immigrants who were granted legal permanent resident status were from Latin America, and a further 36 percent were from Asia (including Russia). Only 9 percent were from Europe.

At the same time, many of the descendants of the earlier generations of immigrants who had come from Europe were moving out of the cores of older metropolitan areas as they established themselves in U.S. society. As a result, the ethnic composition of cities has changed. Cambodian, Colombian, Cuban, Dominican, Haitian, Jamaican, Korean, and Vietnamese neighborhoods are now elements of the social mosaic of many

URBAN VIEW 3

Boomerang Generation: Y Us?

In a world that is run by media, the issue of unemployment in college graduates was addressed in an episode of *Greek* on ABC Family that aired on January 31, 2011. In season 4, episode 5, entitled "Home Coming and Going," Asleigh Howard, the sorority girl and recent graduate of the fictional CRU, struggles to face the harsh reality of leaving the comfortable bubble of college and entering into the working world ... Best friend and fellow sorority sister, Casey Cartwright, urges Ashleigh to think of what's best for her right now and settle for any paying job. But Ashleigh is not yet ready to give up the dream of a career ... We are a generation that has been told all our lives that we can do whatever we want when we grow up, as long as we work hard and study harder ... So we go to college, work hard, and now have nothing to show for it ... This painfully true insight is what is forcing so many Millennials to give up their dreams, move back home with their parents, and settle for any job with a paycheck.[14]

After World War II, young people left the army, started jobs and families, and created the Baby Boom generation. Now many young people who graduate from college are unsuccessful in their attempt to start a career in a tough job market, and saddled with high student loan and credit card debt, are forced to move back into their parents' homes, creating the Boomerang generation. In 2010, the Pew Research Center reported that 37 percent of 18 to 29 year olds were either unemployed or out of the labor force.[15] College seniors who graduated from college in 2009 carried an average student loan debt of $24,000.[16] It should come as no great surprise then when a survey by the AFL-CIO finds that one in three young workers (those younger than 35) were living at home with their parents.[17]

U.S. cities, replacing or supplementing the older ethnic residential concentrations of past waves of immigrants. There are some significant differences, however, in the destinations of these new immigrants. New York has traditionally been the most powerful single magnet, the metropolitan area being the place of residence of the largest number of legal permanent resident immigrants each year. Other important magnets now include metropolitan areas on the West Coast, Southwest, and Southeast—places that historically attracted relatively few immigrants. The impact of recent immigration, measured in terms of immigrant arrivals as a percentage of the total population, is now greatest in Los Angeles and Miami, while New York, Washington, D.C., Chicago, and Boston are the only traditional destinations among the ten metropolitan areas with the highest rates of immigration (the others being San Francisco, Houston, Dallas, and Atlanta). In short, recent streams of immigration are highly differentiated and have enormous impacts.

ECONOMIC CRISIS, RESTRUCTURING, AND NEW METROPOLITAN FORM (1973–PRESENT)

Economic Crisis and Urban Distress (1973–1983)

The period of postwar economic prosperity came to an abrupt end at the beginning of the 1970s in the United States and Western Europe. In October 1973 an Arab-Israeli war led Arab countries to impose an embargo on shipments of oil to the United States and other Western countries, and, soon afterward, OPEC quadrupled oil prices. The subsequent shock to the economic system reinforced a number of long-term structural economic problems, with the result that the U.S. economy was plunged into an episode of **stagflation** (falling demand and rising inflation).

During the 1960s productivity in U.S. manufacturing had averaged between 2 and 3 percent growth each year. By the late 1970s productivity had fallen to less than 1 percent annual growth. At the end of the 1970s the average U.S. family had only 7 percent more real purchasing power than a decade before. Between 1970 and 1983 the average weekly wages of U.S. workers fell, in real terms, from $375 to $365. Between the late 1970s and the mid-1990s, the average weekly wages for production workers continued to fall, in real terms. Unemployment, having remained steady at around 4.5 percent until the early 1970s, reached 8.5 percent in 1974 and a peak of almost 10 percent in 1982. In 1971 the U.S. economy recorded its first trade deficit of the twentieth century; since then it has only twice shown a positive balance of trade with the rest of the world, in 1973 and 1975. By 1983 the United States had slipped below Switzerland, Japan, Norway, Sweden, and Finland in the international league table of *per capita* **gross domestic product (GDP)**.

The fourfold rise in oil prices in 1973 as a result of the OPEC cartel is widely believed to have been responsible for initiating this downturn. The truth, however, is that the sudden rise in energy prices only brought to a head a series of longer-term trends that were already working against the U.S. economy. Falling demand and rising inflation (stagflation) were aggravated by both the increasing penetration of Japanese and European companies into American markets and increased competition from **newly industrializing economies** or **NIEs** (such as Taiwan, Mexico, Hong Kong, and South Korea) with significantly lower labor costs. At the same time, there was a shift to new **technology systems**. As with previous periods of transformation, many of the new industries, based on new technologies and exploiting new resources, needed new settings. *Cities that had been the chief settings for "old" industries, meanwhile, experienced the brunt of economic and social dislocation.*

The result was a period of economic crisis that reverberated around the urban system while corporate America attempted to come to grips with the new situation. Within a decade corporate restructuring and redeployment had accelerated the decentralization of jobs from the older industrial neighborhoods of central cities (particularly in the Manufacturing Belt) and had

URBAN VIEW 4

Contemporary European Urbanization[18]

For many visitors to Europe, the most striking aspect of the older parts of European cities is the general absence of skyscraper offices and high-rise apartments. Cities were developed long before reinforced steel construction and the elevator made high-rises feasible. Building codes—to minimize the spread of fire—maintained building heights between three and five stories during the industrial period. Paris fixed the building height at 65 feet in 1795, and other large cities introduced height restrictions in the nineteenth century. Still regulated today, high-rises are found only in redevelopment areas or on land at the edge of the city, like La Défense in Paris (Figure 17). Skyscrapers have also been built in the central financial districts of some of the very largest cities, including London. In the former socialist parts of Europe, there was no private ownership of land, and so no urban land market. Until recently, the tallest buildings were usually Communist Party and state administrative buildings, massive "Houses of the People," and TV towers.

A glance at a map of Europe shows that each country developed its own national capital city and urban system. Historically, rural-to-urban migration was the most important component of urban growth, especially during the industrial period. This form of internal migration within Europe, though, has largely ceased. And as birth rates have fallen considerably in recent decades, European cities are among the slowest growing in the world.

Even so, cities in Europe have begun to expand outward and coalesce into megapolitan regions in tandem with the transportation and communications infrastructure. Europe now contains about 50 megapolitan regions with more than a million inhabitants. The Rhine-Ruhr megapolitan region in Germany has a diameter of about 70 miles and runs from Düsseldorf and Duisburg in the west to Dortmund in the east. Of similar diameter is the Randstad, a densely populated horseshoe-shaped region in the Netherlands that runs from Utrecht and Amsterdam in the north through The Hague and Rotterdam in the west, to Dordrecht in the southeast.

Richard Cummins/Superstock

FIGURE 17 At the edge of the otherwise low-rise center of Paris with its Arc de Triomphe, La Défense, with its modern arch, is a purpose-built edge city designed to serve as a business district for the French capital city.

Only 60 miles apart, these two megapolitan regions may eventually coalesce to become a dominant European metropolitan core. The metropolitan regions between London and Newcastle form another area of extensive urbanization in England.

European urban areas are part of networks of cities that operate at a number of spatial scales. Since 1989, cities on either side of the former Iron Curtain have become more interconnected. Increasing European Union (EU) economic and political integration has influenced the development of the European urban system. For example, the removal of national barriers to trade within the EU, with the internationalization of the European economy, has encouraged population increases along certain border regions. Urban growth zones straddle the boundaries between the Netherlands and Germany, Italy and Switzerland, and the southern Rhine regions of France, Germany, and Switzerland.

European cities are linked through trade and other mechanisms to major urban areas throughout the global economy. Many of the cities within the EU form part of an international trading bloc—an evolving Eurozone—that contains nearly half a billion people with a combined gross national income greater than that of the United States.

A select group of cities contain the headquarters of major international agencies, many of which were founded after World War II to promote economic, political, or military cooperation. Geneva is the main European center for the United Nations. Paris is the headquarters for the Organization for Economic Cooperation and Development (OECD) and the European Space Agency. Vienna is the headquarters for the Organization of Petroleum Exporting Countries (OPEC). Important decision-making functions are located in the EU's "capital cities": Brussels (with many of the bureaucratic structures, including the Council and European Commission), Strasbourg (the Parliament), and Luxembourg (Court of Justice and Court of Auditors). Brussels is also the headquarters of NATO. The major centers of international banking and finance in Europe historically have been London and Paris, but now include Frankfurt, Zürich, and Luxembourg. Frankfurt hosts the Bundesbank, Germany's influential central bank, as well as the European Central Bank that manages the euro, making it the financial capital of the EU.

London is the only European city to have grown in status to become a **world city** on a par with New York. London and Paris contain the headquarters of some of the most powerful transnational corporations in the world. London contains about 17 of the 500 largest global companies (58% of the United Kingdom's total), including BP, HSBC, and Lloyds. Paris has even more—25 of these companies (64% of France's total), including AXA, Christian Dior, and Vivendi. Rome contains Vatican City, the seat of the Roman Catholic Church. Milan and Paris are major European centers of fashion and design, while London is the premier insurance center. Accessibility via the latest transportation and communications technologies allows some cities to strengthen their international positions. The high-speed train network reinforces the dominance of London, Paris, Brussels, Amsterdam, and Cologne. The cities with the busiest airports are London, Paris, Frankfurt, and Amsterdam.

stimulated the growth of new production space in Sunbelt cities, where land, labor, and local taxes were all cheaper.

The political repercussions of recession, meanwhile, led to a decisive shift away from the Keynesian approach to macroeconomic management and to a significant change in attitudes toward metropolitan management. Although the deepening economic recession after 1973 accentuated the vulnerability of more and more households, recession also made it difficult, both politically and economically, to finance the welfare programs that had expanded dramatically in the 1960s. Taxpayer "revolts," led by Proposition 13 in California in 1978 and Proposition 2½ in Massachusetts in 1980, were followed by electoral victories by politicians who argued that not only had Keynesianism generated unreasonably high levels of taxation, a bloated class of government workers, and disincentives for ordinary people to work and save, but it also may have fostered soft attitudes toward problem groups in society. It was part of an ideological shift from the egalitarian liberalism of the mid-twentieth century to a neoliberalism that was first marked by the government of Margaret Thatcher in the United Kingdom (1979–1990) and the presidency of Ronald Reagan (1981–1989) in the United States.

Jamie Peck and Adam Tickell have characterized the process as a combination of *"roll-back" neoliberalism and "roll-out" neoliberalism.* Roll-backs have meant deregulating finance and industry, dismantling public housing programs, privatizing public space, cutbacks in redistributive welfare programs, shedding many of the traditional roles of central and local governments as mediators and regulators, curbs on the power and influence of labor unions and government agencies, and reducing investment in the physical infrastructure of roads, bridges, and public utilities.

Roll-out neoliberalism has involved establishing public-private partnerships, encouraging central city gentrification, creating free-trade zones, enterprise zones and other deregulated spaces, asserting the principle of "highest and best use" for land-use planning decisions, and privatizing government services. Brenner and Theodore suggest that the implicit goal of neoliberalism at the metropolitan scale has been "to mobilize city space as an arena both for market-oriented economic growth and for elite consumption practices."[19]

The effect has been to "hollow out" the capacity of the central governments while forcing municipal governments to become increasingly entrepreneurial in pursuit of jobs and revenues; pro-business in terms of their expenditures; and oriented to the kinds of planning that keep property values high. Similarly, urban planning practice has become estranged from theory and divorced from any broad sense of the public interest: pragmatically tuned to economic and political constraints rather than committed to change through progressive visions. Public-private partnerships have become the standard vehicle for achieving change, replacing the strategic role of planning with piecemeal dealmaking.

For their part, many larger companies responded to the stagflation crisis by reorganizing their production processes, eliminating the duplication of activities among existing facilities, rearranging the division of tasks among them, closing down or trimming back activities in high-cost locations, investing in new facilities in low-cost locations (often overseas), and diversifying into nonmanufacturing activities. The economic sector that was hit hardest by the crisis was the traditional manufacturing sector, and the cities that were most impacted were specialized manufacturing centers. Already suffering from post-World War II regional decentralization, the Manufacturing Belt fell into an accelerated decline that has been characterized as **deindustrialization** (Figure 18). In Youngstown, Ohio, which became emblematic of deindustrialization, the closure of the Campbell Steel Works in 1977 eliminated more than 10,000 jobs overnight. During the 1970s, Detroit lost more than 166,000 jobs (a loss of nearly 30 percent of its 1970 employment base), while nearby Flint lost nearly 16,000 jobs (a 23 percent loss).

Deindustrialization on this scale involved a mutually reinforcing series of problems: a sort of cumulative causation in reverse

Alexander Mul/Superstock

FIGURE 18 Padlocked gates symbolize deindustrialization. The photograph shows a Ford Motor Company plant in Wixom, Michigan, that was closed in 2007.

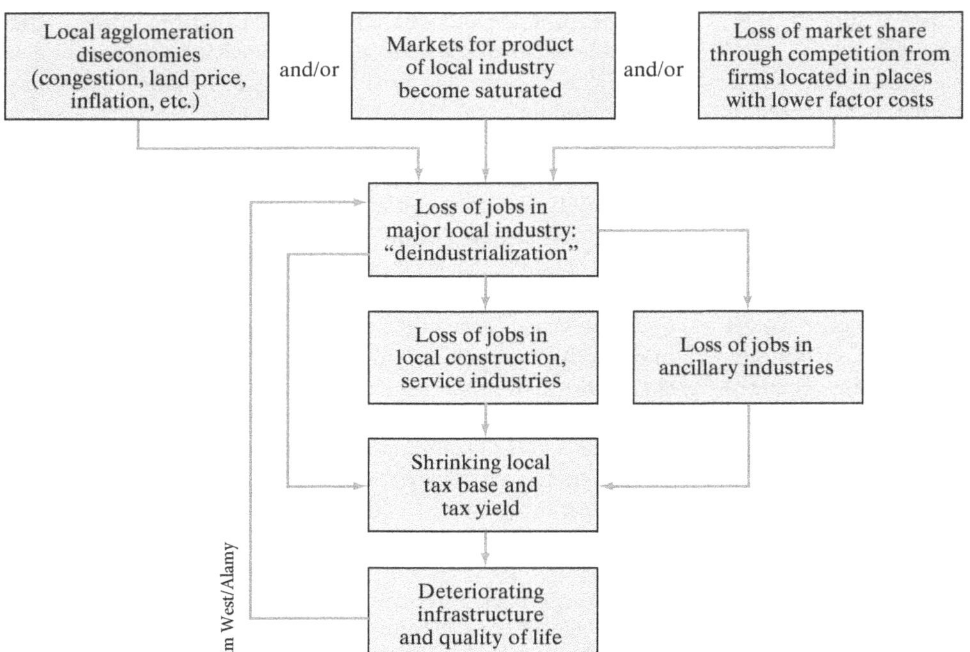

FIGURE 19 Deindustrialization and the downward spiral of economic decline.

(Figure 19). Plant closures led to job losses in ancillary industries, and these in turn led to recession in retailing and personal services. Unemployment and **underemployment** led to lower incomes and increased levels of poverty, which in turn led to outmigration. Decreased prosperity within communities came to be seen in aging housing and infrastructure and was reflected in weakened tax bases, which meant that city governments were unable to maintain or improve public services or amenities. In some cases the loss of tax revenues, combined with increasing needs in terms of infrastructure maintenance and the provision of services and amenities, led to fiscal crises that threatened to bankrupt cities. All this amounted to a discouraging environment for investment, at least until wages and land prices were driven so low that some kind of **comparative advantage** might be recaptured. Meanwhile, many cities and the people who lived in them became acutely distressed.

It would be misleading, however, to portray the decade from 1973 to 1982 as a period of unmitigated economic gloom. Some sectors of the economy were able to prosper, while the quadrupling of oil prices in 1973 and the doubling of oil prices in 1979 enabled oil and petrochemical companies to generate vast profits. Yet, with much of the economy unable to offer attractive prospects for investment, the result was a classic **overaccumulation crisis**, with surplus labor, surplus money capital, and idle productive capacity existing side by side.

Economic Restructuring and New Metropolitan Form (1983–Present)

As in previous periods of economic and urban development, new technologies have been critically important in facilitating the restructuring of the U.S. and European economies. Since the 1980s three kinds of "permissive" or **enabling technologies** have been significant:

- *Production process technologies* such as electronically controlled assembly lines, automated machine tools, robotics, and computerized sewing systems have increased the separability and flexibility of production processes. It is now easier for companies to take advantage of spatial variations in the costs of labor and land.
- *Transaction technologies*, and computer-based just-in-time inventory control systems in particular, have also increased companies' locational and organizational flexibility, allowing materials, components, or information to be purchased as and when needed and eliminating the necessity for large buffer stocks of parts.
- *Circulation technologies* such as communications satellites, fiber-optic networks, microwave communications, e-mail, and wide-bodied jets have reduced the time and cost of distribution, bringing a wider geographic market within the range of an increasing range of business activities.

As larger corporations have exploited the locational and organizational flexibility made possible by these new circulation and transaction technologies, so new **divisions of labor** have evolved among metropolitan areas nationally and internationally. It is important to emphasize that the "time-space compression" of these new technologies has, paradoxically, heightened the importance of geography. The reduction of spatial barriers has had the effect of greatly magnifying the significance of small differences between local labor and land markets, because new technologies enable these differences to be quickly (if temporarily) exploited. As a result, the urban system is becoming a continuously variable geometry of labor, capital, management, production, and consumption. Meanwhile, the *pace* of change has been accelerating, raising some important questions about

URBAN VIEW 5
Australian Cities

Most people in Australia live within a relatively narrow—125-mile wide—band along the east, southeast, and southwest coasts, which includes its largest cities—Sydney, Adelaide, Melbourne, Brisbane, and Perth (Figure 20).[21] The beach has a special place in Australian people's identity and recreational activities that include sunbathing and surfing. Away from the coast, settlement is very sparse, especially in the desert areas of the outback.[22]

Although people and businesses have been decentralizing to the suburbs of the largest cities since the 1950s, Australian central cities have never suffered the same level of disinvestment, infrastructure deterioration, and racial segregation as those in the United States. Meanwhile, the recent economic restructuring toward business and professional services and other services such as tourism has tended to reinforce the corporate primacy of Australian CBDs.

The relative lack of urban freeway construction and peripheral ring roads has worked against the development of **edge cities** in Australia. Likewise, despite suburbanization, the radial orientation of the transportation networks has helped maintain central city vitality and facilitated recent urban revitalization. In contrast to the United States, stricter planning regulations in Australian cities have also helped restrict the development of edge cities. State government planning agencies in Australia have tightly managed land release for urban expansion and for the staging of corridor development. In addition, by specifying the location and assigning land for "district centers," state governments have curbed speculative business development beyond the edge of the built-up area.

That is not to say that economic restructuring has not had an impact on Australian cities. In contrast to the experience in the United States, however, the brunt of deindustrialization in Australia has been quite selective in that it has most adversely affected the postwar industrial suburbs. Particularly hard hit have been the middle and outer suburbs in Sydney, Melbourne, and Adelaide that formerly contained both the traditional manufacturing jobs and the public housing developments that were built for factory workers and their families. This public housing, and tracts of low-rent accommodation surrounding public resettlement hostels, now contain a significant proportion of recent immigrants, many of whom arrived in Australia from Southeast Asia as "boat people" and political refugees. Nevertheless, the level of ethnic residential segregation in Australia is not as extreme as in U.S. central cities such as Detroit and Milwaukee.

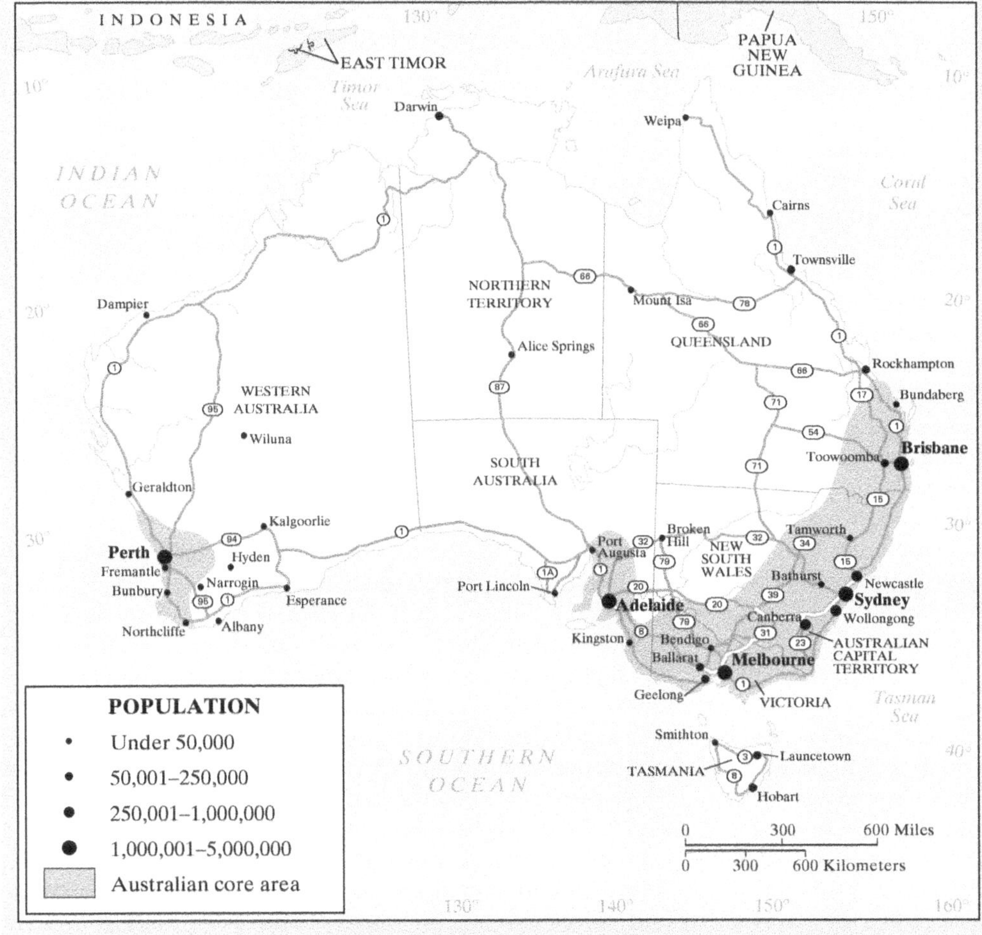

FIGURE 20 Urban Australia, including its largest cities, Sydney, Adelaide, Melbourne, Brisbane, and Perth. Most people in Australia live near the coast and the beach.

(continued)

URBAN VIEW 5
Australian Cities (*continued*)

Like the United States, however, Australia has experienced a polarization of occupations and incomes that has been associated with economic restructuring. Unlike the trend in U.S. cities, however, the growth of business and professional services jobs in Australian cities has been associated with a particularly strong centralization of wealth. High levels of housing reinvestment associated with the gentrification of the inner suburbs have been a feature of Australian cities (Figure 21).

Since the 1970s, as Australian cities have become increasingly integrated into regional and global networks of cities, the national and state governments have adopted neoliberal strategies to revitalize the old industrial landscapes and waterfront areas of the city centers. The national Better Cities program, for example, incorporated strategic investment in urban infrastructure and other projects with the goal of enabling cities like Sydney to capture a greater share of the financial flows in the Pacific region and move up in the league table of world cities. As in the case of the neoliberal policies in the United States and United Kingdom, however, the Better Cities public-private partnerships have promoted upscale central city revitalization projects—luxury downtown apartments and condominiums, and festival marketplaces—at the expense of affordable housing.

FIGURE 21 Gentrification in The Rocks, Sydney. The Rocks is the site of Australia's first European settlement. High levels of housing reinvestment associated with gentrification have been a feature in central parts of Australian cities.

what might happen to the social stability of cities, civic loyalties, and people's sense of place.

One of the most important aspects of the new intermetropolitan division of labor concerns the location of professional and business services whose growth has been a fundamental part of the shift to advanced and now globalized capitalism. Although much of this growth occurred in larger metropolitan areas, the greatest proportional growth has been in midsized metropolitan areas. Much of this growth has been localized and specialized. Boston, for example, has become a major center for computer and data-processing services, engineering services, and R&D laboratories. Washington, D.C., has a concentration of jobs in management consulting services; San Jose in personnel supply services, R&D laboratories, and computer and data-processing services; Raleigh–Durham, Austin, and Orlando in service jobs related to high-technology industries; and Huntsville, Colorado Springs, and Norfolk in R&D labs and private professional and business services related to military activity.[20]

World Cities

Whereas the dominance of large traditional manufacturing centers in the United States and Western Europe characterized the urban systems from the turn of the twentieth century to the 1950s the pivotal role of metropolitan areas specializing in service industries, particularly producer services, characterizes these urban systems under globalized capitalism. With globalized capitalism, some cities (not necessarily the very largest) have become so closely integrated within the global economic system that they are known as "world cities." The term **world city** originated in 1915 with Patrick Geddes, a leading figure in the evolution of city planning. Geddes used it to highlight the primacy of those few cities in which a disproportionate amount of the world's business was conducted.

The fact that the term world city predates contemporary globalization processes points to the changing nature of spatial interdependence. In the first phases of capitalism, the key roles of world cities involved the organization of trade and the implementation of colonial, imperial, and geopolitical strategies. Among the world cities of the seventeenth century were London, Amsterdam, Antwerp, Genoa, Lisbon, and Venice. In the eighteenth century, Paris, Rome, and Vienna became world cities, while Antwerp and Genoa became less influential. With the onset of the Industrial Revolution, the likes of Berlin, Chicago, and Manchester became world cities, while Venice and Lisbon became less prominent.

Since the mid-1970s, the key roles of world cities have been concerned less with the orchestration of trade and the deployment of imperial power, and more with transnational corporate organization; international banking and finance; fashion, design, and the media; and the work of international agencies. World cities have become the sites of extraordinary concentrations of activities associated with organizing finance and investment and creating and managing flows of information and cultural products that collectively underpin the economic and cultural globalization of the world, including processes of neocolonialism and postcolonialism.

World cities provide an interface between the global and the local. They contain the economic, cultural, and institutional apparatus that channels national and provincial resources into the global economy, and that transmits the impulses of globalization back to national and provincial centers. Cities that have been most caught up in these processes—London and New York, in particular—have been called "global cities" by sociologist Saskia Sassen. Central to Sassen's argument are the agglomerative tendencies and dynamics of advanced business services and design services that result in the distinctive clusters and districts that are the locus of influence and innovation (Figure 22). These clusters are reinforced by some important aspects of sociality: personal interaction with clients and after-work drinking and dining in settings where key professionals can learn about new opportunities, review one another's products and practices, and keep abreast of the internal politics of one another's companies.

But the system of world cities is not just a collection of mini-Londons and little New Yorks. All world cities have mixtures of cutting edge economic functions that need not just be advanced producer services. Also, as geographer Peter Taylor has pointed out, world cities exist in networks that are complex and multilayered. One set of networks, for example, is constituted by the infrastructures of airline, telephonic, and Internet systems. Another is constituted by the relations among and between companies conducting business with a global reach; and another still is constituted by the social as well as economic relations within the distinctive clusters and business districts that characterize world cities. In an increasingly interconnected world, many "ordinary" cities have developed certain attributes of world- or global-cityness: settings for innovative production clusters in the fields of information and communications technology, medical engineering, biotechnology, the media industry, and for minor concentrations of transnational corporate organization; international banking and finance; fashion, design, and the media; and international nonprofit and government agencies.

Using data on the office networks of global service firms in accountancy, advertising, banking/finance, insurance, law, and management consultancy, Taylor and his colleagues have identified overall levels of integration within the world city network, resulting in a classification of top-level world cities (Table 1). London and New York are tightly inter-related with one another and both are significantly more integrated in the overall world city network than any other city. Hong Kong has emerged as the third most highly integrated city. Note that, apart from New York, Chicago and Los Angeles are the only United States cities listed in Table 1. In the U.S., cities have always exhibited lower levels of integration in the world city system than might be expected, mainly because of the very large domestic market for advanced business services. This means that foreign companies find it difficult to penetrate the U.S. market and tend to represent clients through just a New York

Wolfgang Kaehler/Corbis Images

FIGURE 22 Office workers in the City of London, England.

TABLE 1 Alpha-level World Cities

Alpha ++	Alpha +	Alpha	Alpha –
London	Hong Kong	Madrid	Warsaw
New York	Paris	Moscow	Sao Paulo
	Singapore	Toronto	Zurich
	Tokyo	Brussels	Amsterdam
	Sydney	Buenos Aires	Mexico City
	Milan	Mumbai	Jakarta
	Shanghai	Kuala Lumpur	Dublin
	Beijing	Chicago	Bangkok
			Taipei
			Istanbul
			Rome
			Lisbon
			Frankfurt
			Stockholm
			Prague
			Vienna
			Budapest
			Athens
			Caracas
			Los Angeles
			Auckland
			Santiago

Source: Taylor et al., *Measuring the World City Network: New Developments and Results*, GaWC Research Bulletin 300, 2009; http://www.lboro.ac.uk/gawc/rb/rb300.html.

office. It also means that U.S. business service firms, with a big domestic market, have less reason to gamble on global expansion.

These world cities have not only experienced profound changes in their economic and demographic profiles but also undergone dramatic transformations in their physical appearance. These include gentrification, branded neighborhoods, large-scale urban regeneration projects, iconic buildings, and "semiotic districts" specializing in goods and services with high semiotic content: flagship stores, megastores, shops-in-shops, high-end restaurants, cafés, art galleries, antique stores, and luxury retail shops.

World cities derive a huge comparative advantage as a result of the image they acquire from the particular concentrations of products, business services, and companies they are associated with: advertising and finance in New York; architecture, insurance, and banking in London; haute couture in Paris; logistics in Singapore; design in Milan; and so on. Favorable images, reinforced and amplified by the media and movies, create entry barriers for competing places, while the wealth generated in successful cities helps them to become thriving settings for high-end consumption, establishing them in turn as global tastemakers. These positive images of world cities relate, of course, to their "front regions": the financial districts, cultural quarters, design districts, entertainment districts, and semiotic districts that are the principal settings for activities with international connections. Less well publicized and documented are the "back regions" of world cities—their gentrified neighborhoods, neobohemias, and "ordinary" neighborhoods—and the associated issues of difference, diversity and inequality.

Globalization and Urban Change

Networked infrastructures of transportation, information, and communications technologies, such as telephone systems, satellite television, computer networks, electronic commerce, and business-to-business Internet services are central to the increasingly important relationship between urbanization and globalization. According to UN-Habitat (the United Nations Centre for Human Settlements),[23] information and communications technologies are intensifying global urbanization in three main ways:

1. They allow specialist urban centers, with their high value-added services and high-tech manufacturing, to extend their powers, markets, and control to ever-more distant regional, national, international, and global spheres of influence.
2. The growing speed, complexity, and riskiness of innovation in a global economy require a concentration of technological infrastructure and an associated knowledgeable technology-oriented culture in order to sustain competitiveness.
3. Demand for information and communications technologies is overwhelmingly driven by the growth of metropolitan markets. World cities especially are disproportionately important in driving innovation and investment in networked infrastructures of information and communications technologies—due to their cultures of modernization, concentrations of capital, relatively high average disposable personal incomes, and concentrations of internationally oriented companies and institutions.

In contrast to the infrastructure networks of earlier technology systems that underpinned previous phases of urbanization, these information and communications technologies are not locally owned, operated, and regulated. Rather, they are designed, financed, and operated by **transnational corporations** to global market standards. Detached from local processes of urban development, these critical networked infrastructures are highly uneven in their impact, and contribute to the so-called **digital divide** because they selectively serve only certain neighborhoods, certain cities, and certain kinds of metropolitan settings. Geographers Stephen Graham and Simon Marvin refer to this tendency as **splintering urbanism**.[24]

Splintering urbanism is characterized by an intense geographical differentiation, with individual cities and parts of cities engaged in different—and rapidly changing—ways in ever-broadening and increasingly complex circuits of economic and technological exchange. Traditional patterns of urbanization have been overwritten by a very new dynamic dominated by enclaves of superconnected people, companies, and institutions, with their increasingly broadband connections to elsewhere via the Internet, mobile phones, iPads, and satellite TVs and their easy access to information services. We can identify several distinctive kinds of urban settings that are the product of splintering urbanism:

- Enclaves of international banking, finance, and business services in world cities and major regional centers. Examples include the business districts in Lower Manhattan, the City of London, Frankfurt, Hong Kong, and Kuala Lumpur.
- Enclaves of Internet and digital multimedia technology development, mostly in world cities in the developed countries. Examples include "Multimedia Gulch" in the SOMA (South of Market Street) district of downtown San Francisco and New York's "Silicon Alley" (just south of 41st Street in Manhattan).
- Technopoles and clusters of high-tech industrial innovation. These have emerged in campus-like suburban settings around world cities in developed countries (as in London, Paris, and Berlin); in new and renewed industrial regions within developed countries (as in southern California, Baden-Württemberg in Germany, and Rhône-Alps in France); and in emerging high-tech production and innovation spaces in NIEs (as in Bangalore, India, and the Multimedia Super Corridor south of Kuala Lumpur, Malaysia).
- Places configured for **foreign direct investment (FDI)** in manufacturing, with customized infrastructure, expedited development approval processes, tax concessions, and, in some cases, exceptions to labor and environmental

regulations. Such places have emerged in economically depressed regions of developed countries (including Northern England and parts of the U.S. Manufacturing Belt) but are mostly found in or near major cities in less developed countries (as in the Brazilian cities of Porto Alegre and Paranà, which have attracted foreign-owned auto plants).

- Enclaves of back-office spaces, data-processing, e-commerce, and call centers. These have emerged in older industrial cities within developed countries (such as Roanoke, USA, and Sunderland, England) and in many cities within NIEs, most notably in India, the Philippines, and the Caribbean.
- Spaces customized as "logistics zones." Airports, ports, **export processing zones**: enclaves in major cities around the world within which the precise and rapid movement of goods, freight, and people are coordinated, managed, and synchronized between various transport modes.

Connected to one another through a complex dynamic of flows, these urban spaces and settings are key elements in the spatial articulation of economic globalization. They are embedded within regions and metropolitan areas whose economic foundations derive from earlier technology systems and whose social and cultural fabric derives from more traditional bases. The result is that the local effects of splintering urbanism are transforming traditional patterns of land use and spatial organization in many parts of the world.

The Polycentric Metropolis

Until the middle of the twentieth century, urban and metropolitan form could safely be conceptualized in terms of the outcomes of processes of competition for land and of ecological processes of congregation and segregation—all pivoting tightly around a dominant central business district and transportation hub. Since then, urban development, now a product of the combination of increased automobility and the blossoming of egalitarian liberalism in the form of massive federal outlays on highway construction and mortgage insurance, developed **urban realms**—semiautonomous subregions bound together through urban freeways—has displaced the

URBAN VIEW 6

Japanese Cities: Tokyo and the Tokaido Megapolitan Region[25]

Japan's phenomenal economic growth in the decades following World War II stimulated dramatic urban development. Between 1950 and 1970 the percentage of people living in cities with a population of 50,000 or more rose from 33 to 64 percent, while the overall urban population reached 72 percent. As the population in cities exploded, the number and size of larger cities increased dramatically. Since 1970 the urban population has continued to increase, but more slowly, reaching 86 percent by 2010.

A distinguishing feature of Japan's urban landscape is the concentration of its major cities into a relatively small portion of its already small land area. The Tokaido megapolitan region contains about 90 million people, about 70 percent of Japan's population. It contains the three urban industrial regions of Keihin'yō (Tokyo–Yokohama), with nearly 35 million people; Keihanshin (Osaka–Kobe–Kyoto), with more than 18 million; and Chūkyō (Nagoya), at nearly 9 million (Figure 23).

Japan's urban system then comprises a capital region centered on the primate city of Tokyo and the other regions elsewhere in Japan. The rapid urban growth from the late 1950s through the early 1970s saw a shift in population from the rural areas to the big cities in general and to Tokyo in particular. Since then, rural-to-urban migration has been increasingly toward Tokyo at the expense of the rest of the country.

The plans for Tokyo's urban development that were drawn up immediately after World War II were not all implemented. Instead, the city experienced largely haphazard growth that produced congestion and a disorganized city layout. In general, urban growth has been concentrated around key subcenters, such as Shibuya, Shinjuku (an edge city that is now the city government headquarters), and along the key transportation arteries (rail and expressway) radiating outward from the Chiyoda District, the old historic core of the Imperial Palace.

In contrast to many U.S. cities, Tokyo's CBD has maintained its corporate primacy. This has occurred despite the decentralization of middle-income workers to peripheral areas in search of affordable housing as land costs spiraled upward in the 1980s in the lead-up to the recession of the 1990s. The workers who commute to the central city each day use one of the busiest and most crowded mass transit systems in the world—the average Tokyo commute is a two-hour journey in each direction.

The March 11, 2011 Sendai 9.0 magnitude earthquake, tsunami, and damaged Fukushima nuclear power plant wreaked havoc on the lives of people living less than 200 miles northeast of Tokyo. Initial estimates are that the cost of the damage to roads, homes, factories, and infrastructure will exceed $300 billion, making it the world's most costly natural disaster. Tokyo's municipal government has emergency response plans for earthquakes that involve evacuation to open spaces like large parks and for tsunamis that involve evacuation to higher ground farther away. The Japanese government also enforces strict building codes in seismic areas and runs public education programs. Nevertheless, had the earthquake hit Tokyo, where the Sendai earthquake was felt, the human and economic toll would have been much worse than in the Sendai area because of the concentration of people and administrative and commercial activities in Japan's capital city and megapolitan region.

(continued)

URBAN VIEW 6
Japanese Cities: Tokyo and the Tokaido Megapolitan Region (*continued*)

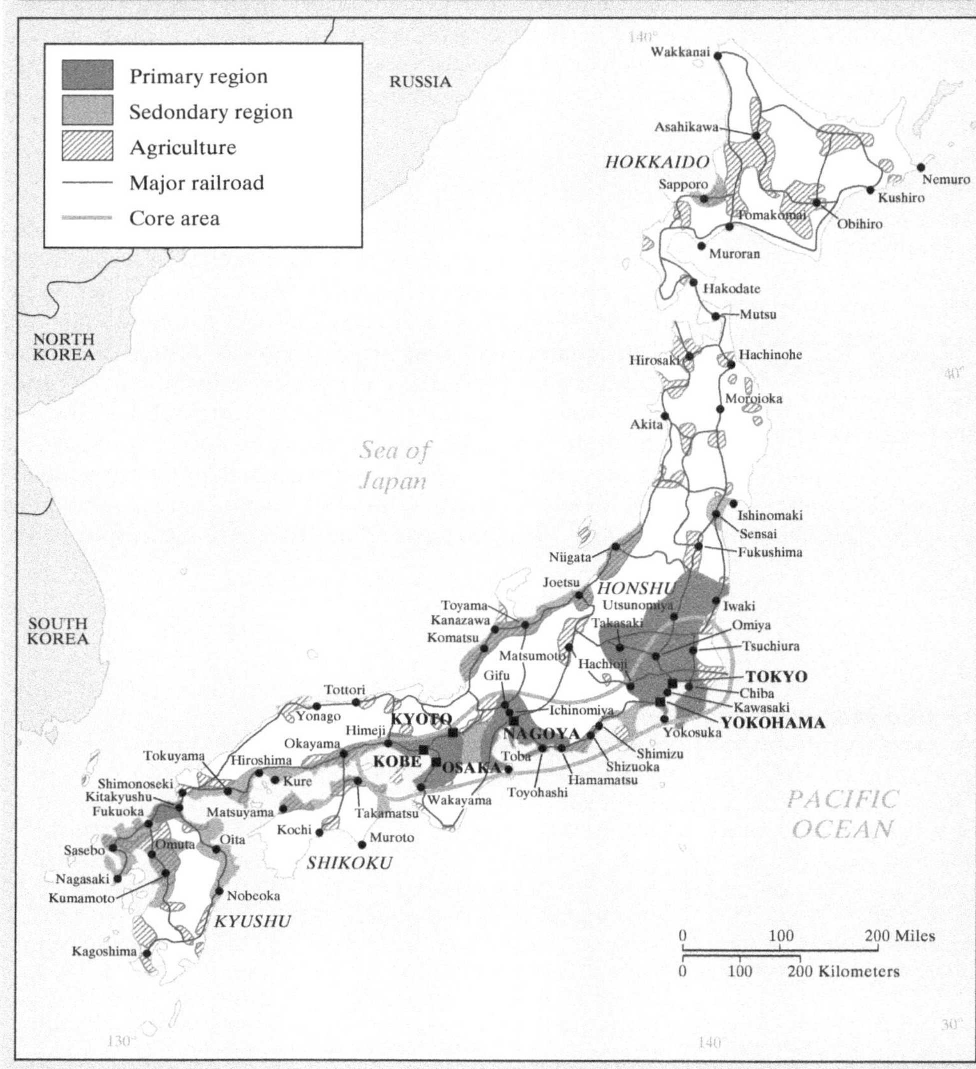

Paul L. Knox

FIGURE 23 Tokaido megapolitan region. A distinguishing feature of Japan's urban landscape is the concentration of its major cities into a relatively small portion of its already small land area. The Tokaido Megalopolis contains about 90 million people, about 70 percent of Japan's population. It comprises three major urban industrial regions centered on Tokyo–Yokohama, Osaka–Kobe–Kyoto, and Nagoya.

traditional core-periphery relationship between central cities and their suburbs. Each urban realm tends to function semi-independently, with a broad mix of land uses and populations of between 175,000 and 250,000. Each realm has retail, commercial, and residential subareas, as well as a commercial and retailing node that functions as a high-order central place for the majority of local residents. As a result, most residents of metropolitan areas had less and less to do with the central core except for occasional trips to major sporting events, large concerts, and so on.

Today, the traditional form of the metropolis has slipped into history. Geographer Pierce Lewis coined the term **galactic metropolis** to capture the disjointed and decentralized urban landscapes that resulted from the splintering urbanism of globalized capitalism. The galactic metropolis is fragmented and multinodal, with mixed densities and unexpected juxtapositions of urban form and function. It is characterized by **edge cities**—suburban hubs of shops and offices that sometimes overshadow the old downtown. Edge cities are nodal concentrations of shopping and office space situated on the fringes of metropolitan areas, typically located on an axis with a major airport, sometimes adjacent to a high-speed train station, always linked to an urban freeway system. Examples include Washington's Dulles corridor (Figure 24), London's Heathrow district, the O'Hare area (Chicago), and Schipol (Amsterdam). The result is a polycentric metropolitan structure that now has variants around the world.

FIGURE 24 Part of the corridor of development that runs between Tysons Corner and Dulles airport, in the Washington D.C. metropolitan area, an example of an edge city.

Geographer Michael Dear notes that:

> It is no longer the center that organizes the urban hinterlands but the hinterlands that determine what remains of the center. The imperatives of fragmentation have become the principal dynamic in contemporary cities . . . . In contemporary urban landscapes, 'city centers' become, in effect, an externality of fragmented urbanism; they are frequently grafted onto the landscape as an afterthought by developers and politicians concerned with identity and tradition. Conventions of 'suburbanization' become redundant in an urban process that bears no relationship to a core-related decentralization.[26]

Edward Soja has offered the term "exopolis" to capture some of the key dimensions of contemporary urbanization, including the growth of edge cities and the increasing importance of external forces associated with globalization. Traditional models of metropolitan structure and traditional concepts—city, suburb, metropolis—are fast becoming examples of what sociologist Ulrich Beck calls "zombie categories," concepts that embody nineteenth- to late-twentieth-century horizons of experience distilled into analytic categories that still mold our perceptions and sometimes blind us to the significance of contemporary change.[27]

The challenges of characterizing the evolving outcomes of urbanization have in fact prompted a great variety of neologisms, including postsuburbia, exurbia, exopolis, boomburbs, cosmoburbs, **stealth cities**, nerdistans, technoburbs, generica, satellite sprawl, and mallcondoville (see the Urban View 7 entitled "From Boomburbs to Bustburbs?"). The term "metroburbia" has emerged to capture the way that residential settings in suburban and exurban areas are thoroughly interspersed with office employment and high-end retailing. The polycentric metropolis, meanwhile, is an encompassing term for the stereotypical urbanized region that has been extended and reshaped to accommodate increasingly complex and extensive patterns of interdependency in networks of half a dozen or more urban realms and as many as fifty nodal centers of different types and sizes, physically separate but functionally networked (Figure 25).[28]

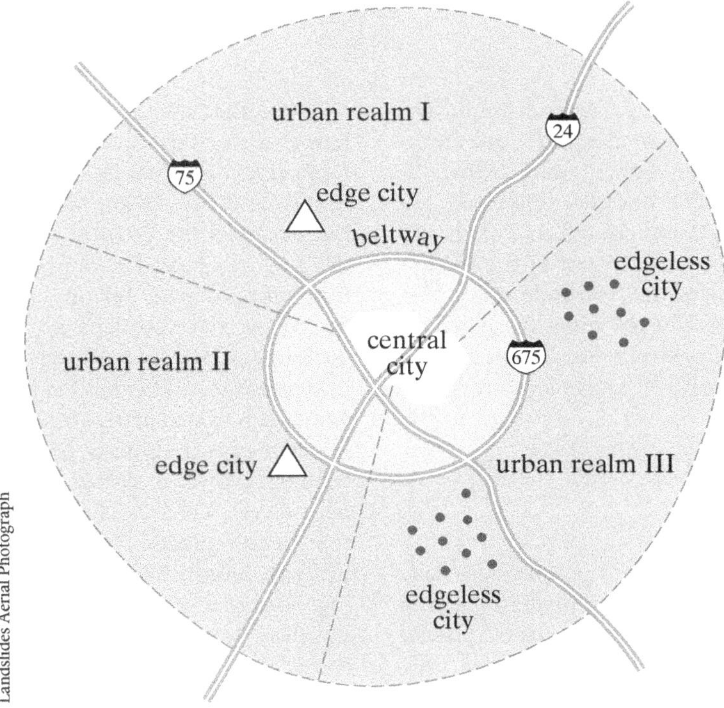

Landslides Aerial Photograph

FIGURE 25 Evolving twentieth-century metropolitan form.

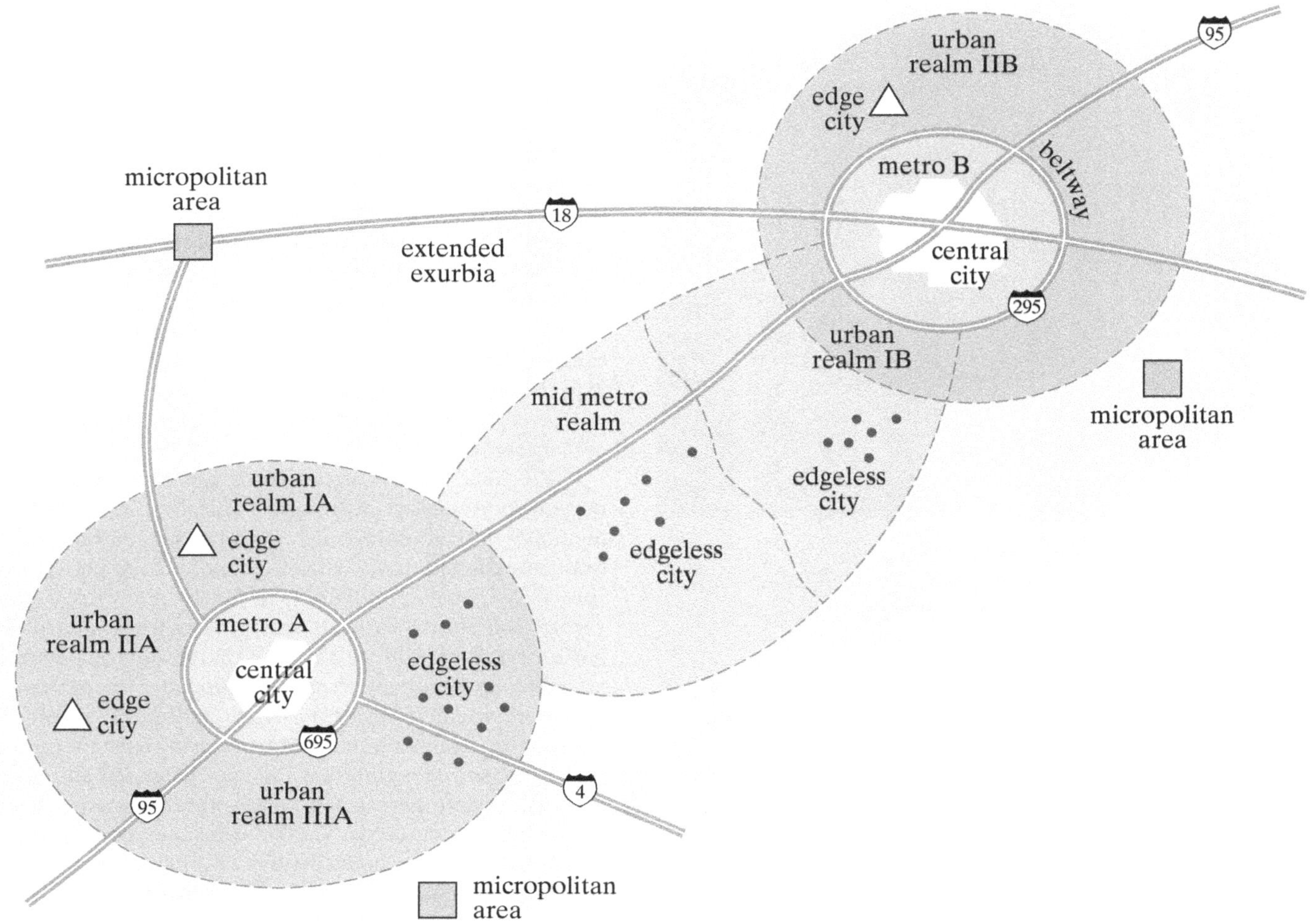

FIGURE 26 Evolving megapolitan form.

Bound together through urban freeways, arterial highways, beltways, and interstates, polycentric metropolises are themselves now beginning to coalesce functionally into "megapolitan" regions that dominate national economies (Figure 26; see also the Urban View 6 entitled "Japanese cities: Tokyo and the Tokaido Megalopolis"). Indeed, the largest of the world's polycentric metropolises have become "100-mile cities"[29]—metropolitan regions that are literally 100 miles or so across, consisting of a loose coalition of central cities, urban realms, edge cities, boomburbs, office parks ("edgeless cities"), and exurbs.

The End of "Suburbia"

Certainly the scale and organization of metropolitan areas have reached the point where urban functions are dispersed across decentralized landscapes that are not suburban in any traditional respect. Meanwhile, many central cities are now "central" only in the most limited geographical sense, their economic and demographic importance having been eclipsed by surrounding urban realms. Hence historian Robert Fishman's announcement of the end of "suburbia."[31]

The other side of the same coin is that central cities are no longer exclusively "urban." Central cities have not only experienced a continued deindustrialization and decentralization of jobs and population but also a selective reinvestment that has reversed the downward spiral of some neighborhoods' decline, conserved and preserved some of the remaining fragments of past periods, and introduced residential and mixed-use developments to some areas that were formerly declining. CBDs have experienced a selective recentralization of economic activity that has brought a renaissance of urbanity and a rush of speculative building.

At the heart of recent land use changes in central cities has been the growth of skilled jobs associated with advanced business services (advertising, banking, insurance, design, etc.) that have replaced the semiskilled manufacturing jobs lost through deindustrialization and the decentralization of retailing. For the moment, we should note that it was the advanced business service companies that filled the office space created by the building booms of the 1980s and 1990s, and it was their employees who fueled the processes of gentrification and historic preservation.

URBAN VIEW 7
From Boomburbs to Bustburbs?[30]

Between the mid-1980s and the housing crisis of the late 2000s, the population of U.S. metropolitan areas increased by double digits, and more than 25 million acres of farmland and open space (an area larger than Indiana) were developed around these metropolitan areas.

During that time, the touchstones of the polycentric metropolis were **boomburbs**, the fastest-growing suburban jurisdictions in the United States, typically located along the interstate beltways that ring large metropolitan areas in the western United States. Boomburbs do not resemble traditional central cities or older satellite cities. Although they possess most elements found in cities, such as housing, retailing, entertainment, and offices, they are not typically patterned in a traditional urban form. Boomburbs almost always lack, for example, a dense business core. As such, they can be seen as distinct from traditional cities not so much in their function but in their low density and loosely configured spatial structure.

Robert Lang of Virginia Tech's Metropolitan Institute defines boomburbs as places with more than 100,000 residents that are not the largest city in their metropolitan areas, and which maintained double-digit rates of population growth in recent decades. While boomburbs may be found throughout the United States, they occur mostly in the Southwest, with almost half in California alone (Figure 27). There are more than 50 boomburbs in the United States, the most populous of which is Mesa, Arizona—bigger than such traditional large cities as Minneapolis, Miami, and St. Louis. Arlington, Texas—the second-biggest boomburb—falls just behind Pittsburgh and just ahead of Cincinnati. Even such smaller boomburbs as Chandler, Arizona, and Henderson, Nevada, surpass older mid-size cities such as Knoxville, Tennessee, Providence, Rhode Island, and Worcester, Massachusetts.

In these boomburbs the scale of the development industry has been such that urbanization occurred in large increments, cutting—and within months filling—swathes of rural land with residential subdivisions, condominium complexes, bleak access roads, strip malls, parking lots, office parks, big-box stores, and, most characteristically, strip development. Beyond these strips lie subdivisions dominated by large-lot, single-family homes. Most of the architecture and urban design is without merit, adding up to what architect Rem Koolhaas has called "Generica." Following the logic of a fast return on investment and flexibility in use, most commercial structures are simple boxes, while economies of scale dictated a cookie-cutter approach for all but the most upscale residential subdivisions—where "monster homes," "starter castles," and "McMansions" took over as the norm.

The recent recession and housing crisis caused decades of double-digit growth to come to a screeching halt for nearly half of the boomburbs in the United States. It may now be necessary to consider replacing the "boom" in the name with "bust" for those boomburbs that have lost population in the last few years. Fifteen—among them Bellevue, Washington, near Seattle; Coral Springs, Florida, near Fort Lauderdale; Fullerton, California, near Los Angeles; and Lakewood, Colorado, near Denver—are ending the decade with less than a 10 percent overall growth in population because of recent losses. "They will drop out of boomburb status," says Lang who goes on to say that this "may signal a real shift in the American landscape where suburbs can no longer assume to be gaining on traditional cities." "The irony is that if they want to keep growing, they must grow as cities, which is diametrically opposite to how they got so big in the first place" concludes Lang.

FIGURE 27 Boomburbs in the United States. This map shows the distribution of the 53 places with more than 100,000 residents in 2000 that were not the largest city in their metropolitan areas and that maintained double-digit rates of population growth in the 1980s and 1990s.

Places and Regions in Global Context: Human Geography 3rd ed., P. Knox, S. Marston, Prentice Hall (2004). Reprinted with permission.

Paul L. Knox

FIGURE 28 South Street Seaport in New York City, one of a number of high-profile "specialty marketplace" projects designed to attract both residents and tourists by utilizing waterfront locations and historic buildings to create settings in which people can shop, promenade, and safely engage in the timeless pursuit of people watching.

Between them these changes transformed many of the landscapes of central cities. Where they were insufficient to rejuvenate the most acute cases of deindustrialization and retail decentralization, **civic entrepreneurialism** stepped in to engineer the appearance of large, set-piece mixed-use developments designed to bolster the city's image and pull in other forms of development. The best-known examples include Baltimore's Harbor Place, Miami's Bayside, Riverwalk in New Orleans, Riverfront in Savannah, Quincy Market in Boston, Pioneer Square in Seattle, and South Street Seaport in New York (Figure 28).

Even some of the largest and most spectacular of these developments have encountered financial difficulties, however, partly because of the inherent high risks of such projects, and partly because intense competition between cities for investment dollars contributed to over-building. Meanwhile, the underlying shift toward a service-based and information-based economy has created numerous land-use conflicts and social tensions in central cities.

FOLLOW UP

Key Terms

- annexation
- balloon-frame construction
- boomburb
- civic entrepreneurialism
- cumulative causation
- edge city
- enabling technologies
- galactic metropolis
- greenbelt cities
- metropolitan consolidation
- regional decentralization
- "roll-back" neoliberalism and "roll-out" neoliberalism
- splintering urbanism
- urban realms
- world cities

Review Activities

1. Construct a graph that shows the changing population since 1950 of the central city and suburban counties of your nearest metropolitan area. Annotate the graph to suggest reasons for the trends that you find.
2. Update your *portfolio*. The material in this chapter lends itself especially well to maps of all kinds. For example, you might prepare a hand-drawn, GIS, or computer cartography map of the city you know best, showing major industrial spaces, shopping centers, shopping strips, suburban downtowns, major highways, and, if it is big enough, urban realms. Population, employment, and unemployment data are available in most libraries and online and provide a good opportunity to devise maps, charts, and graphs that update and amplify the changing experience of different cities. The same topics are also the subject of frequent reports on the websites of various government agencies (the U.S. Bureau of the Census and the U.S. Department of Labor, in particular). Magazines like *American Demographics* are good sources of up-to-date information on population changes. You should also monitor major newspapers and business magazines for stories that relate to the topics you cover in your basic reading. Try to give the portfolio a balanced content as well as a distinctive flavor.
3. Research the employment structure of the city or town in the United States that you live in or that most interests you. Use census data available from the U.S. Census Bureau's website to construct an industry-by-industry profile of employment and summarize the data in a graph or table. What do the data suggest in relation to the city's function within the urban hierarchy?
4. Consider the extent to which your daily life depends on economic linkages between cities around the country and around the world. Look, for example, at the labels on all the clothes you are wearing. How many different places were involved in their production? How did they get to you? How much was a product of your local economy?

Log in to **www.mygeoscienceplace.com** for self-study quizzes, *MapMaster* layered thematic and place name interactive maps, *Urban View* Google Earth™ tours, key resources and suggested readings, related websites, "In the News" RSS feeds, and additional references and resources to enhance your study of urban systems and cities in transition.

NOTES

1. Based on K. T. Jackson, *Crabgrass Frontier: The Suburbanization of the United States* (New York: Oxford University Press, 1985); E. Schram, "The Pig Stand: The First Fast Food Drive-in," *U.S. News and World Report*, 2005.
2. Jackson, Crabgrass Frontier, 264.
3. J. Borchert, "Major Control Points in American Economic Geography," *Annals of the Association of American Geographers* 68 (1978): 230.
4. Peter Hall, *Cities of Tomorrow*, 3rd ed. (New York: Basil Blackwell, 2002), 316–319.
5. Robert Beauregard, *When America Became Suburban* (Minneapolis: University of Minnesota Press, 2006), 6.
6. A. Scott, *Metropolis: From the Division of Labor to Urban Form* (Berkeley, Calif.: University of California Press, 1989).
7. J. A. Casazza et al., *Shopping Center Development Handbook* (Washington, D.C.: Urban Land Institute, 1985), 16.
8. Pierre Filion. "Growth and Decline in the Canadian Urban System," *GeoJournal* 74 (2009): 403–13.
9. Ibid.
10. Gary Sands, "Prosperity and the New Economy in Canada's Major City Regions," *GeoJournal* 74 (2009): 1–14.
11. Filion, *GeoJournal*.
12. J. K. Galbraith, *The Affluent Society*, 4th ed. (Boston: Houghton Mifflin, 1984).
13. P. Harvey, "As College Graduates Hit the Workforce, So Do More Entitlement-Minded Workers," University of New Hampshire website, May 17, 2010 (*http://www.unh.edu/news/cj_nr/2010/may/lw17gen-y.cfm*).
14. K. Eckert, "The Boomerang Generation Goes Greek," *Suite101.com*, March 5, 2011 (*http://www.suite101.com/content/the-boomerang-generation-a355926*).
15. Pew Research Center, *Millenials: Confident, Connected, Open to Change* (Washington, D.C.: Pew Research Center, 2010).
16. The Project on Student Debt (*http://www.projectonstudentdebt.org/*).
17. AFL-CIO, Young Workers: A Lost Generation (Washington, D.C.: AFL-CIO, 2009).
18. See L. McCarthy and C. Johnson, "Cities of Europe," in *Cities of the World: World Regional Urban Development*, eds. S. Brunn, M. Hays-Mitchell, and D. Zeigler, 4th ed. (Lanham, Md.: Rowman & Littlefield, 2011).
19. N. Brenner and N. Theodore, "Cities and the Geography of 'Actually Existing Neoliberalism,'" in *Spaces of Neoliberalism. Urban Restructuring in North America and Western Europe*, eds. N. Brenner and N. Theodore, (Oxford: Blackwell, 2002), 21.
20. B. Ó hUallacháin and N. Reid, "The Location and Growth of Business and Professional Services in American Metropolitan Areas, 1976–1986," *Annals of the Association of American Geographers* 81 (1991): 254–70.
21. R. J. Stimson and S. Baum, "Cities of Australia and the Pacific Islands," in *Cities of the World: World Regional Urban Development*, eds. S. D. Brunn, J. F. Williams, and D. J. Zeigler, 3rd ed. (Lanham, Md.: Rowman & Littlefield, 2003), 456–88.
22. B. Badcock, "The Imprint of the Post-Fordist Transition on Australian Cities," in *Globalizing Cities: A New Spatial Order?*,

eds. P. Marcuse and R. van Kempen (Oxford: Blackwell, 2000), 225.

23. UN-Habitat (United Nations Centre for Human Settlements), *Cities in a Globalizing World: Global Report on Human Settlements 2001* (London: Earthscan Publications, 2001), 6.
24. S. Graham and S. Marvin, *Splintering Urbanism* (New York: Routledge, 2001).
25. The discussion of Japanese cities draws heavily on J. F. Williams and K. W. Chan, "Cities of East Asia," in *Cities of the World: World Regional Urban Development*, eds. S. D. Brunn, J. D. Williams and D. J. Ziegler, 3rd ed. (Lanham, Md: Rowman & Littlefield, 2003), 412–55.
26. Michael Dear, "Comparative Urbanism," *Urban Geography* 26 (2005):248.
27. Ulrich Beck, W. Bonss, and C. Lau, "The Theory of Reflexive Modernization," *Theory, Culture, and Society* 20 (2003):1–33
28. Robert Lang and Paul Knox, "The New Metropolis: Rethinking Megalopolis," *Regional Studies*, 43, 2009, 789–802.
29. D. Sudjic, *The 100-mile City* (New York: Andre Deutsch, 1999).
30. Robert E. Lang and Jennifer B. LeFurgy, *Boomburbs: The Rise of America's Accidental Cities* (Washington, D.C.: Brookings Institution Press, 2007); Haya El Nasser, "Housing Bust Halts Growing Suburbs," *USA Today*, November 20, 2009.
31. R. Fishman, *Bourgeois Utopias* (New York: Basic Books, 1987).

Urbanization in the Less Developed Countries

From Chapter 5 of *Urbanization: An Introduction to Urban Geography*. Third Edition. Paul L. Knox, Linda McCarthy.

Jochen Tack/Photolibrary, Inc.

Urbanization in the Less Developed Countries

From a geographical perspective, the most significant aspect of recent world urbanization is the dramatic difference in trends and projections between the more developed countries like the United States, United Kingdom, Japan, and Australia and the less developed countries of Latin America, Africa, and Asia. In 1950 just under 60 percent of the people who lived in cities were concentrated in the more developed countries. Since then the world's urban population has increased nearly fivefold, with the bulk of the growth in the less developed countries. Economic development and industrialization in Europe depended greatly on the exploitation of people and regions elsewhere. Inevitably, the international division of labor that underpins this relationship fundamentally influenced the patterns and processes of urbanization in the less developed countries. Although variations in internal and external factors produced different urbanization experiences in each country, a pressing problem today for people in many less developed countries is a process of overurbanization in which cities are growing more rapidly than jobs and housing.

LEARNING OUTCOMES

After reading this chapter, you should be able to:

- Describe current urbanization trends and projections for developed countries versus less developed countries.
- Explain how demographic change, rural-to-urban migration, and exceptionally high rates of natural increase combine to promote urban growth in less developed countries.
- Describe the motivations that drive people and families in less developed countries to migrate from farms and villages to cities.
- Compare and contrast the developmental approach to modernization with the dependency theory perspective and its conceptualization of underdevelopment.
- Explain the concept of "overurbanization" and understand what this phenomenon means for people living in the cities of less developed countries.

CHAPTER PREVIEW

This chapter examines urbanization in the less developed countries in global and historical context. A look at current trends shows urban population increasing at almost twice the general population growth rate in the less developed countries.[1] The factors promoting this kind of urban growth vary within and between the different parts of the world. In sharp contrast to the experience of the world's more developed countries, where urbanization was largely an outcome of economic growth, urbanization in the less developed countries has resulted from demographic growth that preceded economic development. We consider some of the attempts made during the last 50 years to disentangle the relationship between urbanization and economic development and to explain "underdevelopment" in the less developed countries.

An entirely new group of cities has grown up since the 1960s around the oil and natural gas fields of countries such as the United Arab Emirates, Kuwait, Saudi Arabia, Bahrain, and Qatar. This photograph shows the skyscrapers and camel race course of Dubai, U.A.E.

URBAN VIEW 1
The Dream of a Better Life as a Garment Girl in Dhaka[2]

By age 23, Mahmuda Akhter had already lived two quite different lives. Until she was 16, she lived in a village in Bangladesh. Married off at 15 in a country where most marriages are still arranged, Mahmuda soon gave birth to a baby girl, Tamina. In the months after Tamina's birth, Mahmuda had difficulty breastfeeding because she suffered from malnutrition, chronic bouts of dysentery, and calcium deficiency. With no money for baby formula, she fed the baby cow's milk and water. The baby cried a lot despite Mahmuda's efforts to comfort her. One night, Tamina suddenly stopped crying. Infant mortality like this is high in poor countries such as Bangladesh where 60 percent of the children in the poorest families are malnourished.

Barely 16, Mahmuda's life changed dramatically when she did what hundreds of thousands of other villagers in Bangladesh do every year, she moved to the capital city, Dhaka, with the dream of a better life. Her husband had already found work there and helped to find her a job ironing in the same garment factory in which he worked as a "polyman," packing shirts in plastic for shipment (Figure 1). The shirts with the "Made in Bangladesh" tag are sold in retail chainstores like the GAP, Old Navy, and Banana Republic.

> The difference between our life in the village and our life now is huge, Mahmuda says. When we lived in the village, we had no income. My husband had no money. We couldn't afford enough to eat. Now I can afford meat every week. I can buy good fish, big fish. I can afford all kinds of food. Now we can plan on having more children, and raise them in Dhaka. My father had a dream that one day I would become a lawyer. I have the same dream. When I have another baby, maybe he or she will grow up to be a lawyer.

Jorgon Schytte/Photolibray, Inc.

FIGURE 1 Young women working in a shirt factory in Dhaka, Bangladesh.

Colonization and the expansion of trade around the world allowed Europeans to influence the world's societies and economies. Although the various regions were at different levels of urbanization on the eve of the European encounters, colonization and the Industrial Revolution created unprecedented concentrations of people in cities that were connected in networks and hierarchies of interdependence around the world. We use a sequence of six phases of colonial urbanization as a framework for examining how this process and its impacts changed over time in different parts of the world. We consider, in turn, mercantile colonialism; industrial colonialism; late colonialism; early independence; neocolonialism; and **globalization** and neoliberalism. This brings us to a pressing problem for people in many cities in the less developed countries today: a process of overurbanization in which cities grow more rapidly than jobs and housing (Urban View 1 entitled "The Dream of a Better Life as a Garment Girl in Dhaka").

URBANIZATION TRENDS AND PROJECTIONS: THE LESS DEVELOPED COUNTRIES IN GLOBAL CONTEXT

In 2007, the world reached an urban milestone: the percentage of people who lived in cities surpassed 50 percent for the first time. In 2010, the world's urban population reached nearly 3.5 billion and it is expected to rise to almost 6.3 billion by 2050.[3] Of the major world regions, the United States and Canada is the most urbanized region, with over 82 percent of the people living in towns and cities. Africa (40 percent) and Asia (42 percent) are the least urbanized (Figure 2).

To put these figures into historical perspective, in 1950 less than 30 percent of the world's population was urbanized. In that year only 75 metropolitan areas had a million or more people, and just 6 topped the five million mark. By 2010, there were 442 metropolitan areas of a million or more, with 54 containing over 5 million people. Looking ahead to 2025, there

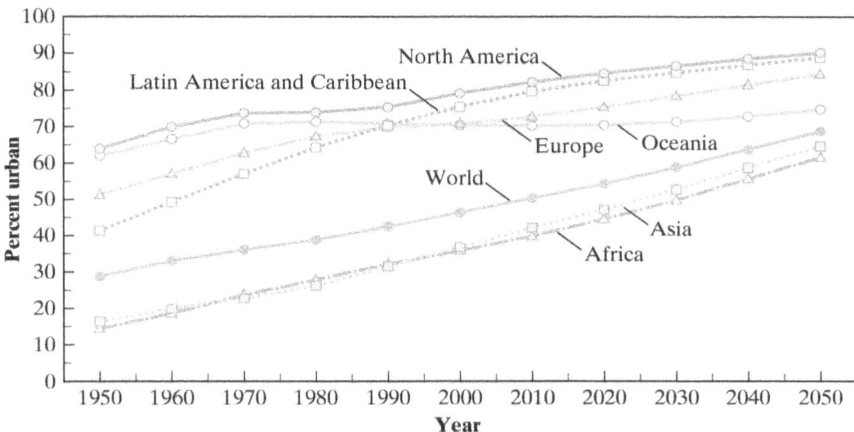

FIGURE 2 Urbanization by major world region.

will be about 547 cities with a population of a million plus, including about 75 with over five million. In fact, the predictions for the near future are for virtually all the world's population growth to occur in urban areas. By 2050, for example, more than 69 percent of the world's population is expected to be urbanized.

From a geographical perspective, the most significant aspect of recent world urbanization is the incredible difference in trends and projections between the more developed regions containing countries like the United States, United Kingdom, Japan, and Australia and the less developed regions comprising the countries of Latin America, Africa, and Asia. In 1950 just under 60 percent of the people who lived in cities was concentrated in the more developed countries. Since then the world's urban population has increased nearly fivefold, with the bulk of the growth in the world's less developed regions. In fact, almost all of the population increase between now and 2050 will be absorbed by the less developed countries, whose urban population is expected to rise from more than 2.5 billion in 2010 to over 5 billion or nearly 83 percent of the world's urban population by 2050 (Figure 3).

Of the world's 25 largest metropolitan areas in 1950, 10 were in Europe and the United States. It took only until 1990 for this number to drop to 5, and by 2025 only 3 of these will be on the list. What is more, the largest metropolitan regions are getting much larger because the number of people living in them is rising (Table 1).

Asia provides some dramatic examples of this trend. From a region of villages, Asia is quickly becoming a region of towns and cities. Between 1950 and 2010, for example, its urban population increased more than seven and a half times to nearly 1.8 billion people. By 2050 almost 65 percent of the people in Asia are expected to be living in urban areas (Figure 2).

Nowhere is the trend toward rapid urbanization more pronounced than in China. For decades the communist government imposed strict controls on where people could live because it feared the transformative and liberating effects of cities. By tying people's jobs, school admission, and even the right to buy food to the places where they were registered to

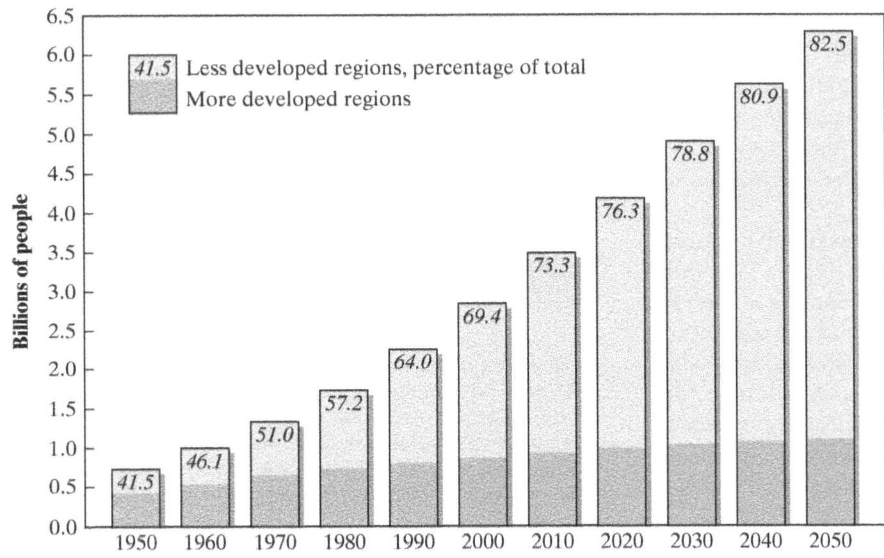

FIGURE 3 Urban population growth.

TABLE 1 Trading Places on the Top 25 List: The World's Largest Metropolitan Areas, Ranked by Population Size (Millions)

1950	Population	1990	Population	2025	Population
New York, USA	12.34	Tokyo, Japan	32.53	Tokyo, Japan	37.09
Tokyo, Japan	11.27	New York, USA	16.09	Delhi, India	28.57
London, UK	8.36	Ciudad de México, Mexico	15.31	Mumbai, India	25.81
Paris, France	6.52	São Paulo, Brazil	14.78	São Paulo, Brazil	21.65
Moskva (Moscow), Russia	5.36	Mumbai, India	12.31	Dhaka, Bangladesh	20.94
Buenos Aires, Argentina	5.10	Osaka, Japan	11.04	Ciudad de México, Mexico	20.71
Chicago, USA	5.00	Kolkata, India	10.89	New York, USA	20.64
Kolkata (Calcutta), India	4.51	Los Angeles, USA	10.88	Kolkata, India	20.11
Shanghai, China	4.30	Seoul, South Korea	10.54	Shanghai, China	20.02
Okaka, Japan	4.15	Buenos Aires, Argentina	10.51	Karachi, Pakistan	18.73
Los Angeles, USA	4.05	Delhi, India	9.73	Lagos, Nigeria	15.81
Berlin, Germany	3.34	Rio de Janeiro, Brazil	9.59	Kinshasa, Dem. Rep. Congo	15.04
Philadelphia, USA	3.13	Paris, France	9.33	Beijing, China	15.02
Rio de Janeiro, Brazil	2.95	Al-Qahirah, Egypt	9.06	Manila, Philippines	14.92
Sankt Peterburg (St. Petersburg), Russia	2.90	Moskva, Russia	8.99	Buenos Aires, Argentina	13.71
Ciudad de México (Mexico City), Mexico	2.88	Jakarta, Indonesia	8.18	Los Angeles, USA	13.68
Mumbai (Bombay), India	2.86	Manila, Philippines	7.97	Al-Qahirah, Egypt	13.53
Detroit, USA	2.77	Shanghai, China	7.82	Rio de Janeiro, Brazil	12.65
Boston, USA	2.55	London, UK	7.65	Istanbul, Turkey	12.11
Al-Qahirah (Cairo), Egypt	2.49	Chicago, USA	7.37	Osaka, Japan	11.37
Tianjin, China	2.47	Karachi, Pakistan	7.15	Shenzhen, China	11.15
Manchester, UK	2.42	Beijing, China	6.79	Chongqing, China	11.07
São Paulo, Brazil	2.33	Dhaka, Bangladesh	6.62	Guangzhou, China	10.96
Birmingham, UK	2.23	Istanbul, Turkey	6.55	Paris, France	10.88
Shenyang, China	2.15	Tehran, Iran	6.36	Jakarta, Indonesia	10.85
Total	**108.43**	**Total**	**264.04**	**Total**	**427.02**

Source: United Nations, *World Urbanization Prospects: The 2009 Revision,* New York: Department of Economic and Social Affairs, Population Division (*http://esa.un.org/unpd/wup/index.htm*).

live, the government made it almost impossible for rural residents to migrate to towns or cities. As recently as 1985, more than 77 percent of Chinese people still lived in the countryside; today the rural population is nearer 50 percent.

China is now rapidly making up for lost time. Having decided that towns and cities can be engines of economic growth within a communist system, the Chinese government not only relaxed enforcement of its residency laws but drafted plans to establish over 430 new cities. Between 1980 and 2010 the number of people living in cities in China more than tripled, from 190 million to 636 million, and the number of cities with a population of three-quarters of a million or more increased from 20 to 133 (Figure 4).

In the world's developed countries, levels of urbanization are high and have been for some time (Figure 5). United Nations figures show that Belgium and Iceland are 90 percent or more urbanized, while Australia, Canada, Denmark, France, Luxembourg, the Netherlands, New Zealand, Spain, Sweden, the United Kingdom, and the United States are all more than 75 percent urban. Despite low *rates* of urbanization, especially when compared to the less developed countries, the urban population in the developed countries is still expected to rise to more than 86 percent by 2050.

Linda McCarthy

FIGURE 4 Guangzhou, in southern China's Pearl River Delta, has experienced phenomenal growth (from 1.0 million people in 1950 to 10.9 million by 2025) and is projected to become one of the 25 largest metropolitan areas in the world by 2025. The construction cranes in CITIC Plaza in the growing Tianhe District of Guangzhou are building new skyscraper offices, apartment complexes, a train station, a metro station, and a sports stadium.

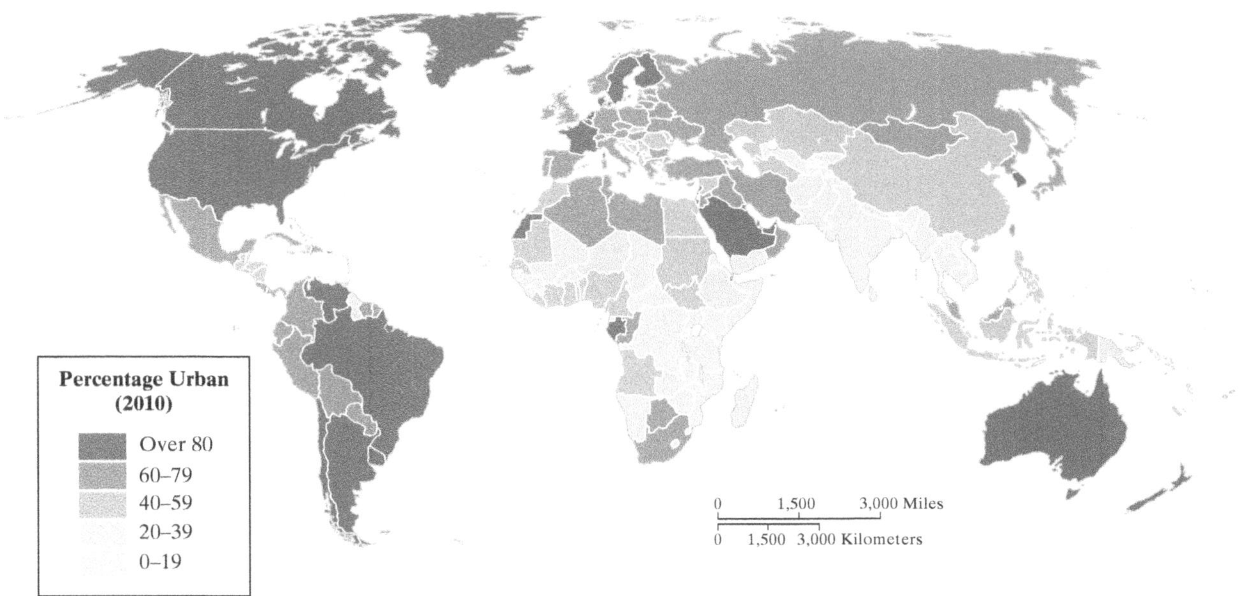

FIGURE 5 World urbanization.

Levels of urbanization are also very high in many of the world's **newly industrializing economies (NIEs)** (Figure 5). Brazil, Mexico, Taiwan, Singapore, and South Korea are all at least 75 percent urbanized. And compared to the developed countries, their rates of urban growth are also high.

In other less developed countries rates of urbanization are even higher. Jakarta, Indonesia, for example, grew from 1.4 million to 9.2 million between 1950 and 2010 and is expected to reach 10.9 million by 2025. Likewise, Lagos, Nigeria, a city of less than 300,000 in 1950, reached 10.6 million in 2010 and is projected to have a metropolitan population of 15.8 million by 2025 (Figure 6). Delhi, Mumbai, and Kolkata (India), São Paulo (Brazil), Dhaka (Bangladesh), Mexico City, and Shanghai (China) are all projected to have metropolitan populations in excess of 20 million by 2025 (Table 1).

Many of the very largest metropolitan areas are growing at annual rates of 4 percent or more each year. To put the

George Esiri/Reuters Pictures

FIGURE 6 Rates of urbanization can be very high in some less developed countries. In Nigeria, for example, Lagos, a city of less than 300,000 in 1950, reached 10.6 million in 2010 and is projected to have a metropolitan population of 15.8 million by 2025.

Marka/Alamy

FIGURE 7 Congestion in central Dhaka, Bangladesh, a rapidly-growing megacity.

situation in numerical terms, metropolitan areas like Dhaka and Delhi are adding up to half a million people to their populations each year: nearly 10,000 every week, even with deaths and out-migrants. It took London 190 years to grow from half a million to 10 million inhabitants; New York took 140 years. More recently, Mexico City, São Paulo, Buenos Aires, Kolkata (Calcutta), Rio de Janeiro, Seoul, and Mumbai all took less than 75 years. Now megacities such as Dhaka, Delhi, and Lagos take about half that time to grow from half a million to 10 million people (Figure 7).

FACTORS PROMOTING URBAN GROWTH

The factors promoting urban growth vary within and between different parts of the world. In sharp contrast to the experience of the world's developed countries, where urbanization was largely an outcome of economic growth, urbanization in the less developed countries has resulted from demographic growth that preceded economic development.

The more rapid decline in death rates compared to birth rates—part of a **Demographic Transition** in the less developed countries—is a fairly recent trend that has generated large increases in population well in advance of any significant levels of industrialization or rural economic development. In rural regions this has produced fast-growing populations in places that face problems with agricultural development. This has been described as *urbanization by implosion*[4] in which in-place population growth is happening, almost unnoticed, in villages and the countryside across vast stretches of rural India, Bangladesh, Pakistan, China, Nigeria, and other less developed countries. These high-density rural regions have population densities that exceed 1,000 persons per square mile—the threshold for defining urban settlements.

Rural-to-urban migration involves impoverished rural residents migrating to the larger towns and cities in search of a better life. These people are driven by the desire for employment and the prospect of access to schools, health clinics, piped water, and the kinds of public facilities and services that are often unavailable in rural regions. Overall, the metropolises in the less developed countries have absorbed almost nine out of ten of the 2.2 billion city dwellers added to the world's population since 1970.

Rural migrants have moved to cities out of desperation and hope, rather than being drawn by actual jobs and opportunities (Urban View 2 entitled "Fleeing the Countryside for Life in the City in Africa"). Because of the disproportionate number of teenagers and young adults in these migration streams, an important additional component of urban growth has followed—high rates of **natural increase** of the population. In most less developed countries the rate of natural increase in cities exceeds that of in-migration. On average, about 60 percent of urban population growth in the less developed countries is due to natural increase.

Political and environmental circumstances can also promote urban growth. Wars in Africa have caused countless refugees to flee to cities. After civil war broke out in Cote d'Ivoire (Ivory Coast) in 2002, hundreds of thousands of people fled areas of conflict for the cities of Abidjan, Grand Bassam, and Yamoussoukro. Refugees from Sudan, Somalia, Congo, and Ethiopia have left the refugee camps in Kenya for its capital, Nairobi. Many people fleeing war-torn areas have refused free humanitarian assistance in crowded and squalid refugee camps and risked persecution and marginalization in cities in the hope of a better life. Instead, many refugees experience exploitation, harassment, prostitution, rape, and dreadful living conditions.[5] In Mauritania, Niger, and other countries

URBAN VIEW 2

Fleeing the Countryside for Life in the City in Africa[6]

For her entire 60 years, Fatima Yadik, a mother of 12 and grandmother to 18, had lived a nomadic life in the northern part of the Central African Republic with her husband, children, and the family's long-horn cattle, which kept them constantly on the move. One day, her camp of Peuhl nomads was attacked by bandits who killed her husband and all the other men and then stole their cattle. Bandits often target Peuhl people because they own livestock.

Terrified, Fatima fled south with her children to the town of Yaloké. After settling in the town, the family survived by collecting and selling firewood. But Fatima realized that for her children to have a future, they needed to go to school. "Until now, none of my children have gone to school," she said, "but now that we no longer have any cattle, they will have to find jobs, and they need to go to school to prepare."

But years of conflict and violence had devastated the already fragile education system in the Central African Republic. Many schools had been looted or damaged, and the teachers had moved to the capital city, Bangui. So a group of Peuhl parents formed a small NGO called Association Mboscuda and, together with other parents, they built the Fraternité School. The school now has more than 600 students, and Mboscuda helps pay for teachers and runs campaigns to encourage Peuhl parents to send their children, especially the girls, to school.

along the southern edge of the Sahara, deforestation and overgrazing in conjunction with government inaction have forced people to move to the cities as the expanding desert has overtaken whole villages.

THEORIES OF URBANIZATION AND ECONOMIC DEVELOPMENT

Historically, an association has existed between urbanization and economic development: Countries with higher levels of urbanization tend to have higher levels of economic development (Figure 8). What is not as clear is the direction of causality—the extent to which economic development promotes urbanization, or urbanization promotes economic development. In the developed countries, although urbanization was largely an outcome of economic development, it was a reciprocal arrangement in which the urbanization that was driven by economic growth in turn stimulated further economic development. The various attempts made during the last 60 years to disentangle this relationship and explain urbanization and "underdevelopment" in the less developed countries can be grouped into three categories.

Modernization Theories: The Developmental Approach

In the 1950s ideas about the development of the less developed countries were based on extrapolations of the European experience. This developmental approach prescribed an economic transition along a continuum of progress from a "traditional" rural society toward a "modern" urban industrialized one. Models, such as Rostow's stages of economic growth, informed the developmental approach.[7] This model shows the five

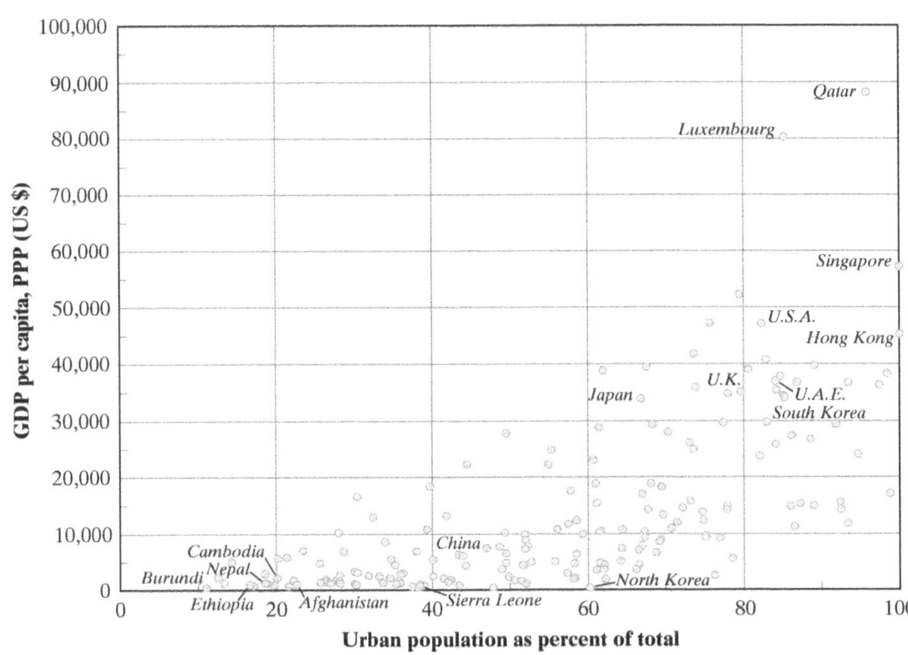

FIGURE 8 Urbanization and economic development, 2009.

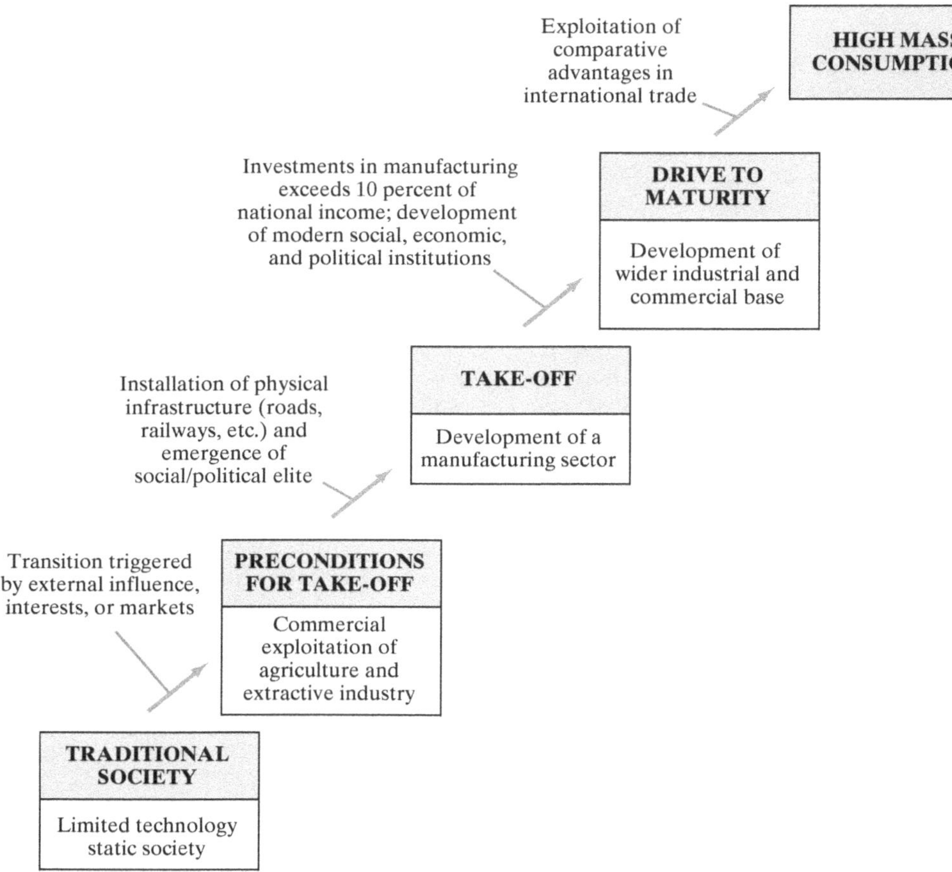

FIGURE 9 Rostow's stages of economic development model.

successive stages through which less developed countries must pass in order to achieve economic convergence with the developed ones (Figure 9).

Similarly, Myrdal's notion of **cumulative causation**[8] saw growth in the less developed regions following the patterns of urbanization experienced by Europe during the Industrial Revolution. Economic growth in one region would trigger strong demand for food, consumer goods, and other manufactures that local producers could not satisfy. This demand would create the opportunity for investors in less developed regions to establish a local capacity to meet the demand; entrepreneurs would take advantage of the cheaper labor and land there. If strong enough, these **spread effects** could enable less developed regions to develop their own upward spiral of cumulative causation.

Myrdal's influential model was followed by others who used a similar logic. Hirshman's model described **trickle-down effects**.[9] Perroux highlighted the importance of the *propulsive industries* that are characteristic of regions with high rates of economic growth,[10] such as the textile industry in England during the Industrial Revolution. As the propulsive industry grows, it attracts other related industries, generating a set of **agglomeration economies**. A **growth pole** is formed and an urban growth center develops. These ideas are shown spatially in Friedmann's core-periphery model.[11] The model shows an urban core of economic advantage and growth, surrounded by nearby agricultural areas that are in the process of development due to their proximity to the core, and a distant stagnant or declining periphery (Figure 10).

Although these kinds of development models have informed public policy and practice in the past, they are now regarded as too simplistic. They perpetuate the myth of "developmentalism," that all countries and regions—despite differences in their political, cultural, technological, and other characteristics—are on the same economic growth trajectory to become "modern" urban industrialized societies. A major weakness of developmentalism is that it fails to appreciate that the prospects for late starters are different from those of places that enjoyed an earlier **initial advantage** free from effective competition and limiting precedents. People in the less developed cities and countries must compete in a crowded field and overcome barriers that were created by the success of some of the early starters in Europe and North America. Most problematic for the modernization theories, though, was the indisputable evidence that the pattern of urbanization in the less developed countries was not following that of the developed countries—urban growth was not producing the expected boost in economic development.

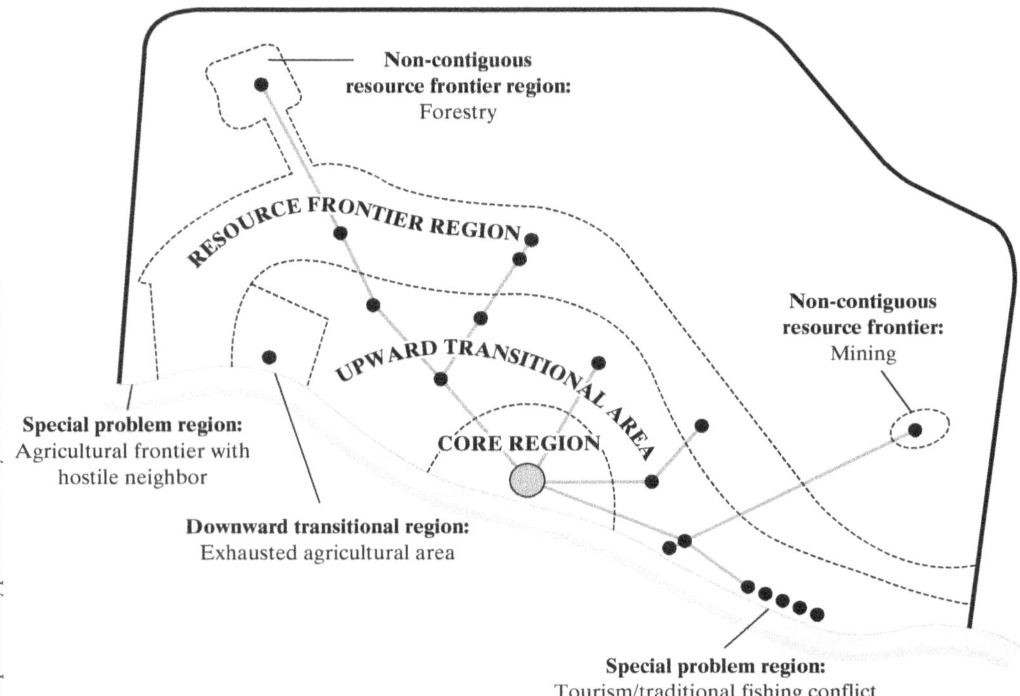

FIGURE 10 Friedmann's core-periphery model.

Urban Bias and *Under*development

In 1977, Michael Lipton coined the term *urban bias* in his book *Why Poor People Stay Poor: Urban Bias in World Development*.[12] Urban bias describes how the urban-based elite who hold power in some less developed countries tend to implement policies that allocate resources for the benefit of cities. By concentrating resources in the urban areas, urbanization rates accelerate and overall national economic development is impaired; urban–rural inequalities intensify because many of the poor people still live in rural areas despite rural-to-urban migration.

Although influential at the time, this notion of an urban–rural divide is now viewed as a simplistic generalization that, at best, describes only those countries experiencing very rapid urbanization. As was the case for modernization theories, the urban bias idea has been faulted for neglecting the international constraints on economic development in the less developed countries.

An avalanche of critical writings argued that the prosperity of the developed countries *depended* on *under*development in other parts of the world. The less developed countries, their role in the world-system already established (and firmly controlled by the economic and military power of the more developed countries), could not "follow" the historical experience of developed countries. In fact, the global system of unequal trade, exploitation of labor, and profit extraction guaranteed that the less developed countries and the people living in them would become more, rather than less, impoverished.

Writers like André Gunder Frank rejected the idea that underdevelopment was the result of geographical isolation or a failure to embrace Western technology, investment, and values. Instead, underdevelopment stemmed directly from the unequal nature of the interrelationships between people in the developed and less developed parts of the world. Figure 11 shows the system as described by Frank:

> as a photograph of the world taken at a point of time, this model of a world metropolis (today the United States) and its governing class, and its national and international satellites and their leaders—national satellites like the Southern states of the United States, and international satellites like São Paulo. Since São Paulo is a national metropolis in its own right, the model consists further of its satellites: the provincial metropolises, like Recife or Belo Horizonte, and their regional and local satellites in turn. That is, taking a photograph of a slice of the world we get a whole chain of metropolises and satellites, which runs from the world metropolis down to the hacienda or rural merchant who are satellites of the local commercial metropolitan center but who in their turn have peasants as their satellites. If we take a photograph of the world as a whole, we get a whole series of such constellations of metropolises and satellites.[13]

Certainly the unequal structure of the world economy and the kinds of monopoly power of a metropolis over its satellites have changed over time—for example, with the switch from **merchant capitalism** to **industrial capitalism** in the nineteenth century or following political independence for former colonies. But the transfer of wealth from satellites to metropolis has continued to fuel growth for people in some places at the expense of others.

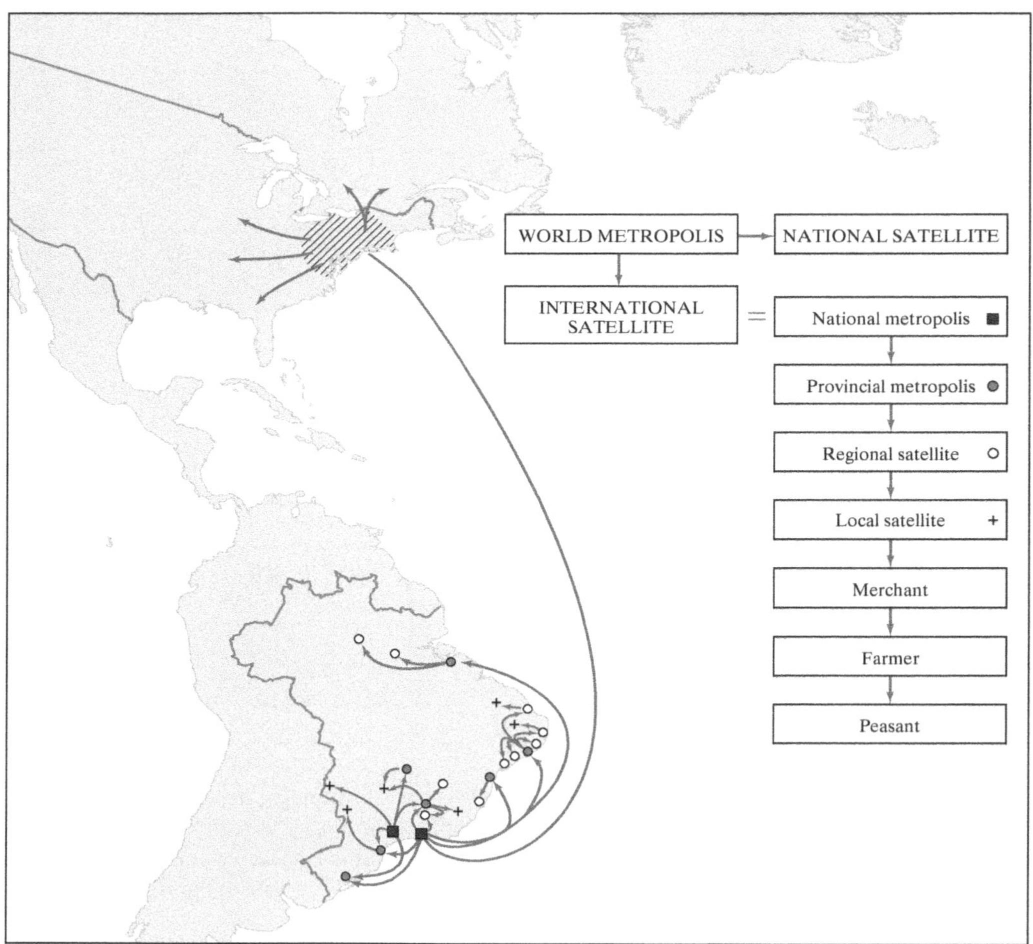

FIGURE 11 The Frank model of dependency.

Frank's approach is an example of *dependency theory*, which has been very influential in explaining global patterns of development and underdevelopment. Dependency theory states, essentially, that development and underdevelopment are the reverse sides of the same global process: independent development is impossible because development in one location requires underdevelopment in another.

Immanuel Wallerstein's *world-system theory*[14] addressed certain criticisms of dependency theory: that dependency theory ignored differences in the characteristics and processes among the less developed countries and focused too much on how these countries are "locked into" a position of dependence. According to the world-system perspective, the entire world economy is an evolving economic system with a hierarchy of countries comprising a core, semi-periphery, and periphery. The more developed core countries take advantage of their dominant position in the world economy to exploit the peripheral and semi-peripheral countries. The semi-peripheral countries (such as NIEs) also exploit peripheral countries (many of which are in Sub-Saharan Africa), while in turn being exploited by core countries. What is important about this conceptualization for the less developed countries is that the composition of this hierarchy is variable, with movement possible in either direction, including from periphery to semi-periphery and semi-periphery to core.

Looking specifically at cities, the model of peripheral urbanization used a political economic approach[15] to extend the dependency and world-system perspectives to the national **urban systems** of less developed countries (Urban View 3 entitled "A Model of Peripheral Urbanization"). This six-stage model describes how the extension of the global economic system to the less developed countries generated a strong process of urbanization.[16] But like dependency theory, this model has been criticized for being too deterministic; it implies that the incorporation of the less developed countries into the world economy makes the associated problems of urban development identified in the model seem inevitable. David A. Smith, for one, has stressed that:

> structural similarity between nations or regions in the global hierarchy may lead to parallel patterns of urban growth. But this will not always be the case . . . The real challenge is to identify the ways in which social relations and outcomes on the local level are linked to macrostructural processes, including those of the global political economy.[17]

URBAN VIEW 3

A Model of Peripheral Urbanization

The model of peripheral urbanization captures how the global system of production and trade generates a strong process of urbanization that affects people in individual cities and systems of cities in the less developed countries.[18]

1. Rural-to-urban migration increases as "traditional" forms of agriculture are disrupted by the introduction of commercial agriculture, by fiscal taxes on the rural population, and by competitive pressures on craft industries, initially from cheap imports and later from products by national manufacturers.
2. Production in the rural areas by national and foreign businesses promotes the development of major transportation and market centers and the rapid expansion of national capitals and major port cities.
3. The growth of manufacturing concentrates production even more within the largest cities, stimulates the expansion of a national government bureaucracy to encourage the process of industrialization, and leads to the concentration of high-income groups in the major centers.
4. Workers move to the largest cities in search of employment, and their labor and spending support further economic expansion.
5. The national government supports industrial expansion and the system is maintained through the provision of physical infrastructure in the main urban centers and social services to selected groups.
6. As development accelerates, private investment begins to spread outward to avoid the rising land prices, labor costs, and traffic congestion in the central city. The government may encourage this process of deconcentration with measures to encourage metropolitan decentralization.

New Models from the Less Developed Countries: Opportunities for Development

Significant departures from earlier thinking characterize contemporary perspectives on urbanization and economic development.[19]

First, there has been a shift away from the dependency theory focus on the international constraints on development in the less developed countries. The emphasis is now on the internal opportunities for development, using models, assumptions, and objectives that have been devised by people within the less developed countries themselves. This shift has been associated with an increasing recognition of the importance of local context for development and of the potential benefits of the economic and cultural interrelationships that already exist within the less developed countries.

Second, the related renewed interest in the role of indigenous social and cultural institutions and in the structure of the economic relations within less developed countries has led to a resurgence of concern for gender divisions in terms of how these affect and are affected by economic development.

Third, the relationship been the environment and development within the context of **sustainable urban development** for the less developed countries and at a global scale is attracting increasing attention. This represents a shift away from conceptualizing development as economic growth and "modernization" until the 1970s, and from then until the mid-1980s of assigning limited importance to the environmental context in dependency theory.

A HISTORICAL PERSPECTIVE ON COLONIAL URBANIZATION

We begin with a brief review of the state of urbanization around the world at the eve of the European encounters. We then focus on colonial urbanization in historical perspective and examine how this process and its impacts changed over time.

Indigenous Urbanization at the Eve of the European Encounters

The Europeans were not the first to create city-based empires. In fact, while Western Europe was in a period of urban stagnation during the **Dark Ages**, urbanization was going strong in other parts of the world. Cities and city systems of great size and political, cultural, economic, and technological importance thrived in what are now the less developed countries of the world *before* the arrival of the Europeans.

Islamic influence and Muslim culture dominated the cities of Southwest Asia from the seventh century C.E. In Africa between the eleventh and sixteenth centuries, Arab influence extended in a band across the north and south of the Sahara and along the east coast. The kingdoms of West Africa, including Mali, and the cities of Timbuktu, Jenne, and Gao, thrived on long-distance trans-Saharan caravan trade (Figure 12). Farther south, cities like Mogadishu, Mombasa, Zanzibar, Bulawayo, and Great Zimbabwe were important centers of commercial and cultural exchange, religion, and learning (Figure 13).

In Asia urban civilizations, such as the Ming Dynasty during the fifteenth and sixteenth centuries in China, flourished. In Southeast Asia the urban centers that developed as early as the first century C.E. were usually either inland sacred cities with ritual rulers, or coastal or riverine centers—some with up to 100,000 inhabitants—that thrived on their long-distance trade connections.

In Mesoamerica the Aztec urban system in the highlands of central Mexico was centered on Tenochtitlán, the capital city from 1325 C.E. The towering, brightly colored temples and palaces really impressed the Spanish when they

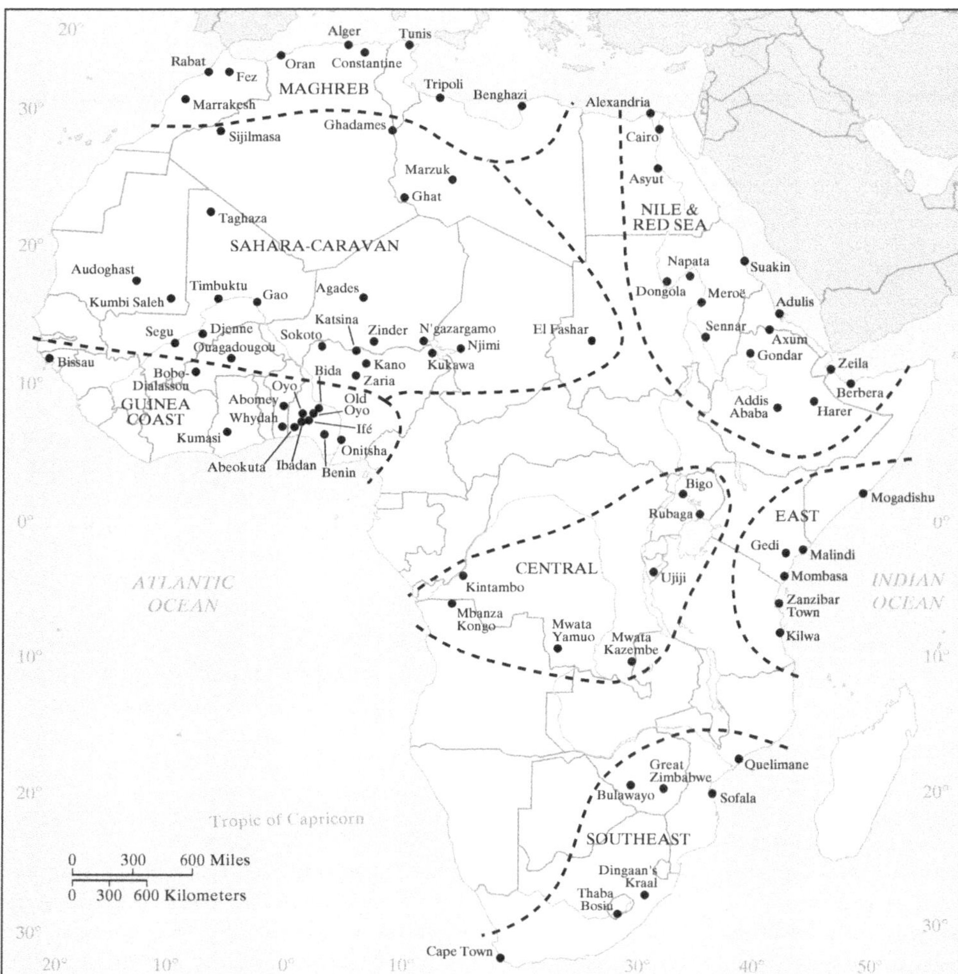

FIGURE 12 Historical centers of urbanization in Africa.

FIGURE 13 Great Zimbabwe, the capital of the Rozvi Mutapa Empire, flourished between the fourth and ninth centuries.

Ian D. Walker/Shutterstock

FIGURE 14 Tenochtitlán (just north of modern-day Mexico City) was the capital of the Aztecs from 1325 C.E. At an estimated 200,000 people, it was one of the largest cities in the world at the time, and its towering, brightly colored palaces and temples, such as the Temple of the Sun, really impressed the Spanish when they arrived in 1519.

arrived in 1519 (Figure 14). With about 200,000 inhabitants, the city was larger than Lisbon or Seville at that time. The Aztecs had a highly stratified society that included a ruler, soldiers, priests, skilled craftsmen such as gold and metalworkers, merchants, and large numbers of agricultural peasants using irrigated farming and terraced fields. Farther south along the Andes between central Chile and Colombia, the Inca Empire reached its height in the fifteenth century C.E. The capital city, Cuzco, had impressive monuments and a population of between 100,000 and 300,000 people. The Inca Empire was organized around a system of 170 administrative centers connected by an elaborate road system that was used to transport the military as well as the agricultural produce from the irrigated and terraced field farming system.

Colonial Urbanization

Economic development and industrialization in the developed countries depended greatly on the exploitation of people and regions elsewhere. Inevitably, the **international division of labor** that this relationship put in place fundamentally influenced the patterns and processes of urbanization in the less developed parts of the world. The colonial powers established **gateway cities** in these regions. In an effort to establish economic and political control over continental interiors, **colonial cities** were deliberately established or developed as centers of administration, political control, and commerce.

One type of colonial city was the new city that was "planted" in a location where no significant urban settlement had previously existed. These cities were laid out expressly to fulfill colonial functions, with ceremonial spaces, offices, and depots for colonial traders, plantation representatives, and government officials; barracks for a garrison of soldiers; and housing for colonists. As these cities grew, housing and commercial land uses were added for the local people who were drawn to the city by employment opportunities as service workers, such as servants, clerks, or porters. Examples of pure colonial cities were the original settlements of Mumbai (Bombay), Kolkata (Calcutta), Ho Chi Minh City (Saigon), Hong Kong, Macao (Figure 15), Jakarta, Manila, and Nairobi.

The other type of colonial city had colonial functions grafted onto an existing settlement to take advantage of a good site and a ready supply of labor. Examples include Mexico City, Shanghai (Figure 16), Tunis, and Delhi. In these cities the colonial imprint is most visible at the center of the city in the formal squares and public spaces, the layout of avenues, and the presence of colonial architecture and monuments. This architecture includes churches, city halls, and railway stations; the palaces of governors and archbishops; and the houses of wealthy traders, colonial administrators, and landowners.

The colonial legacy can also be seen in the planning and building regulations of many cities. Colonial planning regulations were usually the same ones used for the colonizing country itself. Because they were based on Western concepts, these regulations were often inappropriate for colonial contexts. Most colonial building codes, for example, used Western models of a small family home in a residential neighborhood at some distance from the workplace. This is at odds with the needs of large, extended families whose members work in a busy domestic economy in family businesses that are traditionally integrated with the residential

Linda McCarthy

FIGURE 15 Macao, an example of a "planted" colonial city with its Portuguese colonial architecture.

setting. Colonial planning, with its **gridiron** street layouts, **land use zoning** regulations that do not allow for mixed land uses, and building codes designed for European climates, ignored the specific needs and cultural preferences of local communities.

King[20] offered a conceptual framework for analyzing different colonial cities at different times within the context of their internal and external relationships. Because the role and function of colonial cities could vary depending on the scale of analysis, it can be helpful to use a number of spatial scales for examining colonial cities:

1. *City:* The city itself—its internal dynamics, functions, and form.

Linda McCarthy

FIGURE 16 The Bund. Shanghai had British "concession" settlement and architecture grafted onto the existing Chinese walled city.

2. *Region:* The city within the context of its immediate region—its regional production systems, trading relations, settlement patterns, transportation networks, and labor movements.
3. *Colonized Society or Territory:* The city in relation to the colonized society or territory—in terms of changes in social stratification, cultural attitudes, and so on.
4. *Metropolitan Power:* The city within the context of its interactions with the colonial metropolitan power—through trade, capital investment, and colonial policies.
5. *Colonial Empire:* The city within the context of its assigned role within the empire—as an administrative center, commercial port, or transportation hub.
6. *World Economic System:* The city and its place in the evolving world economy and global system of cities.

Internal and external factors produced a different urbanization experience for every country with a colonial history. Timing was important—colonial control was imposed at, and for very different, periods of time in various parts of the world (Figure 17). Regional differences in ethnicity and culture, kinds and levels of urbanization, social, political, and economic systems, environmental conditions, and level of technology all influenced how colonial urbanization played itself out and impacted people in these cities. The motivation for colonial expansion into a particular part of the world was also a factor.

Simon[21] theorized about the determinants and evolving nature of the relationships between the colonizers and the colonized people. He identified ten general determinants of colonial and post-colonial urban form:

1. The motives for colonization—for example, trade (mercantilism), agricultural settlement, or strategic acquisition.
2. The nature of precolonial settlement—for example, isolated villages or permanent urban centers.
3. The nature of imperial or colonial settlement—for example, imperial control required military security with little if any permanent settlement; colonialism could involve significant levels of permanent settlement.
4. Relationships between the colonizers and the indigenous population—for example, extermination (Australia and the United States), assimilation (Hispanic America after the initial conquest), or some intermediate relationship (much of Africa).
5. The presence or absence of indigenous towns—for example, where such centers did exist, they were

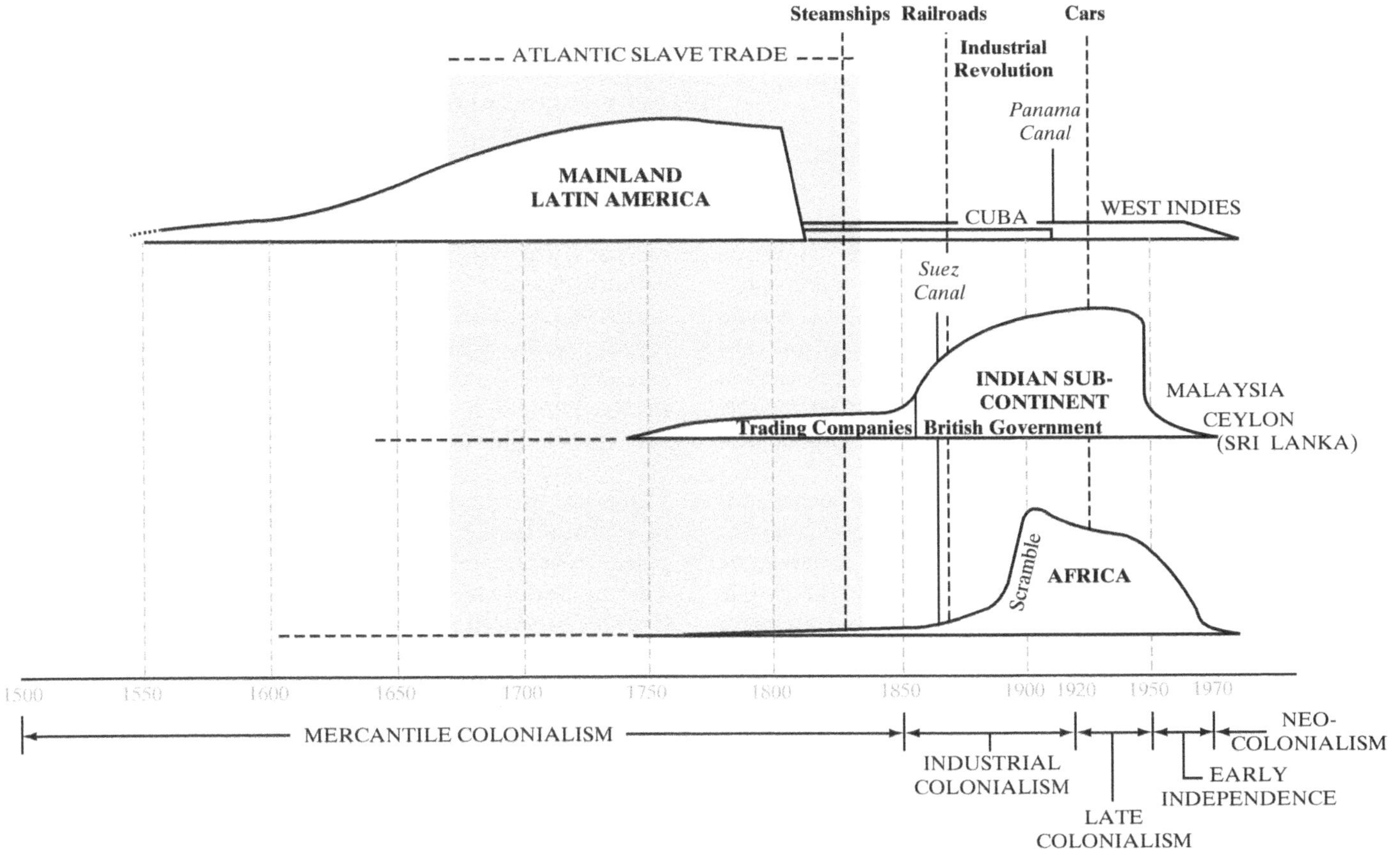

FIGURE 17 Comparative periods of colonial rule and phases of colonial urbanization.

destroyed, ignored, added to, or incorporated within a new planned city. Where there were no preexisting centers, new colonial cities were established, sometimes for the colonists alone, sometimes for colonists and indigenous people in separate sections of the city, and sometimes for all groups without formalized segregation.

6. The nature of the anticolonial struggle, the means by which independence was ultimately gained, and the degree to which the new leadership identified with existing administrative centers, especially the capital city, that symbolized both the colonial past and the achievement of liberation and independence.
7. The extent to which the ex-colonial elite retained economic dominance or were supplemented or replaced by skilled expatriates.
8. Policies pursued by the new national elite with respect to national integration, ethnic and class conflict, and the nature of the country's insertion into the world economy.
9. The functioning of the economy, including government policies affecting the various sectors: private, public, and informal sectors.
10. The extent of urban legislative change under capitalist expansion policies or some model of socialist centralization.

Although not every region had the same experience, Drakakis-Smith's sequence of five phases of colonial urbanization[22] with the addition of a phase of globalization and neoliberalism is a useful framework for examining how this process and its impacts changed over time in different parts of the world (Figure 17).

Mercantile Colonialism

Individual entrepreneurs made initial forays outside Europe in search of riches like gold and silver.[23] Later, attention turned to commodities that were valued within the European trading system, such as spices, silk, and sugar. There was no extensive overseas European settlement because mercantile colonialism was based on private companies rather than state enterprises. These companies could afford to locate only a few permanent representatives within the existing coastal centers. As a result, the local trading and collection networks were retained and incorporated into the new European trading systems. The nature of mercantile colonialism did vary in response to the different local contexts. In Latin America the earliest contacts were incredibly devastating for the people and cities of the Aztec and Inca empires. This contrasts with the experience in Asia where the Chinese Empire did not permit direct European contact in the trade of valuable commodities of Chinese origin.

As profits increased the Europeans began to establish a more extensive presence. Company representatives used troops to take control of local trading and collection networks and to protect warehouses. Later, the demand for commodities of dependable quality forced European companies to become involved in the production process itself. The various East India Companies that operated out of European states, such as England, France, and Holland, were prominent in this process.

Overall, the mercantile colonial period had only a limited impact on individual cities and systems of cities. Europeans were usually confined to small areas of existing cities that were already organized into ethnic or occupational districts. Colonial vernacular architectural styles were European in function but local in design and materials. No new urban hierarchies were created, and settlements of purely colonial origin, such as Lima, Manila, or Cape Town, were the exceptions.

By about 1800, there was diminishing European interest in overseas business activity, which resulted in a transitional period in the nature of colonial urbanization (and independence for much of Latin America). The Napoleonic wars had tied up many of the adventurers and some venture capital in Europe. The shift from trade to production had increased the cost of colonial activities for individual companies, forcing some, most notably the British, French, and Dutch East India Companies, into liquidation and government takeover. Greater profits could be made from the Industrial Revolution in Europe.

Industrial Colonialism

By the 1870s, European investment was again flowing overseas in response to the enormous demand for raw materials and food for the growing urban workforce of the Industrial Revolution in Europe. With government involvement needed to acquire territory and organize production, colonial influence began to have a profound impact on cities and urban systems in Asia and Africa.

Within colonial cities, functional and residential segregation intensified. Although manufacturing was limited to prevent competition with European exports, a large commercial and service sector served the trading and consumer needs of the colonial power. Functional specialization was based on class and ethnicity. Europeans and their institutions dominated foreign trade, expatriate non-Europeans controlled local assembly and distribution, and the indigenous people were involved only in local production, and then under expatriate supervision. This functional specialization reinforced earlier ethnic and occupational residential segregation. Many European districts were separated from non-European ones by physical barriers such as railway lines, parade grounds, police barracks, and racecourses. When combined with the existing social stratification of the indigenous communities, this produced extremely complex patterns of residential segregation.

Industrial colonialism so affected the urban systems in some Asian countries like Malaysia that new urban hierarchies were created. Africa saw a general reorientation of urban economic activity from the interior trading routes to the new coastal ports. Control of production and distribution gave these cities a crucial role in an evolving world economy and international division of labor that supported the growth of the European industrial economy in the nineteenth and early twentieth centuries. But the concentration of economic and political power in certain cities at the expense of others established the foundations for urban primacy that we see today in many less developed countries.

Late Colonialism

World War I and World War II, and an intervening economic recession in Europe, caused erratic demand for **primary products** from the colonies. In an effort to ensure continued profitability, European interests in most regions sought to improve efficiency through **economies of scale**—land reforms and mechanization—that forced out smaller producers and landholders. This fueled rapid rural-to-urban migration that generated more workers than the slower growth in domestic service and factory jobs in the cities could accommodate. A planning process dominated by Europeans allowed uncontrolled **squatter settlements** to develop; at the same time, the high point of colonial planning and architecture was producing new and redesigned European districts based on the *Garden City* concept and impressive institutional structures like city halls, universities, and banks in the downtown areas (Figure 18). Accelerated migration into colonial cities of blue- and white-collar workers attempting to escape the recession in Europe accompanied the rural-to-urban migration. This expatriate influx made it increasingly difficult for the slowly growing group of educated indigenous residents to break into middle-income occupations in either administration or commerce.

FIGURE 18 City Hall and downtown square in Cape Town, South Africa, an example of colonial architecture and urban design, with Table Mountain in the distance.

Early Independence

The 1950s and 1960s saw independence spread rapidly throughout most of Asia and Africa. After the colonial powers departed, there was an influx of indigenous people into the cities in search of jobs in the administrative and commercial sectors from which they had been excluded. During the early years of independence, these job opportunities were limited because of the continuation of European control of commercial companies and the sluggish demand for primary products from a shattered postwar Europe.

Ironically, a major problem in Europe was a shortage of unskilled labor. As a result, unemployed workers from the former **colonial cities** migrated to Western Europe in search of work. Initially, the migrants came from former colonies to former colonial powers (such as Indians to Britain or Algerians to France), but this migration soon spread to many other poor countries, especially those around the Mediterranean Sea (for example, Turkey). The workers were abundant, nonunionized, low-wage, and, being easily threatened with deportation, compliant. During the 1950s and 1960s these migrant workers represented a lucrative bonus for European industrialists and their governments and economies. The governments of the sending countries encouraged this labor migration because it helped slow urban population growth, increase foreign exchange revenues through the remittances sent home, and, they hoped, train some of their workers.

The build-up of workers in Europe was rapid and highly concentrated. By the late 1960s West Germany and France together had some six million foreign workers, concentrated in the industrial cities in the most menial jobs. The sending countries gained few benefits relative to their losses: large numbers of their younger and most trainable workers gone; few migrants receiving useful skills or training; and the remittances used mostly by the returning migrants to finance small consumer businesses in the larger cities, accentuating already serious urban problems. By the 1970s a growing recession in Europe was forcing most countries to tighten their labor immigration laws and reduce the inflow of migrant workers.

During the period of early independence, the economic situation in most less developed countries saw little improvement. Expatriate companies and the same commercial and trading relations from the colonial period continued to dominate these new countries—they exported inexpensive primary products and imported expensive manufactured ones.

In fact, since the period of early independence in many countries, extensive restructuring of the inherited colonial transportation and urban networks has not yet occurred. Many of the rail lines in Africa, for example, still have the predominantly coastal to interior emphasis that fit the exploitation needs of the colonial powers but is less suited to the functioning of an independent country. This has been the experience even in countries where new capital cities were deliberately established in the interior in an effort to promote a more balanced urban system and shed the negative associations that people had for the colonial administrative centers on the coast (for example, Dodoma replaced Dar-es-Salaam in Tanzania, Abuja replaced Lagos in Nigeria, Yamoussoukro replaced Abidjan in Côte d'Ivoire, Brasilia replaced Rio de Janeiro in Brazil, and Islamabad replaced Karachi in Pakistan).

Socially within cities, the major change during early independence was the emergence of a large segment of poor people who were unable to secure paid employment and so could not afford adequate housing, education, and healthcare for themselves and their families. These households looked inward for their survival to create an **informal sector** in which meager incomes are earned and then spent in a wide variety of illegal and quasi-legal economic activities (Figure 19).

Neocolonialism and the New International Division of Labor

During the late 1960s and early 1970s there was a dramatic change in the way that workers from the less developed countries were integrated into the world economic system. The **new international division of labor** involved **transnational corporations** from the developed countries shifting the labor-intensive parts of the production process to cities in less developed countries. Despite independence, many less developed countries are said to have experienced neocolonialism due to transnational corporations from former colonial powers taking advantage of the workers and exploiting the resources of former colonies. This was done for a number of reasons. First, the cost of production had risen in the cities of the developed countries due to the rising cost of wages, rents, and imported raw materials, combined with declining productivity and increasing environmental regulations. Second, steady rural-to-urban migration kept the cost of labor down in the cities of the less developed countries. Third, a large informal sector, representing a reserve army of workers, depressed demands for wage increases. Fourth, advances in technology allowed the separation of production from management. E-mail, satellite links, and **containerization** made it possible for the labor-intensive parts of the production process to be located in the cities of less developed countries (Figure 20) at the same time that the head offices of companies could remain in the largest cities of the developed countries. And finally, international agencies and national governments supported the new international division of labor based on the assumption that the new jobs created in the growing cities in the less developed countries could help promote economic and political stability.

The new international division of labor has had a varied and complex impact on the cities in the less developed countries. First, rapid economic growth has been highly selective. Only a relatively small number of NIEs, like South Korea and Taiwan, initially experienced rapid industrial growth. Following rising labor costs in these early NIEs, transnational corporations looked for new supplies of cheap labor, and new industrial producers emerged, including China. Second, cities have received the bulk of this **foreign direct investment (FDI)**. This has encouraged more people to join the rural-to-urban migration flow and made existing urban problems worse. Third, the effect on social class formation has been considerable. The new waged workforce is relatively conservative, while the informal sector has continued to grow, raising concerns about urban instability. Fourth, in contrast to the early years of independence, women have been incorporated into the urban workforce at unprecedented rates.

Hubertus Kanus/Photo Researchers, Inc.

FIGURE 19 A woman selling a range of everyday items is part of the informal sector of the economy in Saigon, Vietnam.

Sean Sprague/Still Pictures/Specialist

FIGURE 20 Container port in Valparaiso, Chile. Automated cranes load and unload cargo containers between the ships and the flat beds of nearby trains and trucks.

Globalization and Neoliberalism

Many of the economic and political changes in which globalization has accelerated during the past few decades have taken place in countless cities in the less developed countries based on **neoliberal ideologies and policies.**[24] A neoliberal agenda—so-called free trade, cuts in government spending on social programs, privatized public services, and reduced regulation of private companies—that became firmly established in developed countries like the United Kingdom and the United States, was soon exported to the rest of the world.

The national government's role has been crucial in the adoption of neoliberal policies in many less developed countries. In an effort to encourage economic growth by attracting foreign direct investment in what is seen as an increasingly competitive global economy, many governments borrowed heavily to modernize their cities, adding new airports, conference centers, and **free trade zones (FTZs)**, also known as **export processing zones** (EPZs)(Figure 21).

Already in debt to international banks and agencies, many governments have had to adopt economic adjustment programs dictated by the **World Bank** or the **International Monetary Fund (IMF)** as a precondition for further loans. An important part of these structural adjustments has been a reinforcement of neoliberal policies involving further reductions in government spending in areas such as social welfare. These cutbacks have impacted poor people in the cities most and have worsened the already existing shortfall in the provision of basic urban services. UN Habitat has strongly criticized the neoliberal policies of national governments and international organizations like the World Bank and IMF within the context of globalization:

> The widespread feeling of insecurity that cities and countries are "falling behind" and are in the grip of vast impersonal economic forces has provided an excuse to do nothing and to allow programmes of social redistribution and improvement to languish. In the end, the growth of inequality has happened because national governments have abdicated their responsibilities to their citizens to promote fairness, redistribution, social justice and stability in favour of a chimera of competitiveness and wealth for the few. It is also the outcome of international organizations that have adopted a dominant neo-liberal philosophy, which has failed to deliver on most of its promises almost everywhere that is has been applied.[25]

Atlantide/agefotostock

FIGURE 21 The entrance to an Export Processing Zone in Guangdong Province in southern China.

OVERURBANIZATION

A problem today for many cities in the less developed countries is a process of **overurbanization** in which cities and their populations grow more rapidly than jobs and housing.

Overurbanization and Megacities

Unprecedented rates of urbanization have been associated with the growth of **megacities**. Megacities can be distinguished by a very obvious characteristic—sheer size. Most have a population of 10 million people or more. Examples of megacities include Delhi, Mumbai (Bombay), São Paulo, Dhaka (Bangladesh), Mexico City, Kolkata (Calcutta), Shanghai, and Karachi (Pakistan) (Figure 22). All have more inhabitants than 100 of the countries that are members of the United Nations.

These megacities are usually characterized by primacy and a high degree of centrality within their national urban systems. **Primacy** and primate cities occur when the population of the largest city in an urban system is disproportionately large in relation to the second-largest and third-largest cities in that system. **Centrality** refers to the functional dominance of cities within an urban system. Cities that have a disproportionately large share of national economic, political, and cultural activities have a high degree of centrality within their urban system. This combination of size and centrality often causes megacities in different countries to have more in common with one another than with the smaller metropolitan areas and cities within their own countries.

Although most megacities do not function as **world cities**, they do perform an important intermediate role between the upper tiers of the system of world cities and the provincial towns and villages. Not only do the megacities link the local and provincial economies into the world economy, but they also represent important points of contact between the formal and informal sectors of the urban economy (Urban View 4 entitled "The Harsh Realities of Life in a Megacity").

Widespread Overurbanization

Megacities and the people in them grab a great deal of attention because of their enormous size and continued growth. But only about 1 in 8 urban residents in cities of 100,000 or more in the less developed countries live in megacities. This means that most

FIGURE 22 Major urban agglomerations of the world.

population increase will be in urban areas of 500,000 people or less. The fact that smaller cities and towns will absorb the bulk of future urban growth may be both good news and bad news. The good news is that it may be easier to respond to growing population pressures because smaller urban areas may have more flexibility in urban decision-making and territorial expansion than megacities. The bad news is that smaller towns and cities tend to have weak planning implementation capabilities and face increasing responsibilities due to a worldwide tendency to decentralize government functions from national to local governments.

This vast urban expansion in developing countries has global implications. Cities are already the locus of nearly all major economic, social, demographic and environmental transformations. What happens in cities of the less developed world in coming years will shape prospects for global economic growth, poverty alleviation, population

URBAN VIEW 4
The Harsh Realities of Life in a Megacity[26]

Adegoke Taylor, a thin, serious thirty-two-year old itinerant trader with a restless gaze, shares an 8' by 10' room in an alley with three other young men. Adegoke moved to Lagos from Ile-Oluji, a Yoruba town about a hundred miles away. He has a degree in mining from a technical college and had the dream of a professional career in the big city. After arriving in Lagos, he went to a nightclub that played juju—pop music infused with Yoruba rhythms—and got home at 2 a.m. "This experience alone makes me believe I have a new life now. Everywhere you look, you see crowds. I was motivated by that. In the village, you're not free at all, and whatever you're going to do today you'll do tomorrow," he said in English, the lingua franca of Lagos.

But it did not take long for Adegoke to realize that none of the few mining jobs that were advertized in the Lagos newspapers were going to be his. "If you are not connected, it's not easy, because there are many more applicants than jobs. The moment you don't have a recognized person saying 'This is my boy, give him a job,' it's very hard. In this country, if you don't belong to the elite, you will find things very, very hard," he said.

Adegoke ended up taking odd jobs: changing money, peddling stationery, and moving heavy loads in a warehouse for a daily wage of the equivalent of US$3. Sometimes he worked for West African traders who came to the markets near the port and needed someone to locate goods. When Adegoke first arrived in Lagos, he stayed with the sister of a childhood friend, and later found cheap accommodation in a shared room for US$7 a month until the building was burned down during ethnic riots. Having lost everything, Adegoke moved to Lagos Island where he pays a much higher rent, US$20 a month.

Adegoke tried to emigrate from Africa but was denied a visa by the U.S. and British embassies in Lagos. There are times when he longs for the tranquility of his hometown, but he never seriously considered returning to the early nights and monotonous days or the prospects of a lifetime as a manual laborer. His future is in Lagos. "There's no escape, except to make it," he says.

stabilization, environmental sustainability and, ultimately, the exercise of human rights.[27]

This close relationship between globalization and urbanization means that the traditional patterns of urban land use and spatial organization in many less developed countries are being transformed. The local outcomes include squatter settlements that often are associated with severe problems of social disorganization and environmental degradation. Nevertheless, the people in many neighborhoods have been able to develop self-help networks that have formed the basis of community within often overwhelmingly poor and crowded cities.

FOLLOW UP

Key Terms

centrality
colonial cities
containerization
Demographic Transition
dependency theory
economies of scale
free trade zones or FTZs (also known as export processing zones (EPZs)
megacities
modernization theories
newly industrializing economies (NIEs)
transnational corporation (TNC)
underdevelopment
urban bias
urbanization by implosion
world-system theory

Review Activities

1. If you have access to a library with DVDs and videos, arrange to watch Episode 3 in the *Americas* video series, produced by WGBH Boston and Central Television Enterprises for Channel 4, UK (1993). This video, *Continent on the Move: Migration and Urbanization*, examines rural-to-urban migration in Mexico within the context of the underlying processes driving this movement of people to the cities and the associated social and economic problems that have overwhelmed the resources of city governments.
2. Use the population figures in Table 1 to make a map that shows the 25 largest metropolitan regions around the world in 2025. Make a list of the most important factors that you think are responsible for the spatial distribution shown on your map. Categorize these factors into those operating mainly at a metropolitan or national scale and those that are triggered by larger global processes. Then think about some of the social, economic, political, and environmental implications of what you have identified for the urban residents and city governments of these metropolitan regions.
3. Pick a large metropolitan region with a colonial history that interests you, perhaps Lagos (Nigeria), São Paulo (Brazil), or Jakarta (Indonesia). Do some research online to find historical information that you can use to fill in some details for your metropolitan region under these headings: Mercantile Colonialism; Industrial Colonialism; Late Colonialism; Early Independence; and Neocolonialism. Think about some of the reasons why what you find out about the history of your metropolitan region might differ from that of other large metropolitan regions in other parts of the less developed world.
4. Work on your *portfolio*. You might consider spending some time going through the United Nations Human Settlements Program (UN-Habitat) website (at *http://www.unchs.org/*) to find supplementary information about some of the processes and outcomes of urbanization in the less developed countries that you are least familiar with. It might also be helpful for you to find maps and data that help you to consider how and why the urban experiences of the less developed countries has been different from those of the developed countries.

Log in to **www.mygeoscienceplace.com** for self-study quizzes, *MapMaster* layered thematic and place name interactive maps, *Urban View* Google Earth™ tours, key resources and suggested readings, related websites, "In the News" RSS feeds, and additional references and resources to enhance your study of urbanization in the less developed countries.

NOTES

1. United Nations, *World Urbanization Prospects: The 2009 Revision* (*http://esa.un.org/unpd/wup/index.htm*); *World Population Prospects: The 2008 Revision* (*http://esa.un.org/unpd/wpp2008/index.htm*). (New York: Department of Economic and Social Affairs, Population Division, 2010; 2009).
2. S. Pyne and E. German, "The Dreams of Dhaka's Garment Girls," *Global Post*, September 8, 2010, *http://www.globalpost.com/*.
3. United Nations, *World Urbanization Prospects: The 2009 Revision*.
4. M. Qadeer, "Urbanization by Implosion," *Habitat International* 28 (2004): 1–12.
5. S. Pavanello, S. Elhawary, and S. Pantuliano, *Hidden and Exposed: Urban Refugees in Nairobi, Kenya*, working paper (London, Humanitarian Policy Group, 2010).
6. B. Stark-Merklein, "In Central African Republic, Newly Settled Nomadic Children go to School," *www.unicef.org/*.
7. W. W. Rostow, *The Stages of Economic Growth: A Non-Communist Manifesto* (Cambridge, UK: Cambridge University Press, 1960).
8. G. Myrdal, *Economic Theory and Underdeveloped Regions* (London: Duckworth, 1957).
9. A. O. Hirschman, *The Strategy of Economic Development* (New Haven, Conn.: Yale University Press, 1958).
10. F. Perroux, "Note Sur la Notion de Pole de Croissance" (1955), in *Development Economics and Policy: Selected Readings*, ed. I. Livingstone, 1979 (London: Allen and Unwin, 1979), 182–87.
11. J. Friedmann, *Regional Development Policy* (Cambridge, Mass.: MIT Press, 1966).
12. M. Lipton, *Why Poor People Stay Poor: Urban Bias in World Development* (Cambridge, Mass.: Harvard University Press, 1977).
13. A. G. Frank, *Capitalism and Underdevelopment in Latin America: Historical Studies in Chile and Brazil* (New York: Monthly Review Press, 1967), 146–47.
14. Wallerstein, *The Politics of the World-Economy*.
15. See section in Chapter 1 on approaches to urban geography.
16. A. Gilbert and J. Gugler, *Cities, Poverty and Development: Urbanization in the Third World*, 2nd ed. (Oxford, UK: Oxford University Press, 1992), 44.
17. Ibid.
18. D. A. Smith, *Third World Cities in Global Perspective: The Political Economy of Uneven Urbanization* (Boulder, Colo.: Westview Press, 1996), 8.
19. Drakakis-Smith's account forms the basis of this discussion. See D. Drakakis-Smith, *Third World Cities*, 2nd ed. (London: Routledge, 2000), 46–49.
20. A. D. King, *Urbanism, Colonialism, and the World Economy: Cultural and Spatial Foundations of the World Urban System* (London: Routledge, 1990).
21. D. Simon, *Cities, Capital and Development: African Cities in the World Economy* (London: Belhaven Press, 1992), 25–33.
22. Drakakis-Smith, *Third World Cities*.
23. Myrdal, *Economic Theory and Underdeveloped Regions*.
24. United Nations Human Settlements (UN Habitat), *Global Report on Human Settlements 2003: The Challenge of Slums* (London: Earthscan, 2003).
25. Ibid., p. 53.
26. This Urban View is based on page 5 in United Nations Population Fund (UNFPA), *State of World Population 2007: Unleashing the Potential of Urban Growth*. (New York: UNFPA, 2007).
27. Ibid., 6.

Urban Form and Land Use in the Less Developed Countries

Rachael Bowes/Alamy

Urban Form and Land Use in the Less Developed Countries

A variety of historic and contemporary processes—including colonialism, rural-to-urban migration, and overurbanization—have shaped the cities of less developed countries. The patterns of land use and functional organization in cities in different regions remain somewhat distinct, reflecting variations in such factors as historical legacies, levels of technology, and environmental and cultural influences. But the local effects of globalization are transforming these historical patterns of urban land use and spatial organization. Economic and cultural globalization is generating new urban landscapes of innovation, economic development, and cultural transformation. A major problem is that the uneven diffusion of the networked infrastructures of information and communications technologies has intensified economic and social inequalities not only between urban residents in the more and less developed parts of the world, but also among city dwellers in the less developed countries themselves.

LEARNING OUTCOMES

After reading this chapter, you should be able to:

- Recognize the factors that shape the internal structure of cities around the world, and understand how and why these factors combine to create different urban forms in each world region.
- Summarize the impacts of a history of colonialism on the cities of Latin America.
- Describe the challenges faced by people living in squatter settlements on the edge of Latin American cities.
- Explain the variety of types of cities that have developed in Africa.
- Demonstrate how Islamic cities provide good examples of cultural values, economic necessities, and environmental conditions being reflected in urban form and land use.

CHAPTER PREVIEW

In the less developed countries, as in other parts of the world, competition among people for territory and location strongly shapes urban form and land use. In general, different categories of land users—commercial and industrial, as well as residential—compete for the most convenient and accessible locations in cities. This chapter examines the various patterns of urban form and land use in Latin American, African, Islamic, South Asian, Southeast Asian, and East Asian cities. We draw on some well-known descriptive urban models—simplifications of reality—to establish useful snapshots of the major differences among these regions. Of course, these rather static generalizations cannot capture all the outcomes

Since the earliest cities, the internal urban structure and behavior of city dwellers have varied from one region of the world to another in response to the influence of factors such as environmental conditions, history, and culture. In this crowded street in Old Delhi, India, for example, a cow, considered sacred by Hindus, sits undisturbed by passersby.

Julian Nieman/Alamy

FIGURE 1 Business people cross the broad pavement from the Bangkok City Tower building on Sathon Tai road in the CBD of Bangkok, Thailand.

of the historic and contemporary social, cultural, economic, political, technological, and environmental processes affecting cities and the people who live in them (Urban View 1 entitled "Life in a Haiti Tent City").

Since the earliest cities, internal urban structure has varied from one region of the world to another in response to the influence of factors such as history, culture, environmental conditions, and the different roles of individual cities within the **world-system**. Even so, the urban form and land use of many cities in the less developed countries bear the imprint of certain general demographic, cultural, economic, and political processes.

Recent demographic processes such as **rural-to-urban migration** have shaped urban form and land use within less developed countries. Overurbanization has led to a large **informal sector** of the economy, as recent migrants and other people who cannot find regularly paid work resort to earning a living in jobs that are not regulated or taxed by government. The resulting pronounced *dualism*, or juxtaposition in geographic space of the formal and informal sectors of the economy, is quite evident in the form and land use of cities.

For example, there has been a "quartering" of cities into spatially partitioned, compartmentalized residential enclaves. Luxury homes and apartment complexes correspond with an often dynamic formal sector that offers well-paid jobs and opportunities in the **central business district (CBD)** (Figure 1); these residential developments contrast sharply with the slums and **squatter settlements** of people working in the informal sector who are disadvantaged by a lack of formal education and training and often rigid **divisions of labor** shaped by gender, race, and ethnicity (Figure 2). In addition, the imprint of colonialism remains evident, often at the center of the city in the formal squares and public spaces, the layout of avenues, and the presence of colonial architecture and monuments. The colonial legacy, combined with the general processes of urban demographic, cultural, economic, and political change, continue to influence the internal structure and land use at urban and metropolitan scales within less developed countries.

Paul L. Knox

FIGURE 2 Children in a favela (squatter settlement) in Rio de Janeiro, Brazil.

URBAN VIEW 1
Life in a Haiti Tent City[1]

A rooster crows at 6 A.M. and people are waking up in the tent city across from the badly damaged presidential palace in the capital, Port-au-Prince (Figure 3). Some women and girls are already waiting in line to use the portable toilet. Many others are lining up with buckets to get some water when the water truck arrives. Some people who are still in their tents are washing their faces in small plastic washbasins.

One day blends into the next. Without jobs, most people just clean their tents, sit around, and chat. If they have any money, they buy some food from one of the street vendors who hawk brown rice, cornmeal, flour, beans, vegetables, dried fish, and salami. Too many people are adjusting to a life that has become far too permanent in tent cities that were never supposed to be anything but temporary.

The outpouring of support from around the world was immediate and generous. The 2010 magnitude 7.0 earthquake that rocked Port-au-Prince and the surrounding area killed 230,000 people, injured another 300,000, made a million more homeless, and left most of the city as rubble. But everyone agrees that, despite a great deal of effort and concern by governments and international organizations, progress in rebuilding the city and moving people into more permanent housing has been far too slow.

After the earthquake, hundreds of thousands of terrified people were moved into tent cities established by the government or NGOs and many more were forced to live in impromptu tent cities that they set up themselves in parks, soccer pitches, and schoolyards because they had nowhere else to go. Their homemade tents are so close together that, from above, they look like a patchwork quilt of plastic tarps and bed sheets. More disturbing, if that is possible, are the growing number of roofs of corrugated iron or wood—a sign that some tent cities are becoming permanent squatter settlements with time.

Some observers have asked why these people are not more angry. The disturbing explanation that some have offered is that after two centuries of poverty, corruption, and political instability, Haiti's poor had been living in such utterly desperate conditions before the earthquake that, however unpalatable to outsiders, the tent cities may actually represent an upgrade in their standard of living. As dreadful as they are, the tent cities are often a source of food rations, clean water, medicine, and some education for the children in the makeshift school tents. But as night falls, and the candles light up the tents, people gather for evening church services, desperately praying for something better for themselves and their children.

Craig Ruttle/Alamy

FIGURE 3 Some of the survivors of the 2010 magnitude 7.0 earthquake that rocked Haiti's Port-au-Prince and surrounding area had to live in a tent city across from the badly damaged presidential palace. The earthquake killed 230,000 people, injured another 300,000, made a million more homeless, and left most of the city as rubble.

PATTERNS OF URBAN FORM AND LAND USE

Latin American Cities

The internal structure of many Latin American towns today shows evidence of a colonial past—always most visible in what is now the urban core. The most numerous were the Spanish colonial towns of the seventeenth century that generally were located and planned in compliance with royal edicts—the so-called "Laws of the Indies"—which contained design features reflecting the legacy of the Roman Empire in Spain itself. This city planning legislation dictated a **gridiron** pattern of square or rectangular blocks subdivided into long narrow lots along similarly narrow streets. A central plaza (market square) was surrounded by the important buildings—the Roman Catholic church, town hall, hospital, governor's palace, and commercial and retail arcades (Figure 4). The Spanish settlers lived near the center of

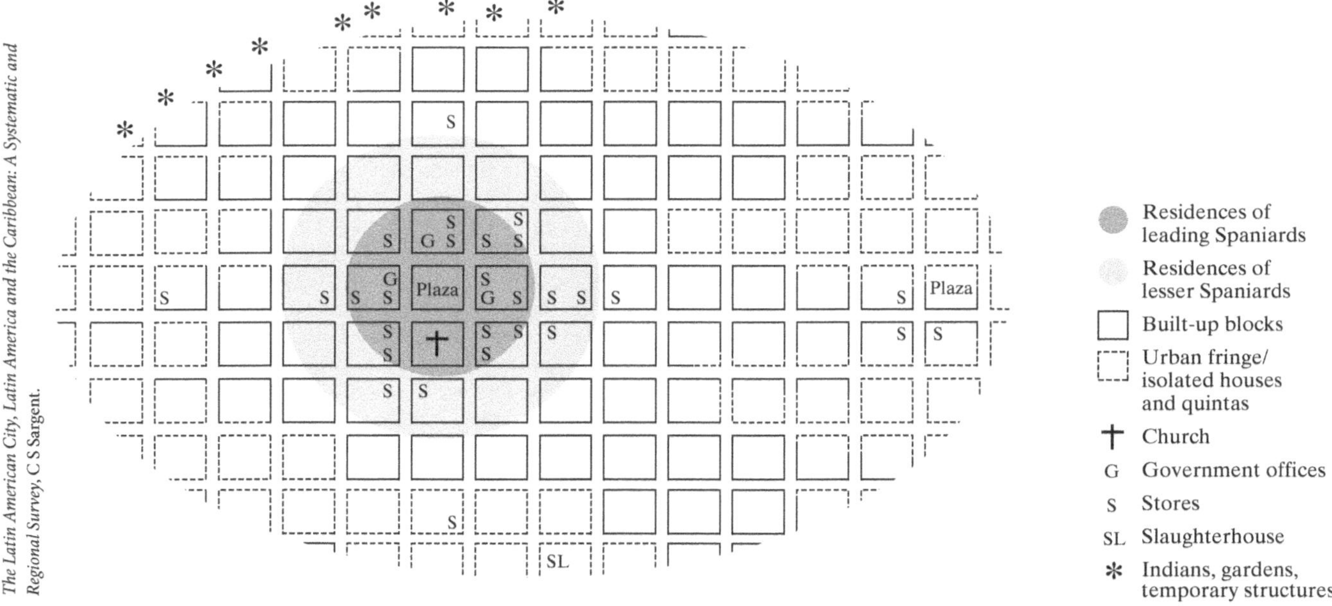

The Latin American City, Latin America and the Caribbean: A Systematic and Regional Survey, C S Sargent.

FIGURE 4 Colonial Spanish town plan conforming to the "Laws of the Indies."

the city, with the wealthy closest to the plaza and residents of middle or lower income in neighborhoods (*barrios*) farther out (Figure 5). The indigenous population was located at the periphery along with undesirable land uses such as slaughterhouses and cemeteries.

As an urban society, the Spanish used towns as social control mechanisms in the pursuit of their three goals: "God, Glory, and Gold." The Roman Catholic Church wanted to convert as many of the indigenous people as possible, and forcibly relocating them to the cities made this task easier. The towns facilitated the expansion of empire, and the indigenous people—as forced laborers in the mines and fields—could more easily be recruited and controlled by the Spanish in an urban setting because the gridiron street pattern allowed them to quickly put down uprisings.[2]

Although the original plaza remains important as a historic or tourist attraction in larger cities and might retain many of its commercial and administrative functions

Jo Stephens/Alamy

FIGURE 5 Cuzco, Peru, with the Church of La Compañia on the central Plaza de Armas.

in smaller towns and cities, contemporary Latin American cities exhibit certain additional characteristic features of urban form and land use. The model in Figure 6 shows a commercial spine and adjacent elite residential sectors surrounded by a series of encircling zones in which residential quality decreases with distance from the central business district (CBD) in what has been described as an **inverse concentric zone pattern.**[3]

The downtown comprises a CBD and a market that reflect how many Latin American downtowns have modern self-contained commercial and retail districts that are now quite separate from the more traditional mixed market districts with their small, street-oriented businesses. The dominance of the CBD reflects a road and public transportation system focused on the downtown and the presence of a large number of relatively affluent people living in central city neighborhoods. An extension of the CBD—a commercial "spine" surrounded by elite residential neighborhoods with tree-lined boulevards and parks—stretches outward along the main transportation artery. This sector has the best urban services and most of the high-end commercial locations outside the CBD: golf courses, restaurants, and office buildings. A suburban shopping center at the periphery of the commercial spine might compete with the downtown. A separate industrial sector, along a railroad or main highway, terminates in a suburban industrial park with modern factories, warehouses, and distribution facilities.

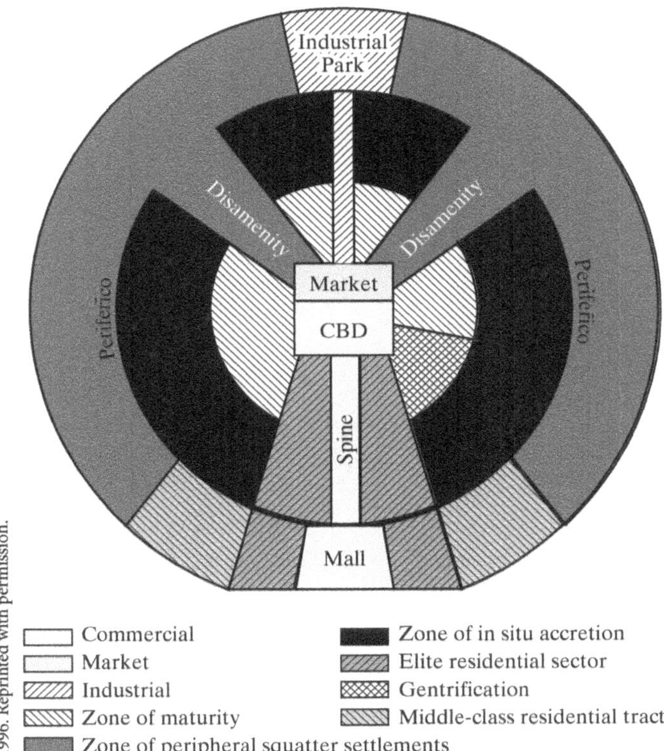

FIGURE 6 Ford's model of Latin American city structure.

"A New and Improved Model of Latin American City Structure," L Ford Ed., *The Geographical Review 86(2)*, 1996. Reprinted with permission.

A zone of maturity surrounding the CBD comprises stable middle-income neighborhoods that are well served by paved streets, lighting, public transportation, schools, water, and sewerage. Newer middle-income suburban housing tracts are located adjacent to the elite sector. There is a small area of **gentrification** in some attractive historic neighborhoods within the zone of maturity near the CBD and elite sectors. Farther out is a zone of *in situ* accretion (literally, in place addition) containing lower-income neighborhoods that show signs of becoming part of the zone of maturity. Because currency fluctuations and devaluations can reduce the value of a bank account overnight, many low-income homeowners instead spend their savings on building materials to enlarge their homes to generate rental income from a lodger. The zone is in a constant state of construction, and many houses have half-finished rooms or second stories.

Around the edge of the built-up area is a zone of peripheral squatter settlements housing impoverished recent immigrants to the city. This zone has the worst quality housing, with most people making shacks from scavenged materials such as timber and corrugated iron. The zone is almost completely without urban services. Sectors of disamenity—along polluted rivers and industrial corridors—run outward from the CBD. Squatter settlements in these sectors tie into those at the periphery. In larger cities a peripheral ring road (*periférico*) connects the shopping mall and industrial park, but development is restricted by the difficulties of expanding infrastructure and upgrading the zone of peripheral squatter settlements.

It is particularly difficult, however, to generalize about urban form and land use in contemporary Latin American cities because land use regulations are weak and often completely disregarded (Figure 7). Figure 8 shows a model of land use that combines typical commercial, residential, and industrial patterns within a broad overview that reflects the general disregard for **land use zoning** regulations in Latin America.[4] Some important features are: locations where the informal sector of the economy dominates; widely dispersed individual commercial establishments, such as small grocery stores, specialty stores (furniture, clothing, housewares), and restaurants, reflecting the fact that most people do not own a car; strip malls along major highways; industrial activities in individual factories throughout the city and low-income housing covering a large part of the city, including slums in less desirable locations near the downtown, especially near older industrial areas.

In contrast to Latin American cities farther south, many towns in northern Mexico have experienced growth stimulated by **foreign direct investment (FDI)** from U.S. companies since the early 1990s and Mexico's entry into the North American Free Trade Agreement (NAFTA) and before that as a result of the Border Industrialization (**Maquiladora**) Program. Since 1965 this program has allowed many U.S. companies to import materials and components duty free into Mexico if the goods then produced by the low-cost Mexican workers are reexported

Paul L. Knox

FIGURE 7 A view of Rio de Janeiro showing the CBD and the unregulated favelas appearing to spill down the mountainside in the left foreground of the photo.

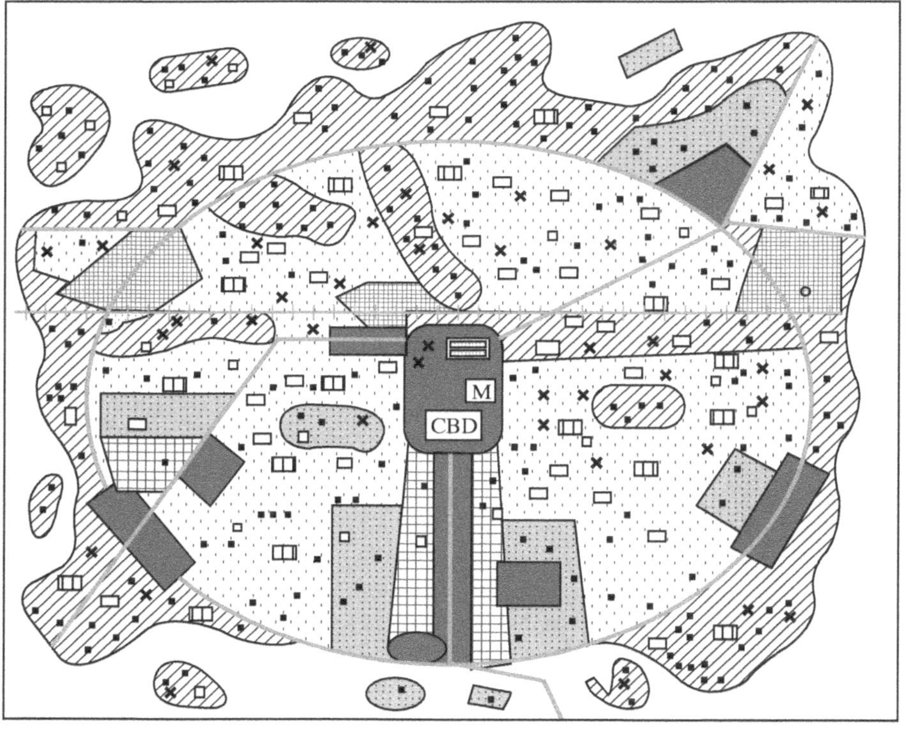

RESIDENTIAL
Elite
Middle-class
Working class
Squatters–slums

INDUSTRIAL
Zones
Individual factories
Small plants

COMMERCIAL
CBD, spine, strips & centers
M Market
Informal economy
Individual stores
Mom and pop
Arterial
Rail line

"Order and Disorder—A Model of Latin American Urban Land Use." Fig. 8, p.28, W Crowley. *Yearbook of the Association of Pacific Coast Geographers 57*. Used with permission.

FIGURE 8 Crowley's model of major land uses in Latin American cities.

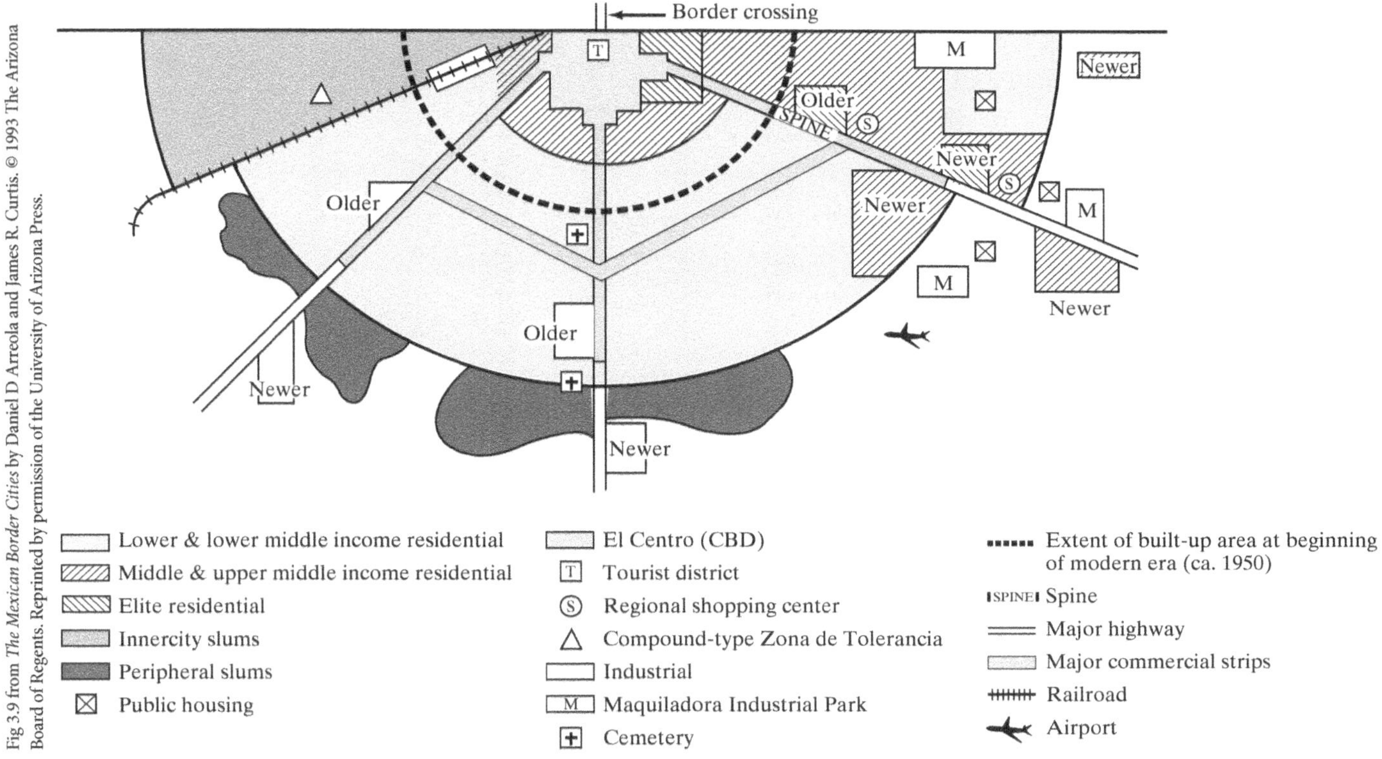

FIGURE 9 Arreola and Curtis's model of urban structure for Mexican border cities.

to the United States. This imprint of **globalization** is captured in Arreola and Curtis's model of the Mexican border city shown in Figure 9.[5] Note some characteristic features: the pre-1950 "traditional" urban core is truncated along its northern edge by the international border; the CBD contains a small tourist district; and, the maquiladoras are close to the U.S. border and the airport.

A distinguishing feature of recent urbanization in parts of Latin America has been the extension of some **megacities** beyond their city and metropolitan boundaries to form **extended metropolitan regions (EMRs)**.[6] In Brazil "polygonized development" reflects a relative decline in the formerly overwhelming dominance of the city of São Paulo as a result of a relatively "concentrated decentralization" pattern of new economic activity that is locally concentrated in surrounding smaller cities and towns (Figure 10).[7] New economic activity has been concentrated in the geographical polygon that now surrounds São Paulo and encompasses Belo Horizonte, Uberlândia, Londrina, Maringá, Porto Alegre, Florianópolis, and São José dos Campos. The main **growth poles** of high technology in Brazil have developed within this new industrial polygon.

A number of factors have led to this expansion of São Paulo into an extended metropolitan region and have worked against a broader deconcentration of new investment throughout Brazil. These include: **agglomeration diseconomies** in the metropolitan area of São Paulo and growing **agglomeration economies** in nearby urban centers; government intervention in the region surrounding São Paulo, including public investment in infrastructure and government subsidies for companies; and the extraordinary regional concentration of income and research resources in São Paulo that has restricted development to the immediate region.

African Cities

Although there have been attempts to create a generalized model of the urban structure of African cities,[8] the varied histories and rapid changes in urban structure since independence in African countries make it more realistic to think in terms of six categories of African cities.[9]

1. *The indigenous (native) city:* Most towns established before major European contact were administrative centers of indigenous power that also had craft and trade functions (Figure 11). These indigenous cities include the Yoruba cities of southwest Nigeria that date back to at least the tenth century. The larger towns were walled and contained central palaces and courtyards. Some, like Ibadan, had populations exceeding 50,000 people by the time Europeans arrived and were profoundly influenced by colonial rule. Farther east, Addis Ababa grew to become the largest indigenous city after Emperor Menelik made it his capital in 1886; although Ethiopia

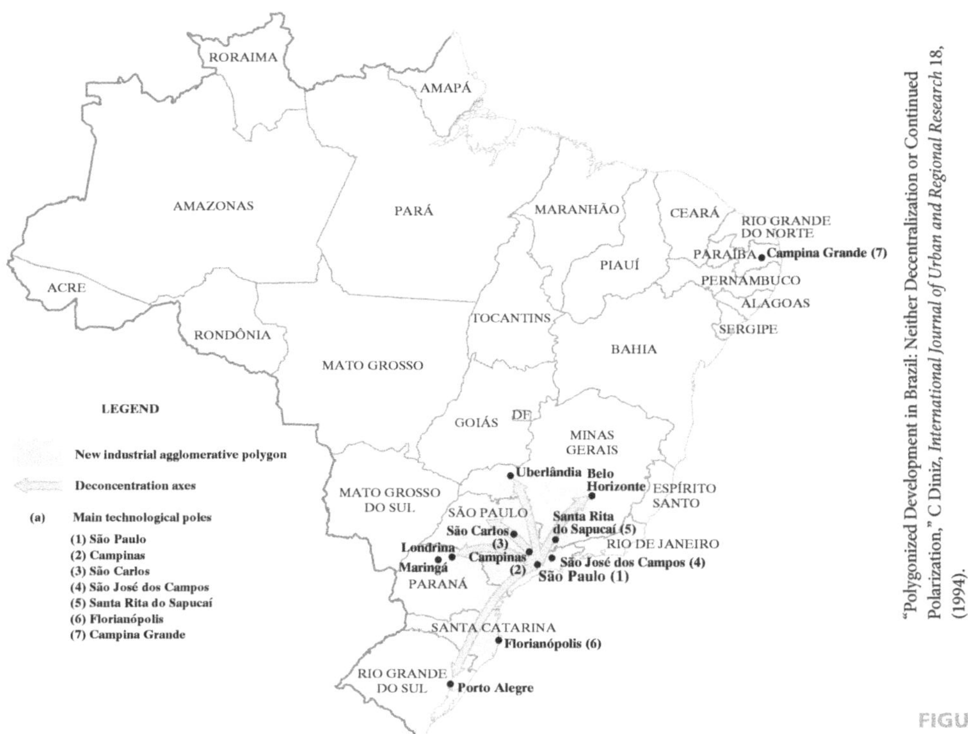

"Polygonized Development in Brazil: Neither Decentralization or Continued Polarization," C Diniz, *International Journal of Urban and Regional Research* 18, (1994).

FIGURE 10 São Paulo's extended metropolitan region.

retained its political independence, the city was still influenced significantly by European contact.

2. *The Islamic city:* Islamic cities in Africa—including Kano in predominantly Muslim northern Nigeria, Dar-es-Salaam in Tanzania, and Merca in Somalia—were founded by African people or invaders who brought Islamic influences (Figure 12). These cities flourished as the capitals of empires, religious centers, and marketplaces at the end of long-distance trans-Saharan caravan routes. They contained features characteristic of Islamic cities in the Middle East and other parts of Asia, including the main covered bazaars or street markets (*suqs*), mosques, a citadel (fortress), and public baths.
3. *The colonial (administrative) city:* Many cities in Africa date from the late nineteenth and early twentieth centuries and are colonial in origin. These cities were founded for administrative and trade purposes rather than as settlements for Europeans. Many, such as Dakar (Senegal) and Freetown (Sierra Leone) (Figure 13), were ports that served as the main points of contact between the colonial powers and the local people. The colonial transportation

Robert Harding Picture Library Ltd/Alamy

FIGURE 11 Harar, Ethiopia, an example of an indigenous African city, is the fourth holiest city of Islam, and contains 82 mosques. This market is at the entrance to the Shoa Gate, one of six gates in the sixteenth century walls of this city.

Saurabh Das/AP Photo

FIGURE 12 The Central Mosque of Kano, Nigeria, the oldest city in West Africa, and in the foreground, horsemen charging to show off their riding skills during the annual Durbar festival.

network of road and railway lines connected the interior regions with these coastal trade centers; this allowed the Europeans to maintain political and military control and to extract minerals and agricultural produce from the interior areas for export. These cities were characterized by quite sharp functional and residential segregation that was imposed by the colonial powers.

4. *The "European" city:* The "European" city is a special category of **colonial city**. It is a true "colonial" city in the original sense of a colony as a place of permanent new settlement. Cities such as Harare (formerly Salisbury) in Zimbabwe (Figure 14), Lusaka in Zambia, and Nairobi in Kenya were established primarily as places for Europeans to live and to provide urban services for permanent European settlers in the surrounding rural areas. They were designed as replicas of towns in Europe and reflected European town planning ideas. Although their main functions were administration and trade, manufacturing to meet the needs of the European settlers was allowed. The Europeans imposed such sharp functional and residential segregation that the Africans who lived and worked in these cities were viewed as temporary residents. The most extreme expression of the "European" city was the apartheid city of South Africa (the Urban View 2 entitled "Fighting Racial Discrimination in South African Cities with Soccer?").
5. *The dual city:* The dual city represents a combination of two or more of the indigenous, Islamic, colonial, or "European" categories of cities. Kano, for example, combines both an ancient walled Muslim city surrounded by a new area of colonial origin. Greater Khartoum comprises the indigenous city of Omdurman on the west bank of the Nile and the colonial city of Khartoum on the opposite bank. These dual cities were interdependent but physically separate—each had a full range of urban functions and each developed with a certain degree of independence.

Gary Schulze

FIGURE 13 The port of Freetown, Sierra Leone, an example of a colonial (administrative) city

National Archives Salisbury

FIGURE 14 Harare (formerly Salisbury), Zimbabwe is an example of a "European" colonial city. This is a photo of Stanley Avenue in the 1930s looking east.

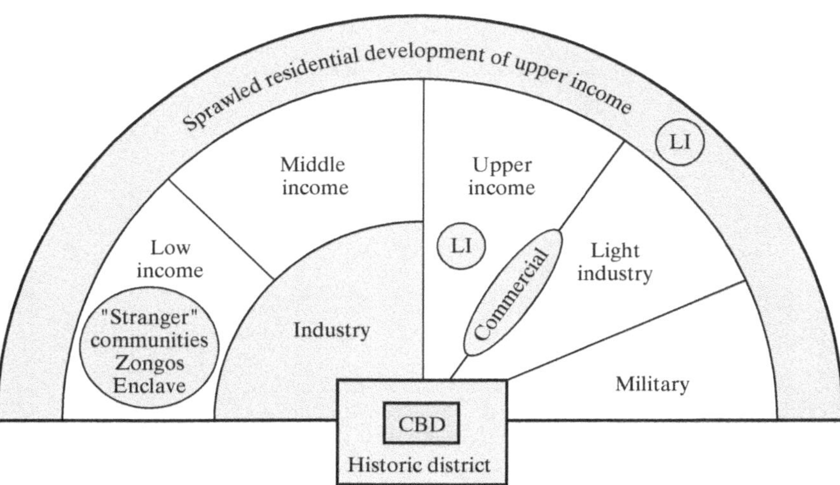

FIGURE 15 A model of a Sub-Saharan African city.

Aryeetey-Attoh, Samuel A, *Geography of Sub-Saharan Africa*, 2nd, © 2003. Printed and Electronically reproduced by permission of Pearson Education, Inc., Upper Saddle River, New Jersey.

6. *The hybrid city:* Indigenous and foreign elements are integrated in the hybrid city. Since independence, most African cities have become hybrids as their indigenous, Islamic, colonial, and "European" elements have become more integrated. Accra, Ghana, is a good example of this kind of city.[10] It was an indigenous city, founded prior to the sixteenth century, that became a colonial administrative center in the late nineteenth century. It combines indigenous, colonial, and modern development. Aryeetey-Attoh's model of the internal structure of Accra shows an irregular pattern of ethnic and religious enclaves superimposed on residential sectors of different income level (Figure 15). Unplanned peripheral sprawl of high-income development accompanied Accra's rapid expansion. This recent trend contrasts with earlier descriptions of an inverse concentric zone pattern for African cities in which the wealthy and middle-income groups congregated close to the government center to avoid the hazards of a time-consuming commute, while the poor occupied the peripheral locations.

A characteristic of many African cities is the existence, within the same municipal boundary, of two enclaves: a poor highly marginalized population and a wealthy expatriate elite.[11] The urban elite is connected to the international economy; poorer people are tied into the informal sector of the urban economy. Greater integration is difficult because the African formal sector has not been able to disengage from its unequal political, economic, and cultural links with other parts of the world and make urbanization work for the benefit of African cities and their residents. It is difficult for people in the informal sector to make the transition to the more prosperous formal sector of the urban economy. Modern technology, which is relatively easily available to those in the formal sector, is virtually inaccessible to most people in the informal sector. This situation presents particular problems for cities that are experiencing **overurbanization**. These include Sub-Saharan Africa's only megacity, Lagos (Nigeria), and other rapidly growing **primate cities**, such as Nairobi (Kenya) (Figure 16), Addis Ababa (Ethiopia), and Harare (Zimbabwe).

Islamic Cities

Islamic cities are found in the Arabian Peninsula and the Middle East—the heartland of the Islamic Empire under the prophet Muhammad (C.E. 570–632)—and in those regions into which Islam spread, including parts of Africa, South-Central Asia, and Indonesia. Many elements of the traditional Islamic city can be found as far away as Seville, Granada, and Córdoba in southern Spain; Kano in northern Nigeria and Dar-es-Salaam in Tanzania; and Davao in the Philippines.

Islamic cities are good examples of how cultural values (Islam), economic necessities (trade), and environmental conditions (dry climate) are reflected in the form and land use of

Images of Africa Photobank/Alamy

FIGURE 16 Nairobi, the capital of Kenya, a rapidly growing African primate city whose skyline is visible in the distance as wildebeest cross a dirt road in Nairobi National Park.

URBAN VIEW 2

Fighting Racial Discrimination in South African Cities with Soccer?

When it was announced that South Africa had been chosen to host the 2010 World Cup, people celebrated and Nelson Mandela could not hold back the tears. "The World Cup will help unify people. If there is one thing in this planet that has the power to bind people, it is soccer," he said. During his 27 years in prison for his anti-apartheid activism, he played soccer with other prisoners in the prison yard.[12]

Apartheid was introduced by the white minority in South Africa with the explicit goal of imposing separate racial development in the "European" city. The 1950 Group Areas Act required the spatial separation of the officially defined racial groups (Whites, Indians, Coloureds, and Blacks). Different areas—separated by clear natural divides such as greenbelts or physical barriers such as railways, roadways, and industrial strips or, if necessary, buffer zones of vacant land—were designated for each group (Figure 17). Ten "homelands" (*Bantustans*) comprising only about 14 percent of the area of the country—mostly barren land—were designated for black people, who comprised 70 percent of the population. The 1952 "Pass Laws" gave black males temporary work permits that allowed them to work in the cities only as needed.

The white minority used apartheid to control the black majority, using fundamentally geographical means to achieve the spatial separation of races within cities, within the countryside, within buildings, in employment, and by marriage. For nearly 40 years, one sport—soccer—thwarted this system and perhaps even foreshadowed the end of apartheid.[13] As Leepile Taunyane, life president of South Africa's Premier Soccer League said: "Soccer played a very crucial role as a form of resistance, never yielding to being divided by government policy." Although blacks were legally limited in where they could go within cities, there were white people who helped black teams find somewhere to play. This was one of the few examples of the beginning of a process that established a foundation for the racial integration that would come with the end of apartheid. Later, in the 1960s, the top white teams wanted to play against black teams to test their abilities. By the late 1970s, as black teams drew more fans, some major South African corporations sponsored matches.

The international pressure that contributed to the end of apartheid included sports boycotts (that prevented South African soccer players from representing their country internationally), a forced withdrawal from the British Commonwealth, trade sanctions, and voluntary investment bans by major transnational corporations. A number of white South Africans were also vocal in their opposition to the system. The early 1990s saw the end of apartheid in South Africa. Nelson Mandela was freed from jail, and President F. W. de Klerk agreed to political power sharing. In 1994, South Africa held the first election in its history in which blacks were allowed to vote. Nelson Mandela was elected the first black president.

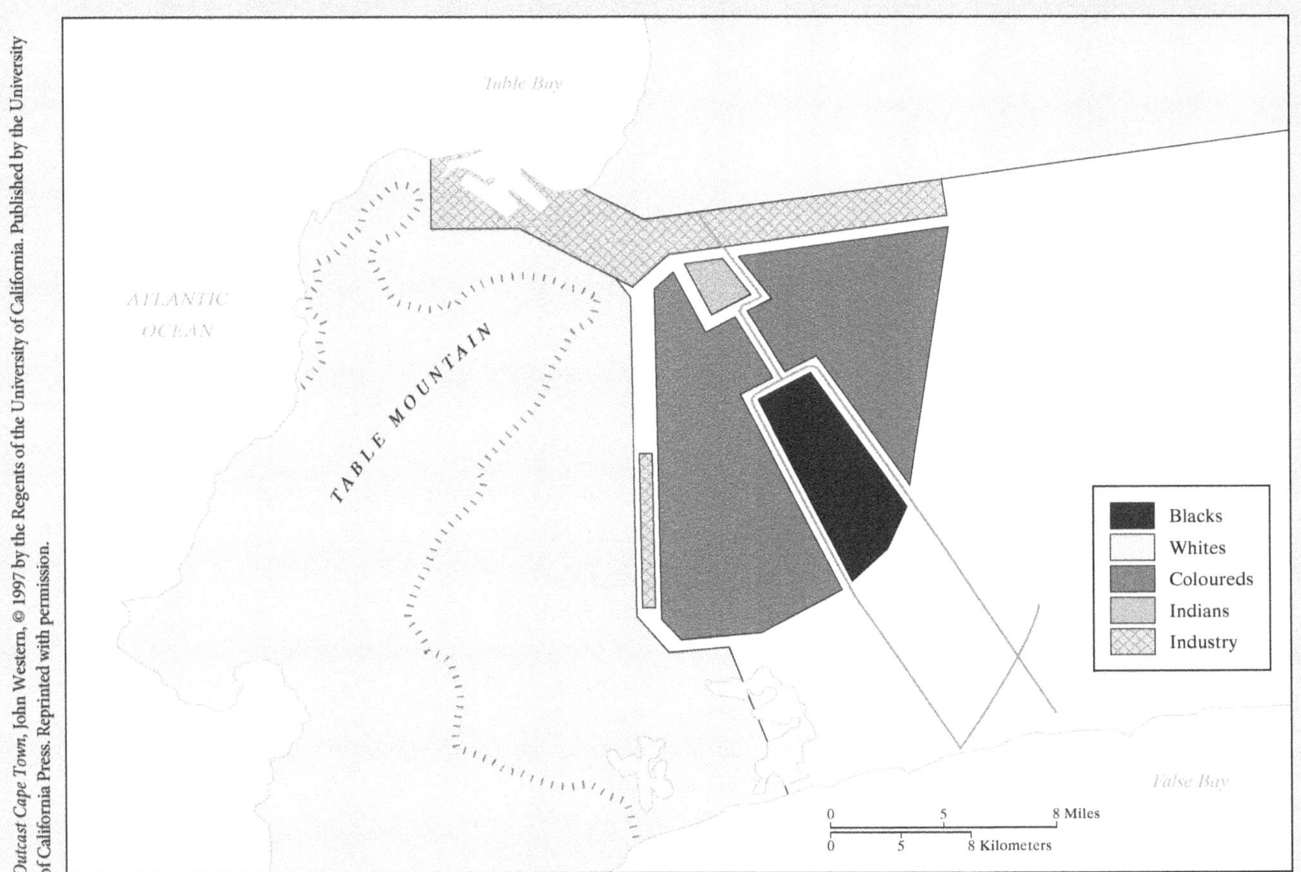

FIGURE 17 Cape Town apartheid city plan.

(continued)

URBAN VIEW 2
Fighting Racial Discrimination in South African Cities with Soccer? (*continued*)

Despite efforts to promote greater integration, the shadow of apartheid planning is expected to remain visible in the geography of South African cities for decades.[14] The white elite still occupy the affluent neighborhoods, while vast numbers of poor blacks live in squatter settlements. Much progress remains to be made in addressing apartheid's legacy and in reducing the deep social, economic, and spatial divisions.

Some observers are optimistic about South Africa's urban future—that planners and politicians can work closely with the communities they serve, attempting to meet their urban services and other needs despite limited public resources. This is why some of the government's beautification strategies in preparation for the World Cup have been so troubling. In Cape Town, massive slum clearance that displaced thousands of poor people and the building of the N2 gateway housing project as "beautiful formal housing opportunities" have been a distressing reminder of the forced removals during apartheid. The locations of eviction and gentrification in the run up to the World Cup (Figure 18) mirror the locations of the coloured and black sectors of Cape Town during apartheid (Figure 17).[15]

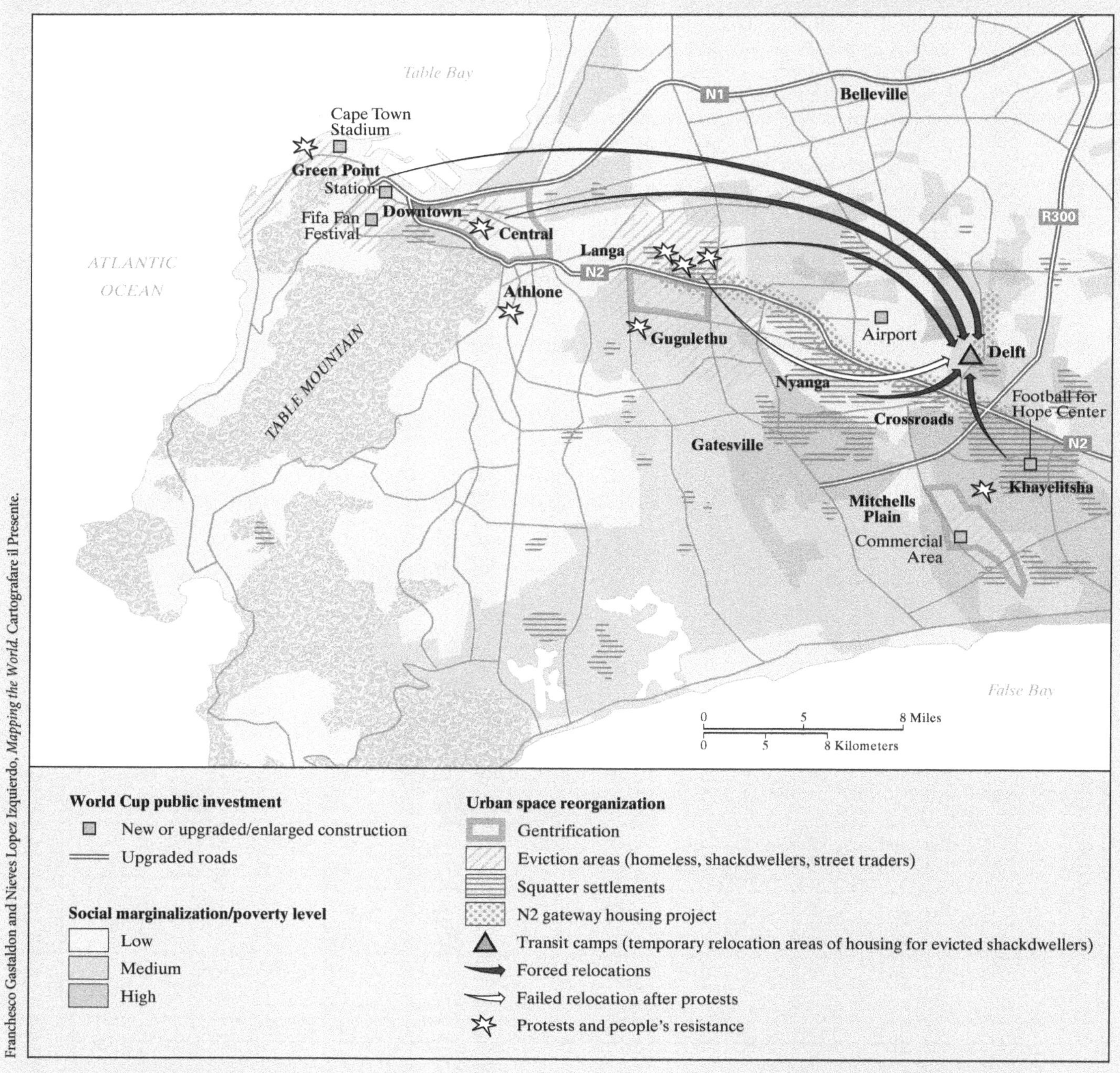

FIGURE 18 Cape Town: World Cup for tourists.

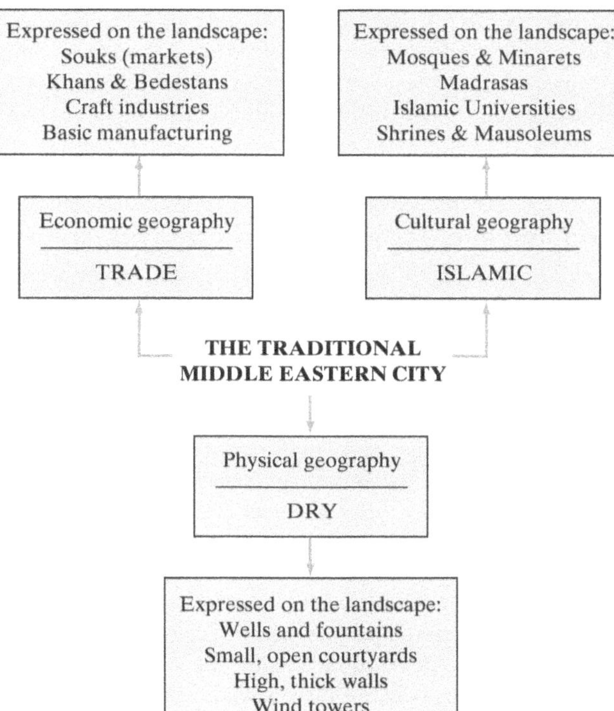

D. J. Zeigler, "Cities of the Greater Middle East," in: S. D. Brunn et al. *Cities of the World: World Regional Urban Development*, 3rd ed., Lanham, Md.: Rowman & Littlefield, 2003, Fig. 7.2, p. 257. Reproduced with permission.

FIGURE 19 The traditional Middle Eastern city.

John Warburton-Lee Photography/Alamy

FIGURE 20 Privacy concerns and climatic conditions are given urban form in the maze of narrow streets in many Islamic cities like Zanzibar, Tanzania.

cities (Figure 19). The fundamentals of the layout and design of the traditional Islamic city are so intimately connected with Islamic cultural values that they are found in the *Qur'an* (the Koran), the holy book of Islam. Although urban growth does not have to conform to any master plan, certain basic principles are intended to ensure Islam's emphasis on personal privacy and virtue, on communal well-being, and on the inner essence of things, rather than on outward appearance.

Privacy is central in the construction of the Islamic city. According to Islamic values, women must be protected from the gaze of unrelated men. Traditionally, doors must not face each other across a minor street, and windows must be small, narrow, and above eye level. Dead-end streets are used to restrict the number of people approaching the homes, and angled entrances prevent intrusive glances (Figure 20). Larger homes are built around courtyards that provide an interior private focus for domestic life (Figure 21).

The rights of others are given strong emphasis. The Qur'an specifies an obligation to neighborly consideration and cooperation that traditionally is interpreted as applying to a minimum radius of 40 houses. In traditional designs parapets surround the roofs to prevent views of the homes of neighbors and drainage channels are steered away from adjacent houses. Because most Islamic cities are in hot, dry climates, these basic principles of urban design have evolved in conjunction with certain practical responses to intense heat and sunlight. Narrow winding streets help to maximize shade, as do latticework on windows, wind towers that draw air into houses to cool them, and the courtyard design of residential areas (Figures 20 and 21). The overall effect is a compact, cellular urban structure within which it is possible to maintain a high degree of privacy.

The traditional Islamic city grew up at desert oasis sites, and prospered from trade connections. The model of the internal structure of the Islamic metropolis in Figure 22 reflects some important features.[16] At the heart of the city is the walled *Kasbah* or citadel (fortress). Its often ornate gates open onto the palace buildings, baths, barracks, small mosque, and shops. Surrounding the Kasbah is the *medina* or old city. The medina usually had a wall with several watchtowers for defense. City gates controlled access, allowing strangers to be scrutinized and merchants to be taxed.

Until the eleventh century, the absence of corporate bodies in a society composed of rulers and subjects meant that there was no need for public buildings and communal meeting places.[17] After that time mosques began to proliferate, and the *Jami*—the city's principal mosque—became the dominant feature of the traditional Islamic city. Located centrally, the mosque complex is not only a center of worship, but also a hub for education and a broad range of welfare functions. As cities grew, new, smaller mosques were built toward the periphery, each out of earshot of the call to prayer of each other.

The covered bazaars or street markets (*suqs*) nearest the Jami typically specialize in the cleanest and most prestigious goods, such as books, perfumes, prayer mats, and modern

Aivar Mikko/Alamy

FIGURE 21 The central courtyards of larger homes in Islamic cities were designed to provide privacy and shade for the family, as in this historic home in Cordoba, Spain.

consumer goods. Those nearer the city gates specialize in bulkier and less valuable goods such as basic foodstuffs, building materials, textiles, leather goods, and pots and pans. Within the suqs, each profession and line of business has its own alley.

The residential population in the old city tended to be concentrated in distinctive quarters or *ahya'*, many of which had gates of their own. There were different quarters for workers with similar occupations, for Jews, Europeans, Christian sects, and ethnic groups, and for people of different village, tribal, or regional origin. Both rich and poor lived together in each quarter.

In the colonial era a new city developed outside the old city walls (Figure 22). The traditional elements of urban form—mosques, public baths, and small shops—were incorporated but augmented with the modern amenities and architectural styles of the colonizers—new government buildings, wide boulevards, hotels, department stores, and corporate offices. Like cities everywhere, contemporary Islamic cities bear the imprint of globalization despite the fact that Islamic culture is self-consciously resistant to many aspects of Western-based culture. The new city that developed outside the old city walls during the colonial era became surrounded by the modern, postcolonial city (Figure 22); as this happened, courtyard homes all but disappeared, their place taken by apartment blocks, with high-income and single-family units increasingly common. The postcolonial city became the zone of international hotels, department chain stores, skyscrapers and office blocks, universities, factories, and highways. The postcolonial city has also become the location for squatter settlements that are now home to recent rural immigrants.

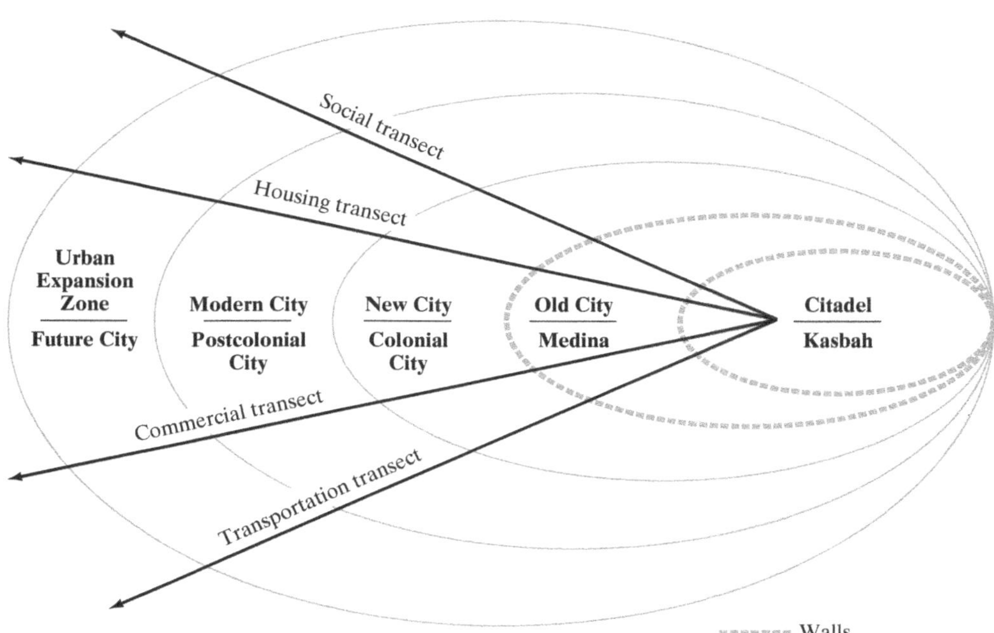

D. J. Zeigler, "Cities of the Greater Middle East," in: S. D. Brunn et al. *Cities of the World: World Regional Urban Development*, 3rd ed., Lanham, Md.: Rowman & Littlefield, 2003, Fig. 7.7, p. 269. Reproduced with permission.

FIGURE 22 Internal structure of the Middle Eastern metropolis.

An urban expansion zone in which small villages are undergoing urbanization has grown up beyond the modern city (Figure 22). New industrial estates, an international airport, modest new housing tracts, or, in the case of countries such as Egypt and Saudi Arabia, new cities, are being added. In the richest countries experiencing growing car ownership, this can also be a zone of uncontrolled urban sprawl.

The contemporary Islamic city has become an interlocking set of zones—the citadel, old city, new city, modern city, and urban expansion zone—that show changes in urban form and function that are patterned in time. A series of different transects extend out from the center of the city through each of these zones (Figure 22).[18]

1. *Social Transect:* The old city has become socially marginalized as wealthy residents have moved out. The modern city is attracting most of the new housing investment and the best urban services. Even foreign tourists stay in hotels in the modern city and make only brief visits to the old city in air-conditioned buses.
2. *Housing Transect:* The old city usually comprises traditional two-story courtyard houses. The modern city contains multistory apartment blocks—now the most characteristic feature of the residential landscape of the contemporary Islamic city.
3. *Commercial Transect:* The old city still contains suqs, traditional industries, and small-scale, family-owned craft businesses. The modern city offers department stores as well as international and national franchise businesses.
4. *Transportation Transect:* The old city contains narrow congested streets of pedestrians, donkey carts, and taxis. The new city reflects the proliferation of private cars and associated amenities, such as gas stations and parking spaces.

A feature of recent urbanization has been the entirely new group of Islamic cities that has grown up around the modern oil and gas fields. These are best exemplified by the cities of the United Arab Emirates, Kuwait, Saudi Arabia, Bahrain, and Qatar (Urban View 3 entitled "Towering Ambition in Persian Gulf Cities and the Global Economic Downturn").

Another characteristic of recent urbanization is the development of increasingly integrated urban regions containing

URBAN VIEW 3

Towering Ambition in Persian Gulf Cities and the Global Economic Downturn

An entirely new group of cities has grown up since the 1960s around the oil and natural gas fields of countries such as the United Arab Emirates, Kuwait, Saudi Arabia, Bahrain, and Qatar.[19] Petrodollars financed this city building. These cities often comprise two zones: a tiny historic core, anchored by a fort and perhaps what remains of a traditional wharf area, whose layout and architectural style provide a source of local identity; surrounding the core is a modern city of glittering high-rise offices, expensive shopping malls, exclusive apartment complexes, sprawling suburbs, and mosques.[20]

Oil wealth has paid for the world's star architects—"starchitects" such as Frank Gehry and Jean Nouvel—to design cities and buildings that blend modern urban design and architecture with traditional themes. The world's tallest skyscraper was officially opened in 2010 in Dubai in the United Arab Emirates. At 160 stories, it can be seen from as far away as 60 miles (Figure 23).[21] But the celebrations and incredible high-tech fireworks display could not disguise the impact of the global economic downturn and property market collapse for Dubai, whose oil wealth is much less than neighboring Abu Dhabi's. The announcement that the building will be known as the Burj Khalifa in honor of the President of the United Arab Emirates and head of Abu Dhabi's ruling family, rather than the Burj Dubai, reflects Abu Dhabi's multi-billion bail-out loans to Dubai.

Until the global recession, the economic magnetism of cities like Dubai attracted unskilled workers from countries such as India, Pakistan, and Bangladesh, and skilled workers

Andre Maslennikov/Still Pictures/Specialist Stock

FIGURE 23 The world's tallest skyscraper, the Burj Khalifa, was opened in 2010 in Dubai, United Arab Emirates. At 160 stories, it can be seen from as far away as 60 miles.

(continued)

URBAN VIEW 3
Towering Ambition in Persian Gulf Cities and the Global Economic Downturn (*continued*)

from Europe, the United States, and other parts of the Arab world. There are many accounts of the exploitation of the unskilled workers and their lives in the overcrowded make-shift camps an hour outside the city.[22] The common story is of young men who were recruited by an employment agent who came to their village and signed them up to work on the construction sites in cities like Dubai. The agent told the young men that they could earn $650 each month for an 8-hour workday. The young men sold their family land and took out loans from local lenders to pay the $3,700 fee for a work visa, which they expected to easily pay off in just six months. After they landed at the airport, their passports were taken by the construction company and they were told that they would be working 14 hour days for $150 a month. Without a passport and money, they could not return home to face their family who depended on them or the local moneylender who would have them thrown in prison. They stayed to work and pay off their loans, but the property market crash has meant that construction companies have slashed jobs and now these workers have no choice but to return home.

Persian Gulf cities are also associated with serious environmental challenges.[23] A surprising and very visible feature of desert cities like Dubai is how much green—lawns, parks, golf courses—there is. Figuratively, the oil is turned into water and the water into green space. But in a desert, the water comes from processing seawater in huge desalination plants, which makes it the most costly water—economically and environmentally—on earth. Dubai has also expanded so quickly that its sewage treatment plants have not been able to keep pace and its tourist beaches are becoming polluted.

Dubai's neighbor, Abu Dhabi, the oil-rich capital of the UAE, is generally viewed as having taken a more cautious approach that has involved diversifying into cultural and environmental sustainability areas. The cultural strategy includes Saadiyat Island that offers high-end commercial, residential, and tourist developments, including a branch of the Louvre and Guggenheim museums. The environmental strategy includes Masdar City, which is designed as the world's first carbon neutral eco-city—quite a contrast to the Burj Khalifa in Dubai.

a number of Islamic cities. These **megalopolises** include one that stretches along a new superhighway across the desert from Cairo to Alexandria and another, anchored by Istanbul and Bursa, that forms Turkey's Marmara megalopolis.

South Asian Cities

Two main forces—colonial and traditional—have influenced the urban form of South Asian cities.[24]

The *colonial-based* city model incorporates land use patterns found in colonial cities in other parts of the world but also contains certain regional characteristics that reflect the colonial functions specific to the Indian subcontinent (Figure 24). The development of colonial cities in South Asia typically involved certain characteristics. The port facility often became the nucleus of the colonial city. The need for trade and military reinforcements required a waterfront location accessible to oceangoing ships. A walled fort nearby functioned as a military outpost. If the port contained factories that processed agricultural raw materials for export, this formed the initial basis for colonial trade. Beyond the fort, the native town—overcrowded, unplanned, inadequately serviced—housed the local workers who serviced the fort and the colonial administration.

A nearby Western-style CBD contained the government, administrative, commercial, retail, and public buildings. The European town grew in a different area or direction from the native town. It was low-density with spacious bungalows, elegant apartment houses, planned boulevards, and good urban services and recreational facilities. Between the fort and European town, an extensive open space (*maidan*) was reserved for military parades and European recreational activities, such

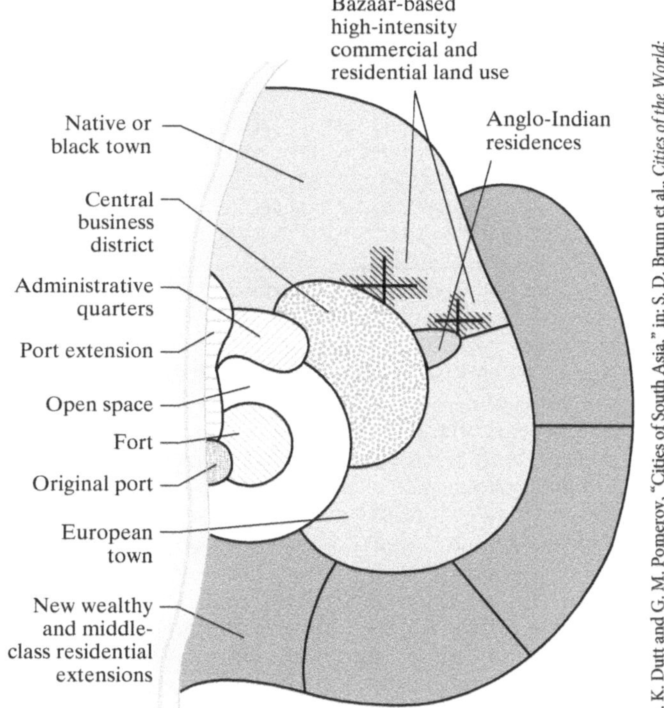

FIGURE 24 The colonial-based city in South Asia.

as cricket and horse racing. Between the native and European town, Anglo-Indian colonies contained the Christian offspring of European-Indian marriages who were never fully accepted by either community. Since independence the colonial city has been extended into reclaimed lowland or undeveloped areas to create new living space, particularly for the native elite.

Bazaar-based cities were part of a thriving **urban system** in South Asia before colonialism, and they contain features that predate the colonial period (Figure 25). Their urban form today reflects their roles as early centers of trade and commerce, administration, or religious pilgrimage. Typically there is a retail business concentration at the main crossroads. Around this intersection—known as a *chowk* in northern India—are the houses of the wealthy and the merchants who live above or behind their shops and warehouses.

The bazaar, or city center, contains a variety of land uses, although commercial activities dominate. With the bulk of family income spent on basic necessities—food, clothing, and shelter—most streets contain retail outlets for foodstuffs and clothes; sidewalk vendors are everywhere. As the bazaar evolved, functional separation of retail and wholesale businesses occurred. There are specific areas for retail traders—textile shops and tailors, grain and bread shops, jewelry stores and pawnshops, and vendors of perishable goods (vegetables, meat, and fish) that people buy daily because most do not own a refrigerator. Public or nonprofit inns provide inexpensive accommodation in smaller towns; hotel accommodation in the larger towns and cities reflects Western influence.

Surrounding this inner core is a zone of wealthy residences, with separate housing for domestic servants, cleaners, shop assistants, and porters. The homes of the poor comprise a third zone, beyond which are the native elite and middle-income groups who settled in what had been the "civil lines" in the colonial period—zones containing civil functions, such as the courthouse, police headquarters, jail, hospital, and public library. Ethnic, religious, linguistic, and caste neighborhoods formed in specific areas depending on the timing of settlement and the availability of land; the **Dalits** ("Untouchables") always live at the periphery of the city, the site of most squatter settlements. As in the colonial cities, many traditional bazaar-based cities were extended as the city grew. Governmental, semigovernmental, cooperative, and private agencies built planned high-income developments, especially for the native elite.

The *planned city* represents a third category of South Asian urban form. Ancient cities that were planned, such as Mohenjo Daro and Harappa, did not survive. Other planned cities from the precolonial, colonial, and independence periods formed the basis for later development. An example of a precolonial planned city is Jaipur, which was founded in 1727 by the Maharajah of Jaipur (Figure 26). A city that was

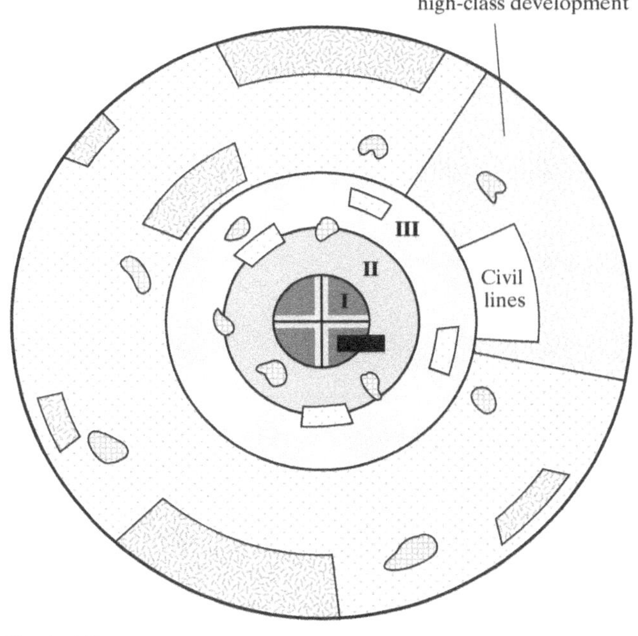

Physical Space

 Bazaar-based traditional city from the precolonial times with rich in zone I and poor in zone III

New postindependence extensions with extensive squatter settlements

+ Chowk, or crossroads

High-intensity commercial and residential land uses

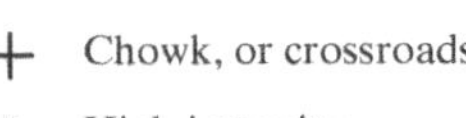 Wholesale market

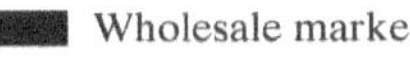 Squatters/slums

Cultural Space

Religious and linguistic clusters and Untouchables

A. K. Dutt and G. M. Pomeroy, "Cities of South Asia," in: S. D. Brunn et al., *Cities of the World: World Regional Urban Development*, 3rd ed. Lanham, Md.: Rowman & Littlefield, 2003, Fig. 9.8, p. 345. Reproduced with permission.

FIGURE 25 The bazaar-based city in South Asia.

Brian A. Vikander/Corbis Images

FIGURE 26 Jaipur, India, an example of a precolonial planned city, which was founded in 1727 by the Maharajah of Jaipur. This is the old section of the city, known as the Pink City, seen from the Jantar Mantar, an astronomical observatory also built by the Maharajah. The City Palace is at center.

Howard Sochurek/Getty Images

FIGURE 27 Jamshedpur, India, an example of a city that was planned in the British colonial period, which was designed as a company town around India's first steel mill. This photo of the Tata Iron and Steel Co. works was taken in 1952.

planned in the British colonial period is Jamshedpur, which was designed as a company town around India's first steel mill (Figure 27).

As they grew, the colonial-based, traditional bazaar-based, and planned cities developed hybrid elements. Colonial functional demands affected the traditional bazaar city layout; traditional elements were added to the colonial cities. Many bazaar and colonial cities have new planned extensions at their peripheries. The original cores of the planned cities were modified over time and surrounded by later unplanned traditional developments and semiplanned modern extensions. The traditional bazaar often thrived alongside the modern CBD, even in former colonial centers like Kolkata (Calcutta) and Mumbai (Bombay) that have since become megacities. In South Asia, megacities such as these have grown so far beyond their city and metropolitan boundaries that they now form extended metropolitan regions.

Although many cities in South Asia are only marginally connected to the world economy, some are beginning to tap into the service employment opportunities associated with foreign direct investment. In India, cities like Bangalore and Hyderabad have focused on providing educated English speakers and good telecommunications infrastructures to attract **transnational corporations** (Urban View 4 entitled "A Day in the Life of a Call Center Worker in India"). As a result, India has become a global center for software development and back-office processing (accounting, medical transcription, payroll management, maintaining legal databases, and processing insurance claims and credit card applications). In a world economy, India has been able to also attract call center business (telemarketing and help desk support) by offering these services for 30 to 40 percent less than countries that have traditionally provided these services (such as Ireland, Canada, Australia, and even Hong Kong and the Philippines). And although the first companies to set up back-office processing or call centers in India were from Europe and the United States (e.g., Swissair, British Airways, General Electric, and American Express), Indian businesses now also offer outsourcing opportunities to transnational corporations (Figure 28).

URBAN VIEW 4
A Day in the Life of a Call Center Worker in India[25]

Parag Arora, 23, works in a call center of a British transnational corporation in Mumbai. He has a BA in economics from Delhi University and earns the equivalent of about US$75 a week.

4 p.m. Parag works nights to coincide with office hours in the United Kingdom, and so wakes up in the afternoon. He shares an apartment with three other call center workers.

5 p.m. Breakfast/lunch: paratha (a type of bread stuffed with vegetables), dhal (lentil soup), *subgi* (a cooked vegetable dish) and *idli* (a fluffy rice cracker dipped in a savory sauce).

5:30 p.m. Parag relaxes with his friends, plays computer games, does some shopping or goes to a movie.

11 p.m. The journey to work on the bus should take 20 minutes, but the streets are often choked with traffic.

12 midnight The ten-hour shift begins with a short team meeting to set targets for the workers before they put on their headsets. Parag answers credit card calls such as ones about payment problems. Workers must aim for an average call-handling time of two and a half minutes—the system also monitors how often customers are put on hold (which implies that the worker needs to ask for help). More than 2,000 people are employed in the call center and about 700 work on any given night.

5 a.m. A half-hour break and a meal in the crowded canteen: similar dishes to lunch (Chinese food is also popular). Workers are permitted another two 15-minute breaks during their shift.

10 a.m. Parag leaves work. The shifts are hard to adjust to, as they change every few weeks; invariably they involve working far into the morning. He works a five-day week but his days off vary.

11 a.m. Bed.

Joerg Boethling/Alamy

FIGURE 28 Indian employees at a call center in Kolkata, India provide service support to international customers.

Southeast Asian Cities

The earliest cities in Southeast Asia originated as a result of the diffusion of Indian and Chinese influences into the region beginning in the first century C.E.[26] Two main types of cities developed: the sacred city and the trading city.

Sacred cities were inland agrarian capitals that were centers of spiritual authority. They contained great temple complexes—such as Angkor Wat in Angkor, Cambodia—for the spiritual rulers. These sacred cities were located and designed according to cosmological principles and were frequently relocated on the advice of court astrologers. The fortunes of these cities rose and fell depending on agricultural productivity and their rulers' military successes.

In contrast, *trading cities* were bustling coastal or river ports that were part of a complex international trading network. The largest contained between 50,000 and 100,000 people. The ethnic elite lived inside the walls of these cities while the native and foreign traders lived outside.

The arrival of Europeans resulted in the establishment of colonial cities and systems of cities in Southeast Asia. The early centuries of European contact produced an embryonic colonial urban network of existing and new **gateway cities** to facilitate control of Asian trade.[27] In the nineteenth century the need for markets and raw materials for the rapidly industrializing countries of Western Europe prompted a shift in colonial emphasis toward Southeast Asia. This fueled the growth of a larger network of colonial urban settlements that included port cities, administrative centers, and mining and market towns, linked together and to areas of production by an extensive transportation system.

The colonial port cities were oriented primarily toward Europe. Virtually all were located on sites of existing settlements—Saigon (Ho Chi Minh City, Vietnam) was built near the Chinese village of Cholon; Singapore was built in the vicinity of a Malay fishing village; and Batavia (Jakarta, Indonesia) was located at the site of an existing trading center.

Colonial cities were typically laid out using a gridiron street plan. The commercial and administrative functions were concentrated at the urban core. The exclusive residential districts for the Europeans were located outside the central "urban villages" (*kampungs*) that contained different ethnic groups. In some cities elements of the precolonial urban form—religious, military, or court functions at the core and the markets and merchants at the periphery—have survived, as in the old walled fort in Manila and the Shwe Dagon Pagoda and Buddhist monasteries in Yangon (Rangoon).[28]

Aside from their considerable size and growth rates, a major characteristic of the larger colonial port cities was the tremendous mixing of economic activities and land uses (Figure 29).[29] The port zone—the center of economic activity in the colonial era—has retained its importance in the postcolonial period.

Instead of a single CBD, there are separate civic and commercial/retail concentrations: a government zone; a Western commercial zone with foreign-owned banks, office towers, department stores, and modern hotels; and one or more high-density "alien" (often ethnic Chinese or Indian) commercial zones, containing small businesses—jewelers, tailors, chemists—in two-story shops and warehouses where the goods are manufactured and sold and where the business owner lives.

In addition to the colonial-era elite residential neighborhoods near the government zone, new high-income suburbs and gated communities have been built to accommodate recent growth. Similarly, in addition to the middle-density

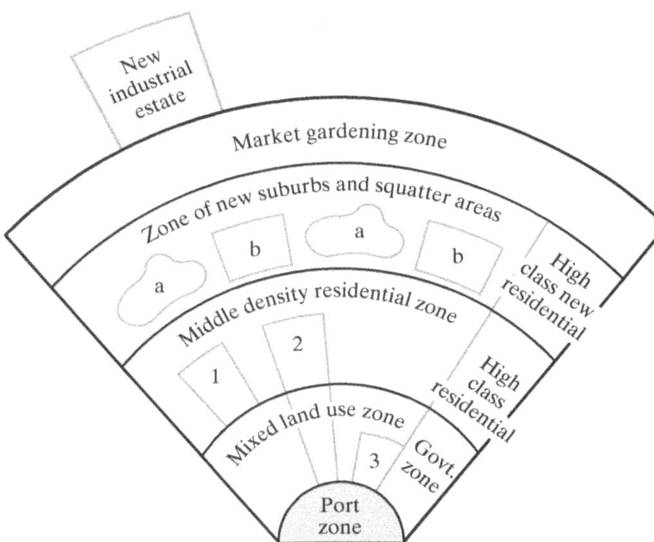

1 = Alien commercial zone
2 = Alien commercial zone
3 = Western commercial zone
a = Squatter areas
b = Suburbs

FIGURE 29 The Southeast Asian city.

The Southeast Asian City: A Social Geography of the Primate Cities of South East Asia, T G McGee, Frederick A. Praeger, Publishers, 1967. Reprinted with permission.

neighborhoods (*kampungs*) of different ethnic groups closer to the urban core, there are now new middle-income suburbs. Squatter settlements are found in zones of disamenity throughout the city (e.g., along polluted rivers) and at the edge of the built-up area. A peripheral zone of intensive market gardening supplies fresh produce to the city's markets. New industrial estates have also been built at the urban periphery to attract transnational corporations and promote job growth.

As in other parts of Asia and in Latin America, a distinguishing feature of recent urbanization in Southeast Asia has been the extension of megacities beyond their city and metropolitan boundaries to form extended metropolitan regions. Megacities like Jakarta, Bangkok, and Manila now lie at the epicenter of extended metropolitan regions.

> Extended metropolitan development tends to produce an amorphous and amoebic-like spatial form, with no set boundaries or geographic extent and long regional peripheries, their radii sometimes stretching 75 to 100 km from the urban core. The entire territory—comprising the central city, the developments within the transportation corridors, the satellite towns and other projects in the peri-urban fringe, and the outer zones—is emerging as a single, economically integrated 'mega-urban region,' or 'extended metropolitan region.' Within this territory are a large number of individual jurisdictions, both urban and rural, each with its own administrative machinery, laws, and regulations. No single authority is responsible for overall planning or management.[30]

The extended metropolitan region of Jakarta is called Jabodetabek, which gets its name by combining the first letters of each jurisdiction that comprises the larger region (Jakarta, Bogor, Depok, Tangerang, and Bekasi) (Figure 30). Jabodetabek contains more than 25 million people and thousands of factories that together represent much of Indonesia's industrial employment.

Surrounding these megacities are "peri-urban" suburban fringes that are within commuting distance of the urban core. Farther out are what McGee calls *desakotas*—whose name comes from combining the Bahasa Indonesian words for village (*desa*) and town (*kota*)—to capture the intensive mix of agricultural and nonagricultural land uses in areas at the urban periphery and along the transportation corridors connecting the major cities (Figure 31).[31] Six features characterize the *desakota* areas:

1. A large population engaged in smallholder rice cultivation
2. An increasing amount of nonagricultural activities in areas that were largely agricultural previously
3. Intense movements of people and goods using relatively cheap transportation, such as two-stroke motorbikes, buses, and trucks
4. Intense mixing of land uses, including agriculture, cottage industries, industrial estates, and suburban residential developments
5. Increasing participation of females in nonagricultural labor
6. "Invisible" or "gray" areas of informal sector activities where government regulations may not apply or are difficult to enforce

The evolution of high-density rural areas of intensive agriculture close to the major urban centers establishes the preconditions for the development of extended metropolitan regions in Asia.[32] These preconditions include sizable pools of labor that are available for relocating industries; significant improvements in transportation and communications that have increased the accessibility of regions close to the urban core and in some cases led to corridor development between two major urban centers; and areas within the extended metropolitan regions that are attractive to decentralizing activities, such as labor-intensive industry and housing.

Jorgen Schytte/Still Pictures/Specialist Stock

FIGURE 30 Jakarta, Indonesia, at the heart of the sprawling Jabodetabek extended metropolitan region.

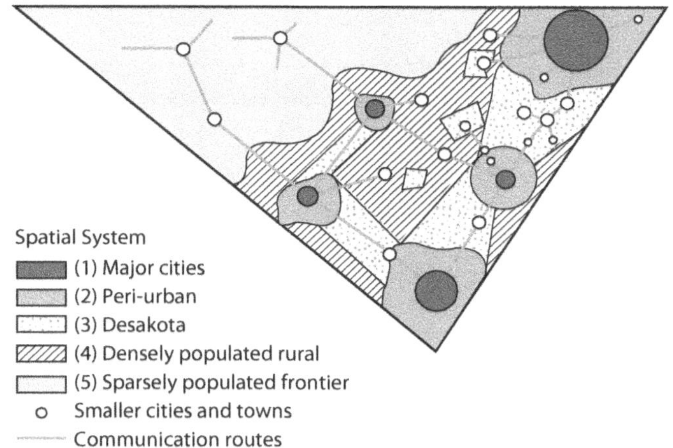

FIGURE 31 Spatial configuration of extended metropolitan regions in Asia.

Although this kind of urbanization appears to repeat the experience of the developed countries during the nineteenth and early twentieth centuries, there are some important differences:[33] The distinction between rural and urban activities is breaking down faster in Asia than it did in the developed countries; considerable advances in technology—facilitating the flow of goods, people, and capital—have allowed extended metropolitan regions to develop; and rapid industrialization and economic growth have focused primarily on the extended metropolitan regions and particularly on the peripheral parts of these regions.

East Asian Cities

Since World War II, the distinctive differences that have emerged in the internal structure of East Asian cities reflect the influence of the socialist tradition in Communist China, North

URBAN VIEW 5
Shanghai, a World City and "Dragon Head" of China's Economy

Shanghai, with an estimated population of 20 million residents in 2010, is China's largest city. It is located on the Huangpu River near the mouth of the Chang Jiang (Yangtze) River (Figure 32).

At the time of the Communist revolution of 1949, Shanghai had a population of over a million, was one of the three largest manufacturing centers in the world, and was the busiest international port in Asia. Its growth was fueled by its role as the primary Asian entrepôt for trade in opium, silk, and tea. The British were the first to develop Shanghai in this way at the end of the Opium Wars in 1842. By 1847 the French had arrived, and by 1895 large sections of the city had been divided into colonial concession zones. The centerpiece of the city was the Bund, the riverfront development of monumental neoclassical buildings containing the major international banks and trading houses.

Linda McCarthy

FIGURE 32 The megacity of Shanghai with the Oriental Pearl Tower and Pudong new area in the foreground on the east of the Huangpu River and the extended metropolitan region stretching off into the distance and smog.

Just as the foreigners had drained the wealth out of the city, now the Communist government officials tapped into its economic strength, but to redress regional imbalances in prosperity across China. Shanghai was heavily taxed—75 percent or more of its revenues went to the national government—with very little reinvested in the city. As a result, Shanghai lost its vibrancy and became rundown despite industrialization and growth under the Communists.

Dramatic changes came in the 1980s when the central government recognized the need for massive investments in Shanghai's infrastructure. Designated as one of 14 "open cities," the Shanghai Economic Zone and several Economic and Technological Development Zones were established as attractive locations for international investment. In addition, about 200 square miles of low-density farm, industrial, and residential land to the east of the Huangpu River was designated as the Pudong New Area (literally, "east of the (Huang)pu (*Pudong Xinchu*)") (Figure 32). Shanghai was to be the "dragon head"—the economic engine that would stimulate development along the entire length of the Chang Jiang valley—the "dragon body."[34]

The official designation of Pudong as a national development project in 1990 was an acknowledgment of the strategic importance of Shanghai for China's overall development. This designation allowed city government officials to offer a comprehensive range of incentives, including tax breaks, to attract domestic and foreign companies and investment and to utilize central government funds for infrastructure and other expenditures.[35] Plans were drawn up for new bridges across the Huangpu River, inner and outer ring roads, subway lines, sewage works, and a second international airport. Pudong is now a multibillion dollar concentration of export-processing industries; office space in Lujiazui that serves as the city's new CBD and as

(continued)

URBAN VIEW 5

Shanghai, a World City and "Dragon Head" of China's Economy (*continued*)

a national financial center; scientific, research and educational facilities; and new residential communities.

Shanghai's phenomenal growth has not only reinforced its status as China's largest city but has also propelled it into the role of a **world city**, a key node in the flows of capital, goods, and information that underpin the global economy. But world cities are cities of extremes, and Shanghai is no exception. It contains places of grinding poverty and vice alongside luxury apartment buildings and upscale department stores, sprawling suburbs of high-rise apartments and gated communities, and the stories of 20 million people.

Four 22-year olds in a punk rock band rehearse in downtown Shanghai behind the metal blast doors of an old subterranean bomb shelter known as 0093. The all-female band members share an apartment in a new downtown high-rise from where they can see across the Huangpu River to the site of what was the contemporary version of the world's fair, *Expo 2010* and the Expo Center, Culture Center, and Houtan Park that remain for people to enjoy after the Expo ended. A total of 190 countries and 56 international organizations exhibited in pavilions designed to amaze and display different aspects of urban development: urban dwellers, urban beings, urban planets, urban footprints, and urban dreams. But although this international Expo attracted more than 73 million visitors, most came from mainland China. A 26-year-old migrant worker can also see the Expo site from where he hangs suspended hundreds of feet above Shanghai's downtown as he helps build another new skyscraper. He passes the glitzy stores along Nanjing Road everyday as he walks to and from the construction site but sends most of his wages home to his family in their rural village. A rat scurries across the room where a 91-year-old grandmother is serving a family meal of *Hong Shao Rou*, a Shanghai favorite of sweet and fatty braised pork belly. She worries that her traditional Shanghai neighborhood or *lilong* has deteriorated and its courtyard homes have become so overcrowded that it will be demolished and its rich sense of community lost. Many old Shanghai neighborhoods have been demolished already to make way for new development, including one nearby to make room for a power switching station that lit up *Expo 2010*.[36]

Korea, and Mongolia versus the capitalist system in South Korea, Taiwan, Hong Kong, and Macao.

The internal structure of the free market cities reflects the high levels of industrialization and urbanization that come with being tied into the world economy. Their urban form and land uses reflect the operation of the capitalist economic system, including private ownership of property, the importance of people's private investment decisions, relatively high standards of living, increasing reliance on the private car, and at times significant stratification by socioeconomic background.[37]

In contrast, the planning of socialist cities in China and Mongolia between the late 1940s and the late 1970s, and still today in North Korea, reflects the Communist ideal of a centralized and highly standardized form of social organization that integrated a classless society with government economic planning based on national and local self-reliance.[38] Changes to the internal structure of cities associated with their transformation to a socialist model after 1949 involved replacing the retail concentration at the core with political-cultural-administrative functions, standardizing housing, and adopting the *self-contained neighborhood unit* concept.

Maoist policies created a peculiar mix of old and new in the urban form and land use of Chinese cities. Dutt et al.'s model of a traditional Chinese city with post-1949 socialist developments captures this mix (Figure 33).[39] The surviving elements of the traditional Chinese city reflect that the original design was based on cosmological principles of urban planning. This dictated a normally square city shape with a surrounding series of walls and moats, three gates in each city wall, and three streets within the walls that ran east–west and north–south. Government buildings and the palace of the imperial household were very important in the traditional Chinese city and were located at the core. Religious and commercial land uses were secondary. Residential land uses were differentiated based on people's social status or occupation.

Post-1949 changes include the addition of a number of long, wide boulevards running in an east–west direction through the city that established the basis for the city street system. Streets were named for revolutionary heroes and events. The streets cut the city into grids of varying sizes, many of which comprise self-contained neighborhood units—large rectangular parcels of housing and other uses surrounded by enclosing walls. The buildings are arranged in rows and columns and divided into subcompounds based on their residential, office, service, or other functions. Most buildings share the same uniform, box-like character. Near the city center there is a huge square for people to attend mass gatherings that contains a monument such as a revolutionary hero's sculpture. Government compounds line both sides of the major boulevards toward the center of the city, along with Communist party committee compounds, revolutionary history exhibition halls, and some commercial buildings. Large factories and factory workers' housing estates are located at the periphery, and the universities are concentrated in one area of the urban outskirts (Figure 33).

In general in East Asia, colonial penetration was more limited than in other parts of Asia. Even so, European and Japanese contact had a strong influence on Chinese port cities like Shanghai and Tianjin. The Treaty of Nanjing in 1842 ceded the island of Hong Kong to Britain and gave the British the right to reside in a small number of Chinese port cities. Subsequent refinements to this treaty—the so-called *Open Door policy*—gave other foreign powers the same rights as the

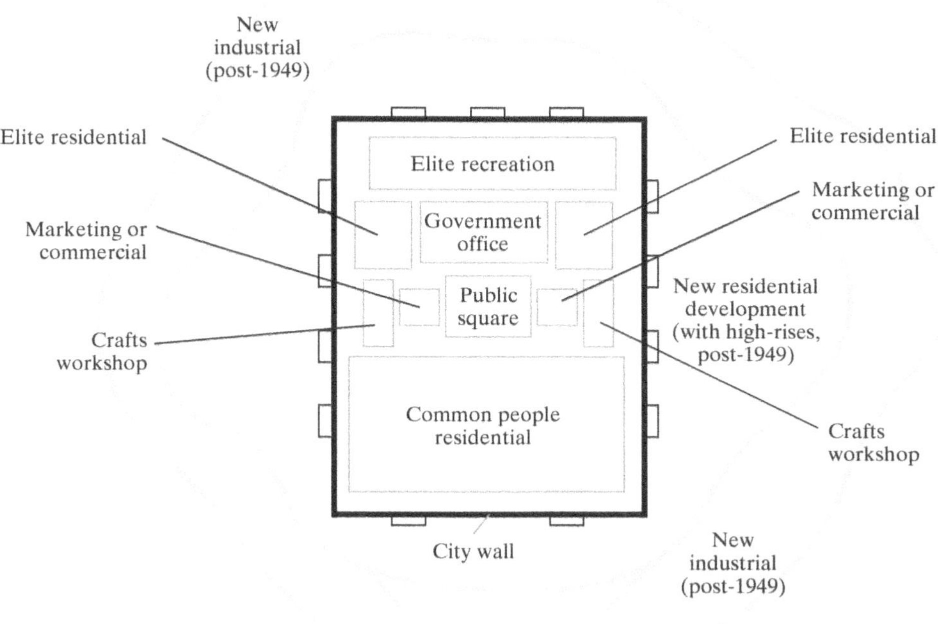

FIGURE 33 Traditional Chinese city model with post-1949 socialist developments.

British. By 1911, after more treaties, about 90 cities were open to foreigners—along the coast, up the Chang Jiang (Yangtze) valley, in northern China, and in Manchuria (Figure 34).[40]

After being opened up to foreigners, these colonial concession-based cities became the largest industrial, commercial, and transportation centers at that time in East Asia.[41] They contained three sections: *a concession zone:* Docks and military bases were established initially along coasts or rivers. Factories and warehouses were built later. With increased trade, commercial zones and elite Western residential districts were added; *a Chinese residential zone:* The majority of the Chinese population lived in the old city that had been designed using traditional Chinese principles of urban planning; and *a buffer zone:* This area became a highly desirable location for the Chinese elite—senior Chinese officials, business people, and high-ranking employees of foreign companies.

FIGURE 34 The legacy of colonialism is visible in the architecture at the center of Guangzhou (then called Canton), which became a British and French concession in 1859–60.

After the defeat of Japan in 1945, the colonial era ended in East Asia except in (British) Hong Kong and (Portuguese) Macao. After that time the internal structure of the colonial concession-based cities were significantly changed as they were transformed to reflect socialist planning principles (Figure 35). The Western residential districts became offices or houses for the party bosses, new port facilities were added as part of industrial expansion, infill neighborhood units were built on available sites in the area surrounding the old city, and high-rise housing developments were constructed at the periphery.

In the reform (post-Mao) era since the late 1970s, free market forces and global capital investment have been allowed increasingly to dictate the course of urban development (Figure 36).[42] As has occurred in cities like Seoul in South Korea and Taipei in Taiwan, the internal structure of the larger cities in China has been modified to include shiny new and revitalized central business districts with modern high-rise offices and luxury hotels. Competition to have the world's tallest building has increased in Asia since the Petronas Towers in Malaysia's Kuala Lumpur took the title from the Sears Tower in Chicago in 1998. The Petronas Towers lost the title in 2003 on completion of the

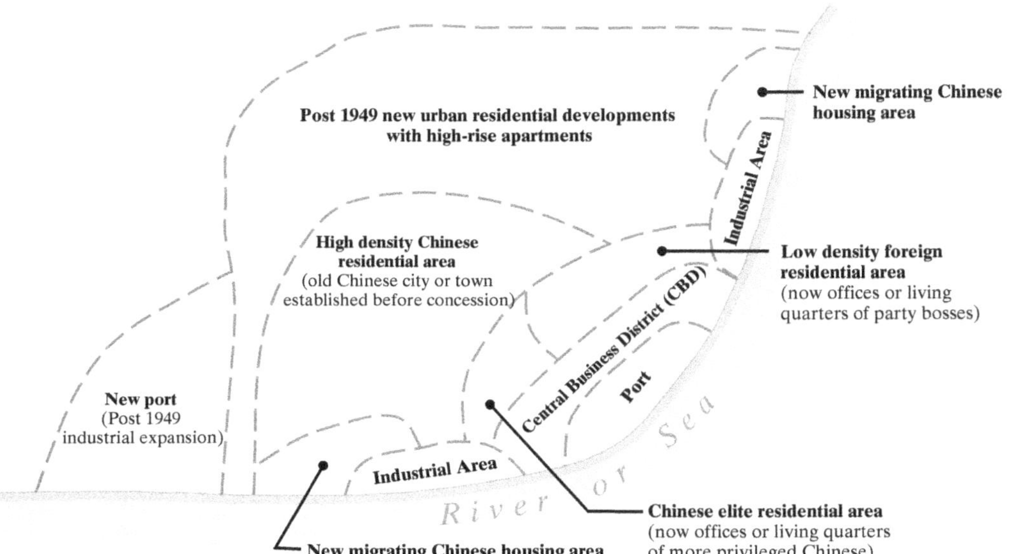

The Asian City: Processes of Development, Characteristics and Planning, 1994, p. 41, A K Dutt et al., Figure 3.7, with kind permission from Springer Science+Business Media B.V.

FIGURE 35 Colonial concession-based Chinese city model.

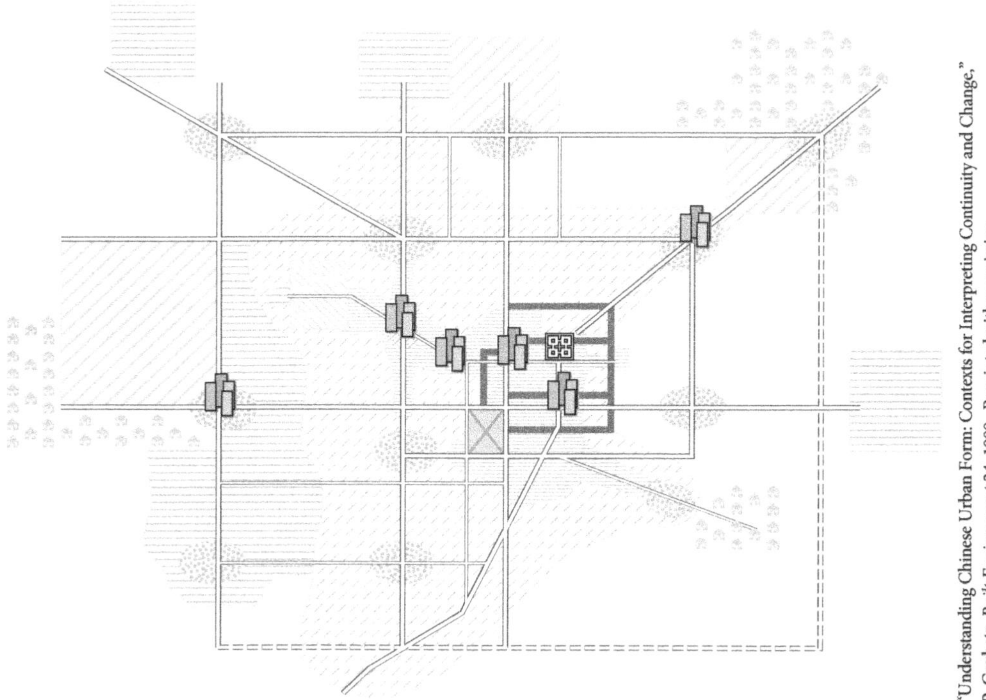

"Understanding Chinese Urban Form: Contexts for Interpreting Continuity and Change," P. Gaubatz, *Built Environment* 24, 1999. Reprinted with permission.

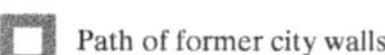

Path of former city walls.

Former administrative center, now restored as a tourist attraction.

Redeveloped commercial districts, residents relocated to new housing on outskirts.

Former treaty port. Some structures remain as housing, others restored for commercial use.

Revitalized central business district, now houses fashionable international and domestic shops and restaurants.

Public square. Concessionaires operate refreshment and entertainment stands on weekends.

Maoist work-unit districts. Many units remain, but industry increasingly separated from housing. Local retail and entertainment centers developed at key crossroads.

Economic development zone.

High-rise hotels, restaurants, entertainment, and office structures.

Villa and expensive housing developments.

New housing developments (mid- and high-rise) for relocated workers.

New commercial centers.

Planned highway.

FIGURE 36 The Great International Chinese City Model.

101-story "Taipei 101" in Taiwan, which lost its title in 2010 to the Burj Khalifa (Figure 23).

Larger cities in China also have seen extensive renovation of older central city commercial and residential districts, some preservation districts in older parts of the urban core that provide a visual link to their presocialist past and are popular with tourists, ring roads to channel some of the much increased traffic around the city, and satellite towns to absorb new population growth.

The imprints of globalization such as these are most visible in the megacities and extended metropolitan regions of East Asia. Megacities like Shanghai and Beijing in China and Seoul in South Korea have experienced enormous transformations in their economies, societies, and internal structure as their government officials and business people have made concerted efforts to take advantage of their increasing integration into the world economy (Urban View 5 entitled "Shanghai, a World City and the 'Dragon Head' of China's Economy"). As in other parts of Asia and in Latin America, the megacities in East Asia are also growing beyond their city and metropolitan boundaries to form extended metropolitan regions (Urban View 6 entitled "Foreign Direct Investment and Regional Development in the Pearl River Delta, the World's Largest Extended Metropolitan Region").

URBAN VIEW 6

Foreign Direct Investment and Regional Development in the Pearl River Delta, the World's Largest Extended Metropolitan Region

Anchored by the major metropolitan centers of Hong Kong, Guangzhou, and Macao, the Pearl River Delta, home to about 120 million people, is the world's largest extended metropolitan region (Figure 37). Since the liberal economic reforms initiated with the post-Mao "Opening and Reform" in 1978, the Chinese government has fostered this region as one of its "engines" of capitalist growth. With this reversal of its antiforeign investment stance, China became a magnet for foreign direct investment, and the Pearl River Delta became the country's most intensive region of foreign direct investment, economic development, and urbanization.

Within the Pearl River Delta, Hong Kong, a British colony until 1997, is a metropolis of more than 7 million people with a world-class financial, service, and logistics base that includes one of the world's busiest container ports and international air cargo handling centers (Figure 38). In recognition of Hong Kong's role as a capitalist economic dynamo, the Chinese government set it up as a Special Administrative District with its own currency, legislature, and legal system, based on the principle of "one country, two systems."

Hong Kong's success in attracting foreign direct investment through incentives such as low corporate taxes prompted the Chinese government to create two of its first Special Economic Zones (SEZs) in nearby Shenzhen and Zhuhai (Figure 37). These were established as **export processing zones** (**EPZs**), also known as **free trade zones** (**FTZs**), that offered cheap labor, land, and tax breaks to attract transnational corporations and their investment, technology, and management practices. Many Hong Kong businesses shifted their labor-intensive manufacturing or contracted out their assembly-line work to Chinese subcontractors in this new "*back factory.*" Hong Kong acted as the "*front shop,*" concentrating on marketing, design, purchasing raw materials, inventory control, management and technical supervision, and financial arrangements.[43] The entire delta region was subsequently designated an Open Economic Area, in which local governments, individual businesses, and farm households enjoy a high degree of independence in economic decision-making. This policy change and the increasing cooperation and coordination between government officials in Hong Kong and Guangdong Province have meant that the so-called

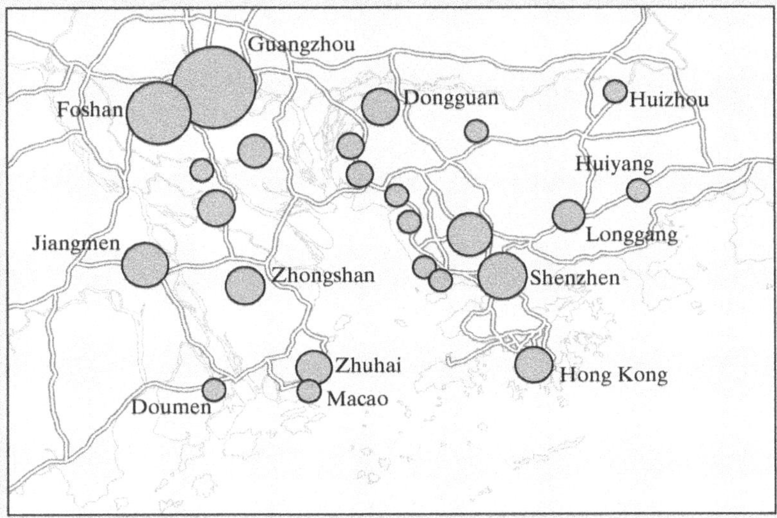

Alain Bertaud, "Spatial Structure of the Pearl River Metropolitan Region" 2001 http://alain-bertaud.com.

FIGURE 37 Anchored by the major metropolitan centers of Hong Kong, Guangzhou, and Macao, the Pearl River Delta, home to about 120 million people, is the world's largest extended metropolitan region.

(continued)

URBAN VIEW 6

Foreign Direct Investment and Regional Development in the Pearl River Delta, the World's Largest Extended Metropolitan Region (*continued*)

Linda McCarthy

FIGURE 38 Hong Kong, a metropolis of more than 7 million people, has a world-class financial, service, and logistics base.

"front shop, back factory" relationship has shifted to more of a partnership model in which manufacturing in Guangdong is promoted certainly, but so too is the services sector.

The relaxation of central government controls also allowed the region's growing rural population to migrate to urban areas in search of assembly-line jobs or to stay and diversify their rice-paddy cultivation into more profitable agricultural activities such as market gardening, livestock husbandry, or fisheries. Economic freedom also allowed rural industrialization, involving mostly low-tech, small-scale, labor-intensive industries.

The triangular area between Hong Kong, Guangzhou, and Macao has emerged as a particularly important economic zone due to its relatively cheap land and labor and the enormous investments by regional and local government officials in the transportation and communications infrastructures to meet the needs of local and international businesspeople. New infrastructure includes major airports, high-speed toll roads, satellite ground stations, port installations, light rail networks, and new water management systems. These, in turn, have attracted business and technology parks, financial centers, and resort complexes in a loose-knit sprawl of urban development. The result is a distinctive extended metropolitan region in which numerous small towns play an increasingly important role in fostering the process of urbanization, with an intense mixture of agricultural and nonagricultural activities and an intimate interaction between urban and rural areas.

At the same time, concerns have been raised about the long-term sustainability of the Pearl River Delta. One issue is that the region's growth is dependent on foreign direct investment, something that can be extremely volatile in its source, volume, and destination over time. Other concerns relate to the pressures of economic development and urbanization on the environment and agriculture in the delta region. The poor quality of the living and working conditions of the millions of low-skilled workers in the factories has attracted international attention. A simple poem "The Factory Girl" offers a personal account of the life of one young female migrant worker in Guandong Province who works 13-hour days with two meal breaks for about US$40 a month.

From the damp, dirty hallway, from the long lines of
the cafeteria,
From the rumble of the machines and the unbearable
factory noise,
The daylight drifts by, the starlight drifts by.
Forever crying on the production line, the factory girls
endure exhaustion and hardship.
Every peaceful, lonely night, the factory girl is bursting
with longing for her village.
And she can hear the sweet call of the mountain goat.
She dreams about the soft, warm bend of her
mother's arm,
And she can smell the sweetness of the old, secluded
garden.[44]

As we saw in this chapter, the close relationship between globalization and urbanization means that traditional patterns of spatial organization are being transformed. The processes involved in urbanization and globalization are inherently problematic for urban dwellers in the less developed countries, and how space and place often play key roles.

FOLLOW UP

Key Terms

agglomeration economies and diseconomies
apartheid
colonial concession zone
desakota
export processing zone or EPZ (also known as free trade zone (FTZ)
extended metropolitan region or EMR
foreign direct investment or FDI
"front shop, back factory" model
growth pole
informal sector of the economy
inverse concentric zone pattern
maquiladora
overurbanization
rural-to-urban migration
squatter settlements

Review Activities

1. If you have time, read a novel that is evocative of urban life in one of the less developed regions of the world. There are hundreds to choose from, including bestsellers about Mumbai in India, such as *Q & A* by Vikas Swarup (New York: Scribner), which was made into the movie, *Slumdog Millionaire*, that won 8 Academy Awards in 2009.
2. Hosting a mega-event like the World Cup soccer tournament in a city like Cape Town in South Africa can be associated with drawbacks as well as benefits for the local people. Pick a mega-event (Olympic games, world expo, etc.) that was hosted by a city in a less developed country that interests you and go online to find out more about the positives and negatives for the city and its residents.
3. Go online to find maps that were produced at different points in time to chronicle the growth of Lagos in Nigeria as it grew to become the first Sub-Saharan African mega-city. Examine the extent and direction of growth and think about some of the most important social, economic, technological, and political processes that drive this growth.
4. Work on your *portfolio*. Make a diagrammatic and annotated entry for each of the models of urban structure in the less developed countries that were covered in this chapter. Make a list of the advantages and disadvantages of using models such as these to help our understanding of the internal structure of cities in the less developed countries.

Log in to **www.mygeoscienceplace.com** for self-study quizzes, *MapMaster* layered thematic and place name interactive maps, *Urban View* Google Earth™ tours, key resources and suggested readings, related websites, "In the News" RSS feeds, and additional references and resources to enhance your study of urban form and land use in the less developed countries.

NOTES

1. M. Landsberg, "A Day in the Life of a Haiti Tent City," *Los Angeles Times*, January 27, 2010; F. Grimm, "In Haiti's Tent Cities, a Return to Normalcy is Unimaginable," *Miami Herald*, July 12, 2010.
2. C. S. Sargent, "The Latin American City," in *Latin America and the Caribbean: A Systematic and Regional Survey*, eds. B. W. Blouet and O. M. Blouet (New York: John Wiley & Sons, Inc., 2002), 167–68.
3. See L. R. Ford, "A New and Improved Model of Latin American City Structure," *The Geographical Review* 86 (1996): 437–40.
4. W. K. Crowley, "Modeling the Latin American City," *Geographical Review* 88 (1998): 127–30; W. K. Crowley, "Order and Disorder—A Model of Latin American Urban Land Use," in *Yearbook of the Association of Pacific Coast Geographers* 57, ed. D. E. Turbeville III (Corvallis, Ore.: Oregon State University Press, 1995), 9–31.
5. D. D. Arreola and J. R. Curtis, *The Mexican Border Cities: Landscape Anatomy and Place Personality* (Tucson, Ariz.: The University of Arizona Press, 1993), 69.
6. M. Hays-Mitchell and B. J. Godfrey, "Cities of South America," in *Cities of the World: World Regional Urban Development*, eds. S. D. Brunn, D. J. Zeigler, and M. Hays-Mitchell, 5th ed. (Lanham, Md.: Rowman & Littlefield, 2011).
7. C. C. Diniz, "Polygonized Development in Brazil: Neither Decentralization nor Continued Polarization," *International Journal of Urban and Regional Research* 18 (1994): 293–314.
8. United Nations, *Urban Land Policies and Land-Use Control Measures: Volume 1. Africa* (New York: UN Department of Economic and Social Affairs, 1973), 10.
9. A. O'Connor, *The African City* (London: Hutchinson & Co., 1983).

10. R. J. Davis, "The Spatial Formation of the South African City," *GeoJournal* Supplementary Issue 2 (1981): 59–72.
11. A. Mehretu and C. Mutambirwa, "Cities of Sub-Saharan Africa," in *Cities of the World: World Regional Urban Development*, eds. S. D. Brunn, D. J. Zeigler, and M. Hays-Mitchell, 5th ed. (Lanham, Md.: Rowman & Littlefield, 2011).
12. D. Crary, "Soccer in South Africa: Long Legacy of Barrier-Busting," in *USA Today*, 6/3/2010.
13. Ibid.
14. A. Lemon, ed., *Homes Apart: South Africa's Segregated Cities* (London: Paul Chapman Publishing Ltd., 1991).
15. C. Newton, "The Reverse Side of the Medal: About the 2010 FIFA World Cup and the Beautification of the N2 in Cape Town," *Urban Forum* 20 (2009): 93–108.
16. D. J. Zeigler, "Cities of the Greater Middle East," in *Cities of the World: World Regional Urban Development*, eds. S. D. Brunn, D. J. Zeigler, and M. Hays-Mitchell, 5th ed. (Lanham, Md.: Rowman & Littlefield, 2011).
17. S. Lowder, *The Geography of Third World Cities* (Totowa, N.J.: Barnes & Noble Books, 1986), 29–31.
18. D. J. Zeigler, *Cities of the World.*
19. Ibid., Zeigler's description of these cities forms the basis for this discussion.
20. Ibid.
21. B. Macintyre, "Towering ambition always comes before a fall," in *Times*, 12/13/2009.
22. Ibid.
23. J. Hari, "The Dark Side of Dubai," in *Independent*, 4/7/2009.
24. A. K. Dutt and G. M. Pomeroy, "Cities of South Asia," in *Cities of the World: World Regional Urban Development*, eds. S. D. Brunn, D. J. Zeigler, and M. Hays-Mitchell, 5th ed. (Lanham, Md.: Rowman & Littlefield, 2011).
25. Based on R. Aspden, "The Bangalore Effect," in *Newstatesman*, January 30, 2006.
26. J. Goss, "Urbanization," in *Southeast Asia: Diversity and Development*, eds. T. R. Leinbach and R. Ulack (Upper Saddle River, N.J.: Prentice Hall, 2000), 110–132.
27. T. G. McGee, *The Southeast Asian City: A Social Geography of the Primate Cities of South East Asia* (New York: Frederick A. Praeger, Publishers, 1967).
28. J. Goss, "Urbanization," in *Southeast Asia: Diversity and Development.*
29. T. G. McGee, *The Southeast Asian City*, 128.
30. T. G. McGee and I. M. Robinson, *The Mega-Urban Regions of Southeast Asia* (Vancouver: UBC Press, 1995), ix–x.
31. T. G. McGee, "The Emergence of Desakota Regions in Asia: Expanding a Hypothesis," in *The Extended Metropolis: Settlement Transition in Asia*, eds. N. Ginsburg, B. Koppel, and T. G. McGee (Honolulu: University of Hawaii Press, 1991), 3–25.
32. See N. Ginsburg, B. Koppel, and T. G. McGee, *The Extended Metropolis.*
33. Ibid.
34. Ibid.
35. C. Olds, *Globalization and Urban Change: Capital, Culture, and Pacific Rim Mega-Projects* (Oxford, UK: Oxford University Press, 2001).
36. B. Larmer, "Shanghai Dreams," *National Geographic Magazine*, March (2010).
37. J. F. Williams and K. W. Chan, "Cities of East Asia," in *Cities of the World: World Regional Urban Development*, eds. S. D. Brunn, D. J. Zeigler, and M. Hays-Mitchell, 5th ed. (Lanham, Md.: Rowman & Littlefield, 2011).
38. C. P. Lo, "Shaping Socialist Chinese Cities: A Model of Form and Land Use," in *China: Urbanization and National Development*, Department of Geography Research Paper No. 196, eds. C-K. Leung and N. Ginsburg (Chicago: University of Chicago, 1980), 130–55.
39. A. K. Dutt, Y. Xie, F. J. Costa, and Z. Yang, "City Forms of China and India in Global Perspective," in *The Asian City: Processes of Development, Characteristics and Planning* (Dordrecht, The Netherlands: Kluwer Academic Publishers, 1994), 41.
40. Ibid.
41. A. K. Dutt, Y. Xie, F. J. Costa, and Z. Yang, "City Forms of China and India in Global Perspective," in *The Asian City: Processes of Development, Characteristics and Planning*, 43.
42. Q. Luo, "Shanghai: The 'Dragon Head' of China's Economy," *Issues & Studies* 33 (1997): 17–32.
43. V. F. S. Sit and C. Yang, "Foreign-Investment-Induced Exo-Urbanisation in the Pearl River Delta, China," *Urban Studies* 34 (1997): 647–77.
44. C. Shuck, "Labor is not a Commodity," *The International Labor Rights Forum*, blog *(http://laborrightsblog.typepad.com/international_labor_right/2008/11/chinese-workers-poem-captures-factory-life.html).*

Urban Problems and Responses in the Less Developed Countries

Shehzad Noorani/Still Pictures/Specialist Stock

Urban Problems and Responses in the Less Developed Countries

Rapid urbanization in many less developed countries has created a host of problems for urban residents and weakened the role of cities as engines of economic growth. Pervasive poverty, inadequate housing, lack of urban services, transportation problems, and environmental degradation all contribute to dreadful living conditions for many urban dwellers. Recent progress has been made in responding to these problems within the context of sustainable urban development. Government involvement is important because this is the point where urban sustainability as a concept overlaps with urban management as a practical process. Government efforts can also help to establish democratic institutions and a participatory planning process that facilitate the involvement of impoverished local people—those disproportionately affected by the problems of urbanization.

LEARNING OUTCOMES

After reading this chapter, you should be able to:

- Explain how and why many of the people living in cities in the less developed countries have become trapped in poverty.
- Identify the main kinds of informal sector activities and the kinds of workers each attracts.
- Appraise the impacts of inadequate housing and a lack of services for people living in cities in the less developed countries.
- Explain the double threat that HIV/AIDS poses for children in Africa.
- Explain what sustainable urban development involves and why it provides a compelling framework for response to the urban problems discussed in this chapter.

CHAPTER PREVIEW

This chapter is not intended to provide an exhaustive inventory of urban problems and the responses to them in the less developed countries. Instead, the goal is to show how the processes involved in urbanization are inherently problematic for urban dwellers and how space and place often play key roles (Urban View 1 entitled "Narrowing the Digital Divide in Africa: Skipping Landlines for Cell Phones"). We focus on five major sets of problems—poverty, inadequate housing, lack of urban services, transportation problems, and environmental degradation.

The close relationship between globalization and urbanization means that traditional patterns of spatial organization are being transformed. The local impacts for the people in these cities are associated with acute problems of social polarization and environmental degradation. The problems are visible in the extensive areas of slum and squatter settlements in **megacities**, high rates of unemployment and underemployment, and a large informal sector of the economy in which people seek economic survival.

If present trends continue, an increasing number of the fastest-growing settlements in less developed countries are likely to face these problems of urbanization. Certainly, cities in both the **newly industrializing economies (NIEs)** and the oil-rich countries face serious difficulties. Islamic culture and urban design principles,

Rapid urbanization in many less developed countries has created a host of problems for urban residents. Pervasive poverty, inadequate housing, lack of urban services, transportation problems, and environmental degradation all contribute to dreadful living conditions for many urban dwellers like this woman who lives in a tent near new luxury highrises in Dhaka, Bangladesh.

URBAN VIEW 1
Narrowing the Digital Divide in Africa: Skipping Landlines for Cell Phones

In Yanguye, South Africa, Bekowe Skhakhane lives very much as her mother did before her. She spends four hours every day getting water from the river, she cooks on an open fire using firewood that she gathers nearby, and she lights her modest hut with candles. In one fundamental way, though, Bekowe's life is very different to her mother's—when she wants to talk with her husband who works 250 miles away in a steel factory in Johannesburg, she calls him on her cell phone.

"It is a necessity" she says, pausing from washing clothes in a plastic bucket as she pulls her pink Nokia from the pocket of her flowered apron. "Buying air time is on my regular grocery list." It costs Bekowe the equivalent of US$1.90 a month for five minutes of precious telephone time on the cell phone that her husband purchased with savings from his factory job. Before the cell tower was erected by Vodacom, an African mobile communications company, Bekowe kept in touch with her husband by letter and waited weeks for mail. The nearest public phone box is 10 miles away and the phone is often broken. Bekowe and other cell phone owners in her neighborhood pay a neighbor, Nsle, the equivalent of 80 cents to charge their cell phones. Nsle keeps a car battery in her home and regularly takes it with her on the bus to the nearest gas station 20 miles away to recharge it.[1]

Many people like Bekowe have helped to make Africa the fastest-growing cell phone market in the world. A 2010 report by the International Telecommunications Union indicated that many more people in Africa use cell phones than in any other part of the world. Whether it is simply making a call, transferring money, or checking the market prices of crops, individual cell phones and "village cell phones" are bridging many divides in Africa. Cell phones are beginning to create all kinds of personal and economic possibilities for millions of Africans despite lack of online access, too few bank branches, inadequate local medical facilities, lack of teachers and books, and bone-rattling roads (Figure 1).[2]

Greatstock Photographic Library/Alamy

FIGURE 1 Young students in South Africa overcome the lack of computers in their school by using a cell phone instead.

for example, have not always been able to cope with the pressures of contemporary rates of urbanization. But this chapter focuses on the poorest countries of Africa, Latin America, and Asia that must confront the most severe urban problems.

Many people have been able to establish self-help networks and organizations—with and without the help of government agencies and nongovernmental organizations (NGOs)—that have formed the basis of community cohesion within the often overwhelmingly poor and crowded cities. There is also increasing recognition that responses to the problems of urbanization must be framed within the context of the concept of sustainable urban development; nevertheless, the intensity of urban problems swamps efforts to respond.

Part of the difficulty arises from the "*globalization paradox*" faced by cities. The residents, business people, and government officials in cities must act as a collective unit to respond to international competition in the world economy; but the people in cities also face growing internal social and economic fragmentation that hampers their capacity to build coalitions, mobilize resources, and develop good governance structures.

During the last decade or so four changes in urban governance have emerged that represent attempts to address this "globalization paradox": decentralization and formal government reforms, participation of local communities in urban policymaking and implementation, multilevel governance and public-private partnerships, and process-oriented and area-based policies. As we will see, although some of these changes are not new, the rationale for introducing them is.

URBAN PROBLEMS

Poverty

The number of people subsisting below the World Bank's *international poverty line* of US$1.25 a day (in 2005 prices) fell from 1.9 billion in 1981 to 1.4 billion today.[3] This reduction in extreme poverty partly reflects international, NGO, and other efforts to reach the United Nations Millennium Development Goal of halving the proportion of people below the international poverty line between 1990 and 2015. But the reduction in extreme poverty has been uneven and millions of people remain trapped in poverty, especially in Sub-Saharan Africa and South Asia. Poverty is also increasingly becoming an urban phenomenon. In countries such as the Democratic Republic of Congo, Zimbabwe, and Haiti, more than 50 percent of the urban populations live below their respective national poverty lines.

Poor people are trapped in poverty for a variety of reasons including a lack of education; bad health; degraded environmental resources; and conflict and mismanagement that waste public expenditures. Rampant corruption by government officials in some countries has led to public funds being siphoned off into Swiss bank accounts by corrupt politicians, while building regulations are not enforced by city inspectors who take bribes for new buildings that later collapse and crush to death whole families. After decades of being forced to live in poverty under oppressive dictators, many people have taken to the streets in popular uprisings that began in the streets of Sidi Bouzid in Tunisia and spread across North Africa and the Middle East.

Within the context of globalization, despite the fact that some terrorist leaders did not grow up in poverty, many of their followers often come from impoverished families in desperately poor countries such as Yemen. A concern of some humanitarian groups is that focusing too narrowly on fighting terrorism neglects the dreadful social and economic conditions of the ordinary people. In Yemen, oil resources are being used up without a plan for a post-oil economy and water resources are so depleted that many people are predicting that Sana'a will be the first capital city in history to go dry.[4]

In less developed countries, a combination of internal and external factors influences the economies of cities and shapes their labor markets. Within national economies, the urban and rural sectors interact through exchanges of people, goods, services, money, and information (Figure 2). Many cities are experiencing unprecedented rates of growth that are driven by rural **"push" factors**—overpopulation and the lack of employment opportunities—rather than urban **"pull" factors**—prospective jobs and a better quality of life. With high rates of **natural increase** and many more people moving to the cities than available jobs, the result is widespread unemployment and underemployment. **Underemployment** occurs when people work less than full time even though they would prefer to work more. Underemployment is difficult to measure accurately, but estimates range from 30 to 50 percent of the employed workforce in the less developed countries.

The urban and rural sectors of the economy, separately and together, are also linked through trade and other exchanges into the world economy (Figure 2). Through the **new international division of labor**, towns and cities in different parts of the less developed world play a key, if often unequal, role in international economic flows that link provincial centers into their national urban network and, ultimately, into the hierarchy of **world cities** and the global economy.

Based on the assumptions underlying modernization theories, in the 1970s many governments adopted **neoliberal policies** to promote urban economic development, including fiscal and other incentives to attract **foreign direct investment**. Public officials were attempting to take advantage of the new international division of labor—to attract some of the transnational corporations that were gravitating toward the greater profit opportunities offered by lower-cost locations.[5]

Cheap labor is certainly an important factor in the locational decision-making of **transnational corporations**. But the demand of the global market for quality products means that other factors—educated and trainable workers, a good-quality transportation network, and reliable services such as

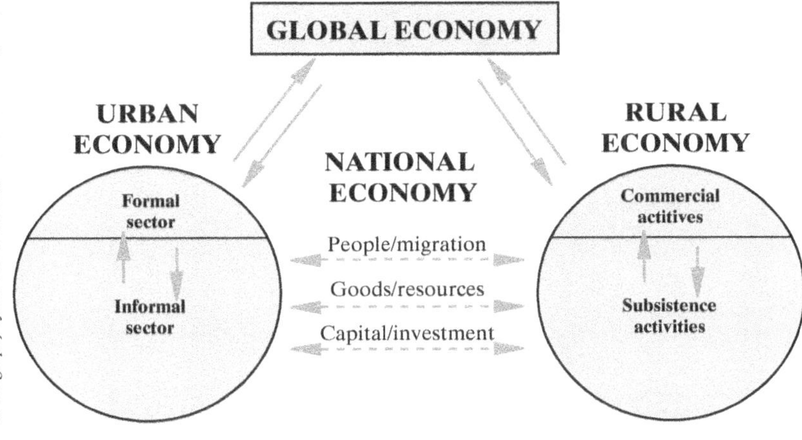

FIGURE 2 Spatial and sectoral interactions involving the urban economies in less developed countries.

water and electricity—are also important. This meant that the poorest countries in Sub-Saharan Africa and elsewhere were bypassed. The relatively few locations that could offer attractive conditions in the 1970s experienced rapid growth, most notably the four so-called Asian Tigers (Hong Kong, Singapore, South Korea, and Taiwan), as well as Mexico and Brazil. Since then, international investment has promoted growth in other countries, most notably China. With time, international investment has shifted to the larger urban centers, reinforcing existing investment patterns and aggravating already serious problems of urban **primacy** and **uneven development**.

Dependence on international capital and global economic trends has introduced instability into the labor market and further fragmented both the urban and the rural sectors of the economy into large informal and smaller formal components (Figure 2). **Overurbanization** has led to a large informal sector because people who cannot find regularly paid work resort to earning a living in jobs that are not regulated by government. The resulting pronounced *dualism*, or juxtaposition in urban space of the formal and informal sectors of the economy, is quite evident in the built environment: Modern high-rise office and apartment complexes and luxury homes correspond with an often dynamic formal sector that offers well-paid jobs and opportunities; these contrast sharply with the *slums* and **squatter settlements** of people working in the informal sector who are disadvantaged by a lack of education, formal training, recognized qualifications, and often rigid **divisions of labor** shaped by gender, race, and ethnicity (Figure 3).[6]

Edgar Clejine/Photolibrary, Inc.

FIGURE 3 A squatter settlement in Lagos, Nigeria.

But the informal sector of the economy is ambiguous—it encompasses wealth and poverty, productivity and efficiency, exploitation and liberation (Urban View 2 entitled "Defying Gender Stereotypes: Las Cholitas in Bolivia").[7] Four main kinds of **informal sector** activities attract different kinds of workers (Figure 4).

1. **Subsistence activities** in which the goods and services, such as clothing and repairs, are primarily for household consumption rather than for profit
2. Small-scale producers and retailers, often self-employed, who produce or sell to earn an income—the typical workers in this category are street traders, artisans, and food vendors
3. Petty capitalists who see greater opportunities for profit in the informal sector where they can avoid taxes and legal employment regulations, such as minimum wages and safe working conditions
4. Criminal and socially undesirable activities, including drug dealing, smuggling, theft, extortion, and prostitution

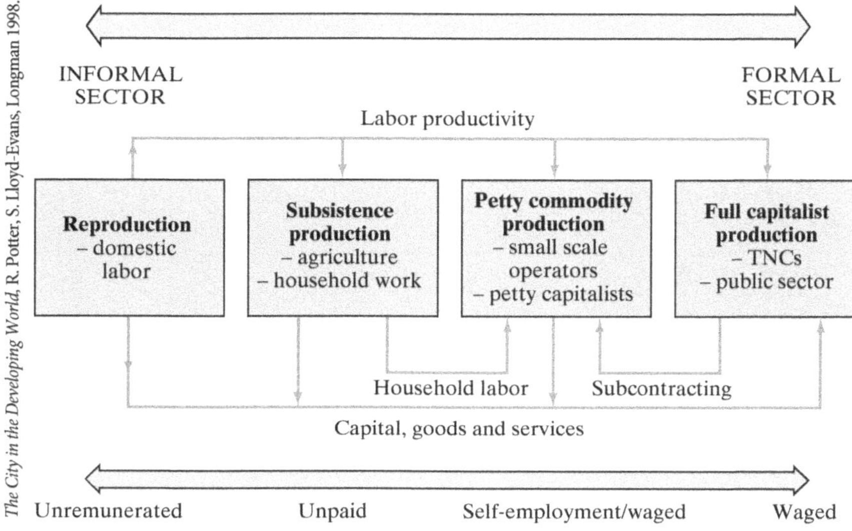

The City in the Developing World, R. Potter, S. Lloyd-Evans, Longman 1998.

FIGURE 4 The informal–formal sector continuum.

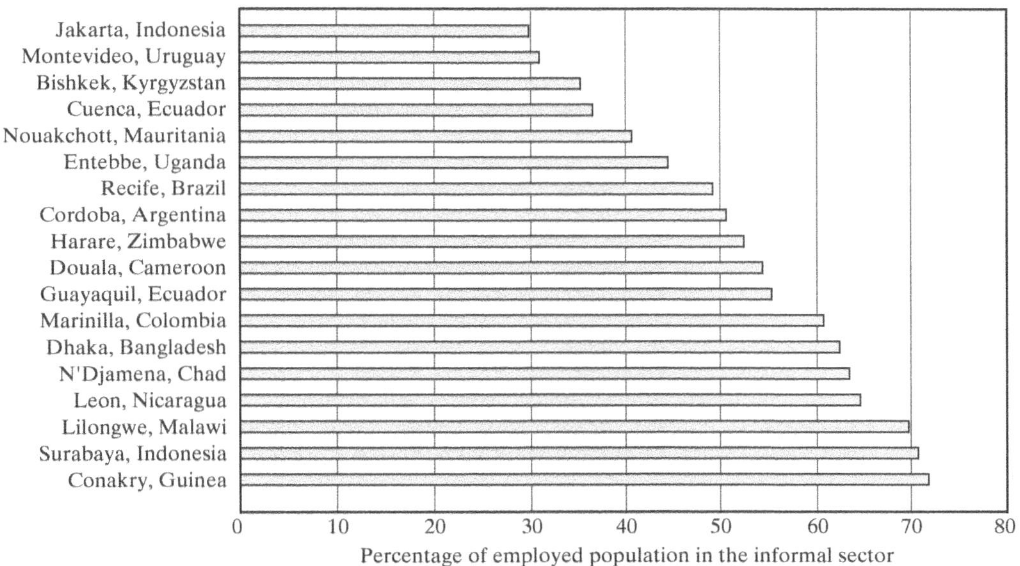

"The State of the World's Cities Report 2001," *United Nations Centre for Human Settlements* (2001). Reprinted with permission.

FIGURE 5 Informal employment in selected cities.

In many cities more than one-third of the population is engaged in the informal sector; in some cities this figure is more than two-thirds (Figure 5). Globally, the share of workers in the informal economy is projected to grow and could exceed two-thirds of the workforce by 2020. Although informal jobs—selling souvenirs, driving pedicabs, or dressmaking—might seem marginal from the point of view of the world economy, they support nearly two billion people around the world.

In many countries the labor force of the unregulated informal sector includes the world's most vulnerable workers—women and children (Figure 6). In environments of extreme poverty, every family member must contribute something. Industries in the formal sector often take advantage of this situation and the fact that labor standards are nearly impossible to enforce in the informal sector. Many companies farm out their production using subcontracting schemes that are based, not in factories, but in home settings that use child workers. Table 1 lists a combination of factors that creates an unfavorable position for women in both the informal and formal sectors of urban economies in the less developed countries.

Urban geographers recognize that the formal and informal sectors are interconnected—the informal sector represents an important resource for the formal sector of urban economies. The people working in the informal sector provide a huge range of cheap goods and services that reduce the cost of living for workers in the formal sector, allowing employers to keep wages low. Although this arrangement does not contribute to urban economic growth or help alleviate poverty, it does keep many companies competitive within the global economic system. For export-oriented businesses, in particular, the informal sector provides a considerable indirect production subsidy. And while this subsidy is often passed on to consumers in the

David Loveall/NewsCom

FIGURE 6 A young child selling chewing gum on the street.

TABLE 1 Factors Producing an Unfavorable Position for Women in the Urban Labor Force in Less Developed Countries

1. Women's reproductive role, particularly in relation to domestic work
2. Colonial introduction of the notion of the male breadwinner and female housewife
3. Orthodox economic perception of women as having relatively low levels of skills, aptitudes, and/or education
4. Women seen as more "passive" workers, less likely to organize and resist exploitation by employers
5. Gender stereotyping of women as dexterous and able to undertake repetitive tasks

Source: R. B. Potter and S. Lloyd-Evans, *The City in the Developing World,* Harlow, UK: Longman, 1998, pp. 166–67.

developed countries in the form of lower prices, the poorest households in the cities of the least developed countries are forced to resort to increasingly drastic strategies for coping with worsening poverty (Table 2).

Inadequate Housing

Social polarization is a dimension of urbanization that is intensified by globalization. The United Nations Human Settlements Programme (UN-Habitat) has identified social polarization as an indirect but crucial determinant of contemporary patterns of segregation of people and land uses around the world—the "quartering" of cities into spatially partitioned, compartmentalized residential enclaves. In less developed countries, these separate "quarters"

URBAN VIEW 2
Defying Gender Stereotypes: Las Cholitas in Bolivia[8]

The women of the Aymara ethnic group in El Alto a suburb of La Paz, Bolivia, still mostly wear their traditional brightly colored shawls pinned with filigree jewelry, multilayered skirts, and trademark black felt bowler hats perched jauntily on their braided hair. The story goes that a clever salesman from Manchester with a shipment of bowler hats started this fashion trend in the nineteenth century.

In the twenty-first century, Bolivia remains very patriarchal and women still lag in literacy and economic opportunities compared to their counterparts in neighboring Argentina and Chile, whose people have already elected female presidents. But in El Alto, some women have experienced a newfound pride and respect—as female wrestlers! Las cholitas are female lucha libre performers who attract a huge paying audience to watch their freestyle wrestling every Sunday night in El Alto's largest public gym (Figure 7). On a good night they can earn US$15—not enough to give up their day job in the fields but enough to help pay the bills and convince their husbands to support their unorthodox second career.

"Bring them on! Bring them on!" the crowd shouts impatiently to the beat of the music. And then the houselights go dim and the master of ceremonies appears and announces their names into his microphone as the curtains to the locker room part and "Amorous Yolanda" and "Evil Claudina" come out to thunderous applause. They wave and twirl their petticoats as they make their way to the wrestling ring. The music suddenly stops and they quickly take off their bowler hats and unpin their shawls. Then they start bashing each other. Claudina grabs Yolanda by her braids, and lifts and hurls her against the ropes. "Watch out" the audience roars. Too late. Claudina, twirling in celebration, trips and falls on her face while Yolanda launches herself through the air and lands on her. The audience goes wild. "She is destroyed forever" the announcer yells hysterically. But "Evil Claudina" will be back to fight "Amorous Yolanda" another day because, in lucha libre, no defeat is ever final.

Johnathan Burkham

FIGURE 7 Las Cholitas. These female wrestlers are defying gender stereotypes in Bolivia.

TABLE 2 Household Strategies for Coping with Worsening Urban Poverty

Increase resources
- Add more household members to the labor force
- Set up new businesses
- Grow own food
- Increase scavenging of items for use or resale; increase foraging for wild foods
- Rent out rooms to tenants

Limit consumption
- Reduce or eliminate consumption of items such as new clothes, meat, or "luxury" food and drink
- Buy cheaper food and secondhand clothes
- Withdraw children from school
- Delay medical treatment
- Postpone repair or replacement of household equipment
- Defer house repairs or improvements
- Reduce social events, including visits to rural homes

Change household composition
- Postpone or stop having children
- Increase household size, with women and married children, to keep wage earners in the family and enable older women to work
- Migration

Source: C. Rakodi. "Poverty lines or household strategies? A review of conceptual issues in the study of household poverty." *Habitat International* 19(4), 1995, p. 418.

include protected enclaves of the wealthy; middle-income neighborhoods; low-income neighborhoods; and ethnic enclaves.

The typology[9] of urban housing supply shown in Figure 8 describes several kinds of housing options in the cities of less developed countries. Private housing (such as owner-occupied, rental, and employer-provided) and public units (built or subsidized by government) comprise a relatively small proportion of the total housing supply. The "popular" housing category (that includes slums and squatter settlements) contains the majority of urban residents—the very poor, the never employed, the permanently unemployed, and the homeless.

The United Nations has estimated that more than one and a half billion of the world's urban residents live in inadequate housing, mostly in the slums and squatter settlements in the less developed countries (Figure 9). More than half of the housing in these countries is substandard, with one-quarter being makeshift structures and more than one-third not complying with local building regulations.[10]

The informal labor market, then, is directly paralleled in the slums and squatter settlements: With so few jobs offering regular wages, most people cannot afford rent or mortgage payments for decent housing. Unemployment, underemployment, and poverty result in overcrowding. In situations where overurbanization has overtaken the supply of cheap housing—outstripping the capacity of builders and governments to provide new and affordable homes—the inevitable outcomes are slums and squatter settlements that offer precarious shelter at best.

Whereas slums are, at least technically, legal permanent homes that have become dreadfully substandard over time, squatter settlements contain nonpermanent makeshift housing that is built illegally on land that people desperate for shelter

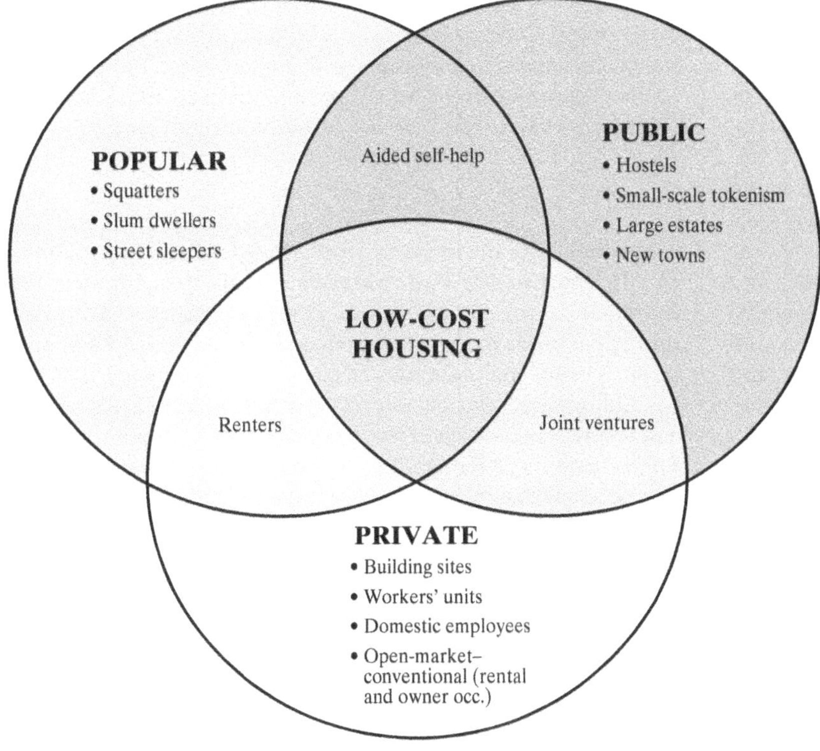

FIGURE 8 Typology of low-cost urban housing supply in less developed countries.

G. Bizzarri/Food and Agriculture Organization of the United Nations

FIGURE 9 A squatter settlement built by poverty-stricken people along a huge water pipe in New Delhi, India.

neither own nor rent. This housing is constructed on unpaved streets in the least desirable locations—derelict sites, poorly drained land, even cemeteries and waste dumps—usually with open sewers and no basic services like electricity or running water. Shacks are cobbled together from any material available—corrugated iron, planks, mud, thatch, tar paper, and cardboard. In Chile squatter settlements are called *callampas*, meaning "mushroom cities"; in Turkey, *gecekindu*—literally, "built after dusk and before dawn." In India they are called *bustees;* in Brazil, *favelas;* and in Argentina, simply *villas miserias*. In the worst cases, overcrowding, inadequate sanitation, and lack of maintenance promote extremely high levels of illness and infant mortality.

Faced with the growth of these squatter settlements, many governments initially responded by eradicating them. Encouraged by Western housing experts and development economists advocating neoliberal policies, many cities tried to stamp out this kind of unplanned urbanization through large-scale eviction and clearance programs. In dozens of cities—including Jakarta (Indonesia), Caracas (Venezuela), Lagos (Nigeria), and Kolkata (India)—hundreds of thousands of people in slums and squatter settlements were ordered out on short notice and their homes were bulldozed to make way for public works, land speculation, luxury housing, urban renewal, and, on occasion, to improve the appearance of cities for visitors and tourists. In preparation for mega events such as the Olympic Games, World Cup soccer tournament, and World Expo, governments have forcibly evicted people to make way for beautification programs. It has been estimated that the Olympic Games has resulted in the displacement of more than 2 million people during the last 20 years. Although disputed by the Chinese government, estimates are that as many as 1.25 million people were displaced in Beijing in preparation for the Olympic Games in 2008.

Most cities cannot afford to build new low-income housing to replace the demolished neighborhoods. Displaced households have no option but to create new squatter settlements elsewhere in the city. The futility of slum clearance has led to a widespread reevaluation of such neoliberal policies; this housing is now seen as a rational response to poverty. Slum and squatter neighborhoods can provide affordable shelter as well as reception areas that might offer supportive community networks and informal employment opportunities for rural migrants. Recognizing the positive functions of this housing and the many self-help improvement efforts by residents, city governments have begun to support them instead of sending in police and municipal workers with bulldozers.[11] But there are limits to community-based efforts, and there are many squatter settlements where self-help organizations do not emerge.

Lack of Urban Services

The influence of the informal labor market is felt in the inability of cities to provide basic services (Figure 10). Because the informal sector yields no tax revenues, there are limited municipal funds for providing adequate health and educational services or for maintaining a clean, safe environment.

Although the number of people without access to an adequate and secure source of water has dropped below 1 billion worldwide for the first time, it remains a sobering fact that a 5-minute shower by a person who lives in the United States will take more water than that used in a whole day by the average person who lives in a slum in a less developed country.[12] Many governments classify the existence of a water tap or standpipe within about 300 feet of a house as "adequate," but this does not guarantee enough water for good health for nearby households. In some cities residents must wait in long lines to fill even one bucket of water because communal

Christine Osborne/Corbis

FIGURE 10 A garbage-strewn alley and water channel in Cairo, Egypt, caused by a lack of basic urban services and illegal dumping.

taps or standpipes often function for only a few hours very day. In 2009, the 1.8 million people in Bhopal, India, were rationed to 30 minutes of water supply every other day because the monsoon rains did not come as expected. The 100,000 people there who live in slums without piped water rely on tankers to deliver water. Fights have broken out and people have been hacked to death by angry neighbors who accused them of stealing water.

Some people tap into the main water supply pipe. Others try to collect water leaking from the ventilation valves of the main pipelines. Many women are forced to get up in the middle of the night to walk over a mile to the nearest pumping station where someone has usually removed some bricks from the base that allows a steady flow of water to run out. Many people have no alternative but to use contaminated water—or at least water whose quality is not guaranteed. Less than half of the households in the less developed countries, many of whom are typically the residents of the most affluent neighborhoods, has water piped into their homes.

Sewage services are just as bad. Only just over half of the people in less developed countries are connected to sewers.[13] The bulk of these sewers discharge their waste untreated into a nearby river, a lake, or the sea. São Paulo has a thousand miles of stinking open sewers; raw sewage from the city's slums drains into the Billings reservoir, a major source of the city's drinking water. Jakarta has no municipal sewage system but there are plans to build one. In the meantime, septic tanks serve about one-quarter of the city's population; others use pit latrines, cesspools, and roadside ditches. Despite the fact that India's 900 million Hindus revere and bathe in many of the country's rivers, especially the Ganges, less than 30 percent of the sewage from its towns and cities running into these rivers is treated. Every day, a shocking 7 billion gallons of untreated wastewater runs into India's rivers; the Yamuna collects up to half a billion gallons of this as it passes through Delhi. A major problem is that India's population is growing faster than new sewage treatment plants are being built.

> If there are, as it is often the case in big cities, no toilets . . . people have to use a bucket that will, under the veil of darkness, be tipped out into a drain or sewer tunnel. Or they have to squat down above a spread piece of paper or plastic bag that afterwards goes to the rubbish dump in which dogs or pigs are rummaging about. These packets of dirt are sometimes called "flying toilets" . . . Today millions of people catch all sorts of diseases through contact with excrements that have been left on fields, paths, streets, or the banks of rivers . . .[14]

In the less developed countries, 30 to 60 percent of all municipal household waste is collected. In Dar es Salaam, for example, only 35 to 40 percent of all garbage and solid waste is collected and removed—the rest is partially recycled informally, tipped into gullies, canals, or rivers, or simply left to rot. Even the household waste that is collected can create problems. Neoliberal policies involving privatized collection has resulted in private garbage collection companies taking advantage of lax municipal oversight to dump in unauthorized sites where the foul stench and health hazards affect nearby residents. In the worst cases, workers are not provided with boots and gloves and must pick through garbage containing the rotting corpses of dogs and cats with their bare hands. When these workers become ill, they do not have healthcare coverage to address the commonly reported symptoms of headaches, coughing and diarrhea, and bacterial infections from the decomposing garbage.

Although dreadful, these problems provide employment opportunities for people in the informal sector. Street vendors, who get their water from private tanker and borehole operators, sell water in cans by the gallon. They typically charge 5 to 10 times the local rate that public water utilities charge affluent residents whose homes have piped water; in some cities the street vendors charge 60 to 100 times as much. Similarly, many cities have informal-sector mechanisms for sewage disposal. In some Asian cities, for example, handcart operators remove human waste at night. The problem with this arrangement is that this waste is often disposed of improperly and eventually pollutes the rivers or lakes from which the urban poor draw their water.

In NIEs such as Singapore, South Korea, and Taiwan, an educated labor force was an important factor in economic growth.[15] Yet access to education remains a problem in many countries, especially for rural residents, because the centers of education are concentrated in cities. Investments in education have benefited middle-income groups more than the poor, and men more than women. The bias against women varies from country to country but is most marked in Islamic cities and in the poorest households.

Urbanization does not necessarily bring improved health because physical well-being and poverty are inextricably intertwined.[16] The United Nations has identified a number of problematic issues specific to urban health.[17]

- Urban health patterns are different from those in rural areas because urban populations are leading the *epidemiological transition:* a shift from communicable to noncommunicable diseases. The World Health Organization has predicted that by the third decade of the twenty-first century, the social and economic changes associated with urbanization will make depression, heart attacks, and traffic accidents the leading healthcare burdens in the cities of less developed countries, as opposed to respiratory diseases, diarrhea, and early childhood conditions now.
- Some communicable health problems, including HIV/AIDS, still dominate urban areas. In fact, urbanization has been paralleled by increasing HIV/AIDS cases in cities, particularly in Sub-Saharan Africa (the Urban View 3 entitled "A Terrible Human Toll: HIV/AIDS in Sub-Saharan African Cities"). Within cities, children are particularly susceptible to communicable diseases, including acute respiratory infections.
- Poor urban residents suffer from the health-related problems of rural areas, such as inadequate diet and poor sanitation, in addition to being exposed to health problems

URBAN VIEW 3

A Terrible Human Toll: HIV/AIDS in Sub-Saharan African Cities

Of the 33 million people worldwide who were estimated to be living with HIV/AIDS at the end of 2008, more than 22 million were in Sub-Saharan Africa.[18] Despite having only about 13 percent of the world's population, this region is home to about 67 percent of the infected people. About 70 percent of all new HIV infections are in Sub-Saharan Africa. AIDS has already claimed the lives of more than 15 million people and orphaned 14 million children there. In the worst-hit countries, such as Swaziland and Botswana, life expectancy has plunged by half, to an unimaginable age of 32 in the case of Swaziland.

Infection rates are high in cities because the concentration of large numbers of people at high density increases the speed of transmission of HIV. Prostitution, multiple sexual partners, and teenage pregnancy are more common in cities than in rural areas. In addition to this devastating human tragedy, the spread of HIV/AIDS adds to already overwhelming economic problems and drains limited public resources in Sub-Saharan African cities in a number of ways:[19]

- Productivity is reduced because AIDS deaths decrease the number of experienced workers, especially those in their most productive years.
- International competitiveness is hurt by higher production costs because of a shortage of skilled workers.
- Employment creation is slower due to lower government revenues and reduced private savings.
- Higher public expenditures are needed for monitoring, prevention, and health care.

Research suggests that there is a link between rapid urbanization amid poverty and the urban character of the HIV/AIDS problem. The African dilemma is particularly troublesome because the percentage of urban poor living below the international poverty line of US$1.25 per day is particularly high. In Tanzania and Nigeria, for example, the urban poverty percentages are 70 and 55 percent respectively. In some cities—Kinshasa, Lagos, Karachi, and Dhaka, to name a few—more than 50 percent of residents live in slums and squatter settlements.

A recent study in Nairobi, Kenya, found that poor slum dwellers start sexual intercourse at an earlier age, have more sexual partners, and are less likely than other city residents to know about or adopt condom use or other preventative measures against contracting HIV/AIDS.[20] Three features of slum conditions help socialize children into very early sexual intercourse: poverty, a social context that is accepting of prostitution, and household arrangements that do not allow privacy, causing children to think that sexual intercourse is an activity that they too can participate in at a very early age.

HIV/AIDS poses a double threat for children in Africa: it makes orphans of them by killing their parents and it infects them too. In Sub-Saharan Africa, nearly half a million children die from AIDS every year; most die before reaching their fifth birthday. For most children, their mother transmits the deadly virus to them at birth or through breast-feeding. Children are also infected through sexual contact. Girls have much higher rates of infection than boys of the same age. Many girls are infected because of the cruel myth that has spread across Africa that sex with a virgin can protect against or cure HIV infection. Some girls are raped. Others agree to sex in exchange for payment, usually of school tuition.[21]

The poor economic conditions in Sub-Saharan Africa and its cities increase the likelihood that women, especially adolescents, will engage in risky sexual behavior for economic survival despite the risk of contracting and spreading HIV/AIDS. Desperation can push some women to rely on sexual relations to supplement the household income to cover the cost of rent, schooling, and other basic necessities; many of these women maximize the number of sexual partners in an effort to increase their economic security.

The Nairobi study highlighted the need to treat people in squatter settlements as a low-income group that is particularly vulnerable to contracting HIV/AIDS because of poor access to health services, family planning facilities, education, and basic amenities due to their relative geographic isolation, low incomes, and illegal or informal residence. The results of the study suggest that it might be difficult to improve the reproductive health status of residents in slums and squatter settlements without also improving their economic and living conditions.

associated with factors specific to cities, such as stress arising from overcrowding and poor working conditions.
- The health burden of the urban poor can only be fully understood within the context of overall inequities within cities.

Transportation Problems

Even though city governments in less developed countries typically spend nearly all of their budgets on transportation infrastructure in a race to keep up with population growth, conditions are bad and rapidly getting worse. Cities in less developed countries have always been congested, but in recent years, congestion has turned into near gridlock (Figure 11). In megacities the availability and use of private cars have increased sharply. Not only are there now more people and traffic, but also the changing spatial organization of these cities has increased the need for transportation. Traditional patterns of land use have been superseded by the agglomerating tendencies inherent in modern industry and the segregating tendencies inherent in contemporary societies. The greatest single change has been the separation of home from work, which has caused a significant increase in commuting for many people.

The general trend has been for increasing automobile dependence and decreasing urban density. Figure 12 shows the relationship between city form and dominant transportation system in three kinds of cities:[22]

1. The traditional "walking" or "pedestrian" city with densities of 250–500 persons per acre.
2. The "transit" city with densities of 175–250 persons per acre.
3. The "automobile dependent" city with densities of 25–50 persons per acre.

The trend away from traditional "walking" cities and toward "automobile dependent" cities in the less developed countries has generated unprecedented traffic problems. Despite some innovative responses, transportation systems in many cities are breaking down, with poorly maintained roads, traffic jams, long delays at intersections, and frequent accidents. Many governments have invested in expensive new freeways and street-widening schemes. Because the new roads tend to focus on city centers—still often the settings for the majority of jobs and most services and amenities—they ultimately fail, emptying vehicles into a congested and chaotic mixture of motorized traffic, bicycles, animal-drawn vehicles, and hand-drawn carts. Beijing, Bangkok, and Mexico City have some of the worst traffic problems in the world (Urban View 4 entitled "How Rationing Can Backfire: The 'Day Without a Car' Regulation in Mexico City"). In Bangkok, the 15-mile trip from Don Muang Airport to the city center can take 90 minutes. Traveling at a speed of ten miles an hour would have been welcomed by many people caught up in what may have been the longest traffic jam in the world so far in 2010: from August 14 to 28, road construction and coal trucks caused thousands of vehicles to be stuck in near total gridlock on China National Highway 110 between Inner Mongolia's border and Beijing. Many drivers moved only half a mile each day and some were stuck in the traffic jam for up to five days. People who lived along the highway took advantage of the stranded drivers and sold them water, instant noodles, and cigarettes at, literally, highway robbery prices. The costs to the national economy of these traffic backups can be enormous. In Jakarta, Indonesia, the world's largest city

Hartmut Schwarzbach/Photolibrary, Inc.

FIGURE 11 Traffic on Ratchadmari Road in Bangkok, Thailand. Roads in the larger cities of less developed countries have always been congested, but in recent years, congestion has turned into near gridlock.

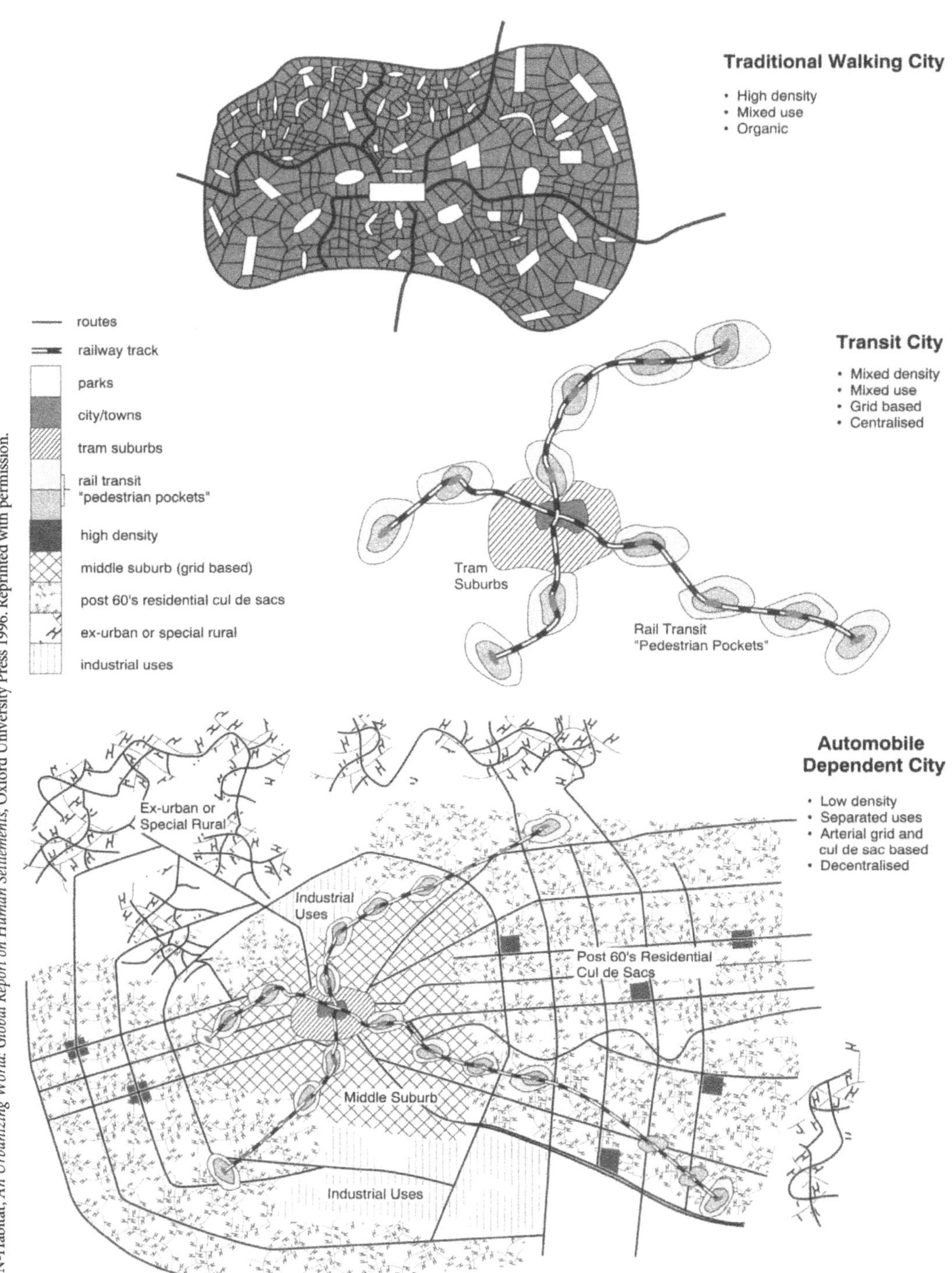

FIGURE 12 Different city forms according to dominant mode of transport.

URBAN VIEW 4

How Rationing Can Backfire: The "Day Without a Car" Regulation in Mexico City[23]

When children in Mexico City paint a picture of their city, they color the sky black. The children are too young to be making a political statement; they are merely painting what they see, a city with one of the worst air pollution problems in the world because of car use.[24] In Mexico City daily traffic backups can extend for miles and it can take 2 to 4 hours to drive across the city (Figure 13).

In an effort to stem the use of cars, Mexico City's government introduced a regulation in 1989 that banned the use of each car on a specific day of the week. The "Day Without a Car" regulation—*Hoy no circula* ("today it does not circulate")—prohibited cars with license plate numbers ending in 1 or 2 from being driven on Thursdays, 3 or 4 from being driven on Wednesdays, and so on.

URBAN VIEW 4

How Rationing Can Backfire: The "Day Without a Car" Regulation in Mexico City (*continued*)

age fotostock/Superstock

FIGURE 13 Avenida de la Reforma, Mexico City, Mexico is clogged with traffic and smog during the afternoon rush hour.

The regulation was controversial. Proponents believed that it was reasonable to expect car owners to contribute to easing traffic congestion and air pollution problems. Those who opposed the regulation argued that it was unfair because some people could avoid the ban more easily than others.

Ultimately, the regulation was counterproductive and a failure—it caused higher congestion and pollution levels because people who could afford a second car purchased one and effectively circumvented the ban on their first vehicle. Total car use and traffic congestion in Mexico City actually increased due to the regulation.[25] Even though a second car was purchased primarily to replace a first car on its driving ban day, total car use increased in households with multiple drivers because a second car was now available. The regulation led to more rather than less pollution because, in addition to higher car use causing more emissions, people tended to buy older second cars with lower technical standards. Although Mexico City introduced emissions testing in 1990 in an attempt to address the problem of higher emissions from older cars, many drivers cheat by bribing testing technicians or bringing another car to the emissions test.

without a rapid transit system, where traffic moves at an average speed of 8 miles per hour, estimates of the annual cost of traffic delays range from US$1.4 billion by the national government to US$4.9 billion by the World Bank.

Environmental Degradation

Due to pressing problems of poverty, poor housing, and inadequate service and transportation infrastructures, cities in less developed countries are unable to devote many resources to environmental issues (Figure 14). Because of the speed of population growth, these problems are escalating rapidly. Industrial and human wastes pile up in lakes and lagoons and pollute long stretches of rivers, estuaries, and coastal zones. Chemicals leaching from uncontrolled dump sites pollute groundwater. The demand for timber and domestic fuels are denuding forests near many cities. This environmental degradation is, of course, directly linked to human health. People living in such environments have much higher rates of diarrhea, respiratory infections, and tuberculosis and much shorter life expectancies than people living in the surrounding rural communities. Children in squatter settlements may be 50 times as likely to die before the age of five as those born in affluent countries.

Air pollution has escalated to hazardous levels in many cities. With the growth of a modern industrial sector and car ownership, but without enforceable pollution and vehicle emissions regulations, tons of lead, sulfur oxides, fluorides, carbon monoxide, nitrogen oxides, petrochemical oxidants, and other toxic chemicals are pumped into the atmosphere every day in large cities. The burning of charcoal, wood, and kerosene for fuel and cooking in low-income neighborhoods also contributes significantly to poor air quality. The World Health Organization has estimated that up to 2 billion people worldwide die prematurely every year as a result of air pollution, and that many more suffer from breathing problems, heart infections, lung infections, and even cancer. A recent study of 18 megacities found that all had at least one major air pollutant at levels exceeding World Health Organization (WHO) guidelines.[26] Based on WHO guidelines, 13 were ranked as poor and 5 were ranked fair. Based on a multi-pollutant index, the most polluted megacities are Dhaka in Bangladesh, Beijing in China, Cairo in Egypt, and Karachi in Pakistan.

Such pollution is not only unpleasant, but dangerous. In Dhaka, it is estimated that every year 15,000 people die prematurely, and several million people suffer from respiratory, heart, and other illnesses attributed to poor air quality.

Kevin R. Morris/Corbis

FIGURE 14 Garbage floats behind the sampans of the Chao Praya River in Bangkok, Thailand.

Proximity to industrial facilities—often because poor people need to live near their place of employment—poses another set of risks. In 1984 the world's worst industrial accident at the Union Carbide chemical factory in Bhopal, India, caused nearly 4,000 deaths and more than 500,000 injuries, mostly among the residents in the nearby squatter settlements. The interrelationships between poor housing conditions, poverty, and urban environmental problems reflect a number of factors:

- A greater incidence of sickness and death is closely associated with housing that is crowded, poorly built, located in unsafe areas (subject to natural and other hazards), and inadequately serviced by water facilities, sewage treatment, and garbage disposal.
- Poor quality housing is often makeshift and temporary, dilapidated, poorly maintained, and fire-prone.
- Low incomes combined with rising land and property prices result in overcrowding and homelessness.
- Insecure and illegal land tenures come with eviction and demolition, which remove any motivation to invest in home improvements.
- The poorest members of society are the most disenfranchised and the least able to articulate their concerns about environmental and other problems.[27]

RESPONSES TO THE PROBLEMS OF URBANIZATION

Sustainable Urban Development

The relationship between urban economic development and environmental conditions has attracted increasing attention within the context of sustainable development. There is now greater recognition that the responses to the problems of urbanization must be framed within the context of the concept of **sustainable urban development**. Responses to the urbanization problems in the less developed countries must address several interconnected components of sustainable urban development (Figure 15):[28]

- *Economy:* The most sustainable economic activities are those that integrate the external role of cities within the regional, national, and world economies with the needs

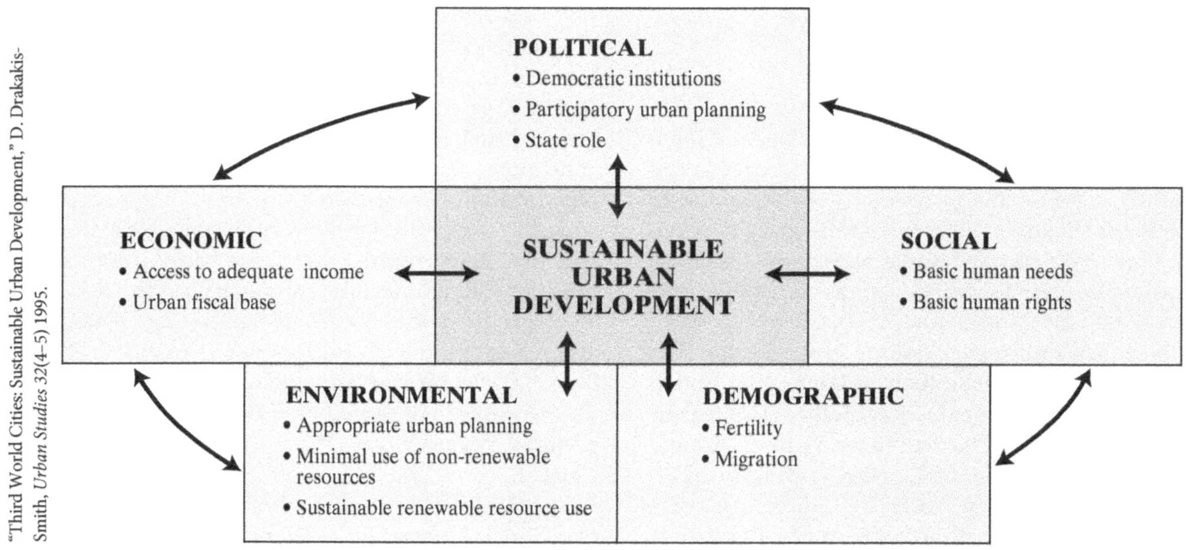

"Third World Cities: Sustainable Urban Development," D. Drakakis-Smith, *Urban Studies* 32(4–5) 1995.

FIGURE 15 The components of sustainable urban development.

of the labor force and the revenue requirements of city governments. Jobs in the informal sector of the economy have a role, but government employment policies are also necessary. Addressing poverty is fundamental not only for the well-being of urban residents but also for the prosperity of the cities themselves.

- *Environment:* The *Brown Agenda* and *Green Agenda*[29] involve addressing urban environmental problems such as pollution and land degradation (Table 3). They are important for poor people in cities in the less developed countries because environmental concerns are not a luxury.[30] Both agendas focus on the complex and unintended side-effects of human activity. The Brown Agenda is concerned more with immediate, localized, and health-related effects. It emphasizes the necessity to reduce the environmental threats to health that are associated with poor sanitary conditions, crowding, inadequate water provision, hazardous air and water pollution, and local solid waste accumulation. The Green Agenda is concerned more with delayed, dispersed, and ecological effects. It emphasizes reducing the impacts of urban-based production, consumption, and waste generation on natural resources and ecosystems at regional and global scales. Prevention of environmental problems is central to both agendas. Both acknowledge the challenge of ensuring that people whose principal motivations are elsewhere should be made to take environmental impacts into account. Both agendas are concerned with equity: the Brown Agenda is concerned more with the burdens on low-income groups in the present and the Green Agenda with burdens likely to affect future generations. Some of these sustainability issues, such as water and air quality, are citywide and are primarily the responsibility of government planning efforts. Other issues, involving renewable and especially nonrenewable resources, can involve individuals and households to a greater extent.
- *Society:* The basic needs of society include a range of issues, from shelter to food to education to healthcare. Government has a role in meeting citywide needs, such as schools and hospitals, as well as in improving human rights. Opportunities for genuine community involvement and self-help programs are also important.
- *Demographic situation:* Exceptionally high rates of natural increase of the population within cities, combined with often massive **rural-to-urban migration**, have not only overwhelmed the resources of city governments, but also intensified and complicated the social and ethnic dimensions of urbanization, threatening the sustainability of much urban development.
- *Political sphere:* The role of government is an important dimension of urban sustainability. Governments can help to establish democratic institutions and a planning process that allows the participation of all members of society, including the poorest people. The political arena is also the point at which urban sustainability, as a concept, overlaps with urban management as a practical process.

TABLE 3 Characteristics of the Brown and the Green Agendas for Sustainable Urban Development

	The "Brown" Environmental Health Agenda	The "Green" Sustainability Agenda
Characteristics of problems high on agenda:		
First order impact	Human health	Ecosystem health
Timing	Immediate	Delayed
Scale	Local	Regional and global
Worst affected	Lower income groups	Future generations
Characteristic attitude toward:		
Nature	Manipulate to serve human needs	Protect and work with
People	Work with	Educate
Environmental Services	Provide more	Use less
Aspects emphasized in relation to:		
Water	Inadequate access and poor quality	Overuse; need to protect water sources
Air	High human exposure to hazardous pollutants	Acid precipitation and greenhouse gas emissions
Solid waste	Inadequate provision for collection and removal	Excessive generation
Land	Inadequate access for low income groups for housing	Loss of natural habitats and agricultural land to urban development
Human waste	Inadequate provision for safely removing fecal material (and waste water) from living environment	Loss of nutrients in sewage and damage to water bodies from its release of sewage into waterways
Typical proponent	Urbanist	Environmentalist

Source: G. McGranahan and D. Satterthwait, "Environmental Health or Ecological Sustainability: Reconciling the brown and green agendas in urban development," in C. Pugh (ed.), *Sustainable Cities in Developing Countries*, London: Earthscan, 2000.

The "Globalization Paradox" and Recent Changes in Urban Governance

The United Nations[31] has drawn attention to the "globalization paradox" faced by people in cities in the less developed countries as they try to achieve sustainable urban planning and management. The increased competition and fragmentation associated with **globalization** and neoliberalism are having contradictory effects on cities. The residents, businesspeople, and government officials of a city must act as a collective unit to respond to international competition and compete effectively in the world economy while being less able to rely on higher levels of government for assistance. At the same time, growing social and economic fragmentation hampers the capacity of people in cities to build coalitions, mobilize resources, and develop good governance structures. This predicament is particularly detrimental when urban change is dramatic and requires enhanced decision-making capacities—as in many of the cities and metropolitan regions in less developed countries.

During the last few decades, four changes in urban governance have emerged as attempts to address the "globalization paradox" in the less developed countries.[32] The relative importance of these changes varies among countries, and they do not necessarily herald the end of more traditional systems of governance. Some of the changes, such as decentralization and formal government reforms, have been around for some time. What is new is the rationale for introducing them.

1. *Decentralization and formal government reforms:* Although certain issues that affect cities and the people who live in them, such as rural poverty and rural-to-urban migration, are best addressed at the national scale, many urban problems are better tackled closer to the source. This principle underlies the trend toward decentralizing power to regional and local levels of government. Decentralization has included strengthening the powers of city government politicians, such as in some Latin American countries.

   Regional programs—such as the Consulting and Training System for Local Development (SACDEL) in Latin America—support the decentralization efforts. Funding and assistance for SACDEL come from the national governments of Latin American countries such as Colombia, Costa Rica, Ecuador, and Peru, as well as the Economic Development Institute of the World Bank and UN-Habitat, to name a few. The program operates from a regional center in San Salvador, El Salvador, and provides training and assistance to public and private training agencies that are charged with enhancing the local capacity of municipal governments in Latin America.[33]

   There have also been attempts at formal government reform at the metropolitan level—including establishing new government structures and agencies with *metropolitan-wide* responsibilities for strategic planning, economic development, urban services, and, more recently, environmental protection. The rationale for these reforms includes long-established goals of efficiently and cost effectively providing urban services and infrastructure. More recent reasons for reform include the need to devise and implement policies that better address metropolitan-wide problems of environmental degradation, social exclusion, crime, and violence.

   The United Nations maintains that strategies that create new governmental forums can increase the voice of marginalized groups, particularly in cities where ethnic minorities are geographically concentrated. The Popular Participation Law that was passed in 1994 in Bolivia, for example, created municipal councils in which representatives from the Quechua (descendants of the Incas) and Aymara (pre-Inca ethnic group) minorities play a role in allocating resources within cities. In India a constitutional amendment was passed in 1993 that reserved one-third of the seats in local government for women. In 2009, the Indian government took a step further and became the first country to reserve 50 percent of these seats for women. Mandating the representation of women in new governmental forums is a good first step; but the empowerment of women in many less developed countries is often still constrained by traditional gender relations.

2. *Participation of local communities in urban policymaking and implementation:* The emphasis is now on the internal opportunities for development, using models, assumptions, and objectives that have been devised within the less developed countries themselves. This shift has been associated with an increasing recognition of the importance of local context for development and growing support for participatory approaches that give citizens a stronger voice in urban policymaking and implementation.

   The greater involvement of urban residents—notably through participation by community-based organizations—is becoming more common in urban planning. By helping to legitimize policymaking, resident participation in devising and implementing policies—by both men and women—can make public policy implementation more efficient and responsive to the needs of the community (Figure 16). This is particularly important in cities where,

Jenny Matthews/Alamy

FIGURE 16 Women from a self-help sanitation committee build latrines in an effort to raise the level of health in their community in Chinyamunyamu, Malawi.

because of the inability of local public institutions to provide basic urban services, people have had to organize themselves to undertake self-help water and sanitation projects (Urban View 5 entitled "Urban Social Movements and the Role of Women: Mahila Milan in Mumbai, India").

When local citizens in Dar es Salaam, Tanzania, found that the government was economically and technically unable to improve the water supply, they looked for affordable ways to do it themselves.[34] Local residents formed community-based organizations with elected members, such as the Kijitonyama Development Community (KIJICO) in 1991 for the 30,000 residents in the more than 5 square mile Kijitonyama neighborhood. Negotiations between KIJICO and the city government with the participation of outside agencies like the World Bank resulted in the residents receiving assistance for a number of projects, including constructing a deep well and a more than 9-mile water distribution network with reservoirs and water pumps with a capacity of nearly 32,000 gallons a day.

Orangi, a community of about 1 million people in Karachi, Pakistan, had no public sanitation system because it was a squatter settlement.[35] In 1980, the community established the Orangi Pilot Project, which helped them to organize into small groups of households surrounding each lane. The Pilot Project helped them to survey the site, draw up plans, and prepare cost estimates. The residents collected money to pay for the sewer installation. As more households became involved, the local government began to provide some financial support. Since then, the residents administered the construction of more than 72,000 latrines and 1.3 million feet of sewer lines.

3. *Multilevel governance and public–private partnerships:* Multilevel governance involves a set of joint practices, including partnership and cofunding. Decentralization has increased the overlap of responsibilities, creating the need to involve different levels of government in urban policymaking. In many cities multilevel governance involves not only public institutions but also private companies, nonprofit agencies, and NGOs. The private sector is often involved because of neoliberal privatization policies, as when the water systems were privatized in many African countries. Community-based organizations may be involved because they have legitimacy in representing the people and first-hand knowledge of local problems. NGOs too can play an important role because they are knowledgeable about program management or because they have assumed certain responsibilities from government. Multilevel governance often comes about for

URBAN VIEW 5

Urban Social Movements and the Role of Women: Mahila Milan in Mumbai, India

In many cities in less developed countries, poor residents are joining self-help efforts to improve their living conditions.[36] When community-based organizations form broad-based associations, the possibility of an **urban social movement** arises. An urban social movement can be defined as a social organization with a territorially based identity that strives for emancipation through collective action. The concept of emancipation here involves satisfying basic needs (including housing and urban services), developing a respectful attitude toward the environment, lack of discrimination (on the basis of race, ethnicity, religion, gender, socioeconomic status, or residential location), and gaining access to urban policymaking and implementation.

Theorists like Manuel Castells believe that a basic characteristic of a true urban social movement is that it seeks societal change.[37] Others, like Schuurman and van Naerssen,[38] argue that the poor in the less developed countries might not be able to achieve societal reform because they cannot always overcome the political and social constraints that limit their efforts. Nevertheless, urban social movements often represent the only mechanism for improving the living conditions of poor people and are becoming more prevalent in the less developed countries. Because women comprise a large percentage of the poor, they have high participation rates in urban social movements in many countries, including India.

In 1985, a Supreme Court ruling gave the Municipal Corporation of Greater Mumbai the authority to demolish pavement dwellings in the city. Pavement dwellings are the most basic kind of informal settlement in megacities. As their name suggests, these shacks are built on the sidewalks and run like ribbons along city streets; this makes them more vulnerable to demolition than larger squatter settlements on less central stretches of vacant land. A group of 500 poor women organized and formed Mahila Milan ("women together" in Hindi) to successfully prevent the demolition of their pavement dwelling homes.[39]

Today, Mahila Milan has grown to become a decentralized network of women's collectives that empowers women to become involved in community and city issues. For example, the women established a crisis credit fund by saving tiny amounts of money on a regular basis so that they could make small emergency loans to women in poor communities for crises like medical injuries where a family needed to borrow money to purchase medicine or to when a wage earner lost a job and the family needed a loan to buy food. Undeterred by illiteracy, they record their savings with different colored squares of paper representing different amounts of money.

Although these poor women cannot save enough for larger loans for upgrading a shelter or buying a home, the women have gained skills and confidence from managing their savings and loans. They expanded their credit scheme with the help of NGOs and established a fund that offers micro-loans for productive enterprises in the informal sector such as vegetable and fruit vending, carpet repair, rag picking, and garbage recycling. Because the women carefully consider the viability of the proposed micro-enterprise, the creditworthiness of the borrower, and the terms of the loan, including interest rate, they have had an almost 100 percent rate of repayment.

practical reasons, but it is increasingly considered a new way of policymaking and implementation.

For example, a partnership of residents, national and local government agencies, and NGOs came together to set up the Masese Women's Self Help Project in 1989 to build 700 homes in the Masese slum in Jinja, Uganda. With the municipal government providing the land and the Project receiving a loan from an NGO through a national government department, the women received secure land tenure and ten-year loans to pay for construction materials. The loans were intended to be provided through a revolving loan fund that would be able to offer future loans once the earlier loans were paid off. More than 300 new homes were built, but twenty years on, only 60 loans had been paid off because of lack of income, which has severely restricted new loans from being made to other residents from the revolving loan fund.[40]

4. *Process-oriented and area-based policies:* Urban governance today involves a system in which a variety of participants work in partnership with government agencies. This complicates policymaking. Policymaking now involves coalitions and compromises; it also requires discussion and debate, which, in turn, depend on appropriate negotiation procedures. This is a new way of reaching decisions in which policy content is, at least in part, a function of the decision-making process itself. These new forms of collective action can no longer take place at the central government level. Area-based regulation has replaced national regulation because of the government's inability to solve problems and address issues at lower levels of government and the need to integrate the diverse elements of effective public policies.

The extent to which these changes in governance address the "globalization paradox" and represent the kind of policymaking and implementation that move cities in the less developed countries toward sustainable urban development can be judged using certain basic criteria.[41] These include the quality of life of urban residents, including levels of poverty, social exclusion, and human rights; the scale of nonrenewable resource use and waste recycling; and the scale and nature of renewable resource use, including freshwater resources.

The cities and governments in the less developed countries—especially the poorest ones—face a daunting set of problems with often limited resources. There is little comfort in the fact that, despite significantly higher levels of prosperity, problems of urbanization persist in the cities of developed countries like the United States.

FOLLOW UP

Key Terms

- Brown Agenda
- decentralization of government power
- digital divide
- dualism
- epidemiological transition
- Green Agenda
- "globalization paradox"
- international poverty line
- natural increase
- push and pull factors
- slums
- subsistence activities
- sustainable urban development
- underemployment
- urban social movement

Review Activities

1. Go to YouTube and watch for a short video from a reputable source that shows how people are attempting to overcome urban problems in Africa, Asia, or Latin America. What surprised you about the problem and how the people tried to overcome it?
2. Go online and collect some digital images that document some of the acute housing, traffic, and pollution problems in the cities of the less developed countries. Display the images with accompanying captions in a short Microsoft PowerPoint presentation©.
3. Because of their rapid growth and high underemployment, it is among the peripheral metropolises of the world—such as Mexico City (Mexico), São Paulo (Brazil), Delhi and Mumbai (India), Dhaka (Pakistan), Jakarta (Indonesia), Karachi (Pakistan), and Manila (the Philippines)—that we can find contenders for the title of "shock city" of the twenty-first century. Using online and library sources, document the most remarkable and disturbing economic, social, and cultural changes for one of these megacities. Think about how and why

the processes and outcomes you identify for this "shock city" in a less developed country today are similar to and different from those in Manchester (UK) and Chicago in the nineteenth century.

4. Work on your *portfolio.* Collect official figures that document some urban problems in less developed countries that particularly interest you. The United Nations (UN-Habitat) has economic and social indicators for selected cities in recent editions of its annual *Global Report on Human Settlements* or *State of the World's Cities* that you can find in the library or on the UN's website (*http://www.unhabitat.org/*).

Log in to **www.mygeoscienceplace.com** for self-study quizzes, *MapMaster* layered thematic and place name interactive maps, *Urban View* Google Earth™ tours, key resources and suggested readings, related websites, "In the News" RSS feeds, and additional references and resources to enhance your study of urban problems and responses in the less developed countries.

NOTES

1. Based on S. LaFraniere, "Cell Phone Frenzy in Africa, World's Top Growth Market," *New York Times*, August 26, 2005; Katine Chronicles blog, *Guardian.co.uk*, January 14, 2010, *http://www.guardian.co.uk/katine/katine-chronicles-blog/2010/jan/14/mobile-phones-africa.*
2. Ibid.
3. United Nations Centre for Human Settlements (UN-Habitat), *State of the World's Cities 2010/2011: Bridging the Urban Divide* (London: Earthscan, 2008).
4. J. Cohen, *Islamist Radicalism in Yemen*, backgrounder, June 29 (Washington, D.C.: Council on Foreign Relations, 2010).
5. D. Drakakis-Smith, *Third World Cities*, 2nd ed. (London: Routledge, 2000).
6. R. B. Potter and S. Lloyd-Evans, *The City in the Developing World* (Harlow, UK: Longman, 1998).
7. Ibid.
8. Based on A. Guillermoprieto, "Bolivia's Wrestlers," *National Geographic*, September 2008; R. Carroll and A. Schipani, "Bolivia: Welcome to Lucha Libre—The Sport for Men making Heroes of Women," *Guardian*, August 30, 2008.
9. Drakakis-Smith, *Third World Cities.*
10. United Nations Centre for Human Settlements (UN-Habitat), *Cities in a Globalizing World.*
11. J. Stewart and P. Balchin, "Community Self-Help and the Homeless Poor in Latin America," *The Journal of the Royal Society for the Promotion of Health* 122 (2002): 99–107.
12. UNICEF and World Health Organization, *Progress on Drinking Water and Sanitation* (UNICEF, New York and WHO, Geneva, 2008).
13. Ibid.
14. M. Black, "Solutions for the Global Waste Problem," *Le Monde Diplomatique* February 12, 2010 (quoted in *Mother Pelican* 6 (2010):1).
15. Drakakis-Smith, *Third World Cities.*
16. Ibid.
17. United Nations Centre for Human Settlements (UN-Habitat), *Cities in a Globalizing World.*
18. See the United Nations website: *http://www. unaids.org.*
19. ILO (International Labor Office), *HIV/AIDS: A Threat to Decent Work, Productivity and Development* (Geneva: ILO, 2000).
20. E. Msiyaphazi Zulu, F. Nii-Amoo Dodoo, and A. Chika-Ezeh, "Sexual Risk-Taking in the Slums of Nairobi, Kenya, 1993–98," *Population Studies* 56 (2002): 311–23.
21. J. Maxwell, "Africa's Lost Generation," *CNNfyi.com.*
22. P. Newman, J. Kenworthy, and P. Vintila, *Housing, Transport and Urban Form* (Canberra, Australia: Institute for Science and Technology Policy, Murdoch University, 1992).
23. See G. S. Eskeland and T. Feyzioglu, "Rationing Can Backfire: The 'Day without a Car' in Mexico City," *The World Bank Economic Review* 11 (1997): 383–408.
24. L. Sweet, "Room to Live," *Healthy Cities for the Urban Century*, IDRC Briefing No. 4.
25. Eskeland and Feyzioglu, *The World Bank Economic Review.*
26. B. R. Gurjar, A. S. Nagpure, and T. P. Singh, "Air Quality in Megacities," in *Encyclopedia of Earth*, C. J. Cleveland (ed.) (Washington, D.C.: Environmental Information Coalition, National Council for Science and the Environment, 2008).
27. S. W. Williams, "'The Brown Agenda': Urban Environmental Problems and Policies in the Developing World," *Geography* 82 (1997): 17–26; World Bank, *Housing: Enabling Markets to Work*, A World Bank Policy Paper (Washington, D.C.: World Bank, 1993).
28. D. Drakakis-Smith, "Third World Cities: Sustainable Urban Development, 1," *Urban Studies* 32 (1995): 659–77.
29. G. McGranahan and D. Satterthwait, "Environmental Health or Ecological Sustainability: Reconciling the Brown and Green Agendas in Urban Development," in C. Pugh (ed.) *Sustainable Cities in Developing Countries* (London: Earthscan, 2000).
30. Massachusetts Institute of Technology's Environmental Strategies for Cities website (*http://web.mit.edu/urbanupgrading/urbanenvironment/*).
31. United Nations Centre for Human Settlements (UN-Habitat), *Cities in a Globalizing World*, 58.
32. Ibid.
33. United Nations Centre for Human Settlements (UN-Habitat), *An Urbanizing World: Global Report on Human Settlements, 1996* (Oxford, UK: Oxford University Press for UN-Habitat, 1996), 331, Box 9.25.
34. United Nations Centre for Human Settlements (UN-Habitat), *Cities in a Globalizing World*, 125, Box 10.4.
35. United Nations Centre for Human Settlements (UN-Habitat), *An Urbanizing World*, 313, Box 9.18.
36. F. Schuurman and T. van Naerssen, *Urban Social Movements in the Third World* (London: Routledge, 1989).
37. M. Castells, *The City and the Grassroots: A Cross-Cultural Theory of Urban Social Movements* (London: Arnold, 1983).
38. Schuurman and Naerssen, *Urban Social Movements in the Third World.*
39. Society for the Promotion of Area Resource Centres (SPARC) in India website: *http://www.sparcindia.org/.*
40. D. Musingo, "Masese Tenants Fail to Pay Loan," *New Vision Online* November 11, 2010 (*http://www.newvision.co.ug/*)
41. United Nations Centre for Human Settlements (UN-Habitat), *An Urbanizing World*, 422.

The Urban Development Process

Kevin Allen/Alamy

The Urban Development Process

It is useful to think of urban development as a process that involves people as decision makers, each with rather different goals and motivations. Although architecture and urban design are important in contributing to the distinctive character of different parts of cities, much of the decision making about *what kind* of structure gets built and *when* and *where* is in the hands not of architects and urban designers but of other "city makers" such as developers and politicians. As these different groups of people interact with one another over specific development issues, they constitute an organizational framework for city building. These frameworks have been called "structures of building provision" and it is through them that the built environment is created and modified. The resulting process of urban development involves a complex ebb and flow of investment, disinvestment, and reinvestment.

LEARNING OUTCOMES

After reading this chapter, you should be able to:

- Explain how the flow of investment in the built environment is influenced by the interaction of rent, market conditions and investment cycles in the broader economy.
- Describe the evolution of real estate industry in the United States and explain how U.S cities have been affected by changes in the organization and dynamics of the industry.
- Account for the roles and linkages between the various people and mediating institutions (such as government agencies and banks) that are involved in the process of actually building the urban environment.
- Recognize the increasingly diverse range of commercial and residential of types of "building products" that are available to customers.

CHAPTER PREVIEW

In this chapter we take a closer look at the processes and some of the principal groups of people involved in the production of the built environment. What makes the built environment especially interesting within the overall context of urbanization is that it reflects, through its very creation, the decisions of *city makers* such as landowners, financiers, developers, builders, politicians, and bureaucrats as well as members of the design professions. Townscapes must be seen as the culmination of land development processes that involve all these key people within the framework of changing metropolitan form, land use, and architecture. Understanding these processes requires us to identify the key people, their motivations and objectives, their interpretations of market demand, and their relationships with one another (Urban View 1 entitled "Two Sides of the Housing Crisis: Skateboarding and Foreclosure"). These, in turn, must be seen within the context of one key precondition for investment in the built environment: the expectation of a satisfactory rate of return on a prospective project. We begin, therefore, with a review of the relationships between property, location, rent, and investment.

What makes the built environment especially interesting within the overall context of urbanization is that it reflects the decisions of city makers such as landowners, financiers, developers, and politicians. In London, U.K., public funding leveraged billions of dollars in private investment so that a formerly derelict dockland area now has a new image—created by the largest single urban redevelopment scheme in the world—centered on the sleek office towers of Canary Wharf.

URBAN VIEW 1
Two Sides of the Housing Crisis: Skateboarding and Foreclosure[1]

On a recent sunny Saturday morning, 23-year-old Josh, who goes by J Mac, was smiling as wide as a *half-pipe* as he did an *ollie* and acrobatically launched himself into the air before landing and skating up the other side of the swimming pool of a foreclosed home in Fresno, California. Skateboarders use real estate tracking websites or satellite images from *Google Earth* to find foreclosed homes with pools. Across the country, swimming pools, once conspicuous indicators of suburban middle-class success, are now stinking symbols of the housing crisis as the foul stagnant water attracts uninvited guests like smelly algae, putrid muck, and scurrying rats. J Mac is uninvited too, but at least he arrived prepared to clean out the pool using his portable gas-powered pump, shovel, and mop. Skateboarders admit that what they are doing is illegal but maintain that they are helping the environment because the standing water in these pools is a rich breeding ground for mosquitoes carrying the West Nile virus. Skateboarders like J Mac also uphold skateboarder rules: no trash or graffiti and never use the houses, only the pools (Figure 1).

Tom and Anne Smith and their two young children have fond memories of living in that house and swimming in that pool. They bought the house with a $200,000 mortgage when Tom was making good money as an Internet marketer. As their equity in the house rose, they borrowed another $100,000 to pay off credit card bills and install the swimming pool. When Tom was suddenly laid off, they could not keep up with the mortgage payments and the bank foreclosed on their home. They moved in with Tom's sister. Her two bedroom apartment is much too small for all of them but they have a *foreclosure* on their credit history and, anyway, cannot afford to rent an apartment of their own. Tom feels depressed and guilty about the foreclosure. "When I took out the mortgage, I committed to making the payments, and then, all of a sudden, we didn't have the money. Then you have to look your kids in the eye and tell them that they have to leave the home and pool that they love so much."

PYMCA/Alamy

FIGURE 1 Skateboarding in the swimming pool of a foreclosed home.

PROPERTY, LOCATION, RENT, AND INVESTMENT

In general terms, investment in the built environment depends on:

- efficiently managing capital by mediating financial institutions, responding to prevailing rates of return in the different circuits of capital investment flows, and
- government intervention, acting on principles of **Keynesian** economic management and principles of social welfare and conflict resolution or neoliberal ideologies and principles of competition and the free market.

We must also recognize that investment in the built environment depends in part on the broader context of other investment opportunities. One dimension of this broader context is related to the role of the built environment in relation to the **circuits of capital investment** flows that maintain the circulation of the economy. One important aspect of this is the role of investment in the built environment as a response to the occasional **overaccumulation crises** of capitalism.

Overaccumulation occurs when three surpluses exist at the same time: surplus labor (unemployed and underemployed people), surplus productive capacity (idle factories and machinery), and surplus capital (because investors with profits, interest, and dividends cannot find enough reasonably safe, profitable investment opportunities in productive businesses). At such times capital seems to "switch" from the primary circuit (investment in manufacturing production) into the secondary circuit (fixed capital assets such as buildings) or the tertiary circuit (science, technology, and social infrastructure like education), to alleviate the consequences of underconsumption.

It was geographer David Harvey who first conceptualized these circuits and pointed to the central role of the built environment and the city-building process in the overall dynamics of capitalism.[2] Harvey has shown not only that investment in the built environment is critical at times of overaccumulation but also that it is consistently important as a precondition for successful capital accumulation. If the built environment is not renewed and extended, the economy will stagnate and social tensions will develop.

A second dimension of the broader context of investment opportunities stems from the overall relationship between the supply of property, prevailing interest rates, and current rates of profit from investment in property. Using the example of investment in housing, we can identify four main ways in which rent interacts with market conditions to affect the flow of capital into the built environment:[3]

1. If interest rates are high relative to profits from house building, there will be a financial disincentive for capital to go into new construction, even if there is a shortage of housing. Continued housing shortages, however, will push rents higher for existing dwellings, which in turn may encourage speculative investment in vacant land. Eventually, continued housing shortages will result in higher profits from construction, and lead to renewed investment in housing.
2. If interest rates are high relative to profits from house building but housing space is plentiful, rents will be depressed and there will be a tendency for disinvestment. This trend will continue until the supply of housing is reduced to the point that rents rise, leading to an eventual upturn in investment in housing.
3. With low interest rates relative to profits from house building and a shortage of housing space, land acquisition and construction will boom until the backlog in demand is met. Typically, construction overshoots this point, leading to overbuilding, high vacancy rates (though not necessarily of the newly constructed housing), lower profits, and lower rents.
4. When interest rates are low relative to profits from house building, there will be a speculative boom even if there is plenty of housing space. This boom may also encourage the upward revaluation (on paper, at least) of existing investments in housing.

A third dimension of the broader context of investment opportunities operates at the intrametropolitan scale and involves the existence of localized property submarkets. Although interest rates do not normally vary within metropolitan areas, other market conditions do. Demand for certain types of residential, commercial, retail, or industrial space varies sharply from one part of the metropolis to another. At the same time, different parts of the metropolis consist of property with different degrees of capital investment ("capitalization"). In some areas, new, highly capitalized property precludes most development initiatives. Around the fringes of the metropolis and in some isolated pockets within the metropolis are parcels of undeveloped property that are open to a variety of initiatives. And in between are older properties that are undercapitalized and devalued by age *where renewed investment could yield higher rents.*

Geographer Neil Smith described this last situation as representing a **rent gap**. We will see the importance of rent gaps in explaining processes of neighborhood change such as **gentrification**. For the moment, however, it is sufficient to note their role in the overall process of **uneven development** that is a fundamental characteristic of capitalist development at any scale. Investment always flows to locations with relative advantages in terms of cost and revenues where, other things being equal, rates of return will be highest. In the case of property development, capital moves away from locations where land and building costs are high to those where they are low; from locations where revenues are low to those where they are high; and from undercapitalized projects (such as older housing or outdated 1960s enclosed shopping malls) to projects that can command higher rents at the same location (such as condominiums or new shopping malls designed to replicate the traditional town center main street). There is, therefore, a constant restlessness to the built environment, as both simultaneous and sequential processes of investment, disinvestment, and reinvestment take place.

Historically, real estate development in U.S. cities was a predominantly local affair, organized on a project-by-project basis by real estate promoters, financiers, or investors and implemented under contract by small, local construction firms. The stabilization of the mortgage market and the setting of national minimum standards for housing financed by the Federal Housing Administration in the 1930s allowed more and more builders to become what Marc Weiss calls "community builders"—developers who design, engineer, finance, construct, and sell buildings on extensive new subdivisions.[4] These community builders were the precursors of the developer-builder companies that now dominate the design and construction of the U.S. metropolitan built environment. It was the community builders of the 1930s and 1940s who pioneered deed restrictions mandating uniform building lines, front and side yards, standards for lot coverage and building size, and minimum construction standards, as well as innovations in landscaping, street layout, and planned provision for retail and office buildings, parks and recreation facilities, and churches and schools. After World War II, it was the community builders who extended these features to developments for middle-income homebuyers, unfolding the "democratic utopia" of post-World War II suburban development. This democratic utopia was also, of course, a developers' utopia, with sustained high levels of demand, plenty of land, relatively cheap capital, weak environmental regulations, and little opposition to development in the form of **NIMBYism** ("Not In My Backyard").

More recently, the development industry has followed the trends of other producer and service industries—with mergers and acquisitions, vertical and horizontal integration, product diversification, the deployment of new technologies, just-in-time delivery, and *niche marketing*—resulting in a much greater market dominance of big, publicly traded companies with complex and sophisticated operations. The profitability of smaller firms has been constrained by the economies of scale and scope enjoyed by these larger firms. Smaller firms also find it harder to deal with the dramatic increase in NIMBYism; with the widespread introduction of impact fees; and with environmental regulations that are now more complex and more strictly enforced.

At the same time, neoliberal reforms that began with the Reagan administration have weakened trade unions, radically altered the system of housing finance, loosened capital markets, and weakened corporate tax law: For the larger firms, it is still a developers' utopia. The housing booms of the 1980s and the late 1990s to mid-2000s that accompanied the emergence of the new, polycentric metropolitan afforded billions in profits for the largest firms. And, although the housing market crashed in 2008 (Urban View 2 entitled "Global Financial Meltdown, Local Disinvestment"), the long-term prospect for developers is rosy: the United States will add approximately twenty-eight million households by 2025, along with about forty-five million new jobs, turbocharging the infinite game of real estate development. Two million homes will need to be constructed each year, and non-residential construction may top three billion square feet annually. Up to $30 trillion will be spent on development between 2000 and 2025. Half the residential structures expected to be in place by 2025 did not exist in 2000. Developers' decisions in orchestrating and delivering all this will not only determine their commercial success but also influence the evolution of the form and appearance of the U.S. cities. The leading edges of metropolitan regions, in particular, are the product of the decisions of independent developers with a "supply-side aesthetic" that is heavily influenced by the market research and production decisions of the largest firms in the home building industry.

URBAN VIEW 2

Global Financial Meltdown, Local Disinvestment[4]

The boom in the "new economy" of the 1980s and 1990s fueled a housing boom, especially at the top end of the market. This was at a time when mortgage interest rates hit their lowest levels in more than four decades. Then, with the collapse of the dot-com speculative bubble in 2000–2001, property markets received a further boost as the built environment became a refuge for capital. The collapse of the dot-com boom made real estate an attractive investment for affluent households: people traded up as fast as they could, aided by a credit industry that became increasingly competitive and increasingly lenient, offering all sorts of mortgage arrangements to help people afford houses that otherwise would be beyond their reach: interest-only mortgages, graduated-payment mortgages, growing-equity mortgages, shared-appreciation mortgages, and step-rate mortgages that supercharged the market and contributed to the housing market bubble of the first half of the 2000s.

Traditionally, banks and savings-and-loan companies had financed their mortgage lending through the deposits they receive from their customers. Together with the involvement of federal agencies, this has always limited the amount of mortgage lending they could do, creating a natural stabilizing effect in the market. With the homebuilding boom of the early 2000s, however, mortgage lenders moved to a new model, selling the mortgages on to bond markets. This made it much easier to fund additional borrowing, but in the neoliberal political climate that had developed it also led to abuses, as banks no longer had an incentive to carefully check out borrowers. The new types of mortgages included "sub-prime" lending to borrowers with poor credit histories and weak documentation of income (also pejoratively referred to as the NINJAs—those with no income, no jobs, and no assets), who were shunned by the "prime" lenders underwritten by federal agencies. They also included "jumbo" mortgages for properties priced in excess of the federal mortgage limit of $417,000. Such business proved extremely profitable for the banks, which earned a fee for each mortgage they sold on. Naturally, they urged mortgage brokers to sell more and more of these mortgages. By 2005, one in five mortgages was sub-prime, and they were particularly popular among recent immigrants trying to buy a home for the first time in the expensive housing markets of big metropolitan areas.

The problem was that these *sub-prime mortgages* were "balloon" mortgages: payments were fixed for two years and then became variable and much higher. Inevitably, this led to defaults, and as the bad loans accumulated, mortgage lenders found themselves, in turn, unable to meet their repayments. The internationalization of finance meant that the first casualty was a British company, Northern Rock. Meanwhile, a wave of foreclosures and repossessions began to sweep across the United States. There were massive levels of defaults of loan repayments and widespread repossession of housing, usually at values far less than the initial loans.

The consequences were experienced at two levels. First, at the macroeconomic level the worldwide banking industry was thrown into crisis. This arose because sub-prime mortgages had been bundled into packages known as CDOs (collateralized debt offerings) or SIVs (structured investment vehicles) and resold to other banks and financial institutions around the world. It had been hoped that this financial reengineering would spread the risks associated with these loans and that lenders could avoid federal limits on their lending-to-capital ratios. However, it appears that few managers in the financial services industries fully understood what they were buying and that various risk assessment agencies were seriously at fault in underestimating the risks associated with these loans. The result was massive losses in many banks around the world, leading to risk aversion and a lack of liquidity (capital) for loans—a so-called "credit crunch." By 2009, the country's biggest homebuilding companies had laid off tens of thousands of workers, and were collectively producing only about one-third of the volume of new homes that they had been producing during the boom of the mid-2000s.

Second, at the micro level, many neighborhoods were blighted by empty homes. The credit crunch had severe impacts on millions of households and many of these were concentrated in particular U.S. regions. As a proportion of all loans offered,

URBAN VIEW 2
Global Financial Meltdown, Local Disinvestment (*continued*)

sub-prime loans were above 40 percent of the total in two belts of U.S. states: in the Midwestern heartland and in the South. In terms of absolute numbers California, Illinois, and Ohio were especially hit by the mortgage crisis; among individual cities, Cleveland was particularly affected by mortgage defaults, repossessions, housing abandonment, and neighborhood decline (Figure 2).

Meanwhile, African Americans have been disproportionately affected. Instead of helping to narrow the wealth and homeownership gap between whites and African Americans, the mortgage crisis helped to strip a great deal of the equity out of African-American neighborhoods. In effect, the credit crunch amounted to a massive redistribution of wealth away from the African-American community in the United States. This is particularly troubling because African-American homeownership had been rising sharply following decades of discriminatory lending and zoning practices that prevented many African Americans from buying a home. The large number of foreclosures of homes and rental properties has resulted in many African-American families losing their homes, savings, and credit, their neighbors seeing the value of their homes plummet, and renters being evicted.

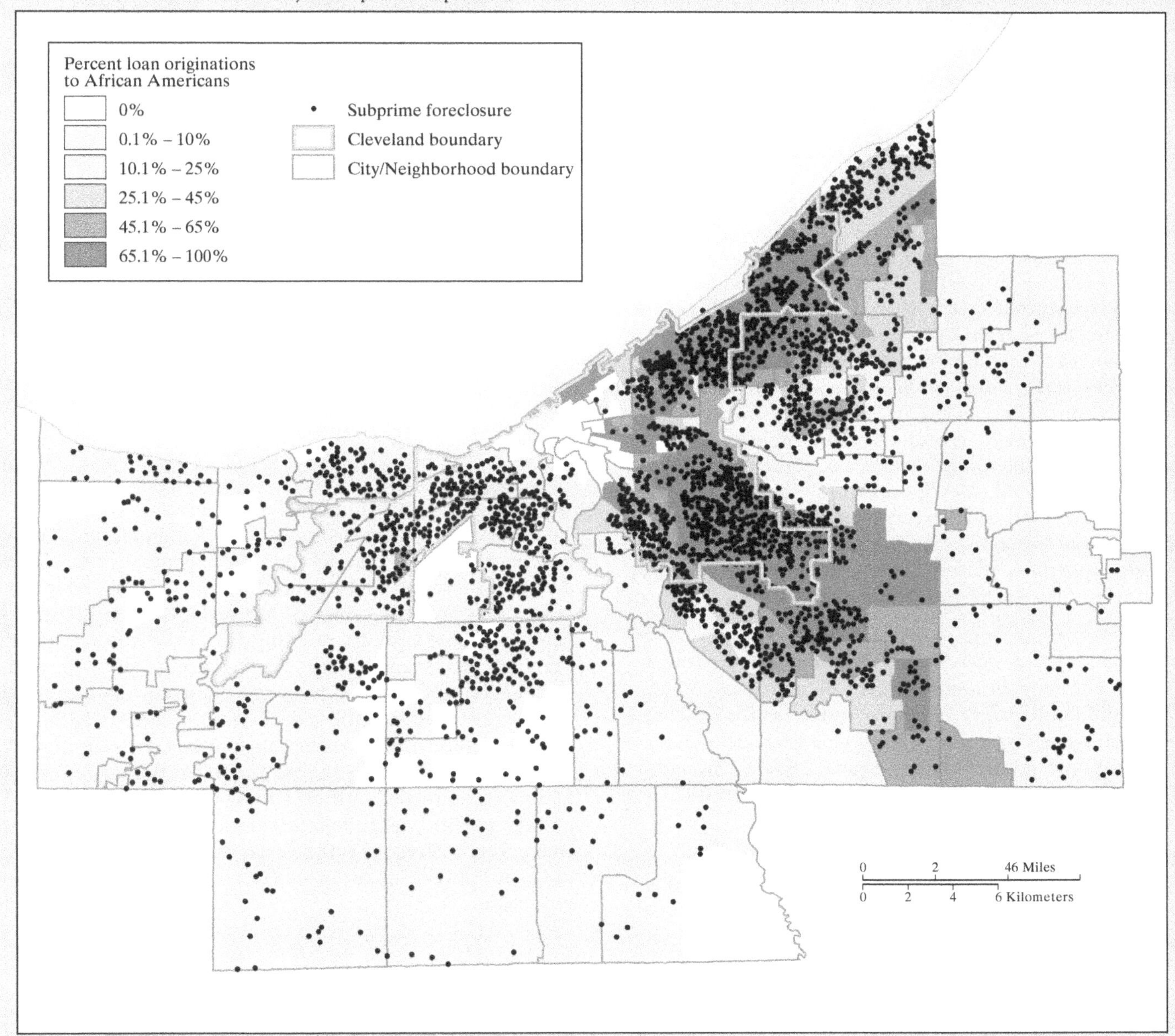

FIGURE 2 Sub-prime mortgage foreclosures in the Cleveland metropolitan area, 2005–2008.

Patterns of Investment in Land and Property

The restlessness of the built environment has been complicated by the fact that some property owners do not behave "rationally" especially when it comes to their home. Rather than treating a property purely as a financial asset, some people may treat it in part as a source of social status, political influence, or "pocket money," and then fail to respond (or respond too late) to changes in market conditions. Investment in land and property varies in terms of both the *purpose* of the investment (the use value of the land or property or its future exchange value) and the *time horizon* in which the investment decision is made (present or future). As Anne Haila explains, "A piece of land can be acquired and a building can be constructed for occupation (use), or for monetary return (exchange) the property yields when rented (annual rent) or when sold (capital gain, defined as the difference between the purchase price and the resale price). An investor acquiring real property and developing it can be oriented towards satisfying a present need or receiving a short-term revenue (present), or can expect to receive benefit or revenue in the long term (future)."[5]

Property as a Financial Asset

For many observers, contemporary urban development processes are dominated by speculators who try to anticipate the change of prices and to sell and buy in favorable market situations. For example, they may try to manipulate the market by lobbying for or against planning permissions. Their source of investment is capital borrowed and gathered from different sources (savings of small investors who buy shares in real estate investment trusts (REITs); capital of firms, financial institutions and public institutions that form joint-stock companies for construction projects). The principal purpose of investment is capital gain. Speculative developers are mainly interested in—and so consider—land as a financial asset. David Harvey has argued that in contemporary society, profits from property development are seen as in principle no different from the returns on investments in bonds, stocks, and shares:

> The money laid out is interest-bearing capital in every case. The land becomes a form of fictitious capital, and the land market functions simply as a particular branch—albeit with some special characteristics—of the circulation of interest-bearing capital. Under such conditions the land is treated as a pure financial asset which is bought and sold for the rent it yields.[6]

As a result, interest-bearing capital circulates through land markets continuously in search of higher future **ground rent**. Although Harvey sees this change as part of the overall evolution of the economy, there are several specific trends that have been important in consolidating the tendency to treat property as a financial asset:[7]

- The **globalization** of the economy has made the property market more international, increasing competition (particularly for office and high-status residential space), driving up rents while increasing the degree of differentiation between rents in different localities, and opening local property markets to a wider circle of investors (including international investors).
- The rapid reorganization of urban form and land use, has increased competition in local property markets, which has increased the pressure for property owners to respond rationally to market conditions.
- The more sophisticated use of advertising and marketing in promoting property investments has helped both to attract a wider pool of investors and to promote the idea of property as an "ordinary" asset or commodity.
- The loosening of planning and development controls has further allowed investors to treat property as an ordinary asset.
- The involvement of new kinds of investors (transnational corporations, pension funds, and so on) and the emergence of new kinds of professionals (real estate managers and property investment analysts) have propagated a more calculating (rent-maximizing) attitude of people in property markets generally.
- Until the global financial meltdown, the deregulation of financial markets had removed many institutional and legal barriers between capital markets that were previously highly segmented. This deregulation had also increased competition within property markets, driven up rents, attracted a wider pool of investors, and facilitated their ability to treat property purely as a financial asset. As a result, there was an overall tendency toward increasing investment in the built environment throughout the postwar period, cyclical fluctuations including late 2000s disinvestment notwithstanding (Figure 3).

THE STRUCTURES OF BUILDING PROVISION

The concept of the *structures of building provision* is based on the observation that, in addition to general principles of demand and supply and universal theories of rent and investment, each building project is the result of the actions of a variety of people and mediating institutions (government agencies, banks, and so on). The creation of the built environment, therefore, must be seen in terms of the functional linkages between specific sets of decision makers and institutions. Table 1 indicates the range of key decision makers, or "city makers," involved in the urban development process. In this section we identify the most important of these city makers, paying particular attention to their motivations and constraints before going on to examine the overall functional linkages between them in the context of land development processes.

City Makers

In any given situation the creation of the built environment is the result of a variety of people, all with their own objectives, motivations, resources, and constraints, and all connected

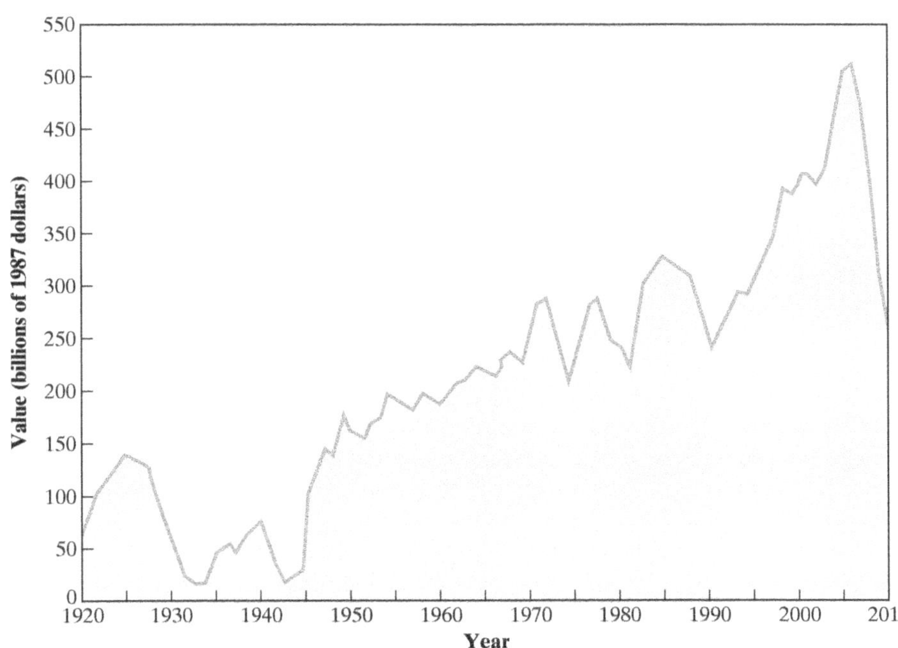

FIGURE 3 The value of new private construction put in place each year in the United States, in constant (1987) dollars.

TABLE 1 Urban Development: Decision Categories and Selected Decision Makers in the United States

1. *Industrial and commercial location decisions*:
   - Executives of industrial companies
   - Executives of commercial companies
2. *Development decisions*:
   - Executives of development companies (developers)
   - Land speculators and landowners
   - Apartment owners and landlords
3. *Financial decisions*:
   - Commercial bankers
   - Executives of savings and loan associations ("thrifts")
   - Executives of insurance companies
   - Executives of mortgage companies
   - Executives of real estate investment trusts (REITs)
4. *Construction decisions*:
   - Builders and developer-builders
   - Executives of architectural and engineering firms
   - Construction subcontractors
5. *Support decisions*:
   - Chamber of Commerce executives
   - Real estate brokers
   - Executives of leasing companies
   - Apartment management firms

Source: J. Feagin and R. Parker, *Building American Cities: The Urban Real Estate Game*, Englewood Cliffs, N.J.: Prentice Hall, 1991, Table 1.1.

with one another in several different ways. The suburban residential development process has been described as "a sort of three-dimensional spider web that can be moved by impact in any corner."[8] The same applies to the development process in the central business district (CBD), to the office development process, and to the process of developing industrial space—though the structure and composition of the web will vary in each case. To complicate things still further, each group of people will involve many individuals. In any city of any size, for example, there will be hundreds of major landowners and dozens of developers and builders. Some of these people will act for themselves within the web of the development process; others will be representing clients, large corporations, or public agencies. Some people may play more than one role at a time. Landowners may be actively involved in subdividing and building, for example, while city government officials may act as both regulators and entrepreneurs. Finally, all the people within the web of the development process must operate within a locally and historically specific context of market conditions and political constraints.

As long as we bear these caveats in mind, it is possible to identify the people that are typically involved in the creation of the built environment. It is also possible to make some generalizations about the roles and objectives of each of the major groups within the web of the development process.[9]

Landowners Landowners stand at the beginning of the chain of events involved in urban development. There are three major types of landowners, each with rather different perspectives:

- The first, *landed estates*, encompasses "old" money, whose ancestors were able to acquire large land holdings at an early stage in the history of development. For this group of landowners profitability is important, but it is often modified by social and historical ties. Decisions about the sale of land are made with an eye to the very long term.
- The second, *industrial land owners*, is dominated by commercial farmers, a group that is crucial to the land

conversion process at the edge of urban areas. Their decision-making typically has to balance short-term financial considerations against longer-term lifestyle considerations. Maintaining an occupation and lifestyle as a farmer often means paying higher taxes that result from the inflation of land values because of proximity to an expanding urban area. The alternative is to capture the rising value of land by selling some (or all) of it to speculators or developers. In response, many state and local governments in the United States have established tax laws that favor agricultural uses.

- The third type is based on *financial ownership*, dominated by property companies and financial institutions such as insurance companies and pension funds. As we will see, the importance of financial institutions has grown rapidly, as savings and profits have been channeled into long-term investments and as large transnational corporations have diversified into property. Property companies, in contrast, are less concerned with the long-term appreciation of assets, focusing on the exploitation of urban land markets for short- and medium-term profits.

Although each of these groups tends to behave in rather different ways within the web of city makers, all of them have influenced the outcome of the city building process in two broad ways: (1) through the size and spatial pattern of parcels of land that are transferred to speculators and developers and (2) through conditions that they may impose on the subsequent nature of development. In terms of the size and spatial pattern of land parcels, much of course depends on the overall pattern of land holdings. The large *ranchos* and mission lands around Los Angeles, for example, have formed the basis of extensive tracts of uniform suburban development, while in eastern cities, where the early pattern of land holdings was fragmented, development has been more piecemeal. Because of the structure of the tax system, however, it is often preferable for landowners to sell smaller parcels over an extended period. As a result, developers often arrange "installment contracts" with a variety of landowners to maintain a sufficient supply of building land. The outcome in terms of spatial patterns of development tends to be one of apparently random urban sprawl.

Because many landowners often sell only part of their holdings at a time, they have a strong interest in what happens to the land they sell. Any change in the use of sold-off parcels is likely to affect the future exchange value of their remaining holdings. In the past it was very common for landowners to sell off parcels of land with contractual provisos—**restrictive covenants**—that limited the nature of subsequent development. Such covenants usually discriminated against people with low incomes or socially undesirable land uses, sometimes in a very explicit way. With changed social attitudes and tougher laws against discrimination, restrictive covenants are somewhat less common but they have by no means disappeared. Rather, the practice has been to frame them obliquely, stipulating minimum lot sizes or low residential densities, for example, to ensure development for more affluent residents.

Speculators Speculators try to buy relatively low-priced land just before it begins to appreciate rapidly in value and to sell it just as it reaches a peak. Sociologists John Logan and Harvey Molotch have identified three very different kinds of speculators:[10]

- The first is the *serendipitous speculator*—someone who has inherited property or who has bought it with a particular use in mind and then finds that it would be more valuable sold or rented for some other use (such as a landlord who can sell a rundown apartment building near the CBD to a developer who wants to build a high-rise office).
- The second is the *active speculator*—the individual who hopes to anticipate changing patterns of land use and land values, buying and selling land accordingly. The prototypical active speculator is a small- or medium-scale investor: individuals (not corporations) who attempt to monitor the investments and disinvestments of bigger investors, using local and online social networks to find out who is going to do what, when, and where.
- The third is the *structural speculator*—the bigger investors who rely not merely on an ability to anticipate changing patterns but who also hope to influence or engineer change for their own benefit. "Their strategy is to create differential rents by influencing the larger arena of decision making that will determine local advantages."[11] They may attempt, for example, to influence the route of a freeway or the location of a rapid transit stop, to change the **land use zoning** map or the master plan, or to encourage public expenditure on particular amenities or services.

Developers The principal role of developers is in deciding on the nature and form of new building projects, platting large parcels of land into smaller lots, installing the infrastructure necessary for a particular use (e.g., streets, sewer and water mains, gas and electric lines), and selling the lots to builders. These activities generally fall under the descriptive label of "subdivision." Many development companies, however, have extended their activities well beyond the business of subdivision to include land assembly and speculation, design, construction, and marketing. Because it is the developers who must decide on the *type* of project to be undertaken on a particular site, they can fairly claim to be the single most important group of city makers.

The Development Process. Figure 4 shows the major steps involved in the overall development process.[12] The preliminary development activities consist of site selection, conceptualizing the project (whether it will be a residential subdivision, private master-planned community, office park, or whatever), and assessing the concept's feasibility. *Site selection and project conceptualization* stand together at the very beginning of the process, each influencing the other. This first step is clearly very important to the outcome of the city-building process, since the developers are inscribing their judgment and interpretation onto the landscape. Other things being equal, developers will

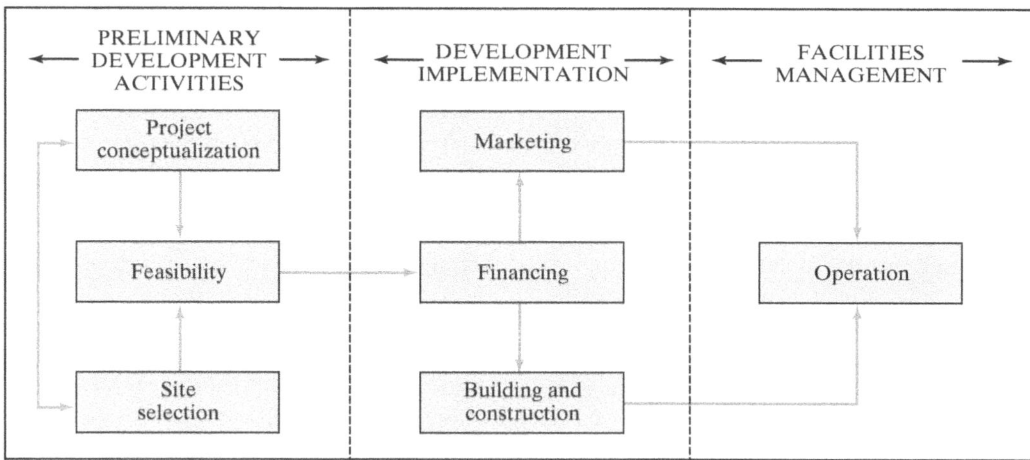

FIGURE 4 The development process.

opt for what is easiest to produce and what is the safest bet in terms of effective demand—the middle of the market. Only a few will have both the nerve to gamble on innovative projects and the ability to persuade financiers and customers that the potential outweighs the risks. In terms of residential development, this conservative approach translates into housing for the "typical" household (or, at least, the developer's idea of the typical household).

Through the 1960s and 1970s this approach resulted in the prevalence of three-bedroom single-family suburban housing (Figure 5), with little provision for atypical households—who were effectively excluded from new suburban tracts. Only in the 1980s, when marketing consultants caught up with social shifts that had made the "typical" household a demographic minority, did developers begin to cater to affluent single people, divorcees, retirees, and "DINKs" (dual-income, no kids), adding luxury condominiums, townhouses, and the like to their standard repertoire.

Among a developer's concerns in the site selection process will be the location, size, and cost of available land, the availability of utilities and municipal services, the possibility of special engineering or construction needs because of the site's physical properties, and the potential reaction of neighboring land users to the project. For most residential development, the decisive factor is site costs, although the nature of the proposed project can introduce important aesthetic considerations:

> Builders constructing lower-priced units are concerned primarily with minimizing costs. They seek to supply basic housing without frills; flat terrain with few trees is ideal. By contrast, builders of higher-priced houses look for sites with natural or social amenities. The costs of building on more rugged terrain and in denser vegetation are higher, but builders invariably find that the value of the completed house is increased by an even greater amount. When units will be sold for a higher price, builders and subdividers may create ponds out of marshes or add to the relief of an otherwise flat area by cutting streets lower and piling excess dirt up on the building sites. Characteristics that are limiting factors to the builder of

Bettmann/Corbis

FIGURE 5 Large-scale development of lower-priced housing depended on the availability of large tracts of cheap land, on which Fordist principles of mass production can be exploited.

lower-priced houses therefore are considered in a positive light in the trade-off calculations of the builders of higher-priced houses.[13]

The final phase of predevelopment activities is determining *feasibility*. Typically, this phase requires coordination with local planners in order to check on compliance with zoning ordinances and legal codes, approaching community leaders in order to gauge reactions to the proposed project, undertaking detailed market analyses, drawing up alternative schematic designs ("schematics"), investigating any special technical issues arising from these schematics, and projecting costs and revenues.

Having completed the preliminary phase, the developer moves into implementation: financing, marketing, design, and construction (Figure 4). *Financing* involves convincing others of the project's feasibility. Typically, the developer, just like the would-be homeowner, must put down part of the cost as cash: the developer's *equity*. The remainder is sought from a bank or from some other backer or consortium of backers—pension funds, insurance companies, and the like—who may themselves require certain changes in the nature of the project. The development industry is highly "leveraged," meaning that the developer's equity often works out to be a much smaller proportion of the overall cost than the homeowner's equity.

Marketing has become increasingly sophisticated as development projects have become larger. In addition to professional market research, this phase involves active promotion of the project early on. If possible, finished space will be sold or leased in advance of construction. In extreme cases, where stakes are very high, developers may buy out the existing leases of prestige customers in order to allow them to move into the project as a way to lure in other tenants.

The *design* and *construction* phase begins with detailed contract drawings being produced from the schematics, followed by a bidding process in which general contractors are invited to bid for various aspects of the engineering, construction, and landscaping. Timing is essential during this phase. Interest has to be paid on construction loans that increase as the project proceeds. Any delays in producing revenue from a project can result in significant losses due to increased interest payments, particularly when delays occur toward the latter stages of construction. Although it usually takes about 10 years for a typical mixed-use project to break even, the survival of a project depends on generating some revenues as quickly as possible simply to avoid foreclosure. Once construction is completed, the developer has to search for and manage tenants, collect rents, generally maintain and administer the project, or sell to new owners. This is the *facilities management* stage of the process (Figure 4).

The actual development process will vary a good deal, of course, depending on the size and nature of the project and the resources and scope of operations of the development company. Many developers specialize in a certain type of project: suburban subdivisions, master-planned communities, office blocks, mixed-use developments, hotels, factories, and so on. And although some developers are active across several regional markets, many specialize in just one metropolitan market.

Builders As we have seen, developers sometimes extend their operations to include building; more often than not, however, developers will subcontract to general building contractors. At the same time, many small and medium-sized building firms will undertake their own speculative land acquisition and development functions. Much depends, as with development companies, on the size and internal organization of the company. Before World War II, about a third of all new houses in the United States were built by small, local, general contractors for their new owners, while another third were built speculatively by operative builders who each averaged between three and five houses per year. Medium- and large-sized community builders accounted for the rest, though no firm produced housing for large regional markets, let alone the national market, and none had the capacity to build more than a few hundred houses a year.

There are still around 80,000 home builders in the United States, most of them tiny, but the biggest builders are rapidly getting bigger, and taking up an increasing share of the market. In 1986, the market share of the 100 largest builders (in terms of new home sales) was 24 percent. At the peak of the most recent housing boom, in 2006, it was 44 percent, with the top ten builders capturing just over 25 percent on their own. It has been the top twenty builders who have taken market share away from the rest. Pulte Homes, the top homebuilder in 1986 with sales of 9,500, was third-largest in 2006, with sales of 41,487. The top builder in 2006 was D.R. Horton, with sales of 53,410 homes in seventy-seven markets across twenty-six states. Between them, the ten largest builders accounted for the sale of almost 296,000 homes in 2006: the equivalent of the entire housing stock of a metropolitan area the size of El Paso, Texas.

The increasing dominance of big firms is due to a combination of factors: access to land and capital, mergers and acquisitions, strategic alliances, geographic diversification, improved production methods, and product innovation. One of the strongest driving forces in the consolidation of the building industry is the importance of access to land—not only land in the right sort of location but also in sufficient quantity for developers to be able to achieve economies of scale. Big firms have a huge advantage because it takes a lot of time to buy land and get the necessary permits to build on it, which means having the personnel for intelligence-gathering, scouting, and handling the fiscal and regulatory paperwork. Big firms also have the upper hand because they have the financial resources to acquire prime parcels of land. A $30 million deal for a piece of land has become commonplace, and a $150 million deal is not unheard of. With deals in this league, most large companies in fact usually purchase an option on land rather than buying it outright. This enables the company to back out of a deal if a municipality refuses to provide the clearances needed to build. The larger the firm, the greater the need to acquire a **land bank** in advance of development, simply to ensure a smooth flow of projects. At any given time, builders like Toll Brothers Inc., Hovnanian

Enterprises, and Pulte Homes control enough land for tens of thousands of houses.

Big publicly traded firms also have access to big capital, and usually at better rates than smaller, private firms that have to rely on banks. Credit-rating agencies have rewarded the success of the top home builders and, coupled with the reduced risk that many builders enjoy by optioning most of their land, instead of owning it, have helped the top builders secure larger credit lines at lower interest rates. The big firms have also become more creative in the ways they finance deals and manage risk, particularly with the use of joint ventures. Vertical and horizontal integration has also extended economies of scale and scope as well as intensifying the structural dominance of large firms within the industry. Some of the largest builders have meanwhile brought parts of their supply chain in-house in an effort to reduce cycle time (the average time it takes to complete a house) and exert more control over the tight labor supply in the construction industry.

The big builders have also deployed new technologies and refined production methods to push down costs and undersell small and medium-sized builders. Toll Brothers, for example, uses prefabricated wall panels and roof-truss systems, shipped from its factories to home sites; Pulte Homes has introduced prefabricated concrete foundation plates instead of site-poured concrete foundations. What is more revolutionary is that the big builders do not really build anything, at least in a technical sense: nearly all the physical work is contracted or subcontracted to electrical, framing, roofing, painting, masonry and plumbing companies, many of which follow the big firms in itinerant fashion from development to development. These contractors do the actual building in accordance with the big firms' signature designs and management guidelines.

Consumers Consumers—households and commercial and industrial businesses—represent the demand side of the development process. Demand for space within the built environment is an important and complicated topic.

One point that needs to be made in the present context, however, is that consumer preferences and consumer behavior must be seen as having developed in a social context "that is fundamentally shaped by markets ruled by capitalists and top managers in industrial and development corporations. . . . The physical structure of production builds barriers and sets limits to individual choices. Moreover, citizen preferences are frequently created or manipulated by powerful investors and their associates working through advertising, public relations, and the mass media."[14]

It should also be stressed that people need not always react individually—as consumers—to the choices created for them by business decision makers and managers. Individual people may also affect the development process through citizen protests over specific development projects, through involvement in pro-growth, no-growth, or slow-growth politics, or through involvement in residents' associations.

Realtors, Financiers, Other Professional Facilitators Realtors, financiers, and other professional agents are essential to the development process as facilitators, intermediaries, and specialized experts. A wide range of professionals is involved, including surveyors, market analysts, advertising executives, lawyers, title insurance agents, appraisers, property managers, engineers, ecologists, and geologists. The most important of these "exchange professionals," however, are mortgage financiers and realtors. Their activities go well beyond the actual *creation* of the built environment to encompass continuing processes of neighborhood change.

Government Agencies The development industry forms a staple of local politics and the focus of a good deal of local policy. Indeed, local economic development in the United States since the 1970s has been associated with an unprecedented form of **growth machine** politics and unprecedented partnerships between development companies and public agencies. City government officials have increasingly shifted to a new culture of **civic entrepreneurialism** that draws heavily on **public-private partnerships** in which public resources and legal powers are joined with private interests in order to undertake development projects. This shift has fostered a speculative and piecemeal approach to the management of cities. Local government officials (with additional funding leveraged from state and federal agencies) have subsidized projects such as downtown shopping malls, festival market places, new stadiums, theme parks, and conference centers because these have been seen as having the capacity to enhance property values and generate retail turnover and employment growth. If successful, the reasoning goes, such projects can cast a beneficial glow over the whole city; but even in the face of poor economic performance they can be regarded as a kind of "loss leader" that may bolster the image of a city and attract other forms of development.

Development is much less of a political issue at the federal level but federal policy can be a crucial factor that influences urban development. As a result, many important development-related decisions are made by politicians and government officials at the federal as well as the local level (Table 2). We will explore several issues where the development process intersects with urban politics and policy. Meanwhile, it is worth noting that the United States is somewhat unusual compared to other countries in the degree to which regulation of the property development process is weak and decentralized—another direct legacy of enshrining private property rights as a civil liberty under U.S. land law. Moreover, many federal responsibilities were decentralized and many spheres of activity involving federal oversight were deregulated in the 1980s as part of the overall retrenchment and restructuring that marked the shift away from Keynesian economic management.

TABLE 2 Urban Development: Decision Categories and Selected Government Decision Makers in the United States

1. *Actions on utility services, building codes, zoning, tax abatements, boosterism, and progrowth agendas:*
   Mayors and city council members
   County governmental officials
   Officials on local zoning and planning commissions
   Officials in special local agencies
2. *Housing, redevelopment, and tax decisions:*
   Members of Congress
   U.S. Department of Housing and Urban Development (HUD) officials
   Local and state government officials

Source: J. Feagin and R. Parker, *Building American Cities: The Urban Real Estate Game,* Englewood Cliffs, N.J.: Prentice Hall, 1991, Table 1.2.

Market Responses of the Development Industry

Like most other industries, the development industry has undergone some radical changes in the past 40 years in response to globalization and structural economic change. In parallel with producers in many other industries, developers and builders have found it advantageous to decrease their emphasis on **Fordist** strategies of mass production-mass consumption in favor of more flexible approaches aimed at exploiting profitable new market niches. Meanwhile, patterns of investment, along with patterns of corporate and professional organization, have become global in scope (Urban View 3 entitled "Urban Development is Less and Less a Local Activity"). In this section we illustrate some of these changes, noting how they have inscribed new elements into the built environment.

URBAN VIEW 3
Urban Development is Less and Less a Local Activity

Investment in the built environment has come increasingly to emphasize large-scale projects, and because of this trend urban development has, in turn, become more international and less and less a local activity. Take, for example, 30 St. Mary Axe, the first tall structure to be built in the heart of heritage-conscious London since the early 1980s and the third-tallest skyscraper in the one-mile-square historic City of London, after the 1980 Tower 42 and 2011 46-story Heron Tower (although all three will be dwarfed by The Pinnacle (also known as The Bishopsgate Tower) when it is completed). The 40-story St. Mary Axe building in London's financial and insurance district is cone-shaped so that the wind will pass easily around it (Figure 6). It has won a number of prestigious architectural awards, including the 2004 Royal Institute of British Architects Stirling Prize. The building has also been featured in movies such as *Harry Potter and the Half Blood Prince.*

Author Series

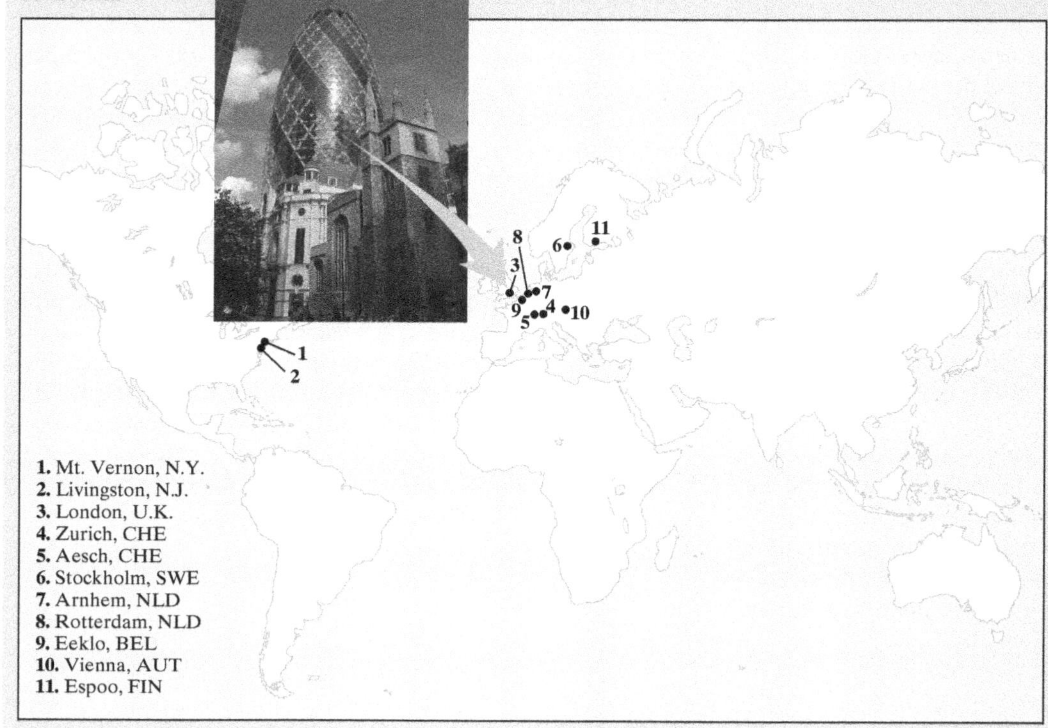

FIGURE 6 30 St. Mary Axe in London, completed in 2004, is a reflection of how the scale of activity in the development industry is often international in its dimensions.

URBAN VIEW 3
Urban Development is Less and Less a Local Activity (*continued*)

The project involved a number of international companies and/or their subsidiaries. The developer and primary occupant is Swiss Re, a Swiss reinsurance company. The architect was Foster and Partners, a British company headed by Sir Norman Foster, whose designs include the Centre Pompidou in Paris and the Hong Kong and Shanghai Bank in Hong Kong. The general contractor was Skanska, a Swedish company. The structural engineers were Ove Arup & Partners, a British company, as was the mechanical and electrical engineer, Hilson Moran. The facade supplier was Schmidlin, a Swiss company, and the facade maintenance system supplier was Lalesse Gevelliften, a Dutch company. The structural steel supplier was Victor Buyck-Hollandia, a joint venture by two companies—Belgian and Dutch. The steel dome was supplied by Waagner Biro, an Austrian company. The elevator supplier was KONE, a Finnish company, and the elevator engineers were Van Deusen & Associates, a company from New Jersey. Universal Builders Supply (UBS), a company from New York, was subcontracted to supply the tower and hoist erectors (Figure 6).The significance of this kind of international dimension to the development industry and its giant corporations lies in their potential for removing much of the debate and control over patterns of development from local arenas of municipal government, from the influence of local "growth machine" coalitions, and from the voices of neighborhood and environmental groups.

Although large national and international development companies often work with local partners in joint ventures to exploit local networks of key contacts, design professionals, and construction companies, the larger, stronger partners tend to learn quickly, setting up local subsidiaries or simply gaining the experience to take on more of the projects themselves. A case in point is the 46-story Heron Tower that involved Skanska, the Swedish company that had been the general contractor for 30 St. Mary Axe. For Heron Tower, though, Skanska removed the need for separate subcontractors—for mechanical, electrical, plumbing, IT, piling, steel decking, suspended ceiling, and engineering work—by providing what the company calls a "total construction and building services solution." Heron Tower still retains an international dimension though due to its U.S. architects, Kohn Pederson Fox Associates (KPF), and British structural engineering company, Ove Arup & Partners.

International is also a good way to describe the larger-than-life U.K. property developer, Gerald Ronson, arguably the most respected developer in Europe and perhaps the last of the great British tycoons. Ronson is infamous for his involvement in the largest U.K. financial scandal of the 1980s: an attempt to manipulate the stock market on a massive scale to inflate the price of Guinness shares and make possible a £2.7 billion takeover bid for Distillers, a Scottish drinks company. Ronson was found guilty (a ruling that was criticized by the European Court of Human Rights) and spent six months in jail. With his company overextended, Ronson barely avoided bankruptcy by securing loans from wealthy international friends including Bill Gates (Microsoft), Rupert Murdock (Australian-born media magnate), and the Sultanate of Oman. Ronson then spent the last few decades rebuilding his company and reputation as evidenced by a favorable decision to allow Heron Tower to be built following a multimillion pound public inquiry. Opposition to the skyscraper came from the British government's English Heritage (officially, the Historic Building and Monuments Commission for English) and Westminster City Council. They argued that the building would have a negative impact on the view of St. Paul's Cathedral from Waterloo Bridge.

New Products The development industry has responded to the need for greater flexibility through the pursuit of *product differentiation* and niche marketing. In the *commercial* sector product differentiation has resulted in a variety of new formats for hotels: luxury/full service, executive conference resorts, extended-stay (with kitchen and laundry facilities en suite), economy-only, and all-suite. Developers of office buildings responded to the changing business climate by producing self-consciously luxurious buildings (Figure 7). Developments for retailing have similarly seen different formats for different market segments: upscale downtown gallerias and malls, for example (Figure 8), and "big box power centers" (community shopping centers anchored by a major tenant such as Wal-Mart, along with one or two other major retailers and a complementary mix of specialty retailers, restaurants, banks, or other **consumer services**). Another significant new "product line" for developers is the specialized mall: a medical mall, for example, that is designed to provide busy, affluent consumers with one-stop shopping that offers physicians, counselors, therapists, medical laboratories, pharmacies, outpatient facilities, fitness centers, health food stores, and cafes.

Developers of business and industrial parks, meanwhile, are offering "flex-space": single-story structures with "designer" frontages, loading docks at the rear, and interior space that can be used for offices, R&D labs, storage, or manufacturing, in any ratio (Figure 9). Old product lines can also be "treated" to enhance flexibility within the market. Business and industrial parks have been repackaged as "planned corporate environments" with built-in daycare facilities, fitness centers, jogging trails, restaurants and convenience stores, lavish interior decor, and lush exterior landscaping.

In the *residential* sector some developers have repositioned themselves away from single-family "starter" homes to build more multifamily projects (that, like business parks, are packaged with services: in this case, security systems, concierge services, exercise facilities, bike trails, and so on) or more expensive homes for the "move-up" market, where the basic product—single-family suburban homes—is differentiated by features such as dramatic master bedroom–bathroom suites, marble floors, and signature landscaping.

At the very top end of the residential market are speculative homes differentiated by the most lavish "designer" features such as elaborate master bedrooms, bathrooms with spa amenities (steam shower, whirlpool tub, and sauna), exercise room, game room, home theater, library, "gourmet" kitchen

Grant Smith/Corbis

FIGURE 7 La Coline office development, Paris, France, is a good example of developers' attempts to capture a distinctive market niche among upscale business and professional services. The building incorporates a wooden bridge that arches across an ornamental pool in a glass-ceilinged atrium.

Linda McCarthy

FIGURE 8 Upscale shopping mall in Las Vegas, Nevada.

Bob Rowan/Corbis

FIGURE 9 "Flexspace": a new product line for developers in the 1980s, this concept combines "designer" office frontages with rear-area loading bays and interior space that can be used as office, industrial, or warehousing space, in any proportion.

Stan Rohrer/Alamy

FIGURE 10 Consumer booms and increased materialism can bring a rash of new speculative building at the top of the residential market, which is normally dominated by custom building such as this.

with butler's pantry, temperature-controlled wine cellar, and integrated multizone air conditioning, media, home appliance, and security systems controlled from touchscreens throughout the house (Figure 10). Most recently, "designer" features have also incorporated environmentally friendly elements such as nontoxic building materials and solar thermal radiant flooring.

Large, privately planned communities have also become popular with developers because they allow more flexibility in design and product type and enable developers to respond quickly to changing market demand (such as the growing number of older Baby Boomers seeking smaller homes with less maintenance).[15] The essential features of these communities are "a definable boundary; a consistent, but not necessarily uniform, character; overall control during the development process by a single development entity; private ownership of recreational amenities; and enforcement of covenants, conditions and restrictions by a master community association."[16] The direct descendants of the planned communities of the 1960s, they are a result of an extreme form of product differentiation and carefully targeted niche marketing. By exploiting new and more flexible land use zoning regulations, developers can put together projects that are attractive to a very profitable sector of the residential market while retaining scope for flexibility in the composition and timing of the development.

Residents of such communities are offered sequestered settings with an extensive package of amenities that typically include a golf course, tennis courts, swimming pools, play areas, jogging courses, an auditorium, exercise rooms, a shopping center, a daycare center, and a security system symbolized by imposing gateways and operated by electronic keyless entry systems. Housing is typically a mixture of expensive single-family houses, upscale townhouses and condominiums, and smaller studios or apartments for young singles or the elderly, all in High Suburban style: mock-Tudor, mock-Georgian, neo-Colonial, Giant Cape Cod, and so on (Figure 11).

Mark Winfrey/Shutterstock

FIGURE 11 Townhouses in a new, private, master-planned community. Note the variety that has been introduced to windows, doors, rooflines, and so on in an attempt to signal the distinctiveness of the project.

The entire ensemble is typically framed in a carefully landscaped setting that might contain a lake stocked with swans or a neoconservationist assemblage of remnant woodland, an artificial wetlands environment, and plantings of wild flowers. The landscape is completed by a parade of joggers in expensive warm-up suits and by busy delivery vans bringing affordable luxuries ordered online from designer clothing and household furnishing websites. The very names of the communities are carefully selected to set a tone of distinction, heritage, and authenticity. Advertising imagery draws on the totemism of golf, equestrianism, and pastoral landscapes, and leaves no doubts about the status and stylishness of the product.

Clearly the importance of design in the built environment is increasing. As one developer put it, "My buildings are a product. They are products like Scotch Tape is a product, or Saran Wrap. The packaging of that product is the first thing that people see. I am selling space and renting space and it has to be in a package that is attractive enough to be financially successful."[17] Such sentiments are by no means new, though they are perhaps felt more keenly nowadays. Design is much more than packaging. It involves language and ideologies that go well beyond the orbits of developers' worlds. It follows that we can "read" these "designer" neighborhoods as the product of our times, the carriers of our society's concern with materialism and social distinction. These themes have emerged very clearly in the residential patterns of U.S. cities in particular.

FOLLOW UP

Key Terms

- active speculator
- circuits of capital investment
- city makers
- foreclosure
- ground rent
- land banks
- niche marketing
- NIMBYism
- overaccumulation crises
- rent gap
- restrictive covenants
- serendipitous speculator
- structural speculator
- structures of building provision
- sub-prime mortgage

Review Activities

1. It is interesting to follow the various stages of the development process for a single project. Two books that do this are *From the Ground Up*, by Douglas Frantz (New York: Henry Holt, 1991) and *Skyscraper*, by Karl Sabbagh (New York: Penguin, 1989).
2. Go online and access one of the *Planet Money* podcasts or blogs by Alex Blumberg and Adam Davidson ("This American Life" and NPR News) at *http://www.npr.org/money/*. "The Giant Pool of Money" show explains in an understandable way how the housing crisis and the turmoil on Wall Street were connected, and why banks made half-million dollar loans to people without jobs or income.
3. Consider the statement that "There is . . . a constant restlessness to the built environment, as both simultaneous and sequential processes of investment, disinvestment, and reinvestment take place." Can you find examples from your nearest town or city that illustrate this?

4. Update your *portfolio* and gather more materials. A very good source in relation to the material covered in this chapter is the online real estate section of any major metropolitan newspaper. Real estate sections usually appear once a week and often contain useful insights and illustrations of the activities and interactions of various city makers. Look for features that illustrate the treatment of property as a financial asset and the responses of the development industry to the opportunities presented by urban change. You may also find data on vacancy rates and on land and property prices that you can use to illustrate aspects of changing urban geography.

Log in to **www.mygeoscienceplace.com** for self-study quizzes, MapMaster layered thematic and place name interactive maps, Urban View Google EarthTM tours, key resources and suggested readings, related websites, "In the News" RSS feeds, and additional references and resources to enhance your study of the urban development process.

NOTES

1. J. McKinley and M. Wollan, "Skaters Jump in as Foreclosures Drain the Pool," *New York Times,* December 28, 2008; J. Scelfo, "After the House is Gone," *New York Times,* October 22, 2008.
2. Harvey first outlined these ideas in a 1975 essay, "The Political Economy of Urbanization in Advanced Capitalist Societies: The Case of the United States," in *The Social Economy of Cities,* eds. G. Gappert and H. Rose (Beverly Hills, Calif.: Sage Publications), 119–163. He has subsequently developed them in a series of books and articles. See "The Urban Process Under Capitalism," *International Journal of Urban and Regional Research* 2 (1978): 101–31; *The Limits to Capital* (Oxford, UK: Basil Blackwell, 1982); and *The Urbanization of Capital* (Oxford, UK: Basil Blackwell, 1985).
3. R. J. King, "Capital Switching and the Role of Ground Rent: 1, Theoretical Problems," *Environment & Planning A* 21 (1989): 450.
4. Based on P. L. Knox and S. Pinch, *Urban Social Geography,* 6th ed. (London: Pearson, 2010), Chapter 6; M. Powell and J. Roberts, "Minorities Affected Most as New York Foreclosures Rise," *New York Times,* May 15, 2009.
5. A. Haila, "Four Types of Investment in Land and Property," *International Journal of Urban and Regional Research* 15 (1991): 348.
6. Harvey, *The Limits to Capital,* 347.
7. See A. Haila, "Land as a Financial Asset: The Theory of Urban Rent as a Mirror of Economic Transformation," *Antipode* 20 (1988): 79–101.
8. E. J. Kaiser and S. Weiss, "Public Policy and the Residential Development Process," *Journal of the American Institute of Planners* 36 (1970): 30–37.
9. The following description is based on sections of T. Baerwald, "The Site Selection Process of Suburban Residential Builders," *Urban Geography* 2 (1981): 339–57; and on P. Knox and S. Pinch, *Urban Social Geography,* 4th ed. (Harlow, UK: Pearson Education, 2000), 180–200.
10. J. Logan and H. Molotch, *Urban Fortunes: The Political Economy of Place* (Berkeley: University of California Press, 1987), 29–31.
11. Ibid., 30.
12. Figure 11.5 and the accompanying discussion are based on the outline of the development process given by Harvey Rabinowitz in *Urban Planning,* eds. A. J. Catanese and J. C. Snyder, 2nd ed. (New York: McGraw-Hill, 1988), 245–49.
13. Baerwald, "The Site Selection Process," 351.
14. J. R. Feagin and R. Parker, *Building American Cities. The Urban Real Estate Game* (Englewood Cliffs, N.J.: Prentice Hall, 1991), 21. Emphasis added.
15. D. R. Suchman, "Housing and Community Development," in *Development Trends 1989,* ed. D. Schwanke (Washington, D.C.: Urban Land Institute, 1990), 35.
16. Ibid., 23.
17. *Architectural Record* (June 1937), 9. Quoted in S. Zukin, "The Postmodern Debate over Urban Form," *Theory, Culture and Society* 5 (1988): 437–38.

How Neighborhoods Change

From Chapter 9 of *Urbanization: An Introduction to Urban Geography*. Third Edition. Paul L. Knox, Linda McCarthy.

Peter Casolino/Alamy

How Neighborhoods Change

In many parts of the world the overall form of larger metropolitan areas has evolved from a relatively straightforward, monocentric structure to a "galactic" sprawl with a polycentric structure—fragmented and multinodal—characterized by surrounding new urban spaces and settings that include edge cities and boomburbs. Embedded in this framework of splintering urbanism is the kaleidoscope of residential neighborhoods, arranged in terms of intersecting cleavages of people's socioeconomic background, household type, ethnicity, and lifestyle. If we look at these neighborhoods closely, we can see that each one is the product of a steady flux of change: arrivals and departures of people, births and deaths, additions and replacements, abandonments and demolitions that collectively carry each neighborhood in a certain direction at a particular speed. In this chapter we examine this dynamism in detail, emphasizing how people operate in housing markets and their residential mobility plays a role in shaping and reshaping neighborhoods.

LEARNING OUTCOMES

After reading this chapter, you should be able to:

- Explain how neighborhood change results from a combination of factors, including physical deterioration of housing stock, obsolescence of housing stock, cohort aging, and change in the composition of inhabitants.
- Recognize the factors that prompt people to invest or to disinvest in a neighborhood.
- Explain the relative lack of public housing in U.S. cities as compared to cities elsewhere.
- Articulate the variety of factors that shape the relocation decisions of people and households.
- Describe the role of key urban professionals in shaping and constraining people's location and relocation decisions.

CHAPTER PREVIEW

This chapter examines how and why neighborhoods change. The first task is to clarify the different components of neighborhood change: the aging of residents, the movement of households into and out of the neighborhood, and the aging of the physical environment. We will see that each of these components exhibits a different periodicity and that the overall effect can be conceptualized in terms of neighborhood life cycles.

Having established these fundamental dynamics, the next task is to review another important dynamic: the changing pattern of people's housing tenure that links broad shifts in the political economy to the context of local housing markets. At this point an important question is raised: Why has public housing for low-income families been so limited in U.S. cities compared to urban housing markets in other developed countries such as the United Kingdom?

If we look at a neighborhood closely, we can see that each one is the product of a steady flux of change: arrivals and departures of people, births and deaths, additions and replacements, abandonments and demolitions that collectively carry each neighborhood in a certain direction at a particular speed. Urban geographers examine this dynamism for cities like New Haven, Connecticut, emphasizing how people operate in housing markets and how their residential mobility plays a role in shaping and reshaping neighborhoods.

Another important issue raised here concerns the existence of spatial submarkets in which the dynamics of neighborhood change are played out. The majority of people in these submarkets comprise of course individual households, whose decisions about where to live and whether or not to move are the subject of the middle part of the chapter. In this section we will see how similarities in the behavior of particular types of households (in terms of people's socioeconomic background, household type, ethnicity, and so on) can be linked in causal terms to processes of neighborhood change.

Yet, although we can consider households' patterns of demand within spatial submarkets to be the fundamental motor of neighborhood change, the actions of key "gatekeepers" such as real estate agents and mortgage financiers also influence the outcomes. As we will see, they are sometimes the agents of social and racial discrimination and bias.

The final section of the chapter draws on all these ideas—neighborhood life cycles, housing submarkets, household behavior, and social gatekeepers—in illustrating the nature of neighborhood transformation that affects people's lives through to gentrification, involving social changes and physical upgrading in older, central city residential neighborhoods.

URBAN VIEW 1

Staying Put Despite the Spiral of Neighborhood Decline: A Mad Hatter?

Gary Witkowski's business, *The Custom Hatter*, one of the last remaining makers of handmade hats in the United States, is staying in the historic Polonia neighborhood despite neighborhood decline (Figure 1). Polonia was the name given to this East Side neighborhood with the heaviest concentration of the 80,000 Polish immigrants who started arriving in Buffalo, N.Y., in the early 1870s. Some of the very long and narrow "Polish cottages" that were built for these new arrivals have survived. With more than one family often living in the 1½-story clapboard cottages, these immigrants had to sleep in shifts. But as devout Roman Catholics, they contributed money to build churches whose spires are visible for miles around. Hundreds of businesses sprang up around the intersection of Broadway and Fillmore Avenues. And when the Broadway Market opened nearby in 1888, this shopping district rivaled the main street of downtown Buffalo.

Today, the Broadway Market has managed to retain some family-owned and -operated businesses that continue the tradition of selling fresh produce that harkens back to the earliest days of the market: black olives from Spain, cheeses from Italy, dates from Africa, smoked salmon from Scotland, and jellied eels from England. But the neighborhood has suffered a spiral of decline. Second- and third-generation Polish-Americans started the downward trend as they moved upward socioeconomically and outward spatially to neighborhoods farther from the downtown, and from there to suburban neighborhoods. The churches lost their congregations, many businesses followed their customers, and the neighborhood became blighted.

AP Photo/Bill Sikes

FIGURE 1 Gary Witkowski's business, *The Custom Hatter*, one of the last remaining makers of handmade hats in the United States, is staying in the historic Polonia neighborhood of Buffalo, New York. Despite neighborhood decline, it remains familiar to him—there is a strong sense of place for this hatter to the stars.

But Gary Witkowski stayed. His website lists over a dozen Hollywood movies for which he made hats during more than 35 years in business, including one worn by Leonardo DiCaprio in *Revolutionary Road* and one for Harrison Ford in *Indiana Jones and the Last Crusade*. He goes by his pseudonym, Gary White, because people told him that his Polish surname might keep away wealthy clients. His store website actually addresses the question: Are you mad as a hatter? He explains that this term comes from the traditional practice of using mercury to put a special finish on hats. This could lead to insanity for hatters who were exposed to this chemical element. Today, mercury is no longer used because of the obvious health hazards. But even so, could Gary still be mad as a hatter? Not really! Despite the changes to the neighborhood, it remains familiar to him—there is a strong sense of place. His store is only several blocks from where he grew up on Lombard Street in a typical East Side Polish and Roman Catholic family.

> From morning through evening during Witkowski's youth, the community was alive. Paperboys perched on corners, selling *The News* and the now-defunct *Courier-Express* to factory workers and passengers catching trains at the local station, the city's largest. Patrons browsed the racks at Sattler's department store at 998 Broadway, just east of Fillmore Avenue and a short walk from where the Custom Hatter sits. Diners enjoyed Friday fish fries at the Broadway Grill, which, like Sattler's, has closed . . . The old neighborhood . . . now exists only in memory. In summer, Witkowski will sometimes stand out on the sidewalk, sanding hats and chatting with visitors who stop to watch . . . He still loves the people. Sometimes, children who live nearby drop in . . . But the Broadway-Fillmore area is a ghost of what it once was, and Witkowski knows that.[1]

NEIGHBORHOOD CHANGE

Several components of neighborhood dynamism can be isolated, although, as Figure 2 suggests, they are all highly interdependent, each influencing—and being influenced by—the others. The most obvious and straightforward aspect of neighborhood change is that of the *physical deterioration* of the housing stock. The rate of physical deterioration of any neighborhood is chiefly a function of two factors: the quality of initial construction and the level of subsequent maintenance by property owners. Both of these are in turn related to the socioeconomic backgrounds and lifestyles of the occupants. Although it is not uncommon for fragments of the urban fabric to last for 100 years or more, 50 to 60 years can be considered to be a reasonable life expectancy in most circumstances in the United States. In general terms, each subdivision or tract of new housing can be thought of as describing a **depreciation curve** over time as it ages (Figure 3). It should be acknowledged, however, that such curves are averages that mask local unevenness in physical deterioration. Housing of the same age and initial quality wears unevenly because of variations in maintenance and improvements by its owners and the localized effects of road works, redevelopment, conversion, abandonment, fire damage, and so on. This unevenness in the rate of depreciation sets up an important precondition for neighborhood social change.

Closely related to physical deterioration is the structural and technological *obsolescence* of the housing stock. Both, in turn, are clearly a function of the needs and expectations of the occupants (and potential occupants) of the housing. Structural obsolescence occurs when the nature of the housing becomes unsuited to people's contemporary needs. Obsolescence does not necessarily bring abandonment or demolition, but it often brings a change of occupants and can lead to a shortened physical life span. A good example is provided by the earliest suburbs of many cities, where housing built before 1910 suddenly became obsolescent for many potential occupants because of the lack of off-street parking and garage space for cars. Meanwhile, the trend away from large families, combined with a sharp rise in the cost of domestic servants, made large, free-standing town mansions something of an anachronism by the 1920s, with the result that many mansions were converted into nonresidential uses. Technological obsolescence occurs when the functional design and equipment of housing and the neighborhood infrastructure become outmoded. Advances in kitchen design and appliances, in heating and cooling systems, and in the addition of swimming pools, bike trails, and community centers are among the more important triggers of such change.

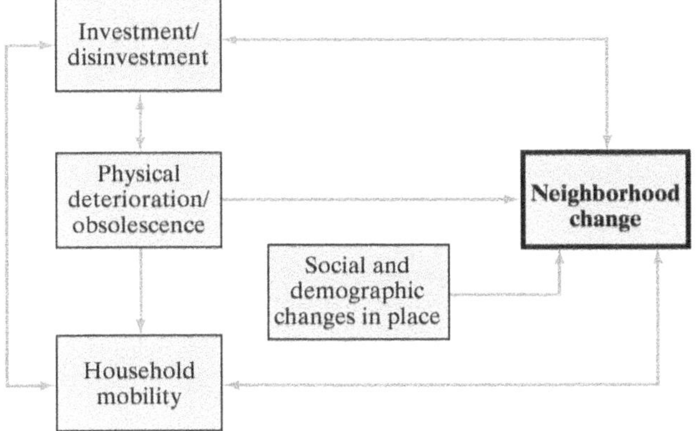

FIGURE 2 The principal determinants of neighborhood change.

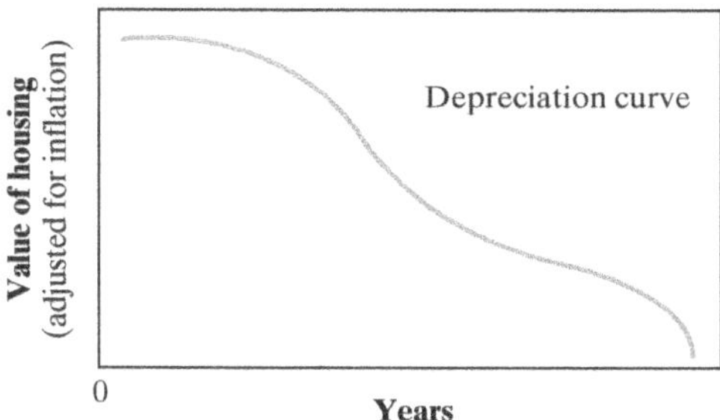

FIGURE 3 A hypothetical depreciation curve for a tract of suburban housing.

Given that most tracts of housing are developed for a specific and relatively homogeneous group of people, the principal point about social demographic *changes in place* (i.e., changes experienced by a community in a particular locality) concerns the gradual aging of the original "colonizing" cohort of people (i.e., the initial group of residents). But certain stages of the household life cycle tend to trigger moves by people to different kinds of accommodation and different parts of the city (Figure 4). Although those households who are unwilling or unable to move may have a life span that approximates that of the housing itself (say 60 years: formed around the age of 25 until death at about 85), the tendency for family life cycle changes to trigger a change of residence means that the periodicity of demographic change tends to be much shorter than that of physical deterioration and obsolescence.

Consequently, successive rounds of **filtering** will bring about a *changing composition* of inhabitants, generally resulting in an influx of younger and slightly less affluent households to the neighborhood. In addition, of course, physical deterioration, structural and technological obsolescence, and the aging of the original "colonizing" cohort of people will collectively have changed the neighborhood to the point where further changes are induced, eventually to the point where people of a significantly different socioeconomic, demographic, ethnic, or lifestyle group move into the neighborhood.

Redevelopment and Reinvestment

All these changes, in turn, translate into changing opportunities for *investment*. Other things being equal, most neighborhoods over time attract a certain amount of investment by homeowners and landlords in renovations and improvements. Where physical deterioration, obsolescence, and social change discourage people from doing this, *disinvestment* may take place—deliberately neglecting routine maintenance, putting homes, apartment buildings, and vacant land on the market, or abandoning them altogether. Other things are not always equal, though.

URBAN VIEW 2
The Household Life Cycle

Figure 4 shows an idealized model of the life cycle of middle-income households, implying that there are recognizable stages, each with a distinctive household composition that is associated with people's particular space needs, specific preferences in terms of accessibility, type of housing tenure, locational setting, and different propensities to move.

According to this scheme, the typical **household life cycle** begins with a brief stage based on single-person households or communal households of young single persons in their late teens or early 20s—like Michael and Emily—who have left home to study or find work. In this stage the physical size of the dwelling is relatively unimportant to these young people, whereas accessibility to jobs and urban amenities is often a major consideration. The first year or two of living together or marriage can represent a continuation of this phase for people like Michael and Emily when they hit their mid-20s. Later, in the event of the birth or adoption of a child, for couples and singles alike, the household's real and perceived needs change dramatically, reversing the relative importance of dwelling space and accessibility and placing a premium on safe, quiet residential environments. Over time, during their 30s and early 40s, a series of adjusting moves will be made as Michael and Emily's household income increases.

By the time their children are of junior-high or middle-school age, access to high-quality neighborhood facilities—especially schools and social amenities—becomes increasingly significant. Meanwhile, the household's earning power is likely to have increased, partly through job promotions for Michael and if Emily returns to the labor force after a period of homemaking. This increased income enables the household to move to a new suburban or exurban home. But in their early 50s, Michael and Emily's divorce and remarriage results in more moves. Some middle-aged adults who experience divorce or lose a job may move in with their elderly, usually widowed parent—offering companionship and assistance in exchange for sharing the parent's home. At the "empty nester" stage, when a household's space requirements decrease considerably, neighborhood ties and sentimental attachments can prevent many couples or single parents from moving to smaller homes; housing professionals have recently begun to pay increasing attention to household desires to remain in a cherished home or valued neighborhood by refitting existing homes to meet changing household configurations. At the same time, the increasing financial independence of older people since the 1960s has been accompanied by increasing residential independence and associated housing options, such as assisted living accommodation or specialized retirement communities (Figure 5).[2] The final residential change for many people like Michael and Emily, however, can come in their late 60s or 70s at the stage when they move to live with a son or daughter or to an assisted living facility.

This model fits with much of the available evidence about conventional residential mobility and urban structure. It should be stressed, however, that it is an idealized model that is *inherently family centered* and *middle income* in orientation. (For people who choose to remain single, for example, different assumptions have to be made. Similarly, purchasing a home cannot be assumed to be an option for low-income households.) The idea of a household life cycle therefore needs to be linked to intersecting cleavages of socioeconomic background, ethnicity, and lifestyle before we can chart the impact of household type on the residential kaleidoscope with any sensitivity.

Richard Smith/Alamy

FIGURE 5 Elderly women enjoy water aerobics in a swimming pool in Sun City, on the edge of the Phoenix metropolitan area, a specialized adult community—in this case, with an active, resort retirement lifestyle—whose more than 40,000 residents make it the largest retirement community in the country.

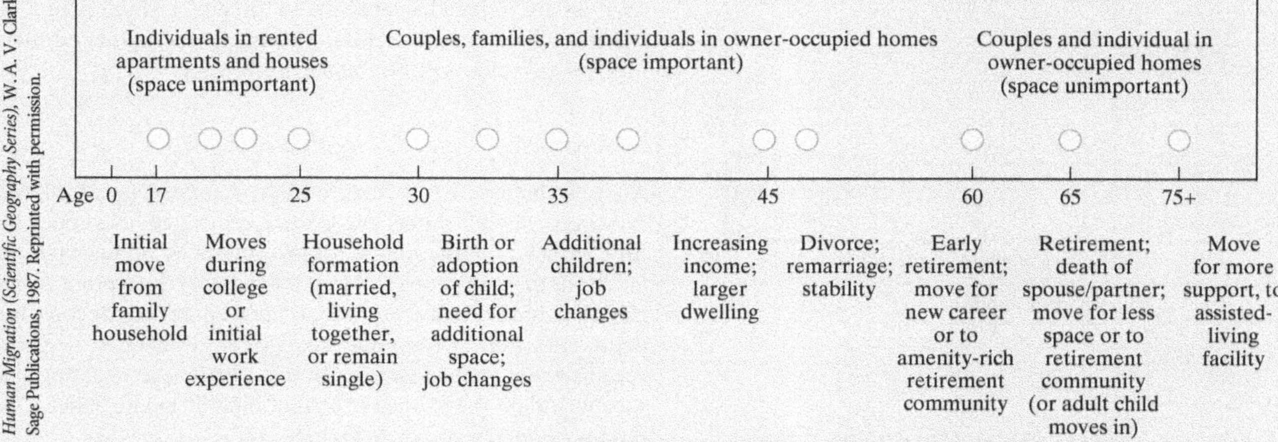

Human Migration (Scientific Geography Series), W. A. V. Clark, Sage Publications, 1987. Reprinted with permission.

FIGURE 4 A household life-cycle model of middle-income residential mobility and housing preferences.

First, differences usually exist between neighborhoods in the rate and nature of change, affecting the landscape of investment opportunities for property developers. For example, a neighborhood may be physically quite sound and socially and demographically very stable—characteristics that ordinarily mean stability without major reinvestment. Such a neighborhood may nevertheless be considered ripe for *redevelopment* or *reinvestment* because of the differences between current rates of return on property in the area and the rates of return anticipated from investment in a change in neighborhood character or land use. Put in crude terms, the anticipated profits from established low-income housing may be significantly lower than the profits anticipated from a new shopping mall in the same location.

Second, neighborhood change also has to be seen in relation to changes in market demand. For example, a neighborhood may have changed to the point where modest investments by property owners are able to capture new or expanding markets through conversions (from rental apartments to upscale condominiums, from industrial loft space to residential lofts, from town mansions to funeral homes, and so on).

Neighborhood Life Cycles

Drawing on these observations about neighborhood change, we can make some tentative generalizations about the typical sequence and interdependent relationships that constitute the neighborhood life cycle. Interactions between physical deterioration, obsolescence, sociodemographic change of people in place, and sociodemographic change of people that is induced by filtering may result in a five-stage life cycle.

1. *Suburbanization.* The beginning of the life cycle, which is characterized by low-density, single-family housing occupied by young families of relatively high socioeconomic background.
2. *In-filling.* Multifamily and rental dwellings are added on vacant lots, increasing the density and decreasing the social and demographic homogeneity of the neighborhood.
3. *Downgrading.* The longest phase of the life cycle. A period of slow but steady deterioration and depreciation in the housing stock, of aging in place, and of increasing population turnover.
4. *Thinning Out.* The beginning of the end: high population turnover bringing social and demographic change; conversion and demolition of some residential units.
5. *Renewal or Rehabilitation and Gentrification.* Renewal ends the neighborhood life cycle abruptly and begins a new one in the form of new tracts of housing, usually at a high density that reflects the neighborhood's (now) relatively central location. In certain cases, rehabilitation and gentrification by new and existing residents can extend the neighborhood life cycle through conversion and reinvestment.

This example shows very clearly how housing "acts both as a determinant and a consequence of neighborhood change."[3] It also helps us to see the residential kaleidoscope as a result in part of the juxtaposition of neighborhoods at different stages of successive life cycles. New townhouses, condominiums, and apartment buildings in central city locations correspond to neighborhoods that have already experienced at least one cycle of structural obsolescence and that have now been redeveloped. Nearby are areas of older housing that have become technologically obsolescent for most households because of their lack of amenities. As a result, successive neighborhoods of this sort have seen the very lowest socioeconomic groups move in.

Where the cycle of physical deterioration has been completed, some neighborhoods stand scarred and partially abandoned; in others, **urban renewal** has made room for public housing projects that have begun a new life cycle of their own. Where the original central city housing was built to a higher standard, the cycle of physical deterioration is still incomplete. Neighborhoods of larger dwellings—built to house prosperous nineteenth-century merchants and industrialists—remain as the basis of residual neighborhoods that have filtered down the socioeconomic ladder to become rooming house areas, with the structural and technical obsolescence of large units being resolved by subdivision and multiple occupancy. Pockets of well-built but smaller dwellings remain as the basis of gentrifying neighborhoods: areas that had slowly filtered down the socioeconomic ladder but that have seen younger, middle-income households move in and invest real capital or sweat equity,[4] or both, in renovating the housing.

Farther out are the city's middle-income suburbs, which, because they date mostly from the 1960s onward, have suffered only moderately from physical deterioration and structural or technological obsolescence. Most still contain a significant proportion of the people who originally moved in, added to over the years by households of similar types. The chief exceptions are the older, innermost suburbs, which have reached the thinning-out stage, with a large number of younger families moving in and the conversion and demolition of some homes to make way for commercial developments and highway improvements. Finally, the newest suburbs, in the outermost reaches of the polycentric metropolis and in occasional pockets of development, represent households in neighborhoods at the beginning of their first life cycle.

HOUSING MARKETS

Any consideration of just how these patterns of residential mobility and neighborhood change come about must take place with reference to the operation of housing markets and the way people treat housing as a commodity. Housing, as geographer David Harvey has observed, "is fixed in geographic space, it changes hands infrequently, it is a commodity which we cannot do without, and it is a form of stored wealth which is subject to speculative activities in the market. . . . In addition, the house has various forms of value to the user and above all it is the point from which the user relates to every other aspect of the urban scene."[5] In market terms, therefore, housing extends well beyond the shelter provided for people by the dwelling itself to include a complex package that is often referred to in terms of housing services. We can identify four main aspects of housing services:[6]

1. Shelter and privacy.
2. Satisfaction and status associated with the size and quality of the dwelling and the prestige of the address.
3. Environmental quality, including both the quality of the physical environment (trees, vistas, parks, sidewalks, bike trails) and the quality of the social environment.
4. Accessibility to places of work, schools, shopping, friends, sport and recreational facilities, and other services and amenities.

The overall utility of these services is generally referred to as the **use value** of housing. Because it depends a great deal on the needs and preferences of particular households, the use value attributed to a particular dwelling will tend to vary for people according to their socioeconomic background, household type, lifestyle, and so on.

The role of housing as a form of stored wealth adds a fifth aspect to housing services:

5. Equity (for owners)—the financial return on an investment in housing (specifically, the difference between the market value of a dwelling and the amount of any outstanding mortgage debt on the property) that is (for owner-occupiers) tax-free. In this context we should note that the equity value of housing, along with that of other real estate investments, rises and falls with property bubbles as they expand and burst. The potential for gaining unearned income through equity increases, together with the use value of a dwelling, will determine its **exchange value** in the marketplace.

At this point we are confronted by the fact that by no means all of the housing market is for people who want to be owner-occupiers. Other forms of tenure, particularly private rentals, account for a large proportion of the stock of available housing in every city; so it is less useful to think in terms of overall urban housing markets than to think in terms of a series of *submarkets*. The changing composition of these housing submarkets is both a product of urbanization processes and a determinant of sociospatial differentiation. **Housing submarkets** can be delimited in terms of dwelling type, price range, and location, but people's *tenure* represents the single most important factor.

Urbanization and the Tenure Transformation

The rise of homeownership has been not only the greatest single aspect of the transformation of urban housing but also one of the most important elements in the social and cultural evolution of American urban life. The ideal of home ownership as a central element of the "American Dream" took root for people in the 1930s. The product of Depression-era politics, the original notion of the American Dream built on the idea of individual freedom, especially the possibility of dramatic upward social mobility through a person's ingenuity and hard work, with the promise that successive generations would enjoy steadily improving economic and social conditions. It did not take long, though, for the ideal of home ownership to be grafted onto this notion, thanks largely to real estate interests. Realtors, in particular, systematically developed and deployed strategies that helped to make "normal" this notion of home and neighborhood so that it was desired by "middle class" U.S. families. As Jeffrey Hornstein notes in his book *Nation of Realtors*, "The conception of real estate brokerage as an occupation-cum-profession depended upon the existence of 'home' as an intellectual and cultural object . . . Thanks in large measure to real estate brokers' cultural and political work, the single-family home on a quarter-acre lot in a low-density suburban development became the 'American Dream,' and the vast majority of Americans bought into it."[7]

From about 1915 through the 1920s, realtors collaborated with various government agencies and civic groups to promote single-family home ownership. Their Own-Your-Own-Home campaign sought to reinforce the idea for people of the home as a privileged consumer durable, worth sacrificing and going into debt for. This mission to make the United States into a land of universal homeownership "was merely the culmination of a long republican tradition linking civic virtue to property ownership. The American republic would be able to save itself from the degenerative ravages of historical time, class struggle, and urban corruption by providing all citizens with a home of their own in healthful, natural surroundings: American civilization would develop in space rather than time."[8]

When the housing market crashed at the onset of the Depression, the National Association of Real Estate Boards was in a position to work closely with President Hoover's White House Conference on Home Building and Home Ownership. In doing so, it secured support for a reduction of taxes on real estate and endorsement for a federal mortgage discount bank to facilitate long-term mortgages. These became key elements in the ambitious programs of the Roosevelt administration's New Deal.

The overall proportion of owner-occupied dwellings in U.S. cities rose from 20 percent in 1920 to 44 percent in 1940, 60 percent in 1960, 66 percent in 1980, and 68 percent by 2010 (Figure 6). There are four principal reasons for this trend:

1. *Increasing affluence among a progressively wider section of society*, which made homeownership possible for a greater proportion of households. Meanwhile, **economies of scale** and other innovations in the development and construction industries kept down the costs of suburban single-family dwellings, reinforcing the basic affordability of home ownership.
2. *Increasing perception of the benefits of homeownership* in terms of
   - establishing social status and fulfilling the American Dream of independence based on property rights.
   - achieving residential segregation and pursuing exclusionary social and political strategies.
   - acquiring financial benefit through equity gains that in turn help to finance upward social mobility on the housing "ladder."
   - fulfilling, in many cases, family-centered lifestyles.

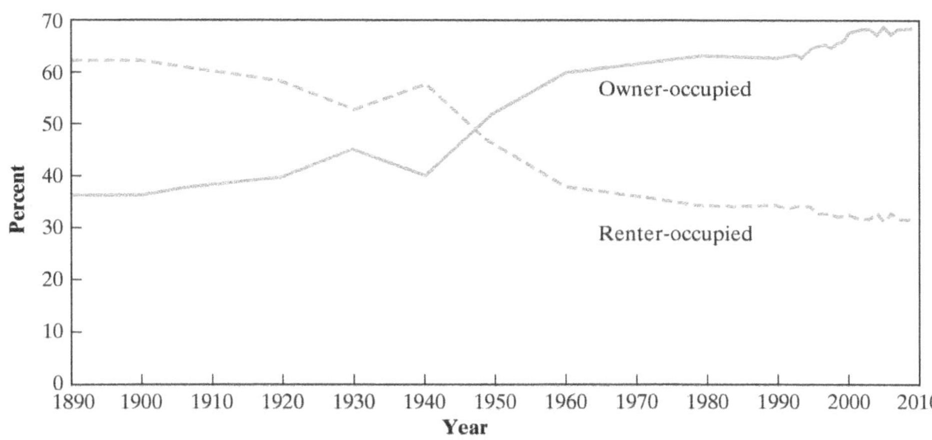

FIGURE 6 Tenure of occupied nonfarm dwelling units in the United States.

3. *Increasing recognition of the economic and political significance of homeownership.* There were several aspects to this factor.
   - The role of homeownership in the circulation of capital: Traditionally, saving for the substantial down payment needed for entry into the homeownership submarket created a pool of capital that financial institutions could use to lend to entrepreneurs to invest in industrial projects; interest payments on existing mortgages, meanwhile, ensure a steady additional flow of (potential) investment capital; and financial institutions can buy and sell the mortgages themselves and use them to leverage further sources of investment capital. Lastly (but by no means least), homeownership promotes the circulation of capital by creating a platform for all kinds of consumption, from furniture and furnishings to yard care equipment, that stimulate cycles of investment/production/consumption/profitability in a variety of industries.
   - The role of homeownership as a regulatory economic mechanism: The size of the housebuilding industry and its **multiplier effects** in relation to the market for consumer goods made it a key element in Keynesian economic management in the past: one of the principal levers that governments have been able to manipulate in boosting or if necessary curbing economic growth.
   - The role of homeownership in promoting social and political stability: The more people there are who have a stake in the private property market, the less likely they are to behave in ways that threaten economic and political stability. The more people there are who carry the significant encumbrance of debt that comes with a mortgage, the more individual and social incentives there are to protect and enhance the exchange value of property, to support employment stability (even at the cost of foregone wage raises), and to adopt a conservative approach to economic and political affairs.

As a result of the increasing recognition of these economic and political aspects of homeownership, *successive governments have actively fostered the growth of homeownership.* In broad terms in the United States, the federal government has done so through two main sets of policies, those aimed at:

- Cultivating and protecting savings and loan institutions involved in financing homeownership. In addition to insuring mortgage loans made by private institutions for home construction or purchase, the institutions themselves received privileges and concessions in terms of corporate law and taxation.
- Encouraging households to purchase rather than rent. Two particular aspects of income tax policy are important here: the inclusion of interest payments on mortgage loans for "qualified" residential property as allowable deductions before tax rates are applied and a generous exemption of equity gains from capital gains tax.

4. *A sharp decline in the profitability of rental units.* This decline is also the result of a combination of several factors:
   - The introduction and enforcement of more rigorous building standards and housing codes, which caused landlords to eliminate the cheapest rental units.
   - The relatively slow rise in the incomes of tenants in privately rented accommodation, which has made it difficult for landlords to raise rents.
   - Rent control legislation, initially introduced to divert investment capital and civilian labor away from housing and into defense industries during World War II.
   - The increasing physical deterioration of the rental stock, especially buildings that had been constructed in the nineteenth century.
   - Taxation policies that have made it unattractive for landlords to maintain or improve their rental properties.
   - The decrease in demand by people for rental accommodation as a result of the government-supported financial advantages associated with homeownership.

URBAN VIEW 3

Cabrini-Green: A Cherished Home in a Place That was a Symbol of Everything Wrong with Public Housing[9]

The final tenant with her four youngest children moved out of the last high-rise public housing unit in Chicago's notorious Cabrini-Green in December 2010. With the help of Hope VI grants from the federal government, all of the mid- and high-rise towers are being demolished and replaced with low-rise mixed-income market-rate apartments and townhouses, with some units reserved for public housing tenants. The former residents of Cabrini-Green have mostly been moved into other public housing units elsewhere in the city or have moved into market-rate apartments made affordable with the help of government housing vouchers.

The last tenant, Annie Ricks, a 54-year old teacher's aide, grew up in a one-room shack in Riverview, Alabama. She moved into Cabrini-Green 21 years ago after a fire destroyed her West Side house. Her daughter, Tasha, now 29, a union construction worker who owns a two-bedroom condo on the South Side, still remembers what life was like after the fire. "We were homeless and walked the streets for months and slept in the homes of relatives and friends, and even for two weeks in the waiting room of Cook County Hospital. There was Mama and seven kids with no place to go. We tried to get into all the different projects but couldn't. Then we got into Cabrini. I guess my mother loves it so much because it's the only thing that got us off the streets."

So what became a symbol of everything that was wrong with public housing was a cherished home for Annie Hicks. And now it is gone. Residents like her feel the loss of their close-knit community. Over the years, in the face of shared hardships, the Cabrini-Green residents organized to pressure the City of Chicago for help with the gang violence and neglect, as well as to protect and support each other. They even went to court to try to ensure that replacement housing would be built before the Hope VI-funded demolition of their high-rises.

The Cabrini-Green high-rises were built during the late 1950s and early 1960s as part of urban renewal efforts to clear a slum known as "Little Hell." Cabrini-Green eventually housed 15,000 people. This public housing project was named after St. Frances Xavier Cabrini, the first U.S. citizen to be canonized by the Roman Catholic Church. A downward spiral was inevitable because the Chicago Housing Authority's budget for maintaining the buildings and addressing crime came from the rents that had to be kept low because of the poverty of the residents. Cabrini-Green attracted national attention in 1981 when a gang war killed 11 residents. In 1992, 7-year-old Dantrell Davis was shot by a stray bullet as he held his mother's hand on his walk to school. In 1997, a 9-year-old girl—known as Girl X—was found in a stairwell of one of the high-rises after being brutally raped, choked, and poisoned with roach poison sprayed down her throat, leaving her blind, paralyzed, and unable to speak.

> There were many other towers of poverty in Chicago. Most gone now. And most of them were places were the flash of a knife or the pop of the gun could on any given day come quicker than at Cabrini Green. But what made Cabrini Green stand out was its proximity to the glitter and the gold prime real estate, the wealth of the surrounding neighborhood. Cabrini Green stood in the shadows of the downtown towers of Chicago. The skyline as seen from Lake Michigan where the power elite counted money, made deals and then went out to play. Cabrini Green stood, since its very inception, in a vise. Continually squeezed, as the surrounding real estate got more and more expensive. Chicago is a city where generations grew up knowing on an almost biological level, that there are certain streets you just don't cross. And the streets around Cabrini got tighter and tighter.[10]

As a result, progressively fewer rental units were built, and increasing numbers of rental units were demolished, abandoned, sold into owner-occupancy, or converted to nonresidential uses. Over time, then, U.S. cities experienced an almost complete reversal of the relative importance of the owner-occupier and rental submarkets.

Public Housing

In U.S. cities public housing is conspicuous by its absence. Cities in many other developed countries contain a substantial amount of "social housing" that has been provided to address acute problems of housing need that have arisen because the private market has been unable to provide decent housing for low-income households and still make a profit. In Europe, for example, social housing represents a relatively large proportion of the housing stock in some countries (35 percent in the Netherlands, 25 percent in Austria, 21 and 20 percent in Denmark and Sweden respectively) to a small proportion in other countries (6 percent in Germany and 4 percent in Hungary (after mass privatization)). In most countries, the percentage of social housing has decreased during the last decade because its provision has not kept pace with the rate at which either private housing has been constructed or social housing has been privatized or demolished.[11]

In the United States, the 1.2 million public housing units represent less than 1 percent of all housing units.[12] An important feature that makes U.S. public housing distinctive, however, is that more of it is provided indirectly through rent subsidies (2.2 million units in the "Housing Choice Voucher" program) to private landlords (instead of being provided directly as public housing units by government housing agencies). In addition, there are another 1.7 million multi-family assisted units (private-owner, project based).

Despite having only these 5.1 million publicly subsidized units (representing just under 4 percent of all housing), U.S. cities have just as much need for public housing as the Netherlands, Austria, Denmark, and Sweden. Although escaping the devastation of wartime bombing (unlike many cities in Europe and Japan), U.S. cities have certainly experienced acute problems of housing need. Slums have persisted within every large city for decade after decade. A succession of studies has shown that about one-third of all families in central cities are in need of basic housing, which they cannot afford,[13] and homelessness has risen alarmingly. It is, therefore, instructive to examine how and why U.S. urbanization has not produced more public housing units.

In fact, there was a time when public housing might have developed into a substantial segment of the housing market. The initial legislation mandating public housing was passed in 1937 in the political climate of the post-Depression New Deal. Although public housing was widely seen as ideologically distasteful (as it had been in most other countries), the depth of the Depression was seen as having created exceptional need, and the program was able to be "sold" politically in terms of employment creation rather than housing subsidy. The program established a federal government presence in urban housing, but it was not until after World War II that a more extensive and ambitious national program was launched. The Housing Act of 1949 linked public housing construction to slum clearance programs and represented a significant commitment to subsidized housing for the poor (Figure 7). Strong opposition to the act was deflected by a coalition of different interest groups, all with different agendas and different assumptions about the main purpose of the legislation. Political liberals (thinking of both the slum clearance and the provision of public housing) saw it as a means of eliminating slums and rehousing the poor, but business interests (thinking mainly of the slum clearance) saw it as a means of increasing central city property values, while many local politicians (equating slum clearance with urban renewal) saw it as a means of bolstering their tax base and luring back some of the more affluent consumers and taxpayers who had moved to the suburbs.

The 1949 Housing Act authorized the construction of 810,000 low-rent units over six years. The overall contribution of this program to the housing stock of U.S. cities has been minimal, however. The 810,000 units called for under the 1949 legislation took 25 years to materialize. Besides, it did so in the form of very poor quality housing that was localized in some of the most economically distressed neighborhoods in cities. Meanwhile, the rate of construction of public housing was swamped by the scale of urban poverty and the rate of physical deterioration of inexpensive private rental units. By the mid-1970s public housing had become synonymous with failure, and the prospect of establishing a substantial program of public housing had vanished, at least for the foreseeable future. Looking back, we can see several reasons why public housing did not take root in the United States:

- An exceptionally strong free-enterprise ethic (Figure 8) in which public housing aroused suspicions of "creeping socialism." In this context it is not insignificant that the crucial period of debate about public housing took place during the notorious McCarthy witch-hunt for Communist sympathizers.
- Racial bias and discrimination that caused many people to be reluctant to support a program whose beneficiaries were likely to be mostly African American or Hispanic.
- Highly organized and well-funded opposition from the National Association of Real Estate Boards, the National Association of Home Builders, the United States Savings and Loan League, the U.S. Chamber of Commerce, the

Corbis

FIGURE 7 Public housing built under the provisions of the 1949 Housing Act. The relaxed atmosphere depicted in this photograph soon deteriorated to a more oppressive one as public housing in the United States became widely regarded as a receptacle for "problem" families.

Library of Congress

FIGURE 8 Free-market propaganda, cleverly wrapped in patriotism and democratic ideals, was organized by industrial, financial, and real estate organizations to minimize the "creeping socialism" of public housing legislation and other liberal urban policies in the United States.

Mortgage Bankers Association of America, and the American Bankers Association.

- Overly restrictive terms and conditions governing the construction and management of public housing. Because of a dollar limit on construction costs (pegged at $2,400 between 1949 and 1965) some tenants were given homes that had no bathroom doors, no toilet seats, and no baths or showers. Strict means-testing for potential tenants was commonly coupled with a rule that the household had to vacate the property as soon as its income exceeded the qualifying minimum by 25 percent or more. As such, public housing was programmed to be no more than a jerry-built haven for the chronically poor. Public housing projects soon came to be stigmatized as "instant slums," their residents stigmatized as economically incompetent and socially undesirable. Many of the residents themselves, feeling **labeled** and with no security of tenure, found no reason to care for their homes or their neighborhood, which helped to fuel the arguments of opponents of public housing, who maintained that it would be a breeding ground for social problems (Figure 9).

In the early 1990s the federal government introduced an improvement program called HOPE VI, designed to revitalize the worst public housing projects, such as Cabrini-Green, into mixed-income developments. Cities were awarded block grants to plan, demolish, and redevelop public housing projects, drawing on the fashionable design principles of New Urbanism. Redeveloped projects were expected to be high-density, low-rise, pedestrian friendly, and transit-accessible. By 2010, when funding for the program ended, more than 250 HOPE VI Revitalization grants had been awarded, totaling more than $6 billion. The principal criticism of the program has been that it did not require one-for-one replacement of the nearly 100,000 demolished housing units, so that there was usually a loss of affordable housing for the poor, with tenants being displaced. The program has also been criticized for being deployed as part of cities' neoliberal strategies, with HOPE VI redevelopments effectively seeding the gentrification of "difficult" neighborhoods.

Before the Hope VI program ended, the Choice Neighborhoods program was launched with a budget of $65 million for 2010. Choice Neighborhoods is intended as a successor to Hope VI that not only continues the focus on improving public housing but also incorporates a broader approach to concentrated poverty by widening the scope of the efforts to the surrounding neighborhood by adding daycare centers, parks, sidewalks and even farmers markets, with related investments in school reform, public transportation, and improved access to jobs.

AP Photo/Beth A. Keiser

FIGURE 9 A self-fulfilling prophecy: underfunding and overly restrictive tenancy regulations led to the failure of many public housing projects, including this one—Cabrini-Green—in the process of demolition in Chicago.

URBAN VIEW 4

Public and Private Housing in European Cities[14]

Apartment living is common in Europe—residents of all income groups own or rent apartments. Apartments are a good land use choice in high-density and compact European cities where space is at a premium and land values are high. Traditionally, instead of growing outwards, cities grew upwards to the limit of a city's height regulations.

The multistory apartment house originated in northern Italy to accommodate the wealthy during the Renaissance. By the early eighteenth century, the apartment house had spread to the larger cities in continental Europe and Scotland. Until the invention of the elevator, social stratification within individual buildings was vertical: Wealthier families occupied the lower floors, while poorer residents lived in smaller apartments above. Horizontal social stratification also developed within apartment blocks. The large expensive units were located in the front of buildings, with small low-rent units facing the rear. By the late eighteenth century, as the Industrial Revolution spurred increasing urbanization, apartment blocks had spread to medium-size cities. Speculators built large-scale standardized tenements for middle-income occupants and barracks for low-income residents.

The two-story terrace of single-family houses is distinctive to England, Wales, and Ireland. This row house tradition can be traced back to efforts to restrict congestion in London in the late 1500s that made it illegal for more than one family to rent a new building. The narrow multistory house with an apartment on each floor, a variation of the row house, is found along the North Sea coast from cities like Rouen and Lille in northern France to Bremen and Hamburg in northern Germany.

The serious housing shortage that started with the Great Depression of the 1930s was exacerbated by the lack of construction and significant destruction during both world wars. The public housing programs that began in Vienna in the early 1920s were stepped up across Western Europe after World War II. Modern architecture and urban design principles were combined with low-cost factory production methods. This resulted in the replacement of the war-damaged historic houses and dilapidated nineteenth-century tenements in the central parts of cities by monotonous high-rise apartment buildings.

In the 1950s and 1960s most governments adopted a policy of metropolitan decentralization of public housing to inexpensive open land at the edge of cities. Modern high-rise apartment blocks were concentrated in large peripheral estates known by their French name—*grands ensembles*. Developments like Park Hill in Sheffield, Sarcelles in northern Paris, and Chorweiler in Cologne contained 1,000 or more apartments at densities as high as 120 residents per acre.

Traditionally, West European governments spent more on public housing construction than on rent subsidies. Because the poorest segment of society cannot afford public housing, less stigma is attached to public housing in Europe compared to North America. The very poorest residents are distributed instead throughout European cities in the very cheapest private rental units. These poor quality dwellings are found in damp basements, attic spaces, and run-down rear apartments. Where a stigma is associated with public housing, it applies only to specific kinds of government-subsidized developments. These include the purpose-built housing projects that replaced the peripheral shantytowns of poor rural immigrants and postwar refugees in Spain, Italy, France, and Germany.

Historically, the severity of need and the political leanings of governments determined the amount of public housing provided within each city and country. The amount was highest in cities with serious housing shortages and liberal municipal governments such as Edinburgh and Glasgow in Scotland where the number of public units grew to well over half the total housing stock. Until the 1970s, public housing comprised 20 to 30 percent of the total in England, France, and Germany, and 10 percent in Italy. Public housing represented only 5 percent or less of the housing in the more affluent and conservative Swiss cities. Since the 1970s, however, the amount and percentage of public housing have declined due to government cost-cutting privatization programs.

In general, the cities that developed under socialism throughout Central and Eastern Europe were less spatially segregated than those that evolved under capitalism. Certainly, mansions—the prewar residences of the social elites—were used for political purposes to house party officials, foreign delegations, or institutes. But housing was viewed as a right, not a commodity; each family was entitled to its own home at reasonable cost.

In the face of the tremendous housing shortfalls following World War II, from the 1950s to the 1970s the Communist governments began massive building programs of *housing estates:* typically 11-story, prefabricated, multifamily apartment blocks. These apartments were small (about 460 to 650 square feet), poorly constructed, and almost universally disliked by residents. Often, housing estates were built in large clusters, sometimes forming massive concrete curtains, usually on land near the edge of cities. Because of this, urban population densities, which are typically higher in Central and Eastern Europe than in North America, could actually *increase* near the urban periphery (Figure 10).

The housing estates were often constructed in groups to form a **neighborhood unit** consisting of three to four apartment blocks arranged into a quadrangle surrounding shops, green space, and play areas for children. The purpose of these units, in keeping with socialist planning philosophy, was to promote greater social interaction; however, the neighborhood unit concept, along with the construction of housing estates, was abandoned by the mid-1980s after governments recognized the futility of trying to achieve social engineering through physical planning.

imagebroker/Superstock

FIGURE 10 Former socialist suburban housing in Dresden, Germany.

RESIDENTIAL MOBILITY AND NEIGHBORHOOD CHANGE

> *Residential mobility is a central facet of urban social geography, for it provides a spatial expression of the link between the individual household and the social structure, between the household's life-world and its biographical situation, between internal culture-building processes and the spatial template of the city.* The residential choices of individual households in aggregate define the social areas of the city. *The rapid transformation of American central cities and the plight of their dwindling fiscal base was caused in part by the massive migration of middle-class families to the suburbs in the 1960s. Similarly, the anticipated revitalization of the downtown areas of some cities in the 1970s . . . was brought about by the reverse migration of childless professionals to "gentrified" central neighborhoods.* But there is a two-way relationship between individual and aggregate levels, *for at the same time the individual's pattern of choices is constrained by the preexisting set of spatial opportunities in the city and the household's own biography—those characteristics of income, stage in the life cycle, ethnic status and lifestyle which will close off certain housing options to it and substantially reduce its range of choice.*[15]

This quotation from the work of social geographer David Ley provides a good description of the overall relationship between residential mobility and urban residential structure. Although people's migration actively creates and remodels the social and demographic profile of neighborhoods, it is also conditioned by the existing pattern of neighborhoods. This relationship is outlined in Figure 11, which emphasizes the effects of household mobility and residential structure on each other. The residential mosaic is seen as a cumulative product of residential mobility, which in turn is seen as a product of housing opportunities and household needs and expectations. These needs and expectations of people, in their turn, are influenced by what Ley calls the household's biography (in terms of income, lifestyle, household type, ethnicity, and so on) and by the individuals in each household's knowledge and perceptions of housing opportunities.

Movers, Stayers, and Neighborhood Change

We can take these relationships seriously only if it can be shown that rates of residential mobility in cities run consistently at reasonably significant levels. In fact, the turnover of addresses in most U.S. cities is strikingly high: About 1 in every 12 metropolitan households moves to a new home each year. Naturally, this figure masks some important differences between metropolitan areas, between central city and suburban areas, between neighborhoods, and between tenure sectors (such as owners versus renters).

These differentials suggest that residential mobility is a selective process, with some neighborhoods being dominated by households with little propensity ever to move. These households have been called *stayers*. Among "stayers," people in households who are home owners, middle-aged and elderly, and with low-to-middle socioeconomic background tend to be overrepresented. A major contribution of the people in these households to the city's residential mosaic is to impose a relatively high degree of stability on certain neighborhoods. *Movers*, whose impact as we have seen, may either reinforce neighborhood composition or initiate change, tend to be people who are younger, renters, and at one extreme or another of the socioeconomic ladder.

Given that people age, adopt new lifestyles, experience changes in income levels, and so on, it follows that stayers can become movers, and vice versa. In addition, research has shown that there is an independent duration-of-residence effect, whereby the longer a household remains in a dwelling the less

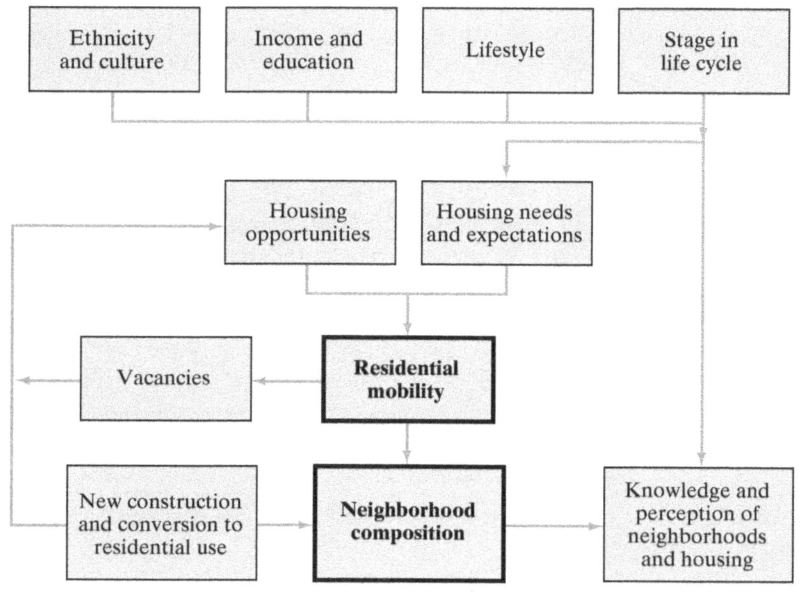

FIGURE 11 Relationships between housing demand, residential mobility, and neighborhood change.

likely it is to move, mainly because of the emotional attachments that people develop toward the home and the immediate neighborhood and because of the social networks established there. Research has also supported the idea that people can be thought of as *locals* or *cosmopolitans*, depending on their natural propensity to develop such attachments.

The Impact of New Arrivals to the City

Another important distinction to make in relation to patterns of residential mobility by people concerns the difference between intraurban moves (i.e., *within* the metropolis) and other kinds of mobility—intermetropolitan migration, inmigration from nonmetropolitan areas, and immigration directly to metropolitan areas. In the typical U.S. metropolis, intermetropolitan migration, inmigration, and immigration account for up to one-third of all moves—substantially less than in the past, when migration and immigration not only fueled urban growth to a much greater extent but also dominated neighborhood dynamics.

The impact of these long-distance moves today can best be divided into two categories: the arrival of low-income migrants and the arrival of middle-income and high-income migrants. Traditionally, flows of low-income migrants and immigrants have mainly focused on central city neighborhoods with inexpensive housing, but with a significant degree of localized differentiation according to ethnicity and place of origin. More recently, some of these flows have been directed to first-tier and second-tier suburbs, where established ethnic communities provide a port of entry for new arrivals. A good example is the "Little Saigon" area in the Clarendon neighborhood of Arlington, Va., a 60-year-old suburb of Washington, D.C. Such flows reflect the continuing importance of **chain migration**, whereby migrants who were encouraged and assisted in their move by friends and relatives from their place of origin encourage other friends or relatives to join them, helping them to find accommodation and jobs when they arrive.

The impact of middle-income and high-income migrants, in contrast, is much less focused in spatial terms. These people do, however, exhibit certain broad regularities in spatial behavior. An occupationally and geographically mobile group of cosmopolitan movers, they are often constrained by limited amounts of time available to search for a home. In addition, because they are willing and able to move in search of better jobs or career advancement, if they do purchase a home rather than rent, they tend to opt for homes that are likely to be relatively easy to resell. Finally, experience may have alerted them to the possibilities of moving into a neighborhood only to find that their own values and lifestyle are out of tune with the established character and social networks of the neighborhood. The overall result is a tendency to select newly built homes in suburban tracts or packaged developments, moving on after a year or two, when they have developed a better sense of the geography of their new city.

Intraurban Moves

Moves by people within metropolitan areas account for at least two-thirds of the total turnover. In addition to the follow-up moves of newly arrived migrants, they include a great variety of adjustments as new households are formed, old ones split up, and existing ones respond to changed circumstances. Most moves take place over relatively short geographical distances—about one-sixth to one-quarter of the diameter of the city. This fact is generally taken to reflect the importance of people's desire to preserve local ties and, in particular, children's ties to friends and schools. Household changes in housing quality and neighborhood status are usually incremental, and there is a marked directional bias to patterns of movement, with many households staying within the same sector or quadrant of the city. Most metropolitan areas are so large that many people prefer to consider first the housing vacancies that occur within the part of the metropolis that they know best.

Analyses of the geographical interrelationships between patterns of residential mobility and neighborhood housing attributes and socioeconomic attributes have also shown that, in addition to these broad generalizations, housing *type* (as measured by owner-occupancy rates, the incidence of single-family dwellings, and the number of rooms per dwelling) rather than housing *quality* (as measured by housing values, housing age, and the number of bathrooms per dwelling) is the chief determinant of intraurban mobility.[16] The same studies have shown that it is family status rather than socioeconomic status that is a function of housing type (and therefore of mobility), thus underscoring the importance of changes in the household life cycle in triggering residential mobility (Figure 4). As we might expect from our knowledge of changing metropolitan form and changing patterns of residential differentiation, however, these relationships have become progressively weaker

URBAN VIEW 5

Neighborhood Stability in West European Cities[17]

West European cities enjoy remarkable neighborhood stability. Europeans move much less frequently than North Americans, and homeowners maintain their characteristic solidly constructed concrete block, brick, and stone housing. As a result, older neighborhoods at or near the center of large cities enjoy remarkably long lives despite suburbanization.

The districts of handsome mansions built by speculative developers for wealthy families in the seventeenth, eighteenth, and early nineteenth centuries remain stable high-income neighborhoods, such as Belgravia, Bloomsbury, and Mayfair in central London (Figure 12). High-income suburban neighborhoods developed typically in the western part of older

(continued)

URBAN VIEW 5
Neighborhood Stability in West European Cities (*continued*)

Paul L. Knox

FIGURE 12 Belgrave Square in Belgravia, central London. Districts of handsome mansions such as this, which were built by speculative developers for wealthy families in the seventeenth, eighteenth, and early nineteenth centuries, remain stable high-income neighborhoods.

industrial cities, upwind of industrial smokestacks and residential chimneys.

Many wealthy residents, in fact, have remained at or near the city center in Western Europe since before the Industrial Revolution. Higher taxes on city land until the late nineteenth century kept the poorest residents and immigrants outside the city walls. Beginning with Paris in the second half of the nineteenth century, this tradition was strengthened by the replacement of areas of slums and former city walls with wide boulevards and imposing apartments.

Since the eighteenth century in Western Europe, however, urban growth has spread to suburban zones and has even enveloped freestanding villages and towns. Yet these separate urban centers became distinct quarters within the expanding city as they maintained their long-established social and economic characteristics, major landmarks, and shopping streets. During the second half of the nineteenth century, **annexations** of groups of these suburban quarters produced distinctive city districts with their own shopping areas and government institutions.

In the last few decades governments in older industrial cities have funded urban renewal projects designed to attract higher-income residents to the revitalized parts of central areas. The success of these large-scale city center redevelopments has given rise to gentrification in the surrounding area. Demand for housing that has the potential to be renovated for higher-income occupants, however, has raised property values in certain areas and pushed out lower-income residents.

since 1950, thus pointing to an increasing complexity in the pattern and logic of residential mobility.

Reasons for Moving

We can begin to understand the complexity of residential mobility a little better by investigating the behavior of individual households, and try to find answers to the question of why families move by distinguishing between voluntary and involuntary moves and between "push" and "pull" factors (Figure 13).[18]

Involuntary moves can account for a significant and surprisingly high proportion of intraurban mobility—between 15 and 25 percent. They are necessitated by events totally beyond the household's control: property demolitions, **eminent domain** proceedings, evictions for rent arrears or defaulting on mortgage repayments, and disasters such as fire or flood. In addition to these purely involuntary moves are somewhat more voluntary moves that are *induced* by unwanted or unforeseen circumstances, such as divorce, ill health, a death in the family, or corporate relocation (a change in the location of the place of work—because of office

MOVE					
Involuntary/Forced	**Voluntary**				
• Demolition • Eminent domain • Eviction • Disaster (flood, fire)	**ADJUSTMENT**			**INDUCED**	
	HOUSING FACTORS • Space • Quality/design • Cost • Tenure change	NEIGHBORHOOD FACTORS • Quality • Physical environment • Social composition • Public services	LOCATION AND ACCESSIBILITY • Workplace • Shopping/school/ amenities • Family/friends	EMPLOYMENT STATUS CHANGE • Job change • Retirement	LIFE CYCLE EFFECTS • Household formation • Change in marital status • Change in household size

Life Cycle and Housing Adjustment as Explanations of Residential Mobility, W. A. V. Clark, J. Onaka, Urban Studies 20 (1983).

FIGURE 13 Reasons for household relocation.

decentralization, for example). On average, such moves account for a further 15 percent or so of all intraurban mobility.

People give a wide variety of reasons for purely *voluntary* moves. Most are adjustment moves that are intended to change the type and quantity of housing. Classic migration theory groups these reasons into **push factors** and **pull factors**. The most frequently cited push factor by far is the lack of sufficient space—or, to be precise, the *feeling* of insufficient space for household needs. Once again, this factor underscores the relevance of changes in the household life cycle. Other frequently cited push factors include the costs of maintenance and repair, structural or technological obsolescence in the dwelling, aspects of the physical environment of the neighborhood (e.g., poor upkeep of neighboring properties, heavy traffic), and characteristics of the social environment (e.g., too many or too two few children, too much or too little street life). The most frequently cited pull factors by far are those associated with a change of employment. Other pull factors include accessibility to shops, amenities, and friends, the attractions of better-quality schools and public services, the chance to switch housing submarkets (usually from renting to owning), and the attractions of particular settings in pursuing particular lifestyles.

Understanding Household Behavior: The Decision to Move

In practice, of course, households are "pushed" and "pulled" by a variety of factors at once. In addition, the combined effect of push and pull factors may not be enough to cause a household to look for alternatives. Even if a search among potential alternatives is begun, a household may not necessarily eventually move: Adjustments can be made to perceived wants and needs, to the dwelling itself, and even to the neighborhood. These possibilities are accommodated in Figure 14, which is based

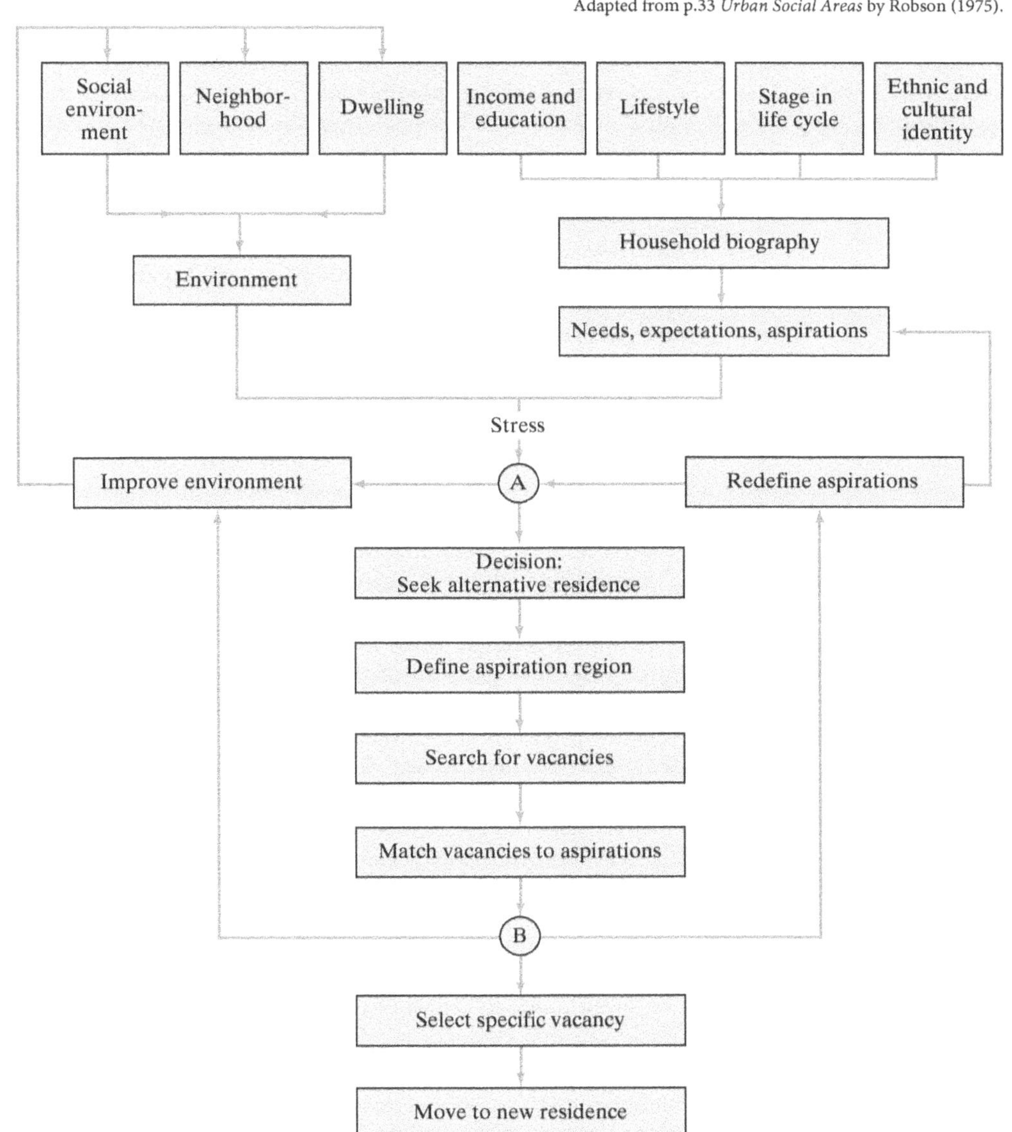

FIGURE 14 A model of the residential location decision process.

on a conceptual model of the household relocation and search process first developed by urban geographers Larry Brown and Eric Moore.[19]

This model recognizes that a household will derive some degree of "utility" from its current dwelling: a product of its own biography and of the qualities and attributes of the dwelling itself and the neighborhood. This utility may not be positive, however. "Stressors," arising from the interaction of a variety of push and pull factors, may reduce the utility of the dwelling so that the household feels forced to do something about the situation (point A on Figure 14).

One course of action is to address directly the shortcomings of the dwelling or neighborhood. Dwellings can be remodeled, extra space can be added, or kitchens can be remodeled. Neighborhood shortcomings can be addressed through joining or forming residents' associations, by confronting troublesome neighbors, or by political action.

If such action is not feasible, or if it is unsuccessful, a second course of action may be pursued: modifying the household itself as a means of coming to terms with existing conditions. Plans to have children may be deferred or abandoned, household lifestyle may be modified or constrained, or, more likely, household aspirations and expectations may be lowered. The older people get, the more adept they become at lowering their aspirations to fit their experience of reality. This, then, is an important explanation for the tendency for older households to become stayers.

Understanding Household Behavior: The Search for Alternative Places to Live

The third course of action is to seek alternative places to live. Figure 14 shows that there are three stages to this course of action: (1) defining the "aspiration region" (based on site characteristics including the attributes desired in a new home and its locational situation), (2) searching for dwellings that fall within the aspiration region, and (3) comparing possible alternatives. The importance of these stages is that *different kinds of households behave differently at each stage, thus reflecting and sustaining the processes of social and residential differentiation.* Households with different biographies and income constraints will set out to look for vacancies with quite different housing goals in mind. In addition, some households will develop a much clearer idea of what they are looking for, depending in part on the personality of household members but mostly on their ability to organize their thoughts (partly a function of education) and on their overall knowledge and experience of housing and neighborhood types.

The search process itself is much more closely linked to household socioeconomic attributes. Typically the search does not cover the whole city, or even whole housing submarkets, but takes place within the household's *awareness space*, which is a product of its *action space* and its *information space*, both of which are closely linked to income, education, and occupational status. Consequently, while some households rely on word-of-mouth within a very localized search space, others systematically pore over newspaper advertisements, while still others base their search on wide-ranging neighborhood visits and online presentations, guided by real estate agents with computerized multiple listings.

The search is not simply a function of the range and quality of information, however. Households differ in the extent to which they face barriers and constraints during the search process. Lack of personal transportation is an obvious barrier for some people that is directly related to socioeconomic status. More important, but less clearly a function of socioeconomic status, is the constraint of time. Each household has to balance (1) the probability of making a better decision after an exhaustive search against (2) the costs—both real and psychological—of doing so. Where time is limited, the anxiety produced by lack of success may result in a modification of the household's aspiration region, a restriction of its search space, or a shift in its use of information sources. The pressure of time may also simply lead to confused or irrational decision making. Then again, the longer the search goes on, the greater the household's knowledge of the housing market, although there comes a point, as in shopping for anything else, where the cost of missed opportunities outweighs the value of additional knowledge.

Having identified possible alternatives, households must make a choice. The better defined the household's aspiration region, the more clear-cut this decision should be, though research has suggested that, having gone through the search process, most households do not refer rigidly to their original formulations. Rather, they are likely to compromise a good deal. In the final decision, neighborhood quality (and in particular the quality of neighborhood schools, for households with children of school age) seems to weigh more heavily than housing quality or accessibility, and interior style and appearance weigh more heavily than exterior style and appearance.[20] Households that are unable to find vacancies that fall within their aspiration region at this point (B in Figure 14) must revert to one of the two other possible courses of action—going back and addressing the shortcomings of their existing dwelling or neighborhood, or redefining their aspirations and needs.

Empirical analyses of people's behavior in large and complex housing markets show that their search patterns tend to be quite restricted. Most households do not, for example, search anywhere near all of the housing submarkets within their affordable price range. Rather, they are likely to attempt to reduce search costs and uncertainty by searching within a single housing submarket. Searches tend to be anchored around familiar community areas and key nodes such as workplace locations, and most households are very persistent in restricting themselves to one or two preferred local areas.[21]

One important factor that is not explicit in Figure 14 is discrimination and bias within housing markets. Most scholars agree that this is an important factor in shaping patterns of residential location, along with affordability, job location, household preferences, and household information. There is, however, considerable debate as to just *how* important discrimination is within the overall context of the residential kaleidoscope.[22] The various *ways* in which bias and discrimination take place are discussed in the next section.

HOUSING MARKET GATEKEEPERS, BIAS, AND DISCRIMINATION

Decisions about where to live and when to move are constrained not only by income, barriers to the search process, and the interdependence of people's actions, but also by the decisions and behavior of key groups who influence the supply of housing and housing finance. Speculators, developers, builders, realtors, and public officials, as well as individual households, can be regarded as "city makers." Yet city making does not end once housing has been built and sold. The continual restlessness of people and capital investment constantly remakes the sociodemographic landscape in response to new needs and opportunities (and in so doing inevitably modifies the physical landscape). At the center of this process are the "exchange professionals" who facilitate residential mobility: realtors, mortgage financiers, insurance agents, appraisers, landlords, and so on.

Although these exchange professionals project themselves as being socially neutral, in certain circumstances they may act as "social gatekeepers," facilitating residential mobility by certain groups of people and financing housing in certain neighborhoods but limiting mobility by other groups of people and suppressing the capital available to other neighborhoods. This gatekeeping is not necessarily the result of conscious attempts to influence the pattern of flows of people and capital. In some circumstances exchange professionals may not even be aware of the gatekeeping consequences of their actions.[23] Given the importance of their roles in the marketplace, however, the dynamics and constraints of their professional environments make a certain amount of bias almost inevitable.

Every profession tends to develop its own distinctive professional ideology or view of the world, and individual professionals are influenced and guided by these perspectives in making their day-to-day decisions.[24] These professional ideologies are the product of a variety of factors, including

- selectivity in recruitment (including the self-selection that draws people with certain backgrounds, interests, and dispositions to certain careers).
- values, attitudes, and priorities imbued through education and training.
- the criteria of professional success sponsored by professional journals, magazines, and websites.
- the reward system and career structure.

Within every profession, however, there is also scope for deliberate discrimination: consciously wielding professional power or influence to deny members of particular groups access to housing in certain neighborhoods or to restrict the flow of finance to certain properties or neighborhoods. This discrimination may be personally motivated, but more likely it is the result of the exploitation of social prejudices for personal (or corporate) financial gain. Civil rights legislation has made such activity illegal in the United States, but most cities still bear the imprint of housing discrimination of the pre-civil rights era. In addition, discrimination remains endemic to housing markets because it is very difficult to police.

Real Estate Agents as Social Gatekeepers[25]

The key to understanding the professional behavior of real estate agents is in their reward system, which is normally a commission based on a percentage of the sale price of a property. Ideally, therefore, real estate agents would like to see a situation where house prices are high *and* there is a high turnover rate. This is by no means the usual situation, however, and in many circumstances high prices work against a high turnover or vice versa. In addition, real estate agents can find themselves working on behalf of both buyers and sellers in the same market, thereby not only dividing their loyalties but also complicating their attitudes toward asking prices. As a result, real estate agents develop a keen sense of the residential kaleidoscope as an opportunity structure that has to be managed and exploited carefully. Neighborhoods with high prices (and therefore high potential commissions) must be monitored for changes that might depress exchange values, neighborhoods with low turnover rates must be carefully monitored for signs of movement, and lower-priced neighborhoods must be monitored for signs of gentrification. It is a short step for some unscrupulous real estate agents (a small minority within their profession, it must be emphasized) to encourage or contrive such changes.

The principal activity in this respect has always been **steering**: keeping like with like, deterring households from moving into neighborhoods occupied by households of a different socioeconomic background, ethnicity, or sexual orientation in order not to jeopardize local prices. In 1955 real estate agents could openly explain steering in this way:

> People often try to get in higher class areas than they'll be accepted in. We just don't show them any houses in those areas. If they insist, we try to talk them out of it one way or another. I've purposely lost many a sale doing just that. It pays in the long run. People in the community respect you for it and they put business your way.[26]

The result was "slammed door" discrimination, mainly against African Americans. In 1968, Title VIII of the Civil Rights Act (Fair Housing Act) made such activity illegal. Discrimination continued, however. A 1989 report by the U.S. Congressional Research Service described how "revolving doors" had replaced "slammed doors":

> The new technique—apparently developed to evade statutory prohibitions against discrimination—consists essentially in deceiving a minority applicant into believing that he or she would be welcome as a tenant or owner, but in withholding information about available housing. . . . In a sales situation, the applicant may be told the seller has already accepted another sales contract or told the owner had changed his mind and taken the dwelling off the market. Everything is done in such a way that the applicant will go away believing that nothing is available and without the slightest awareness that he or she has been discriminated against. The black or Hispanic applicant has been led in and then conducted out, so to speak.[27]

In 2009, U.S. Department of Housing and Urban Development (HUD) and HUD-certified state and local agencies that enforce laws equivalent to the Fair Housing Act received 10,242 complaints. The most common basis of complaints was disability discrimination (44 percent) followed by race discrimination (20 percent).[28] The number of housing discrimination complaints typically exceeds 10,000 but certainly a large number of complaints are not reported. By some estimates as many as four million instances of housing discrimination occur each year in the United States.[29]

Every ten years, HUD commissions a "paired testing" housing discrimination study. Having set new guidance on housing discrimination against members of the LGBT community in 2010, the next HUD study will for the first time address LGBT discrimination as well as the usual focus on discrimination against minorities. For the 2000 HUD study, 4,600 pairs of testers—one minority and the other white—posed as otherwise identical home seekers and visited real estate or rental agents to inquire about the availability of advertised housing units that were for sale or rent in 23 metropolitan areas across the United States.[30] This methodology is designed to provide direct evidence of differences in the treatment experienced by minorities and whites in their search for housing and a way to track the extent to which housing discrimination is changing over time.

Comparing the results of the 2000 and 1989 studies of metropolitan home sales markets shows that there was a decline in the level of discrimination experienced by blacks and Hispanics seeking to buy a home (Table 1). Between 1989 and 2000, however, steering based on neighborhood racial composition increased for black homebuyers—by 10 percent in the case of homes recommended and by almost 8 percent in the case of the homes that were inspected. So although black homebuyers faced less discrimination overall in 2000, blacks and whites were apt to be recommended and shown homes in different neighborhoods. Blacks were more likely to be steered to neighborhoods that were predominantly black compared to the neighborhoods recommended to comparable white homebuyers. Hispanic homebuyers, in particular, faced significant discrimination from real estate agents when it came to help in obtaining financing. Between 1989 and 2000, the overall incidence of white-favored treatment on financing assistance (offers of help with financing, lender recommendations, and discussion of down payment requirements) rose from 33.3 percent to 38.6 percent.

The results of the 2000 study of metropolitan rental markets showed that there had been a modest decrease in discrimination toward blacks seeking to rent a housing unit. Hispanic renters, however, did not experience this downward trend (Table 1). Hispanics, for example, were more likely to be quoted a higher rent for the same housing unit than their white counterparts in 2000 compared to 1989. In addition, Hispanic renters were more likely to experience discrimination in their housing search than African-American renters in 2000.

There is evidence that housing discrimination has increased due to online advertising.[31] The rise in complaints has been driven by discrimination in online advertising for rental properties, with most of the ads discriminating against families with children, which is illegal under the federal Fair Housing Act. Discrimination against minorities in online rental listings on Craigslist is also being increasingly documented. This is an issue because three-quarters of all renters use the Internet to search for vacancies.[32]

Of course, housing discrimination is not confined to the United States. In Sweden, for example, researchers carried out a housing discrimination study based on online rental listings and found ethnic and gender discrimination there too.[33] Three fictitious renters with distinctive sounding ethnic and gender names were created. These potential renters applied for vacant rental apartments that were advertized online by landlords. The male with the Arabic name received many fewer responses and apartment showings than the male with the Swedish name. The female with the Swedish name had the least difficulty finding an apartment online.

TABLE 1 Discrimination by Real Estate and Rental Agents against Minorities in U.S. Metropolitan Housing Markets, 1989 and 2000

	1989	2000	Examples of discrimination
Homebuyers			
Black	29.0%	17.0%	White homebuyers were more likely to be able to inspect available homes, to be shown homes in predominantly white neighborhoods, and to receive more information/assistance in financing as well as more encouragement.
Hispanic	26.8%	19.7%	White homebuyers were more likely to receive information/assistance in financing and to be shown homes in non-Hispanic neighborhoods.
Renters			
Black	26.4%	21.6%	Whites were more likely to receive information about available housing units, had more opportunities to inspect available units, and were more likely to be offered rental incentives.
Hispanic	23.7%	25.7%	Whites were more likely to receive information about available housing units, had more opportunities to inspect available units, and were more often quoted a lower rent for the advertised unit than similarly qualified Hispanic renters.

Source: The Urban Land Institute, *Discrimination in Metropolitan Housing Markets: National Results from Phase 1 HDS 2000,* Washington, D.C.: U.S. Department of Housing and Urban Development, 2002.

In the real estate market, steering and "revolving door" tactics are not the only kinds of intervention pursued by unscrupulous real estate agents. Local submarkets can be manipulated by **block busting**, whereby prices are deliberately driven down, temporarily, allowing real estate agents or their associates to buy up as many properties as possible before restoring equanimity to the market and then selling to a new group of purchasers. Prices can be driven down in various ways: by steering minority households *toward* a lower-income white neighborhood, for example, supplemented by scare tactics designed to hasten "white flight"—posting bogus FOR SALE signs, or hustling for listings. In extreme cases outsiders may be hired to commit petty acts of vandalism that are intended to signal the social deterioration of the neighborhood. Alternatively, neighborhood price decline can be "seeded" by purchasing houses only to leave them vacant and neglected. As residents see the neighborhood beginning to slip, more will put their homes on the market. When a sufficient number of dwellings has been acquired, the real estate agents or their associates will hope to derive a large profit, either by selling them one by one to members of an incoming group of people or by selling them all to a single developer seeking a large lot for a big project.

Mortgage Finance Managers as Social Gatekeepers

Discrimination against minorities in the metropolitan home sales market may also arise in subsequent contacts involving applications for loans. Bank officers and savings and loan association managers tend to operate within decision-making frameworks that are fairly closely circumscribed by company policy, nationally established money market rates, and federal law. Nevertheless, these lenders still exercise a considerable amount of discretion in allocating mortgages, so that they are able to play a pivotal role in deciding "who lives where, how much new housing gets built, and whether neighborhoods survive."[34]

The chief allegiance of commercial lending institutions is not to borrowers but to investors. Funds are therefore not to be risked on loans to vulnerable households, for unconventional properties (because they may be difficult to sell if the borrower defaults), or for properties in unpromising neighborhoods (also difficult to resell in the event of a default). Risk minimization means attaching a good deal of weight to creditworthiness, with special emphasis on income stability. This practice tends to favor salaried middle-income households while making it more difficult for wage earners, the self-employed, and female-headed households to secure a mortgage. Risk minimization also involves categorizing applicants and properties in terms of stereotypes. Because mortgage finance officers traditionally have tended to be predominantly white, male, middle-income and married with children, such stereotyping tends to work to the disadvantage of minority groups, female-headed households, other nontraditional household types, and people with unconventional lifestyles.

The easiest means of stereotyping properties is on the basis of neighborhood quality, drawing a line around "high-risk" neighborhoods and using this as a basis for determining loans. This practice, known as **redlining**, also results in a bias against minorities, female-headed households, and other vulnerable groups, because they tend to be localized in high-risk neighborhoods. A further problem with redlining is that it becomes a self-fulfilling prophecy, as neighborhoods starved of mortgage funds become progressively more run-down and so increasingly risky. We can summarize the consequences of redlining in terms of seven stages:[35]

1. The cheaper housing typical of redlined areas is effectively placed beyond the means of many lower-income households, because with regular mortgages difficult or impossible to obtain, loans can be secured only by means of higher down payments, higher interest rates, and shorter loan maturities.
2. The worst dwellings, denied mortgage loans even from more expensive sources, are left vacant and neglected, thus initiating a spiral of neighborhood decline that is difficult to stop.
3. The neighborhood declines to the point where even home improvement loans are difficult to obtain, leading to rapid physical deterioration and, in some cases, abandonment.
4. As conditions worsen, property insurance becomes expensive and difficult to get. Neighborhood businesses begin to fail, taking essential services and important sources of cash flow out of the neighborhood.
5. In the face of accelerating decline in property values, those who are able to do so begin to move away, leaving behind a residual population dominated by elderly and disadvantaged households.
6. The city treasury feels the results of neighborhood decline as property tax revenues decline while the demand for public services increases.
7. The entire city is drained of so much revenue that public services have to be cut back across the whole city or tax rates have to be increased, or both.

The severity of these problems, coupled with the fact that it was most often minority neighborhoods that were the worst casualties, led to a series of laws designed to eradicate redlining in the United States. These laws included the Civil Rights Act (federal Fair Housing Act) of 1968, the Home Mortgage Disclosure Act of 1975, the Equal Credit Opportunity Act of 1976, the Community Reinvestment Act of 1977, and the Fair Housing and Equal Opportunity Act of 1989. Nevertheless, many cities still carry the blight of neighborhood decline initiated (or at least reinforced) by redlining.

There is evidence that housing discrimination in home loans has increased due to the housing crisis.[36] As the number of foreclosures increased, private fair housing centers around the United States have seen more cases of discrimination against minorities in mortgage lending than before the housing crisis. In addition, recent studies continue to document discrimination.[37] A 2010 study in the *Federal Reserve Bulletin* reported that in 2009 the rejection rate for conventional home loans for African-American borrowers (32.3 percent) was about 2.5 times higher than the rejection rate for white borrowers (13.1 percent), while the rejection rate for Hispanic borrowers (25.6 percent) was about twice as high.[38]

A 2009 study by the nonpartisan Center for American Progress of data from 14 large lending institutions found that 17.8 percent of white borrowers were given higher-priced mortgages (defined by the Federal Reserve as having an annual percentage rate at least 3 points higher than a Treasury security of the same maturity) when they borrowed from large banks in 2006, compared to 30.9 percent of Hispanics and a staggering 41.5 percent of African Americans. Disparities in lending remain even after accounting for differences in creditworthiness or income. High-income African-American and Hispanic borrowers were about three times more likely (32.1 and 29.1 percent respectively) than high-income white borrowers (10.5 percent) to be given higher-priced loans.[39] Another study in 2007 in the *Federal Reserve Bulletin* reported that 53.7 percent of African Americans and 46.6 percent of Hispanics were given subprime loans compared to about 17.7 percent of whites.[40] While some portion of this overall racial gap may be explained by careful underwriting that takes into account differing household incomes, credit scores, and loan-to-value ratios, the enormity of the gap and the fact that studies continue to find that the gap persists across the same income levels of minorities and whites raise important questions about the lending practices of financial institutions.

Evidence of discrimination in lending also comes from other sources that document how some mortgage finance managers targeted minority communities for risky, high-cost loans. In a 2008 lawsuit, *Mayor and City of Baltimore v. Wells Fargo Bank*, for example, affidavits filed by two former employees of the bank showed that the lender:

- targeted for subprime loans African-American communities (but not white ones) in Baltimore and Prince George's County, Maryland
- provided significant financial incentives to bank employees for steering African-American borrowers into subprime loans despite these borrowers actually being qualified for prime mortgages
- used black churches as platforms for selling subprime loans by sending African-American bank employees to make presentations there.[41]

The rejection rates for federally backed (**Federal Housing Administration [FHA]**, Veterans Administration [VA], and the Farmers Home Administration [FmHA]) mortgages show similar patterns of disparities. This situation raises concerns because FHA loans have comprised a disproportionate, and since the housing bubble burst, rising, share of the financing used by African-American and Hispanic homebuyers. In 2008 federally backed loans accounted for 51 percent of all home purchase loans to African-American applicants and 45 percent of home purchase loans to Hispanics (compared to 27 percent for white applicants) (the lower 2001 percentages were 42, 38, and 19 respectively).[42] In addition, in the lead up to the housing crisis, HUD had identified abuse of the FHA program by lenders and sellers who had tried to take advantage of first-time homebuyers and have used FHA loans to carry out "property flipping" scams. Property flipping involves purchasing distressed properties at a negligible price, and then, after minimal cosmetic or even no work being done to the property, selling it at a price far above what it is worth. Victims of property flipping are often unsuspecting low-income, minority first-time homebuyers.

Insurance Agents as Social Gatekeepers

Discrimination against minorities in the metropolitan home sales market may also arise in contacts involving applications for home insurance. Discrimination in underwriting may seem less noteworthy than discrimination in lending, but its impact is no less profound. A lender will not provide a mortgage unless a potential homebuyer first obtains property insurance. The insurance policy is required because it minimizes the risk of financial loss to the lender in the event that the house is damaged or destroyed.

Households who experience problems of insurance availability tend to be located within central city neighborhoods, often with high concentrations of minority residents. Certain risk factors—such as older wood frame homes, electrical and heating systems that have not been updated, and higher theft rates—are greater for central cities than for suburbs.[43] A study of loss costs in eight major metropolitan areas in the United States published by the Insurance Research Council found that the frequency of claims, the size of the claims, and consequently industry costs per insured home were higher for urban than suburban policy holders.[44] At the same time, anecdotal and quantitative evidence points to the existence of discrimination on the part of property insurers based on racial stereotyping.

Systematically ascertaining the extent of discrimination is difficult due to the limitations of current data availability—the insurance industry is not subject to the federal disclosure requirements that apply to home lenders. Nevertheless, a statistical analysis of 33 U.S. metropolitan areas by the National Association of Insurance Commissioners, a trade association of state law enforcement officials who regulate the insurance industry, found that the racial composition of the neighborhood remained significant in the number and cost of policies after controlling for risk factors covering loss experience and other demographic factors.[45] Paired testing of major insurers in nine cities by the National Fair Housing Alliance found evidence of illegal discrimination in the shares of tests: Chicago (83 percent), Atlanta (67 percent), Toledo (62 percent), Milwaukee (58 percent), Louisville (56 percent), Cincinnati (44 percent), Los Angeles (44 percent), Akron (37 percent), and Memphis (32 percent).[46] Fair housing organizations have filed a series of lawsuits and administrative complaints against some of the country's largest insurers (and lenders) that have resulted in settlements in the millions of dollars. Recent progress in addressing discrimination in underwriting include voluntary educational, mentoring, and outreach initiatives on the part of insurance companies and the establishment in 1994 of the National Insurance Task Force (NITF), renamed the NeighborWorks Insurance Alliance in 2004, comprising insurance companies, government regulators, and community groups, which has the mission of developing partnerships between the insurance industry and community groups.[47]

URBAN VIEW 6
Hoxton's Serial Transformations[48]

Like many central city neighborhoods in large metropolitan areas, Hoxton has gone through a series of transformations in response to structural economic change, to cycles of investment and disinvestment, to its changing situation within the evolving overall framework of the city's infrastructure and spatial organization, and to the movement of different groups of people into and out of the district. In the late seventeenth century Hoxton, just beyond London's city walls, was seen by property developers and London gentry as far enough away from the dirt and squalor of the city but close enough for a short commute to work. It was duly transformed into a superior sort of district, a phase that is marked by a few surviving Georgian residences and by its formal street pattern featuring Charles Square and Hoxton Square.

The Industrial Revolution brought another transformation. By the midpoint of the nineteenth century Hoxton had become a working-class neighborhood, the center of the London furniture industry, with workshops and warehouses surrounded by hastily erected housing for a rapidly increasing population of migrants and immigrants. By the end of the century, Hoxton had become run-down, its population notorious for poverty and squalor. For much of the twentieth century, despite the elimination of the worst of the slum housing and the provision of social housing, Hoxton remained a classic central city problem area, its manufacturing base sharply diminished and its population trapped in a cycle of deprivation. Hoxton is still one of the most underprivileged and run-down areas of London, with poor schools, high levels of unemployment, and a built environment dominated by obsolescent and semi-derelict structures. Yet it has famously undergone a double transformation since the mid-1980s as the processes of gentrification have flickered across the landscape.

The trigger for the gentrification of Hoxton was that it acquired a neo-bohemian aspect as aspiring artists and young designers and musicians found inexpensive live/work spaces in the district's industrial lofts and buildings toward the end of the 1980s. The streets of Hoxton began to be used for experimental shows and the buzz attracted galleries and dealers. Soon, the neighborhood's atmosphere began to attract an avant-garde of graphic design firms, independent music labels and studios, interior design firms, photographic studios and galleries, architecture firms and new media companies who colonized the old furniture workshops of the district. Disused and obsolescent workshops and warehouses were renovated as offices, galleries, and bookstore/cafés. Loft spaces were renovated as apartments.

By moving into Hoxton, designers could signal their trendy commitment to themselves and to others, including clients. They had distinct ideas about what living like a designer should entail, and this in turn impacted the atmosphere of the district. Hoxton's signature identity was "fashion rebel": scruffy clothes and daft haircuts. For men, the "uniform" was vintage Levi jeans paired with T-shirts bearing the names of obscure record labels; the "look" was completed by a haircut that became known as the Hoxton Fin. The Hoxton-girl look was deliberately trashy, featuring Blondie T-shirts, plastic jewelry, and pixie ankle boots.

Several projects funded through government central city regeneration programs helped to reinforce this phase of gentrification, including basic refits of warehouses and the remodeling of an old electricity generating station. Meanwhile, neoliberal government policies—fewer controls on development, a reduction in social support services for working class families, and the privatization of the public housing stock—contributed to a gradual displacement of the existing multicultural working class community. In their place came young "creatives" of all sorts, attracted by the low rents, the neo-bohemian atmosphere of the district, and the gritty authenticity of the built environment. Abandoned warehouses began to be converted into lofts and pubs; and clubs, including a pioneering gay club, the *London Apprentice*, opened around Hoxton Square. Media representations of the district, ignoring its less glamorous aspects and the displacement of disadvantaged households, helped to promote Hoxton as the exemplar of an edgy and innovative cultural quarter: *Time Magazine*, for example, referred to Hoxton as one of the "coolest places on the planet" in 1996.

This strengthened the trendy reputation of Hoxton but made it unaffordable for the neo-bohemians at the heart of its initial transformation, because most of them never had the capital or the creditworthiness to purchase their property and protect themselves from rent increases. As property prices began to rise, many of them left for cheaper space farther east. In their place has come a more affluent cohort of gentrifiers with the capital and creditworthiness to purchase and renovate loft apartments and to patronize an increased number of galleries, upscale cafes, noodle bars, sushi restaurants, and Italian delicatessens. Near enough (less than a mile) to the expanding fringe of London's banking and financial quarter, Hoxton has also developed an embryonic night-time economy, its restaurants, clubs, and bars catering increasingly to affluent office workers. As a result, a second transformation has occurred, tipping Hoxton from a district of cultural production to one of cultural consumption; and from an avant-garde neo-bohemia to just another gentrifying district (Figure 15).

Grant Smith/Photolibrary, Inc.

FIGURE 15 Near the expanding fringe of London's banking and financial quarter, Hoxton has developed an embryonic night-time economy; its clubs, bars, and restaurants (like Jamie Oliver's *Fifteen*) cater increasingly to affluent office workers. As a result, a transformation has occurred, tipping Hoxton from a district of cultural production to one of cultural consumption; and from an avant-garde neo-bohemia to just another gentrifying district.

PUTTING IT ALL TOGETHER: THE EXAMPLE OF GENTRIFICATION

It should not be surprising—given what we have seen in previous sections about the differential physical deterioration of neighborhoods, fluctuations in the flow of investment capital, variability in household behavior, changing institutional constraints, and the influence of key gatekeepers—that neighborhood change is not easy to predict. For the same reasons, explanations of changes that *have* taken place are difficult to establish and are often hotly contested among academics, city planners, policy makers, and politicians.

Figure 16 brings together the major elements influencing neighborhood change in general terms (and fits within the structuralist approach). Here, households are seen as one of a number of interdependent groups of people whose actions are set within the context of the financial climate, legal frameworks, public policies, institutional practice, and professional ideology. What Figure 16 cannot capture, however, are the relative importance and specific interactions of particular elements in relation to any given type or case of neighborhood change. Consequently, it is possible to "explain" neighborhood change in different ways, depending on the emphasis given to different elements. In this section we examine the case of gentrification (Figure 17) as a particular type of neighborhood change, showing how various elements depicted in Figure 16 can be interpreted and how they in turn are connected to broader shifts in the trajectory of economic, social, cultural, political, and urban change.

Gentrification is simultaneously

> . . . a physical, economic, social and cultural phenomenon. [It] commonly involves the invasion by middle-class or higher-income groups of previously working-class neighborhoods or multi-occupied "twilight areas" and the replacement or displacement of many of the original occupants. It involves the physical renovation or rehabilitation of what was frequently a highly deteriorated housing stock and its upgrading to meet the requirements of its new owners. In the process, housing in the areas affected, both renovated and unrenovated, undergoes a significant price appreciation. Such a process of neighborhood transition commonly involves a degree of tenure transformation from renting to owning.[49]

Although gentrification has been most pronounced in **world cities** like New York and London and in regional **nodal centers** that have evolved from older urban cores, such as Minneapolis and Philadelphia in the United States or Manchester and Glasgow in the United Kingdom, by the early twenty-first century it has become virtually global in its incidence.[50] Research on gentrification in the United States suggests that it is increasing and involves between 1 and 5 percent of urban households and the displacement of an estimated 900,000 households each year.[51]

> Displacement imposes substantial hardships on some classes of displacees, particularly lower-income households and the elderly. Although some displacees report finding similar or improved dwelling units and neighborhoods, a substantial number report a deterioration in post-move dwelling units and/or neighborhood quality. Rents almost always increase, modestly for some households, substantially for others. Lower-income outmovers are particularly hit, finding the least satisfactory alternative units and neighborhoods and facing the highest proportional shelter-cost increases. For elderly displacees, the neighborhood studies show particular hardships.[52]

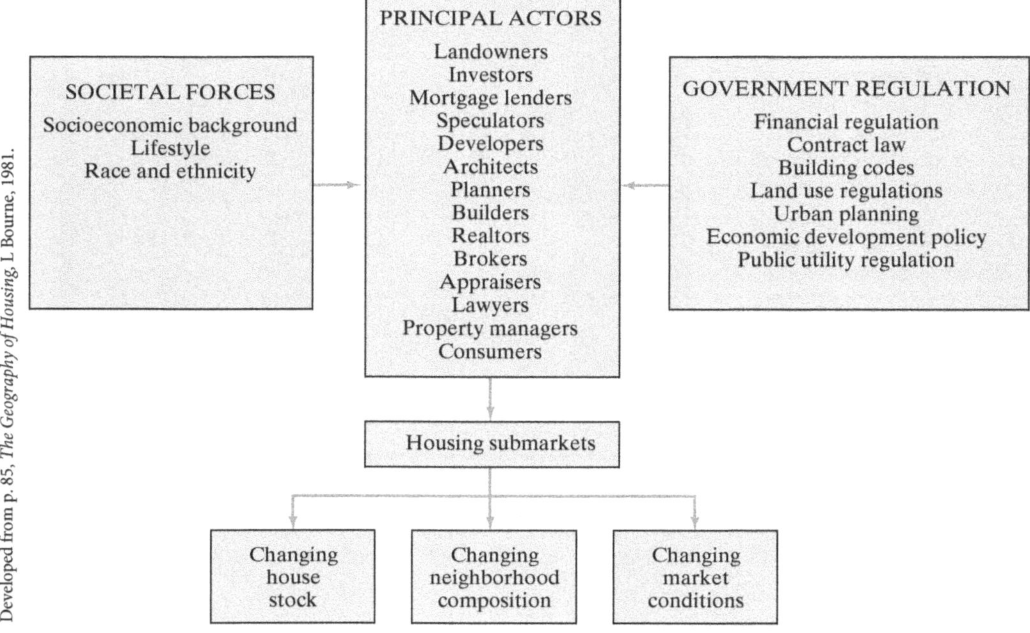

Developed from p. 85, *The Geography of Housing*, L Bourne, 1981.

FIGURE 16 Actors and institutions affecting neighborhood change.

FIGURE 17 Houses in a gentrified neighborhood of the Richmond district of San Francisco—part of the streetcar suburb illustrated in an earlier chapter.

The significance of gentrification really lies in its qualitative, symbolic, and ideological implications for urban change. Gentrification involves dramatic changes in neighborhood character, with a good chance for social conflict. Because it creates improvements to the built environment, encourages new retail activity, and results in the expansion of the local tax base without necessarily drawing heavily on public funds, it has become an important symbol and prospect for urban change for ideological conservatives. Because it fosters capital accumulation, caters to the consumption patterns of higher-income groups, and results in the displacement of vulnerable and disadvantaged households, it has become emblematic of urban restructuring and a portent of urban change for ideological liberals. Because it can be seen as the product of a wide variety of factors, it has become the focus of theoretical debates contrasting the effects of production and consumption, supply and demand, capital and culture, and gender and socioeconomic background in accounting for neighborhood change.

The scope of these debates is important, partly because it shows how the same causal factors can be seen very differently in relation to one another, but most importantly because it shows how the elements depicted in Figure 16 are in turn connected to broader contextual changes in the trajectory of economic, social, cultural, and political life. The scope of the debates has been framed around two main sets of ideas. The first, associated in particular with the work of David Ley,[53] emphasizes the importance of occupational, social, and cultural shifts in influencing patterns of demand. This *humanistic* argument gives priority to human agency and consumer preferences and can be paraphrased as follows. The increased pool of professional, administrative, managerial, and technical workers, together with the politicization of middle-income interest groups and the emergence of postmodern cultural sensibilities, has generated an emerging group of potential gentrifiers. Because many of these potential gentrifiers are employed in central city settings, and because their postmodern sensibilities lead them to reject modern homes in downtown condominiums or suburban tracts in favor of settings with some history, human scale, and ethnic and architectural diversity, they are attracted to older central city neighborhoods. Once established in sufficient numbers, gentrifiers are able to consolidate their lifestyle through local political influence, enhancing their neighborhood and attracting still more gentrifiers by electing representatives who "deliver" better security, environmental improvements, and the preservation of historic buildings. Real estate agents and mortgage finance managers further reinforce the process by exploiting the trend while commercial investors reinforce the process through investments such as in upscale clothing boutiques, chic wine bars, and bookstores with coffee shops. The most important people in this perspective, however, are the individual households who initiate and sustain the demand for "gentrified" settings.

In contrast, the alternative interpretation of gentrification, associated mainly with the work of Neil Smith,[54] sees the most important people as real estate agents and developers. This *structuralist* argument gives priority to the process of capitalist economic development and, in particular, moves by capital to arrest declining rates of profit and can be paraphrased as follows. The suburbanization of economic activity and households has steadily distorted the classical land value gradient that fell steadily from the CBD, to the periphery of the city. In particular, land values in central city neighborhoods have fallen relative to the CBD and suburban nodes, creating a "valley" in the land value gradient. This valley intensified after World War II, resulting in the "devalorization" of central city neighborhoods, a situation in which the rent from land uses allocated under the market conditions of earlier (preautomobile) times was significantly less than the **ground rent** that could be obtained under new uses. This situation is the **rent gap**, the fundamental precondition for gentrification, which is then initiated by three types of developers: (1) professional developers who purchase property, redevelop it, and resell for profit; (2) occupier-developers, who buy and remodel property and inhabit it after completion; and (3) landlord developers, who rent property to tenants after rehabilitating it. The role of gentrifying households, meanwhile, is interpreted less in terms of individual consumer preferences and more in terms of socioeconomic relationships, intertwined with the dynamics of culture and politics. To varying degrees in different cities, gentrification has been incorporated into the **neoliberal policies** of city governments. Gentrification is therefore a back-to-the-city move by private-sector capital that is directly facilitated by public-sector policies. Smith sees this move—together with deregulation, **privatization**, and the other neoliberal reforms of the 1980s and 1990s—as a new form of revenge by the powerful in society for the moral and economic decline of city life following the social reforms of the 1960s (hence the term **revanchist city**—the French word *revanche*, meaning revenge).

Drawing on both sets of ideas, three conditions can be posited as being necessary to explain the occurrence of gentrification:[55]

1. A pool of potential gentrifiers. This pool can be traced to the production and concentration of key factions of professional, administrative, managerial, and technical workers in major cities around the world. It is argued that it is a product of the restructured social and spatial **division of labor** associated with the onset of **advanced capitalism**.
2. A supply of potentially gentrifiable central city property (the rent gap theory). But the existence of a rent gap does not necessarily lead to gentrification:

   > Without the existence of a pool of potential gentrifiers and available mortgage finance, gentrification will not occur however great the rent gap and however great the desire of developers to make it happen. And where appropriate housing stock does not exist in sufficient quantity, as for example in cities such as Dallas, Phoenix and other new southern and western U.S. cities, gentrification may be very limited.[56]

3. A degree of effective demand for central city property from potential gentrifiers. Preferences for central city settings

   > depend on both the growth of service class job opportunities downtown, and on demographic and lifestyle changes which have seen large numbers of women enter the labor force and growing numbers of both single households and dual career childless couples. For these groups, with a high disposable income, inner-city locations offer proximity to employment and to restaurants, arts and other facilities.[57]

Finally, it should be noted that other theorists put the emphasis elsewhere. Liz Bondi and Alan Warde, for example, both emphasize the *interaction of class and gender* in understanding gentrification.[58] From this perspective, the location of dual-earner households in the central city is a solution to problems of access to work and home and of combining paid and unpaid labor for married middle-income women and men in well-paid career jobs. This development, of course, is related to other broad sociodemographic shifts, including the restructuring of family life that has been reflected in the postponement of child bearing, decreased family size, and closer spacing of children. These trends result in small, affluent households prepared to pay high prices for sought-after housing because they benefit more from the reduction in commuting costs associated with inner-urban residential locations than do those with only one adult working in the city center.

Based on their work on gay involvement in gentrification, writers such as Larry Knopp and Mickey Lauria have introduced sexuality to the equation.[59] They argue that the decline in traditional manufacturing employment in many U.S. cities, combined with the increase in administrative, managerial, and **producer services** jobs in the CBD, drew many first-generation and second-generation openly gay and lesbian people to central city neighborhoods. For a variety of reasons, the LGBT community (especially gay men) have traditionally been disproportionately represented in these types of jobs. The economic draw of these jobs, combined with the heterosexism and homophobia experienced in suburban "family" neighborhoods, has made central cities very attractive. In addition, the inexpensive and renovatable housing stock in depressed neighborhoods gave members of the LGBT community (and again, particularly gay men, because of their higher earning power as male wage earners) the opportunity to develop a territorial and economic base for establishing a political voice and for developing community resources.

Writers such as Sharon Zukin and Rosalyn Deutsche, meanwhile, have emphasized the role of the avant-garde and the contextual shift toward a "society of the spectacle" in which stylish materialism is increasingly important.[60] In larger metropolitan areas, they argue, the artistic avant-garde have become less high-culture and elitist and more important as cultural intermediaries. Politicians, speculators, and developers have come to see avant-garde art and culture as a crucial element in any new project. Meanwhile the avant-garde themselves, with a lifestyle that focuses very much on identity, appearance, presentation of self, fashion design, decor, and symbolism, have been critically important not only in pioneering the "resettlement" of rundown, low-rent areas but also in simultaneously providing these areas with the "designer" touch of radical chic necessary for them to be seen in a new light by the newly affluent and aestheticized professional service workers.

All writers on gentrification acknowledge that both economic and cultural processes are at work, so the crucial issue is *which factor is most important*. This might seem like some esoteric academic debate, but it has important implications, such as for neighborhood planning by city governments and for political action by community groups and individuals adversely affected by gentrification. If the forces of capital are seen as overwhelmingly dominant, as in the structuralist explanations, then human agency can achieve relatively little without wholesale reforms of the operation of capital markets. If, however, more importance is given to the autonomous role of cultural movements, as in the humanistic arguments, then these can influence the nature of capitalist development itself. What is clear from the research, however, is that the relative importance of economic and cultural factors varies in different cities. In Berlin, geopolitics has been a factor, the reunification of the country in 1990 amplifying rent gaps and inducing dramatic changes in household mobility. In Mexico City, gentrification is not nearly as highly capitalized or widespread as in New York City. In the Caribbean, the increasing interrelationships between gentrification and global capital investment tend to filter through the tourist industry, giving gentrification there its own distinctive flavor.[61]

FOLLOW UP

Key Terms

- action space
- awareness space
- block busting
- chain migration
- depreciation curve
- eminent domain
- exchange value
- gentrification
- household versus neighborhood life cycles
- information space
- redlining
- revanchist city
- spatial submarkets
- steering
- use value

Review Activities

1. Consider your own neighborhood or one that you know. How does it fit within the conceptualization of neighborhood life cycles? What signs of change can you see in the neighborhood? How is this change related to changes in nearby neighborhoods and to changes in the rest of the city?
2. List the residential moves that your family (or one that you know) has made within a city. How can these moves be explained (1) individually and (2) collectively?
3. Academic journals sometimes contain a series of articles that represent a back-and-forth academic debate about some important topic—such as one about gentrification between Chris Hamnett and Neil Smith. In this case Hamnett's initial paper reviewed the major theories of gentrification, Smith's commentary took issue with some of Hamnett's arguments (and to being characterized as a "blind man" whose theory only perceived part of the "elephant" of gentrification), and Hamnett then responded to Smith. Use your library's website to get a copy of these three articles and see what you think—do you tend to agree more with Hamnett's arguments or Smith's or do you find yourself somewhere in between? The articles are in the *Transactions of the Institute of British Geographers*. You should read them in chronological order: (1) Hamnett, "The blind men and the elephant: the explanation of gentrification" 16(1991): 173–89; (2) Smith, "Blind man's buff, or Hamnett's philosophical individualism in search of gentrification" 17(1992): 110–15; and (3) Hamnett, "Gentrifiers or lemmings? A response to Neil Smith" 17(1992): 116–19.
4. Update your *portfolio*. Remember to keep a note of any questions that are raised by your reading. Now that you have covered more material, you may also find it useful to review earlier sections of your portfolio and perhaps add new material or observations.

Log in to **www.mygeoscienceplace.com** for self-study quizzes, *MapMaster* layered thematic and place name interactive maps, *Urban View* Google Earth™ tours, key resources and suggested readings, related websites, "In the News" RSS feeds, and additional references and resources to enhance your study of how neighborhoods change.

NOTES

1. C. Hsu, "Rise and Ruin: Making Hats for Hollywood on Buffalo's East Side," *The Buffalo Story Project*, blog posted April 17, 2010 (*http://buffalostoryproject.com/2010/04/17/rise-and-ruin/*); The Custom Hatter website (*http://www.custom-hatter.com/*); Forgotten Buffalo website (*http://www.forgottenbuffalo.com/buffalospoloniahistory.html*).
2. M. Farnsworth Riche, *The Implications of Changing U.S. Demographics for Housing Choice and Location in Cities* (Washington, D.C.: The Brookings Institution, 2001).
3. L. Bourne, *The Geography of Housing* (London: Edward Arnold, 1981), 23.
4. Sweat equity is a term generally used to describe an increase in the value of residential property that results from householders' unpaid labor in remodeling and refurbishing.
5. D. W. Harvey, *Society, the City and the Space-Economy of Urbanism* (Washington, D.C.: Association of American Geographers, College Resource Paper no. 18, 1972), 16.
6. Bourne, *The Geography of Housing*, 14–15.
7. Jeffrey M. Hornstein, *A Nation of Realtors. A Cultural History of the Twentieth-Century American Middle Class* (Durham, NC: Duke University Press, 2005), 7.
8. Ibid., p. 121.
9. D. Terry, "The Final Farewell at Cabrini-Green," *Chicago News Cooperative*, December 9, 2010; E. Schmall, "Last Resident of Chicago's Cabrini-Green faces Uncertain Future," *AOL News*, December 9, 2010 (*http://www.aolnews.com*).
10. Chicago Guy, "Cabrini-Green's Last Tenant," *Open Salon*, December 10, 2010 (*http://open.salon.com/blog/chicago_guy/2010/12/10/cabrini_greens_last_tenant*).
11. C. Whitehead and K. Scanlon, *Social Housing in Europe* (London: London School of Economics, 2007).
12. HUD (U.S. Department of Housing and Urban Development), *A Picture of Subsidized Households—2008* (Washington, D.C.: Office of Policy Development and Research, 2008).
13. J. S. Fuerst, ed., *Public Housing in Europe and America* (London: Croom Helm, 1974).
14. This discussion is based on L. McCarthy and C. Johnson, "Cities of Europe," in *Cities of the World: World Regional Urban Development*, eds. S. Brunn, M. Hays-Mitchell, and D. Zeigler, 5th ed. (Lanham, Md.: Rowman & Littlefield, 2011).
15. D. Ley, *A Social Geography of the City* (New York: Harper & Row, 1983), 238–39. Emphases added.
16. M. T. Cadwallader, "A Unified Model of Urban Housing Patterns, Social Patterns and Residential Mobility," *Urban Geography* 2 (1981): 115–30.
17. This discussion is based on L. McCarthy and C. Johnson, "Cities of Europe," in *Cities of the World: World Regional Urban Development*, eds. S. Brunn, M. Hays-Mitchell, and D. Zeigler, 5th ed. (Lanham, Md.: Rowman & Littlefield, 2011).
18. P. H. Rossi, *Why Families Move: A Study in the Social Psychology of Urban Residential Mobility* (New York: Free Press, 1955).
19. L. A. Brown and E. G. Moore, "The Intra-Urban Migration Process: A Perspective," *Geografiska Annaler* 52B (1970): 1–13.
20. E. W. Butler et al. *Moving Behavior and Residential Choice: A National Survey*. National Cooperative Highway Research Program Report no. 81 (Washington, D.C.: Highway Research Board, 1969).
21. See, for example, J. O. Huff, "Geographic Regularities in Residential Search Behavior," *Annals of the Association of American Geographers* 76 (1986): 208–27.
22. See, for example, W. A. V. Clark, "Residential Segregation in American Cities: A Review and Interpretation," *Population Research and Policy Review* 5 (1986): 95–127; G. Galster, "Residential Segregation in American Cities: A Further Response to Clark," *Population Research and Policy Review* 8 (1989): 181–92; and W. A. V. Clark, "Residential Segregation in American Cities: Common Ground and Differences in Interpretation," *Population Research and Policy Review* 8 (1989): 193–97.
23. See, for example, S. Kim and G. D. Squires, "The Color of Money and the People Who Lend It," *Journal of Housing Research* 9 (1998): 271–84, which found that the likelihood of home loan approval for black or Hispanic applicants increases as the proportion of black or Hispanic employees, and particularly administrative and professional workers, increases at a lending institution. This finding is consistent with the "cultural affinity" hypothesis: When more employees at a lending institution are from a particular community, they may be better placed to obtain information from marginal applicants and approve applications that do not clearly meet all the objective indicators. At the same time, other reasons for the relationship could be the minority loan officer's knowledge of, or sympathy with, fair lending concerns, the influence of minority coworkers on white loan officers, and self-selection.
24. R. E. Pahl, *Whose City?* 2nd ed. (Harmondsworth, UK: Penguin, 1975).
25. The term "social gatekeeper" was first used in a study of real estate agents in New Haven, Conn., in 1955 by R. Palmer in his Yale University Ph.D. dissertation, *Realtors as Social Gatekeepers*.
26. R. Palmer, *Realtors as Social Gatekeepers*. Ph.D. diss., Yale University, 1955, 77.
27. P. M. Downing and L. Gladstone, *Segregation and Discrimination in Housing: A Review of Selected Studies and Legislation*, Congressional Research Report 89-317 (Washington, D.C.: The Library of Congress, 1989), 25–26.
28. U.S. Department of Housing and Urban Development, *The State of Fair Housing: Annual Report on Fair Housing FY 2009* (Washington, D.C.: HUD, 2010).
29. National Commission on Fair Housing and Equal Opportunity, *The Future of Fair Housing* (Washington, D.C.: National Commission on Fair Housing and Equal Opportunity, 2008).
30. The 2000 study results are available in U.S. Department of Housing and Urban Development (HUD), *Discrimination in Metropolitan Housing Markets: National Results from Phase 1 of the Housing Discrimination Study*, Washington, D.C.: HUD, 2002 (*http://www.huduser.org/portal/publications/hsgfin/phase1.html*). The comparability of the 2000 data with the 1989 *Housing Discrimination Study* results is discussed in Annex 5 of the 2002 study.
31. National Fair Housing Alliance, *2009 Fair Housing Trends Report* (Washington, D.C.: NFHA, 2009).
32. See, for example, A. G. Carpusor and W. E. Loges, "Rental Discrimination and Ethnicity in Names," *Journal of Applied Social Psychology* 36 (2006): 934–52; S. Friedman, G. D. Squires, and C. Galvan, *Cybersegregation in Boston and Dallas: Is Neil a More Desirable Tenant than Tyrone or Jorge?* (University of Albany, SUNY: Department of Sociology, 2010) (*http://media2.myfoxboston.com/pdf/discriminationstudy.pdf*).
33. A. Ahmed and M. Hammarstedt, "Discrimination in the Rental Housing Market: A Field Experiment on the Internet," *Journal of Urban Economics* 64 (2008): 362–72.
34. M. E. Stone, "Housing, Mortgage Lending, and the Contradictions of Capitalism," in *Marxism and the Metropolis*, eds. W. K. Tabb and L. Sawers (New York: Oxford University Press, 1978), 190.

35. J. Darden, "Lending Practices and Policies Affecting the American Metropolitan System," in *The American Metropolitan System: Present and Future*, eds. S. D. Brunn and J. O. Wheeler (New York: Winston, 1980), 93–110.
36. National Fair Housing Alliance, *2009 Fair Housing Trends Report* (Washington, D.C.: NFHA, 2009).
37. National Fair Housing Alliance, *2010 Fair Housing Trends Report* (Washington, D.C.: NFHA, 2010).
38. R. B. Avery, N. Bhutta, K. P. Brevoort, and G. B. Canner, "The 2009 HMDA Data: The Mortgage Market in a Time of Low Interest Rates and Economic Distress," *Federal Reserve Bulletin*, December 2010.
39. A. Jakabovics and J. Chapman, *Unequal Opportunity Lenders? Analyzing Racial Disparities in Big Banks' Higher-Priced Lending* (Washington, D.C.: Center for American Progress, 2009).
40. R. B. Avery, K. P. Brevoort, and G. B. Canner, "The 2006 HMDA Data," *Federal Reserve Bulletin*, December 2007.
41. National Fair Housing Alliance, 2010, p. 38.
42. R. B. Avery, N. Bhutta, K. P. Brevoort, G. B. Canner, and C. N. Gibbs, "The 2008 HMDA Data," *Federal Reserve Bulletin*, December 2009.
43. For a review of racial discrimination in underwriting in U.S. cities, see G. D. Squires, "Racial Profiling, Insurance Style: Insurance Redlining and the Uneven Development of Metropolitan Areas," *Journal of Urban Affairs* 25 (2003): 391–410.
44. Insurance Research Council, *Homeowners Loss Patterns in Eight Cities* (Wheaton, Ill.: IRC, 1997).
45. R. W. Klein, "Availability and Affordability Problems in Urban Homeowners Insurance Markets," in *Insurance Redlining: Disinvestment, Reinvestment, and the Evolving Role of Financial Institutions*, ed. G. D. Squires (Washington, D.C.: The Urban Institute Press, 1997), 43–82.
46. S. L. Smith and C. Cloud, "Documenting Discrimination by Homeowners Insurance Companies through Testing," in *Insurance Redlining: Disinvestment, Reinvestment, and the Evolving Role of Financial Institutions*, ed. G. D. Squires (Washington, D.C.: The Urban Institute Press, 1997), 97–117.
47. T. C. Pittman, "Rejoinder to Racial Profiling, Insurance Style: A Spirited Defense of the Insurance Industry," *Journal of Urban Affairs* 25 (2003): 411–22.
48. J. Cartner-Morley, "Where Have All The Cool People Gone?" *Guardian*, November 21, 2003 (*http://www.guardian.co.uk/lifeandstyle/2003/nov/21/fashion1/*).
49. C. Hamnett, "Gentrification and Residential Location Theory: A Review and Assessment," in *Geography and the Urban Environment*, eds. D. T. Herbert and R. J. Johnston (Chichester, UK: Wiley, 1984), 284.
50. N. Smith, "New Globalism, New Urbanism: Gentrification as Global Urban Strategy," *Antipode* 34 (2002): 427–50.
51. For reviews of this research, see N. Smith and P. Williams, eds., *Gentrification of the City* (Boston: Allen and Unwin, 1986); and the special issue of *Urban Studies* 12 (2003) devoted to gentrification.
52. R. T. LeGates and C. Hartman, "The Anatomy of Displacement in the United States," in *Gentrification of the City*, eds. N. Smith and P. Williams (Boston: Allen and Unwin, 1986), 197.
53. D. Ley, *The New Middle Class and the Remaking of the Central City* (Oxford, UK: Oxford University Press, 1996).
54. See N. Smith, *The New Urban Frontier: Gentrification and the Revanchist City* (London: Routledge, 1996); and N. Smith, "New Globalism, New Urbanism: Gentrification as Global Urban Strategy," *Antipode* 34 (2002): 427–50.
55. C. Hamnett, "The Blind Men and the Elephant: The Explanation of Gentrification," *Transactions of the Institute of British Geographers* 16 (1991): 173–89.
56. Ibid., 186.
57. Ibid., 187.
58. L. Bondi, "Gender Divisions and Gentrification: A Critique," *Transactions of the Institute of British Geographers* 16 (1991): 190–98; A. Warde, "Gentrification as Consumption: Issues of Class and Gender," *Society and Space* 9 (1991): 223–32.
59. See L. Knopp, "Sexuality and the Spatial Dynamics of Capitalism," *Environment and Planning D: Society and Space* 10 (1992): 651–69; M. Lauria and L. Knopp, "Towards an Analysis of the Role of Gay Communities in the Urban Renaissance,"*Urban Geography* 6 (1985): 152–69.
60. S. Zukin, "The Postmodern Debate over Urban Form," *Theory, Culture and Society* 5 (1988): 431–46; R. Deutsche, "Uneven Development: Public Art in New York City," *October* 47 (1988): 3–53.
61. N. Smith, "*New Globalism, New Urbanism,*" 440.

The Politics of Change: Urbanization and Urban Governance

From Chapter 10 of *Urbanization: An Introduction to Urban Geography*. Third Edition. Paul L. Knox, Linda McCarthy.

Rob Casey/Alamy

The Politics of Change: Urbanization and Urban Governance

In this chapter we examine urban governance and the politics of changing urban geographies. The emphasis is on the *relationships between urbanization and the changing roles of urban government, and on the changing politics of urban development*. Our concern is with the generalizations that can be made about the influence of urban change on urban governance and vice versa rather than with the details of governmental organization, the structures of political participation, or the political careers of influential urban leaders. Similarly, we look beyond "symbolic politics" (the "hot" issues that provide easy copy for newspaper editors while satisfying the need for political figures to stay in the public eye)[1] to focus on long-term politics that determine "who gets what, when, and how."[2]

LEARNING OUTCOMES

After reading this chapter, you should be able to:

- Be familiar with the principal phases in the evolution of urban governance and how they relate to patterns of industrialization and urbanization and to the shifting economic and political power of various social groups.
- Discuss how the emergence of neoliberal ideologies has affected perceptions and strategies of planners, local politicians, developers and civil society groups.
- Discuss the related problems of fiscal squeeze and fiscal retrenchment in central cities and explain the privatization of many city services.
- Illustrate the consequences of the increasing entrepreneurialism of city governments.
- Compare and contrast the major theoretical models of political power in the city and their characterization of the nature of power, governance and conflicts in local politics.

CHAPTER PREVIEW

The purposes of this chapter are (1) to show how, as the economic base of cities evolved, the fortunes of different groups of people changed and cities themselves generated new problems and challenges; and (2) how, as cities became larger and more complex, the scope of urban government broadened to include the regulation and provision of an increasingly wide range of infrastructural elements, goods, and services for residents—all of which had a direct and sometimes profound effect on the social geography as well as the built environment of cities.

We will also see that the combination of sociospatial change, changing urban problems, and the changing scope of urban government resulted in a succession of different types of power holders; different, that is, in terms of these people's motivations and objectives. The ethos and orientations of urban governance, reflecting these changes, in turn fostered further changes in the nature and direction of urban development. In the broadest terms, it is possible to recognize distinctive phases in the evolution of urban governance and urban politics that reflect the overall evolution of cities and urban systems. In the first part of this chapter, we summarize the

Historically, as cities such as San Francisco, California, became larger and more complex, the scope of urban government broadened to include the regulation and provision of an increasingly wide range of urban infrastructure, goods, and services for residents—all of which had a direct and sometimes profound effect on the social geography as well as the built environment of cities.

history of urban governance and politics in the United States in terms of the periods of development described earlier.

In the second part of the chapter, we focus on the entrepreneurial politics and neoliberalism that have shaped the United States since the mid-1970s. In the concluding section of this chapter, we turn to conceptual and theoretical perspectives on urban governance and politics to answer three broad questions raised by the changes we have reviewed: (1) How is power structured within cities and how does its structure change with the changing dynamics of urbanization? (2) How can we interpret the role of central governments in relation to urban development? and (3) How can we interpret local conflicts in relation to patterns and processes of urban change?

First, though, we should clarify what is meant by urban governance. Urban govern*ance* is not the same as urban govern*ment*. Governance is a concept that recognizes that power exists outside as well as inside the formal institutions of government. Most definitions of governance include three main groups: government, the private sector, and civil society. Urban governance emphasizes "process." It recognizes that any decision-making that affects cities is based on the complex interrelationships among many individuals and groups who have different goals and priorities. It is the reconciliation of these competing goals and priorities that lies at the heart of the concept of urban governance. The United Nations Human Settlements Program (UN-HABITAT) has proposed the following definition:

> Urban governance is the sum of the many ways individuals and institutions, public and private, plan and manage the common affairs of the city. It is a continuing process through which conflicting or diverse interests may be accommodated and cooperative action can be taken. It includes formal institutions as well as informal arrangements and the social capital of citizens.[3]

URBAN VIEW 1
The Disneyfication of Times Square

Appropriately enough, Times Square in New York contains the largest Disney store in North America. The Disney connection is apt because Times Square itself has been Disneyfied. Disneyfication is a term used to describe a place that has been made more marketable and attractive in a very sanitized way by removing unpleasant aspects, including even historical ones. The term is usually used in a derogatory way because it captures how a place may end up resembling a sterile theme park version of its original authentic self.

For Times Square, that was back in the early decades of the twentieth century when it was a lively cultural hub of theaters, music halls, and ritzy hotels. It was a wonderful place for people to come together and celebrate events like World Series wins. But Times Square went downhill during the Great Depression in the 1930s and afterwards gained a dreadful reputation as a dangerous hub not only for criminal activity involving pimps and prostitutes on street corners, and drug dealers and addicts shooting up in the streets, but also for sleezy businesses like sex shops and peep shows. After Rudolph Giuliani became mayor of New York in 1994, he did his part to help clean up Times Square by closing down the seedy businesses, forcing out drug dealers and other criminal elements, and encouraging more upscale tourist-friendly establishments.

One of the few businesses that survived the cleanup of Times Square is Jimmy's Corner, a bar on West 44th Street. Its owner, 80-year-old Jimmy Glenn, was a boxing trainer who also owned the now long gone Times Square Gym on the second floor of a building on 42nd Street where boxers like Muhammad Ali, Joe Frazier, and Mike Tyson trained. Now, he says of Times Square: "it's like a pinball machine out there." But he does not miss the drug addicts, porn shops, and criminals. "Everybody loves Times Square now," he says.[4]

Although Times Square is a cleaner and safer place for visitors, its Disneyfication has come with a loss of authenticity. An online review by a customer of Jimmy's Corner captures a certain nostalgia for the past: "They really don't make them like this anymore. If I ran the landmarks and preservation committee, be assured I'd give this place the historic status it richly deserves so as to keep at least one tiny part of Times Square good. Not hip, not cool. Just good."[5] Nearby neighborhood residents, including those in Hell's Kitchen, have seen negative impacts after the Disneyfication of Times Square, including gentrification. "We're like a big plum, and all the real estate people are salivating, and if we don't watch it we'll become the next Columbus Avenue," said Marisa, a 51-year-old actress who has lived in Hell's Kitchen for decades.[6]

But the fact is that Times Square today is economically incredibly successful, attested to by its huge aniated neon and LED advertisements and giant curved seven-story NASDAQ sign (Figure 1). It is home to corporate giants like Morgan Stanley, Viacom, and Condé Nast. Disney is only one of many retailers, including Aeropostale, Forever 21, and American Eagle Outfitters, that have opened glittering stores in the hopes of making customers of some of the estimated 1.5 million people who pass through Times Square everyday. But this freewheeling, dazzling economic success story is a bit of a paradox. As Tim Tompkins, President of the Times Square Alliance, puts it: "The irony is that this place represents in many ways the epitome of free-market capitalism, but its transformation is due more to government intervention than just about any other development in the country."[7]

Anthony Bayford

FIGURE 1 Times Square. Tourists enjoy the safe and clean Disneyfied Times Square with its huge aniated neon and LED advertisements and abundant shopping and entertainment opportunities.

THE HISTORY OF URBAN GOVERNANCE

As economic, social, cultural, technological, and other changes have shaped one another and simultaneously shaped and re-shaped urban geographies, they have also been involved as both cause and effect in the broad sweep of changes in urban governance. In this section, we describe the main features of five distinctive periods:

1. Laissez-faire and economic liberalism (1790–1840)
2. Municipal socialism and the rise of machine politics (1840–1875)
3. Boosterism and the politics of reform (1875–1920)
4. Egalitarian liberalism and *metropolitan fragmentation* (1920–1945)
5. Cities as growth machines and service providers (1945–1973)

Laissez-faire and Economic Liberalism (1790–1840)

The earliest phase of the evolution of urban governance and politics, roughly coincident with the era of the mercantile city, was dominated by the doctrine of economic liberalism, a laissez-faire philosophy that rested on the assumption that the maximum public benefit would result from unfettered market forces. This phase was also the "Age of the Common Man," with rapid demographic growth and an expanded electoral franchise combining to wrest urban political power from the exclusive control of the literate and the wealthy. Because of the pervasive ideology of free enterprise, urban government was weak and disorganized. And because of the pervasive opportunism in political life, the system was also endemically corrupt, with victorious politicians routinely handing out jobs and rewards to friends and supporters. The result was that urban governance or politics rarely affected urban development in a direct or beneficial way. The indirect consequences were that cities became increasingly perplexing, their overall economic growth being accompanied by social polarization and mounting disorder, disease, and congestion.

The basic framework of local government in most of the United States dates from this period, as counties, municipalities, and school districts were set up to assist state governments in carrying out their responsibilities. Municipalities were established to provide essential local services for people. School districts were established as *independent* units because of a strong conviction that the public education of children was important enough to society to warrant its economic and political freedom from other government units. A notable exception to this pattern of local government was the township, which originated during the colonial period and is still prevalent in New England. In addition to the principle of elected representatives guiding local affairs according to their interpretation of the "public good," an important feature of the New England township system is the element of participatory democracy for residents provided by town meetings (Figure 2).

In the early nineteenth century local government was quietly dominated by a small pool of "natural" leaders drawn from "established" local families. Few of these people regarded their role as extending beyond providing a reasonably sanitary and lawful environment for urban residents. Local governments covered their expenditures, such as they were, directly from revenues from property taxes, fines, and fees. The main disadvantage of this system was that capital expenditures for costly undertakings such as roads, sewers, and bridges were limited to the amount that could be covered by annual tax levies. It was a disadvantage that became acutely apparent to people as the pace of urbanization quickened. It was solved by **incorporation**, an act that changed urban government from a passive, regulatory activity to an active agent in economic development.

Jim Evans/Associated Press

FIGURE 2 New England town meeting.

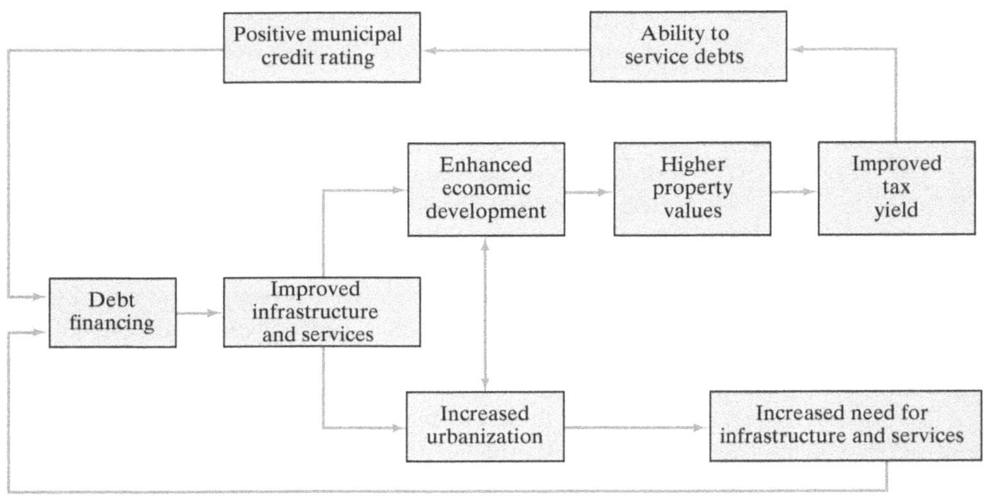

FIGURE 3 The interdependence between urbanization, economic development, and debt financing.

What was so important about incorporation was that it conferred limited liability and, therefore, the opportunity for **debt financing** through issuing bonds. Bond repayments for incorporated places were relatively secure: They could reasonably be expected to be met from property taxes, revenues from which could be expected to increase as a result of the growth stimulated by the new infrastructure and improved services financed by bond issues (Figure 3). The mercantile cities of the period were all incorporated to allow them to compete with one another as central places. *As a result, U.S. urban governance and politics came to be founded on a fragile web of debt financing, economic growth, and increasing property values.*

Municipal Socialism and the Rise of Machine Politics (1840–1875)

The second phase in the evolution of urban governance and urban politics, roughly coincident with the era of the early industrial city, saw the erosion of laissez-faire ideology, an increase in the range and strength of local government powers, and an increase in expenditures (paid for through an expansion of debt financing), all in response to the continuing threats posed by distinctively urban problems: fire, disease, and mob violence. This was also the age when **machine politics** emerged, with charismatic leaders controlling hierarchical (and often corrupt) political organizations that drew on working class support for more paternalistic urban governance (Figure 4).

Florida Center for Instructional Technology (FCIT)

FIGURE 4 This cartoon depicts New York corruption with Boss Tweed holding onto the reins of the Democratic Party prior to the 1872 election.

Crowding and uncontrolled development had resulted in city residents being vulnerable to fires and epidemics. The free market, meanwhile, had created an impoverished workforce whose living conditions were so bad that rebellion and disorder emerged as a constant threat and an occasional reality. The combination was sufficient to convince a majority of people of the need for **municipal socialism**—local government intervention in the marketplace to impose standards and ensure the provision of key services and basic amenities for people (what is now sometimes referred to as **collective consumption**). Governance thus became involved with efforts (1) to moralize the lives of citizens through regulating the physical and social environment, and (2) to develop a framework of welfare provision that would systematically change the conditions that caused individual impoverishment and fostered social disorder.

This period of radical change attracted a variety of new, self-made entrepreneurs—merchants, manufacturers, building contractors, and real estate speculators—to vie for political control in order to steer the expanded range of public policies in directions suited to their own interests. In this volatile and rather confused phase of urban politics, it was increasingly apparent that the key to success was *organization*. Political parties and trade unions became increasingly important, and they in turn fostered the emergence of new kinds of political leaders whose careers were based on close attention to the concerns of their constituents. In larger cities with sizable working-class populations, organization began to take the form of "machines" that delivered votes in return for the patronage jobs and benefits that could be granted by successful politicians.[8]

Machine politics helped to ensure that patronage and corruption became firmly embedded in urban governance. Paradoxically, the corruption of the machine system helped in some ways to accelerate the modernization of cities. Machine bosses were astute enough to appreciate the general economic benefits of major infrastructure improvements, as well as the personal kudos that could be derived from being associated with them. In more pragmatic terms, the more housing codes, building regulations, and municipal ordinances that were passed to regulate the emerging infrastructure, the more opportunities there were to solicit bribes in return for not enforcing them. It should be noted, though, that Southern cities were for the most part an exception to this trend toward organized machine politics. Because Southern cities lagged in terms of industrialization, did not attract large numbers of immigrants, and did not allow African Americans to vote, their working-class vote could not be delivered in the same way as in the industrial centers of the Northeast.

Boosterism and the Politics of Reform (1875–1920)

The era of the industrial city saw a struggle between the interests of ethnic and working-class groups and those of the new urban elite. Downtown business and professional interests organized themselves to restore urban rule to the "the better elements" of society and to shift the emphasis of urban government to activities that might attract investors—**boosterism**. The expanding middle class, meanwhile, made its presence felt in the latter part of this period through the politics of the Progressive Era (1895–1920), lending its support to the reform movement that was aimed at reducing corruption, ridding urban governance of machine politics, and establishing the basis for a more progressive approach to urban development.

Corruption became the symbolic issue that the new urban elite chose in seeking to undermine the legitimacy of the political machines. It was no coincidence that this period also saw the characterization of immigrant neighborhoods (the power base of the machines) as breeding grounds for social problems. Editors, novelists, and reformers competed with each other to draw attention to the corrupting moral influence of the "Great Unwashed" and to emphasize the theme of Anglo-Saxon superiority.

Meanwhile, the economic recessions of the 1870s and 1890s changed the context of urban politics decisively, stirring the urban elite to form alliances that would eventually replace the corrupt machines and install representatives who would improve the business climate of the city for new investment. The leadership of every town and city, big and small, soon came to realize that, under the dramatic spatial reorganization involved in forging a modern industrial economy, their economic future depended on outdoing their competitors in attracting investors in manufacturing and business. What is particularly significant about this period for the subsequent evolution of U.S. cities is that the public interest came to be defined more in terms of aggregate economic benefits than in terms of equity, social justice, civil rights, democracy, or community well-being.

Against this background, the common theme of reform emerged in the 1870s and 1880s. Religious and moral reform focused on temperance as the way to combat gambling, prostitution, and social malaise (Figure 5). Educational reform focused on school attendance, administrative reorganization, and revised curricula as the way to imbue the urban masses with standards of citizenship, democracy, and diligence. Housing and sanitary reformers saw improved physical conditions as a prerequisite for moral and social well-being and as a necessary step toward improved efficiency and productivity in the workplace.

The Progressive Era From the mid-1890s to 1920 the reform movement dominated local politics nationwide. This was the Progressive Era. The Progressive movement was reinforced by several concurrent trends. The need for expertise in municipal government was increasingly apparent with every technological innovation and every increment in the size and scope of public activities. City governments had to regulate the laying of pipes and cables and the setting up of streetcar systems. They needed inspectors to regulate and inspect new building methods, elevators, and gas and electricity supplies. They needed accountants who could handle increasingly complex city budgets and administrative officers who could coordinate and professionalize civic boosterism. The people in these growing occupations sought to establish and legitimize themselves with professional associations and trade magazines, and they realized from

Corbis Images

FIGURE 5 Women in Madison, Minnesota, campaign for the prohibition of alcohol. Minnesotans ratified national prohibition in 1919 and only repealed the state's dry laws in 1934, a year after the 21st amendment was passed.

the start that it was in their own best interests to support the ideals of the Progressive movement. Their voices were joined by those from another emergent group: social scientists and their professional organizations, who saw in the Progressive movement a ready audience for the careful documentation and analysis of every facet of urban life. At the same time, falling paper prices and technical advances in printing allowed a great expansion in the circulation of newspapers and popular magazines. Although the ideals of the Progressive movement did not often make for good copy, scandals, exposés, and shocking revelations about slum life did. The consequent crusades had the incidental effect of publicizing the objectives of Progressive reform.

By the early years of the twentieth century, the Progressive movement had achieved some tangible success. Galveston, Texas, became the first city to be run on the business model when an emergency situation after a hurricane prompted the state legislature to replace the city council and mayor with a commission of business leaders, each of whom was responsible for administering a different branch of municipal affairs. This form of government was subsequently adopted on a permanent basis. By 1915 more than 450 towns and cities had dopted the commission format of urban governance (with the mayor usually being the commissioner who is responsible for public safety). Nonpartisan, citywide balloting was meanwhile instituted in a number of cities—Akron, Boston, Detroit, Los Angeles, and Nashville being among the first.

In 1908 the city manager form of government was introduced in Staunton, Virginia. After Dayton, Ohio, adopted it in 1914, the city manager form of government spread rapidly. The elected council and the chief elected official (e.g., mayor) are responsible for making policy, and the city manager, who is appointed by the council, has full responsibility for the day-to-day operations of the government. About half of all cities in the United States now operate with a city manager.

The most widespread reform between the mid-1890s and 1920, however, was the introduction of the civil service model, which embodied the principles of specialization and scientific management that were central to the ethos of the industrial era. By 1920 some form of civil service was in operation in nearly all the larger and more industrialized cities. The chief holdouts and exceptions, as in so many other aspects of urban governance and politics, were Southern cities.

Annexation In keeping with the spirit of boosterism and efficiency that pervaded the period, the geography of cities was transformed by an acceleration of **annexation**—the addition of unincorporated land into cities. Boosterism was one important motive for annexation. With the advent of streetcar systems and new developments in home construction techniques, city populations had begun to decentralize, and city pride dictated that these people be recaptured.

Progressive ideals also played a part. Annexation, it was argued, created the framework for **economies of scale** in government—economies that would benefit both the suburbs and the central city as the costs of infrastructure provision and bureaucratic expertise were spread over a larger population. It also helped the Progressive cause that the addition of middle-class voters from outlying districts would weaken the potential strength of machine-style politics. Even more compelling (for the annexing governments, at least) was the importance of capturing the tax base represented by the growing suburbs. Annexation enabled cities to maintain, perhaps even strengthen, the delicate web of economic growth, increasing property values, and debt financing (Figure 3). Finally, real estate interests and development-related entrepreneurs (e.g., builders and their suppliers) were strong advocates of annexation because it enhanced opportunities for speculative development. Annexation of undeveloped land was particularly attractive to them, because it carried the implicit promise that this land would eventually be equipped with all the utilities and amenities of the city.

Suburbanites, of course, recognized that they had to pay higher taxes once annexed. They also had to share schools and municipal services with city residents, which sometimes led to conflict. As suburban speculative development increased in sophistication (with improved packages of amenities) and suburban self-consciousness became stronger, opposition to annexation increased. Citizen participation in such decisions was still some way off, however, and state legislatures generally took the view that no small territory should be allowed to impede metropolitan development. It would not be long, however, before state legislatures began to retreat from forced annexation in the face of suburbia's mounting electoral power and political influence.

Egalitarian Liberalism and Metropolitan Fragmentation (1920–1945)

Between 1920 and 1945 the suburbanization associated with the mass production of cars gave rise to the incorporation of suburban jurisdictions that quickly developed a politics of their own: exclusionary and competitive. An important corollary of this shift was that central cities lost not only a large section of better-educated middle-class voters and taxpayers but also a significant component of the pool of potential political leaders. In the early 1930s the Depression helped to generate a climate of opinion that was much more favorable to governmental provision of goods and services for people, resulting in an extension in the roles and legal powers of both federal and local governments and in a shift in the nature of urban politics.

Beginning in the 1920s, the pace of suburbanization quickened. The combination of cheap, mass-produced cars and federal mortgage insurance policies resulted in a major shift in metropolitan spatial dynamics. For the first time, population growth in the suburbs outpaced that of the central cities in both relative and absolute terms. Middle-class suburbanites sought not only to escape from proximity to the slums of central cities but also to escape from the tax burdens of central cities and to establish a distinctive setting for governance and politics in which middle-class values and preferences might flourish.

Rather than waiting to fight annexation, the people in many suburban communities took the preemptive strategy of petitioning for **suburban incorporation** for themselves (not necessarily as cities, since incorporation as a village was sufficient to protect their reputation, status, and independence). At the same time, state legislatures, recognizing the changing spatial distribution of voters, became increasingly reluctant to alienate suburbanites by approving annexation. Gradually, the prevailing legal view came to be that annexation should be voluntary, with the clear support of the residents of the affected area. The result was that central cities came to be encircled by a ring of independent and often hostile governments (Figure 6). Geopolitically, the U.S.'s metropolitan areas were fragmented, and that *metropolitan fragmentation undercut the ability of central city governments to deal with urban change and development even as it intensified sociospatial segregation.*

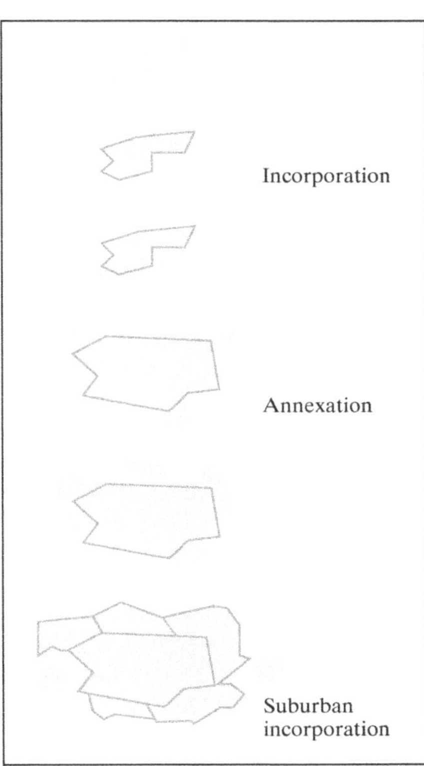

FIGURE 6 The process of metropolitan growth and political fragmentation.

Metropolitan fragmentation diminished the ability of central city governments to deal with urban change and development in several ways:

- The suburbanization of the middle classes reduced the pool of experienced and educated political leaders and reformers. Suburbanites with an impulse to become involved in civic affairs now did so in suburban jurisdictions, channeling their efforts into establishing and preserving a rather narrow and sectionalized version of the public interest.
- The loss of a large percentage of the middle classes, meanwhile, not only directly weakened the central city's tax base but also confined it by putting future property taxes out of reach. This loss, of course, made debt financing more difficult, which impaired the ability of central city governments to provide safe and efficient settings for economic development. Without room for new development, central cities were left with a steadily aging environment requiring increasing levels of maintenance and expensive fire protection.
- At the same time, the loss of middle-class people and the continuing influx of low-income migrants and immigrants forced central city governments to face even

greater burdens in terms of the provision of welfare-related services and social control functions.

- Central city governments continued to be responsible for providing metropolitan-wide amenities such as galleries and museums and for expenditures on roads, parking space, utilities, and policing incurred by suburban commuters and shoppers.

All these changes and trends put central city governments in an unavoidable **fiscal squeeze** (Figure 7) as revenue capacity fell while expenditure demands increased. The issues and conflicts associated with this squeeze have shaped the politics and geography of central cities ever since.

The second consequence of metropolitan fragmentation, *intensified sociospatial segregation*, was a result of the political independence of the middle classes, which enabled them to erect barriers to keep out lower-income people. The reasons for wanting to exclude lower-income households had to do with the perceived threat to property values and to the lifestyles and the kind of moral order the middle-class suburbanites sought to achieve in their schools and communities.

The chief instrument at their disposal was **exclusionary zoning**. By carefully framing their **land use zoning** ordinances, suburban jurisdictions were able to make it difficult or impossible for "undesirables" to move in. The zoning tactics invented in San Francisco to discriminate against the Chinese and refined in New York in 1916 to discriminate against undesirable factories were soon deployed by suburban communities to exclude undesirable social groups and unwanted land uses. By the spring of 1918, New York had become a place of pilgrimage for citizens and officials wanting to find out how zoning worked.[9] In 1924 the U.S. Department of Commerce drafted a model zoning law as a basic instrument that could be used across the country, and in 1926 the Supreme Court gave its approval to zoning in the landmark case of *Euclid v. Ambler*. The decision was significant because it recognized land use zoning as a proper use of the **police powers** of local governments. But the Court actually went further in its judgment, explicitly ruling that it was legitimate for local governments to protect the character of neighborhoods of single-family dwellings against the threat of lower property values and negative externalities associated with apartment housing, commercial activities, and industrial land use. Reformers and planners endorsed this view because they felt that zoning provided a means of controlling urban growth and fostering efficient patterns of land use. But the chief motivation for introducing zoning ordinances in suburban jurisdictions was to keep out undesirable social groups by artificially raising the cost of housing. Normal practice quickly came to involve limits on the amount of new construction, minimum lot sizes, maximum densities, and bans on apartment houses. *Such barriers have underpinned metropolitan sociospatial fragmentation ever since.*

The political independence of middle-class suburbs also contributed to the third consequence of metropolitan fragmentation, the *fragmentation of metropolitan political life*. As political scientist Ken Newton has observed, social groups can confront each other when they are in the same political arena, but this possibility is reduced when they are separated into different jurisdictions. "Political differences are easier to express when groups occupy the same political system and share the same political institutions, but this is more difficult when the groups are divided by political boundaries and do not contest the same elections, do not fight for control of the same elected offices, do not contest public polities for the same political units, or do not argue about the same municipal budgets."[10] This attenuation of democracy means in turn that community politics tends to be low-key, while the politics of the whole metropolitan area are often notable for their absence. The balkanization of the municipalities within metropolitan areas means that it is difficult to make, or even think about, area-wide decisions for area-wide problems. The result is

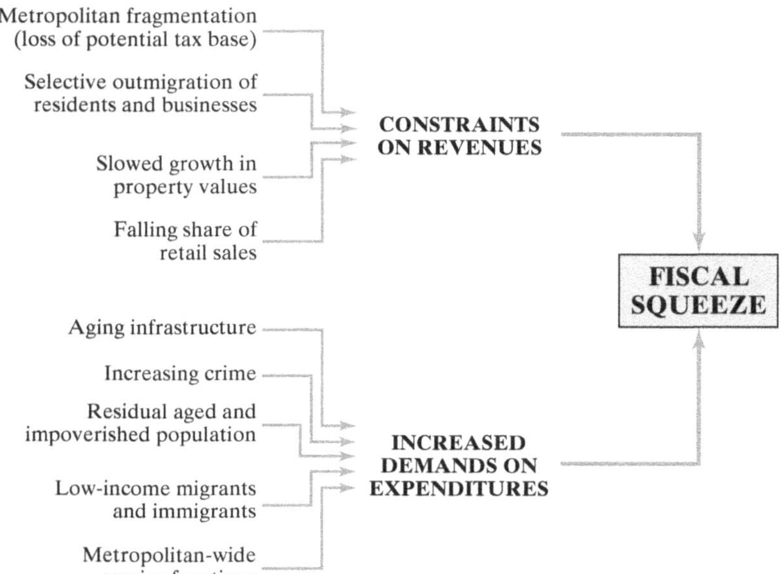

FIGURE 7 The fiscal squeeze on central city governments.

that small issues rule the day for want of a political structure that could handle anything larger. Metropolitan political fragmentation has imposed a legacy of competition rather than cooperation between neighboring governments. Not only do suburban jurisdictions compete with central cities for land users with a high tax yield and low demands for public services, they also compete with one another. In addition to land use zoning, local governments seek to gain a competitive edge through their tax packages and the "bundles" of services that they offer to businesses and residents. The result has been termed **fiscal mercantilism**, an allusion to the fierce and unrelenting competition between trading nation-states in the sixteenth and seventeenth centuries.

The New Deal The economic distress of the Depression in the early 1930s brought about new federal roles that not only restructured the relationship between the federal government and the country's cities but also brought about a realignment of politics at the metropolitan level. What was so significant about the changes ushered in by the New Deal was their *scale*. After the New Deal, cities found themselves the beneficiaries of heavy federal expenditure on a series of programs, and this fiscal relationship, in turn, brought about a close interdependence between federal and local politics that lasted for almost 50 years.

These developments were attributable in part to the geography of distress during the Depression years. Cities were certainly hit hard, and the more specialized industrial cities experienced acute economic and social distress. Many city governments attempted individually to provide relief, expand public employment, and initiate public works programs. But economic recession had caused a slump in property values, which triggered a precipitous decline in city revenues from property taxes. Faced with the prospect of municipal bankruptcy, city governments desperately petitioned the federal government directly for help.

There were two fundamental reasons for the particularly close relationship between cities and the federal government during the Depression:

1. The federal government's acceptance of responsibility for national economic growth, the maintenance of aggregate demand, and the minimization of unemployment made cities important targets for **Keynesian** strategies of economic management because of the potential **multiplier effects** of public expenditures in metropolitan settings.
2. Urbanization had made metropolitan votes much more important to the composition of the Electoral College that determines presidential elections.

Several dimensions of the New Deal were of particular importance to the political economy of urbanization. One was federal spending on public works and infrastructure, part of a broader package of measures to stimulate economic growth. A second was federal expenditure to provide relief from poverty and unemployment, bolster personal incomes, and defuse urban social unrest. A third was federal expenditure aimed at clearing slums and resettling slum dwellers. A fourth was federal support for home mortgage insurance, another part of the strategy to stimulate economic growth. Together, these policies and programs not only had a significant impact on patterns of urban development, but they also changed the pattern of intergovernmental relations and fostered new political alignments within cities; both of these shifts have left their mark on the trajectory and outcomes of U.S. urbanization.

Meanwhile, the New Deal also fostered a new coalition *within* cities: that between liberal reformers and the blue-collar constituencies that had previously supported the political machines. Political scientist John Mollenkopf has argued persuasively that this alliance represented a **progrowth coalition** that came to be established across most of the country and that formed the basis of modern party politics:

> Domestic urban development programs . . . became the principal means through which the modern Democratic party was created. Progrowth coalition building thus became a central feature of national as well as local politics. National politicians and the federal government became important actors in local politics, and this involvement, in turn, integrated local politicians, program administrators, and program beneficiaries into a new national political framework.[11]

Cities as Growth Machines and Service Providers (1945–1973)

One of the immediate carryovers of New Deal politics was a slum clearance and public housing program. The funding for urban renewal programs provided an ideal platform for the development of progrowth coalitions. During the 1950s downtown revitalization became the rallying cry for civic leaders around the country. The coalitions varied in character from city to city, but they typically included business elements (developers, bankers, and financiers), blue-collar interests (labor unions), liberal interests (planners and welfare agencies), and representatives of both local government (city managers and political leaders) and the federal government (urban renewal executives). It soon became clear that the most important people were the investors who had the economic power to organize the coalitions into would-be **growth machines**[12] that might restore prosperity to the central cities. Growth machines could not flourish simply on the basis of an alliance between business interests and local politicians, however. The urban renewal projects that were the preoccupation of urban governance and politics during the 1950s were carried forward within a web of interdependent interests (Figure 8).

By the early 1960s it had become increasingly apparent that growth machines were producing results that were not entirely beneficial. Although most members of the growth machines were happy, an increasing number of citizens began to feel that the benefits of growth were being channeled unevenly and that their cities were being stripped of their identity as blank Modernist offices, apartment blocks,

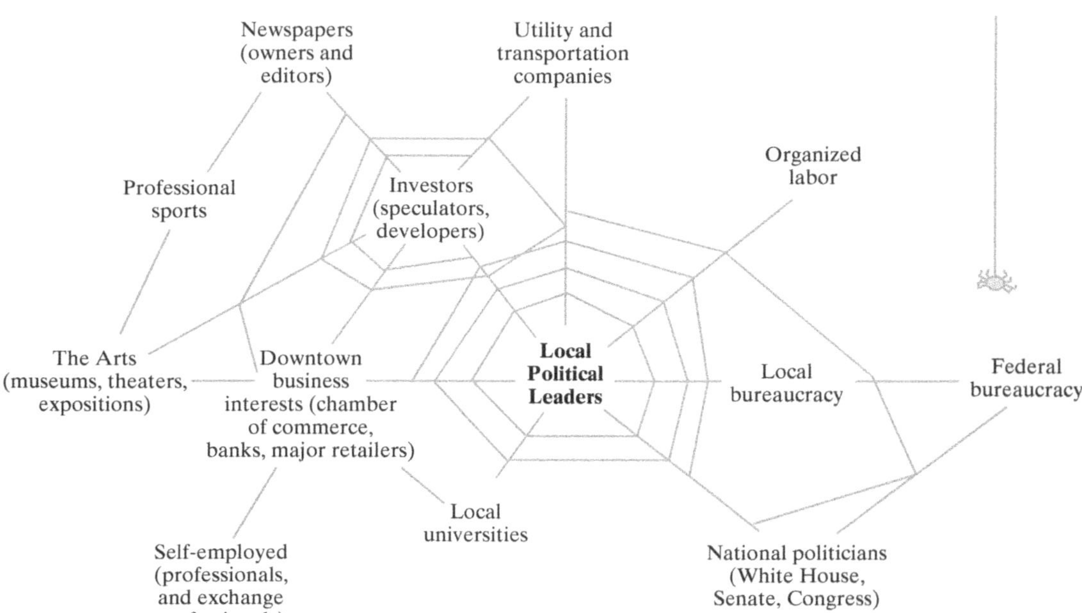

FIGURE 8 The web of growth machine politics.

and mixed-use developments came to take the place of distinctive old settings. The biggest losers were the residents of neighborhoods that were officially designated as blighted (Figure 9). In particular, the lack of *affordable* units in renewal projects effectively dismantled whole communities; their members were scattered across the city to make room for upscale housing and commercial land uses. By removing the structure of social and emotional support provided by the neighborhood and by forcing people to rebuild their lives separately among strangers elsewhere, urban renewal imposed serious psychological costs on low-income households. At the same time, displaced families typically faced a steep increase in rents because of their forced move upmarket.

Community disruption such as this eventually provoked a new kind of militant activism strong enough to change the dynamics of urban politics. Although renewal agencies had been careful to enlist support from established neighborhood organizations—indeed, they had to in order to qualify for federal assistance—they found increasing opposition from *new* organizations and coalitions of grassroots protest groups. Remember too that by the 1960s the Baby Boom generation, with its rebellious counterculture of iconoclastic politics and a collectivist approach to the public interest, was coming

Minneapolis Public Libraries (Special Collections Department)

FIGURE 9 Downtown Minneapolis in 1962. Urban renewal during the 1950s and 1960s involved the demolition of about 200 buildings across 25 blocks, representing about 40 percent of the downtown area. This photo shows the Guaranty Loan Building being razed and what were intended to be temporary surface parking lots on the sites of buildings that had already been demolished.

into young adulthood. Finding themselves outside the framework of both formal politics and the web of interdependent interests that constituted growth machines, community activists and their supporters came to represent the vanguard of **urban social movements** that epitomized the "people power" that permeated the urban politics of the 1960s. Their methods were chiefly those of door-to-door campaigns, community-wide meetings, and militant but (mostly) peaceful demonstrations. They were eventually successful in turning the tide of public opinion against renewal programs. Bolstered by books such as Jane Jacobs's *The Death and Life of Great American Cities*,[13] antirenewal, slow-growth, community-oriented, and environmentally sensitive perspectives increasingly came to be incorporated into urban politics.

Community groups gained representation on renewal agencies; accommodations began to be made; and a few outright victories were achieved, such as the abandonment of construction on Milwaukee's Park East freeway which threatened to cut off the lakefront (Urban View 2 entitled "Milwaukee Demolishes the 'Freeway to Nowhere'"). **NIMBYism** (Not In My Backyard-ism) was born (though, subsequently, it was in middle-income suburbs rather than central city neighborhoods that NIMBYism and the more extreme BANANAism [Build Absolutely Nothing Anywhere Near Anything] came of age). Equally important, the interaction between community organizations and city officials that took place over redevelopment issues resulted in much stronger local leadership networks.

Black Power and Black Politics Because of the social geography of cities, black communities and black businesses were typically the chief casualties of urban renewal programs. Indeed, the removal of black communities from the immediate environs of CBDs was very often an implicit goal of growth machines, because the key participants were acutely aware of the need to protect property values and to maintain an attractive environment for upscale retailing and leisure aimed at a predominantly white clientele. But African-American community activism over renewal projects has to be seen against the context of wider challenges to discrimination and segregation.

URBAN VIEW 2
Milwaukee Demolishes the "Freeway to Nowhere"[14]

"Taking down a freeway without help from an earthquake is remarkable," said one of the judges from the Congress for the New Urbanism who awarded the City of Milwaukee's Park East Redevelopment Plan a Charter Award in 2003. What was commonly referred to as the "Park East Freeway," however, was not really a freeway at all. It was the remnant of an abandoned plan to encircle Milwaukee's downtown with freeways. About half of the Park East Freeway had been built when citizen protest in the 1970s halted construction of the connecting highway segments, including one that was planned to run along the lakefront. This left the Park East Freeway as a spur—a half-mile of elevated roadway—that actually limited access to the downtown. The underutilization of the freeway and the potential redevelopment opportunities of the land beneath and surrounding this freeway spur prompted the city of Milwaukee to demolish it and adopt a tax increment finance district to provide financing for the infrastructure needed to reconnect the land with the local street grid (Figure 10).

Linda McCarthy

FIGURE 10 The Park East "Freeway to Nowhere"—the remnant of an abandoned plan to encircle Milwaukee's downtown with freeways in the 1970s—in the process of being removed in 2003 (at right); the 22-acre former Pabst Brewing Co. headquarters (behind) is being redeveloped as The Brewery, a residential, commercial, and educational complex.

Urbanization had exposed African Americans to the promises and frustrations of mass consumer society and to the American Dream of opportunity and social mobility. Urbanization had also brought enough African Americans to the cities for collective action to be effective. In the early 1960s this action took the form of nonviolent protest (Figure 11). The model had been the boycott of the bus company in Montgomery, Alabama, inspired by the personal protest of Rosa Parks in 1955 and sustained for a year by Dr. Martin Luther King, Jr. The tactics that led the Supreme Court to eventually overturn Alabama's segregation laws inspired boycotts, sit-ins, and demonstrations elsewhere. In 1964, after President John F. Kennedy's assassination, President Lyndon Johnson was finally successful in getting Congress to pass the Civil Rights Act, which outlawed racial discrimination. The following year, Congress authorized federal agents to register voters who had been illegally kept off the electoral rolls in parts of the South.

The year 1964 also marked the beginning of ghetto riots. The stress of ghetto life, exacerbated by rapidly rising but unfulfilled expectations in African-American communities and by antiblack activism (provoked by white fears about losing hegemony over schools, housing, and jobs), was sparked into violence and disorder by incidents of (white) police aggressiveness. In 1965, Malcolm X, a leader of the militant and separatist Black Muslims, was assassinated. In 1968, Dr. King was assassinated, as was Robert Kennedy, a presidential candidate who as Attorney General in his brother's administration had spearheaded the White House drive for civil rights legislation.

The effect on black politics was decisive. Integration through nonviolent protest was displaced by self-advancement in the context of collective "black power." Racial pride, inspired by the emergence of independent black countries in Africa, became the basis for political, social, and cultural solidarity. African-American leaders such as the Reverend Jesse Jackson organized aggressive boycotts to persuade white-owned businesses to hire African Americans and to do business with black-owned businesses. At the same time, they organized overt registration drives to maximize black power at the ballot box. Black culture, black history, and black issues demanded—and received—respect and attention as black voters and community leaders made their presence felt in urban politics.

The Struggle for Social Justice and Spatial Equity

Black politics was not the only source of change and reform, however. Even as the riots and civil disorder of the 1960s were forging the bases of a new black politics, the New Deal coalition between urban blue-collar classes and liberal reformers was reaching maturity. President Johnson initiated a "War on Poverty" in an attempt to build the "Great Society" implicit in the New Deal vision. In 1966, Johnson created the U.S. Department of Housing and Urban Development (HUD), which was a landmark in urban affairs because of the subsequent proliferation of national urban policies and the expansion of federal aid to cities. By 1969 over 500 federal grant programs were targeted for cities, with total annual appropriations amounting to $14 billion. At the beginning of the decade, there had been only 44 such programs, with appropriations amounting to less than $4 billion.[15]

The political mood fostered by these initiatives had the effect of intensifying certain long-standing issues related to metropolitan governance. One of these issues was quintessentially geographic in nature: the difficulty of matching the boundaries of electoral units to the shifting population distributions inherent in metropolitan growth—the **malapportionment** problem. Unless there is an at-large voting system, variations in the number of voters between electoral districts represent an erosion of democracy. To take a simple example: The influence of a single voter in a constituency of 2,000 voters is proportionally much greater than that of another voter who happens to live in a constituency of 3,500 voters. This situation is an example of malapportionment. It arises almost inevitably as urbanization redistributes people across the map of electoral districts from year to year; it can also arise quite deliberately if the legislative bodies that control electoral boundaries are able to create larger-than-average

Everett Collection, Inc./Alamy

FIGURE 11 Non-violent civil rights protesters on the freedom march between Selma and Montgomery, Alabama in 1965.

constituencies in areas where opposing political groups are known to have the support of a majority of voters.

The most flagrant cases of deliberate malapportionment were in fact at the level of congressional districts and state senatorial and assembly districts, where the bias was decidedly in favor of rural districts and against metropolitan areas. In a series of decisions between 1962 and 1965 the Supreme Court ruled against the practice of malapportionment, beginning a "*reapportionment* revolution" that was based on the strict criterion that the number of voters in each constituency should vary by no more than a half a percent. In addition to representing an improvement in the natural justice of the situation, these rulings had the effect of adding decisively to the importance of the urban voter in national and statewide politics.

The Supreme Court's ruling did not apply to local jurisdictions, however, with the result that large variations (as much as 30 percent in Atlanta, Chicago, Philadelphia, and St. Louis) persisted in the size of the electorate from one city ward to another. This variation effectively disenfranchises large numbers of registered voters. If, as is often the case, the malapportioned group involves the central city poor, the problem assumes more serious proportions. Policies affecting rent control and waste collection and questions such as the location of undesirable land uses and facilities or the imposition of a commuter tax will be decided in favor of suburban communities.[16]

A related geopolitical issue is that of **gerrymandering**. This occurs where a specific group or political party deliberately manipulates the spatial configuration of election boundaries in relation to known concentrations of supporters or opponents. It is quite possible to satisfy fairly strict apportionment criteria but still bias the outcome of election results by carefully drawing election boundaries in relation to the social ecology of the city. The problem is that it is very difficult to establish conclusive proof of gerrymandering. There is strong circumstantial evidence, however, of gerrymandering against African-American communities. Research suggests that in cities with relatively small African-American neighborhoods the most common type of gerrymandering has involved cracked districts (in which opposition voting strength is fragmented, leaving the voters neutralized as a minority in a large number of seats). In contrast, in cities with African-American neighborhoods that would be extensive enough to warrant two or more elected representatives, it has been common for the electoral map to minimize the number of African-American majority districts, with the rest of the African-American electorate scattered across predominantly white districts; the U.S. Congressional districts of New York and Chicago, the state senatorial districts of Milwaukee and Philadelphia, and the city council districts of Atlanta have been cited as examples of these in the past.[17]

Meanwhile, the spirit of local autonomy that characterized the 1960s accelerated the creation of **special districts**, resulting in a confusing overlay of metropolitan jurisdictions. Special districts, such as port authorities, were seen as an attractive solution to a wide range of problems because they were not subject to the statutory limitations on financial or legal powers that apply to municipalities. In particular, a community can increase its debt or tax revenue by creating an additional layer of government for a specific purpose such as fire protection or sewage treatment. Special districts also have the potential advantage of being customized to correspond closely to specific functional areas and, therefore, of being finely tuned to local social organization and participation. Between 1942 and 1972 the number of nonschool special districts in the United States increased from 6,299 to 23,885. By 1977 the largest 35 metropolitan areas in the United States each had an average of 293 separate jurisdictions of one kind or another (many of them with very narrow functions, such as the one created to provide for the St. Louis Zoo).[18]

The increasing partitioning and layering of metropolitan space created more problems than it solved, however. Taking a metropolitan-wide view, it was increasingly apparent to many people that local autonomy and administrative convenience were more than offset by fiscal imbalances, inefficient and inequitable distributions of public services, the creation of complex, competing, and expensive bureaucracies, and the juxtaposition of conflicting policies. Opportunities for economies of scale and coordination in areas such as transportation, planning, water supply, housing, and public health were forfeited. The sheer proliferation of governmental jurisdictions made it difficult for citizens to keep up with the politics and policies that affected them, leaving the control of many special districts to a self-selected knot of individuals and special interest groups.

Together with the continuing political fragmentation at a metropolitan scale associated with the incorporation of suburban communities, these problems make it increasingly difficult to sustain the notion of *metropolitan* governance and management. There are in fact several ways in which reforms might recapture the fit between governmental structures and socioeconomic reality.[19] These reforms include transferring key functions to a higher tier of government, instituting intergovernmental agreements where a number of jurisdictions jointly provide certain services, and consolidating two or more jurisdictions into a unitary, multifunction authority. In general such reforms have been tentative and exceptional: There are simply too many vested interests in the status quo.

ENTREPRENEURIAL POLITICS AND NEOLIBERALISM (1973–PRESENT)

In the most recent phase in the evolution of urban governance and politics, economic restructuring and spatial reorganization have prompted major changes in the nature of urban governance and politics. Aggravated by the **stagflation** of the 1970s, central city decline reached the point where governments faced mounting expenditures on everything from infrastructure provision and repair to welfare services—while at the same time they were losing the capacity to raise money through taxes and bond issues. The result was a "fiscal crisis" that first became apparent in New York in 1975. With pressure on public spending, the quality of public services, public goods, and physical infrastructures inevitably deteriorated, which in turn added even more pressure for those with money to spend it privately. People's concern to have their children attend "good" schools intensified demand for housing in upscale developments with their own community pre-schools and elementary schools. Increasing numbers of people began to buy private security systems, enroll their children in private extra-curricular lessons and activities, and spend time at the mall rather than the local playground. It is only human nature that people paying for private services will tend to resent paying for public services that they feel they no longer need. Also resentful of continued spending on socially and geographically redistributive programs, they began to support the view of certain policy experts and politicians who were asserting that government at all levels had grown too big and too expensive. As a result, **neoliberal ideologies** began to permeate urban politics. Neoliberal ideologies are predicated on a minimalist role for government, assuming the desirability of free markets as the ideal condition not only for economic organization, but also for political and social life.

Labor-market "flexibility" became the new conventional wisdom. The proponents of neoliberalism regard policies designed to redistribute resources to disadvantaged districts or cities as necessitating excessive taxation of wealthy people, thereby discouraging entrepreneurial leadership, reducing investment capital, and undermining productivity. If necessary, social goals and regulatory standards have to be sacrificed, it is argued, to ensure that businesspeople have the maximum latitude for profitability. The rising tide of economic development, the argument goes, will then float all boats, urban and rural, central city and suburban. If the tide does not happen to rise high enough in any given location, people can "vote with their feet" and move to more prosperous cities or regions of the country.

So, rather than focus on metropolitan problems such as poverty, environmental degradation, and traffic congestion, urban governance became more concerned with providing a "good business climate" that might attract investment. As geographer David Harvey noted, urban governance shifted decisively from management to entrepreneurialism.

The 1970s and 1980s also saw the rise of a strong libertarian element in U.S. political thought. Individual liberty and the sanctity of private property ownership were the cornerstones of a neoliberal ideology that sought to undermine the rationale for land use zoning. In the political climate of the 1980s, popular support for progressive notions of scientific management had ebbed away, to be replaced by a greater emphasis on individual self-expression and a new resistance to central authority. In populist politics as well as in law and economics journals, the traditional arguments for land use zoning—abating nuisances, creating efficiencies in planning for the common good—were increasingly dismissed. Instead, zoning was portrayed as an inequitable redistribution of the rights to develop land. The argument began to form among libertarian theorists that municipal zoning amounts to a "taking" of property owners' development rights that should not be allowed, or that should be compensated financially.

The overall effect has been to "hollow out" the capacity of central government while forcing municipal governments to become increasingly entrepreneurial in pursuit of jobs and revenues; increasingly pro-business in terms of their expenditures; and increasingly oriented to the kind of planning that keeps property values high. Neil Brenner and Nik Theodore suggest that the implicit goal of neoliberalization at the metropolitan scale has been "to mobilize city space as an arena both for market-oriented economic growth and for elite consumption practices."[20] As a result, planning practice has become estranged from theory and divorced from any broad sense of the public interest. Urban policy has become pragmatically tuned to economic and political constraints rather than being committed to change through progressive visions. **Public-private partnerships** have become the standard vehicle for achieving change, replacing the strategic role of planning with piecemeal dealmaking. Planning has become increasingly geared to the needs of producers and the wants of consumers and less concerned with overarching notions of rationality or criteria of public good.

Inevitably, the ascendance of neoliberalism has affected the tenor and vitality of civil society. The meaning of civil society has changed over time but is generally understood to involve all the main elements of society outside of government—the "parapolitical" elements that serve as mediating agencies between individuals and the formal machinery of institutional politics. These include business organizations, trades unions, homeowners' associations, and voluntary groups of all kinds, such as charities and conservation societies. Although relatively few such organizations are explicitly "political" in nature, many of them are politi*cized* inasmuch as they occasionally pursue group activities or campaigns through the medium of government. Indeed, there is a school of thought among political scientists that argues that private groups are highly influential in raising and defining issues for public debate.[21] According to this school of thought,

URBAN VIEW 3

Tax Increment Financing (TIF)[22]

> To its advocates, tax increment financing is the greatest thing to happen to cities since paved streets . . . As the Denver-based Front Range Economic Strategy Center recently noted, 'Tax increment financing subsidies are like a development credit card for the city: Buy a project now and pay it off with future revenue.' But as anyone who has a credit card knows, the temptation to misuse it can be strong.[23]

As the federal government reduced its role in funding urban development beginning in the 1970s, and as neoliberal ideologies took hold, entrepreneurial city governments turned increasingly to **tax increment financing**. TIF is a mechanism used by cities to finance redevelopment efforts that is directly tied to the success of those efforts. If an area in a city can be made more attractive to private developers and new development occurs, the tax revenue collected from that area would be expected to rise. Tax increment financing taps into any increase in tax revenues by using the tax "increment" (the difference between the taxes after redevelopment and the expected taxes without redevelopment) to finance the improvements and other activities that stimulated the redevelopment to occur in the first place.

URBAN VIEW 3
Tax Increment Financing (TIF) (*continued*)

In U.S. states with statutes authorizing the use of TIF, originally in areas that were "blighted" but increasingly anywhere that will generate significant economic benefits, the first step for a city is to ascertain the property tax revenue that is being collected in a particular area before redevelopment. Following feasibility studies and cost-benefit analyses, the city establishes a tax increment district (TID) for a specified period (often 20 years), formulates a development plan, enters into development agreements with private developers, and borrows money (through loans or by selling bonds) to use in various ways to improve the development prospects of that district: property acquisition, site preparation, loans to developers and new businesses, capital improvements such as new roads and street lights, and new services such as improved street cleaning and security patrols. As redevelopment occurs, tax revenues increase, and the tax above the preredevelopment property tax revenue is used to pay off the city's loans or bonds.

Although tax increment financing sounds very attractive in theory—the local government does not lose any revenue because it is assumed that the tax increment would not have occurred without the redevelopment efforts financed by that increment—it is not without potential drawbacks. Obviously, cities can find themselves in trouble if the redevelopment does not produce the estimated increment (that has been earmarked to pay off the loans or bonds). In situations where the TIF involves a general cap at pre-TIF levels on property valuations or tax assessments, other public entities that normally receive property tax revenues—school districts, special districts, the county—do not receive their part of the tax increment for the life of the TID or until the loans or bonds are paid off, even in situations where the redevelopment creates additional demands for their services. If TIF is used unnecessarily—in areas where redevelopment would have occurred in the absence of a TID—some or all of the tax increment represents revenues that the local government would have collected anyway and that now instead cuts into general revenue because it has to go to pay off the loans or bonds. There may also be a lack of "transparency" for ordinary citizens—because the TIF bonds are not "general obligation" bonds, voter approval is not required. This has led to concerns about high levels of TIF debt due to an overdependence on TIF by cities. But "TIF is like any other tool, useful or harmful according to how it's used . . . [like a colander] . . . 'Use it as a rain hat and it doesn't work very well. Use it to drain spaghetti and it works just fine.'"[24]

politicians and officials tend to back off until it is clear what the alignment of groups on any particular issue will be and whether any official decision-making will be required. In essence, this gives municipal governments the role of umpiring the struggle among private and partial interests, leaving them to decide the outcome of major issues in all but a formal sense.

Fiscal Crisis

As we have seen, metropolitan fragmentation left many cities with a legacy of chronic fiscal problems as a result of the fiscal squeeze—between falling revenues and increasing demands on municipal expenditures (Figure 7). After World War II this pressure intensified, particularly in the older industrial cities of the Northeast. Freeways and metropolitan sprawl drew out even more of their middle classes and employers, and with them went their tax dollars. Only a relatively small portion of the properties they left behind was replaced in urban renewal programs by newer, bigger, or more-luxurious structures. The remainder continued to deteriorate, the worst falling into a **spiral of decay** and abandonment (Figure 12). The aggregate yields of property taxes fell accordingly. Sales taxes also fell, the suburbanization of the middle-income groups having drsawn retailers along in their wake. During the 1960s retail sales in larger central cities, when adjusted for inflation, fell by between 15 and 30 percent. Meanwhile, the demand for central city services increased at an unprecedented pace. In addition to the overall trend toward providing an increased range of services that was inherent to the progrowth coalition's welfare state, central cities faced increasing expenditures for several reasons:

- Their original infrastructure of roads, sewers, water mains, bridges, and so on had reached the end of its designed

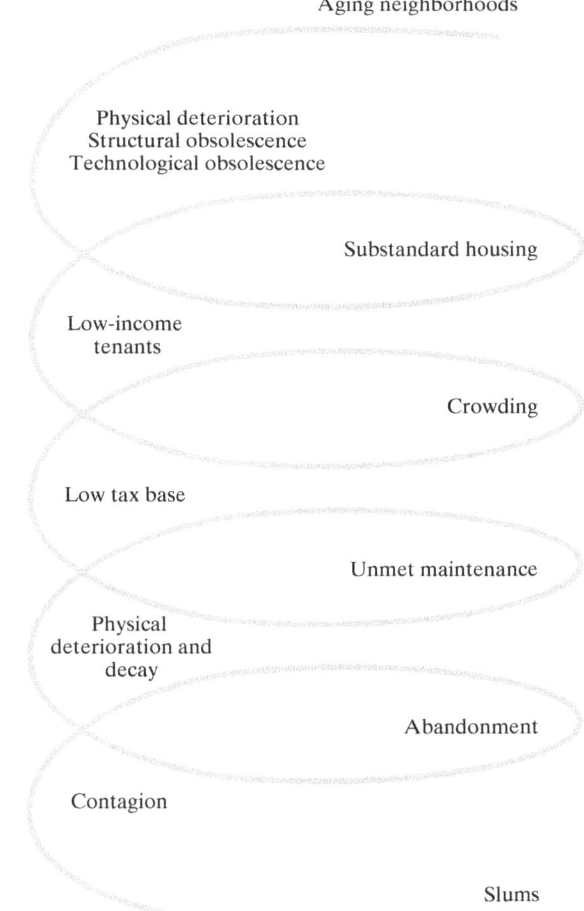

FIGURE 12 The spiral of neighborhood decay.

life of 75 to 100 years; repairs and replacements began to impose an increasing burden on city expenditures.

- Rising crime rates induced corresponding increases in spending on policing, incarceration, and the courts.
- Rising proportions of immigrants, the lone elderly, single-parent families, and the unemployed in older central city neighborhoods required increased expenditure on specialized public services and amenities.
- The costs of federal welfare programs began to be passed on to local governments. Federal Medicaid, for example, provided free medical care for welfare-dependent people, but state governments were required to contribute 50 percent of the costs. It did not take long for states to begin to pass along part of this burden to their cities.

The fiscal problems of U.S. cities in the 1970s were rooted, then, in a long-standing and intensifying fiscal squeeze. But the fiscal squeeze on its own did not add up to crisis. What pushed the fiscal squeeze to the edge of crisis was the transition that had taken place in the restructuring of metropolitan economic geography and, with this restructuring, a shift in the political economy of central cities:

> On the one hand, new economic growth in the central cities did not provide sufficient employment and income benefits to the central city's residents. Industrial jobs were taken by suburbanized union workers. Construction work was dominated by restrictive craft unions. And the new office economy was drawing on the better educated, better heeled suburban workforce. Industrial investments were now part of vast multilocational networks of plants, thus weakening the local multipliers from local plant investments. This export of the income benefits of local economic growth meant a continuous reservoir of poor, structurally unemployed people who turned to city governments for jobs and services.
>
> On the other hand, the rising office economy of the central city required a restructuring of urban space to move people and information most efficiently. This required a massive investment in public capital for mass transit, parking, and urban renewal, [as well as] the more traditional forms of infrastructure.[25]

These infrastructure investments were often effectively insulated from conflict and political debate by the new forms of administration and financing that had evolved with the rise of growth machines: autonomous special districts, banker committees, and new forms of revenue and tax increment finance bond issues. As a result, two spheres of city expenditures emerged: one oriented to constructing the new infrastructure necessary for profitable private development, the other to providing services and public employment for the city's residents. These two worlds—of social wage and social capital—were structurally separated, with the first governed by electoral politics and the excesses of patronage and the second housed in bureaucratic agencies dominated by public administrators who survived by their efficiency.[26]

URBAN VIEW 4

The Fiscal Squeeze and U.S. Central Cities

New York City was the first to reach the crisis point of fiscal distress. Between 1965 and 1970 the city's budget had doubled, as had the number of people on its welfare rolls. By 1975 its accumulated operating deficit was $2.5 billion and it faced an unfunded budget gap that could not be filled by borrowing. In the end, New York was bailed out not by the federal government but by the New York State Legislature, which turned over control of the city's finances to an Emergency Financial Control Board, headed by the governor and dominated by representatives of the major banks. It also created the Municipal Assistance Corporation (which inevitably came to be referred to as Big MAC) to lend New York City the funds it needed to service its debts. These institutions imposed a number of austerity measures that were to foreshadow the subsequent retrenchment of other local governments.

Between 1975 and 1978 New York laid off more than 60,000 public employees, initiated salary rollbacks and wage freezes, established user fees for some municipal services, reduced budget allocations, and consolidated or eliminated several municipal departments. Meanwhile, banking interests, appalled at the scale of the problem, persuaded a reluctant federal government to make loans and guarantees available. It did so, making it very clear that the sole reason was to avoid city bankruptcies that could trigger a national banking crisis. This development was important because it *refocused the public interest once more in terms of economics, leaving the interpretation of urban affairs to the representatives of banks, real estate investors, and corporate financiers.*

New York's experience was meanwhile repeated, although in somewhat less dramatic fashion, in a number of cities, including Baltimore, Boston, Cleveland, Detroit, Philadelphia, and St. Louis. A combination of bailouts, retrenchment, and a national economic recovery put an end to the crisis, but the **fiscal squeeze** has persisted, leaving central cities vulnerable.[27] In 1991, for example, Bridgeport, Connecticut, was threatened with bankruptcy, as was Chelsea (in the Boston metropolitan area). In Philadelphia the city treasury was so short of cash that weekly meetings were convened to decide which bills could be paid. About 25 years after its first crisis, New York was again facing a growing deficit and another crisis, even though its role as a provider of services had by then become secondary to its role as facilitator of economic development. This time, the city's enormous

URBAN VIEW 4

The Fiscal Squeeze and U.S. Central Cities (*continued*)

budget gap—around $6 billion in 2002—was partly the result of simultaneous events outside the city's control—the September 11 attacks, the decline of the stock market, the national recession, and the problems and scandals of the financial industry that is so crucial to New York's economy. Nevertheless, the federal government was no more inclined to help (other than 9/11 reparations) than it was in 1975. The City closed the budget gap with drastic measures such as an 18.5 percent property tax hike and enjoyed economic recovery until the recession hit in late 2007. Since then, cities across the country have struggled to overcome the fiscal squeeze and fund even the most basic of services. New York City's budget gaps are projected to be $3.6 billion, $6 billion, and $6.6 billion for 2012, 2013, and 2014 respectively. Further cutbacks will adversely affect the City's already trimmed down budgets for construction and maintenance of City-owned buildings, pensions, and health benefits for City workers, and City support for groups that shelter the homeless, investigate domestic abuse, and provide families with daycare.[28]

Fiscal Retrenchment and Neoliberalism

The fiscal well-being of central cities was not the only aspect of urban governance and politics to be affected by the economic and metropolitan restructuring that began in the mid-1970s. The political economy as a whole was altered beyond recognition within just a few years. Partly in response to longer-term trends in the economy and society and partly in response to the jarring economic and fiscal experiences of the mid-1970s, a new conservatism—neoliberalism—emerged in local politics. This was a politics that was based on freeing up the economy for recovery by reducing government regulation and control and lowering taxes by cutting back on expenditures for public services. This neoliberalism was widely supported by middle-class voters who were seeing their incomes stagnating after years of steady growth and who had come to believe in the argument that all they had been given in return for their tax dollars was a bloated group of meddlesome and inefficient bureaucrats and an army of welfare "freeloaders." Neoliberalism was also, of course, supported by the entire spectrum of the business community, which naturally saw lower taxes and less regulation as catalysts for renewed profitability.

So a new political alliance was born, between private business interests and the middle income groups. It was an alliance that met relatively little opposition. The circumstances of its creation had damaged its would-be opponents. Working-class solidarity, and organized labor in particular, had been seriously weakened by deindustrialization and the associated changes in economic geography. Corporate restructuring in an ever more centralized and internationalized economy increasingly outmaneuvered workers and left them to operate in local arenas, while companies were able to switch production and investment between cities, regions, and countries.[29] The progressive element of the middle income groups had been demoralized by the obvious failures of welfare provision and Modernist technocratic reform to combat crime, poverty, and environmental degradation. It had also been weakened by the defections of people who no longer felt as progressive once they personally began to feel the pinch of economic recession.

The first and most striking evidence of the new alliance was manifest in "taxpayer revolts." In California, Proposition 13 cut property taxes by 60 percent. In Massachusetts, Proposition $2^1/_2$ limited property taxes to 2.5 percent of assessed real estate value. Similar proposals were passed in Idaho and Nevada. In Arizona, Hawaii, Michigan, and Texas, voters approved bills that severely limited public spending, and everywhere there was increasing reluctance to sanction bond issues for municipal development. In 1980, Ronald Reagan, who had championed this new conservatism as governor of California, was elected president, marking the success of the new alliance at the national level.

The result was that cities were forced to reevaluate their roles, regardless of whether they were in fiscal trouble. The outcome of this reevaluation was a general **fiscal retrenchment** in service provision. In Oakland, for example, the city government responded to the neoliberalism imposed by Proposition 13 by closing a fire station and four branch libraries; reducing the police department's budget for criminal investigations, park maintenance, after-school recreation programs, library and museum hours, and street sweeping and maintenance; and eliminating over 100 administrative positions.[30] By the mid-1980s, many older central cities had cut the number of municipal employees by 10 to 15 percent. Boston and Cincinnati had cut their workforce by more than 20 percent, while St. Louis had eliminated more than 40 percent of its employees.

Other trends accompanied this sort of retrenchment as it was replicated across the country. Cost efficiency in service delivery and public administration came to be emphasized at the expense of need, spatial equity, and social justice. Voluntarism came to be an important means of maintaining service provision, particularly in relation to neighborhood security, elementary education, and library services. But voluntarism is an uneven phenomenon, and for the most part it only served to underline the increasing metropolitan sociospatial polarization and fragmentation.[31] Other aspects of retrenchment included cost-recovery programs, coproduction arrangements (where the public sector works with nonprofit organizations), and "intensification" (increasing labor productivity through managerial and organizational changes). Perhaps the most important single trend in the United States, however, was that of the privatization of public services (as opposed to that privatization that involves selling off public assets, which is more common in other countries), because it reflected so clearly the new ethos of urban governance.

The Privatized City[32]

In 1988 the Report of the President's Commission on Privatization concluded that **privatization** "may well be seen by future historians as one of the most important developments in American political and economic life in the late 20th century."[33] Because the commission was reporting to the Reagan administration, its conclusions are perhaps not surprising. There is, nevertheless, some substance to support the conclusion. The Council of State Governments has estimated that whereas under $30 billion of services were furnished annually by the private sector in the 1970s, the figure had risen to over $80 billion by the early 1980s and $150 billion by the 1990s, when the Local Government Center's database included over 35,000 specific instances of state or local privatization.

These data reflect a trend that has to be seen in terms of the perceived benefits for both the public and the private sector. The particular advantages of privatization as seen by city governments are:

- reducing direct municipal outlays.
- sharing financial risks with the private sector.
- accessing skilled staff not available in the public sector.
- reducing costs for taxpayers through private-sector efficiencies such as savings in construction costs and time, operational productivity, and economies of scale.
- maintaining service levels with no increases in tax rates or user fees.
- improving service quality.
- increasing flexibility from the reduction in bureaucratic complexity and procedures.

For private companies, privatization has provided new markets, new outlets for investment capital, and new entrepreneurial opportunities. Timing is of critical importance here. It was no coincidence that private capital was available for public-private partnerships just as cities were facing retrenchment and fiscal stress: Both were the product of a phase of overaccumulation. *In short, changing circumstances in international, national, and real estate markets brought the private sector to the public sector as much as ideological and fiscal shifts brought the public sector to the private sector.*

In the United States privatization has been most pronounced in relation to infrastructure projects. Both federal and local governments have become increasingly aware of the extent of infrastructure decay and deficit and its implications for metropolitan economic development. Privatization has been seen as a means of funding projects that might otherwise have been left on the drawing board. Overall, the provision of basic public works facilities accounts for the greatest dollar volume of privatization activity. The most important projects have been large-scale schemes for wastewater and sewage handling, water provision, resource recovery, and waste-to-energy plants. During the 1970s, when the U.S. Environmental Protection Agency (EPA) increased water quality standards, many cities turned to the private sector to operate and maintain such plants. With the introduction in 1981 of tax incentives for businesses involved in privatization, the interest of major engineering companies intensified. By 1985, 15 municipalities had opted for privatized wastewater projects. By 1988 one-third of the local governments in a national survey had privatized some roads, bridges, or tunnels.[34] Almost as many had privatized water supplies, nearly one-quarter had privatized water treatment facilities, and almost one-fifth had privatized municipal buildings or garages.

The privatization of services has since accelerated to reach a significant level. Surveys of local governments suggest that more than 80 percent of U.S. cities now use some form of privatization in their service provision. The use of contracting by local governments varies significantly depending on the service, however. More than 50 percent of local governments contract out commercial waste collection, vehicle towing and storage, legal services, and the operation of hospitals, daycare facilities, and homeless shelters.[35] Other services that are contracted privately on a significant scale include domestic waste and recycling collection, bus and ambulance operations and maintenance, street repair, street lighting, data processing, landscaping, and vehicle fleet repair. Concerns about quality and public accountability make contracting out other kinds of government services more difficult. Less than 5 percent of local governments contract out crime prevention, police and fire communications, fire prevention and suppression, traffic control and parking enforcement, sanitary inspection, or prison and jail services.

Meanwhile, the retrenchment of public-sector activities and the unwillingness and inability of urban governments to take responsibility for new or expanded service needs caused another kind of privatization to emerge: the provision of services by the private sector *without contracting or formally cooperating with the public sector*. The most striking example of private-sector services is in the area of security: private security "officers" patrol shopping malls and office buildings, and private security forces protect upscale residential developments.

Some serious questions have been raised concerning the effectiveness and the desirability of privatization. In particular, it remains to be seen whether private contractors will be both able and willing to keep costs down. The obvious danger, of course, is replacing a public monopoly with a private monopoly. There are a number of causes for concern about privatization, including:[36]

1. "Low-balling," in which contractors initially bid lower than their actual costs to secure a contract and then, after the city disbands its own delivery system, raise the price to recover their initial costs and establish a monopoly.
2. Disruption of service resulting from bankruptcy or labor disputes. Clauses can be included in privatization contracts that hold the contractor liable for the remainder of the contract term, but the problem remains as to how cities might deal with disruption of service if they

themselves have relinquished the capacity to provide the service.

3. The replacement of "primary" jobs in the public sector by "secondary" jobs in the private sector, with women and minorities in particular experiencing the effects of less security, lower pay, and fewer benefits. The effects of privatization on career civil servants could include career disruption and dislocation and reduced morale and productivity.
4. The risk of corruption associated with the interdependence of leading bureaucrats, politicians, and local contractors.
5. The emphasis on cost efficiency implicit in privatization; not a bad thing in itself, but cause for concern in that it reinforces the idea of a rather narrowly defined public interest.

Privatopia The politics of privatism are perhaps most starkly obvious in the master-planned speculative developments of suburbia that are controlled by homeowner associations (also known as common interest communities) that effectively operate as private governments. At least half of all housing currently on the market in the fifty largest metropolitan areas and nearly all new residential development in California, Florida, New York, Texas, and suburban Washington, D.C., is subject to mandatory governance by a homeowners' association. In Arizona, Pima County alone has more than 800 associations representing about 100,000 homeowners. They have been characterized as fragmented "privatopias," in which "the dominant ideology is privatism; where contract law is the supreme authority; where property rights and property values are the focus of community life; and where homogeneity, exclusiveness, and exclusion are the foundation of social organization."[37] Privatopias are premium spaces designed to accommodate affluent households in enclaves that are legally controlled by "servitude regimes" (i.e., covenants, controls, and restrictions that regulate both the physical environment and social comportment). Servitude regimes are typically designed by developers not only to preserve landscaping and maintain the integrity of urban design but also to control the details of residents' homes and their personal behavior. Developers thus become benevolent dictators, imposing their framework on the landscapes and communities of master-planned suburbs.

For consumers, these servitude regimes offer a means of narrowing uncertainty, protecting equity values, and, above all, establishing the physical framework for the material consumption that constitutes their lifestyle. Little is left to chance, with covenants, controls, and restrictions specifying what is and what is not allowed in terms of garden fences, decks, hot tubs, and clotheslines, the color of doors and mailboxes, and so on. Most ban all signs except for real estate placards, and restrict what kind of vehicles can be parked outside, even in driveways; some even prescribe how long garage doors can be left open, the type of furniture that can be seen through front windows, the color of Christmas tree lights, and the maximum length of stay for guests. Most limit the number and types of pets that residents may keep, as well as the kinds of activities that are allowed in gardens, driveways, streets, and public spaces, and whether any sort of business can be conducted from the home. Under U.S. law, servitude regimes have to be taken very seriously, because if homeowner associations do not enforce them to the letter they can be accused of being arbitrary and capricious. Homeowner associations have the power to levy fines to bring transgressors into compliance. They also have the power to obtain a lien on the property of recalcitrant homeowners and can even threaten foreclosure.

As a result, the master-planned developments of contemporary suburbia are culturally hermetic spaces, "purified" arenas of social reproduction, characterized by ostentatious consumption and social segregation. Private master-planned communities have an internal politics characterized by NIMBYism and growth/slow-growth/no-growth disputes. Sociologist Setha Low suggests that common interest communities, with weak social ties and diffuse interpersonal associations among homogenous populations, promote "moral minimalism"—a reluctance to get personally involved in any kind of political dispute. Only when residents can be assured that someone else will bear the burden of moral authority, enabling them to remain anonymous and uninvolved, are they likely to participate in any kind of local politics.[38]

NIMBYism, Smart Growth, and the Geopolitics of Suburbia

NIMBY ("Not In My Backyard") squabbles are a chronic aspect of local politics throughout suburbia. Suburban homeowners are especially conservative because what is typically their single largest asset—their home—is always linked to the fate of their neighborhood. The result is that they tend always to vote in local politics in the interest of resisting change that might affect their neighborhood. When changes to the physical or social fabric of a neighborhood are perceived by more than a handful of residents to threaten property values—or, more euphemistically, the "character" of a neighborhood—the flashpoints for NIMBY squabbles become politicized. Voters, faced with the prospect of some unwanted change, are typically quick off the mark with placards, leaflets, and mobilization e-mails with upper-case phrases and multiple exclamation points for emphasis. At issue, almost always, are property values and the closely associated concerns of aesthetics and social exclusion.

The appetite for bigness and bling that has led to the insertion of big new homes—"McMansions"—in place of older homes in established neighborhoods, for example, has led not only to NIMBY bickering among neighbors but also to political campaigns aimed at changes in zoning ordinances. Thus, for example, Chevy Chase, Maryland, an upscale suburb of Washington, D.C., imposed a six-month moratorium on home construction in 2005 to make time to examine how to deal with the proliferation of oversized single-family houses on "scrapeoff" sites. In 2006, the Los Angeles city council passed an ordinance that limits home size on lots of 8,000 sq. ft. or less in the Sunlund-Tujunga area to 2,400 sq. ft. or a floor-to-area ratio (FAR—really the house footprint-to-lot ratio) of 40 percent, whichever is greater. New Canaan, Connecticut, enacted regulations in 2005 that limit the height of new houses, while Austin, Texas, has introduced FAR limits of 40 percent on new housing, a maximum of 2,500 sq. ft. for replacement construction, and a maximum of 20 percent of existing floor space for additions.

Social and demographic change produces similar reactions. In Manassas, Virginia, thirty-five miles southwest of downtown Washington, D.C., the cost of housing has meant that some extended families decided to share space in single-family homes. Many of these families included immigrants working in Manassas and unable to find affordable housing any other way within reasonable commuting distance. This was entirely legal until neighbors began to complain to city hall in Manassas. The city's rather hysterical response was to set up an "overcrowding hotline," while the mayor sent letters to the governor of Virginia asking him to declare a state of emergency. When this failed, the city of Manassas decided to redefine its definition of a "family," adopting a zoning ordinance that essentially restricts households to immediate relatives, even when the total number of persons in the household is below the city's occupancy limit.[39]

The rural-urban fringe, meanwhile, has become a land use battleground, where "developers, long-term landowners, quick-buck land speculators, politicians and realtors are matched against other long-term landowners, politicians, environmentalists, and newcomers who want to keep their communities attractive and fiscally manageable."[40] Planning theorist Dolores Hayden has noted the paradox that successful NIMBY campaigns against new developments can sometimes accelerate suburban sprawl, pushing unwanted development farther out into greenfields.[41] This, of course, is another reason for a regional approach to planning and development.

Zoning Struggles Much of the political struggle in suburbia turns on how to manipulate local zoning laws to protect or achieve some sort of advantage. For affluent neighborhoods in areas designated for low-density residential development, zoning acts as an invisible wall, keeping out undesirable households and land uses. Only those who already own property in the municipality have standing in court to bring suit against the zoning codes, so outsiders are unable to effect change. As a result, as James and Nancy Duncan noted in their study of Bedford, New York, zoning "plays an active structuring role in grounding the practice of an aestheticized way of life in a place. It attempts to maintain sufficient social homogeneity within a territorially bounded and (relatively) defensible space in order to achieve a collective sense of place and landscape."[42] At the same time, low-density zoning pushes up house prices, "zoning out" key workers with modest salaries—teachers, nurses, emergency services workers, and so on—and exacerbating metropolitan transportation problems.

From the perspective of local municipalities, exclusionary zoning is an important tool in competing with other jurisdictions for fiscal health. The goal is to use zoning, along with other land use planning tools, to keep out burdensome populations (with high levels of needs but low tax capacity) and noxious industries while attracting affluent and self-sufficient populations and clean economic activities that pay their workforce well. For the elected boards of suburban municipalities, the need to balance the books through carefully targeted growth must be set against voters' preference for low taxes and their sensitivity to their property values. The result is a struggle between growth-machine coalitions on one side and slow-growth or no-growth coalitions on the other.

Loudoun County, Virginia, provides a classic example. Rapid growth during the 1990s increased the county's population from 86,000 to 170,000: a boom time for local builders, developers, and other members of the local Chamber of Commerce but alarming for both residents and incomers who were witnessing a dramatic escalation of traffic congestion and increasing signs of stress in the county's ability to provide the kind of schooling and public services that they had come to expect. Led by a grassroots citizens group, Voters to Stop Sprawl, voters installed a slow-growth set of representatives in the 1999 elections. The new leadership promptly rezoned large swaths of undeveloped land, effectively placing the western two-thirds of the county off limits to conventional subdivisions, with developers generally limited to no more than one home per twenty acres in northwest Loudoun County (or one house per ten acres in cluster subdivisions with lots of surrounding green space). Progrowth activists challenged the rezoning in Virginia's Supreme Court, which eventually ruled in their favor on a technicality: that the county had not clearly defined in its public notices the boundaries of land to be rezoned. Meanwhile, a new crop of progrowth officials took power in Loudoun after elections in 2004, underwritten by campaign contributions from a growth-machine coalition. One of the first rules the new county board adopted was to take away the agenda-setting function from its chairman, a holdover slow-growth advocate, and place it into the hands of the vice-chairman, a progrowther.

"Smart" Growth An alternative to the continual and unappetizing politics of this sort has emerged as a kind of "third way" politics: have your cake and eat it too. It is the doctrine of "smart" growth. Smart growth is progrowth, but only when it is relatively compact and steered toward strategically designated locales with adequate infrastructure. It is, in other words, a stealthy euphemism for old-fashioned planning of the sort that cannot be openly described as such in the context of a neoliberal political economy. Smart growth adheres to the principles of preserving public goods, minimizing adverse land use impacts, maximizing positive land use impacts, minimizing public fiscal costs, and maximizing social equity. It has been gratefully embraced by an embattled planning profession that has been in retreat to neoliberalism.

A national coalition, Smart Growth America, has attracted dozens of member organizations, including the American

Farmland Trust, the American Planning Association, the Congress for The New Urbanism, 1000 Friends of Connecticut, 10,000 (not to be outdone) Friends of Pennsylvania, the Sierra Club, and the Trust for Public Land. A Smart Growth Network, coordinated by the Environmental Protection Agency's Division of Development, Community and Environment, in conjunction with several non-profit and government organizations, including the International City/County Management Association and the Sustainable Communities Network, has developed a set of ten basic goals (referred to as "principles") for smart growth: mixed land uses; taking advantage of compact (i.e., higher-density) neighborhood design; creating housing opportunities and choices; creating walkable communities; fostering distinctive communities with a strong sense of place; preserving open space, farmland, and critical environmental areas; strengthening and directing development toward existing communities; providing a variety of transportation choices; making development decisions predictable, fair, and cost-effective; and encouraging community and stakeholder collaboration in developer decisions.

Most prominently, Parris Glendening, governor of Maryland between 1995 and 2003, made smart growth the center of his electoral platform. In office, he appointed a cabinet secretary to oversee development policy, pulling together the state agencies—transportation, housing, environmental quality—that had anything to do with growth. Glendening insisted that the state itself take the lead in smart-growth-type policies, locating state agency offices only in downtowns and town centers. He redirected state funding from highways to transit and to infrastructure in higher-density settings, and championed a policy under which developers paid for water, sewer, and other infrastructure in undeveloped areas, while developers building in designated higher-density areas benefited from streamlined approval processes and reduced fees.

Naturally, policies like these were a rude and unwanted provocation to progrowth interests. Libertarian and neoliberal think-tanks like the Heritage Foundation and the Reason Foundation cranked out essays warning of the anti-American implications of smart-growth "abuses" (sometimes portrayed as elitist, sometimes as socialistic), damage to free-market mechanisms, and constraints on individual choice; they provided lobbyists with lists of worrisome key points to use in speaking to legislators, and fed talking points to op-ed writers in local newspapers. Developers simply stepped up their campaign contributions to progrowth candidates in local elections. But, to the dismay of smart-growthers, the most effective challenges to their policies came from citizens themselves, in classic NIMBY responses. In Maryland, for example, residential and retail projects around Metro stations—considered ideal locations for smart growth because they would encourage the use of mass transit—were stopped cold or scaled back because of neighborhood opposition. Even projects in designated residential smart-growth areas have run into local opposition. Maple Lawn Farms, a 508-acre site in Howard County three miles south of Columbia, Maryland, midway between the converging metropolitan areas of Baltimore and Washington, and with a six-lane highway running along one side and subdivisions wrapped around the other three sides, is a classic smart-growth site. But neighbors—including the former chairman of the Howard County chapter of the Sierra Club!—objected, contending that they already faced crowded roads and schools and needed to preserve the remaining open space in the area.[43] After thirty-two public hearings, construction began in Maple Lawn Farms in 2004. Instead of the initial proposed density of three homes per acre—fairly typical for U.S. suburbs but on the low side for a smart-growth project—the final approved density was 2.2 homes per acre. By the time Glendening left office in 2003, his office had concluded that the rate at which farm and forest land was being developed in Maryland had not slowed appreciably.

Civic Entrepreneurialism and the Politics of Image

After a decade of economic and metropolitan restructuring, by the early 1980s the national economy had begun to recover from the stagflation crisis. The economic restructuring since the early 1980s has been associated with a much more flexible economic geography. Civic leaders and local business interests were quick to appreciate this flexibility. They were also astute enough to appreciate that the globalization of the economy made it harder for the federal government to stage-manage patterns of investment. *It was therefore left to local governments to negotiate with large corporate investors and to attempt to stimulate and attract private enterprise by creating the right conditions for profitable investment.*

City governments, in short, had to become entrepreneurial. The cornerstone of this new ethos of governance has been the public-private partnership (not to be confused with the privatization of projects and services).[44] Public-private partnerships can take various forms, but they are often managed by quasi-public development organizations known variously as Development Authorities, Economic Development Corporations, and Local Development Corporations:

> These institutions legitimate and organize an overt alliance between the local state and particular capital factions under the rubric "partnership." They usually implement their decisions without referenda or legislative approval of specific projects. They have a great deal of discretion over the use of public funds and the granting of tax concessions.[45]

In seeking to create the preconditions for economic growth, public-private partnerships have subsidized private development through a variety of mechanisms: tax abatements, tax-exempt industrial revenue bonds, lease financing, sales tax exemptions, and tax "holidays" (whereby businesses are exempted from local taxes for a number of years as a reward for local investment). Partnerships may also involve using public capital as risk capital in true joint ventures, exercising public powers of **eminent domain** in assembling land, constructing public infrastructure customized to the needs of private development, rewriting city ordinances to accommodate private development, and accessing federal funds for urban development.

Ironically, the privatization of urban governance and politics was fostered in this final respect by the last fling of the Democratic progrowth coalition between liberal reformers and blue-collar interests. Under the Carter administration in 1978 the federal government introduced the Urban Development Action Grant (UDAG) program, mainly in the hope of easing fiscal stress by stimulating central city economic development. The UDAG program made federal funds available to distressed cities, allowing wide discretion over their use so long as they leveraged private investment that would create new jobs and taxes. With UDAG funds, cities could enter into partnerships with private developers and leverage millions of investment dollars. In spite of the neoliberalism that characterized the subsequent Reagan and first Bush administrations, the UDAG program was maintained with significant levels of funding into the late 1980s. The Reagan and two subsequent Bush administrations also did what they could to facilitate the privatization of urban governance by reforming federal tax laws and relaxing various regulatory functions. Civil rights, labor laws, occupational health and safety, and environmental protection were all enforced less aggressively.

Strategies for Urban Economic Development We can recognize four basic approaches to entrepreneurial governance in a global economy, each of which involves heavy reliance on public-private partnership and a high level of local boosterism and image-making.[46]

1. *Attempts to maintain the attractiveness of cities as settings for modern production and manufacturing.* This approach typically involves an extension of traditional investments in the kind of physical and social infrastructure that is important to modern industry: everything from roads and bridges to high-tech industrial parks and schools with specialized curricula. It also involves the provision of substantial packages of customized inducements such as tax breaks, subsidies, and purpose-built infrastructure. Such packages often pit one city against another in bidding wars for internationally "mobile" investment in major new developments by **transnational corporations**.
2. *Attempts to capitalize on federal government expenditures.* The federal government is a major employer and contractor. There are some Sunbelt cities whose economic base is dominated by federal outlays, particularly for the military and aerospace. With the "militarization" of the economy during the 1980s and the "war on terror" after September 11, 2001, the flow of federal outlays can represent a particularly important catalyst for urban development in conjunction with the localized multiplier effects from government contractors and subcontractors. Although the pattern of such flows is largely a function of existing geographies of defense and aerospace industries and the geography of congressional pork barreling, local boosters look to public-private partnerships to enhance the infrastructure in appropriate ways—with research parks and research universities, for example.
3. *Attempts to capture or retain the key command and control functions in corporate management, government, and financial and business services.* This approach typically involves the provision of some rather expensive urban infrastructure: airports, international communications networks, convention centers, hotels, and so on. It also requires cities to enter into public-private partnerships to ensure an adequate supply of office space of appropriate quality, together with surrounding upscale amenities.

   City governments themselves have been chiefly responsible, however, for the proliferation of one of the key symbols of status as a **command and control center**. Convention and exhibition centers bring business and professional visitors who not only fill hotel rooms and generate trade (and taxes) in shops and restaurants but also are exposed to the city and its opportunities for business. Public-private partnerships all over the country—from Gary, Indiana, to New York City—compete with one another with their large and sumptuous convention centers.
4. *Attempts to improve the attractiveness of cities as places of consumption.* This aspect of entrepreneurialism is also important in reinforcing the other three. In U.S. society consumption and materialism add up to "quality of life," something that not only generates jobs, incomes, and tax revenues by itself but also enhances the prospects of being able to secure investments in any activity that involves well-paid workers—whether in production activities, government research and contracting, or management and business services.

   It is this aspect of entrepreneurialism, too, that has been most obvious in the changing landscapes of cities. Public-private partnerships have fostered the entire range of settings that are now seen as essential for the proper transaction of the intensified materialism of U.S. society: cultural "anchors" such as museums and renovated turn-of-the century theaters and 1920s movie palaces, for instance. Examples include the restoration of the defunct Woodward Avenue entertainment district in Detroit, which contains the restored 1928 Fox Theatre; Comerica Park, the Detroit Tigers baseball stadium; and Ford Field, the Detroit Lions football stadium. Another example is the creation of a cultural district featuring a performing arts center that houses the Pittsburgh Symphony Orchestra in the renovated shell of a 1927 movie palace in Pittsburgh's Golden Triangle.[47]

   Another seemingly essential component for cities seeking to consolidate their status as top-level **central places** is the major-league sports franchise with top-of-the-line spectator facilities. Just as stadiums for professional sports had sometimes been key elements of renewal schemes, so entrepreneurial city governments saw them as magnets for tourists, regional visitors, and

FIGURE 13 Fans enjoying a baseball game at Oriole Park at Camden Yards in Baltimore: a ballpark built and operated with the help of city funds.

AP Photo/John Mummert

potential investors with a choosy workforce or clientele to please. Thus (to take just a few examples): Minneapolis, faced with the possible loss of the (baseball) Twins and (football) Vikings in the late 1970s, worked with the team owners and floated a special bond to fund a stadium in the downtown area. Baltimore, having lost the (football) Colts to Indianapolis, agreed to subsidize a new ballpark for the (baseball) Orioles (Figure 13) and relinquish the city's stake in incomes from ballpark concessions.

Large mall complexes, gallerias, and festival marketplaces are a third important component. The archetypes of these developments were Boston's Quincy Market, opened in 1976, and Philadelphia's Gallery at Market East, opened in 1977. No city of any size could do without some similar form of development. Well-known examples include Baltimore's Harbor Place (Figure 14), Riverwalk in New Orleans, Pioneer Square in Seattle, and South Street Seaport in New York. Usually the product of

FIGURE 14 Harbor Place, Baltimore. Part of the "carnival mask" of urbanization put in place through civic entrepreneurialism and public-private partnership.

Ian Dagnall/Alamy

some form of public-private partnership, these set-piece developments create focal settings for integrated packages of upscale offices, tourist shops, "impulse" retailing, restaurants, concert halls, and art galleries. They are important assets for entrepreneurial cities because of their scale and their consequent ability to stage—or merely to *be*—the spectacular. They are described by David Harvey as the "carnival mask" of contemporary urbanization, their spectacular spaces being a means to attract capital and people (of the right sort) while diverting attention from the continuing problems of urban decay and social deprivation in nearby (but unseen) neighborhoods (this work fits within the poststructuralist approaches).[48] These developments are settings for "events" (such as concerts, ethnic festivals, and outdoor exhibits) and carefully planned "animation" (based on farmers' markets, street entertainment, and the like), all subsidized by the publicity budget of the city or the developer, or both.

With neoliberalism established as an ideological "commonsense," it was a short step to what geographer Neil Smith has called revanchism, and what sociologist Sharon Zukin has memorably described as "pacification by 'cappuccino:'" reclaiming urban spaces from low-income communities and low-profit settings through urban policy and planning in the cause of a "good business climate." Cindi Katz, in her discussion of the public-private Grand Central Partnership in New York City, provides an example of some of the ways that such an agenda is pursued, and shows how particular groups of people and their activities are removed from view in the interests of ensuring an "orderly," "clean," and "safe" public space. The project, she notes (1998: 42):

> resulted in the removal of all kinds of people from Grand Central and its environs, suggesting that they have, at best, unequal rights to the city. Many of these people, among them the shoeshiners and retailers, earned a modest living in the station and caused no harm to others, but their presence did not seem to fit the new image for Grand Central which includes Michael Jordan's expensive steak house taking up a quarter of the mezzanine, a Godiva chocolate shop, and the redundant but inevitable Starbucks coffee stall. Their lot, like so many aspiring middle and working class people in contemporary New York seems of no moment to the architects of the neoliberal city—witness the enduring assaults on the City University, the Giuliani administration's 1998 harassment of taxi drivers and restrictions on street vendors, and the searingly high rents for even the smallest commercial spaces. Yet, the texture of the city—its very driving force and unique quality—will be lost if such groups of people have no place in it. If Grand Central becomes as much of an ordered contrivance as a Disney production and its commercial attractions no different from any upscale mall, the Partnership's "operation" may be considered a success but the patient will be dead.[49]

All four approaches to entrepreneurial governance have fostered a *dealmaking ethos* that is reflected in some radical changes in the nature of urban planning and policy outcomes. These include the emergence of more flexible approaches to land use zoning and the encouragement of historic preservation and **gentrification**. It is important to note at this point that the roots of land use zoning, historic preservation, and gentrification are to be found in the political economy of entrepreneurial cities.

The Politics of Packaging The increasing entrepreneurialism of urban governance has made rebuilding, repackaging, and rebranding the urban landscape a common priority among large cities. Flagship cultural sites, conference centers, big mixed-use developments, warehouse conversions, waterfront redevelopments, heritage sites, and major sports and entertainment complexes have appeared in many cities. Geared toward consumption rather than production, these developments are designed to provide a new economic infrastructure suited to the needs of a post-industrial economy: business services, entertainment and leisure facilities, and tourist attractions. They are, invariably, closely woven into the narratives of city branding. They are also, invariably, the product of "growth-machine" coalitions of local real estate, finance, and construction interests that seek to propagate an ideology of growth and consumption as well as engaging in tactical politics around local government land-use regulation, policy, and decision-making and in pursuing public-private partnerships.

An early example in the context of the neoliberal era comes from New York City, where a coalition of élites first came together to forge a new image for the city in response to the deep fiscal crisis of the 1970s. In addition to the fiscal fragility of the city government, New York was characterized at that time by civil unrest, blackouts, strikes, neighborhood abandonment, graffiti-covered subways, and soaring crime. Building on initiatives by entrepreneurs in media, real estate, and tourism—notably the "Big Apple" branding exercise by the Association for a Better New York and the efforts of the newly founded *New York* magazine to portray the city as a cool place for young urbanites to live, work, and shop—Mayor John V. Lindsay organized one of the first public-private partnerships, aimed simply at sprucing up the city in time for the Bicentennial celebrations of 1976: cleaning taxicabs, sweeping the streets and handing out golden apple lapel pins at airports and train stations. In 1977 the New York State Department of Commerce started the **I ♥ NEW YORK** campaign, made world famous by the simple graphic designed by Milton Glaser. Miriam Greenberg, in *Branding New York*, argues that this was the beginning not only of an image makeover, but also of a systematic strategy of restructuring political and economic relations in the city toward a more business- and tourist-friendly environment.

The urban politics associated with entrepreneurial governance and the politics of packaging has been characterized by the emergence of a distinctive kind of political leadership characterized as "Entrepreneurial Mayors." These include in the 1990s, Mayor Rudy Giuliani who pursued an aggressive law enforcement and deterrent strategy, with highly publicized crackdowns on relatively minor offences such as graffiti and turnstile jumping on the theory that this would send a message that order would be maintained, thereby enhancing the quality of life for the city's middle classes. His successor, Michael Bloomberg, has continued the rebuilding, repackaging, and rebranding of the urban landscape, supporting the construction of new sports stadiums, pushing to attract and support huge events like the Republican National Convention and New York Fashion Week, establishing the office of a chief marketing officer for the city, and marketing the city under a new slogan, "The World's Second Home."

City life was in vogue again; industrial decay could be repackaged as romantic; downtown could be reinscribed with its former glamor; unexceptional neighborhoods could be "reclaimed" for the city's tax base by yuppies sold on the idea of a "frontier" urban experience.[50] Projects like Detroit's Renaissance Center (Figure 15) were successfully portrayed as part of the regeneration of the economic foundations and social fabric of once-dead central cities. The power of their imagery, reinforced by themes in advertising for everyday products from jeans to cars, must be credited with a substantial (but immeasurable) contribution to the recentralization of higher-order retail and service activity in central cities.

Linda McCarthy

FIGURE 15 The Renaissance Center, Detroit.

PERSPECTIVES ON GOVERNANCE, POLITICS, AND URBAN CHANGE

Having reviewed the evolving relationships between urban change, urban governance, and urban politics, we are in a position to summarize the most important conceptual and theoretical perspectives. We draw here, of course, on some of the substance of political science, though we will touch on only a small portion of that discipline's conceptual and theoretical content. The urban experience that we have reviewed raises three broad questions. One concerns *power*: How is power structured within the urban arena and how does its structure change with the changing dynamics of urbanization? The second concerns *governance*: How can we interpret the role of local government—the "local state" in theoretical parlance—in relation to urban development? The third concerns the *issues* that drive local politics: How can we interpret local conflict in relation to patterns and processes of urban change?

The Structure of Local Power

We should recognize at the outset that in practice different kinds of power structures exist in different cities and that these variations have a great deal to do with variations in cities' economic base and demographic composition. The focus here, however, is on generalized patterns and structures of power. At this level of abstraction, one of the classic models is the **elitist model**.[51] In elitist cities a handful of people who are at the top of a stable pyramid of power make nearly all the important decisions. These individuals, drawn mainly from business and industrial circles, constitute a strongly entrenched and select group: Without their support or consent, little of significance is ever accomplished. Elected officials are not so much the puppets of this leadership as the "understructure" of power, dependent on the patronage and tolerance of powerful individuals, many of whom keep a low public profile. The elitist model assumes that the legitimacy of this leadership is ceded from below rather than imposed from above. That is, the passivity of the bulk of the people in a city is a reflection of their voluntary acceptance of a relationship in which they indulge the leadership's domination in return for that leadership's pursuit of the public interest. The ballot box provides a register for any serious abuse of power.

A critical question raised by this perspective is whether such a consensus over the public interest is consciously and freely agreed to by the bulk of the people or whether it is the product of the systematic suffocation of opposition. The purposeful suppression of opposition is the basis of the **neo-elitist model**, which suggests that there are three main ways in which potential opposition to the preferences of powerful elites are defused:[52]

1. The failure of disgruntled groups of people to press their demands, either because they anticipate a retaliatory response from the powerful elite or because they believe it would be futile.
2. The refusal of those people in powerful positions to respond meaningfully to the political demands of less powerful groups. Response can effectively be denied, for example, by establishing committees or inquiries that take a long time to reach a conclusion.
3. The "mobilization of bias." This involves the manipulation by the elite of the values, beliefs, and opinions of the general public. So, for example, demands for change by some people may be denied legitimacy by being branded a threat to the freedom of the individual or to economic development. In this context, influence over the mass media is particularly important. Given such influence, the elite group is able to restrict public consideration to those matters considered "safe" and, as a result, the demands or grievances of some sections of the community may remain muted.

An entirely different model of local power structure is the **pluralist model**, which posits that power is dispersed, with different kinds of interests dominant at different times over different issues.[53] In this model the business elites that have been found to be in control in some cities are only one among several "power clusters" comprising diverse, autonomous, and nonhierarchical groups. The structure of power is essentially competitive, drawing on a wide range of participants and ensuring a fundamental element of democracy through the need for elites to acquire mass loyalty. This competition keeps society in rough equilibrium, with workers and unions acting as a counterbalance to business people, consumers offsetting the power of retailers, tenants constraining the power of landlords, and so on. These groups of people, additionally, have overlapping memberships, which promotes intergroup contact and makes for tolerance and moderation in local politics.

This pluralist model can be used to describe a *generalized political life cycle* of U.S. cities. In this life cycle there are four successive "dynasties." The first, roughly coincident with the era of the mercantile city, was dominated by an oligarchy of "Patricians." Then, during the era of the early industrial city, the traditional elite of landowners and commercial leaders was pushed aside by the entrepreneurial leaders of immigrant communities—the bosses and their machines. The third "dynasty" flourished in the Reform era at the beginning of the twentieth century, when new middle-class and business interests captured the power base. The fourth "dynasty" saw the rise of the professional politician, the "Ex-Plebe," as a coalition builder whose job it is to take advantage of the natural pluralism of U.S. urban politics.

The **corporatist model** of local power structures elevates government itself to the role of a key player in the power game.[54] In this model the basic framework of power rests on a symbiotic relationship between private organizations (labor unions, community groups, business leadership clubs) and various arms of local government. Key organizations become incorporated into the formal decision-making process (via representation on committees, appointments to boards, licensing, subsidies, and franchises), and city governments delegate a certain amount of authority in return for cooperation and support (moderating members' demands, contributing to public campaigns). The corporatist model is one that sees society in terms of segmented socioeconomic organizations that are taken under the wing of professional politicians and technocrats who are then able to expand and consolidate the scope of their power and authority.

A final model derives from the socioeconomic fragmentation of contemporary metropolitan areas in the United States. It follows from the sociocultural fragmentation and metropolitan spatial restructuring since the 1980s, which sometimes made the pluralism that had formed the basis of progrowth coalitions vulnerable to unstructured, multilateral conflicts in which many different groups fought continuously with one another over a broad range of issues. The rupturing of working-class interests along cleavages of race and ethnicity; the division of middle-class interests between growth, no-growth, and slow-growth perspectives; and the split between downtown business interests and suburban developers, between big business and small businesses, and so on, have resulted in a **hyperpluralistic** situation in which unstable power relations are reflected in unstructured and multilateral conflict. Power is exercised over narrow areas and for limited amounts of time by a variety of special interest groups who go their own way in seeking narrow gains and who are less restrained in their conduct and less likely to accommodate one another than under the pluralist model. Coalitions are essential to the hyperpluralist model, but they are short-lived and ad hoc.

These models should be seen for what they are: crude portrayals of real situations in which power relations are complex and fluid. They are constructs, or "ideal types," that help us conceptualize certain aspects of urban dynamics. In this context, another useful concept is that of **urban regimes**—the idea of power structures that rest on slowly changing coalitions of dominant groups and interests that are represented by city officials (both elected and appointed) who sustain both their own power and that of the coalition by ensuring a variety of particular benefits or policy outcomes for the key groups involved.[55]Certainly urban regimes can involve a good deal of change as mayors and business executives come and go, and economic conditions change. Urban governance in cities may be in transition between different types of regimes. But the tendency has been toward a pluralist or hyperpluralist model due to the gradual reduction in the political base of urban government (decline of political parties, competition from suburban communities) and in business concentration (diversified economies, "mobile" capital).[56] This trend has placed a much greater responsibility upon city mayors for coalition building within urban regimes—thus helping to explain the emergence of "Entrepreneurial Mayors".

The Role of the Local State

Each of the models of power structures that we have described has an implicit interpretation of the role of the local state in relation to urban development.

From an elitist perspective the state is seen as an institution that can be manipulated, influenced, diverted, or somehow orchestrated to conform to the central interests of those in the top levels of economic and social life. For those at the lower levels, therefore, the state can be an oppressor, the instrument of business interests.

From a pluralist or hyperpluralist perspective, the local state is seen as a passive institution that registers and responds to diverse demands. It cannot be captured or manipulated for long; it is a broker and a mediator rather than an oppressor.

From a corporatist perspective, the state is seen as a much more important institution: the only means of sustaining linkages between the divisions among the various levels within society. It is a common means to a multiplicity of ends. In this view, then, the local state can be characterized as an orchestrator and director.

Another school of thought emphasizes the autonomy of the technocrats and bureaucrats in the local civil service. This view originated with sociologist Max Weber's observations about the "dictatorship of the official" that accompanied nineteenth-century modernization. In this view urbanization had become so complex that elected officials had to rely increasingly on the expertise of civil servants. Taken a step further, the local state is seen to be controlled, on a day-to-day basis, by civil servants whose professional ideology and departmental allegiances are crucial in determining a good deal of the shape of local government activity. This interpretation can be characterized as a **managerialist model** of urban governance, in which civil servants effectively become social gatekeepers by mediating policy implementation.

These interpretations are all somewhat narrow, however, in that they do not take into account the relationships between the local state and the national state, nor do they address the broader issues of the role of the local state in relation to the overall trajectory of economic and social change. For such perspectives we have to turn to a *structuralist* approach.[57] From this perspective the local state is seen as a relatively autonomous adjunct of the national state, with both acting in response to the prevailing balance of class interests in society. In broad terms, the local state safeguards the interests of big business while "buying off" the working classes through reformist strategies. In practice, the activities of the local state are interpreted as serving one of three key roles in relation to the economy and society:

1. Enhancing and sustaining private production and capital accumulation through
   a. the provision of infrastructure.
   b. easing the spatial aspects of the reorganization of production—through the planning process, for example, or through urban renewal projects.
   c. investment in human capital—through public schools and technical training programs.
   d. "demand orchestration," which, through public works contracts, for instance, brings stability and security to markets that might otherwise be volatile and unpredictable.
2. Reproducing labor power through collective consumption by
   a. ameliorating the living conditions of the poor (e.g., subsidized public housing and transportation).
   b. enhancing social and cultural amenities (e.g., public libraries, parks, art galleries).
3. Maintaining order and social cohesion through
   a. police services.
   b. welfare programs and social services.
   c. "agencies of legitimation" such as local schools and citizen participation schemes.

Patterns of Local Conflict

However the role of the local state is interpreted, a certain amount of local conflict is inevitable—not just among the principal protagonists in the overall struggle for power but also among individuals, social groups, and communities as they react to the changes imposed by urbanization as mediated through the institutions of government. It is this kind of conflict that is in question here, the stuff of local politics—both formal and informal—from day to day. At this level most conflict centers on the public regulation of privately initiated change and in the manner and quality of provision of various services and amenities for residents. It will come as no surprise to find that those people with the greatest stake in a particular activity or "turf" are the ones who become most involved. Although no formal theories deal with the patterns of such conflict, some useful generalizations can be made using a model of locational conflict.

The model assumes a classic **concentric zone** pattern of land use and **social ecology**. Conflict is generated by change: the changing dynamics of the urban economy, changes associated with the aging of people and places, and changes in people's values and expectations. More specifically, three kinds of sociospatial change generate locally focused conflicts:

- Metropolitan decentralization (new suburban and exurban development).
- Physical restructuring (land use change and redevelopment in central city areas).
- Changing neighborhood composition (through residential mobility, aging, etc.).

Given this scenario, we can see that different settings are associated with different kinds of conflicts. **CBDs**, for example, are the location of conflict among people involving redevelopment, historic preservation, and transportation; established residential areas are striking for their relative absence of conflict, except over schooling; and suburban and exurban residential areas provide the settings for every kind of conflict except those involving historic preservation and redevelopment. We can also see how specific types of conflict vary in intensity across the metropolitan area. Conflict over transportation, for example, tends to be localized in two settings: in areas of suburban and exurban expansion and in the neighborhoods adjacent to the reorganizing city center. Conflict over "cultural" issues (e.g., the encroachment of one ethnic group on the territory of another), in contrast, tends to decline steadily toward the suburbs from a peak in the **zone in transition**.

FOLLOW UP

Key Terms

- annexation
- debt financing
- exclusionary zoning
- gerrymandering
- growth machines
- incorporation
- malapportionment
- metropolitan fragmentation
- municipal socialism
- NIMBYism
- privatization
- public-private partnerships
- reapportionment
- special districts
- urban regimes

Review Activities

1. Consider the way in which the idea of the "public interest" changed for people as cities themselves changed. How would you yourself define the public interest?
2. It is noted that "Central cities came to be encircled by a ring of independent and often hostile governments." Can you find evidence of this sort of situation today? If so, what sorts of conflicts are involved between the different jurisdictions?
3. Can you identify any examples of entrepreneurial governance in your city or town? Your local government may have an economic development department with a website that publicizes its efforts. Search the business section of local newspapers that are available online that contains news stories about entrepreneurial governance efforts. What kinds of projects (including high-profile ones such as a convention center or sports stadium) have been developed, and what can you find out about the initiators and supporters of these projects?
4. Add to your *portfolio*. The material covered in this chapter can be illustrated best with firsthand accounts from city newspapers. If you have online access to archived editions of newspapers, find examples of political cartoons that illustrate topics such as machine politics, annexation, or reform. In current newspapers the online business section contains news stories that cover city council meetings, routine announcements about zoning and highway repairs, and perpetual discussions about local economic development. What examples can you find, and how are they related to the context and concepts discussed in this chapter?

Log in to **www.mygeoscienceplace.com** for self-study quizzes, *MapMaster* layered thematic and place name interactive maps, *Urban View* Google Earth™ tours, key resources and suggested readings, related websites, "In the News" RSS feeds, and additional references and resources to enhance your study of the politics of change associated with urbanization and urban governance.

NOTES

1. M. Edelman, *The Symbolic Uses of Politics* (Urbana, Ill.: University of Illinois Press, 1985).
2. H. Lasswell, *Politics: Who Gets What, When, How* (New York: McGraw-Hill, 1936).
3. United Nations Human Settlements Programme, *The Global Campaign on Urban Governance: Principles of Good Urban Governance* (New York: UN-Habitat, 2002) (*http://www.unhabitat.org/campaigns/governance/Principles.asp*).
4. C. V. Bagli, "After 30 Years, Times Square Rebirth is Complete," *New York Times,* December 3, 2010.
5. *Yelp* online review of Jimmy's Corner by j.a. from Brooklyn, New York: *http://www.yelp.com/biz/jimmys-corner-new-york.*
6. J. Berger, "Hell's Kitchen, Swept Out and Remodeled," *New York Times* March 19, 2006.
7. Bagli, *New York Times.*
8. This evolution drew on a long history of patronage and corruption in American urban politics. The organizational influences of early trade union and socialist politics were also important. See, for example, M. Shefter, "The Emergence of the Political Machine," in *Theoretical Perspectives on Urban Politics*, eds. W. D. Hawley et al. (Englewood Cliffs, N.J.: Prentice Hall, 1976).
9. S. I. Toll, *Zoned America* (New York: Grossman, 1969).
10. Newton, K. 1978, "Conflict Avoidance and Conflict Suppression," in *Urbanization and Conflict in Market Societies*, ed. Kevin Cox (London: Methuen), 84.
11. J. H. Mollenkopf, *The Contested City* (Princeton, N.J.: Princeton University Press, 1983), 3.
12. Ibid., 55.
13. J. Jacobs, *The Death and Life of Great American Cities* (New York: Random House, 1961).
14. This discussion is based on information in the City of Milwaukee's Department of City Development website (*http://www.mkedcd.org/parkeast/*).
15. H. P. Chudacoff and J. E. Smith, *The Evolution of American Urban Society*, 5th ed. (Upper Saddle River, N.J.: Prentice Hall, 2000), 286.
16. J. O'Loughlin, "Malapportionment and Gerrymandering in the Ghetto," in *Urban Policymaking and Metropolitan Dynamics*, ed. J. S. Adams (Cambridge, Mass.: Ballinger, 1976), 539–65.
17. Ibid., 540.
18. J. C. Bollens and H. J. Schmandt, *The Metropolis*, 4th ed. (New York: Harper & Row, 1982).
19. R. D. Honey, "Metropolitan Governance," in *Urban Policymaking and Metropolitan Dynamics*, ed. J. S. Adams (Cambridge, Mass: Ballinger, 1976), 425–62.
20. N. Brenner and N. Theodore, "Cities and the Geography of 'Actually Existing Neoliberalism'," in *Spaces of Neoliberalism. Urban Restructuring in North America and Western Europe*, eds. N. Brenner and N. Theodore (Oxford: Blackwell, 2002), 21.
21. E. Banfield and J. Q. Wilson, *City Politics* (Cambridge, MA: Harvard University Press, 1963).
22. See American Planning Association, *Growing Smart: Legislative Guidebook* (Washington, D.C.: APA, 2002); and the State and Local Government Law website of Washington University Law School for American Planning Association, *Growing Smart Project, Model Tax Increment Financing Statute: Commentary: Tax Increment Financing* (*http://ls.wustl.edu/statelocal/tifstat.htm*).
23. J. Krohe Jr. "At the Tipping Point: Has Tax Increment Financing Become Too Much of a Good Thing?" *Planning: The Magazine of the American Planning Association* 2007 March: 20.
24. Ibid., 25.
25. R. Friedland, "Central City Fiscal Stress: The Public Costs of Private Growth," *International Journal of Urban and Regional Research* 5 (1981): 370–371.
26. Ibid., 371.
27. H. F. Ladd and J. Yinger, *America's Ailing Cities: Fiscal Health and the Design of Urban Policy* (Baltimore, Md.: Johns Hopkins University Press, 1991).
28. W. Alden, "Budget Woes Threaten Vital Services in New York City," *Huffington Post*, January 7, 2011; D. W. Chen and M. Barbaro, "In Fiscal Crisis, Mayor Considers Raising Property Tax 7 Percent," *New York Times*, September 22, 2008; H. Goldman, "New York City Budget Deficits May Be Larger Than Mayor Predicted, Liu Says," *Bloomberg News*, December 15, 2010.
29. F. F. Piven and R. A. Cloward, *Regulating the Poor: The Functions of Public Welfare*, updated ed. (New York: Vintage Books, 1993).
30. C. H. Levine et al., *The Politics of Retrenchment: How Local Governments Manage Fiscal Stress* (Beverly Hills, Calif.: Sage, 1981).
31. For a detailed examination of voluntarism, see J. Wolch, *The Shadow State: Government and the Voluntary Sector in Transition* (New York: The Foundation Center, 1990).
32. This section is based on P. L. Knox, "Public-Private Cooperation: A Review of the Experience in the U.S.," *Cities* 5 (1988): 340–46.
33. President's Commission on Privatization, *Privatization: Toward More Effective Government* (Washington, D.C.: U.S. Government Printing Office, 1988), 251.
34. R. W. Poole, Jr. and P. E. Fixler, Jr., "Privatization of Public-Sector Services in Practice: Experience and Potential,"*Journal of Policy Analysis and Management* 6 (1987): 612–25.
35. E. R. Gerber, C. K. Hall, and J. R. Hines, Jr., *Privatization: Issues in Local and State Service Provision* (Ann Arbor, Mich.: Center for Local, State, and Urban Policy, University of Michigan, 2004) (*http://www.closup.umich.edu/research/reports/pr-1-privatization.pdf*).
36. J. Hanrahan, *Passing the Bucks: The Contracting Out of Public Services* (Washington, D.C.: American Federation of State, Federal, and Municipal Employees, 1983).
37. E. McKenzie, *Privatopia. Homeowner Associations and the Rise of Residential Private Government* (New Haven, CT: Yale University Press, 1994) 177.
38. S. Low, *Behind the Gates. Life, Security, and the Pursuit of Happiness in Fortress America* (New York: Routledge, 2003).
39. P. L. Knox, *Metroburbia USA* (New Brunswick, NJ: Rutgers University Press, 2008), Chapter 6.
40. T. Daniels, *When City and Country Collide: Managing Growth in the Metropolitan Fringe* (Washington, DC: Island Press, 1999), xiv.

41. D. Hayden, *Building Suburbia. Green Fields and Urban Growth,* 1820–2000 (New York: Pantheon, 2003).
42. J. Duncan and N. Duncan, *Landscapes of Privilege. The Politics of the Aesthetic in an American Suburb* (New York: Routledge, 2004), 85.
43. P. Whoriskey, "Planners' Brains vs. Public's Brawn," *Washington Post* (August 10, 2004), A01.
44. D. W. Harvey, "From Managerialism to Entrepreneurialism: The Transformation of Urban Governance in Late Capitalism," *Geografiska Annaler* 71B (1989): 3–17.
45. P. F. Smith, *City, State and Market: The Political Economy of Urban Society* (New York: Blackwell, 1988), 210.
46. Harvey, "From Managerialism to Entrepreneurialism."
47. These examples, together with several of those that follow, are detailed in J. Teaford, *The Rough Road to Urban Renaissance* (Baltimore, Md.: Johns Hopkins University Press, 1990).
48. D. W. Harvey, *The Condition of Postmodernity* (Oxford, UK: Blackwell, 1989).
49. Katz, C. "Excavating the Hidden City of Social Reproduction: A Commentary," *City and Society* 10 (1998): 43.
50. See N. Smith, "New City, New Frontier: The Lower East Side as Wild, Wild, West"; and M. C. Boyer, "Cities for Sale: Merchandising History at South Street Seaport," in *Variations on a Theme Park: The New American City and the End of Public Space*, ed. M. Sorkin (New York: Hill and Wang, 1992), 61–93 and 181–204, respectively.
51. F. Hunter, *Community Power Structures: A Study of Decision Makers* (Chapel Hill, N.C.: University of North Carolina Press, 1953).
52. P. Bachrach and M. Baratz, *Power and Poverty* (New York: Oxford University Press, 1970).
53. R. Dahl, *Who Governs?* (New Haven, Conn.: Yale University Press, 1961).
54. See P. Schmitter, *Comparative Political Studies* (Beverly Hills, Calif.: Sage, 1977); and H. V. Savitch, *Post-Industrial Cities: Politics and Planning in New York, Paris and London* (Princeton, N.J.: Princeton University Press, 1988).
55. C. N. Stone, *Regime Politics: Governing Atlanta, 1946–1988* (Lawrence, Kans.: University Press of Kansas, 1989); A. E. G. Jonas and D. Wilson, eds., *The Urban Growth Machine: Critical Perspectives, Two Decades Later* (Albany, N.Y.: State University of New York Press, 1999).
56. H. V. Savitch and J. C. Thomas, "Conclusion: The End of Millennium Big City Politics," in *Big City Politics in Transition*, eds. H. V. Savich and J. C. Thomas (Newbury Park, Calif.: Sage, 1991), 248.
57. See P. Saunders, *Urban Politics: A Sociological Interpretation* (London: Hutchinson, 1979); and J. O'Connor, *The Fiscal Crisis of the State* (New York: St. Martin's Press, 1973).

Urban Policy and Planning

From Chapter 11 of *Urbanization: An Introduction to Urban Geography*. Third Edition. Paul L. Knox, Linda McCarthy.

Corbis Images

Urban Policy and Planning

The geography of contemporary cities is imprinted with layer after layer of the outcomes—both intentional and unintentional—of policies and plans. It took a long time for the idea of managing urban change and planning city development to become accepted by most people, but in the process there emerged a series of ideas and precedents that left their mark on cities. Each successive phase of economic development and urban change brought new challenges—along with new interpretations of old problems—that brought innovations in policymaking and planning. These innovations are significant because of their legacy for today's policymakers and planners. The lessons derived from past responses to urban challenges condition our whole approach to urban policy and planning. Even the institutional apparatus that we deploy in urban policymaking and planning is part of this legacy. Like all other aspects of urbanization, it must be understood in relation to continuous, interdependent processes of sociospatial change.

LEARNING OUTCOMES

After reading this chapter, you should be able to:

- Compare and contrast the views of the early (19th century) planning pioneers in Europe and the United States.
- Explain the key elements of the Progressive Era and how notions of charity, volunteerism, and paternalism blended with planning strategies to deal with living conditions in the slums.
- Understand the role of the New Deal and its impact on U.S. cities.
- Describe the role of the Supreme Court and the federal government in the evolution of post World War II planning strategies in the U.S.
- Describe some of the major European and U.S. city initiatives to promote greater urban sustainability.

CHAPTER PREVIEW

The objective of this chapter is to highlight urban policy and planning as both products of urban change and modifiers of urban dynamics. Having already reviewed the interactions between urban change, governance, and politics, we can now appreciate the main issues and turning points. The beginnings of urban policy and planning were rooted in a long, slow struggle with the central paradox of urbanization: the unwanted physical and social side effects for people of the very cities that were the crucibles of free-enterprise competitive capitalism. The end of the beginning phase was marked by the severe crisis of the Great Depression and the consequent moderation of free-enterprise principles through government intervention. A relatively brief "golden era" for urban policy and planning followed World War II. Subsequently, the rise of neoliberalism has challenged the idea that cities could be successfully planned and managed. The pressures of economic restructuring across fragmented metropolitan space helped put an end to the golden era of urban policy and planning.

It took a long time for the idea of managing urban change and planning city development to become accepted by most people, but in the process there emerged a series of ideas and precedents that left their mark on cities. Each successive phase of economic development and urban change brought new challenges—along with new interpretations of old problems—that brought innovations in policymaking and planning. These innovations are significant because of their legacy for today's policymakers and planners.

This is not to deny the very considerable importance of urban policymaking and planning in contemporary metropolitan regions. Rather, as we will see, they have, for the time being at least, become fragmented and divorced from any broad sense of the public interest.

THE ROOTS OF URBAN POLICY AND PLANNING

The roots of modern Western urban policy and planning can be traced to the Renaissance and Baroque periods in Europe (between the fourteenth and eighteenth centuries), when artists and intellectuals dreamed of ideal cities and rich and powerful rulers used urban design to produce extravagant symbols of wealth, power, and destiny. Inspired by the classical art forms of ancient Greece and Rome, Renaissance urban planners sought to recast cities in a deliberate attempt to show off the power and the glory of the state and the church. Spreading slowly from its origins in Italy in the fourteenth century, Renaissance urban planning had diffused to most of the larger cities of Europe by the seventeenth century.

Dramatic advances in military ordnance (cannon and artillery) during the Renaissance brought a surge of planned redevelopment that featured impressive fortifications, geometric-shaped strongholds, and an extensive *glacis militaire*—a sloping, clear zone of fire that offered no cover for attacking forces. Inside new walls, cities were recast according to a new aesthetic of grand design—geometrical plans, streetscapes, fancy palaces, and gardens that emphasized views of dramatic perspectives. These developments were often of such a scale that they effectively fixed the layout of cities well into the eighteenth and even into the nineteenth century, when walls or glacis, or both, by then obsolete, eventually made way for urban redevelopment in the form of parks, railway lines, or beltways.

As societies and economies became more complex with the transition to **competitive capitalism**, national rulers and city leaders looked to urban planning and policy to impose order, safety, and efficiency, as well as to symbolize the new seats of power and authority. One of the most important early precedents was the comprehensive program of urban redevelopment that Baron Georges Haussmann carried out in Paris between 1853 and 1870.

Haussmann's ideas were widely influential and extensively copied until the emergence of the Modern movement early in the twentieth century, with its idea that buildings and cities should be designed and run like machines and that urban design, planning, and policy should not simply reflect dominant social and cultural values but, rather, help to create a new moral and social order.

URBAN VIEW 1
Competition at all Costs: Civic Entrepreneurialism Taken Too Far?

You know something is different when a politician speaks out against offering public subsidies to attract companies to his city. But that is exactly what the mayor of Kansas City, Missouri, Mark Funkhouser, an Independent, who refers to himself as a capitalist but also a realist, has done.

> As Mayor, I've assembled a strong series of initiatives designed, at their heart, to build jobs instead of buildings, to build jobs instead of failing bars and restaurants, to build jobs instead of assembling taxpayer-funded giveaways to fatten the bottom lines of large corporations. It's been a painful process. It has cost me some friends, hasn't won me many new ones and made me some flat-out enemies. But it had to be done.[1]

The political and locational situation of Kansas City provides the context for the Mayor's position. His city straddles a state line: Kansas City is in Missouri, but suburban growth has occurred largely outside its city limits, on the other side of the state line, in Kansas. With the economic downturn of the late 2000s, the mayor points out that:

> What had been an informal economic competition between Kansas City and its suburbs became a dangerous game of hunter-prey when Kansas made economic poaching the official state sport with the passage of its Promoting Employment Across Kansas (PEAK) legislation. This legislation provided massive subsidies to companies to move a few miles across the state line from Missouri to Kansas.[2]

The mayor argues that public subsidies that encourage a company to merely relocate across the state line from Kansas City to a suburban community do not create any new jobs for area residents. The company relocation also comes at significant cost—subsidies like grants and tax breaks—that can adversely impact government spending to improve a locality's competitiveness, such as for schools, parks, and roads. And the Mayor argues that the

> unwritten policy of Government-By-Giveaway Economics hasn't hurt only the City of Kansas City. It has left Kansas City a weak metropolitan competitor on the national and global stage.[3]

Bob Regnier, chairman of the Greater Kansas City Chamber of Commerce in 2008, supports the mayor's view that competition for companies by offering tax breaks is out of control, leaving less public funding available for city infrastructure and services:

> Meanwhile, existing businesses (sometimes direct competitors) are paying full taxes and picking up the tab for the relocating entity. Kansas companies are now demanding parity and are threatening to move to Missouri . . . Ironically, the companies we're all trying to relocate and retain are attracted here by our quality lifestyle and the amenities of a world class city: an international airport, professional football and baseball teams, a great zoo, a vibrant arts scene.[4]

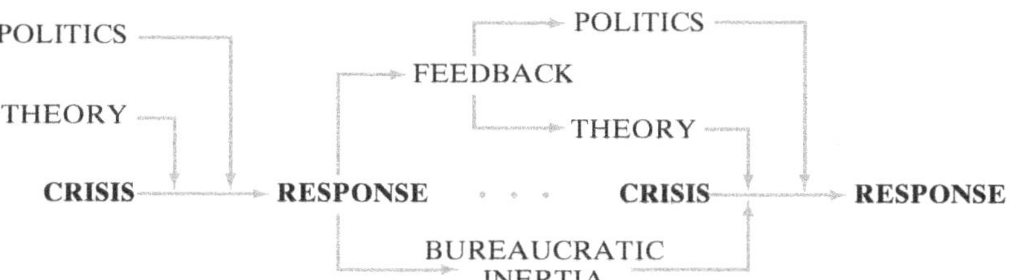

FIGURE 1 A crisis-response model of policymaking and planning.

The history of urban planning and policy since the inception of the Modern movement is best read as a story of successive crises and responses (Figure 1) rather than as the rise and fall of a movement or as a succession of ideas and experiments. Each perceived crisis—economic, social, political, and environmental—has elicited a response that has modified both the theory and practice of urban policy and planning. Successive responses have been shaped not only by the nature of the immediate crisis but also by the experience of previous reactions to earlier crises—the *feedback loop*.

Themes and Perspectives

Running through this story of crisis and response in urban policy and planning is a continuing struggle not only with the central paradox of urbanization (cities as economic necessities but potential social and environmental evils) but also with the tensions and contradictions inherent in a property-owning democracy. And in the United States the primacy accorded under law to private property ownership makes urban policy and planning much more difficult than in many other countries. Although the sanctity of private property rights rigorously protects specific rights of certain individuals (i.e., owners of real estate), cities are less orderly than they might be for many businesses, less convenient than they could be for most residents, and less healthy (in the widest possible sense) than they should be for everyone.

A second source of tension and contradiction for city planners and policymakers centers on the need, on the one hand, to exert some kind of public control over urban space (to make cities more efficient, convenient, and healthy) and, on the other, to avoid stifling capitalist enterprise with too much regulation.

The history of urban policy and planning begins with the increasing expression of *rationalism* in its various manifestations: concerns with efficiency, order, goals and objectives, cost effectiveness, and, above all, the idea of "progress." But as the story gains momentum and complexity, we will describe the interactions between rationalism and several other -isms: the **environmental determinism** of most early social science, the antiurbanism of highbrow culture, the elitism of pioneer urban designers, the managerialism of technocrats, the pragmatism and entrepreneurialism of local politicians, and so on. The result is a complex legacy of thought and practice that continues to shape and modify urban space.

Finally, it is important not to stereotype urban policy and planning as antibusiness or anticapital. True, their history shows that they developed initially in response to criticisms of the laissez-faire approach to economic and urban development. But as planning historian Richard Fogelsong has suggested, it should not be inferred from this that urban planning was anticapitalist in its effects. Although urban planning and policy emerged in response to forces that were endogenous to capitalism, the interventions of planners and policymakers served to stabilize the process of urban development and to mitigate the effects of market forces in ways that contributed to the overall maintenance of the capitalist system.[5] It was the middle classes and business interests that led governments to acquire the power to protect public health, safety, and welfare through planning regulations, public works, and policies.

THE BEGINNING: PHILANTHROPY AND REFORM

There is no need here to reiterate in detail the miserable conditions of nineteenth-century urbanization. In Europe the lives of city dwellers were famously described by Charles Dickens and catalogued by Charles Booth and Friedrich Engels.[6] Physical conditions for people in U.S. cities were equally miserable:

> Pigs still served as garbage scavengers. There were miles of crude paving, usually of cobble, granite or wood. Sewers were infrequent and flush toilets rare. Wooden construction, still prevailing in the centers of most cities, made tenements a firetrap. In consequence, waterworks were designed more for the needs of firefighting than from any concept of pure water or public health needs.[7]

It was abundantly clear that these conditions not only affected the people who lived in the crowded slums and industrial districts but also were a serious threat to everyone else. Fire, disease, and mob violence spilled all too frequently and easily out of the slums for anyone to feel secure. Employers were unhappy that their employees were sickly (and therefore less productive) and fearful of mob protest turning into mob rule (as revolutionary events in Bohemia, France, Germany, Hungary, and Switzerland in the 1840s showed all too clearly). Property owners were unhappy at being perpetually vulnerable to fires. The respectable middle classes were worried about the breakdown of moral order and the contaminating effects of crime and drunkenness. Everyone was unhappy at being constantly exposed to life-threatening diseases such as influenza, tuberculosis, whooping cough, and scarlet fever and to epidemics of smallpox, cholera, typhoid, and even bubonic plague.

Early European Traditions

These concerns were the preconditions for a sense of crisis that was mobilized in Europe early in the nineteenth century. In Britain the first real focus of concern came from the efforts of one person: Edwin Chadwick, first secretary of the Poor Law Board. His campaigning led to the Royal Commission on the Health of Towns. The publication of the commission's report in 1842 prompted progressive-minded liberals to form themselves into voluntary associations such as the Health of Towns Association, the Association for Promoting Cleanliness Amongst the Poor, the Society for Improving the Condition of the Labouring Classes, and the Metropolitan Association for Improving the Dwellings of the Industrious Classes. At first they limited themselves to discussion and passing resolutions, but soon they began to lead by example in constructing model housing: demonstration projects that sought to show how decent housing could be built for working-class households with affordable rents that still yielded a profit.

The problem was that the profit margins were far too modest to attract much serious attention from investors or developers, and the urban crisis intensified. By the 1860s the liberal response was dominated by wealthy philanthropists such as George Peabody and the Guinness family, who were willing and able to build worker housing that yielded profits that were, for the period, strikingly low: typically around 5 percent. On this basis the Peabody Trust alone built more than 20,000 dwellings in London between 1860 and 1890, some of which are still in use (Figure 2).

Such philanthropy did not affect the poorest of the poor (who could not afford the rents), nor did it make significant inroads into the ever-worsening conditions of industrial cities; but it did help to bring some attention in polite society to the issue of how to address the conundrum of industrial cities: sustaining environments in which investors could make a profit in both the labor market and the housing market without leaving large numbers of households so poor that their slum neighborhoods threatened the health and security of everyone else.

Paul L. Knox

FIGURE 2 Peabody Trust housing, near Covent Garden, London.

Ebenezer Howard and Garden Cities The broader Victorian cultural response to industrialization and its urban problems was a Romanticism that dominated literature, poetry, art, and architecture. Central to this Romanticism was an idealized view of the countryside and rural life. When this Romanticism combined with the strong sense of paternalism and noblesse oblige (responsibility that comes with privilege) that the Victorian bourgeoisie had inherited from the European aristocracy, it led to the idea of planned communities that might combine industrial production with the virtues of arcadian environments. Early in the nineteenth century socialist industrialist Robert Owen had built a model industrial community—New Lanark, in Scotland—with a strict paternalistic regime. By the mid-nineteenth century, paternalism, tempered by philanthropy, produced a flurry of model communities. Among these were the company towns of Saltaire, England, built by textile manufacturer Titus Salt, and Margarethenhöhe in Germany, built by steel magnate Alfred Krupp, and the larger and more ambitious company suburbs of Bourneville, England, built by chocolate maker George Cadbury, and Port Sunlight, England, built by W. H. Lever, whose products included Sunlight Soap.

In 1902, Ebenezer Howard published *Garden Cities of Tomorrow*, drawing on the experience and the idealism of these and other projects and adding a rationale that appealed strongly to the Romantic ideals of the time: Garden Cities could be planned and designed so as to combine the best of the countryside (nature, beauty, tranquility) with the best of the city (jobs, amenities, society) while avoiding the worst of both (Figure 3). Howard's goal was harmonious, self-governing communities that would grow into cities of manageable size.

These new self-contained towns would be located outside the commuting range of existing cities. They would be big enough to provide residents with jobs so that they did not have to commute to other urban centers, but not so big as to have the social and environmental problems of the large industrial cities of the time. Although not achieved, Howard's idea was that a number of Garden Cities would form a polycentric **urban system** that together would be the functional equivalent of an existing congested metropolis.

Howard's ideal plan limited each Garden City to 30,000 people on 6,000 acres (about 9 square miles). The built-up area was to be about 1,000 acres (1½ miles in diameter), at the center of which were public gardens surrounded by civic buildings—the town hall, courthouse, library, museums, and hospital—easily accessible by radial boulevards (Figure 4). This walking city would contain a concentric zone of housing and also a separate but accessible zone of factories and warehouses connected by a circumferential rail line. The built-up area was to be surrounded by a permanent green belt that restricted urban sprawl and offered recreational opportunities for residents while at the same time protecting agriculture.

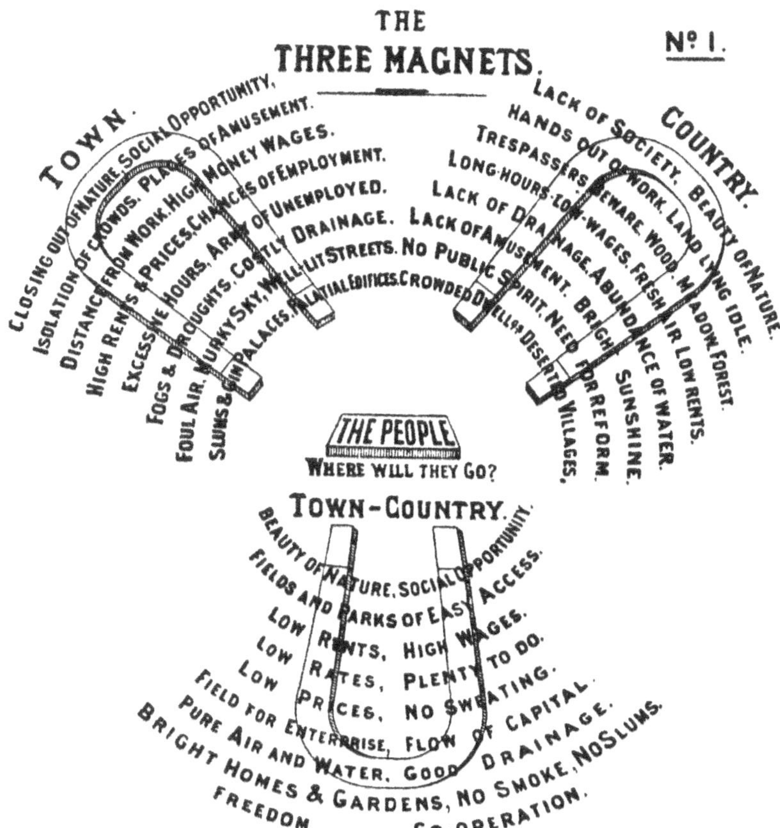

FIGURE 3 Ebenezer Howard's argument for Garden Cities.

Howard invested his own money in the concept, cofounding the Garden City Pioneer Company, which developed Letchworth (18 miles north of London), the first full Garden City (Figure 5). Using Howard's schematic plans, the company laid out roads, parks, and factory sites and invited private developers to build housing (within carefully regulated standards) on prepared sites. The scheme was supported by liberal reformers because it involved utopian ideals that invoked pastoralism and social order, by practical reformers because it involved land use control and centralized direction, and by conservatives because it gave private business more scope to develop real estate. Other Garden Cities were built, including Hellerau near Dresden, Germany; Floreal near Brussels, Belgium; and Tiepolo, in Italy. Although Howard's logic was based on stand-alone settlements, the appeal of his urban design principles also gave rise to *Garden Suburbs*. Examples include Wythenshaw, a suburb of Manchester, England; Hampstead Garden Suburb in northwest London; Romerstadt near Frankfurt, Germany; Hirzbrunnen in Basle, Switzerland; and Milanino outside Milan, Italy.

The popularity of the Garden City concept led to its inclusion as an important element in the formative stages of U.S. city planning. Sunnyside Gardens and Radburn, the planned communities sponsored by the Regional Planning Association of America, were among the first examples, and they were highly influential. It was a U.S. sociologist-planner, however, who developed Howard's principles into the concept of the **neighborhood unit**—a concept that was to become one of the most widely adopted city planning devices throughout the world. Clarence Perry worked for the New York–based Russell Sage Foundation and lived in one of the foundation's model railroad suburbs, Forest Hills Gardens, which had been built in 1911. Perry was strongly influenced by the ideas of Jane Addams and the settlement house movement, seeing the neighborhood as the key setting in which to foster moral order and community spirit. Living in Forest Hills Gardens underlined for Perry the importance of physical design in neighborhood life. The result was a blueprint for urban design (Figure 6), a neighborhood unit bounded by roads carrying heavier flows of automobile traffic and centered on a local elementary school, local shops, and community institutions.

Patrick Geddes and Scientific Planning In the early years of the twentieth century, when scientists was setting discovery after discovery before an increasingly appreciative world, the social sciences in general and the study of urbanization in particular were still in their infancy. In this context the voice of Patrick Geddes, professor of biology at the University of Dundee, Scotland, was persuasive. As a scientist, he commanded considerable respect, and he was able to speak out on social issues without being pigeon-holed as a liberal do-gooder. He was an active campaigner for housing reform and had good connections in all the relevant societies. He was fascinated by cities and appalled by what he saw. He likened them to grease

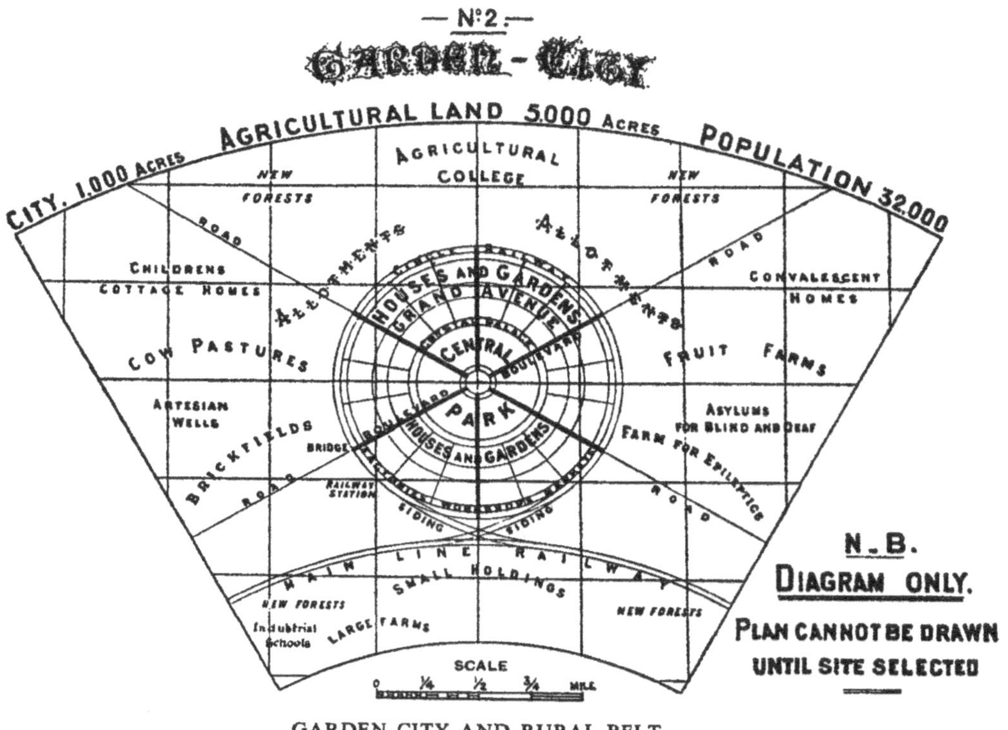

FIGURE 4 Ebenezer Howard's plan for a Garden City.

stains and to coral reefs, growing organically with little order or purpose. In this respect he was a product of his time, intuitively antiurban. More than anything, he believed that cities needed to be managed, just as a farmer might manage fields of crops or herds of animals. The bad had to be eradicated for the good to prosper. There had to be a plan for growth and development, which in turn implied an inventory of present resources.

It was this idea of survey to create an inventory that was Geddes's first major contribution to his amateur interest. He had his own model for undertaking such surveys in his Outlook Tower in Edinburgh, which contained a camera obscura and a collection of photographs of urban life. Influenced by the writings of French sociologist Frederic Le Play, Geddes believed that the information gathered from urban surveys

Linda McCarthy

FIGURE 5 Letchworth, the first Garden City, just north of London. The central public gardens, surrounded by civic buildings such as the town hall and public library, still attract residents today who like to picnic in the sun or play a little soccer.

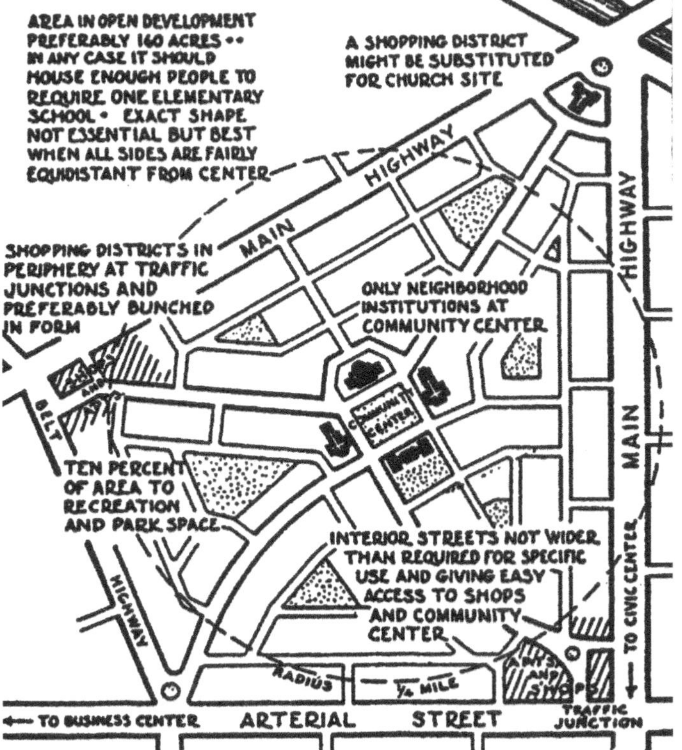

FIGURE 6 Clarence Perry's neighborhood unit concept.

should clarify the availability of resources, the nature of human responses to the physical environment, and the nature of the local cultural surroundings.

Following the writings of French geographer Vidal de la Blache, Geddes believed that this inventory should be undertaken in the context of a city's regional framework. This idea was his second major contribution to planning. The region, he argued in his 1915 book *Cities in Evolution*, had to be the basis for the reconstruction of economic, social, and political life. Cities and regions needed each other; they had to be planned and managed together, particularly, he noted, in view of the decentralizing forces of the new "neotechnic" technologies of electrical power and the internal combustion engine.

United States: Jacob Riis and the Tenement Commissions

In the United States the dangers of unplanned, uncontrolled urbanization were first highlighted in a book published by Jacob Riis in 1890. In his classic *How the Other Half Lives*, Riis graphically described the squalor and depravity of the lives of people in New York's tenement slums—*terra incognita* to many "respectable" families—while adding fuel to popular prejudices against cities in general and immigrants in particular.[8] Though by no means the first example of a sense of crisis over the quality of life of many people in U.S. cities, it provides a convenient starting point—earlier crises had failed to elicit any decisive response. Throughout the nineteenth century there had been bursts of hand-wringing and pamphleteering by liberal reformers and do-gooders, but with relatively little effect either on mass opinion or on government officials. By the time Riis published his book, informed opinion had reached a critical point. It was clear that urban housing markets, left to themselves, were not going to achieve any kind of natural balance; nor did city governments seem able to address the problem of the slums, even where city bosses and political machines owed their position to the support of immigrants and low-income voters.

In response to the attention focused by Riis on crowded tenement life, two successive *Tenement House Commissions* were set up, the first in 1894 and the second in 1900. Both confirmed what people already knew: The slums were haunts of vice, the root of urban degeneracy, and a threat to moral order. The first commission's recommendations—for the introduction of bylaws that would prevent overbuilding—were quickly derailed by speculative developers.

The response of the second commission was a defining moment for U.S. city policymaking and planning, even though it is not usually celebrated as such in the history of planning. In contrast to parallel commissions in Europe, the 1900 commission came down decisively against direct public intervention in the housing market, arguing that it would bring about a ponderous bureaucracy, intensified political patronage, and the discouragement of private capital, all merely to "better the living conditions of a favored few."[9] The best answer, the commission asserted, was tighter physical regulation of private developers through codified *building standards*.

But although building standards might prevent the embarrassment of "instant" slums, it still left the poor without access to decent, sanitary housing at an affordable rent. It was consequently left to volunteers, philanthropists, and visionaries to address the fundamental problems of poverty, ignorance, and neighborhood decline and to propose the nature of change. As a result, the imprint of the upper middle classes came to be firmly stamped on the reforms of the Progressive Era—reforms that were to be woven into the fabric of twentieth-century urban policymaking and planning. Initially, this imprint could be seen in three interrelated movements: settlement houses, urban parks, and the City Beautiful movement.

Progressive Era Reforms

Middle-class people's response to urban malaise was initially expressed through voluntarism and private philanthropy (Figure 7). By the end of the nineteenth century, New York City alone boasted over 1,300 organized forms of philanthropic aid.[10] Some of these charitable institutions were based on ethnic lines and so exhibited a certain degree of spatial and functional coherence. But in general charity was fragmented, uncoordinated, and directed almost entirely toward individuals and families. At the same time, it was increasingly clear to those people who actually worked among the poor that the slums were a collective problem, involving issues that extended to the sociogeographic contexts of neighborhood, school, and factory. This growing realization led to an influential movement—the settlement

Charles F. Bretsman/Library of Congress

FIGURE 7 Urban policy as charity—for a long time the only option, and then only for the "deserving" poor.

house movement—based on concepts of neighborhood and community as the focus for charity.

Settlement Houses The people who supported the settlement house movement believed that by living among the poor people they hoped to serve they could develop a fellowship, learn about these people's needs, and more effectively socialize the poor into middle-class standards of family life and educational achievement. The movement was modeled on a residence called Toynbee Hall, which has been set up in the slums of London, England, in 1884. One of the first settlement houses in the United States was the Neighborhood Guild (later renamed the University Settlement), established in a New York slum tenement by Stanton Coit in 1886; another well-known early example was Hull House (Figure 8), established by Jane Addams in Chicago.

The idea proved to be just what the eager volunteers—mainly women—were looking for. They organized continuing education for school dropouts, summer camps and neighborhood playgrounds to expose children to the virtues of nature, day nurseries for working women, educational programs to save "fallen" women, and clubs for the elderly. Volunteers also undertook social surveys in order to strengthen their arguments for reform, and they campaigned for the prohibition of alcohol. In 1891 there were 6 settlement houses in the United States; by the turn of the century there were more than 100; and by 1910 there were over 400.

It was no coincidence that the social theorists of the time emphasized the dangers of moral disorder and **social disorganization**. Pioneer sociologist Robert Park endorsed the idea of neighborhood centers that would help to mold people's character and instill discipline and control in community life. "The purpose of social settlements," he pointed out in an essay published in 1915, was nothing less than "to reconstruct city life."[11]

Unfortunately, the settlement house volunteers' sense of love, sacrifice, and service was too often vulnerable to the corruption of condescension, paternalism, and evangelical zealotry. Novelist Sinclair Lewis saw them as "upholding a standard of tight-smiling prissiness."[12] Other charity workers accused them of gush and sentimentality and became worried that inexperienced charity workers were themselves becoming a threat to the emergence of an appropriate moral order.

Corbis Images

FIGURE 8 Hull House, a settlement house in Chicago opened by Jane Addams (1860–1935), a leader of the settlement house movement and co-winner of the 1931 Nobel Peace Prize.

Library of Congress

FIGURE 9 A settlement house playground in the early 1900s.

In response, societies were set up in many cities to organize and coordinate charitable activities and to train relief workers to conform to strict sets of rules and record-keeping practices that were designed to make caregiving more effective. This training involved, among other things, learning to spot and weed out the "deserving" from the "undeserving" poor. This distinction, a legacy that has survived to the present day, permeated not only the ideology of professional planners and policymakers but also conventional social wisdom.

The Park Movement Another Progressive Era legacy, much more tangible than our attitudes to charity and welfare, is the network of parks and public open spaces enjoyed by people in cities. The park movement was rooted in the Arcadian Classicism inspired by pastoral ideals and the concept of nature as a spiritual wellspring. Early manifestations of these ideals were in the "rural" cemeteries of the early nineteenth century and in the mid-century creation of New York's Central Park under the direction of Frederick Law Olmsted, Sr.

The reform movement of the late nineteenth century drew on these same ideals, adopting the design motifs and intellectual underpinnings of Arcadian Classicism to cultivate feelings of honesty, beauty, wholesomeness, cleanliness, and natural order among the laboring classes. Parks and public playgrounds (Figure 9) were one early step in this endeavor. Jane Addams opened one of the first real children's playgrounds in 1893 at her Chicago settlement house, and the example turned out to be the beginning of a trend. For the next 20 years, U.S. cities were inscribed with all kinds of parks, playgrounds, and public open spaces, not just for children, but for everyone.

During this time, the rationale for park building quickly moved away from the relief they provided from the smells and diseases of the slums to the advantages they afforded in terms of recreation (for the masses), aesthetic pleasure (for the middle classes), and enhanced property values (for the upper-middle classes).

But for the reformers the decisive attribute of parks and public open spaces was their potential for spreading civilized values and social order within urban society (Figure 10). By bringing the laboring classes into contact not only with the spiritual energy of nature but also with the enlightened manners and comportment of other classes, parks could become a kind of universal moral force, a source of democratic and fraternal feelings: "It was believed that parks would breed a desire for beauty and order, spreading a benign, tranquilizing influence over their surroundings."[13]

Corbis Images

FIGURE 10 Public parks such as this one (Lincoln Park) in Los Angeles were a deliberate attempt to promote "civilized values" and social order.

Endorsed by leading intellectuals like Andrew Jackson Downing, Walt Whitman, and Herman Melville, the park movement rapidly gained momentum. By the turn of the century large parks had been established in Baltimore, Boston, Chicago, Cleveland, New York, Philadelphia, St. Louis, San Francisco, and Washington, D.C. New York had 11 parks of 100 acres or more. Boston had initiated the first *planned* metropolitan system of parks: an "emerald necklace" of parks and parkway links. Smaller parks proliferated in cities everywhere: ornamental parks, zoological parks, parks for strolling, boating, lunching, skating and team sports, waterfront parks, downtown vest-pocket parks (Figure 11), and neighborhood parks.

The overall effect was significant for several reasons. The success of the park movement consolidated the ethos of *paternalism* in urban policymaking and planning. It established environmental determinism as a strong element in city planning. It brought about real change in the shape and pattern of daily life. And, last but not least, it established the infant art of city planning as a semiautonomous activity: Park boards and commissions were typically given responsibility for supervising the design of parks, financing them, building and maintaining them, and enacting regulations governing their use. These boards and commissions were the forerunners of city planning commissions that were appointed early in the twentieth century and of the **special districts** that compounded metropolitan fragmentation.

The City Beautiful Movement The same logic that carried the park movement forward also sustained the City Beautiful movement that emerged at the end of the nineteenth century. The Beaux Arts style exploited by Daniel Burnham in the Chicago Exposition of 1893 was right in line with reformers' desire to extend the civilizing influence of the park movement for people to the broader frame of reference provided by the city's built environment. The neoclassical vocabulary preferred by Burnham and other City Beautiful designers brought a strong element of *conservatism* to the whole enterprise. The symbols and motifs of Beaux Arts architecture and monuments not only suggested a link with the great European classical cities of the past, such as Athens and Rome, but also helped to legitimize the U.S.'s Anglo-Saxon ruling classes and institutions at a time of massive immigration and profound socioeconomic change.

At the same time, the broad boulevards and radiating road networks that framed City Beautiful projects were welcomed by civic boosters as providing an orderly physical framework for economic development and by landowners whose property values escalated in anticipation of the implied redevelopment of large tracts of central city land.

The City Beautiful movement also spurred the development of *city planning commissions* and created an awareness of the need for technical experts to prepare plans. For the most part this task fell to architects, who brought with them the egotistical grandiosity of their profession. "Make no little plans," advised Daniel Burnham. His own plan for Chicago, published in 1909, was a magnificently illustrated volume that was truly metropolitan in scope, with proposals for a system of regional ring roads connected to downtown Chicago by a series of radial highways and **parkways**, for over 60,000 acres of monumental parkland, and for a series of parks, marinas, and developments along the lakeshore. It was a scheme for expansion, a framework for speculation, a design to embellish the city, to

Linda McCarthy

FIGURE 11 Bryant Park, New York, a downtown vest-pocket park of a mere nine acres.

flatter its leaders, and to impress its inhabitants. *It did nothing, however, to address the fundamental problems of slum housing and social malaise.*

THE CITY PRACTICAL

It was at this point, early in the twentieth century, that liberal intellectuals in the United States began to take a broader view of urban problems, influenced in part by the ideas of Patrick Geddes on scientific approaches to city planning. The first National Conference on City Planning was held in Washington, D.C., in 1909. Most of the people attending the conference were government officials or representatives of charitable agencies; a few were architects, landscape architects, or engineers. What they had in common was a conviction that the process of city building was occurring piecemeal through speculative real estate deals without the guidance or benefit of any kind of public policy to protect or enhance the long-term interests of the community.

There was strong agreement that some form of government policymaking concerning urban development was necessary—preferably the sort that was proactive rather than reactive. There was, however, no consensus on the shape of such policies. As Richard Fogelsong has observed, "These planners were searching for a directive system, a method of decision making—a system they called 'planning' without knowing precisely what they meant by the term."[14]

Having slowly inched away from laissez-faire urbanization, struggled with housing reform, rushed ahead with parks, and flirted with monumentalism and neoclassical aesthetics, the urban policy and planning community now found itself a clear objective: the city as an efficient and disciplined spatial framework for both economic and social life, the "City Practical." This perspective required a shift of focus, away from the unwanted consequences of urbanization (poverty, slums, and disorder) and toward the role of cities as generators of prosperity and enlightenment for the people living in them:

> A change in orientation began to occur that eventually would bring a reevaluation of the potency of environmental reform and the beginning of an idea that the American City might be disciplined by the progressive development of human knowledge, state regulatory mechanisms, and public welfare provisions.[15]

Achieving those goals required, first of all, working toward a more economical and efficient system of land use and transportation and a carefully coordinated system of infrastructure provision. Though costly, this aim found widespread support among the business community. It was no coincidence that the notion of the City Practical coincided with the introduction of **Fordism** and **Taylorism**, and local business leaders were quick to appreciate the indirect benefits that this kind of planning would have on profitability.

While city planning commissions busied themselves with implementing the City Practical, intellectuals with an interest in urban policy and planning began to address the implications of the regional economic reorganization and metropolitan decentralization that were beginning to reshape both the U.S. urban system and the form of individual cities. It was in this context that Lewis Mumford came to prominence, articulating very effectively the ideas of Geddes and Howard and helping to draw attention to the need to address urban policy and planning at a larger scale: the alternative being "more and more of worse and worse." Mumford was instrumental in the formation of the Regional Planning Association of America (RPAA) in 1923. Although the direct achievements of the RPAA in the planned communities of Sunnyside Gardens and Radburn were relatively modest, the indirect impact of the RPAA was profound. The ideas promulgated by its members in the 1920s were to be a strong influence on the New Deal that gave policymaking and planning a much more central role in the economic and social life of the country.

THE NEW DEAL

The New Deal, President Roosevelt's response to the **stagflation** crisis of the early 1930s, gave urban policy and planning its modern shape. The New Deal did *not* represent a coherent or integrated policy or planning framework. In the scramble for national economic recovery, programs were often ad hoc and sometimes contradictory. Nevertheless, the New Deal moved the country toward a social democracy that sought to manage the national economy, defuse urban social unrest, and create an environment for coordinating production and consumption.

The New Deal era was a time of unprecedented experimentation in policymaking and planning. From the tentative conceptual, technical, and practical achievements of a few reformers and pioneering professionals, policymaking and planning became a central feature of the political economy of urbanization.[16] Indeed, policymaking and planning became integral to the entire political economy. The National Planning Board, established in 1933 (and renamed the National Resources Committee in 1935), brought the rationality of scientific, survey-based planning to everything from public works improvement to education, unemployment, health, old-age insurance, regional development, and technological assessment.

Urban policy and planning were specifically shaped by several elements of the New Deal package:

- The Civil Works Administration (CWA) and Public Works Administration (PWA) made grants to local jurisdictions for public works projects such as building roads (Figure 12), bridges, airports, and public buildings and sewer and water projects that, it was hoped, would counter the Depression directly by creating jobs and indirectly by improving the infrastructure for economic development.
- The Works Progress Administration (WPA) directly employed workers through a network of local offices in order to achieve a more immediate effect on unemployment rates. Its efforts went mainly into constructing highways and streets, water and sewer systems, and recreational facilities.

Corbis Images

FIGURE 12 A road construction project in Manhattan funded by the Works Progress Administration (WPA).

- The Federal Emergency Relief Administration (FERA) made direct grants to states for unemployment relief and matching loans in order to stimulate public works projects, the majority of which were urban-related.
- The **Federal Housing Administration (FHA),** created under the National Housing Act of 1934, succeeded the Home Owners Loan Corporation as a means of providing federal mortgage insurance in order to prevent mortgage defaults and creating jobs by encouraging private residential construction.
- The Emergency Housing Division of the PWA, which became the U.S. Housing Authority through the U.S. Housing Act of 1937, undertook slum clearance and public housing construction in order to relieve the acute social distress of big-city slum neighborhoods.
- The **Resettlement Administration,** in its brief lifetime, was the most powerful and revolutionary planning authority ever created in the United States. Under the leadership of Rexford Guy Tugwell, RPAA ideas were vigorously pursued through a program of **greenbelt cities** that, it was hoped, would attract enough city dwellers to enable central city slums to be turned into parks. Political opposition limited the scale of the program and brought a premature end to the Resettlement Administration itself.

Although the New Deal established strong government intervention as accepted rather than exceptional practice, public intervention was really rather piecemeal. The imperatives of economic management and social relief increasingly blurred the vision of the City Practical. The New Deal resulted in fragmentation and differentiation rather than a legacy of comprehensive planning. Planning and policy administration became separated from implementation, spatial policies were increasingly separated from social policies, and short-term policies were pursued without reference to long-term goals. It was by default, therefore, that policies initially drafted in relation to economic management and employment relief were to contribute to a Golden Age of urban policymaking and planning after World War II.

POLICY AND PLANNING FOR RENEWAL AND GROWTH (1945–1973)

For roughly three decades after World War II, urban policymaking and planning were central elements of Western political economies. It was an era of technocratic policymaking and planning, whose practitioners enjoyed more public confidence than ever before or since in their ability to re-create cities as equitable and efficient settings for people, commerce, and industry. In Europe the emphasis was on postwar reconstruction and economic renewal. In the United States the emphasis was on economic growth. In both, urban policymaking and planning, urban governance, and urban development all presupposed one another in the rationalization of space required by Fordist systems of production.

Europe: Planning For Renewal

The context for professionalized planning in post–World War II Europe was heavily influenced by three sociopolitical movements with little in common except for a belief in the need for planning. One of these was dominated by paternalistic and idealistic liberal urban reformers who carried the heritage of Geddes and Howard. Another was dominated by conservative rural preservationists who were concerned about the encroachment of urbanization and industrialization on the limited amount of farmland and the associated heritage of picturesque landscapes. The third was dominated by the political representatives of industrial regions whose economic base had been severely damaged in the war and was now threatened by overseas competition.

One of the most effective advocates for professionalized planning was Patrick Abercrombie, who had affiliations with all three movements in the United Kingdom. Abercrombie's main concern was the unprecedented rate of urban sprawl. Because of the postwar economic recession, building materials and manual labor were cheap, and house prices were more affordable in relation to wages than at any time before or since. The consequent rate of suburban sprawl was unprecedented, with alarming losses of agricultural land, increased traffic congestion, and longer journeys to work for commuters. Highlighting these issues, Abercrombie advocated the idea of managing urban development. His strategy, embodied in his highly influential Plan for London (Figure 13), involved three main elements: severe restrictions on sprawl in a "Green Belt" surrounding the city, slum clearance and industrial regeneration in central areas, and the decentralization of industry and households to a number of New Towns located well beyond the Green Belt.

Abercrombie's plan involved decentralizing more than a million people to eight New Towns outside London. It was implemented by the radical Labor government that came to power in 1947. Well suited to the technocratic Modernism of Le Corbusier, Abercrombie's ideas quickly became widely influential. In Scotland new towns were built at East Kilbride and Cumbernauld to accommodate decentralization from Glasgow. In France growth was redirected to five new towns near Paris and to Lille-East outside Lille. In the Netherlands, overspill population was channeled to new towns such as Zoetermeer near The Hague and Bijlmermeer outside Amsterdam. New towns were also built to decentralize population from Copenhagen in Denmark and Stockholm in Sweden.

In Germany overspill population was directed to new towns such as Märkische Viertel near Berlin, Perlach outside Munich, and Nordurestadt near Frankfurt. The new towns of Wulfen and Marl were built to accommodate planned population growth in an area of the northern Ruhr that was designated for mining industry expansion. In contrast, some new towns that were built to stimulate growth in declining or remote regions, such as Cwmbran in South Wales, Glenrothes in Scotland, and Aycliffe and Peterlee in northeast England, were not successful despite government subsidies for new residents and businesses.

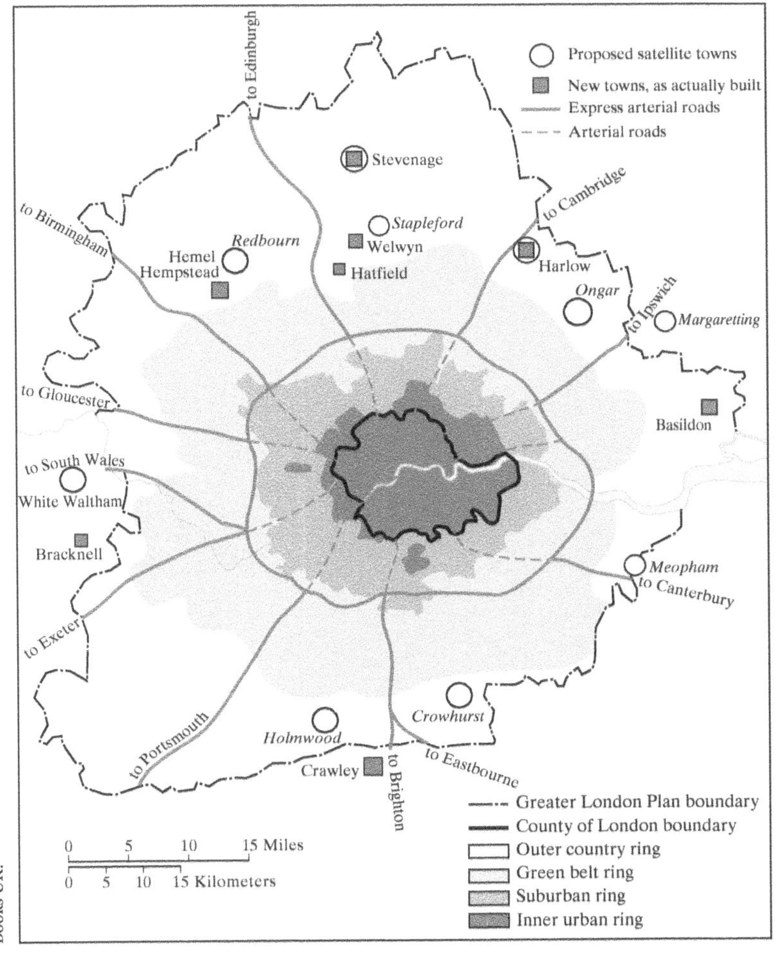

FIGURE 13 The Abercrombie Plan for London.

URBAN VIEW 2

The Visible Legacy of Urban Policy and Planning in European Cities[17]

Although no two cities are identical, the strong legacy of urban policy and planning is visible as a set of common characteristics that make European cities distinctive.

High Density and Compact Form: The constraints of the city walls kept the density of development high during the Middle Ages. A number of factors perpetuated this compact, densely built-up form that is now characteristic of large cities in Europe. A long tradition of planning that restricted low-density urban sprawl dates back to the application of strict city building regulations in the earliest suburbs. The compact urban form also reflects the relatively late introduction and widespread use of cars, as well as high gasoline prices.

Complex Street Pattern: The narrow streets and alleys of the medieval core in many cities evolved in the preautomobile era. Outside the walled town during the medieval period, suburban areas grew around the long-distance roads that radiated outward from the city gates. Influenced by Haussmann's ideas in the nineteenth century, cities like Munich, Marseille, and Madrid made radial or tangential boulevards the axes of their planned suburbs.

Town Squares: The town square, the heart of the Greek, Roman, and medieval towns, has often survived as an important open space. Some medieval town squares boast a continuous tradition of regular open-air markets. The large open square, typical of socialist cities in Central and Eastern Europe, was used for political rallies. Today, like their Western European counterparts, many central squares and their historic buildings have been adapted to economic and social change and contain modern commercial functions such as tourist offices and restaurants.

Major Landmarks: The historic landmarks in West European city centers have traditionally been symbols of religious, political, military, educational, and cultural identity. Many cathedrals, churches, and statues continue to serve their original purpose. Some town halls, royal palaces, and artisan guildhalls have been converted into libraries, art galleries, and museums. Medieval castles and city walls have become tourist attractions. Today, of course, the major landmarks are expressions of economic power—no longer city halls, but the offices of **transnational corporations**; not cathedrals, but sports stadiums. In Central and Eastern Europe the hallmarks of socialist cities were the massive buildings in "wedding cake" style, red stars, and "heroic" statues. Since the late 1980s, socialist political symbols have been replaced by billboards advertising the trappings of consumer culture.

Low-Rise Skylines: For North American visitors, the most striking aspect of the older parts of West European cities is the general absence of skyscraper offices and high-rise apartments (Figure 14). The urban core was developed long before reinforced steel construction and the elevator made high-rises feasible. Building codes designed to minimize the spread of fire maintained building heights between three and five stories during the industrial period. Paris fixed the building height at 65 feet in 1795, while other large cities introduced height restrictions in the nineteenth century. Still regulated today, high-rises are found only in redevelopment areas or on land at the periphery of the city, like La Défense in Paris. Skyscrapers have also been built in the central commercial and financial districts of some of the very largest cities, including London.

In socialist parts of Europe there was no private ownership of land and so no urban land market. With few transnational corporations doing business in these countries, socialist cities were usually devoid of tall commercial buildings that mark the **CBD** (central business district). Until recently, the tallest buildings were usually Communist Party and state administrative buildings, massive "Houses of the People," and TV towers.

Bustling Downtowns: The high density and compact nature of European cities create downtowns that bustle with activity. The vitality of the city center is reinforced by the widespread use of public transportation systems (buses, trains, subways) that converge on the core.

In larger cities distinct functions dominate particular districts. Institutional districts house government offices and universities. Financial and office districts contain banks and insurance companies. A pedestrianized retail zone leads to the train

Anthony Bayford

FIGURE 14 Amsterdam's low-rise skyline. A distinctive aspect of the older parts of West European cities is the general absence of skyscraper offices and high-rise apartments.

URBAN VIEW 2

The Visible Legacy of Urban Policy and Planning in European Cities (*continued*)

station. Cultural districts offer museums and art galleries. Entertainment areas include theater and red-light districts. Many downtown buildings have multiple uses. Apartments are found above shops, offices, and restaurants. Large department stores, such as Harrods in London and Kaufhaus des Westens in Berlin, are prominent features in most European downtowns. Modern downtown malls include Westfield London and Palace Flora in Prague. Suburban malls are becoming prevalent. Many coastal or riverine cities have also refurbished old port and industrial buildings to house mixed-use waterfront redevelopments like Kīpsala in Riga and HafenCity in Hamburg. Other cities have renovated obsolete historic structures, such as London's Covent Garden, as festival marketplaces with specialized shops, restaurants, and street performers.

The United States: Planning for Growth

After World War II there was a "closure" in the public policy dynamic in the United States that can best be understood as a product of a "class accord" between big business and organized labor, an accord that was regulated by the federal government.[18] It was in many ways the high point of Fordist capitalism. The foundation of the accord was the deal tacitly struck between employers and labor unions: higher pay and improved working conditions in return for management-driven improvements in productivity.

This accord seemed quite natural in the context of the sustained economic boom of the postwar period. The policy objectives stemming from the accord were: (1) to consolidate the economic growth on which the accord depended and (2) to redistribute the benefits of this growth at the margins, quelling (or, if possible, preempting) the social and spatial unevenness of the overall growth.

We have already encountered several of the principal outcomes of this approach. We noted the impact of the interstate highway system in facilitating the development of the urban system as the framework for an integrated national economy. We also saw how federal mortgage insurance carried over from the New Deal to underpin managed economic growth and development in the form of postwar suburban expansion. The **pro-growth coalition** generated by New Deal politics carried over into the planned renewal of the fabric and economic base of central cities, how rapid economic development and increasing affluence induced massive increases in spending on **collective consumption** (education and other public services), and how increasing expectations and the backlash from persistent exclusion and inequality resulted in even greater increases in spending on welfare and income support.

In this section we round out the picture of urban policy and planning between 1945 and 1973 by summarizing the main policy initiatives of the period. First, however, we must acknowledge the role of the courts in establishing some of the principles on which the new policy dynamic was based.

The Courts and Urban Policy in the United States

The U.S. Supreme Court was a major factor in translating the social and political currents of the postwar period into landmark decisions that had important effects in framing urban policy and planning. These landmark decisions were based largely on interpretations of the Fourteenth Amendment, which reads, in part:

> No State shall make or enforce any law which shall abridge the privileges or immunities of citizens of the United States; nor shall any State deprive any person of life, liberty, or property, without due process of law; nor deny to any person within its jurisdiction the equal protection of the laws.

This constitutional provision directly affects cities, and city governments as legal agents of their state governments. The pressures of postwar economic, social, and metropolitan change, and in particular the racial conflicts of the 1960s, brought a stream of cases to the Court. Three issues were of particular importance in shaping the urban policy dynamic of the period: school desegregation, open housing, and voting rights.

School Desegregation The landmark decision affecting school desegregation was the 1954 case of *Brown v. Board of Education* in Topeka, Kansas, in which the Supreme Court ruled that segregated schools were unconstitutional. This decision left a certain ambiguity, however, in that the ruling merely outlawed the use of government power or funds that resulted in segregation; it did not require integration. This ambiguity was clarified in the 1968 case of *Green v. County School Board of New Kent County*, in which the Court ruled that school districts must achieve as much racial mixing as possible and that districts that had not been integrated because of past practices must take affirmative action to achieve balanced racial enrollments. This ruling, in turn, required judicial decision makers to confront the realities of U.S. urban geography. The result was the adoption of school busing plans to secure balanced enrollments (*Swann v. Charlotte Mecklenberg Board of Education*, 1971). For the most part, however, the Court has shied away from decisions that would ensure metropolitan-wide desegregation.

Restrictive Covenants The fundamental housing policy case was *Shelley v. Kraemer*, in which the Supreme Court ruled in 1948 that racially **restrictive covenants** represented a violation of the Fourteenth Amendment. The Court broadened this decision in 1968 in the case of *Jones v. Mayer* to ensure the freedom for people of any race "to buy whatever a white can buy, live wherever a white can live."

The Court also found itself confronting issues raised by public housing and **exclusionary zoning**, though its decisions in this area have been much less rigid in their effect on the practice of local policy and planning. In *NAACP v. Mt. Laurel* in 1972, the Court struck down an exclusionary zoning ordinance, arguing that every municipality has an obligation to provide a reasonable share of needed areawide low-income housing. In the 1977 case of *Metropolitan Housing Development Corp. v. Arlington Heights* in Chicago, the Court decided that exclusionary zoning with discriminatory effects was not, *a priori*, illegal; only where discriminatory intent could be shown were such ordinances to be outlawed. This decision, of course, left huge scope for interpretation and effectively allowed a continuation of metropolitan social polarization.

Civil Rights We have already noted the role of the Supreme Court in initiating the "reapportionment revolution" of the 1960s. The landmark reapportionment case was *Baker v. Carr* in 1962. After the passage of civil rights legislation in the form of the 1965 Federal Voting Rights Act, the Court confirmed the principle of equality in the weight of each person's vote in balloting in the specific case of racial differences (*Allen v. State Board of Elections*, 1969). Another aspect of electoral geography was addressed in 1973, when in the case of *White v. Regester* the Court ruled that multimember districts (where more than one legislator represents the same geographic area) violated the Fourteenth Amendment by effectively diluting the minority vote. In 1986 the Court strengthened minority voting rights by interpreting vote dilution in a broader context than the mere geometry and arithmetic of electoral subdivisions, also taking into account the past record of local voting patterns and minority success or failure in gaining representation.

Federal Policy Initiatives

In the decades following World War II, there was a sharp increase in government expenditures on social objectives, as embodied in the United States, for example, in the New Frontier of the Kennedy administration and the War on Poverty and Great Society of the Johnson administration. In practice, most of these federal expenditures took the form of grants-in-aid to state and local governments. During the 1960s and the 1970s, party political differences over domestic policy and planning were largely confined to politicians disagreeing over the distribution of these grants and the conditions attached to them: Democrats sought to perpetuate the progrowth accord between big business and organized labor by channeling grants-in-aid toward central cities and blue-collar communities, while Republicans sought to channel them more toward rural and suburban areas and to Sunbelt cities. Among the many grant programs and policy initiatives that addressed urban issues were the following:

- The Housing Act of 1949, which was the baseline legislation for assistance to city governments for clearing blighted areas, assembling land for redevelopment, and building public housing.
- The Housing Act of 1959, which, among other things, provided federal support for preparing comprehensive plans at a metropolitan level.
- The Federal Aid Highway Act of 1962, which mandated urban transportation planning as a condition for receiving federal funds in urban areas. As a result, it provided a major stimulus to local land use and transportation planning.
- The Economic Opportunity Act of 1964, which created the Office of Economic Opportunity (OEO) as a vehicle for neighborhood-based economic development in central cities. Only 25 percent of OEO funding went to public agencies; the rest was channeled directly to neighborhood groups and private and nonprofit organizations (universities, churches, civil rights groups, settlement houses, family services agencies, United Way organizers) engaged in projects that were designed to assist the people who were casualties of economic and metropolitan restructuring.
- Project Head Start, initiated in 1965 in an attempt to break the central city **cycle of poverty** by funding educational programs designed to improve cognitive development and social, emotional, and physical well-being among younger low-income children.
- The Demonstration Cities and Metropolitan Development Act of 1966, an attempt to help concentrate and coordinate federal, state, and local public and private efforts to improve the quality of urban life and combat deprivation and disadvantage. As such, it was very much a product of the intellectual climate of the time, an attack on "place poverty" rather than "people poverty" through "area-based positive discrimination." In this respect, the Model Cities program was strongly influenced by the concepts of *spirals of neighborhood decay* (Figure 10.12) and localized *cycles of poverty*.
- The National Environmental Policy Act of 1969 and the Environmental Quality Improvement Act of 1970 required federal agencies to adopt a systematic, interdisciplinary approach to planning and policymaking affecting the environment. It also required the preparation of an Environmental Impact Statement (EIS) containing information on the likely environment impacts of proposed projects. The Clean Air Act of 1970 created the Environmental Protection Agency (EPA) and empowered it to set ambient air quality standards.
- The Urban Mass Transportation Assistance Act of 1970 provided the first long-term federal commitment to financing urban mass transit, and the 1970 Federal Aid Highway Act significantly increased the influence of local jurisdictions in highway planning. The Urban Mass Transportation Assistance Act of 1974 authorized for the first time the use of federal funds for transit operating costs.

There is no doubt that these federal policies significantly affected the trajectory of urbanization. Although they did not always have the effects that were planned, their very existence inevitably changed both the pattern of development and the social, political, and economic context in which subsequent change took place. They were, in other words, yet another manifestation of the **sociospatial dialectic**.

This is not to say, however, that any dimension of this phase of urbanization could be seen as having been successfully brought under the spell of rational, planned change or as having been successfully managed because of the policymaking apparatus. Apart from the sheer complexity of the task, the structure of national and local governance works against the chances of success, especially in the United States, where the pork barrel effect tends to dilute and distort the intended effects of policy, and the grant-in-aid format makes innovative policies and plans unlikely to succeed because of the number of layers of government and the variety of outside participants involved.

Evangelical Bureaucrats

Throughout the United States and Western Europe, the postwar period between 1950 and 1975 was characterized by a proliferation of government programs for housing, urban renewal, land use zoning, transportation planning, environmental quality, and comprehensive planning projects. All of these provided jobs for planners and enhanced the profession's visibility and growth. Planners became indispensable intermediaries between various kinds of projects and various layers of government and so there was unprecedented growth in the number of urban planners, in their collective power as a profession, and in their self-confidence concerning the possibility of delivering better, safer, nicer, and more efficient cities.

Higher education shifted into gear: The number of planning programs in the United States increased from about 20 in the mid-1950s to around 90 by the end of the 1970s, with more than 20 programs offering doctoral degrees in planning. The output of formally qualified planners increased over the same period from about 100 per year to over 1,500 per year. At the top of the profession, a small but powerful and influential group of can-do, will-do planners and renewal executives emerged: Robert Moses, whose power base in New York had already been established by World War II; Edward J. Logue, who worked for Richard Lee's administration in New Haven, Connecticut; Edmund N. Bacon (Figure 15) in Philadelphia (actor Kevin Bacon's father); Dave Loeks in Minneapolis–St. Paul; and William Ryan Drew in Milwaukee. These were the men who seemed chosen to bring the bold and inspirational ideas of Howard, Geddes, Corbusier, Wright, and Stein to fruition in a golden age of technocratic planning on a truly heroic scale.

There was an evangelical spirit to the entire profession. Cities *should* be better places. They *could* be. Planners shared a professional orientation that was a product of both the long, slow beginnings and the heady, explosive growth of the postwar era. So they carried with them, through self-selection and through formal education, strong elements of liberal idealism, of utopianism, of environmental determinism and design determinism, a penchant for sweeping, futuristic solutions, and a concern for the aesthetics of the urban environment.

Courtesy of The Ed Bacon Foundation

FIGURE 15 Edmund Bacon, Director of Planning for Philadelphia in the 1960s (right), with staff members Irving Wasserman (left) and R. Damon Childs, examining a model of I. M. Pei's Society Hill Towers.

Equipped with the latest developments in social science research and theory—the languages and toolkits of behavioral theory, regional economics, regional science, quantitative geography, systems analysis, and transportation modeling—the stage was set for a golden age of planning on a truly heroic scale. Cities everywhere were recast through strategic plans that the planners based on the modernizing principles of strict separation of land uses, slum clearance, large-scale civic and commercial renewal projects, urban expressways and, in Europe, new towns and public housing schemes.

Too often, though, the result was a complex of slab-like buildings and parking structures, often in the Brutalist style of reinforced concrete, with pedestrians segregated from traffic on elevated walkways and windswept flights of stairs. The desirability and effectiveness of planned modernity had been challenged as early as 1961 by Jane Jacobs in her book *The Death and Life of Great American Cities*. Jacobs reasoned that planning had taken away the life and vitality of cities, tearing out their sclerotic hearts only to replace them with a "great blight of Dullness" in the form of high-rise apartment blocks. Adherence by planners to the dogma of land use segregation, she pointed out, had resulted in the loss of vitality and serendipity in urban life. Left to planners, she argued, city landscapes ". . . will be spacious, parklike, and uncrowded. They will feature long green vistas. They will be stable and symmetrical and orderly. They will be clean, impressive, and monumental. They will have all the attributes of a well-kept, dignified cemetery."[19] Within little more than a decade, public confidence had been sapped by a series of lengthy, costly, and ultimately abortive development sagas, and by the whiff of corruption surrounding many of the larger redevelopment schemes.

So, even in their finest hour, planners were forced to watch themselves fail. Their evangelism and environmental determinism led them to get bogged down in "bureaucratic offensives" like urban renewal and highway construction, to the point where the communities whose lives they had hoped to improve were angry and afraid. Before they knew it, their rationality and their predilection for efficient and tidy land use patterns had led them to become social gatekeepers.

NEOLIBERAL POLICY AND PLANNING

The neoliberal era was firmly established after 1973. By 1978 the Carter administration had abandoned the idea of an urban policy framework for the United States. The first *National Urban Policy Report*, published in 1978, promoted the disengagement of the federal government from urban affairs. President Carter himself, in his State of the Union address that year, reflected the new attitude very clearly. "Government," he said, "cannot solve our problems. It can't set our goals. It cannot define our vision. Government cannot eliminate poverty, or provide a bountiful economy or reduce inflation *or save our cities*."[20]

It remained only for the apparatus of urban policy and planning to be dismantled. The Reagan administration enthusiastically undertook this task. The objective was now to free private enterprise as much as possible from the constraints of government at every level. Cities were no longer seen as the appropriate scale at which to focus initiatives to promote economic growth. A major government report, *Urban America in the Eighties*, spelled out the logic. Free-enterprise markets are the best means of allocating urban land uses and making investment decisions that maximize the benefits for everyone. Government aid to distressed cities and neighborhoods, the report argued, only hampered the long-term efficiency of free-enterprise markets.[21]

The neoliberal ideologies that have dominated the political economy of Western countries since the mid-1970s are predicated on a minimalist role for the state, assuming the desirability of free markets as the ideal condition not only for economic organization, but also for political and social life. This kind of thinking has led to a dramatic cutback in government aid and a corresponding increase in deregulation and **privatization**. Cuts in federal outlays for urban and regional development programs between 1981 and 1988 amounted to nearly 60 percent. Studies of the urban impacts of these reductions have established, to nobody's surprise, that the outcome was an increase in social polarization and fiscal stress, with the most adverse effects being felt by people in larger, older central cities. Indeed, the increased inequality soon reached the point where it was described in terms of a "new class war."[22] Planning and urban design, deflected from issues involving regulation or social expenditure, have been increasingly pushed toward contributing to artful fragments of upscale suburbia as a means of sustaining professional identity and credibility. Instead of developing and implementing strategic plans, planning departments were reduced to rubber-stamping the subdivisions and mixed-use "town centers" proposed by property developers who were surfing the credit boom of the 2000s. All planners could reasonably hope for was to make developments more artful (through themed design), or at least less artless (through "smart" growth).

The Property Rights Movement The free-enterprise, neoliberal climate encouraged landowners and developers to challenge the fundamental power of governments to protect public health, safety, and welfare through planning regulations and public works that are in the public interest. In the United States this power was always constrained by the Fifth and Fourteenth Amendments to the Constitution, which bar public authorities from "taking" private property for public use without just compensation.

Until the ascendance of neoliberalism, the courts had tended to support the public interest across a broad range of community issues, especially when they concerned health, safety, and welfare issues and were based on technical analysis and a publicly adopted comprehensive plan. Beginning in the 1980s, these rulings have been challenged with increasing vigor by a property rights movement that has also pressed for legislation that would provide compensation for landowners for *any* loss of value resulting from *any* government actions that restrict or curtail the use of their property. Through the 1980s and 1990s the Supreme Court decided several cases in favor of property owners against state agencies (*First English Evangelical Lutheran Church v. County of Los Angeles*, 1987; *Nolan v. California Coastal Commission*, 1987; *Lucas v. South Carolina Coastal Council*, 1992; *Dolan v. City of Tigard*, 1994), although development moratoria were validated in the case of *Tahoe Sierra Preservation Council v. Tahoe Regional Planning Agency* in 2002. Meanwhile, the property rights movement continues to sponsor the introduction in state legislatures of "takings" bills that would, if passed, exert an enormously inhibiting effect on regulatory agencies at all levels of government.

Planning as Dealmaking

The neoliberal ideologies that have flourished since the mid-1970s have undermined the professional identity and credibility of urban policymakers and planners. As a result, urban policymaking and planning have become fragmented, pragmatically tuned to economic and political constraints, and oriented to stability rather than being committed to change through comprehensive plans. Practice has become estranged from theory and divorced from any broad sense of the public interest. It has become increasingly geared to the needs of producers and the wants of consumers and less concerned with overarching notions of rationality or criteria of public good.

Unfortunately, this loss of faith has occurred at a time when urban restructuring portends a particularly urgent need for some form of overall policy framework. Urban areas, as we have seen in earlier chapters, are having to accommodate a wholly new mixture of industry and employment within their polycentric metropolitan form. Economic transformation and social polarization would seem to call for a comprehensive approach to managed change. Yet the principal aim of policy and planning is now job creation rather than producing the city practical, the city beautiful, or, indeed, any kind of *city* at all.

URBAN VIEW 3
Kelo v. City of New London Eminent Domain Lawsuit

> Susette Kelo dreamed of owning a home that looked out over the water. She purchased and lovingly restored her little pink house where the Thames River meets the Long Island Sound in 1997, and had enjoyed the great view from its windows . . . The richness and vibrancy of this neighborhood reflected the American ideal of community and the dream of homeownership.[23]

After Pfizer accepted an incentive package in 1998 that included paying one-fifth of its property taxes for the first 10 years, the pharmaceutical company built a new plant in New London, Connecticut, that opened in 2001. City officials created the New London Development Corporation to purchase the neighborhood of 115 homes nearby and sell the land to a commercial developer with plans for a hotel, retail outlets, and condominiums that would generate higher tax revenues. Susette Kelo and 14 other residents resisted, and so the City instigated eminent domain proceedings to force them to sell. The legal battle reached the U.S. Supreme Court that in 2005 made one of its most controversial rulings—in favor of the City of London. Justice Sandra Day O'Connor wrote in her dissent:

> Any property may now be taken for the benefit of another private party, but the fallout from this decision will not be random. The beneficiaries are likely to be those citizens with disproportionate influence and power in the political process, including large corporations and development firms. As for the victims, the government now has license to transfer property from those with fewer resources to those with more. The Founders cannot have intended this perverse result.

The result was perverse in many ways. The mayor at the time said that she had received 4,000 e-mailed death threats. The City eventually agreed to move Susette Kelo's house to another location and to provide substantial additional compensation to the other homeowners. But their neighborhood was gone. The cost to the City and state for the purchase and razing of the homes was nearly $80 million. The notoriety of the legal case dampened interest and with the economic downturn, the developer was not able to obtain financing and had to abandon the project.[24]

In 2009, two years before the tax breaks for Pfizer were about to expire, the company announced that it would be leaving New London within two years. After hearing the announcement, Michael Cristofaro surveyed the cleared land that had been his parents' neighborhood before it became the epicenter of this high-profile battle over eminent domain. "Look what they did. They stole our home for economic development," he said. "It was all for Pfizer, and now they get up and walk away."[25]

Susette Kelo is still bitter about the loss of her home, which she later sold for a dollar to Avner Gregory, a preservationist. Gregory moved the Kelo house across town where he intends it to stand as a bright-pink symbol of the abuse of eminent domain. Susette Kelo takes comfort from the fact that after the Supreme Court ruling, 9 states passed legislation to limit the use of eminent domain while 43 others passed stronger property rights reform.

One policy development that has carried this dealmaking ethos forward is the idea of **enterprise zones**. Based on the work of geographer Peter Hall and first advanced in Britain, this idea involves designating specific areas in cities and making them attractive to businesses by combining as many concessions as possible—tax breaks, subsidized factory space, and so on—with a relaxation of government controls and regulations. More generally, urban regeneration has become dependent on entrepreneurial public-private partnerships, rather than strategic city planning. Planning has become entrepreneurial and planners have become dealmakers rather than regulators.

URBAN VIEW 4
Urban Regeneration in London's Docklands[26]

Between the late eighteenth and the mid-twentieth centuries, London's docklands became the trading heart of Britain's empire. Occupying the north bank of the River Thames for several miles downstream of Tower Bridge, the docksides and wharves developed into distinctive settings (Figure 16). Immediately downstream from Tower Bridge were elegant multistory Georgian warehouses for high-value goods linked to London's commodity trade as the imperial capital: ivory, teas, furs, tobacco, plant and flower oils, spices, and other exotic imports. Farther downstream were facilities for handling and storing bulkier cargoes, such as tropical fruit and vegetables, coal, cattle feed, chemicals, cement, and paper. Still farther downstream, newer, larger docks and huge, refrigerated warehouses were added later, providing a modern infrastructure for global trade. This commercial activity depended on a huge labor force of dockworkers, who were crowded with their families into cheap and often substandard housing in the neighborhoods surrounding the docks.

Relatively little of this urban setting and community now remains. The docklands fell into a steep decline in the 1960s and 1970s as a result of labor problems; competition from Rotterdam and other European ports; and the construction of container port facilities farther downstream. Employment fell from 30,000 in its heyday in the 1950s to 2,000 by 1980. The disused and derelict docks, so near the center of London, represented both an embarrassment to the national government and a huge potential property asset. So in 1979 the national government, taking up an idea by geographer Peter Hall, established a small experimental enterprise zone in the heart of the docklands

(continued)

URBAN VIEW 4
Urban Regeneration in London's Docklands (*continued*)

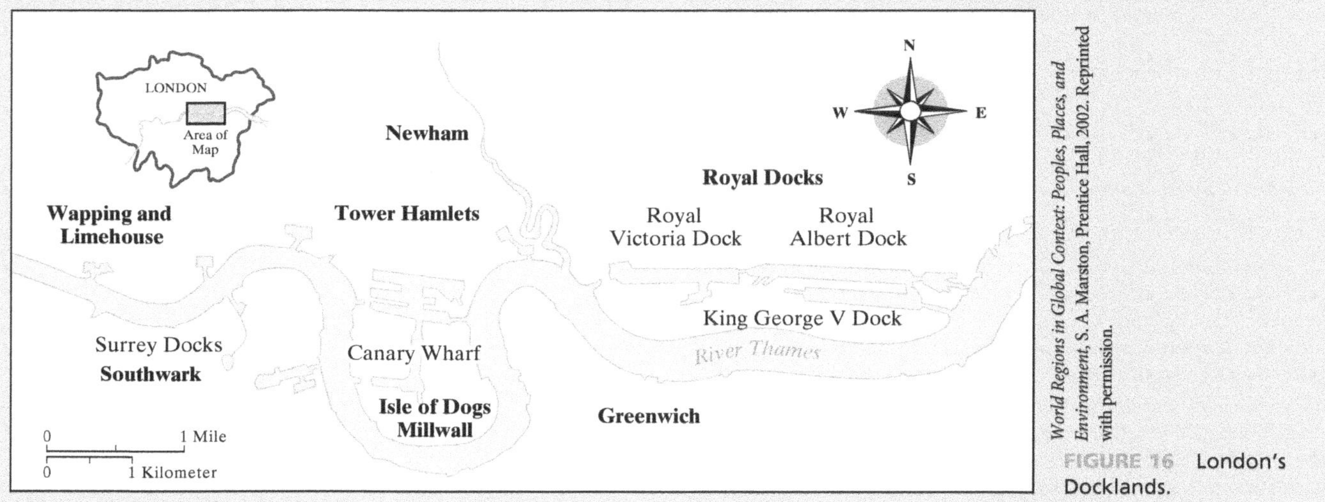

World Regions in Global Context: Peoples, Places, and Environment, S. A. Marston, Prentice Hall, 2002. Reprinted with permission.

FIGURE 16 London's Docklands.

in an effort to attract new businesses to the area by suspending certain taxes and regulations. The following year, the government created the London Docklands Development Corporation (LDDC) and charged it with planning the economic regeneration of an eight-and-a-half square mile area of the docklands. This urban development corporation was given substantial resources and powers that national government ministers believed were necessary to regenerate such a large area in a relatively short time. The LDDC took over the planning powers for the area from the London boroughs of Tower Hamlets, Newham, and Southwark and was given the power to acquire land through eminent domain if necessary.

It was at this point that London's financial markets began to respond to globalization. Automation and information-based financial dealings required large, modern office units, flexible in plan, with deep floor plates to accommodate state-of-the-art technology in suspended ceilings and underfloor cabling. In the heart of the medieval city, London's central financial district had little such space and few opportunities to create it. Then in 1985 the world's largest property development company, Canadian-based Olympia and York, put together an ambitious redevelopment scheme for the docklands that would provide several million square feet of new office space. Although the redevelopment scheme ran into financial difficulties for a while (and Olympia and York fell into disastrous debt), regeneration efforts continued.

Between 1981 and 1998, at which point the government closed down the LDDC, national grants of £1.86 billion were spent on land acquisition and reclamation (including environmental cleanup) and the provision of utilities and transportation infrastructure (including new roads and a driverless light rail system). This public funding leveraged £7.7 billion in private investment so that today London's dockland area has a new image—created by the largest single urban redevelopment scheme in the world. Its new image centers on the sleek office towers of Canary Wharf and its associated office complexes that are populated by day by the office workers in international financial and publishing companies (Figure 17). The new development is surrounded by cleaned-up waterfronts and restored warehouses that have been converted into expensive condominiums and retail complexes. The new business and professional services jobs, high-end retail establishments, and upscale housing, however, do not address the needs of the remaining long-time residents—many of whom are unemployed dockworkers.

Linda McCarthy

FIGURE 17 The docklands development in Canary Wharf, London, across the Thames River from the Chapel Stephen Lawrence Gallery, University of Greenwich.

Mixed-Use Developments and Cluster Zoning Perhaps the single most important aspect of recent changes in planning practice has been the way that planners have accommodated developers' need for flexibility through new approaches to land use zoning. Once the cornerstone of city planning practice, single-purpose "Euclidian" zoning (based on the landmark case *Euclid v. Ambler*) is now seen as wasteful, monotonous, and overly rigid. Planners and planning committees now see mixed-use zoning, which is critical for developers wanting to produce large set-piece projects like the Westlake Center in Seattle (Figure 18), as a means of achieving everything from enhancing a city's tax base to initiating urban revitalization and increasing ridership on public transit systems. In downtown areas mixed-use zoning is being combined with incentive deals in which developers are granted additional height or density allowances in exchange for specified building features (e.g., elaborate facades or building tops) or facilities such as daycare centers, residential space, and space for services that might help restore variety and vitality to downtown districts.

In suburban jurisdictions the solution to the rigidity of Euclidian zoning has been cluster zoning, in which regulations are applied to an entire parcel of land (a Planned Unit Development, or PUD) rather than to individual building lots. With a PUD, developers can calculate densities and profits on a project-wide basis, allowing the clustering of buildings to make room for open spaces (such as golf courses) or to preserve attractive site features (such as ponds or old barns). Cluster zoning facilitates a blend of residential and nonresidential elements and a mixture of housing types that can be adjusted as property sales dictate. For developers, PUDs offer **economies of scale** plus **economies of scope** for product diversity and flexibility, all within a predictable regulatory framework. For planners, cluster zoning offers the prospect of high tax-yield development with services and amenities provided at no cost to the taxpayer.

Linda McCarthy

FIGURE 18 The Westlake Center in Seattle, a mixed-use development (MXD) that incorporates retail, office, hotel, parking, and cultural amenities in a carefully planned package where each element is designed to complement the others.

URBAN VIEW 5
Competitive Regionalism[27]

Entrepreneurial city governments have become the norm across the United States. Based on neoliberal policies, individual localities compete for private companies and their investment for goals such as jobs and tax base. Yet this competition has drawbacks, including the enormous public resources diverted (e.g., as grants) or lost (e.g., as tax breaks) to incentives for companies; instances of fewer gains for an urban economy than projected; the inequitable distribution of the costs and benefits for the people within cities; and poorer cities paying more, with the costs outweighing the benefits for some. Because only one city can win, this competition is zero sum at a regional or national scale if it merely relocates investment between places at public expense. If public funds are diverted from education, technology, and the transportation and communications infrastructures, cities become less competitive nationally and internationally.

Aware of these drawbacks, many government officials agree that eliminating wasteful competition is rational. Yet, acting individually, each local government can feel forced to offer incentives to companies because there is no certainty that others will not. Theories of cooperation (e.g., iterated game theory and the prisoners' dilemma game) provide insights into this collective inaction problem. In a simple two-person game, each player chooses to cooperate or defect. The best outcome for both is if they both cooperate. The worst outcome is if they both defect. Yet each opts to defect because neither can guarantee that the other will cooperate.

Regional cooperation—*competitive regionalism*—may offer an alternative to this prisoner's dilemma at a metropolitan scale. The argument is that

> America's economy should now be seen as a *common market of metropolitan-based local economic regions*. These regions are indeed strongly interdependent, but they also compete with each other and with the rest of the world . . . the new leadership coalitions and networks recognize that the geographic focus of their efforts has to be

(continued)

URBAN VIEW 5
Competitive Regionalism (*continued*)

> the metropolis as a whole, not just the central city or suburbs independently.[28]

Competitive regionalism has the potential to better market an area for investment and reduce wasteful competition, help a metropolitan region to mobilize its strengths and address socio-economic divisions that can weaken its chances of success, and find a profitable niche in the international economy.

> Around the country, there is a dawning awareness that regional approaches to development—and particularly, cooperation between central cities and their suburbs—may be vital to economic survival. . . . This is not . . . the same thing as the movement to regionalize service delivery, or yet another round of drum-beating for metropolitan government. Rather, cities and their suburbs are trying to sort out ways in which a variety of public and private entities can work on behalf of the economic development of an entire region, instead of in competition with one another.[29]

An example from the United States is the Metro Denver Economic Development Corporation (Metro Denver EDC), a public-private partnership—an affiliate of the Chamber of Commerce that includes 70 cities, counties, and economic development organizations—that promotes economic development while reducing wasteful competition for private-sector investment among the localities across the entire Denver metropolitan area. "Our philosophy toward bringing business here is unique. Each of the Metro Denver EDC's economic development partner organizations is committed to the economic vitality of the entire region. As an ambassador for the area, each is ready and able to communicate the benefits of Metro Denver first and individual communities second."[30]

An example from Europe is the Urban Community of Lille Métropole (Lille Métropole Communauté Urbaine (LMCU)) in France—a network of 85 towns and villages with shared regional decision making and investment in metropolitan Lille. LMCU invests in transit, water, and sewage projects, as well as urban renewal, and sporting and cultural facilities, in an effort to improve the regional quality of life while promoting economic development and new corporate investment in the region. The advantages for individual localities include channeling the public funds not provided as incentives into more productive activities, such as improving labor force skills. While providing fewer unnecessary incentives would benefit individual localities, reducing company relocations that occur purely as a result of public subsidies would benefit the national economy by minimizing this unproductive use of public and company funds.

This solution, however, may merely shift the scale from a contest between localities to one between regions. Unless regional **competitive advantage** is promoted through enhancing conditions for business, and economic specialization occurs between regions, competing as a region for mobile investment eliminates competition only between the jurisdictions *within* particular regions—the battle for investment would continue *between* regions. As in the private marketplace, competition and cooperation are not mutually exclusive—cooperation through strategic alliance is still competition.[31] Jonas found this in Southern California where the "beggar-thy-city-neighbor" competition of the 1980s became "beggar-thy-region-neighbor" in the 1990s.[32] The drawbacks of competition at a local scale can be replicated at regional scales. Intensified competition between regions will not necessarily produce new investment or a more productive economy at a national scale.

PLANNING FOR HEALTHY AND LIVABLE CITIES

Not surprisingly, there have been some important counter-movements to neoliberalism and the consequent attenuation of Modernist planning. Broadly speaking, liberal ideals now find expression in a variety of movements and initiatives that are concerned with planning for healthy and livable cities. One example is the Healthy Cities Project, launched in Europe in 1987 by the European Office of the World Health Organization (WHO). The goal of the WHO European Healthy Cities network is to promote a good overall quality of life (in terms of people's social and economic conditions) as well as physical environments that promote good health. The core strategies of Healthy Cities are set out in Health21 and Local Agenda 21 (WHO's main policy frameworks), the Zagreb Declaration of Healthy Cities, and the Charter of European Towns and Cities Towards Sustainability (Aalborg Charter). These documents aim to engage local governments across Europe in the creation of healthy urban settings through a variety of processes of political commitment, institutional change, and innovative action. Their focus is on addressing environmental justice, the social dimensions of sustainability, community empowerment, and urban planning.

More than 90 European cities and towns have been designated as WHO Healthy Cities. They are linked through national, regional, metropolitan, and thematic healthy cities networks. Cities participating in these networks have developed and implemented a wide range of initiatives, including urban environmental health profiles and city health plans, community development programs, and programs that address the needs of vulnerable groups.

Sustainability and Green Urbanism

The Healthy Cities network and efforts by city governments to cut their greenhouse gas emissions are part of a broader movement concerned with people's urban environments. Planning for **sustainable urban development** and **green urbanism** involves facilitating major lifestyle changes (such as walking, bicycling, and reductions in material consumption), the preservation and restoration of the natural environment, and the application of new technologies (such as public transit, district heating, and green building and design).

As with the healthy cities idea, some of the most innovative examples of sustainability and green urbanism are found in Europe. The European Union has strongly endorsed

URBAN VIEW 6

Cities Take Environmental Sustainability Efforts into Their Own Hands

Large cities around the world—from New York and Mexico City to Shanghai and Tokyo—are taking action and cooperating to tackle climate change by reducing greenhouse gas emissions through a range of energy efficiency and clean energy programs. Many of these cities are more populous than some small countries but traditionally have carried no weight at the United Nations Climate Change talks.[33] Now their mayors are speaking up. Recent gatherings include the World Mayors Summit on Climate following the UN talks in Cancun in November 2010 and the annual C40 Climate Summit of the Mayors of the world's large cities (in São Paolo in 2011).

The mayors say that they feel a greater urgency about climate change than their national government leaders. This is partly because when floods, droughts, or torrential rains devastate a city, it is the mayor who must stand before the citizens and offer a response. Urban areas consume up to 60 percent of global energy production and emit 70 percent of greenhouse gases.[34] They contain more than half the world's population and are projected to have more than 69 percent of its people by 2050. The mayors control policies on transportation, water and waste management, street lighting, and the energy efficiency of buildings. They argue that for these reasons they need to play a strategic role in tackling climate change.

The City of Los Angeles' climate action plan, for example, has one of the most aggressive goals of any large U.S. city: reducing greenhouse gas emissions to 35 percent below its 1990 levels by 2030. Efforts range from actions that affect only municipal facilities, such as harnessing wind power for electricity production, energy efficiency retrofits in City buildings, converting the City's fleet vehicles to cleaner and more efficient models, replacing streetlights with efficient LED lamps, and reducing water consumption, to efforts to facilitate changes in businesses and the community, such as rebates for the purchase of energy-efficient appliances.[35]

sustainability through its legislation and directives—in particular its 10-year Environmental Action Program. Most European countries have prepared national sustainability strategies, and many local governments have implemented a variety of environmentally sensitive policies and strategies. Among these are the pedestrianization of central city streets; restrictions on car access; traffic "calming" (slowing) by modifying street patterns; land use planning for compact urban development; urban greenways and forests; increased bicycle use through public bicycle sharing systems, and bicycle paths with priority access to schools, shopping areas, and green space; ecovillages of environmentally conscious residents; and integrated rail, tram, metro, and bus systems.

One of the most innovative approaches to sustainable urban development has taken place on the southern edge of Freiburg im Breisgau, a university town in southwestern Germany. There is a long-standing tradition of sustainable urban development in the region, and when the city purchased an old barracks in the Quartier Vauban in 1994 it decided to convert the barracks into a flagship environmental and social project. An ambitious set of progressive objectives included an emphasis on car-free living, small lots with preferential allocation to cooperative building projects, prohibiting detached houses and buildings exceeding four storys, extensive use of ecological building materials and solar energy, a diversity of building shapes, and strict standards for domestic energy consumption (Figure 19).

Building lots were sold to small cooperatives of owner-occupiers (*Baugruppen*), each comprising between 3 and 21 households. These co-housing groups were responsible for the detailed building design of their property, accommodating their specific needs and aspirations in a common plan, and frequently pursuing additional environmental and social objectives. There is dedicated accommodation, for example, for guests who come to visit Vauban residents, that is designed to avoid the need for building large apartments with rarely used spare bedrooms. Many of the units have solar collectors, and the entire district is connected to a heating grid with a co-generation plant that runs on wood chips.

Paul L. Knox

FIGURE 19 The Quartier Vauban, in Freiburg, Germany.

Perhaps the most radical of the district's policies are those related to car-free living. Every street in the Quartier Vauban except for the main axis is a "play street," or *Spielstrasse*. These streets often function as extended front porches, and are frequently used as places for neighbors to socialize. Some streets are designated as Bicycle Streets, allowing one-way access for slowly driven vehicles and two-way access for bicycles; in addition, there is a well-developed network of pedestrian-only paths. Although cars are allowed in the Quartier Vauban, their use and ownership is sharply restricted. The speed limit on the district's central spine, which also carries the tram line that connects the district with Freiburg's city center, is 20 miles per hour, but elsewhere it is walking speed (3 miles per hour). There is a ban on parking on private property: cars may enter the residential streets but only for pick-up and delivery. Residents with cars are required to purchase or lease a parking space in one of the perimeter multistory community parking garages. Visitors, too, are expected to park their car in a parking garage and pay for the privilege as they would in a downtown car park. Approximately 40 percent of all households in Quartier Vauban have decided not to own a car at all. They are able to join a car-sharing organization (whose vehicles are also parked in the peripheral garages), whose members also receive a free pass for all public transportation within Freiburg and a 50 percent reduction on rail tickets. If a family decides to purchase a car, they are required to make a one-time payment equivalent to $20,000 for the construction of a parking space as well as monthly rental fees and property taxes for the parking space.

Most U.S. towns and cities have made no such ambitious attempts at sustainable planning or green urbanism. Portland, Oregon, and the Minneapolis-St. Paul, Minnesota, are notable exceptions to the general trend of uncontrolled sprawling, auto-dependent urban development that consumes land at a rate that far exceeds the rate of population growth (Urban View 7 entitled "Sustainable Metropolitan Planning Efforts in the United States"). Relative to European cities especially, most U.S. cities and metropolitan areas have high carbon dioxide emissions, draw in excessively large amounts of energy and resources, produce large amounts of waste, destroy large amounts of sensitive habitat and productive farmland, generate high levels of traffic congestion, and impose high costs of infrastructure provision.

Metropolitan Governance and Planning

The political fragmentation of metropolitan areas that has been associated with **suburban incorporation** in the United States has made it difficult to sustain the notion of *metropolitan* governance and planning. The advantages of metropolitan-wide

URBAN VIEW 7
Sustainable Metropolitan Planning Efforts in the United States

By the mid-1960s the Twin Cities (Minneapolis-St. Paul) metropolitan area faced serious urban challenges: inadequately treated sewage released into public waters and contaminating groundwater supplies; the privately owned regional bus system deteriorating due to rising fares, declining ridership, and an aging fleet; urban sprawl threatening recreational and agricultural activities; and growing fiscal disparities among communities making it difficult for some localities to raise enough taxes to provide essential public services. With 186 different city and township governments across seven counties, the metropolitan area was ill equipped to deal with problems that transcended these local boundaries.[36]

The Minnesota legislature responded in 1967 by creating the Twin Cities Metropolitan Council, with members appointed by the governor, to plan and coordinate development in the seven-county metropolitan area. The legislature also created regional wastewater and transit systems and enacted tax-base sharing legislation to reduce fiscal disparities among communities.

The Metropolitan Council's *Regional Development Framework* has the goals of preserving the urban core; phasing new development outside the core based on population, household, and employment forecasts, consistent with the region's capacity to provide services; expanding regional facilities—such as water treatment plants—on a planned schedule; and protecting some outlying land for agricultural, rural, and recreational uses.

The Metropolitan Council established a metropolitan urban service area (MUSA) within which development and redevelopment are encouraged (Figure 20). Improvements in the regional system of sewers, transportation, parks, and airports are made to meet the needs of people living inside this urban growth boundary. A low-density rural service area was established to preserve agriculture. The framework includes a fully developed urban area, a developing urban area, freestanding growth centers, rural centers, a commercial agricultural area, and a general rural use area. Each local government is required to develop a detailed land use plan that is consistent with this framework. Although the MUSA has been extended incrementally over time to accommodate growth and has helped restrict uncontrolled urban sprawl, some new development has leap-frogged into adjacent counties—particularly along the interstate highways—outside the seven-county area.

The Twin Cities Fiscal Disparities Act of 1971 introduced metropolitan tax-base sharing to provide a way for all communities to share in the region's growth. This diverts 40 percent of the increase in commercial and industrial property taxes into a central fund that is redistributed based on each locality's population and the ratio between the locality's per capita valuation of property and that of the metropolitan area. This tax-base sharing has lowered fiscal disparities across the metropolitan area. Tax revenues generated by the expansion of commercial and industrial property (mostly in suburban communities such as Bloomington that contains the Mall of America) have gone disproportionately to localities (at the urban core such as Minneapolis and St. Paul) that have lower tax bases but a high demand for services.

Metro in Portland, Ore., has the distinction of being the only metropolitan government in the United States that is directly elected by the region's voters. This metropolitan council covers the three counties and 25 cities of the Portland metropolitan area and has responsibility for protecting open space and parks, planning for land use and transportation, and managing waste disposal and recycling.[37] Like the Twin Cities Metropolitan Council, Metro manages an urban growth boundary that restricts uncontrolled urban sprawl and encourages development and redevelopment in its existing neighborhoods. Metro's transit-oriented development program has helped facilitate increased travel choices so that more than 90 percent of the region's residents now live within a half mile of transit.

URBAN VIEW 7
Sustainable Metropolitan Planning Efforts in the United States (*continued*)

Since I moved to downtown Portland 16 months ago, I have rarely used my car. I get to most of the places I want to go via foot or bicycle and only occasionally need to combine transit into my trips. The bicycle and pedestrian facilities that Portland is establishing downtown and in other core area neighborhoods enable me to live the kind of lifestyle that contributes not only to my health but that of the planet (Mary Vogel, Portland resident).[38]

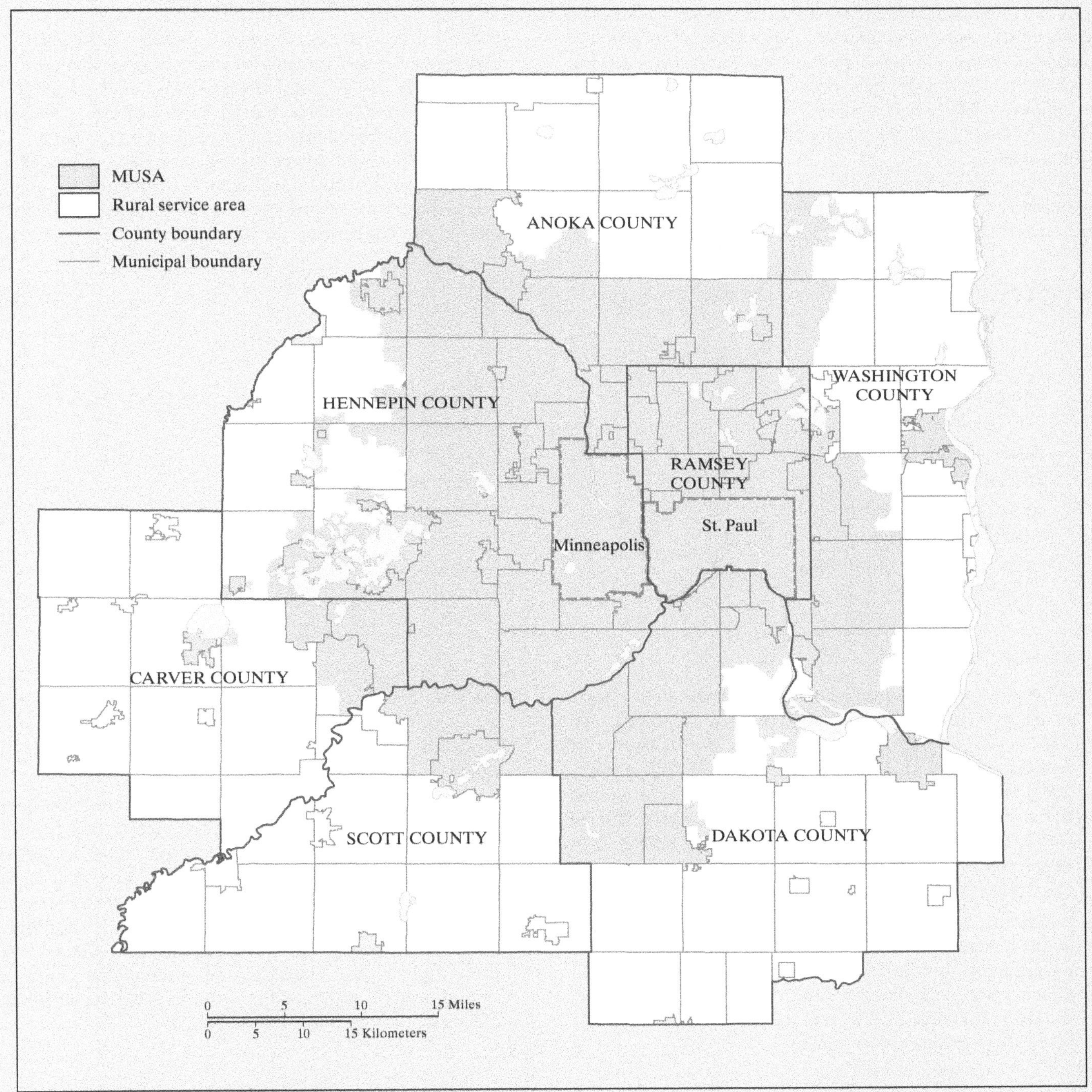

FIGURE 20 The metropolitan urban service area (MUSA) and rural service area to the year 2030 for the Twin Cities of Minneapolis-St. Paul metropolitan area.

planning—including economies of scale in service provision and coordinated infrastructure planning—have led to some governance reforms.[39] Given the strong tradition of local government control over urban planning and the many vested interests in the status quo, however, these reforms have been limited and exceptional.

Examples include when two or more jurisdictions are consolidated into a unitary government as in the city-county consolidation in Indianapolis that created Unigov in 1969; or intergovernmental agreements where one locality provides a service (such as water) to others in the metropolitan area; or when a group of jurisdictions jointly provides certain metropolitan-wide functions or services (such as fire and emergency response); or the transferring of key functions to a higher (metropolitan) level of government (Urban View 7 entitled "Sustainable Metropolitan Planning Efforts in the United States").

Compared to Europe, the more limited concern with planning for healthy and livable cities in the United States suggests that if people in the U.S. once expected too much of urban policy and planning, they now perhaps expect too little. Although we have learned that urbanization is a broad, deep, and complex collection of interrelationships that we cannot hope to completely "shape" or "control," we have also seen that reform is possible and that urban outcomes can be successfully modified. But what is clear from this and previous chapters is that reform has to come from society at large rather than from the policy and design professions.

As Sociologist Janet Abu-Lughod argues:

> If we are to get better cities, if we are to deserve better cities, then we cannot wait patiently for a technological or administrative "fix" . . . We must reach a deeper understanding of the forces and processes that have shaped them . . . Such forces are well beyond the control of planners and administrators, although they are the collective outcomes of all our attitudes, institutions, and social structures. To get changed cities—to change our cities into a better image of society—we must change the society itself.[40]

FOLLOW UP

Key Terms

building standards
City Beautiful movement
cluster zoning
collective consumption
competitive regionalism
enterprise zones (EZs)
environmental determinism
Garden Cities
garden suburbs
green urbanism
neighborhood unit concept
Park Movement
scientific planning
settlement houses
Tenement House Commissions

Review Activities

1. When you have time, read a novel that captures some aspects of urban policy and planning that were covered in this chapter. *The Devil in the White City: Murder, Magic, and Madness at the Fair that Changed America*, by Erik Larson (New York: Crown, 2003) brings the Chicago of 1893 to life as it interweaves the true stories of two men—Daniel Burhham, associated with the City Beautiful Movement and the architect of the 1893 Chicago Exposition and H. H. Holmes, a cunning serial killer who used this world's fair to lure his victims to their deaths.
2. Go online and search newspaper reports to find a story about a high-profile civic entrepreneurial case of local government subsidies, such as tax breaks, being offered to attract a large corporation to a particular city. After reading the story, list some of the pros and cons of offering public subsidies to private corporations.
3. Do you think that there should be a single planning agency that covers entire individual metropolitan areas in the United States? What sort of case might be made for or against this kind of metropolitan planning?
4. Work on your *portfolio*. With the background you have now acquired in urban geography, you should be able to look at your own town or city and "read" it. What examples can you find of federal policies that have been "written into" the urban landscape? Can you find examples of the nature or pattern of urban development having been influenced by local plans or policies? Are there aspects of your city's development that could or should have been planned in order to achieve a different outcome?

Log in to **www.mygeoscienceplace.com** for self-study quizzes, *MapMaster* layered thematic and place name interactive maps, *Urban View* Google Earth™ tours, key resources and suggested readings, related websites, "In the News" RSS feeds, and additional references and resources to enhance your study of urban policy and planning.

NOTES

1. M. Funkhouser, *Kansas City's Economic Future: Creating jobs by replacing government meddling with organic growth* (*http://www.reelectmayorfunkhouser.com/docs/economicfuture.asp*), 2.
2. Ibid., 10.
3. Ibid., 5.
4. R. Regnier, "Incentives War Helps No One," *Kansas City Star,* December 7, 2010.
5. R. Fogelsong, *Planning the Capitalist City* (Princeton, N.J.: Princeton University Press, 1986).
6. Charles Dickens found the subject matter for novels like *Oliver Twist* in the teeming slums of London (see the U.N.'s description of the appalling living conditions in London and New York in the 1800s in "A Tale of Dickens's London" in *The State of the World's Cities Report, 2001* (Nairobi: UN-Habitat) (*http://www.unchs.org/Istanbul+5/20-27.pdf*); Charles Booth, a philanthropic Liverpool shipowner, produced the prominent 17-volume *Life and Labour of the People in London (1886–1903)* to highlight the plight of the poor (for more information about Booth, see the Charles Booth Online Archive at the London School of Economics Library (*http://booth.lse.ac.uk*); Fredrich Engels, shocked by the living conditions of the poor, wrote in 1844 the influential *Condition of the Working Classes in England.*
7. M. C. Boyer, *Dreaming the Rational City* (Cambridge, Mass.: MIT Press, 1983), 18.
8. J. A. Riis, *How the Other Half Lives: Studies among the Tenements of New York* (New York: Scribner's, 1890).
9. R. W. DeForest and L. Veiller, eds., *The Tenement House Problem. Including the Report of the New York State Tenement House Commission* (New York: Macmillan, 1903). Quoted in P. Hall, *Cities of Tomorrow: An Intellectual History of Urban Planning and Design in the Twentieth Century*, 3rd ed. (New York: Basil Blackwell, 2002), 39.
10. Boyer, *Dreaming the Rational City*, 27.
11. R. E. Park, "The City: Suggestions for the Investigation of Human Behavior in the City Environment," *American Journal of Sociology* 20 (1915): 580.
12. Quoted in Hall, *Cities of Tomorrow*, 42.
13. Boyer, *Dreaming the Rational City*, 39.
14. Fogelsong, *Planning the Capitalist City*, 4.
15. Boyer, *Dreaming the Rational City*, 60.
16. J. Hancock, "The New Deal and American Planning: The 1930s," in *Two Centuries of American Planning*, ed. D. Schaffer (Baltimore, Md.: Johns Hopkins University Press, 1988), 197–230.
17. This discussion is based on L. McCarthy and C. Johnson, "Cities of Europe," in *Cities of the World: World Regional Urban Development*, eds. S. Brunn, M. Hays-Mitchell, and D. Zeigler, 5th ed. (Lanham, Md.: Rowman & Littlefield, 2011).
18. R. Florida and A. Jonas, "U.S. Urban Policy: The Postwar State and Capitalist Regulation," *Antipode* 23 (1991): 349–84.
19. J. Jacobs, "Downtown is for People," in *The Exploding Metropolis,* ed. W. Whyte, Jr. (Garden City, NY: Doubleday, 1958), 157.
20. January 19, 1978 State of the Union address by President Carter, Audio/Video Archive on The American Presidency Project website (*http://www.presidency.ucsb.edu/*). Emphasis added.
21. President's Commission for a National Agenda for the Eighties, Panel on Policies and Prospects, *Urban America in the Eighties: Perspectives and Prospects* (Washington, D.C.: U.S. Government Printing Office, 1980).
22. F. F. Piven and R. A. Cloward, "The New Class War in the United States," in *Cities in Recession*, ed. I. Szelenyi (Beverly Hills, Calif.: Sage, 1984), 26–45.
23. Institute for Justice, *Cases: Kelo v. New London, Lawsuit Challenging Eminent Domain Abuse in New London, Connecticut, http://www.ij.org/.*
24. J. Tuccille, "Susette Kelo's Revenge: New London regrets eminent domain fiasco," *Civil Liberties Examiner.com* (*http://www.examiner.com/*).
25. P. McGeehan, "Pfizer to Leave City That Won Land-Use Case," *New York Times,* November 12, 2009.
26. Harvey, "From Managerialism to Entrepreneurialism," 13.
27. This discussion is based on L. McCarthy, "The Good of the Many Outweighs the Good of the One: Regional Cooperation Instead of Individual Competition in the USA and Western Europe?" *Journal of Planning Education and Research* 23 (2003): 140–52.
28. H. G. Cisneros, *Urban Entrepreneurialism and National Economic Growth* (Washington, D.C.: United States Department of Housing and Urban Development, 1995), 3, 19.
29. R. Gurwitt, "The Painful Truth about Cities and Suburbs: They Need Each Other," *Governing* 5 (1992): 56.
30. Metro Denver Economic Development Corporation (Metro Denver EDC) website: *http://www.metrodenver.org/about-metro-denver-edc/.*
31. E. Goetz and T. Kayser, "Competition and Cooperation in Economic Development: A Study of the Twin Cities Metropolitan Area," *Economic Development Quarterly* 7 (1993): 63–78.
32. A. E. G. Jonas, "Regulating Suburban Politics: 'Suburban-Defense Transition,' Institutional Capacities, and Territorial Reorganization in Southern California," in *Reconstructing Urban Regime Theory: Regulating Urban Politics in a Global Economy*, ed. M. Lauria (Thousand Oaks, Calif.: Sage, 1997), 222.
33. T. Johnson, "As Nations Dither on Climate Change, Big Cities Step Up." *McClatchy News Service,* November 23, 2010.
34. Ibid.
35. City of Los Angeles, *Green LA: An Action Plan to Lead the Nation in Fighting Global Warming*, May 2007 (*http://www.ci.la.ca.us/ead/pdf/GreenLA_CAP_2007.pdf*).
36. This discussion is based on information in the Metropolitan Council's website (*http://www.metrocouncil.org/*).
37. This discussion is based on information in Metro's website (*http://www.metro-region.org/*).
38. U.S. Environmental Protection Agency, *2010 Award for Smart Growth Achievement* (*http://www.epa.gov/smartgrowth/awards/sg_awards_publication_2010.htm#policies_reg*).
39. See, for example, M. Orfield, *Metropolitics: A Regional Agenda for Community and Stability* (Washington, D.C.: The Brookings Institution Press, 1997); and D. Rusk, *Cities without Suburbs* (Washington, D.C.: Woodrow Wilson Press, 1993).
40. J. Abu-Lughod, *Changing Cities* (New York: Harper Collins, 1991), 379.

The Residential Kaleidoscope

iStockphoto

The Residential Kaleidoscope

In contrast to European cities, one of the most powerful images of urbanization in the United States is of a system of "melting pots." U.S. cities, according to this view, have absorbed a fluctuating stream of migrants and immigrants, processing people with a tremendous diversity of backgrounds into a pluralistic society with a common language and a shared ideology of democratic free enterprise. The degree to which this image is reality is a matter for some debate. It is clear, however, that a great deal of the dynamism of cities is related to the processes through which new arrivals interact with one another and with existing groups—striving for economic and social success while dealing with their own cultural identity. In these processes people sort themselves—and are sorted—into different parts of the urban fabric, resulting in mosaic patterns of neighborhood differentiation (Figure 1). Through simultaneous processes of segregation, assimilation, and resegregation, different parts of the city and different types of housing come to be characterized, for a time at least, by households of different socioeconomic status, families of different size and structure, and people of different racial and ethnic backgrounds.

LEARNING OUTCOMES

After reading this chapter, you should be able to:

- Understand how the relationships between social distance and physical distance influence how people interact with each other.
- Identify the major factors influencing residential segregation.
- Evaluate the major contributions of the Chicago School of Human Ecology to our understanding of residential ecology.
- Identify the key trends in economics, demographics and ethnicity that have shaped urban social ecology since the 1970s.
- Explain how segmented lifestyle groupings and the increased emphasis on consumption is being mapped onto residential patterns.

CHAPTER PREVIEW

In this chapter we review the "classic" arrangement of residential areas in U.S. and European cities before going on to consider the rearrangements that have taken place as a result of economic reorganization and social change since the mid-1970s. We begin with an examination of the principal features of urban social interaction and residential segregation. In this context the central issues for urban residents are the relationships between physical distance, social distance, and patterns of social interaction. As we will see, there are several very good reasons why physical distance and social distance continue to act as mutually reinforcing aspects of social interaction and residential segregation, not the least of which is territoriality.

Given these relationships, we will then see how and why the cornerstones of neighborhood differentiation in U.S. cities are based on residential segregation in terms of people's social status, household types, ethnicity, and lifestyles. In contrast, in many continental European cities ethnicity is generally not an independent

In general, the cornerstones of neighborhood differentiation in U.S. cities are based on residential segregation in terms of people's social status, household types, ethnicity, and lifestyles. These children out riding their bicycles in their neighborhood both come from middle-income families and nice suburban homes.

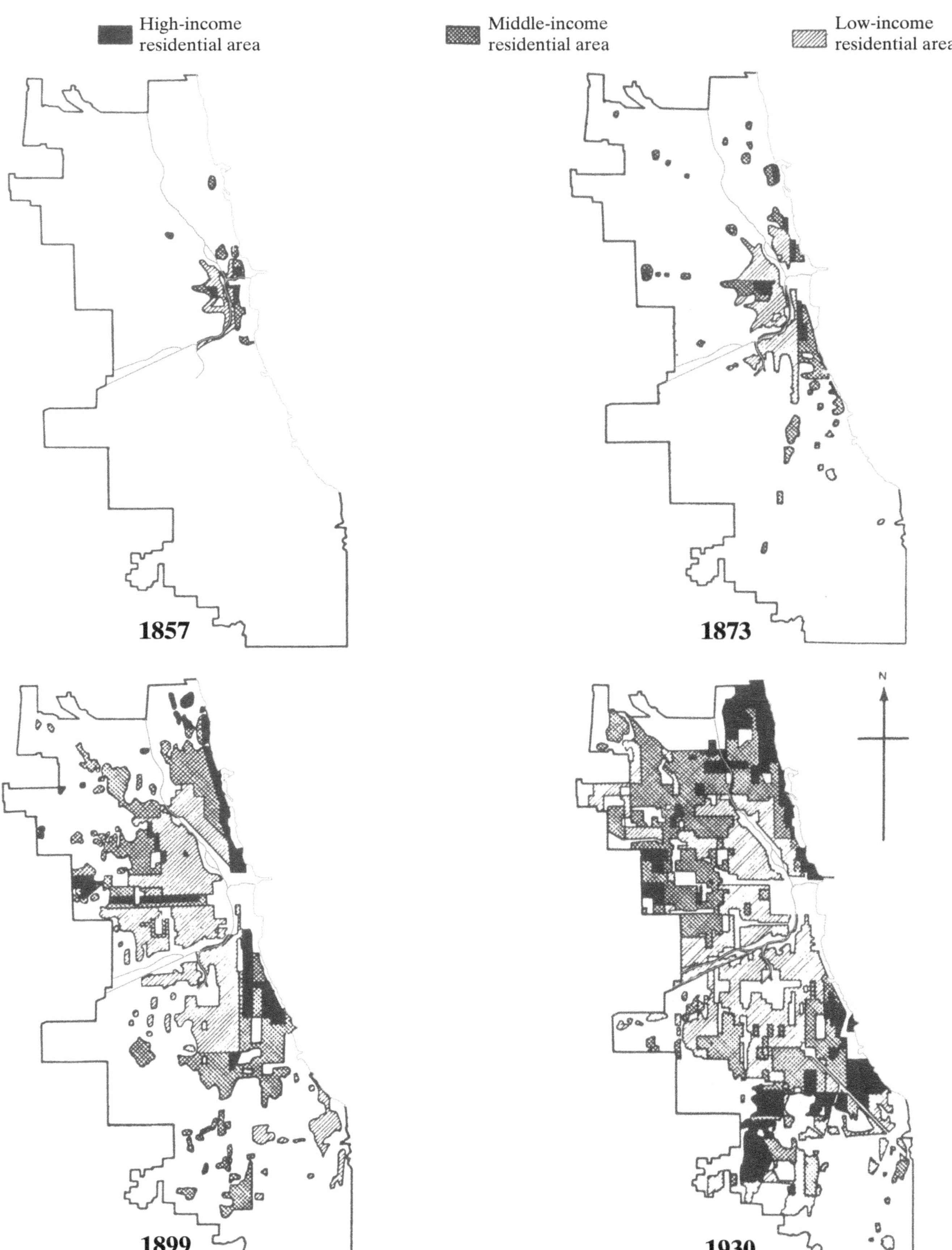

FIGURE 1 The evolution of socioeconomic areas in Chicago, 1857–1930.

URBAN VIEW 1

The French Ghetto Beat of a Muslim Rapper[1]

French rapper, Abd al Malik, was born in France to Congolese parents and grew up in the Neuhof public housing project—a *banlieue*—in the suburbs of Strasburg. This maze of anonymous graffiti-covered high-rise apartments is home to more than 5,000 people whose mini-satellite dishes are as likely to be tuned to an Arabic-language TV network as to a French one. Although Malik now lives in Paris where he has become a successful rap artist and author, his family and most of his rap group, the New African Poets, or N.A.P., still live in Neuhof.

When Malik visits, he and two members of N.A.P., his brother Bilal and friend Mohamed, drive through the public housing projects with the car speakers blaring out tracks from their CDs. One track is about how the 9/11 terrorist attacks made Malik feel ashamed to be Muslim. "Neither fundamentalism nor extremism; Me, I don't mix politics and faith," his voice raps as some young men hurry over to welcome him. They put their hands on their chests as a sign of respect. They recognize that his music comes from authentic experiences of deprivation, criminal activity, redemption, and religious activism.

> Malik pointed out the spot where his friend Fouad was stabbed to death during a brawl. He stopped outside the window where, as a boy, he watched the heroin dealers fleeing police through syringe-filled gangways—a swirl of faces that would one day populate his lyrics. He saw the ground-floor apartment that once housed the mosque where he worshipped after converting from Catholicism to Islam. Malik, Bilal, and Mohamed are grand frères—"big brothers"—now, and they use their prestige to set a good example . . . [But] France is full of tinderbox Neuhofs . . . products of a clash between France's stratified, secular mainstream and immigrant diasporas that are predominantly Arab, African, and Muslim.[2]

Yet this is also where many people see French rap and hip hop at their best, challenging not only artistic but also social and cultural complacency and misconceptions. In the opinion of 20syl, the MC/producer for progressive live hip hop band, Hocus Pocus and DJ in the legendary C2C Crew:

> I see it as a voice for minority and forgotten peoples. It is also a good way to question myself and the world I'm living in. Musically and graphically, it's freedom, you have no rules, no boundaries.[3]

dimension, partly because of the absence of substantial ethnic minorities (at least compared to the United States). Having established these "classic" dimensions and patterns of residential differentiation, we take a closer look at recent changes that reflect the new opportunities and constraints for people of the changing political economy and changing metropolitan form. Here, we will see how new social groups, new kinds of household organization, and new lifestyle orientations have been imprinted on the social map.

SOCIAL INTERACTION AND RESIDENTIAL SEGREGATION

The residential mosaic is what gives definition to cities. It is what gives character and life, flesh and blood to the skeleton of metropolitan form and land use. In particular, the residential mosaic is the framework for (and product of):

- patterns of friendship, relationships, marriage, and social interaction among people;
- the reproduction of distinctive life worlds and communities; and
- local politics.

Unlike a real mosaic, however, urban residential patterns are not set in concrete. Although they do tend to reflect the inertia of self-perpetuating and self-reinforcing patterns, they also change, in the long run, in response to new economic, demographic, social, cultural, and political conditions that influence people's choices and constraints. So a better (though still imperfect) metaphor is provided by the kaleidoscope, with its constant rearrangements of different fragments. Mapping these arrangements and rearrangements is a fundamental task of geographic analysis, providing descriptive models that are valuable in generating and testing hypotheses and theories concerning urbanization processes.

Patterns of urban residential differentiation stem, ultimately, from the dynamics of urban social interaction that develop among people within the structural frameworks of socioeconomic background, demographics, and culture (Figure 2). Inevitably, these dynamics are complex and multidimensional. Although we can think of most social interaction as being sustained by people's affinity for others with similar socioeconomic resources and cultural values ("people like us"), the

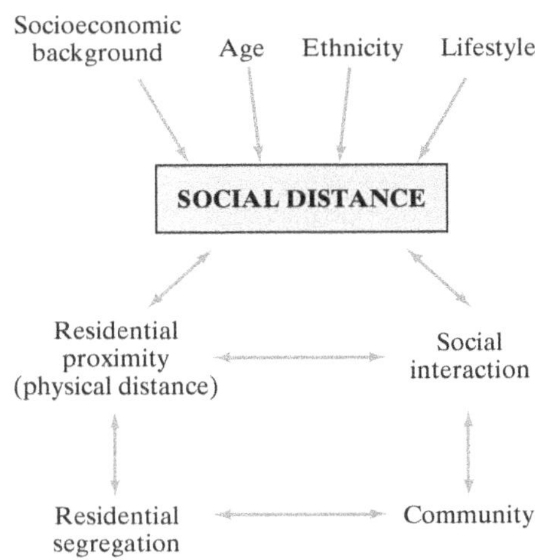

FIGURE 2 Social interaction and residential segregation.

sources of this affinity—lifestyle, age, ethnicity, and so on—are interdependent and constantly shifting.

It is often helpful to distinguish between social interaction based on people's primary and secondary relationships:

- *Primary relationships* include those between family—based on ties of loyalty and duty—and those between personal friends—based on ties of attraction and mutual interest.
- *Secondary relationships* are more purposive, involving individuals who get together for particular reasons. They are often divided into those in which there is some intrinsic satisfaction ("expressive" interaction) for people and those in which the interaction is merely a means of achieving some common goal ("instrumental" interaction). Expressive interaction is typically organized around voluntary associations of various kinds: sports, hobby and social clubs, and volunteer groups. Instrumental interaction, in contrast, normally takes place within the frameworks of business associations, trade unions, political parties, and pressure groups. Such a division reveals only part of the story, however, since a great deal of social interaction can be seen to be both expressive *and* instrumental: interaction among people that takes place within the framework of ethnic, religious, and cultural organizations, for example.

The degree of social interaction between particular groups of people in the city is a function of **social distance**, which can be conceptualized in terms of people's attitudes toward other groups. Short social distances are reflected in people's willingness to think of members of other groups as potential marriage or life partners, and progressively greater social distance is reflected in willingness to have members of another group as friends, neighbors, colleagues, fellow citizens, and (most distant) visitors from another city or country. This scale points to the close relationship for people between social distance and physical distance. The less the social distance, the greater the probable physical proximity between people—their *residential proximity*. Conversely, the greater the residential proximity, the greater the likelihood of social interaction.

This relationship between social interaction, social distance, and physical distance is not as simple as it may appear. The effects of social and physical distance are closely interwoven and difficult to isolate. In addition, the influences of physical distance have been rapidly diminishing in the "shrinking world" of modern technology and mass communications.[4] Improvements in personal mobility and the advent of computers, the Internet, and smart phones have released people from neighborhood ties. The middle- and upper-income people, who inhabit urban worlds without finite geographical borders, have become "cosmopolites," for whom distance is "elastic." Meanwhile, even "localities" have found themselves less bound to residential proximity, it is suggested, as better social services and increased economic security have made local support systems less important and as greater differentiation in the rhythms of daily life has made it more difficult for increasing numbers of people to participate in neighborhood life.[5]

Against such arguments we must recognize that distinctive residential neighborhoods still exist. Neighbors continue to supply much of the raw material for social life for many social groups, while for relatively immobile groups (such as the poor and the very elderly) the neighborhood represents virtually the only opportunity for social interaction. Even the more mobile are susceptible to chance local encounters and the social interaction that may follow; and most householders will establish some contact with neighbors from the purely instrumental point of view of mutual security.

There are in any case several good reasons why social distance and physical distance should continue to act as mutually reinforcing aspects of social interaction and residential segregation. Geographer Ron Johnston, noting that physical distance restricts any social interaction involving face-to-face contact for the simple reason that traveling takes time and costs money, makes the basic case as follows:

> Lasting social contacts are based on common interests. These may be very specialized—a particular hobby, perhaps—but many are based on interests, notably the problems of child rearing, which reflect common values and lifestyles. These commonalities, in turn, reflect incomes, occupations and educational backgrounds. The more similar people are on these criteria, the more likely they are to have in common and the more mutual benefits they are likely to receive from frequent social contact—both formal (local societies, churches, etc.) and informal (coffee mornings, cocktail parties, bridge clubs, etc.). . . . [C]ongregation and segregation are obvious means to this end, ensuring that one's neighbors are potentially valuable social contacts, even if that potential is never realized.[6]

Sociologist Gerald Suttles has identified additional reasons for the persistence of residential segregation.[7] They include:

- Minimizing conflict between social groups with different values and attitudes.
- Maximizing the political voice and influence resulting from spatial clustering.
- Greater degree of social control (i.e., collective self-control and self-policing) that is possible with relatively homogeneous residential groupings.

Territoriality

Inherent to Suttles's interpretation of residential segregation is the idea of **territoriality**: the tendency for particular groups of people to attempt to establish some form of control, dominance, or exclusivity within a localized area. Group territoriality depends primarily on the logic of people using space as a focus and symbol for group membership and identity and as a means of regulating social interaction.[8] In many ways it can be seen as the product of the transition from the rigid social order of the preindustrial city to the competitiveness and social mobility of industrial cities. In preindustrial society the relative stability of society and the rigidity of group membership enabled people effectively to maintain social distance through appearance and comportment—what sociologist Lyn Lofland, coining a dreadful piece of jargon, called "appearential ordering."[9]

In the more rapidly changing environment of the industrial city, with its constant arrivals of strangers from other places, appearances could be deceptive. The city became a stage; people could pose and masquerade, their background and social credentials unchallenged in the turmoil. Meanwhile, the more intense economic competition, the greater variety of people attracted to cities, and the emergence of new occupational groupings caused sensitivities to social distance to become even greater. With appearance no longer a reliable guide, spatial ordering—group territoriality—provided a means for people to establish and maintain social distance. The outcome can be seen most clearly in the demarcation of the "turf" of street gangs, in the segregation of ethnic groups antagonistic to one another (Figure 3), and in the exclusivity of wealthy residential neighborhoods. The very fluidity of modern urban social and economic life, however, makes it very difficult for people to maintain territoriality at such extreme levels.

The Foundations of Residential Segregation

Residential segregation is based primarily on four interrelated dimensions of society—social status, household type, ethnicity, and lifestyle—each of which influences people's perceptions of social distance. Before going on to examine the patterns and intensity of urban residential segregation, therefore, we first need to develop some understanding of the nature of these social cleavages.

Social Status

The structure and dynamics of social status are central not only to social interaction and residential segregation but also to the broader sweep of urban politics and urban change. At a purely empirical level, someone's social status can be interpreted in terms of educational qualifications, occupation, and income.

At a more conceptual level, however, a person's social status is usually taken to involve additional commonalities of values and culture, and it is more appropriate to talk in terms of social *class*. For most theorists these commonalities are rooted in economic organization and structure. Karl Marx, for example, saw class in terms of the fundamental division between the owners and controllers of property and capital (the bourgeoisie (Figure 4)) and those who must sell their labor, however skilled (the proletariat). Max Weber, the second great theorist of social relations and social structure in modern urban settings, argued for a more refined categorization of class, based on the same criteria of control and ownership of property and capital but also recognizing differences (particularly within the huge proletariat) based on people's marketable skills and the consequent differences in access to consumer goods, living conditions, and personal experiences (Figure 5). The development of the ideas of Marx and Weber on social class by contemporary social theorists has resulted in some important distinctions that must be summarized very briefly here.[10]

First is the concept of **class structure**, which refers to the formal categorization of class positions in a society at any given time. It is based on the positions people hold within the division of labor and the framework of economic organization. The broad categories of class structure are cast in terms of fairly heterogeneous groups: the "middle class," for example, consisting of a great variety of occupations. Narrower categories within the class structure ("the professions," for example) are sometimes referred to as **class factions**.

Yet people are not always conscious of the contours of this formal categorization. Rather, "People *experience* class as it is built into their lives in particular ways and come to realize the

Linda McCarthy

FIGURE 3 Belfast, Northern Ireland. This political mural is one of many that promoted loyalist (to the British Crown) political beliefs by glorifying paramilitary groups such as the Ulster Young Militants (UYM), the youth wing of the Ulster Defense Association (UDA), which used the name Ulster Freedom Fighters (UFF) when it chose to claim responsibility for a terrorist attack on republican targets and terrorist groups.

Linda McCarthy

FIGURE 4 Many larger cities in the United States have retained elite residential neighborhoods in central locations around enclaves of town mansions built by wealthy local industrialists in the nineteenth century. This example is from Milwaukee, Wisconsin.

force of class in and through the immediate circumstances they can experience and understand directly."[11] It is these experiences that constitute the process of **class formation**, resulting in conscious collectivities of people. Although many of these experiences are framed around class structure and class factions, many are the result of other dimensions of what geographer David Harvey has termed **class structuration**. According to Harvey, these include (1) the division of labor that determines the formal class structure, (2) institutional barriers to social mobility, (3) the system of authority, and (4) the dominant consumption patterns of a particular time and place.[12]

Several factors are of direct relevance to patterns of social status, social interaction, and residential segregation in today's cities, including local patterns of marriage or of people living together, and the social processes of socialization and stereotyping. By far the most important factor in reinforcing class structure, formation, and structuration is the educational system. Education is an important determinant of the skills that decide a person's starting position in the division of labor. The spatial organization of the U.S. educational system is particularly important here. Most children attend a neighborhood school that is run by a locally elected school board

Linda McCarthy

FIGURE 5 Segregation by socioeconomic status is an attribute of "gentrified" neighborhoods—older residential areas that have been "invaded" and renovated by affluent households seeking the amenities and lifestyle of central-city locations. In Georgetown (Washington, D.C.), gentrification is visible as upscale neighborhoods of older town houses.

that largely depends on local funding, usually through local property taxes. It follows that there are significant differences between school districts and between individual schools in the quality and type of education provided. Consequently, school catchment areas have become an important element in a self-perpetuating cycle of class structuration and residential segregation. "Good" schools will tend to be found in localities with high property values and with people with high average levels of educational achievement who tend to elect progressive school boards. The success of these schools is a strong attraction for educated, affluent households with school-age children. As a result, affluent, middle-class neighborhoods with good schools tend to stay that way, constantly attracting a supply of members of this class faction.

Economic competition, played out in the housing market, is the basic mechanism that ensures this stability, but it is supplemented by a local politics that is directed toward maintaining the quality of educational facilities and the exclusion of families whose finances and attitudes might threaten the educational environment. Given this stability, the school setting is an important element of **social reproduction**, providing (in this case) the skills that are the passport to higher-paid, higher-status occupations. At the other extreme, underfunded schools serving disadvantaged neighborhoods are a factor in the social reproduction of low-income households.

Marriage patterns also involve a circular chain of social reproduction and residential segregation. Studies of how most people pick the person they marry (excluding those who use online dating sites) have shown that a high proportion of marriages are between couples who lived within one or two miles of each other, with greater distances encompassing progressively fewer marriages.[13] This **distance-decay effect** reflects the role of spatial segregation in ensuring social reproduction through marriage: People tend to marry their equals in social status, and because neighbors tend to be social equals, they tend to find marriage partners from the neighborhood.

Family, school, and neighborhood settings are all important to the overall process of *socialization* by which people are introduced to group norms, attitudes, values, dress codes, speech styles, and standards of comportment. This process produces a local convergence that not only reinforces the process of social reproduction but also helps to give identity and character to the kaleidoscope of residential areas within cities. The process is so strong and pervasive that even adult newcomers to a neighborhood often seem to conform, after a while, to local norms. This phenomenon, known as the **neighborhood effect**, is difficult for researchers to track and quantify, but there is evidence that it operates in a variety of ways, influencing lifestyle preferences and voting patterns, for example.

Finally, we should note that these self-reinforcing patterns of social interaction and residential segregation are further consolidated as a result of the *stereotyping* that is itself a consequence of spatial segregation. Because many people do not regularly mix with members of other social groups, they tend to create models of them—stereotypes—that are based on incomplete, second-hand knowledge and that often overemphasize certain attributes. When people exaggerate differences between groups, it increases perceived social distances and sustains the imperative for residential segregation.

Household Type

Households are the basic unit of residential organization. As such, they are very important to an understanding of residential segregation. In the United States, for example, although the traditional suburban household—consisting of the nuclear family unit: mother, father, and children—has by no means disappeared, other forms of household organization are growing in number. The single-parent family has grown to 17 percent of all households, while 27 percent of households consist of a person living alone. Similar changes are occurring in the societies in most other developed countries in response to the social, economic, demographic, and technological processes. Other common household types in U.S. cities are childless (or child-free) young heterosexual couples, gay couples, "empty-nesters" whose grown children have left home, families consisting of divorced and remarried parents and the children from their former marriages, and small groups of people, college students for example, living communally (and not necessarily involving sexual relations).

Members of these different household types tend to share certain attributes, particularly in relation to housing needs and preferences. In addition to the necessity to match household size to housing space, each household type tends to have very specific housing needs and preferences. These similarities in residential needs and preferences lead to spatial congregation and greater within-group interaction, which adds a further dimension to residential segregation.

URBAN VIEW 2

The Social Construction of Race[14]

A dilemma facing urban researchers who want to map and analyze residential segregation and economic inequalities between different groups over time is that the most reliable sources of long-term data, such as the U.S. Census, categorize people by "race." Yet genetic research has not found any significant biological basis for differentiating people into separate racial groups. Rather, race is a **social construction** that artificially categorizes people based on characteristics that include physical appearance (particularly skin color), ancestral heritage, and cultural history.

Historically, nineteenth-century Europeans classified the people in their colonies into an artificially created hierarchy of categories that placed themselves at the top. This social

(continued)

URBAN VIEW 2
The Social Construction of Race (*continued*)

construction inevitably classified some people as inferior and often translated into the economic and social exploitation of these people. In the United States before the reforms associated with the civil rights movement, the social construction of race resulted in legal construction of laws that sought to separate people and treat them differently based on race.

Yet if race is a social construction, it does not explain a person's behavior or economic situation. Similarly, racism—and resultant discrimination—is not a uniform or invariable condition of human nature but, rather, consists of sets of attitudes that are rooted in societal perceptions as they relate to political culture and economic circumstances.

Consequently, the interaction of political culture and economic circumstances is particularly important to an understanding of residential segregation in societies, such as the United States and the United Kingdom, where institutional discrimination carries racism into the housing delivery system. Despite civil rights reforms in these countries, institutional discrimination continues to permeate the legal framework, government policies (relating to, among others, urban renewal, public housing, and suburban development), urban **land use zoning** ordinances, and the practices of builders, landlords, bankers, insurance companies, appraisers, and real estate agents.

The impersonal web of exclusionary practices that results from this institutional discrimination has reinforced the racism and discrimination of individuals to the point where segregated housing has led to *de facto* segregated schools, shopping areas, and recreational facilities. This spatial segregation, in turn, serves to socially reproduce racism and to sustain economic inequalities between different groups of people within cities.

One useful way of looking at the relationship between household types and residential segregation is through the process of the **household life cycle**. As people enter successive stages of the household life cycle, this tends to trigger moves to different kinds of accommodation and different parts of the city.

Ethnicity

The term *ethnicity* covers any group of people that is primarily characterized by attributes of race, religion, nationality, or culture. Implicit in the use of the term is the idea that such groups are minorities whose presence in the city stems from a past or continuing stream of in-migration. In this sense, important ethnic groups in U.S. cities include African Americans, Chinese (Figure 6), Italians, Jews, Mexicans, Puerto Ricans, and Vietnamese. The term charter group is sometimes used to describe the host society, the matrix in which ethnic groups find themselves. The charter group itself may not be ethnically homogeneous, but it is usually dominated, culturally, if not always numerically, by a particular combination of race, religion, and national origin—as in the white, Anglo-Saxon charter group of many U.S. cities (Figure 7). What is clear enough from the available evidence, however, is that most minority groups tend to be highly segregated from the charter group. In addition, this segregation has been shown to be greater than might be anticipated from the socioeconomic status of the groups concerned. In other words, the low socioeconomic status of

Linda McCarthy

FIGURE 6 San Francisco's Chinatown is an area that is completely and intensely ethnic in terms of residence, businesses, cultural institutions, and street life.

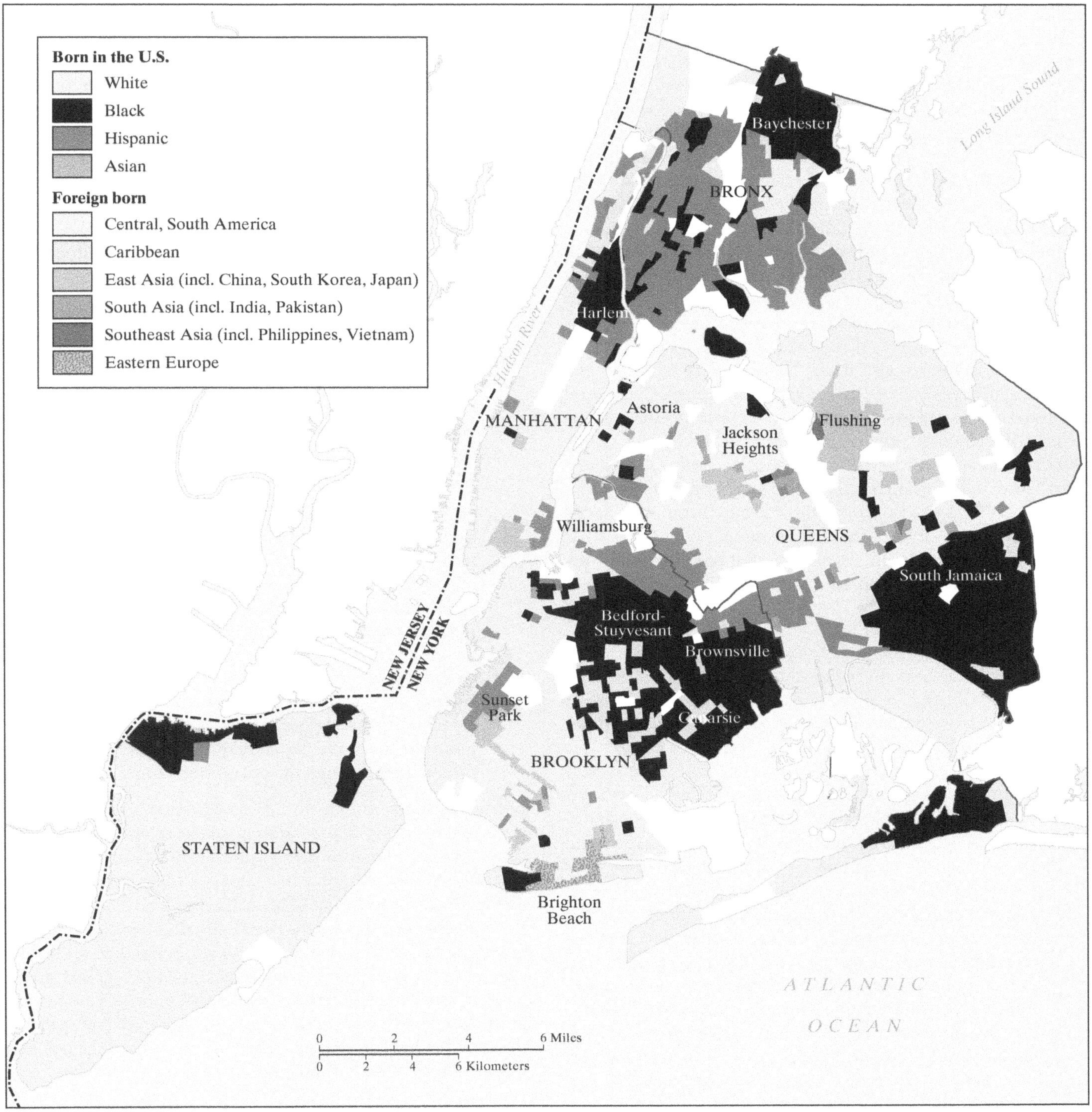

FIGURE 7 Ethnicity in New York, 2009. Despite the image of a melting pot, note how little overlap there is between the white—charter group—of people who were born in the U.S. and the major concentrations of U.S.-born Hispanic, Asian, and African-American people and the foreign-born immigrants.

certain minority groups can only partially explain their high levels of residential segregation. The extent to which ethnic groups are spatially segregated from the charter group and each other varies a great deal, depending on the degree to which the process of assimilation occurs and the amount of time that the process has been operating. Assimilation, however, is not simply the process of one culture being absorbed into another; both the charter and ethnic minority cultures are changed by assimilation through the creation of new hybrid forms of identity. In overall terms, however, the degree of assimilation is a function of the social distance between the charter group and a particular ethnic group. It is, therefore, possible to think of ethnic groups as distinctive class factions, subject to circular chains of social reproduction and residential segregation through socialization in family, school, and neighborhood settings and through stereotyping by other groups.

As a result, the rate and degree of assimilation of a minority group, and the spatial patterns of residential segregation, can be seen to depend on the intensifying effects of (1) external factors, including charter group prejudice, institutional discrimination in the housing market, and the structural effects of often low s ocioeconomic status, and (2) internal group cohesiveness associated with the desire of ethnic group members to maintain cultural identity. In fact, there may be differences not only in the degree and speed of the overall assimilation of different ethnic groups but also in the behavioral assimilation and structural assimilation of particular ethnic groups. **Behavioral assimilation** occurs when members of a minority ethnic group acquire the language, norms, and values of the charter group, thus becoming *acculturated* to the mainstream life of the city. **Structural assimilation**, on the other hand, refers to the diffusion of members of a minority ethnic group through the social and occupational strata of the charter group.

Given this distinction, we can see that **congregation**—the residential clustering of an ethnic minority through choice—fulfills several functions:[15]

- *Defensive functions*, where the existence of a territorial "heartland" helps to reduce the isolation and vulnerability of members of a particular ethnic minority. This is particularly important when charter group discrimination is widespread and intense.
- *Support functions*, where ethnic enclaves serve as a "port of entry" to the city for new migrants and a haven for longer-term residents. By clustering together in a mutually supportive haven, members of the group are able to avoid the hostility and rejection of the charter group (and of other ethnic groups), exchanging insecurity and anxiety for familiarity and strength. The existence of ethnic institutions such as places of worship can be a particularly important source of both practical and spiritual support. In addition, most ethnic groups develop localized, informal self-help networks and welfare organizations. Finally, the existence of ethnic enclaves provides protected niches for ethnic enterprise (both legal and illegal) providing an expression of group solidarity as well as a means of social and economic advancement for successful entrepreneurs and a route for workers to bypass the charter-group dominated labor market.
- *Cultural preservation functions*, where ethnic residential segregation helps to preserve and promote a distinctive cultural heritage. The distance-decay effect on the selection of marriage partners is particularly important here. In addition, the ethnic institutions and businesses supported by residential clustering are important elements of cultural solidarity. Territorial clustering is particularly important to ethnic groups that observe religious precepts relating to dietary laws, the preparation of food, and attendance at prayer and religious ceremonies.
- *"Attack" functions*, where the ethnic neighborhood serves as a base for action. This action is usually both peaceful and legitimate. Spatial concentrations often enable ethnic groups to gain representation within the institutional framework of urban politics, for example. Sometimes, however, ethnic heartlands are used as a convenient base for insurrectionary groups and urban guerrillas, who are able to "disappear" within the camouflage of a distinctive cultural milieu, protected by a silence resulting from a mixture of sympathy and intimidation.

We can also see that ethnic residential segregation might be expressed differently, depending on the importance of these functions combined with internal group cohesiveness and on the importance of charter group attitudes combined with institutional discrimination and structural effects associated with socioeconomic status. Three main types of ethnic segregation and congregation have been identified: colonies, enclaves, and ghettos.[16]

- *Colonies* are the result of situations in which a particular area of a city serves as a port-of-entry for an immigrant (or migrant) ethnic group. It is a temporary phenomenon, a base from which ethnic group members are culturally assimilated and spatially dispersed.
- *Enclaves* are ethnic concentrations that exist over several generations mainly because their inhabitants choose to congregate for functional reasons.
- *Ghettos* are ethnic concentrations that exist over several generations mainly because of the constraints of charter group attitudes and discrimination, which are often institutionalized through the operation of housing markets.

Ethnic segregation can be measured empirically in several ways. After a long debate among social scientists, the most generally used measure is the *index of dissimilarity*, a simple but effective statistic that is analogous to the Gini index of inequality that is widely used in economics. The *index of dissimilarity* is used to compare the spatial distribution of two groups of people (blacks and whites, for example) and provides the answer to the question: What percentage of one group would have to move to a different territorial area in order to achieve a proportional distribution of the groups within each area that is the same as their proportional distribution within the city as a whole? The formula for the index (D) can be written as:

$$D = \frac{1}{2}\sum_{i=1}^{k} | x_i - y_i |$$

where x_i and y_i represent the percentages of group members and others, respectively, in area i, the summation being over all the k territorial areas within the city. This formula, multiplied by 100, yields an index with a theoretical range of values from 0 (no segregation) to 100 (complete segregation). The territorial units of measurement are usually census tracts, sometimes city blocks. It should be noted, however, that the index is very sensitive to variations in the size of the areal units (the smaller the mesh of areal units, the higher the resultant index of dissimilarity). A similar measure, the *index of segregation*, can be used to compare the spatial distribution of one group with *all* other groups (e.g., Asians and all other races). Both the index of segregation and the index of dissimilarity are measures of *absolute* levels of segregation. That is, they quantify the discrepancy between actual spatial patterns and those that would exist if groups were randomly intermixed. As a result, they do not capture differences in the proportional size of the groups. An index of dissimilarity of 90 between Asians and whites in a city may seem striking, but its significance is diminished if we learn that less than 1 percent of the city's population is Asian while 96 percent are white. Where researchers need to gauge the relative mixture of two groups across a set of territorial areas, they use a different index, the *index of isolation* (sometimes called the index of dominance).[17]

For our purposes, the most widely used index, the index of dissimilarity, provides an effective measure of the degree of residential segregation in U.S. cities. There is plenty of evidence of the residential segregation of ethnic groups. Indexes of dissimilarity across large metropolitan areas at the census tract level in the United States range between 30 and 60 for most ethnic groups and the white population, with Hispanics averaging 46 and Asians 45. A striking exception to this generalization is the very high degree of residential segregation experienced by African Americans. At the census tract level, indexes of dissimilarity between the African-American and the white population in large U.S. metropolitan areas average about 58.[18]

These data suggest that *the idea of U.S. cities as melting pots may be something of a myth*. Segregation is highest in the largest metropolitan areas, with the top 5 most segregated metropolitan areas for African Americans located in the Northeast-Midwest "Rust Belt" (Table 1).

Certainly, the residential segregation of African Americans is currently at its lowest point since roughly 1920. And although there remains a large number of "hypersegregated" metropolitan areas, the 2000s continued a trend toward decreasing segregation that began in the 1970s.[19] Between 1980 and 1990, black segregation across all large metropolitan area census tracts declined from 73 to 68; by 2009 it had fallen to 58.0. In 2009, states in the West comprised the most integrated region of the United States (e.g., Nevada 46.5), followed by states in the South (e.g., Mississippi 47.7). Some of the traditionally more rapidly growing metropolitan areas in the West and South—such as Las Vegas and Raleigh-Durham—have relatively low and declining levels of segregation (38.8 and 40.6 respectively in 2009).[20]

These changes have been associated with the efforts of the civil rights movement since the 1960s and government action against discrimination in housing and lending. The most recent 10-year study by the Urban Institute on behalf of the U.S. Department of Housing and Urban Development (HUD) found that 21.6 percent of black renters and 17.0 percent of black homebuyers encountered some form of housing discrimination in 2000. These figures represent a decline in discrimination since the previous HUD study in 1989 (from 26.4 percent for black renters and 29.0 percent for black homebuyers).[21] Meanwhile, the growth of black middle-income groups is reflected in census data that show an increase from 15 percent in 1960 to 38 percent in 2009 of blacks living in the suburbs.

At the same time, however, the HUD study showed that **steering** on the basis of neighborhood racial composition increased significantly for black homebuyers between 1989 and 2000. Blacks were more likely to be steered by real estate agents to neighborhoods that were predominantly black compared to the neighborhoods recommended to comparable white homebuyers. Meanwhile, the many blacks who have not achieved middle-income status—and cannot afford to buy a home—remain concentrated in impoverished central city neighborhoods.

Lifestyle

The intersecting cleavages of social status, household type, and ethnicity, together with people's experience and their personal preferences and aspirations, lead to a variety of urban lifestyles. People pursuing similar lifestyles often cluster together, partly in order to fulfill the lifestyle itself and partly because affinity to people with similar lifestyle preferences tends to narrow perceived social distance, which promotes the self-perpetuating

TABLE 1 Changes in Black Residential Segregation* in the Most Segregated Large U.S. Metropolitan Areas as Measured by the Index of Dissimilarity, 1980–2010

		Index of dissimilarity				Percent change	
2010 rank	Metropolitan area	1980	1990	2000	2010	2000–2010	1980–2010
1	Milwaukee, Wis.	83.9	82.6	81.8	81.5	−0.3	−2.4
2	New York, N.Y.	81.2	81.3	81.0	78.0	−3.0	−3.2
3	Chicago, Ill.	87.8	83.8	79.7	76.4	−3.3	−11.4
4	Detroit, Mich.	87.4	87.4	84.6	72.3	−12.3	−15.1
5	Cleveland, Ohio	85.4	82.4	76.8	74.1	−2.7	−11.3

**Segregation: Blacks versus White Non-Hispanics.*

Source: J. Iceland and D. H. Weinberg, *Racial and Ethnic Residential Segregation in the United States: 1980–2000*, Washington, D.C.: U.S. Department of Commerce, 2002, Tables 5-4 and 5-5; William H. Frey *Analysis of 2010 U.S. Census data* (*http://www.psc.isr.umich.edu/dis/census/segregation2010.html*).

Ron Buskirk/Alamy

FIGURE 8 Residential segregation according to lifestyle preferences has been made possible by the development of specialized, amenity-rich suburban communities such as The Villages of Lady Lake in Central Florida.

chain of social reproduction and residential segregation. The "classic" U.S. lifestyles identified by sociologist Wendell Bell[22] were based on three stereotypes:

- *Familists*, who are home-centered people and who tend to spend much of their spare time with their children. As a result, they are attracted to residential settings such as suburban neighborhoods that are dominated by other familists and are close to schools and parks.
- *Careerists*, who are people attracted to prestige residential settings that are conveniently located in relation to workplaces or transportation nodes.
- *Consumerists*, whose preference is for the material benefits and amenities of cities, leading to a tendency for these people to congregate in central locations close to clubs, theaters, art galleries, restaurants, and so on; or in amenity-rich planned suburban settings (Figure 8).

It is questionable, however, whether individuals or households can realistically be categorized this way. It is perhaps more accurate to think of most of them as subscribing to all three sets of preferences, but to different degrees and at different points in their life cycle (Figure 9). In addition, changing class structure, changing household types, changing social

AP Images

FIGURE 9 "Keynesian" suburbs like this one facilitated residential segregation by socioeconomic background (middle income), household type (married couples with young children), ethnicity (white), and lifestyle (family-centered). This aerial photo shows part of Levittown in 1948, shortly after this mass-produced suburb was completed on Long Island farmland just 25 miles east of Manhattan.

values, and increasing affluence have resulted in numerous additions to these "classic" lifestyles: gay lifestyles, recreation-oriented and sport-oriented lifestyles, "singles" lifestyles (e.g., metrosexuals: straight urban males who are in touch with their feminine side and not afraid to show it by cooking, having good grooming habits, and dressing in style), and the lifestyle of the active retirement community, for example. At the same time, it must be acknowledged that there remains a section of the population for whom "lifestyle" is experienced largely through TV and the Internet, the home and the neighborhood being viewed more as havens from the outside world rather than as settings for the enactment of a favored lifestyle.

URBAN VIEW 3
Social Exclusion and Migrant Workers in European Cities

The rebuilding of Western Europe's urban infrastructure and industry after World War II generated a strong demand for workers, particularly in the cities of more prosperous countries like France, Germany, and the United Kingdom. In the 1950s and 1960s, rural-to-urban migration fueled growth, especially in the largest cities. In addition, tens of thousands of foreign-born migrant workers were brought in to fill low-wage assembly-line and service-sector jobs that the more skilled domestic labor force would not take. These "guest" workers came from the countries of Mediterranean Europe, including Italy, Greece, Portugal, and Spain, and from former European colonies. What was then West Germany attracted migrants from Turkey, Yugoslavia, Italy, and Greece; France brought in workers from Algeria, Tunisia, Morocco, Spain, and Portugal; while Britain drew on Commonwealth citizens from the Caribbean, India, Pakistan, and from Ireland.

The migrants were intended as temporary workers—a ready stream of labor that could be turned on and off as needed. The volume of incoming migrants, in fact, paralleled the highs and lows of the business cycle, with a peak in the mid-1960s and a trough by the early 1970s recession. By the mid-1970s, there were nearly 4 million foreign workers in West Germany, over 3.5 million in France, 1.5 million in the United Kingdom, nearly 1 million in Switzerland, and about half a million each in Belgium, the Netherlands, and Sweden. The onset of a deep economic recession in 1973, however, brought a dramatic check to the flow of migrants. Most governments imposed immigration restrictions. Some countries offered financial incentives to encourage return migration. Inducements were effective only in regions of severe unemployment, such as the Ruhr. Otherwise, the incentives were not enough to motivate migrants to return home to countries where employment opportunities were bleaker than in Europe.

In the European Union today, 4 percent—about 20 million—of the people were born outside its member countries. More than one-third of the immigrants in France are concentrated in the Paris metropolitan region where they represent over 15 percent of the population. Foreign-born residents comprise 15 to 25 percent of the population in German cities like Frankfurt, Stuttgart, and Munich. More than half the population of Amsterdam is non-Dutch, being strongly represented by individuals from Morocco, Turkey, and Ghana. Despite the lack of comparable data for immigrants in cities across Europe, their presence is evident in the cultural diversity reflected in the names of stores and restaurants in some neighborhoods and even in the list of most popular baby names for cities like Brussels (Table 2).

TABLE 2 Top 10 Boys' and Girls' Names in the Brussels Metropolitan Area, 2007

	Boys' Names	Girls' Names
1.	Mohamed	Lina
2.	Adam	Sarah
3.	Rayan	Aya
4.	Nathan	Yasmine
5.	Gabriel	Rania
6.	Amine	Sara
7.	Ayoub	Salma
8.	Mehdi	Imane
9.	Lucas	Ines
10.	Anas	Clara

Source: The Telegraph newspaper, *http://www.telegraph.co.uk*

These immigrants typically live in poor quality suburban high-rise apartments (*banlieue*) or central city enclaves left vacant through suburbanization. Each enclave is dominated by a particular ethnic group. The enclaves in Frankfurt and Vienna contain mostly Turks, while in Paris and Marseille, the different enclaves house Algerians or Tunisians. In large British cities, in contrast, there is significant mixing of different ethnic groups. Within each neighborhood, however, the ethnic groups are highly segregated from each other. And although there are large numbers of Asians and West Indians, the foreign-born population represents only 15 to 20 percent of the population within most neighborhoods.

In addition to outright discrimination, the labor and housing markets help create central city enclaves. Low wages force immigrants to rent lodgings in deteriorating central city locations. The internal cohesiveness of the ethnic groups also contributes to residential segregation. Existing residents are more likely to share information about vacancies in their neighborhood with members of their own ethnic group.

Europe's aging population coupled with demand for low-wage labor means that jobs will continue to be filled by immigrants. But there has been a backlash by some people, often xenophobic and racist, against newcomers and even some second and third generation residents. Most countries have immigration restrictions. The contradiction between rising demand for low-wage labor and an unwillingness to accept non-EU nationals has proven quite costly and dangerous, as the weeks of rioting in the largely Muslim working class suburbs of Paris amply demonstrated in 2005.

INTERPRETATIONS OF RESIDENTIAL ECOLOGY

Contemporary patterns of residential segregation have developed on the basis of residential neighborhoods established in earlier growth phases, and the principal dimensions of urban residential structure were forged in the early decades of the twentieth century. The emergence of the Industrial City brought about a radical transformation of urban space, resulting in a broad framework of *sectors* and *zones* of specialized land use. Within the older, central districts of cities, unprecedented numbers of migrants and immigrants from a wide spectrum of ethnic origins struggled to establish themselves. In the newer suburbs, meanwhile, the newly expanded middle-income groups sought to pursue the new trend of domesticity in quiet, secluded neighborhoods. Observing the resultant residential sorting and resorting, sociologists from the University of Chicago (some of the most notable being Robert Park, Ernest Burgess, and Roderick McKenzie) developed a theory of residential segregation and a model of urban residential structure that became benchmarks of urban theory.

The Chicago School: Human Ecology

Robert Park, having worked in Chicago's neighborhoods as a journalist before becoming chairman of the Department of Sociology at the University of Chicago, was struck by the distinctiveness of different neighborhoods: "A mosaic of little worlds that touch but do not interpenetrate."[23] Each of these little worlds, suggested Park and his colleagues, could be thought of as an *ecological* unit, a particular mix of people that had come to dominate a particular niche in the urban fabric. This interpretation led them to adopt a view of the city as a kind of social organism, with social interaction governed by a "struggle for existence."

This kind of biological analogy provided the Chicago School with an attractive general framework within which to set their detailed studies of the "natural histories" and "social worlds" of different groups of people. Just as in plant and animal communities, they concluded, order in human populations must emerge through the operation of "natural" process such as impersonal competition for territory and dominance. It should be borne in mind that these ideas were the product of an era when the appeal of **neoclassical economics** (with its emphasis on unfettered competition) was powerful, when the influence of Darwinism was relatively strong, and when the social sciences were struggling to establish a "scientific" respectability. Plant and animal science provided a rich source of ecological concepts and a graphic terminology for these researchers to use in portraying the social geography of the city.

The Chicago School saw social interaction as an expression of symbiosis in an overall social context that was dominated by competition for living space. The result was a series of *natural areas* within which different groups of people were dominant. The Chicago School produced some very detailed and painstakingly researched studies of some of these "natural areas." One of the classics was Harvey Zorbaugh's *The Gold Coast and the Slum*,[24] a study of distinctive "natural areas" in Chicago's Near North Side. These included the "Gold Coast," a neighborhood of wealthy residents along the lakeshore, and a slum area containing clusters of migrants and immigrants (Figures 10 and 11). Zorbaugh showed how the personalities of these

Corbis Images

FIGURE 10 Chicago's "Gold Coast" in the 1930s. A "natural area" investigated by the Chicago School researchers.

Lee Russell/Library of Congress

FIGURE 11 A Chicago slum in the 1930s. Another kind of "natural area" investigated by the Chicago School researchers.

different areas were related to their physical attributes—the "habitat" they offered—as well as to the attributes and ways of life of the people who lived in them.

The Human Ecologists of the Chicago School, however, did not see "natural areas" such as these as being permanently fixed. As the numerical strength and competitive (market) power of different groups altered and the relative attractiveness and suitability of different settings changed (through the physical and demographic changes imposed by the occupancy of particular groups of people with distinctive demographic mixes and ways of life), the ecological processes of *invasion* and *succession* (Figure 12) brought modifications to the pattern of "natural areas." Meanwhile, in addition to the biotic dimension of social organizations (based on impersonal competition for territory and resources), Park and his colleagues recognized a cultural dimension, shaped by the consensus of social values.

Ernest Burgess brought these concepts together in his famous **concentric zone model** of residential differentiation and neighborhood change in Chicago.[25] Having delineated the location and extent of some 75 of Chicago's natural areas, Burgess interpreted them as falling into a series of concentric zones (Figure 13). This interpretation was reinforced by the conviction that as the city grew in population, economic competition among people and an increasingly specialized division of labor would lead to the complementary ecological processes of *centralization* and *decentralization*. Centralization occurred through the operation of **agglomeration economies** while decentralization occurred as businesses and economic functions

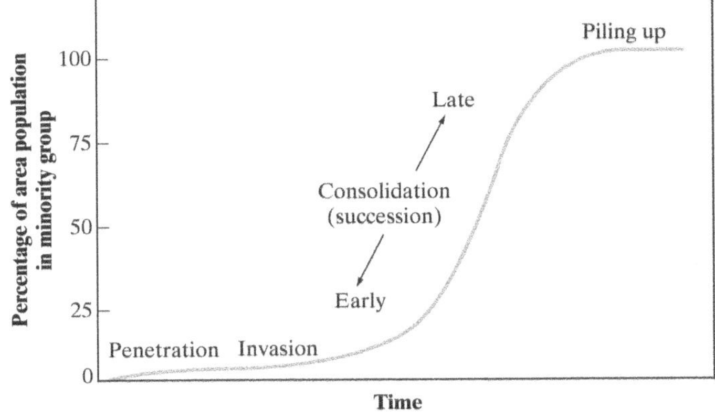

Urban Residential Patterns, R. J. Johnston, Bell 1971. Reprinted with permission.

FIGURE 12 In Human Ecology, the process of "invasion" and "succession" involves typical stages according to the percentage of an area's population comprising the "invading" group.

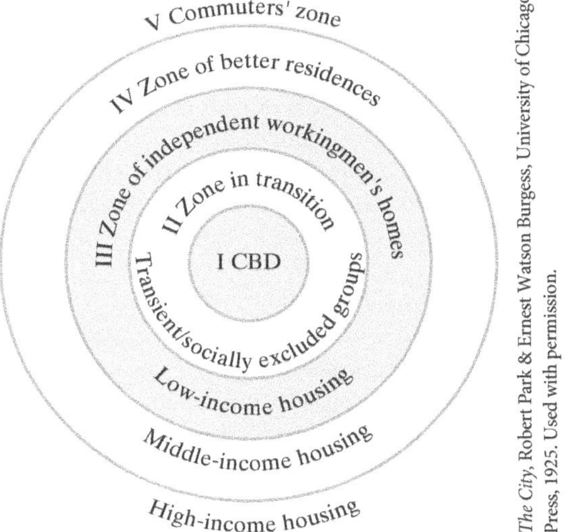

The City, Robert Park & Ernest Watson Burgess, University of Chicago Press, 1925. Used with permission.

FIGURE 13 The ecological model of urban structure, derived from studies of Chicago in the 1920s and 1930s.

that lost out in central city competition removed themselves to peripheral settings.

The innermost zone, the **central business district (CBD)**, was above all the expression of the economic forces of centralization. It was surrounded by a **zone in transition**, comprising an inner belt of factories and warehouses surrounded by deteriorating residential neighborhoods. This deterioration was largely the product, according to Burgess, of the "invasion" of older residential areas by the burgeoning factories and warehouses generated by agglomeration around the CBD. This invasion and deterioration prompted existing residents to leave, which provided housing opportunities for low-income migrants and immigrants. Surrounding the zone in transition was the *zone of independent workingmen's homes*, dominated by second-generation working class migrants and immigrants who had moved upward socioeconomically to the point where they had been successfully able to move outward spatially and "invade" inner suburbs and consolidate themselves there in ethnic communities. Those who made it into the middle classes, or who were born into the middle class, lived in the *zone of better residences*, dominated by single-family homes with spacious yards; or in the *commuters' zone*, small towns and villages with a dormitory function but with little industry or employment of their own.

Although Burgess saw these broad zones as reflections of centralization, decentralization, and the differential economic competitive power of broad groups of people within society, he saw the "natural areas" that made up each zone (Chinatown, Little Sicily, and so on within the zone in transition, for example) as reflections of symbiotic relationships forged on the basis of people's language, culture, and race.

Criticisms of Human Ecology The Chicago School became so influential and the concentric zone model so popular as a device for describing and explaining urban structure that it has become a benchmark in urban studies. It is also important to recognize, however, that the concentric zone model was derived from a particular set of circumstances: a city (Chicago) with a single dominant economic core and a metropolitan area that was growing rapidly through continuous streams of migration and immigration by people that produced a very heterogeneous population. It follows that the model can be applied only to cities where such circumstances exist. It is also important to know that the Chicago School version of *Human Ecology* later came under heavy criticism.

Some of the most stringent criticisms were directed at the relative neglect of what Park and his colleagues had identified (but failed to explore in sufficient detail) as the "cultural" dimension of social organization. Walter Firey, for example, criticized the Chicago School's approach because it overlooked the roles of sentiment and symbolism in people's behavior. He pointed to the evidence of social patterns in Boston where, although there were "vague concentric patterns," it was clear that the persistence of the status and social characteristics of distinctive neighborhoods such as Beacon Hill, the Common, and the Italian North End could be attributed in large part to the "irrational" and "sentimental" values attached to them by particular sections of the population.[26] Firey's point was that social values could—and often did—override impersonal economic competition as the basis for social interaction and residential segregation. Other critics of the Chicago School pointed to the failure of its general structural concepts (such as the "natural area" and concentric zonation) to hold up under comparative examination. The most telling criticisms, however, were those that addressed the biotic analogies on which ecological theory had been developed. With similar analogies being used in an entirely different context to justify the geopolitics of Hitler's Nazi regime, social theories based on biotic analogies were suddenly revealed to be dangerously simplistic.

Factorial Ecology As a result, ecological ideas had to be reformulated, eliminating the crude mechanistic and biotic analogies. Following the early lead of sociologists such as Amos Hawley and Leo Schnore,[27] many analysts now find it useful to think simply in terms of an urban "ecology" consisting of discrete territories (such as "social areas" or "neighborhood types") containing relatively homogeneous populations with distinctive socioeconomic characteristics. One important development prompted by this reformulation was the exploration of the notion of urban ecology as a conceptual framework within which to portray the residential structure of the city through statistical analyses of socioeconomic data. "*Factorial ecology*," involving the use of the multivariate statistical technique of factor analysis, became one of the most widely used quantitative techniques for tackling the complex question of measuring urban sociospatial differentiation (and fits within the spatial analysis approach).

Studies of the factorial ecology of U.S. cities have indicated that the most important dimensions of residential differentiation are, in descending order of importance, people's socioeconomic status, family status or life cycle, and ethnicity—with spatial expressions associated respectively with sectoral, zonal, and clustered differentiation. One of the striking aspects of these findings, beginning with a rash of early studies, was the consistency of these dimensions and their associated spatial patterns from one city to another, often over several decades and despite variations in the input variables used in the statistical analyses. Geographer Robert Murdie suggested that socioeconomic status, family status, and ethnicity represent the main dimensions of social space that, when superimposed on the city's physical template, serve to identify areas of social homogeneity (Figure 14).

With the dimensions of Murdie's idealized model being seen less clearly in cities today, some researchers have suggested that these "classic" spatial patterns are changing. Cities are undergoing fundamental economic transformations that have been accompanied by social, demographic, and technological changes. Advances in telecommunications, for example, have already removed some of the traditional frictions of space, not only for households but also for businesses. This is not to suggest that residential differentiation and segregation

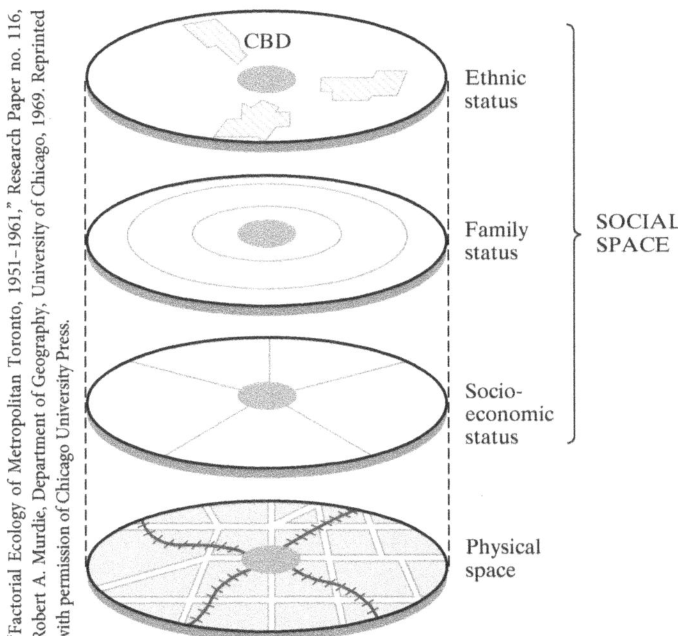

"Factorial Ecology of Metropolitan Toronto, 1951–1961," Research Paper no. 116, Robert A. Murdie, Department of Geography, University of Chicago, 1969. Reprinted with permission of Chicago University Press.

FIGURE 14 An idealized model of the residential structure of U.S. cities, based on the study of factorial ecology.

are disappearing or changing abruptly; but the recent economic and other changes that are affecting cities and the people in them are being manifested in more complex ways and at a finer level of resolution than the "classic" sectors, zones, and clusters that have been associated traditionally with socioeconomic status, family status, and ethnicity.

RECENT CHANGES TO THE FOUNDATIONS OF RESIDENTIAL SEGREGATION

The economic restructuring since the mid 1970s has triggered a number of changes that have resulted in significant rearrangements of urban residential structure. At the heart of these rearrangements are three processes. The first is the *occupational polarization* of workers that has resulted from sectoral shifts in the economy combined with the effects of corporate reorganization and redeployment, advances in robotics and automation, and the increasing participation of women in the labor force. The effect has been an increase in the number of higher-paid jobs (in producer services, high-tech manufacturing, and the media), an increase in the number of low-paid jobs (in routine clerical positions, retail

URBAN VIEW 4
Residential and Economic Structure in European Cities[28]

The classic U.S. models of urban residential structure do not apply well to the diverse social patterns within European cities. A model with concentric circles of increasing socioeconomic status with distance from the core is most applicable to British cities. Mediterranean cities exhibit the **inverse concentric zone pattern** found in Latin America. In Mediterranean Europe the elite concentrate in central areas near major transportation arteries, while the very poor can be found in inadequately serviced parts of the periphery. In Europe it is the demographic pattern that fits a model of concentric circles because the number of people per household usually increases with distance from the city center.

A **sector model** corresponds with the pattern of socioeconomic status where sectors containing different income groups radiate out from the city center. This social pattern reflects the historic preference of the wealthy to locate in pleasant axial areas, such as along monumental boulevards or upwind of pollution sources. The poorer residents were left with unattractive linear sectors of land along railway lines or zones of heavy industry. The multiple nuclei model fits the pattern of ethnic differentiation—different ethnic groups are concentrated either in nodes within the central city or in high-rise public housing projects at the periphery.

Figure 15 shows a model of Northwestern European city structure. The preindustrial core contains the market square and historic structures such as a medieval cathedral and town hall. Apartment buildings contain upper-income and middle-income residents above small street-level shops and offices. The narrow winding streets extend out about a third of a mile. A few wider streets radiate out from the square and form a pedestrianized corridor that runs to the train station and contains national and international department stores, restaurants, and hotels. Skyscrapers are concentrated in the commercial and financial district. New downtown shopping malls or festival marketplaces are in refurbished historic buildings. Some of the old industrial and port areas contain new retail, commercial, and residential waterfront developments. Wealthy mansions grace the western part of the core.

Encircling the core are some zones in transition found in the concentric zone model. The area of the former city wall is a circular zone of nineteenth-century redevelopment. Upper-income residents have bought and gentrified some of the deteriorated housing, while other sections provide low-rent accommodation for students and poor immigrants.

Surrounding this area is another zone in transition—an old industrial zone with disused railway lines. In the 1950s and 1960s, new industrial plants, such as light engineering and food processing, replaced derelict old factories and warehousing. Low-income renters and owners live in the run-down nineteenth-century housing. Some houses have been refurbished or replaced with new units. Certain neighborhoods are quite distinctive because they house foreign immigrants who often live above their exotically painted stores and restaurants.

As in the concentric zone model, beyond this inner area is a zone of "workingmen's homes." This is a stable lower middle-income zone dating from the first half of the twentieth century. These **streetcar suburbs** contain apartment blocks and houses without garages and are anchored by a small shopping area, community center, library, and school. Outside this area are middle-income automobile suburbs, containing apartments and single-family houses with garages, that correspond with the zone of better residences in the concentric zone model. Farther out are nodes containing the most exclusive neighborhoods, as might be expected in the **multiple-nuclei model**.

(continued)

URBAN VIEW 4

Residential and Economic Structure in European Cities (*continued*)

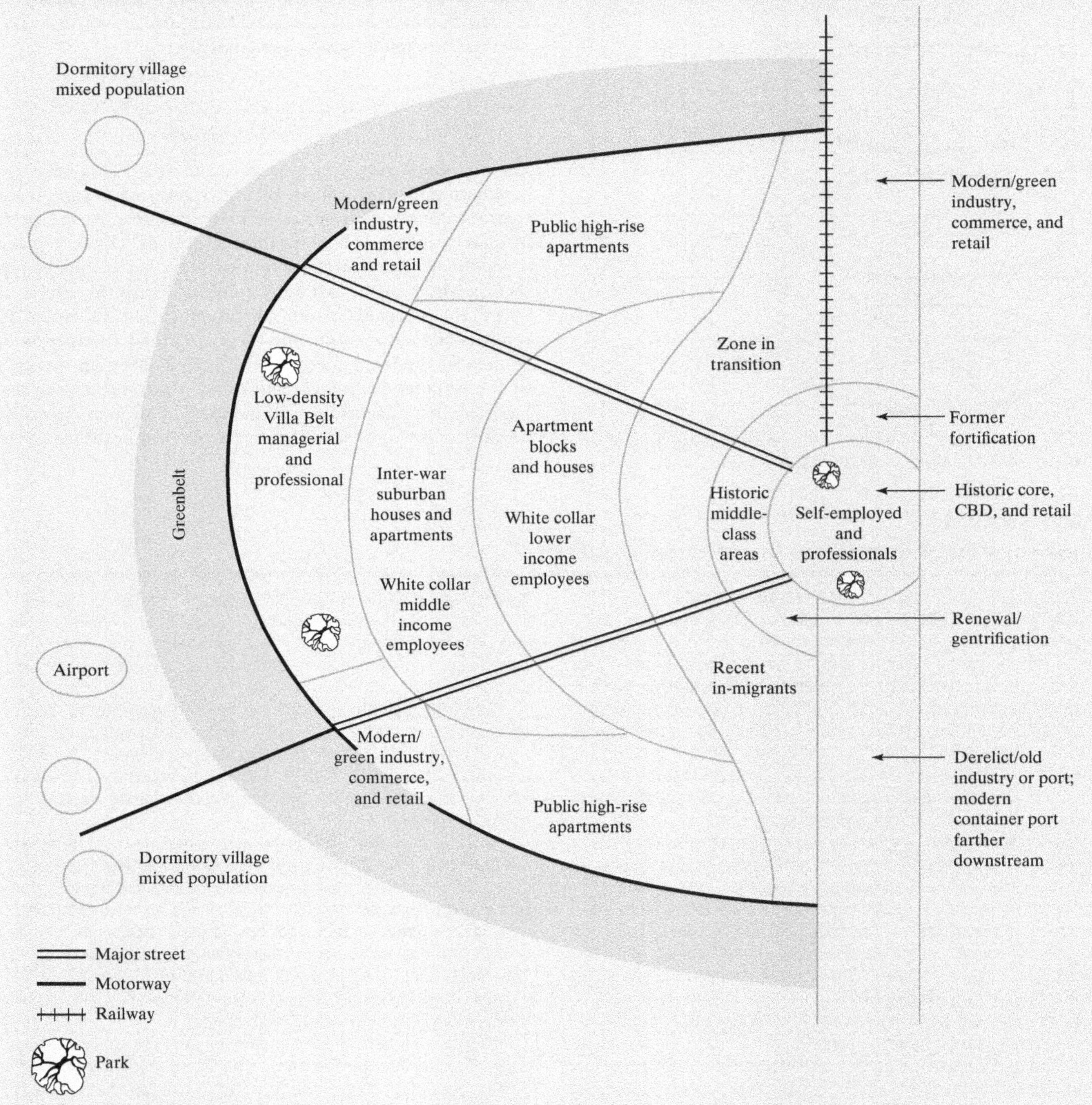

FIGURE 15 Model of Northwest European city structure.

L., McCarthy and D. Danta, "Cities of Europe," in *Cities of the World: World Regional Urban Development*, eds. S. D. Brunn, J. F. Williams, and D. J. Zeigler. Lanham, Md.: Rowman & Littlefield, 2003, 3rd ed., Fig 5.23, p. 198.

The estates of public high-rise apartments and new private middle-income and lower-middle-income "starter" homes at the urban periphery that lack basic amenities like shops and banks also correspond with the multiple-nuclei model. The periphery also contains commercial and industrial nodes, containing shopping malls, business and science parks, and high technology manufacturing.

Beginning in the early twentieth century, cities like London delineated a greenbelt at the edge of the built-up area within which development was prohibited. The greenbelt was intended to prevent urban sprawl and provide recreational space. Commuters live outside the greenbelt in dormitory villages and small towns that correspond with the commuters' zone in the concentric zone model. The airport and related activities such as hotels and modern manufacturing are located farther out on a major freeway.

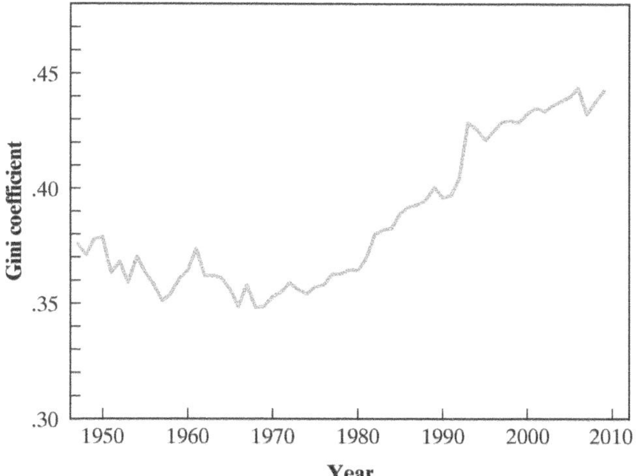

FIGURE 16 Annual changes in family income inequality in the United States. The degree of inequality is measured by the Gini coefficient, which has a range of 0 to 100, with higher values representing greater inequality.

sales, fast food operations, and so on), and a decrease in middle-income jobs (skilled traditional manufacturing). These trends are reflected in the increasing inequality in family incomes in the United States since the 1960s (Figure 16). The new class factions that emerged from this polarization at both ends of the socioeconomic ladder soon left their imprint on urban residential structure, while the thinning ranks of the traditional middle-income groups contributed to the breakup of solid middle-income suburban neighborhoods that previously had been the cornerstones of urban residential structure.

The second set of processes stems from the experience of the *Baby Boom generation*, whose early cohorts found themselves having to cope with the transition from the social awareness, ethnic consciousness, and brotherly love of the counterculture years to the harsh economic realities of labor and housing markets that were suddenly depressed by the effects of **stagflation**. People's attitudes and lifestyles changed, and residential preferences changed with them. Changed attitudes toward work, home, and family, meanwhile, were also reflected in changing patterns of household organization (increases in the proportion of single-person households, single-parent families and dual-income households with no children, for example), and these also came to be reflected in patterns of residential differentiation.

The third set of processes stems from the experience of *ethnic minorities*. The arrival in U.S. cities of new streams of immigrants has recharged the ethnic dimension of residential structure, recalibrated the scales of social distance, and led to the spatial reorganization of ethnic neighborhoods. It has done so differentially, however, because of the differential pattern of origins and destinations. Legal immigration since the 1960s has not only been higher than at any time since the 1920s but it has also been very different in character from previous streams, with fewer Europeans and more people from Latin America, the Caribbean, and Asia. These new immigrants have also been highly localized in their impact—Mexicans locating disproportionately in Los Angeles, New York, Chicago, Miami, and Houston, for example; Puerto Ricans, Dominicans, Jamaicans, Chinese, and Indians in New York; Cubans, with growing numbers of Salvadorans and Nicaraguans, in Miami; Iranians and Salvadorans in Los Angeles; and Filipinos and Chinese in San Francisco.

Meanwhile, the experience of longer-established ethnic minorities has been closely tied to occupational polarization and to the sociocultural changes associated with the Baby Boom generation. For some of these people, these occupational shifts and changed values facilitated behavioral or structural assimilation or both; for others, they resulted in social isolation and economic vulnerability. Black suburbanization, for example, took off in the 1960s with the growth of black middle-income groups (Figure 17). Between 1960 and 2009, the percentage of blacks living in U.S. suburbs rose from 15 to 38 percent (Figure 18).[29] The metropolitan areas with the highest suburban black percentages are in Southern metropolitan areas such as Atlanta.[30] The higher concentration of blacks in the suburbs has not always resulted in higher levels of racial integration however. Suburban blacks in Atlanta, for example, tend to live in predominantly black suburban communities. And, for the United States as a whole, just over half of all blacks still live in central cities—often concentrated in impoverished neighborhoods—compared with more than a quarter of all whites.

New Divisions of Labor, New Household Types, and New Lifestyles

At the upper end of the socioeconomic ladder, the most significant changes have been associated with the growth of producer services occupations (involved in nonmaterial products): management consultants, designers, marketing experts, and so on. These occupations have expanded since the 1960s in the United States to the point where four occupations associated with finance and insurance now comprise 96 percent or more of that employment in metropolitan areas: actuaries, insurance appraisers of auto damage, financial analysts, and brokerage clerks (Figure 19). New York has the highest employment of all four occupations, with nearly one of every five financial analysts and one of every four brokerage clerks located in this metropolitan area. Five life, physical, and social science occupations are found almost entirely in metropolitan areas: political scientists; biochemists and biophysicists; medical scientists (except epidemiologists); sociologists; and industrial-organizational psychologists. Several occupations associated with air travel and public transportation are also concentrated almost exclusively in metropolitan areas. Other occupations concentrated in metropolitan areas include IT occupations, such as semiconductor processors and computer hardware engineers.[31]

At the other end of the socioeconomic ladder, the most significant changes have involved the creation of a *secondary*

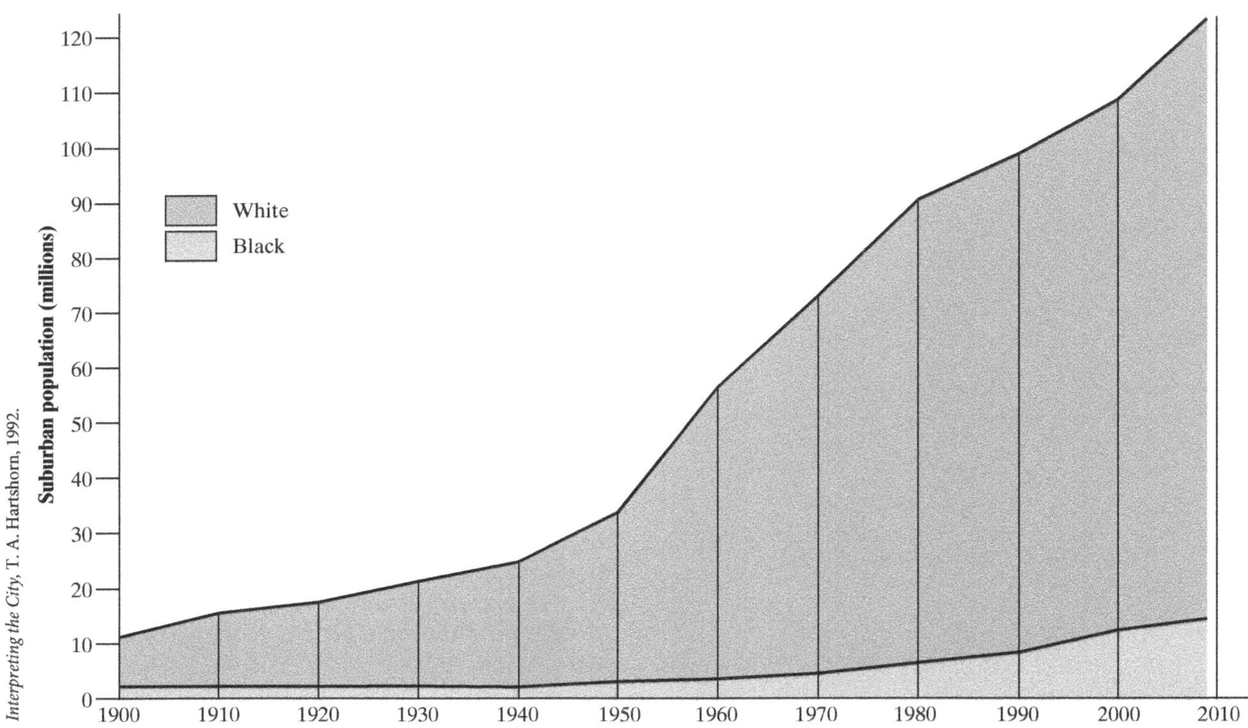

FIGURE 17 Black–white suburbanization trends in the United States.

labor market characterized by workers with low wages, poor working conditions, little or no job security, no benefits, and few prospects of promotion. This labor market has been the product of (1) the relative decrease in metropolitan labor markets of semiskilled and skilled traditional manufacturing jobs (the effect of regional and international decentralization associated with globalization and the **new international division of labor**) combined with (2) new production technologies involving computers and robotics and (3) the growth of unskilled service-sector jobs (such as flipping hamburgers, waiting on tables, and stocking shelves).

The effects for workers of these underlying trends have been compounded by some of the strategies associated with flexible production systems. It has been estimated that at least one in every five jobs in the United States today is "flexible," in that it involves some form of "contingent"[32] temporary, part-time, or independent subcontracting employment arrangement.[33] Elaborate systems of subcontracting, for example, have driven down the wage rates of workers in the many small companies that bid for contract work from corporate giants. **Flexible production systems** in general and subcontracting in particular have also encouraged the use of temporary and part-time labor which, from the employer's point of view, has twin advantages. Employers can fine-tune their workforce according to market conditions, and they can realize substantial savings on the costs of pensions, health insurance, and unemployment insurance fund contributions (which must be paid for permanent, full-time workers). In contrast, such strategies leave the workers vulnerable to unemployment and poverty.

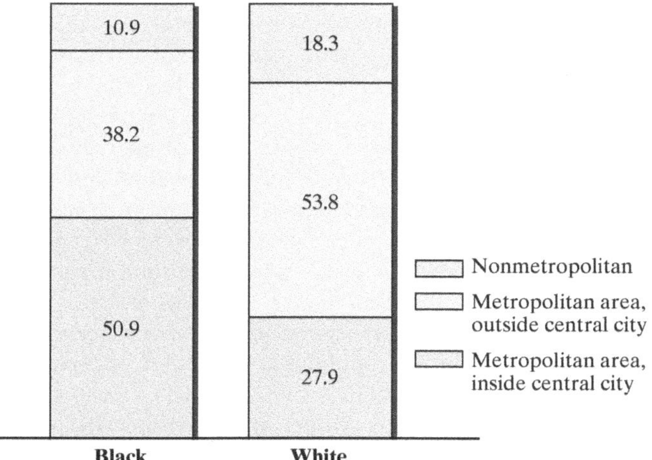

FIGURE 18 U.S. metropolitan and nonmetropolitan residence by race, 2009.

New Roles for Women A particularly important aspect of these trends has been that a large number of part-time and temporary jobs have been filled by *women*. Whereas only about 36 percent of women of working age were actively engaged in

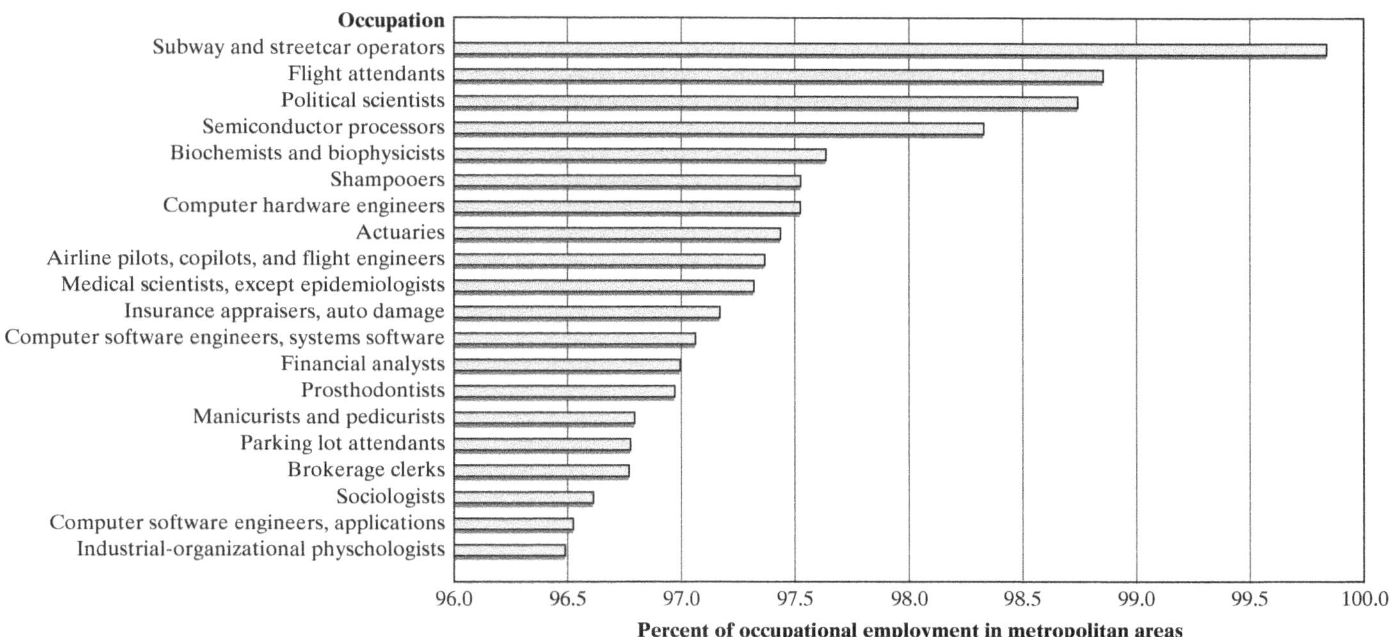

FIGURE 19 Occupations with the highest concentration of employment in metropolitan areas in 2009.

URBAN VIEW 5

The Ethnoburb—A New Suburban Ethnic Settlement

In the large metropolitan areas of the United States for the first time, more than half of all racial and ethnic groups live in the suburbs. Of course, this statistic masks significant variations both in the residential patterns of different minority groups and among different metropolitan areas. Variations in the suburban residential patterns of minorities reflect a variety of conditions that differ across metropolitan areas. These include city-suburban disparities in housing availability and costs, racial and ethnic discrimination, the relative mix and socioeconomic status of an area's minority groups, and the historic development of minority communities as shaped by the specific migration flows and residential patterns of the minority groups in each area.[34]

Compared to blacks or Hispanics, Asians are more likely to live in the suburbs than in central cities. Beginning in the 1960s, upwardly mobile Asians moved out of central city enclaves in search of bigger houses, nicer neighborhoods, and better school districts. Meanwhile, the more recent Asian immigrants with higher educational attainment and professional jobs moved directly into suburban areas.

In the largest metropolitan areas in 2008, 62 percent of Asians were suburban residents (compared to 51 percent of blacks and 59 percent of Hispanics). The majority of Asians are concentrated in a relatively small number of metropolitan areas. Those with the highest Asian suburban percentages are on the West Coast and in Hawaii. Despite their relatively small absolute numbers, Asians comprise 20 percent of the suburban populations in the San Francisco metropolitan area, 28 percent in San Jose, and 35 percent in Honolulu.[35]

And although in the past, Chinatowns in major cities have symbolized the Asian presence in the United States, there are now as many as five distinct Asian national-origin groups in addition to the Chinese—Filipinos, Japanese, Indians, Koreans, and Vietnamese—along with residents from other Asian countries such as Iran and Pakistan. Despite this diversity, the different groups are highly localized in their impact. The New York metropolitan area, for example, has the largest number of Chinese and Indians and the second largest share of Koreans. Los Angeles has the most Filipinos and Koreans and the second largest number of Chinese and Japanese.[36] Meanwhile, the traditional Asian central city ethnic enclaves no longer house the majority of Asian residents. Overall, about half of all Asians lived in the suburbs of metropolitan areas of all sizes in 2009.[37]

Geographer Wei Lei has proposed a new model of ethnic suburban settlement in large U.S. metropolitan areas. In recent decades, under the influence of international geopolitical and global economic restructuring, shifting national immigration and trade policies, and local demographic, economic and political changes, a new type of ethnic area—the **ethnoburb**—has emerged (Figure 20).

Ethnoburbs are suburban ethnic clusters of residential areas and business districts. They are characterized by vibrant ethnic economies that depend on the presence of large numbers of local ethnic minority consumers; they also have strong ties to the global economy that reflects their role as outposts in the international economic system through business transactions, capital circulation, and flows of entrepreneurs and other workers. Ethnoburbs are multiethnic communities in which one ethnic minority group has a significant concentration but does not necessarily comprise a majority. Ethnoburbs function as a settlement type that replicates some features of an ethnic enclave and some

(continued)

URBAN VIEW 5
The Ethnoburb—A New Suburban Ethnic Settlement (*continued*)

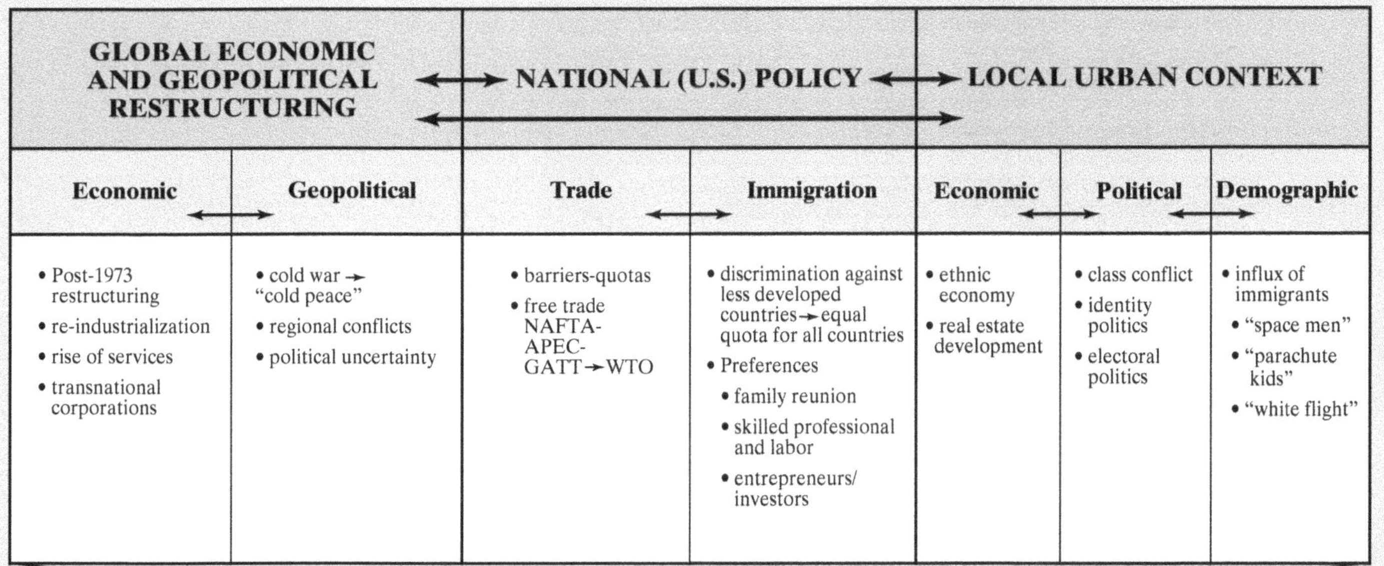

FIGURE 20 The formation of an ethnoburb.

"Anatomy of a New Ethnic Settlement: The Chinese Ethnoburb in Los Angeles," *Urban Studies* 35 (March) (1998).

features of a suburb lacking a single ethnic identity. Ethnoburbs coexist alongside the traditional central city ethnic enclaves.[38]

Wei argues that the clear ethnic imprint on the suburban residential and business landscape sets the ethnoburb apart from the typical U.S. suburb, in which members of ethnic minorities are more dispersed among the white charter group. Her case study analyses focused on the largest suburban Chinese concentration in the United States—the San Gabriel Valley in the eastern suburban part of Los Angeles County.[39] This Chinese ethnoburb evolved in the 1960s in the city of Monterey Park, a suburban bedroom community that was very accessible by the freeway system to the CBD and Chinatown. Since then, the suburbanization of Chinese people and businesses has spread not only throughout this western part of the San Gabriel Valley but also into the east, to suburban areas like Diamond Bar, Hacienda Heights, Rowland Heights, and Walnut.

the labor force in 1960, this number has now reached about 60 percent.[40] This shift includes significant numbers of women who have been able to enter the workforce primarily as a result of changed social attitudes in the wake of the so-called Women's Liberation Movement of the 1960s. It is underpinned, however, by several other trends. Many companies have deliberately recruited women as a flexible source of relatively low-wage and poorly organized (in terms of union membership) labor. Many women have felt compelled to join the labor force in order to maintain household spending power in the face of price inflation. In 1983, wives' incomes represented just 29 percent of total family income; by 2008, this figure had risen to 36 percent. And many more households are now headed by women who must work in order to support dependent children. In 1984, 52 percent of mothers with children under the age of 6, and 68 percent of mothers with children ages 6–17, were in the labor force. These percentages have now risen to 64 and 78 percent respectively.[41]

New Patterns of Household Formation Although the number of conventional two-parent families remained relatively stable, at about 23 million, married-couple families with children as a percentage of all households has fallen from 40 percent in 1970 to just over 20 percent. Meanwhile, single-mother families increased from 3 million to more than 8 million (while the number of single-father families grew from 393,000 to 2.5 million). Overall in 2009, female-headed households with children comprised 25 percent of all families with children.[42]

For whites, the growth in divorce among couples with children was a major reason for the increase in single-parent households. The more liberal attitudes fostered by the counterculture movement of the 1960s and early 1970s and the general increase in female participation in the labor force since then, which has enabled more women to contemplate divorce, have meant that white single-mother families are more likely to result from marital disruption (45 percent were divorced or separated in 2008) than an out-of-wedlock birth (34 percent were never married in 2008).[43]

The increase in single-parent households of African Americans has been particularly striking, and the main reason was not increased divorce rates—black single mothers are least likely to be divorced (20 percent)—but an increase in the numbers of unwed teenage mothers. This increase was less a product of any sustained increased incidence of pregnancy among African-American teenage women and more a consequence of the demise of the "shotgun wedding"—black single mothers are most likely to be never married (62 percent)—in the face of changing social attitudes and, in particular, the diminishing probability that marriage would represent economic support or security (due to, for example, higher college graduation rates among African American women). It was young African-American men whose wages, as a group, grew most slowly, and it was African-American men in general who bore the brunt of the economic and spatial restructuring since the mid-1970s. Figure 21 shows very clearly the somewhat erratic but nevertheless decisive decline in the ratio of African-American to white median family incomes between 1970 and 1993. And despite significant improvement until 2000, the subsequent decline in the ratio to 0.61 by 2009—reflecting a median family income of $38,409 for African-American and $62,545 for white families—captures the continued relatively lower levels of prosperity of many African-American households.

The pattern of household types was further diversified as counterculture sexual liberation movements fostered the proliferation of households of unmarried heterosexual and homosexual couples and of communal households (including small groups of single people who have left home to study or find work). Meanwhile, increased longevity (among whites, at least) contributed to growing numbers of two-person elderly households; and declining fertility levels after the mid-1960s revolution in birth control technology increased the numbers of two-person young-adult households. In addition, the Baby Boomers maintained a conspicuous influence in the mix of household types over the years. Initially, they were the explanation for an increasing number of larger families with young children, and, most recently, for an increase in the number of households consisting of one or more retirees. *This diversification of household types has diminished the utility of the notion of the family life cycle as a basis for understanding residential segregation.* Rather, it is more appropriate to think in terms of a household life cycle consisting of a more broadly defined set of stages, each triggered by events such as entering and leaving school, entering the labor market, marriage, divorce, job change, and retirement, but with many different paths from childhood to retirement.

The diversification of household types has also resulted in large tracts of the city being dominated now by nonfamily households. In Milwaukee, Wisconsin, for example, there is an easterly north-south axis through the heart of the metropolitan area with a concentration of households consisting of unrelated people living together. These households are concentrated in multifamily areas, mostly near the universities (to the northeast of the CBD surrounding the University of Wisconsin–Milwaukee and just west of the CBD surrounding Marquette University) (Figure 22). What is important about all these changes in the present context is that they have altered not only the composition of urban populations that represent the raw material for residential differentiation but that they have also, through their interactions with one another, modified the very foundations of social distance and neighborhood segregation. In particular, they have contributed to (1) increased residential segregation in terms of people's material culture and lifestyle and (2) the spatial isolation of "new" groups of vulnerable and disadvantaged households.

Increased Materialism and New Lifestyles With the intersection of the maturing Baby Boom generation and the growth in high-paying **producer services** occupations (financial analysts, management consultants, marketing experts, and so on), the cultural landscape changed significantly. A much greater emphasis by people on materialism and style came about,

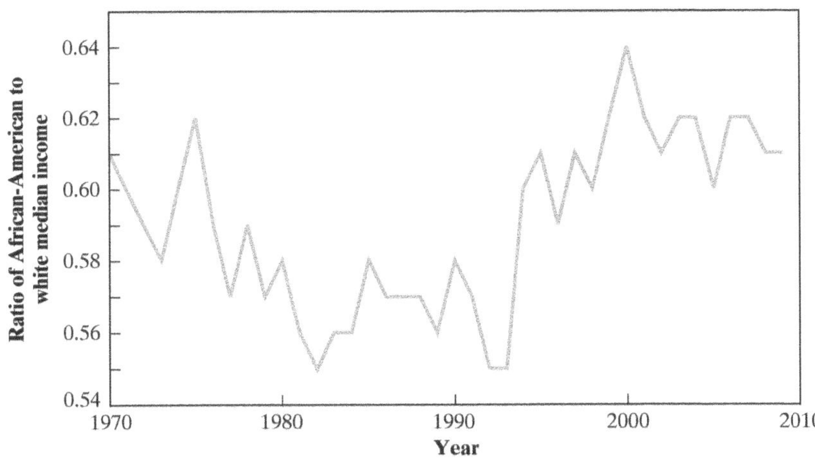

FIGURE 21 Ratio of African-American to white median family income in the United States.

Maryjo Walicki/MCT/Newscom

FIGURE 22 Many students, like these University of Wisconsin-Milwaukee undergraduates, live in nonfamily households while attending university.

only a few years after confident predictions of a pronounced societal trend of declining materialism.[44] The Baby Boomers, whose formative experience was the postwar economic boom, provided the main basis for these predictions. Their rebellion against the apparent complacency of industrial progress and affluence was channeled into a countercultural movement with a collectivist approach to the exploration of freedom and self-realization through a variety of movements, including radical politics, drugs, and sexual liberation. The high-water mark of this "alienation generation"[45] was 1968, the year of sit-ins, student worker alliances, protest marches, general strikes, civil disorder, and riots. The failure of these events produced a generation characterized by materialism and consumerism with a pluralism of taste: a sociocultural environment in which the emphasis was not so much on ownership and consumption per se but on the possession of particular *combinations* of things and the *style* of consumption.

Meanwhile, the space-time compression of new communication technologies and the globalization and homogenization of consumer culture fostered the perceived need for distinctiveness and identity: "*This society eliminates geographical distance only to reap distance internally in the form of spectacular separation.*"[46] In this *society of the spectacle*, where emphasis is on appearances, the symbolic properties of urban settings and people's material possessions have assumed unprecedented importance. Social distinctions, previously marked by people's ownership of particular kinds of consumer goods, now have to be established via the symbolism of "aestheticized" commodities.

In addition, maturing Baby Boomers found themselves flooding labor and housing markets just as the economy was experiencing the worst recession since the 1930s. Wages stood still while food and house prices ballooned. Here is one personal testament:

> In our 20s, my friends and I hardly cared. We finished college (paid for primarily by our parents), ate tofu, and hung Indian bedspreads in rented apartments. We were young: it was a lark. We scorned consumerism. But in our 30s, as we married or got sick of having apartments sold out from under us, we wanted nice things, we wanted houses.[47]

Yet economic circumstances did not permit a smooth transition to materialism, even for college-educated middle-income individuals:

> My friends dressed and ate well, but despite our expensive educations, most had only one or two elements of the dream we had all laughed at in our 20s and now could not attain. We had to choose between kids, houses and time. Those with new cars had no houses; those with houses, no children; those with children, no houses. A few lucky supercouples—a lawyer married to a doctor, say, with an annual combined income of more than $100,000—had everything but time.[48]

Unable to fulfill the American Dream, people saved less, borrowed more, deferred parenthood, comforted themselves with the luxuries that were advertised in glossy magazines as symbols of style and distinctiveness, and generally surrendered to the hedonism of lives infused with extravagant details. The signs—literally—of this new materialism were everywhere in the 1980s: bumper stickers advising SHOP TIL YOU DROP or HE WHO DIES WITH THE MOST TOYS WINS; designer labels—Prada, Dolce & Gabbana, Louis Vuitton, Hugo Boss—deliberately left on clothes and accessories.

The culture of stylish materialism that developed also brought new patterns of social behavior. Barbara Ehrenreich argues that a cornerstone of the "yuppie strategy" was the determination of both men and women to find proven wage-earners as potential partners to marry or live with. The result, she suggests, was the consolidation of a class faction characterized by high educational status and by very high household incomes. The importance of finding suitably qualified partners had, in turn, intensified the potency of material signifiers:

> . . . since bank accounts and resumés are not visible attributes, a myriad of other cues were required to sort the good prospects from the losers. Upscale spending patterns created the cultural space in which the financially

> well matched could find each other—far from the burger eaters and Bud drinkers and those unfortunate enough to wear unnatural fibers.[49]

We can see how these phenomena translate into the urban landscape in a number of ways. Because, for instance, style and distinction are so important to people's consumption (and even to the very experience of shopping), the *retail* landscape became polarized. To avoid going out of business, department stores, for example, were faced with the choice of specializing at either end of the market—upscale for the wealthy or discount for the poor and thrifty. Bloomingdale's and Neiman-Marcus are at one extreme and Wal-Mart and dollar stores are at the other. Undifferentiated chains, such as Korvette's and Gimbel's, with a middle-income consumer market, were forced to close. Sears and JC Penney anxiously attempted to "reposition" themselves to survive in the increasingly segmented market.[50]

The *residential* landscape began to reflect the increased importance of materialism and stylish distinction in a variety of ways, including the **gentrification** of increasing numbers of central city neighborhoods through the physical rehabilitation and social change associated with "invading" young professionals and the development of *private, master-planned communities* as suburban settings for recreation-oriented and consumption-oriented lifestyles. Increasingly, as different lifestyles have proliferated within the various socioeconomic groups, household types, and ethnic groups, developers have provided a greater variety of residential settings (condominium apartments, townhouses, retirement communities, master-planned communities), "packaging" them with keynote design themes and key lifestyle amenities. Thanks to the time-space compression of technological advances such as cable TV and cheaper and more effective telecommunications such as the Internet and smart phones, the resultant lifestyles are not localized and idiosyncratic but, rather, *replicated across the whole country*. Thanks to specialized relocation consultant services and the emergence of nationwide realtor networks such as ERA, Century 21, Coldwell Banker, and Red Carpet, people who can afford to can easily find the kind of lifestyle setting they are interested in and easily move from one to another. The result has been described as a *mosaic culture* that takes the form of "archipelagos" of similar lifestyle communities stretching from coast to coast, with "islands" in every major metropolitan area.[51]

URBAN VIEW 6
"Inconspicuous Consumption?"

Besides the decennial census, the U.S. Bureau of Labor Statistics Consumer Expenditure Survey (CE)[52] is perhaps the most important survey that provides researchers and businesses with insights into consumer behavior. The survey reveals that during the 1990s and 2000s most people in the United States were not the "conspicuous consumers" that economist Thorstein Veblen described more than a century ago in *The Theory of the Leisure Class* (1899)—a label, in fact, that better fits 1980s spending and consumption patterns. Since then, Veblen would be hard-pressed to find the abundance of material goods whose purchase he criticized as a primitive form of snobbery and self-doubt. Instead, as journalist Michael Weiss, who wrote the book *The Clustered World: How We Live, What We Buy, and What It All Means About Who We Are*,[53] saw it, there was "less flash and more cash for life's big and little necessities."[54] Many bargain-hunting consumers were looking for opportunities to buy more for less. Certainly, household spending rose until 2009[55] but even before the economic downturn affected people's expenditures many consumers were spending a larger share of their budgets on health care, housing, and transportation and less on new clothes, jewelry, eating out, and entertainment.

Demographic trends played an important role in influencing consumer spending The Baby Boomers who began reaching age 45 in 1991—the beginning of the peak earning years in the United States—were most influential. The large number of these older Boomers contributed to an increase in spending on health insurance and prescription drugs (more aches and pains), higher education (more kids in college), and housing (more homeownership).[56]

The category of spending that experienced the greatest decline since the 1980s was clothing.

> Among younger age groups, the drop in clothing expenditure reflects more casual attitudes toward fashion . . . fashionistas deemed it acceptable to mix and match pricey fashions with discount rags. . . . It became OK to have a Gucci bag with a $20 raincoat from Target. . . . You could get your T-shirts at the Gap and cashmere sweaters from Bergdorf. . . . The bottom line for such eclecticism was lower clothing bills. That trend also reflects a values shift in which people are less brand-conscious and more time-pressed. Today, working couples who have less time to shop are looking for convenience and value when they go shopping. . . . Boomers don't care what pair of khakis they buy as long as it has the best price. . . . As for the Gen Ys, they'd rather have the coolest electronics than the coolest clothes. . . . Such changing priorities have helped discounters like Target and Wal-mart, while hurting specialty stores like Gap and Abercrombie & Fitch. . . . Apparel used to be about image. Now it's about function.[57]

During the coming decades, of course, that thinking may change as a result of the complicated interplay within cities of economic conditions such as the continued fallout of the financial crisis of the late 2000s and socioeconomic status, household types, ethnicity, and lifestyle. In 2011 the oldest members of the Baby Boom generation reached 65 and are enjoying longer life expectancies than any previous generation. Although it is not certain how future consumption and housing choices will differ from today's, what is certain is that the sheer size of the Baby Boomers as a demographic group means that their choices will continue to translate visibly into the urban landscape.

Social Polarization and Spatial Segregation

The "new economy," based on digital technologies, economic and cultural globalization, and the growth of employment in finance, insurance, real estate, biomedical and dot-com industries, has produced a "winner-take-all" society, with the gap between wealthy households and the poor increasing sharply. A distinctive feature of the new economy is that higher-income earners have emerged in occupations that have only a weakly established social status. A "new bourgeoisie" has emerged consisting of "symbolic analysts:" economists, financial analysts, management consultants, personnel experts, designers, marketing experts, purchasers, and so on. They have been joined by a "new petit bourgeoisie" dominated by well-paid junior commercial executives, engineers, skilled high-technicians, medical and social service personnel, and people directly involved in cultural production: authors, editors, radio and TV producers and presenters, magazine journalists, and the like. Together, the new bourgeoisie and new petit bourgeoisie now dominate the top of the economy, with salaries commensurate with their increasing economic influence. In 1975, the most affluent 20 percent of U.S. households accounted for 43.6 percent of total (pre-tax) household income, while the top five percent accounted for 16.5 percent. By 2009, the numbers were 50.3 percent and 21.7 percent respectively. The average income of the top 20 percent of households in 2009 was $170,844, up from the 1975 average of $109,028 in constant (2009) dollars.[58] The average income of the top five percent meanwhile had jumped to $295,388. These are the households that set an unobtainable ever-rising standard for everyone else of relentless lifestyles of competitive consumption and grab-it-yourself materialism. Instructed by glossy lifestyle magazines like *Architectural Digest, Best Life, Cosmopolitan, Country Living, GQ, House Beautiful, Martha Stewart's Living, Pool & Spa Living, Stuff, Success,* and *Trump Magazine*, America's upper-income classes have developed chronic cases of "luxury fever" and "affluenza."[59] Their lifestyles are increasingly played out in neighborhoods that are highly segregated from the rest of the city's residents. Robert Reich points to the "secession of the successful":

> In many cities and towns, the wealthy have in effect withdrawn their dollars from the support of public spaces and institutions shared by all and dedicated the savings to their own private services. As public parks and playgrounds deteriorate, there is a proliferation of private health clubs, golf clubs, tennis clubs, skating clubs and every other type of recreational association in which costs are shared among members. Condominiums and the omnipresent residential communities dun their members to undertake work that financially strapped local governments can no longer afford to do well—maintaining roads, mending sidewalks, pruning trees, repairing street lights, cleaning swimming pools, paying for lifeguards and, notably, hiring security guards to protect life and property.[60]

For many people, it seems, secession from the city leads to isolation and loneliness. Robert Putnam captured the isolation of contemporary suburbia in his influential book, *Bowling Alone*. Putnam argues that people's social connectedness and civic engagement in the United States reached a peak between 1945 and 1965 and then declined precipitously. On average, Putnam found, involvement in political and community events declined by about 25 percent between 1973 and 2000. This includes declines in presidential voting as well as declines in local civic engagement such as volunteering to serve on town committees, local organizations, and clubs. Levels of basic socializing among people also dropped, as did the general level of mutual trust. The number of people in the United States attending club meetings declined by 58 percent; family dinners declined 33 percent, and having friends over declined by 45 percent.

Putnam ascribes these changes to a combination of factors, including generational change, the pressures of maintaining dual-career households, and the growth of electronic entertainment: multi-channel cable and satellite TV, DVDs, electronic games, the Internet. Another significant factor, he concludes, is suburbanization. The spread of people's daily lives across work, shopping, and errands in the decentralized and fragmented metropolis tends to pull apart any community boundedness that may exist. Ever-longer commutes mean that the amount of time spent driving alone has reduced not only the amount of time available for socializing and civic engagement but also people's energy and inclination to do so. Putnam also suggests that the fragmentation of the leading edges of suburbia into "enclaves segregated by race, class, education, life stage, and so on" undermines rather than sustains community. The homogeneity of these enclaves, he argues, reduces the local disputes and issues that draw neighbors into public contact.[61]

At the bottom end of the socioeconomic ladder, the intersection of occupational shifts, immigration streams, and the cultural and demographic dynamics associated with the Baby Boom generation has resulted in some dramatic changes in the composition of vulnerable and disadvantaged households. Since the 1970, for example, the percentage of both single-parent families with children and low income families have increased, the poverty rate and unemployment rate have both increased, the percentage of immigrants has increased, while the percentage of people aged 25 or older who graduated from high school has decreased (Table 3). At the same time, an increasing proportion of families are *multiply disadvantaged*,[62] and were especially hard hit by the late 2000s downturn in the economy.

Although the 1990s saw some slight improvement, the changes in the composition of vulnerable and disadvantaged households within U.S. metropolitan areas since the 1960s have not been encouraging (Table 3). The trends that contributed to these changes (occupational shifts, corporate reorganization and redeployment, and so on) have also contributed to shifts within the **urban system**, so that there has

TABLE 3 Selected Socioeconomic and Demographic Indicators (Percent of Total) for U.S. Metropolitan Areas, 1970–2009

	1970	1980	1990	2000	2009
Single parent families with children	13.2	19.9	23.5	28.5	29.9
Poverty rate*	11.4	11.5	12.1	11.8	14.3
Low income families (national lowest 20% income bracket)**	17.0	18.2	17.8	19.1	19.0
Persons aged 25 or more who graduated from high school**	31.7	34.5	28.7	26.9	23.4
Unemployment rate	4.2	6.2	6.1	5.7	9.8
Immigrants (foreign born population)	5.7	7.4	9.5	13.0	14.3

*Figures are for 1969, 1979, 1989, and 1999.

**Figure is for 2008 (not 2009).

Source: U.S. Department of Housing and Urban Development, *State of the Cities Data Systems (SOCDS)*. (Available at: *http://www.huduser.org/portal/datasets/socds.html*) for 1970–2000 data; U.S. Census Bureau's American Community Survey, 2009 (*http://www.census.gov/acs/www/*); U.S. Census Bureau Current Population Survey (CPS), Annual Social and Economic (ASEC) Supplement, Table HINC-05 (*http://www.census.gov/hhes/www/cpstables/032009/hhinc/new05_000.htm*).

been a great deal of variability *among* and *within* metropolitan areas in the changing composition of their vulnerable and disadvantaged populations.

An important aspect of these changes is the **feminization of poverty**, particularly among minority populations (Figure 23). Female-headed households (with no spouse present) represented just under 33 percent of all households below the poverty level in 2009, while the percentage for African American and Hispanic female-headed families was much higher at about 40 percent. As a result, one in five children in the United States are now being raised in poverty. A factor in this trend has been the relatively high teenage pregnancy rate in minority communities where economic restructuring had left high levels of unemployment. Other contributory factors include the persistence of a gender gap in wages, inequitable division of assets at the time of divorce, and the ineffective implementation of child support agreements.

It is important to bear in mind that neoliberalism since the mid-1970s has resulted in policy shifts that raised taxes and

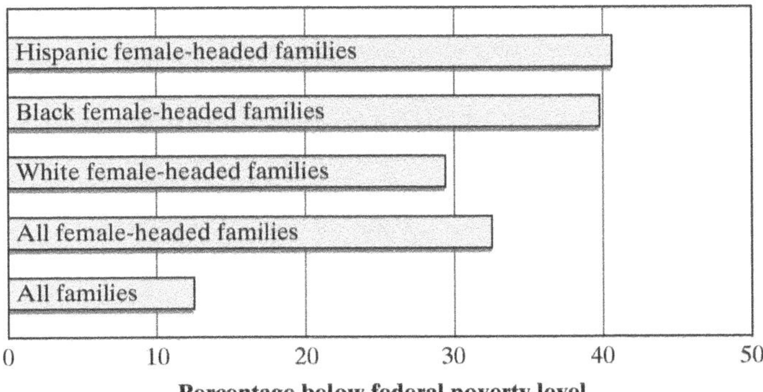

FIGURE 23 Families below the federal poverty level, United States, 2009.

dismantled much of the welfare safety net. Between 1979 and 2006 the average after-tax income of the bottom fifth of households increased by 10.7 percent (compared to 86.5 percent for the top fifth). But during that time the share of after-tax income of the bottom fifth of households fell from 6.8 percent to 4.7 percent (while that of the top fifth rose from 42.4 percent to 52.1 percent) (Figure 24). Unemployment benefits have been reduced, employment and training programs cut, AFDC (Aid to Families with Dependent Children) benefits held down and then replaced in 1996 with the more restrictive Temporary Assistance for Needy Families (TANF), eligibility requirements for work disability benefits tightened, and federal aid to economically distressed cities and neighborhoods virtually abolished.

Congress has amended the Fair Labor Standards Act several times to increase the federal hourly minimum wage—from $2.00 in 1974 to $7.25 in 2010. But because the minimum wage is not indexed for inflation—and during that time, the consumer price index rose by over 218 percent—the value of the increase in the minimum wage has been lost. In fact, the real (inflation-adjusted) value of the minimum wage is now *far less* than it was in 1974.

Sociologist-planner Manuel Castells argues that the result has been the increasing exclusion of the most marginal and vulnerable groups in society,[63] while sociologist William J. Wilson has advanced the concept of an *underclass* of people who have become socially isolated and economically detached from the labor force: mostly African Americans, with a disproportionate share of female-headed households that have become dependent on welfare benefits.[64]

Most striking among the landscapes of the excluded are *impacted ghettos*—spatially isolated concentrations of the very poor (Figure 25), usually (though not always) dominated by African Americans and often denuded of community leaders and containing very high concentrations of

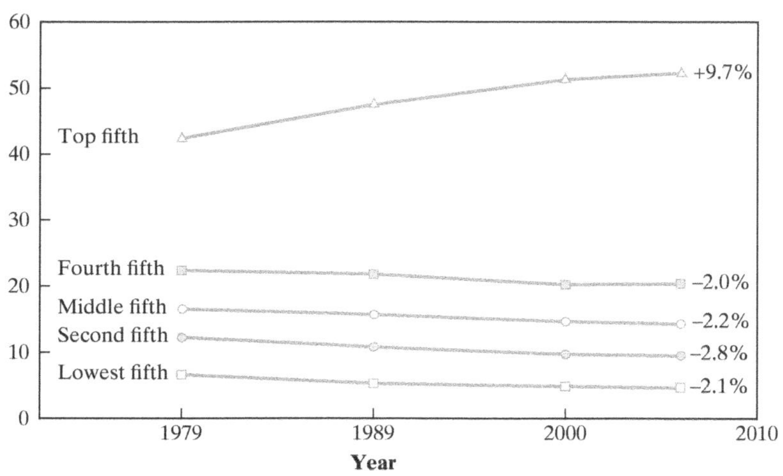

FIGURE 24 Shares of after-tax income of U.S. households, by income group.

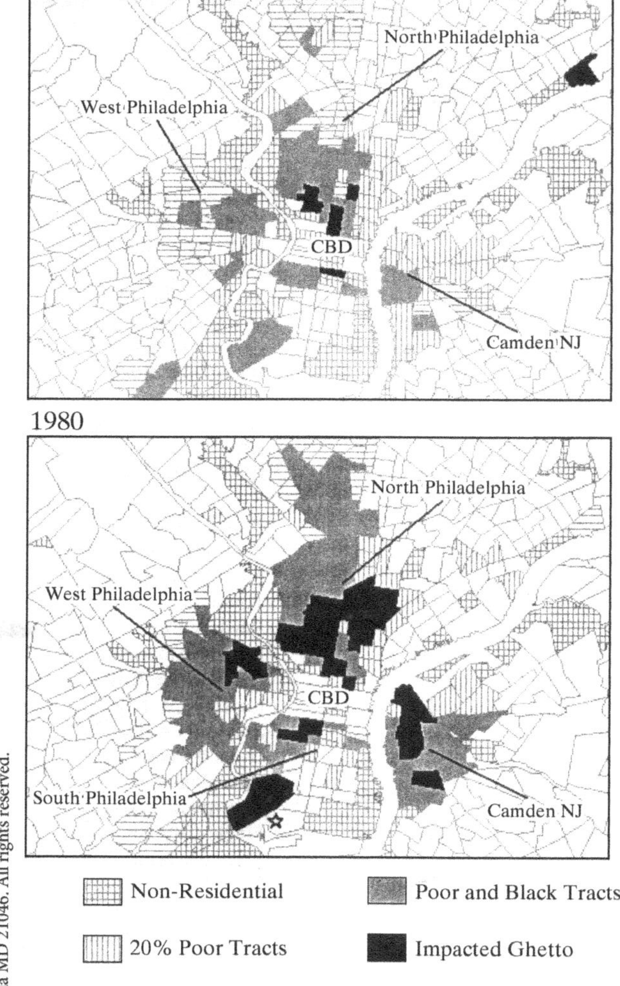

FIGURE 25 The growth of the "impacted" ghetto in Philadelphia.

female-headed households struggling to survive in downgraded environments that also serve as refuges for criminals. Less visible, but more decisively excluded, are the "landscapes of despair" inhabited by the homeless:[65] microspaces that range from the vest-pocket parks of downtown Los Angeles, the city square in San Francisco, and the federal area of the District of Columbia to the anonymous alleyways and park benches of every big city.

THE NEW RESIDENTIAL MOSAIC: "LIFESTYLE" COMMUNITIES

> The reorganization of world markets has expanded the consumption functions of mature urban economies, creating new jobs and new spaces of consumption. Many of these jobs are low-paying jobs in stores, restaurants, hotels and domestic and personal services. While many of the new consumption spaces rely on a high level of skill and knowledge, and provide cultural products of beauty, originality and complexity, others are standardized, trivial and oriented toward predictability and profit. At the same time, individual men and women express their complex social identities by combining markers of gender, ethnicity, social class and—for want of a better word—cultural style. Many of these markers are created in, and diffused from, cities: on the streets, in advertising offices and photography studios, on MTV. Many of the people who create these markers live in cities, too. They are artists, new media designers, feminists, gays, single parents and immigrants . . . urban cultural diversity holds a curious and yet wondrously creative mirror to the paradox of polarization: while cities become more like other places, they continue to attract the extremes of poor, migrant and footloose urban populations and the very rich. Their ability to forge 'urban' lifestyles continues to be the city's most important product.[66]

URBAN VIEW 7

GIS Marketing Applications Help Starbucks to Brew up Better Locational Analyses[67]

Like other companies in the retail and food business, Starbucks Coffee Co. uses GIS—Geographic Information Systems—in its locational analysis when determining where to locate a new store (Figure 26). In CBDs, Starbucks aims to take advantage of high volumes of pedestrian traffic. Figure 27 shows a GIS map of walk times in the vicinity of one coffee shop. This kind of analysis—based on sales forecasting data—helps retail outlets like Starbucks to explore the potential of specific locations so that they can make sound site-selection decisions. Maps like this one can help site-selection teams reduce the risks inherent in attempting to identify a good location for a new store by improving the chances that the new location will cater to a high-volume and responsive customer base.

In addition to incorporating existing government-generated socioeconomic statistics, such as census data, ZIP codes, highway data, and city street maps, maps such as this can also contain data on sales and customer characteristics that the company generates from so-called store intercept studies (which some people are concerned about because of the potential for invasion of privacy through the coupling of different sorts of data sets). Through the application of GIS technologies, companies like Starbucks are also able to uncover spatial patterns, such as residence and consumer spending, that they can then use to help implement more effective marketing campaigns.

Dr Wilfried Bahnmüller/photolibrary

FIGURE 26 High volumes of pedestrian traffic are important to the success of coffee shops because coffee drinkers tend to stop in on an impulse, as at this street cafe on Kohlmarkt Square, 1st district, Vienna, Austria.

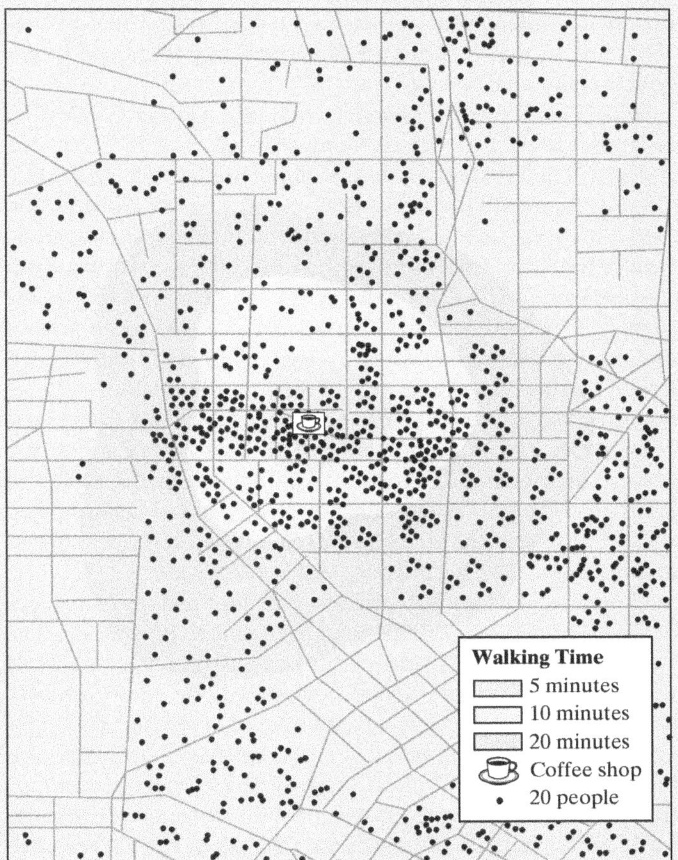

Places and Regions in Global Context: Human Geography 3rd ed., P. Knox, S. Marston, Prentice Hall (2004). Reprinted with permission.

FIGURE 27 Walk times to different trade areas in a CBD—maps like this help Starbucks to understand the hinterland area from which an individual coffee shop is likely to draw its customers.

In societies where traditional institutions and class markers are being eclipsed by fragmented and segmented lifestyle groupings and where individuals have much greater expectations than ever before about expressing and creating their own individual identity, the consumption of everything from houses to furniture and clothes is central. Houses, neighborhoods, interior design, clothes, gadgets, food—everything—is now loaded with meaning. These meanings are shared among people in social groups and become key markers of status, lifestyle, and identity. "Consumption is not simply an act of purchase in pursuit of needs and wants but a social process in which different groups relate to specific goods and artifacts in complex ways, deploying symbolic languages of exclusion, entitlement, and distinction. More importantly, different patterns of consumption have become embedded in a variety of lifestyles."[68] This differentiation makes for a socio-cultural environment in which many people put an emphasis not so much on ownership and consumption *per se* but on the possession of particular *combinations* of things and the *style* of consumption. As a result, the social ecology of the city is a fragmented mosaic, differentiated into neighborhoods that reflect the segmented lifestyle groupings and class factions of contemporary society.

Mapping and analyzing this mosaic requires detailed information about the patterns of people's consumption that are so important to status and identity for many U.S. consumers. Consumer research consultants, using multivariate statistical methods similar to those previously used by urban geographers to tease out the spatial structure of cities, have identified the matrix of contemporary household types in the United States according to their distinctive consumption patterns and their typical residential settings.

The best-known and most comprehensive classification of landscapes is the Nielsen Company's Claritas PRIZM® NE system.[69] It is based on block-level population data from the U.S. Census, merged with consumption and lifestyle information from list-based records for nearly 200 million households, along with detailed profiles of more than 890,000 households from sources such as R.L. Polk's new car buyers and Simmons lifestyle surveys. The PRIZM system crunches these data into fourteen major socio-geographic groups: Urban Uptown, Midtown Mix, Urban Cores, Elite Suburbs, Affluentials, Middleburbs, Inner Suburbs, Second City Society, City Centers, Micro-City Blues, Landed Gentry, Country Comfort, Middle America, and Rustic Living. Each of these, in turn, is subdivided, giving a total of 66 distinctive market segments, each numbered according to socioeconomic rank (taking into account characteristics such as income, education, occupation, and home value).

For example, Elite Suburbs accounted for just over 5 percent of all U.S. households in 2009: roughly 6 million households. The 1.5 million households in the most exclusive market segment within this group (and ranked number 1 in the entire country because they have the most opulent standard of living) are Upper Crust households, dominated by empty-nesting couples between the ages of 45 and 64. Typical aspects of this segment's consumption profile include expenditures of $3,000 or more on foreign travel each year, shopping at Saks Fifth Ave., subscriptions to *Conde Nast Traveler* magazine, watching the Golf Channel, belonging to a country club, and driving a Mercedes SL Class. Blue Blood Estates (ranked number 2 overall) account for a further almost one million households (about 1 percent of the total) and are characterized by married couples with children, college degrees, a significant percentage of Asian Americans and six-figure incomes earned by business executives, managers and professionals living in million-dollar homes with manicured lawns, driving high-end cars like the BMW 750, reading *Architectural Digest* and *Scientific American,* and playing tennis at exclusive private clubs. Movers & Shakers (number 3; 1.9 million households: 1.6 percent of the total in 2009) represent the U.S.'s up-and-coming business class who live in a wealthy suburban world of dual-income couples who are highly educated, typically between the ages of 35 and 54, often without children. Typical aspects of their segment's consumption profile include downhill skiing, eating at Bertucci's, reading *Inc.* magazine, and driving Land Rover Range Rovers. The Winner's Circle segment (number 6; 1.3 million households; 1.1 percent) is characterized by affluent younger households: mostly 35- to 54-year-old couples with large families in new-money subdivisions. Surrounding their homes are the signs of upscale living: recreational parks, golf courses, and upscale malls. With a median income of over $100,000, Winner's Circle residents are big spenders who like to travel, ski, go out to eat, shop at clothing boutiques like Ann Taylor and drive a Mercedes GL Class. The residential mosaic revealed by market segment analysis of the most affluent in society reflects the "secession of the successful" into lifestyle enclaves that are accommodated in master-planned developments and upscale subdivisions. It is, in simple terms, a reflection of demand and supply, mediated by focus groups and advertising. On the demand side, there is an implicit expectation on the part of wealthy consumers that secession brings with it a sense of community or communality through voluntary segregation of like-minded households. On the supply side it is explicit: "community" is a key sales feature in the consumer housing industry, promised as part of every package of amenities by developers.

Nevertheless, there is more to the secession of the successful than the quest for lifestyle communities. Richard Sennett has emphasized the importance, in contemporary society, of avoidance of exposure to "otherness," noting how cities are organized either around spaces limited to (and carefully orchestrating) consumption; or spaces limited to (and carefully orchestrating) experience. "'Exposure' more connotes the likelihood of being hurt than of being stimulated. . . . What is characteristic of our city-building is to wall off the differences between people, assuming that these differences are more likely to be mutually threatening than mutually stimulating. What we make in the urban realm are therefore bland, neutralizing spaces, spaces that remove the threat of social contact . . ."[70] What Sennett is talking about here has more to do with exclusionary social segregation than the propagation of lifestyle and the quest for community. Participating in this voluntary segregation, he argues, people draw themselves a picture of who they are, an image that excludes anything that might convey a feeling of difference or dissonance: a "purification ritual" that produces a myth of community solidarity.

This desire for avoidance by some people of exposure to "otherness" highlights a potential drawback of market segment analysis; it can result in stereotypical and polarized descriptions of different groups of people within metropolitan areas. By comparing, for example, the labels for some of the different market segments, the contrast is clear between labels like "Upper Crust" and "Blue Blood Estates" versus "Shotguns & Pickups" and "Back Country Folks." This kind of market segment analysis is particularly problematical if it promotes stereotyping that causes some neighborhoods to be labeled as poor prospects for certain kinds of economic or other activities (as in **redlining**, where mortgage-lending agencies refused to lend in certain areas). The concern is that the Claritas PRIZM® NE "You Are Where You Live" assumption may become a self-fulfilling prophecy if companies target customers based on the market segment descriptions.

The gap between "urban legends"—such as, "no one works in central city neighborhoods"—and urban realities is wide, and often the private marketing data that people rely on to provide accurate neighborhood information actually exacerbate these myths. These marketing data have several problems. They are often based on infrequently updated Census information that undercounts central city residents; they make misleading generalizations (e.g., crime statistics are not based on the number of crimes committed in the neighborhood but estimated from the "type" of people living there); and they fail to review local data for trend analysis, thereby missing many positive developments in cities. The data have serious repercussions for cities, influencing the location and product decisions of businesses; government policies on welfare, housing, and bus routes; and even prospective students' choice of university.[71]

The main point to repeat here is that lifestyle communities are complex and multidimensional—and not amenable to all-encompassing descriptions. The actual diversity within particular city blocks and even within individual households, of course, mark an important break from the idea of a generalized pattern of residential differentiation as depicted in Figure 14; but they follow from the regional and functional differentiation that is inherent to the economic and demographic trends that we have seen are influencing the cities and metropolitan areas in the United States.

FOLLOW UP

Key Terms

assimilation
charter group
class formation
class structure
congregation
ethnoburbs
factorial ecology
feminization of poverty
Human Ecology
index of dissimilarity
lifestyle communities
neighborhood effect
social distance
socialization
stereotyping

Review Activities

1. Go online to the U.S. Census Bureau's website and use its American FactFinder search tool to gather some socioeconomic and demographic indicators for Detroit, Los Angeles, and Phoenix. Use your library or surf the Web to try to identify particular trends and factors that can help you to explain some of the differences and similarities that you identified among these three urban areas.
2. Go online to the Nielsen Claritas PRIZM® NE site at: *http://www.claritas.com/MyBestSegments/Default.jsp*, navigate to the zip code lookup page by clicking the "Zip Code Lookup" link at the bottom right, and enter your zip code. What market segment groups are included in your zip code? Knowing your own neighborhood, how accurate do you think the descriptions are of the different market segments that are listed for your zip code? What are some of the problems associated with urban analyses based on market segment analysis that may result in stereotypical characterizations of residents and neighborhoods? What may be missed by concentrating on people's income, consumption patterns, and material possessions? What kinds of attributes do *you* think are most important in describing people and their neighborhoods?
3. Consider the observation that "The city became a stage; people could pose and masquerade, their background and social credentials unchallenged in the turmoil". Is this aspect of cities exciting or threatening to you (or perhaps both)? Why?
4. Work on your *portfolio*. The topics covered in this chapter provide a good opportunity for you to personalize your work by writing about your own experience in relation to territoriality, discrimination, lifestyles, and neighborhoods, and relating it to the concepts and generalizations that you have read about. You will also be able to use features and news stories from local newspapers that you can access online to illustrate different aspects of residential segregation. Remember too that you can summarize, quote, reproduce, or analyze material from the suggested reading and online sources. Look for maps and charts that illustrate key points.

Log in to **www.mygeoscienceplace.com** for self-study quizzes, *MapMaster* layered thematic and place name interactive maps, *Urban View* Google Earth™ tours, key resources and suggested readings, related websites, "In the News" RSS feeds, and additional references and resources to enhance your study of the residential kaleidoscope.

NOTES

1. S. Rotella, "A Muslim Rapper's French Ghetto Beat: Rising Star Shuns Extremist Message," *Boston Globe*, October 2, 2005.
2. Ibid.
3. S. Day, "French Hip Hop Narrates a Generation," *Deft Magazine*, May 31, 2010.
4. Melvin Webber first advanced this argument in the early 1960s. See his essay, "The Urban Place and the Nonplace Urban Realm," in *Explorations into Urban Structure*, M. M. Webber et al. (Philadelphia, Pa.: University of Pennsylvania Press, 1964), 79–153.
5. S. Keller, *The Urban Neighborhood: A Sociological Perspective* (New York: Random House, 1968).
6. R. J. Johnston, *The American Urban System: A Geographical Perspective* (New York: St. Martin's Press, 1982).
7. G. Suttles, *The Social Construction of Communities* (Chicago, Ill.: University of Chicago Press, 1972).
8. For a discussion of human territoriality from a geographer's perspective, see R. Sack, "Human Territoriality: A Theory," *Annals of the Association of American Geographers* 73 (1983): 55–74.
9. L. H. Lofland, *A World of Strangers* (New York: Basic Books, 1973).
10. For a detailed discussion of class, see A. Giddens, *The Class Structure of the Advanced Societies*, 2nd ed. (New York: Harper & Row, 1981).
11. R. Walker, "Class, Division of Labor, and Employment in Space," in *Social Relations and Spatial Structures*, eds. D. Gregory and J. Urry (London: Macmillan, 1985), 187. Emphasis added.
12. D. W. Harvey, "Class Structure in a Capitalist Society and the Theory of Residential Differentiation," in *Process in Physical and Human Geography*, eds. R. Peel, M. Chisholm, and P. Haggett (London: Heinemann, 1975).
13. See R. L. Morrill and F. R. Pitts, "Marriage, Migration, and the Mean Information Field," *Annals of the Association of American Geographers* 57 (1967): 401–22.
14. This discussion is based on P. Knox and S. Pinch, *Urban Social Geography*, 4th ed. (Harlow, UK: Pearson Education, 2000), 228–29.
15. See F. W. Boal, "Ethnic Residential Segregation," in *Social Areas in Cities*, vol. 1, eds. D. T. Herbert and R. J. Johnston (New York: Wiley, 1976).
16. Ibid., 109–110.
17. See D. Timms, "Quantitative Techniques in Urban Social Geography," in *Frontiers in Geographical Teaching*, eds. R. Chorley and P. Haggett (London: Methuen, 1965), 239–265.
18. See William H. Frey analysis of 2009 American Community Survey (*http://censusscope.org/ACS/Segregation.html*).
19. E. L. Glaeser and J. L. Vigdor, *Racial Segregation in the 2000 Census: Promising News* (Washington, D.C.: The Brookings Institution, 2001).
20. Frey, 2009 American Community Survey analysis.
21. U.S. Department of Housing and Urban Development (HUD), *Discrimination in Metropolitan Housing Markets: National Results from Phase 1 of the Housing Discrimination Study (HDS)* (Washington, D.C.: HUD, 2002) (*http:www.huduser.org/publications/hsgfin/phase1.html*). This study used "paired testing," in which 4,600 sets of paired individuals—one minority and the other white—posed as otherwise identical homeseekers and visited real estate or rental agents to inquire about the availability of advertised housing units across 23 U.S. metropolitan areas. This methodology is designed to provide direct evidence of differences in the treatment experienced by minorities and whites in their search for housing.
22. W. Bell, "Social Choice, Lifestyles, and a Theory of Social Choice," in *The Suburban Community*, ed. W. Dobriner (New York: Putnam, 1958).
23. R. E. Park, "The City: Suggestions for the Investigation of Human Behavior in an Urban Environment," *American Journal of Sociology* 20 (1916): 608. The following description of human ecology draws on my summary in *Urban Social Geography*, 2nd ed. (London: Longman, 1987), 59–63. For more-detailed discussions, see A. H. Hawley, *Human Ecology* (New York: Ronald Press, 1950); and J. N. Entrikin. "Robert Park's Human Ecology and Human Geography," *Annals of the Association of American Geographers* 70 (1980): 43–58.
24. H. Zorbaugh, *The Gold Coast and The Slum* (Chicago: Chicago University Press, 1929).
25. E. W. Burgess, "The Growth of the City: An Introduction to a Research Project," in *The City*, eds. R. E. Park et al. (Chicago, Ill.: University of Chicago Press, 1925).
26. W. Firey, "Sentiment and Symbolism as Ecological Variables," *American Sociological Review* 10 (1945): 140–48.
27. A. Hawley, *Human Ecology: A Theory of Community Structure* (New York: Ronald Press, 1950); L. F. Schnore, *The Urban Scene* (New York: Free Press, 1965).
28. This discussion is based on L. McCarthy and C. Johnson, "Cities of Europe," in *Cities of the World: World Regional Urban Development*, eds. S. Brunn, M. Hays-Mitchell, and D. Zeigler, 4th ed. (Lanham, Md.: Rowman & Littlefield, 2011).
29. U.S. Census Bureau's American Community Survey 2009 data release for metropolitan statistical areas (*http://www.census.gov/acs/www/*).
30. W. H. Frey, *America's Regional Demographics in the '00s Decade: The Role of Seniors, Boomers, and New Minorities* (Washington, D.C.: The Brookings Institution, 2006).
31. Bureau of Labor Statistics, *Chart Book: Occupational Employment and Wages, May 2009* (Washington, D.C.: Office of Employment Statistics, 2009).
32. The Bureau of Labor Statistics defines "contingent" work as any job situation in which a worker does not have an explicit or implicit contract for long-term employment. These workers include independent contractors, on-call workers, and those working for temporary help services.
33. Bureau of Labor Statistics, *Contingent and Alternative Employment Arrangements, February 2005* (Washington, D.C.: United States Department of Labor, 2005).
34. W. H. Frey et al., *The State of Metropolitan America: On the Front Lines of Demographic Transformation* (Washington, D.C.: The Brookings Institution, 2010).
35. U.S. Census Bureau's American Community Survey 2009 data release (*http://www.census.gov/acs/www/*).
36. J. R. Logan, *From Many Shores: Asians in Census 2000* (Albany, N.Y.: University of Albany, Lewis Mumford Center for Comparative Urban and Regional Research, 2001) (*http://www.s4.brown.edu/cen2000/AsianPop/AsianReport/AsianDownload.pdf*).
37. U.S. Census Bureau's American Community Survey 2009 data release for metropolitan statistical areas (*http://www.census.gov/acs/www/*).
38. W. Li, *Ethnoburb: The New Ethnic Community in Urban America* (Honolulu, HI: University of Hawaii Press, 2009).

39. W. Li, "Los Angeles's Chinese *Ethnoburb*: From Ethnic Service Center to Global Economy Outpost," *Urban Geography* 19 (1998): 502–17.
40. U.S. Congress Joint Economic Committee, *Women and the Economy 2010: 25 Years of Progress but Challenges Remain* (Washington, D.C.: U.S. Congress, 2010).
41. Ibid.
42. U.S. Census Bureau's American Community Survey 2009 data release for metropolitan statistical areas (*http://www.census.gov/acs/www/*) Table S1101; U.S. Congress Joint Economic Committee, *Women and the Economy.*
43. Mather, M. *U.S. Children in Single Mother Families* (Washington, D.C.: Population Reference Bureau, 2010).
44. This section draws on P. L. Knox, ed., *The Restless Urban Landscape* (Englewood Cliffs, N.J.: Prentice Hall, 1993), 1–34.
45. A. Heller, "Existentialism, Alienation, Postmodernism: Cultural Movements as Vehicles of Change in Patterns of Everyday Life," in *Postmodern Conditions*, eds. A. Milner et al. (New York: Berg, 1990), 1–12.
46. G. Debord, *The Society of the Spectacle* (New York: Zone Books, 1994), para. 167.
47. K. Butler, "Paté Poverty," *Utne Reader* (Sept.–Oct., 1989), 77.
48. Ibid., 74, 77.
49. B. Ehrenreich, *Fear of Falling* (New York: Pantheon, 1989), 229.
50. Ibid., 228.
51. B. J. L. Berry, *The Human Consequences of Urbanization* (New York: St. Martin's Press, 1973).
52. U.S. Bureau of Labor Statistics (BLS) *Consumer Expenditure Survey* (CE) (*http://www.bls.gov/cex/*).
53. M. J. Weiss, *The Clustered World: How We Live, What We Buy, and What It All Means About Who We Are* (Boston: Little, Brown and Company, 2000).
54. M. J. Weiss, "Inconspicuous Consumption," *American Demographics*, 24 (2002): 31–32.
55. U.S. Bureau of Labor Statistics (BLS) *Consumer Expenditure Survey* (CE) showing a decline of 2.8 percent in average annual expenditures per consumer unit following an increase of 1.7 percent in 2008.
56. Weiss, *The Clustered World*, 32.
57. Ibid., 37.
58. C. DeNavas-Walt, B. D. Proctor, and J. C. Smith, *Income, Poverty, and Health Insurance Coverage in the United States: 2009* (Washington, D.C.: U.S. Census Bureau U.S. Department of Commerce, 2010).
59. Robert H. Frank, *Luxury Fever* (Princeton, NJ: Princeton University Press, 2000); John De Graaf, David Wann, and Thomas H. Naylor, *Affluenza. The All-Consuming Epidemic* (San Francisco: Berrett Koehler, 2001).
60. Robert Reich, "Secession of the Successful," *New York Times Magazine* (Jan. 20, 1991), p. 17.
61. Robert Putnam, *Bowling Alone: The Collapse and Revival of American Community* (New York: Simon & Schuster, 2000.
62. Experiencing two or more of the following: poverty-level incomes, unemployment, no telephone, incomplete kitchen or plumbing facilities, and crowded living conditions. See P. Knox and R. Rohr-Zanker, *Economic Change, Demographic Change, and the Composition and Distribution of Vulnerable and Disadvantaged Households in the United States* (Washington, D.C.: U.S. Department of Commerce, Economic Development Administration, 1989).
63. M. Castells, *The Informational City* (Cambridge, Mass.: Blackwell, 1990).
64. W. J. Wilson, *The Truly Disadvantaged* (Chicago, Ill.: University of Chicago Press, 1987).
65. M. Dear and J. Wolch, *Landscapes of Despair* (Princeton, N.J.: Princeton University Press, 1987).
66. S. Zukin, "Urban Lifestyles: Diversity and Standardization in Spaces of Consumption," *Urban Studies* 35 (1998): 835–37.
67. This discussion is based on an industry interview, "Location Analysis Tools Help Starbucks Brew Up New Ideas," *Business Geographics* 8 (October 2000): 32–34; and P. L. Knox and S. A. Marston, *Places and Regions in Global Context: Human Geography*, 3rd ed. (Upper Saddle River, N.J.: Pearson Education, Inc., 2004), 94–95.
68. A. Heller, "Existentialism, Alienation, Postmodernism: Cultural Movements as Vehicles of Change in Patterns of Everyday Life," in *Postmodern Conditions*, ed. A. Milner, P. Thompson, and C. Worth (New York: Berg, 1990), 1–13.
69. This Claritas information available at: *http://www.claritas.com/MyBestSegments/Default.jsp?ID=30&SubID=&pageName=Segment%2BLook-up*.
70. Richard Sennett, *The Conscience of the Eye: The Design and Social Life of Cities* (New York: Knopf, 1990), p. xii.
71. J. Pawasarat and L. M. Quinn, *Exposing Urban Legends: The Real Purchasing Power of Central City Neighborhoods* (Washington, D.C.: The Brookings Institution) (*http://www.brookings.edu/es/urban/pawasarat.pdf*), vi.

Urbanization, Urban Life, and Urban Spaces

Paul L. Knox

Urbanization, Urban Life, and Urban Spaces

Urban spaces are created by people, and they draw their character from the people that inhabit them. As people live and work in urban spaces, they gradually impose themselves on their environment, modifying and adjusting it as best they can, to suit their needs and express their values. Yet at the same time people themselves gradually accommodate both to their physical environment and to the people around them. So there is a continuous two-way process, a sociospatial dialectic, in which people create and modify urban spaces while at the same time being partly conditioned by the spaces in which they live and work. Neighborhoods and communities are created, maintained, and modified; the values, attitudes, and behavior of their inhabitants, meanwhile, cannot help but be influenced by their surroundings and by the values, attitudes, and behavior of the people around them. In addition, the ongoing processes of urbanization contribute economic, demographic, social, and cultural forces that must be accommodated within the sociospatial dialectic.

LEARNING OUTCOMES

After reading this chapter, you should be able to:

- Understand the ways in which people create and modify urban spaces while at the same time being partly conditioned by the spaces in which they live and work.
- Evaluate the ideas set out by Louis Wirth in his essay, "Urbanism as a Way of Life."
- Explain how people organize their concepts of space and place via the construction of mental maps.
- Describe how people assign meaning to places based on how they interact with them.
- Recognize the formal and informal ways that gender coding is blended into our everyday experience of life.

CHAPTER PREVIEW

In this chapter we explore the complex relationships between urban settings and people's individual and social behavior. The objective is to establish the main features of the sociospatial dialectic in which people create and shape their environment while at the same time being partly conditioned by it. An additional objective is to show how this two-way relationship has changed as cities themselves have changed.

We will see that there is a long history of ideas about the effects of urbanization on people's behavior. We review the most important of these ideas, focusing in particular on the concept of the "moral order" of urban society and on the very influential theory that urban settings give rise to distinctive types of social interaction and ways of life. This review leads to a closer examination of the issues in the context of different kinds of urban settings. We will see, for example, why we might expect quite different outcomes for people in low-income, central city neighborhoods than in affluent suburbs.

The relationships between society and space are further explored in relation to the concept of "community." This exploration, in turn, leads to an examination of territoriality and of the implications of the ways in which people perceive various elements of their urban environments. All this background, in turn,

The catalytic qualities of urban spaces in generating and accommodating many of the innovative and expressive qualities of individual and social life are widely recognized. Much depends on the eye of the beholder, as in this photograph taken in Vancouver, Canada.

URBAN VIEW 1
The Writing on the Wall for Territoriality? Taggers Respect Community Mural Painted in "Neutral" Colors

The Salvation Army South Los Angeles Center . . . serves one of the most troubled urban regions in Southern California. Offering services in three independent programs, the Center supplies living assistance to low-income families through Family Services, growth and learning development for infants to five-year olds through Child Care, and academic, artistic and athletic activities for students of all ages through its Youth, Adult and Senior Programs.[1]

In the neighborhood around this Salvation Army Center, it seems that every wall and storefront, including the mini-mart's Virgin of Guadalupe mural across the street, have been spraypainted by local taggers (Figure 1). Their work forms a spatial communications network throughout the neighborhood, including announcing which gang has taken over part of another gang's turf and which gangs have allied with each other.[2] But the large mural painted on the side of the Salvation Army Center continues to be spared. This despite the fact that the Center is located right along the territorial boundary—South Central Avenue—of two of the city's most violent gangs, Crips on one side and Bloods on the other.

Police say it's the result of an informal truce among area gangs to spare the center, which provides social services to hundreds of needy families . . . Police are surprised by what seems to be a concerted effort by local gangs to keep the center from harm. They attribute the phenomenon to a sense of ownership the community has taken for the facilities.[3]

Ignoring warnings from neighbors that the art would be vandalized, the director of the center, with the help of a corporate sponsor and some local volunteers, commissioned a local artist to paint the mural in 2007. It shows two boys playing basketball, an adult reading to a child, and a student proudly holding a school diploma. The director of the center says that the mural has raised awareness about the social services offered, and that members of competing gangs play basketball in the gym and work out together in the weight room. "They walk in the building, they go neutral," he says.[4] Territoriality is temporarily put on hold by gang members as they briefly experience a sense of community.

Tim Zinnemann/Corbis Images

FIGURE 1 A tagger marks his gang's territory at night. In neighborhoods where high residential densities bring different ethnic and other groups into close proximity, territoriality can be expressed, as here, through graffiti.

helps in understanding the importance of people's everyday "lifeworlds." We will see how these lifeworlds are established by people through fundamental geographical parameters of individual "time-space" routines that help to structure the very fabric of society.

We conclude by examining one particular dimension of the sociospatial dialectic that tends to be obscured by emphases on specific social groups and specific kinds of settings: women's spaces and women's places in cities. With this theme we will also see how long-term changes in the political economy of urbanization have conditioned sociospatial relationships among people.

SOCIAL LIFE IN CITIES

Rural life is usually portrayed as "natural"—that is, close to nature. City life and city spaces are portrayed as "artificial." Rural society is portrayed as stable and neighborly; urban society is portrayed as volatile and individualistic. Rurality is associated with tradition and familiarity; cities are associated with novelty and variety. Often, the experience of living in cities is characterized in rather negative terms. Art, literature, public opinion, and social theory have tended to portray cities as "necessary evils," places that offer people economic opportunity and accessibility to a broad spectrum of amenities but that are somehow "unnatural"—their sheer size and density leading to personal and social stress. We have already seen how Jeffersonian notions of the virtues of pastoral life, Thoreau's emphasis on the importance of nature, and Turner's celebration of the frontier experience have exerted a powerful influence on U.S. culture. Surveys of U.S. literature and intellectual thought have shown how city spaces are often depicted as arenas of conflict and incubators of loneliness, isolation, and deviance[5]—themes that continue to be reflected in movies, TV series, and music.

At the same time, the catalytic qualities of urban spaces in generating and accommodating many of the innovative and expressive qualities of individual and social life are widely recognized. Much depends on the eye of the beholder. For some people rural settings provide human scale and a comfortable pace, while cities present garish sights and overwhelming and frenetic experiences. For others, rural settings are dull and suffocating, while cities offer exciting sights and experiences.

Making sense of this ambiguity and ambivalence has been a preoccupation of theorists ever since the urban transformation that produced the early industrial city. Amid slowly changing patterns of land use and residential segregation, each district and neighborhood provides a setting in which particular social and cultural groups of people act out their daily lives; there is an intimate relationship between space and society. People create and change urban spaces according to their needs, but at the same time the physical and sociocultural attributes of particular settings partly shape and constrain their behavior—both individually and collectively. So the relationship between urban space and urban society can be characterized as a continuous **sociospatial dialectic**[6] in which each shapes and reshapes the other.

We must also recognize that, within the sociospatial dialectic of overall relationships between space and society, people create individual *lifeworlds* for themselves and for one another: ways of being, ways of seeing, and ways of knowing that are shaped by personal experience through daily routines in time and place.[7] Because of the general tendency for sociospatial segregation, urban spaces—districts and neighborhoods—often generate widely shared ("intersubjective"), taken-for-granted worlds with common horizons. But sociospatial segregation is never absolute, and some people will inevitably construe their lifeworlds in uniquely personal ways. As a result, the relationships between space, society, and ways of life are complex and contingent. As we will see, however, early and influential theorists emphasized the effects of the strangeness, artificiality, and individualism of urban settings on people's behavior: a deterministic perspective that has had a profound effect on ideas in urban geography, urban sociology, and all the related disciplines.

THEORETICAL INTERPRETATIONS OF URBAN LIFE

Our theories about urban life stem from the work of European social philosophers such as Ferdinand Tönnies (1855–1935), Emile Durkheim (1858–1917), and Georg Simmel (1858–1918), who tried to understand the social and psychological implications of the shocking changes brought about by the unprecedented urbanization associated with the Industrial Revolution. At the heart of their analyses was an association between the scale of society and the nature of social organization. Their argument was: in preindustrial society, small, fairly homogeneous populations contain people who know one another, perform the same kind of work, and have the same kind of interests; they consequently tend to look, think, and behave alike, reflecting a consensus of values and norms of behavior (the rules and conventions of proper and permissible behavior); in contrast, the inhabitants of large cities find themselves living at unprecedented densities and organized along new lines as a result of economic specialization.

In such a setting there is, inevitably, contact not only with more people but also with more *kinds* of people. But close relationships with family and friends are less easily sustained because of the fragmentation of life: the factory whistle and the office clock impose rigid new constraints on people's use of time, while the functional separation of land uses results in the spatial separation of home, workplace, and leisure activities. Family life becomes more constrained, and extended family systems have to give way to compact, nuclear families. At the same time, social differentiation brings about a divergence of lifestyles, values, and aspirations, which weakens social consensus and cohesion and threatens to disrupt social order. This divergence, in turn, leads to attempts to adopt "rational" approaches to social organization, to the proliferation of formal social controls, and, where these attempts are unsuccessful, to an increase in social disorganization and deviant behavior.

The "Moral Order" of City Life

The earliest theorists of urban life were, like all of us, products of their time. In wrestling with the implications of the emergence of huge concentrations of people, one of their central concerns was with the "moral order" of urban populations: the norms of behavior and patterns of social interaction that were clearly different (and, for many, disturbingly different) from the well-worn conventions, rigid social hierarchies, and close-knit groups of rural and small-town life. It was in this context that Ferdinand Tönnies, one of the founders of sociology, established the framework for theories of urban life by conceptualizing patterns of human association in terms of a continuum marked by polar extremes.[8] At one end was the concept of ***Gemeinschaft*** (translated as "community"), in which

- the basic unit of organization is the family.
- social relationships among people are characterized by depth, continuity, cohesion, and fulfillment.
- people are bound together in caring, familial ways.
- controls over individual behavior are exerted through the informal discipline of family and neighbors.

At the other extreme was the concept of ***Gesellschaft*** (translated as "society"), in which

- social relationships among people are founded on the rationality, efficiency, and contractual obligations stemming from patterns of economic organization.
- a large proportion of social interaction by people tends to be short-lived and superficial.
- people are bound together by formal ties to institutions and organizations.
- controls over individual behavior are exerted through impersonal, institutionalized codes of behavior.

Although Tönnies did not equate *Gemeinschaft* with rural life and *Gesellschaft* with urban life in a simple or rigid way, the inference was clear: industrial urbanization brings a transition in the very fabric of social life. The same assumptions were elaborated by Emile Durkheim (Figure 2), another of the founders of sociology.[9] His observations on the changing basis of the bonds sustaining human groups led to the idea of two forms of social "solidarity." The first, based on similarities between people, he called **mechanical solidarity**. The second, based on the existence of differences between people stemming

Corbis Images

FIGURE 2 Emile Durkheim (1858–1917), one of the founding fathers of social science. His work provided important insights about urban life and its effect on individual and group behavior.

URBAN VIEW 2
"Sex and the City": Prostitution[10]

Cities have often provided opportunities to transgress moral codes, and one important manifestation of this has been prostitution—the granting of sexual favors in exchange for monetary reward (usually, but not exclusively, by women for men). This is often called the world's "oldest profession," but the actual term "prostitute" came into prominence only in the eighteenth century. In older societies sexual favors outside marriage were often granted by women who were courtesans, mistresses, or slaves. The crucial point is that these women were often known to those procuring the sexual favors.

With the development of large cities, prostitution changed in character in that the women and their clients frequently did not know each other. The reason for this change is fairly obvious; in small-scale agrarian societies, people were much more likely to be familiar with each other, whereas in cities there was a much greater chance of anonymity. In addition, economic destitution in the early industrial cities meant that prostitution was often the only effective means for some women to earn an income.

Prostitution was rife in many nineteenth-century cities. In London in the mid-nineteenth century, it has been estimated that there were 8,000 or more women making a living through prostitution.[11] Because it was against the law for women to stand in one place to attract clients, most solicited trade by walking the streets—as when Jack the Ripper was targeting prostitutes in London in 1888. Men of all social backgrounds used prostitutes, and shops in the Strand and Haymarket areas of London advertised "beds to let," often by the hour.

Timeline of the Early Hours of Friday, August 31, 1888, when Polly Nichols, Jack the Ripper's First Victim, was Murdered[12]

12.30 A.M. Polly left The Frying Pan pub where she had been drinking and returned to her lodging house.

1.30 A.M. At the lodging house, Polly was told to leave because she did not have enough money for the fee for the night ("doss money" of four pence). Polly left but asked that a bed be reserved for her. "Never Mind! I'll soon get my doss money. See what a jolly bonnet I've got now" she said as she showed off her new straw bonnet trimmed with black velvet.

2.30 A.M. Polly met Emily Holland and said that she had had her doss money three times already that day but had spent it on gin. She told Emily that she was going to try to solicit for trade one more time and then return to the lodging house. The services of a destitute and alcoholic prostitute like Polly could be had for two or three pence or a stale loaf of bread. The price of a large glass of gin was three pence.

3.15 A.M. Two police officers patrolling Buck's Row reported nothing unusual.

3.45 A.M. Polly's body was found by Charles Cross and Robert Paul who did not want to be late for work and so arranged Polly's skirts to "give her some decency" and then alerted the first police officer they met on their way to work. In the meantime, police constable John Neil found Polly's body and called for Dr. Llewellyn. The doctor pronounced Polly dead "but a few minutes."

URBAN VIEW 2

"Sex and the City": Prostitution (*continued*)

The inquest testimony recorded that Polly's throat had been violently slit, and that she also had deep wounds two to three inches from the left side of her abdomen, as well as several incisions running across her abdomen. She had a bruise running along the lower part of her jaw on the right side of her face that might have been caused by a blow from a fist or pressure from a thumb. She also had a circular bruise on the left side of her face that might have been inflicted by the pressure of the killer's fingers.

Despite the danger of physical assault from clients and the high risk of contracting sexually transmitted diseases, prostitutes could earn far more on the streets than they could through work in the poorly paid industries of the time. But a double standard operated—although prostitutes were frequently persecuted, the men who used their services were not. Prostitution, therefore, reflects patriarchal gender relations (the inequalities in power relations between men and women).[13]

Prostitution has continued to be a source of much conflict in the twentieth and twenty-first centuries. In the United States and the United Kingdom, for example, prostitution is found in distressed central city neighborhoods. Many of the residents object to the women seeking "custom" on their streets. This has led to community campaigns to expose "curb crawlers" (those who drive through these residential areas looking for prostitutes). Such campaigns can involve cooperation between the police and local community groups, although the result is often that prostitution is merely displaced elsewhere in the city.

Prostitution illustrates well the importance of recognizing the different "voices" in the city. It seems clear, for example, that most of the women are coerced into prostitution through economic disadvantage and often experience considerable physical and psychological harm from both their pimps and clients. We should recognize at this point that the legal and social restrictions surrounding prostitution are full of hypocrisy and contradiction. In the United Kingdom, although prostitution is legal in principle, there are prohibitions on soliciting in the streets or living off "immoral earnings." As Duncan notes, widespread condemnations of prostitutes as being exploited by men or suffering from "false consciousness" exclude prostitutes from the freedom to control their own bodies in safe conditions free from police harassment.[14] It is little wonder then that the discourses surrounding prostitution also display ambiguity and contradiction: portraying women as either victims of sex-hungry predatory males or autonomous providers of social services.

from specialized economic roles, he called **organic solidarity**. This term was based on a biotic analogy, the idea being that in a complex organism (modern, urban society) every single "organ" (social group) is mutually dependent on the rest.

Anomie and Deviant Behavior Durkheim believed that the transition from mechanical to organic solidarity as the dominant form of social organization was a direct result of three factors: the sheer number of people in urban settings, the density of urban residential settings (remember that he was writing just after the introduction of the streetcar), and the dynamism of cities, sustained by improved transportation and communication systems, which put larger and larger numbers of different sorts of people in contact with one another. It is important to note that Durkheim did not see this idea of the "dynamic density" of urban life as inherently bad. The loss of the mechanical solidarity provided by family and neighborhood would, he believed, be compensated for by the growth of organic solidarity provided by the workplace. He did, however, recognize that there could be circumstances in which the regulation and peaceful coexistence of organic solidarity would break down, leading to a social condition of **anomie**. The term literally means "normlessness," but it is more accurate to think of situations in which the norms of personal and social behavior are so weak or muddled that some people become socially isolated, confused, or uncertain about how to behave, while others are easily able to challenge or ignore social conventions. One of the consequences of such circumstances is an increase in **deviant behavior**, one aspect of which Durkheim himself wrote about in his 1897 book *Suicide*.[15]

In contrast to the emphasis given by Tönnies and Durkheim to the contextual effects of urban life on social organization, Georg Simmel believed that all social relationships and all forms of social organization stem from basic human drives, interests, and psychic states. As a result, according to Simmel, people enter into relationships in pursuit of a variety of objectives: survival, status, power, erotic and spiritual instincts, and so on. People's lives, he concluded, are acted out amid tensions and conflicts between their individual creativity and the various institutionalized forms of social interaction and social organization that have evolved to fulfill generalized human needs. In this context, Simmel saw urbanization as both constraining and liberating for people. The constraints to self-fulfillment and creativity, he argued, had two main dimensions.

One dimension was the intensification of objectivity and rationalization by people that accompanied the evolution of an industrialized, urbanized society. In particular, the emergence of a highly developed money economy creates conditions that foster changes in the way people are able to relate to one another. This, he argued, rewards intellectual over emotional development and threatens to dry up the wellsprings of the mind and self.

The other dimension of modern urban society that threatened to constrain self-fulfillment and creativity concerns the intensification of external stimuli. These include the bombardment of the psyche by the sheer numbers of people and the fast pace of city life (dynamic density again) and the bewildering plurality of styles accommodated by city life—compounded by their transitory and illusory nature.

The overall effect of these two constraining dimensions of urbanization, Simmel suggested, was the emergence of a metropolitan "way of life." In Simmel's 1905 essay, "The Metropolis and Mental Life,"[16] he pointed to several key attributes of this way of life. Blasé attitudes and antipathy toward others emerge as people respond to the *psychic overload* generated by the unexpected, discontinuous, and rapidly changing stimuli and situations encountered in the city. At the same time, reserve and privacy emerge in response to people's need for contemplation and inner reflection. But these defenses by people against overstimulation make it difficult for others to express their subjective personalities, so that a third distinctive component of the metropolitan way of life is fostered: heightened and exaggerated forms of display, from dress to architecture.

Liberating Aspects of Urban Life It is here that we can see the liberating side of urbanization for people: The "lonely crowd" created by dynamic density of city life *allows* reserve and privacy to flourish; the plurality of styles accommodated by city life gives full scope to the variety of human creativity. The aestheticization of everyday life had been in fact a central preoccupation of some nineteenth-century theorists and commentators. Following the novelist Honoré de Balzac (1799–1850), they pointed to the emergence of artistic countercultures, the *bohème* and *avant-gardes* of Paris, Berlin, and Vienna as intermediaries in stimulating, formulating, and disseminating a new sense of aesthetics and consumption. According to this perspective, urban life is infused with an intoxicating and constantly changing flow of commodities and sensations.[17] Department stores and shopping arcades become elements in a dream world of consumption; city streets become stages upon which people can act out multiple roles; crowds become audiences for the *flâneur*—the stroller wanting to see and be seen, the dandy "who makes of his body, his behavior, his feelings and passions, his very existence, a work of art."[18]

When theorists turned their attention to the urban experience in the mushrooming cities of the United States, however, they paid little attention to the liberating dimensions of urban life and the individuality of human creativity. Instead, it was the negative aspects of urban life that came to be writ large. This emphasis was in large measure due to the influence of Louis Wirth (1897–1952), a German immigrant who had studied under Simmel and then came to study sociology at the University of Chicago. His particular distillation of ideas in an essay entitled "Urbanism as a Way of Life," published in 1938,[19] was to become one of the single most influential treatises in urban sociology, environmental psychology, and urban geography—even to the point of straitjacketing many people's ideas about urban life.

Urbanism as a Way of Life

Writing from the same perspective as Tönnies and Durkheim, drawing heavily but selectively on the ideas of Simmel, and strongly influenced by colleagues in the Chicago School of Human Ecology, Wirth developed a deductive theory that was based on what he saw as the three fundamental attributes of urbanization:

1. The increased number of people.
2. People's increased (physical) density of living.
3. The increased heterogeneity of the population.

Each of these attributes, he reasoned, gives rise to a slightly different cluster of outcomes. Together, the outcomes are mutually reinforcing, generating a distinctive web of attitudes, behavior, and social organization by people that characterize **urbanism**: a way of life that is an inevitable consequence of urbanization. Although Wirth's theory involved a complex interweaving of hypothesized relationships, the outcomes can be summarized in terms of changes that affect individuals and changes that affect society at large.

Wirth's deductions led him to the conclusion that the combined influence of increased size, density, and heterogeneity of urban populations would affect individual behavior because people had somehow to cope with a great number and variety of physical and social stimuli. The overall result, he argued, was that city life leads people to become more withdrawn, emotionally buffered to the point where they are naturally aloof, brusque, and impersonal in their dealings with others. Also, the general loosening of interpersonal bonds through this adaptive behavior tends to leave people *unrestrained*, and so liable to engage in egocentric and unconventional behaviors. People are also left *unsupported* in times of crisis, leaving them vulnerable to neurosis, alcoholism, suicide, or some form of deviant behavior.

At the same time, Wirth drew a parallel picture of social change associated with the increased size, density, and heterogeneity of urban populations. Economic competition and the division of labor provide the basis of spatial differentiation that is reinforced by impulses of social attraction and social avoidance; this creates a mosaic of social worlds and results not only in the segregation of people by different income groups, ethnic groups, and household types but also the *fragmentation of social life* between home, school, workplace, friends, and family. Wirth accepted that this segregation and fragmentation tend to encourage a general increase in tolerance. At the same time, though, the fact that people have to divide their time and attention among highly differentiated people and places weakens the support and control provided by primary groups such as family, friends, and neighbors. In addition, this trend is reinforced by the weakening of social norms that is a consequence of the divergent interests, lifestyles, and cultures associated with the increasing heterogeneity of urban populations.

The overall societal response is to replace the support and controls formerly provided by primary social groups like family and friends with "rational" and impersonal procedures and institutions like welfare agencies and criminal codes supported by the police. According to Wirth, however, this kind of rational order can never adequately replace the communal order based on consensus and the moral strength of small primary groups of people. As a result, urbanism brings a loosening of the social fabric, allowing egocentric and unconventional behavior by people to flourish and allowing personal crises to snowball into social problems affecting everyone.

In this way, Wirthian theory points to the inevitability of situations and settings in which city life—or, to be precise, life for certain people in certain parts of the city—breaks the bounds of convention and normality, often to the disadvantage not only of the people and neighborhoods involved but also of urban society as a whole. This aspect of urban life can be developed in several ways. The idea of psychological overload resulting from complex or unfamiliar environments has been investigated by environmental psychologists and was popularized by Alvin Toffler in his best-selling book *Future Shock*.[20] Toffler used the notion of "future shock" to describe the outcome of the need for people in modern (urban) settings to "scoop up and process" more and more information. Scientific studies of people's response to this kind of overstimulation have shown that several different strategies are common. One involves eliminating from perception the most unwelcome aspects of reality. In its most extreme form, this can result in some person creating a mythological world—either a world in their heads or in a virtual world with other online players—that becomes an unhealthy substitute for the real world. In such a world this kind of person may consider odd behavior to be "normal." Others, of course, all too easily view this person as just another of the city's eccentrics.

This analogy points to another, more common strategy: managing several distinct roles or identities at once. According to Wirthian theory, this strategy is characteristic of urbanism because of the geographical and functional separation of the "audiences" to which a person's different roles are addressed, including family, neighbors, and coworkers. People are consequently able to present very different "selves" in different social contexts. The city, with its wide choice of different settings, roles, and identities, becomes a magic theater (Figure 3). Unfortunately, the anonymity afforded by the ease of slipping from one role to another not only helps to relieve the stresses and strains of urban life but also facilitates deviant and criminal behavior by some people. In addition, the strain of having to sustain different and perhaps conflicting roles or "selves" over a prolonged period may, for some people, contribute to mental illness or to deviant behavior.

The most widespread strategy for many people, however, is simply to withdraw, to become aloof, and to mind their own business. The social psychologist Stanley Milgram documented various manifestations of this strategy,[21] the most striking of which is the collective paralysis of moral order reflected in the lack of bystander intervention when something clearly antisocial, deviant, or unlawful is happening. The most widespread examples are those involving rudeness, petty theft, or vandalism where bystanders ignore what is going on. The behavior of the people who witness and ignore such behavior is itself deviant in a way; but the significance of this from the point of view of Wirthian theory is in the way that it can encourage the spread of more serious forms of deviancy by eroding social responsibility and social control. Of special interest to many geographers and sociologists is the clear variability in the amount and type of deviant behavior from one neighborhood to another. Deploying Wirthian ideas, an explanation can be traced to the much greater size, density, and heterogeneity of the populations of certain areas, particularly those that are low-rent neighborhoods receiving the brunt of low-income migrants and immigrants. Here, the effects of urbanization go far beyond the generalized notion of a way of life characterized by aloofness, withdrawal, egocentric behaviors, and the substitution of impersonal and institutionalized controls for self-control and social constraints. In these neighborhoods, the Wirthian argument goes, the combined stresses of size, density, and heterogeneity lead not only to increases in anomie and deviant behavior but also to a more widespread breakdown of moral order, social solidarity, and social control: a condition known as **social disorganization**.

Spencer Grant/PhotoEdit

FIGURE 3 Actors in the "magic theater" of the city. Two punkers argue with an evangelist on a sidewalk in Camden Town, London.

The Public and Private Worlds of City Life

Although such interpretations of urban life have proven popular and remarkably durable, they are only partially supported by the evidence of scientific analysis. They are also vulnerable to criticism on the grounds that human beings are in fact more flexible and inventive than allowed for in the deterministic cause-and-effect logic of Wirthian theory. Experiments designed to measure people's willingness to be helpful and cooperative, together with studies of certain categories of crime and interpersonal conflict, tend to support the Wirthian scenario: successively larger towns

Wally Santana/AP Images

FIGURE 4 The sidewalk is usually a neutral zone in urban space, part of the "public world" shared by people of different backgrounds.

and cities exhibit less helpful and more conflictual and deviant behavior by people.[22] At the same time, studies that have compared the number or quality of friendships and personal relations have found no difference between different-sized settlements; while experiments focusing on people's psychological states have found that stress and alienation are just as prevalent, if not more so, in rural and small-town communities.

One way of reconciling the apparent inconsistency of these findings is to reexamine the notion of urban settings, recognizing a distinction for people between the public and the private spheres of life (Figure 4). The public spheres are characterized by situations in which people are strangers, situations that require a special etiquette: reserved, careful, nonintrusive. People must be—or at least appear to be—indifferent to others. But this is *situational behavior, not a psychological state*. From this perspective a person who lives in a city does not lose the capacity for deep, long-lasting, multi-faceted relationships but instead . . . *gains* the capacity for surface, fleeting, restricted relationships.[23] Sociologist Claude Fischer has drawn on this distinction, suggesting that urbanism can be regarded as involving fear, distrust, estrangement, and alienation from "other people" in the wider community but not from family, friends, neighbors, or associates at work or school.[24] In Wirthian terminology this view means that urbanism can accommodate both moral order and social disorganization.

Changing Metropolitan Form and New Forms of Urbanism

It should be remembered that Wirth's observations and theories about people's behavior were based on a very specific phase of urbanization: one fueled by European immigration, driven by industrialization, and only just beginning to feel the effects of low-density decentralization made possible by buses, trucks, and cars. The emergence of a new metropolitan form, characterized by polarized central city cores surrounded by sprawling, low-density **urban realms**, raises the possibility of a new form of urbanism based on new ways of life—urban and suburban certainly, but now in a polycentric metropolis.

The changes brought about by suburbanization, by the intensification of economic differentiation, and by the proliferation of different lifestyles have made it increasingly difficult and unrealistic to think in terms of entire urban or metropolitan populations being shaped in their behavior by variations in the size, density, and heterogeneity of their surroundings. Meanwhile, it has become increasingly apparent that the ecological parameters of social status, household type, ethnicity, and lifestyle operate as powerful determinants not only of patterns of marriage and relationships, friendship, and social interaction among people but also of personality development, sociopolitical attitudes, and cultural values. From this perspective the urban mosaic sustains many different "ways of life," while the significance of "urbanism" is relegated to the notion of the overall variety implied by these relationships.

Urban Villages One of the first writers to recognize the links between ecological areas in cities and people's ways of life was sociologist Herbert Gans. Gans emphasized the cohesion and intimacy of distinctive social worlds based on ethnicity, family, neighborhood, occupation, and lifestyle, strongly rejecting the idea that urban life in any way diminishes these social worlds for people. His conclusions were based on an intensive study of the Italian community in the West End of Boston. So intense

URBAN VIEW 3
Homosexuality and the City[25]

Homosexuality involves emotional and sexual attraction between people of the same sex. Homosexuality has existed in all cultures, with sodomy long prohibited by religious teaching, although not marked out as a particular homosexual act. It was only in the mid-nineteenth century that the term homosexuality was devised, which then identified homosexuals as a distinct and separate group. The association of the term with persecution in the twentieth century led many to prefer alternative terms, such as "gay" or "queer" (the usually pejorative word "queer" being deliberately used in an ironic way to acknowledge the repressive character of social constructions of homosexuality).

In general, cities have provided greater anonymity and tolerance of alternative lifestyles compared with the hostility toward lesbian, gay, bisexual, and transgender people (LGBT) manifest by residents in rural communities. Since the 1960s and the civil rights movement, there have been profound changes in the nature of homosexual spaces within cities that reflect broader social and political changes in society. Together with the race riots, student protests, and anti-Vietnam War demonstrations of the 1960s in the United States, there were a number of gay protests. The most famous of these took place in New York in 1969 when gays and lesbians rioted after a police raid on the Stonewall Inn. Following this, the Gay Liberation Front (GLF) was formed in New York in 1971 to make vocal gay demands for equality. Arguably, there is now greater tolerance of homosexual activity by "straight" segments of society, although the extent of this tolerance should not be exaggerated—the ongoing political struggles to make gay marriages and civil unions legal in the United States being a case in point. Nevertheless, the hidden and covert character of gay activity in some areas has been replaced by distinctive residential districts composed of substantial proportions of gay people in which gay lifestyles are explicitly displayed.

Weightman was among the first to draw attention to areas in U.S. cities with distinctive gay lifestyles.[26] He noted that gays were playing a leading role in the process of **gentrification** in some central city neighborhoods, often displacing poorer residents. Without doubt, the most famous of these residential districts is the Castro district of San Francisco. The origins of this district can be traced back to World War II. Gays and lesbians serving in the armed forces were often discharged in San Francisco and preferred to stay in the city instead of returning to face the prejudice of their home communities. Another well known gay space is the Marigny district of New Orleans. As in San Francisco, this district is located in a culturally mixed, relatively tolerant area. Knopp[27] described how property developers and speculators took advantage of the gay population's demand for property in this area. In particular, they exploited the **rent gap** by artificially inflating the values of properties through bribing private appraisers. Paradoxically, the resultant influx of middle-income gays caused many of the existing middle-income gays in the area to become concerned over the preservation of historically important areas in the city rather than any broader issues affecting the gay community.

Male homosexuality has received much more analysis in urban geography than lesbianism, partly because gay residential areas are generally more visible and easily mapped and studied. Women in general have fewer financial resources than men, and they also face the threat of male violence. Lesbians therefore share a desire for relatively inexpensive housing as well as a concern with personal safety. These factors, combined with the pressures of a predominantly heterosexual society, mean that lesbian residential areas are less overt than gay spaces. Nevertheless, lesbian spaces in cities have been mapped and analyzed, including housing cooperatives and bars, restaurants, nightclubs, cinemas, and bookshops that cater to lesbians.[28]

Such mapping of gay spaces in cities has extended our knowledge, but these studies have also been criticized in recent years. In particular, it has been argued that the work on *gay enclaves* have been limited in their contribution to understanding the processes that lead to this spatial clustering.[29] To begin with, most of these areas cannot be thought of as exclusively gay. In addition, many LGBT people live outside these parts of the city. The crucial point is that such a focus on gay spaces tends to conceptualize them as different when, in reality, *all* spaces in cities are socially constructed in a sexualized manner. For example, research has focused on the processes through which certain public spaces, such as hotels and restaurants, become constructed as heterosexual.[30] Even the domestic environment in the home is constructed around a heterosexual-based notion of sexual reproduction.[31]

It has been suggested that the spatial concentration of LGBT people would, as in the case of ethnic minorities, provide a base for political mobilization against repression and discrimination. However, many have questioned the efficacy of such a separatist strategy. It has been argued that such gay enclaves have helped to maintain the notion of gay lifestyles as separate, different, deviant, and sinful. In the wake of HIV and AIDS, however, such areas assumed a new role as focal points for support networks and healthcare services.

Yet another change in LGBT spaces in recent years has been the commercialization of these areas as entrepreneurs have sought to exploit the high incomes of some gay households. For example, the Soho area of London has developed into a thriving commercial scene in what has been described as the world's Pink Capital. Clubs, bars, and shops catering explicitly to the LGBT community have heightened the visibility of these lifestyles but have tended to cater to specific types of LGBT people—the young and wealthy—with older and poorer LGBT people more likely to be excluded. So a new form of economic division has been imposed on divisions based on sexuality.

was the cohesion among the residents there (and so different from the predictions of Wirthian theory) that Gans titled his book *The Urban Villagers*.[32]

Studies of "urban villages" in other cities have found social interaction among residents to be cast in terms of socioeconomic status as often as race or ethnicity; but the common denominators are the attributes of older, central city neighborhoods with a long history of immobility. These attributes give us a strong clue as to the necessary conditions for the emergence of a "village" type of social world. It is the

Corbis Images

FIGURE 5 "Village" life in the city. Shared spaces and shared experiences with familiar faces can promote a villagelike community atmosphere.

relative immobility—personal, occupational, and residential—of lower socioeconomic groups (of all ethnic backgrounds) that is the key. Immobility results in a strengthening of "vertical bonds" of kinship and "horizontal bonds" of friendship among people.

The typically high degree of residential proximity between family members traditionally in low-income areas like Boston's Italian neighborhood not only made for a greater intensity of interaction among family members but also facilitated the important role of the matriarch in providing practical support for family life—looking after grandchildren, for example—and in passing on attitudes, information, beliefs, and norms of behavior. Primary social interaction between friends is also reinforced by the residential proximity that results from immobility. Relationships formed among children at school are carried over into street life and, later on, relationships (Figures 5 and 6). Finally, the shared and repeated experience of hard times can promote the development of close-knit and overlapping social networks in low-income areas. With everyone "in the same boat" and little population turnover, residents can develop a mutuality of feeling and purpose that can become the mainspring of the social institutions, way of life, and "community spirit" associated with urban villages.

This community spirit can be a brittle phenomenon, however. The stresses and tensions that come with economic insecurity and deprivation underlie the mutuality of urban villages. Studies of life in urban village-type neighborhoods have described a good deal of conflict and disorder, and the main cause seems to be the absolute lack of space. Crowding leads to noise problems, insufficient play space, and inadequate laundry facilities and is associated with personal stress and fatigue. Children especially are likely to suffer from the psychological effects of the lack of privacy that not having enough space imposes. The links between crowding, stress, and deviant behavior are difficult to establish scientifically, however, particularly because so many other potential causal factors are present.[33]

Such stresses and tensions would seem to point to the operation of Wirthian processes. But Herbert Gans suggested that urban villagers develop a *subculture* that short-circuits Wirthian determinism. Gans defines a subculture as

> an organized set of related responses that had developed out of people's efforts to cope with the opportunities, incentives and rewards, as well as the deprivations, prohibitions and pressures which the natural environment and society—that complex of co-existing and competing subcultures—offer to them.[34]

Laimute Druskis/Pearson Education, Inc.

FIGURE 6 Street play can be an important dimension of neighborhood social life.

It follows that, rather than thinking of urbanism as a way of life, we should think of urbanism as fostering distinctive ways of life for different social/demographic/economic/ spatial groups. Claude Fischer took this argument a step further, suggesting that these ways of life, or subcultures, will be intensified by the conflict and competition among people that is inherent to urbanization and that new subcultures will spin off as new social groups (generated by the arrival of new migrants and immigrants and by the evolution of new class factions) reach the "critical mass" required to sustain cohesive social networks.[35] Although Fisher did not address the spatial framework associated with these networks, others have pointed out that their continued existence depends to a large extent on avoiding conflict with other subcultures. Conflict can be avoided by the development of implicit behavioral limits: a sort of "social contract" between subcultures. But it seems to be best avoided by means of spatial segregation. So conflict avoidance by people provides a further dynamic that contributes to the development of urban villages.

The idea of subcultures and ways of life fits conveniently with the long-standing concept of **cultural transmission**, in which distinctive norms of behavior, sometimes norms that would be seen as different by society at large, are passed from one generation to another within a particular community. This concept was given prominence by sociologists of the Chicago School in the 1930s and 1940s, who were particularly interested in those "natural areas" of the city in which the transmission of cultural values and attitudes from one generation to another seemed to foster criminal and delinquent behavior.[36]

Also relevant here is the concept of **neighborhood effects**, in which people tend to conform to what they perceive as local norms of behavior in order to gain or maintain the respect of their local peer group. Evidence from studies of voting behavior and attitudes toward education, as well as studies of criminality and delinquency, tends to support the idea that people act and think in ways that are not always in their best interests objectively; instead they conform to the opinions, both real and perceived, of the residents in their neighborhood who they interact with on a regular basis. One example of such neighborhood effects is the paradoxical syndrome of "suburban poverty." This is a term used to describe the result of neighborhood effects that impose middle-income consumption patterns on some incoming households whose incomes are not quite enough to service a new mortgage *and* "keep up with the Joneses" but who still feel obliged to conform with their neighbors' levels and patterns of consumption.

Suburban Communality, "Habitus," and Contemporary Lifestyles The first interpretations of suburban life, following the traditions of Tönnies, Durkheim, Simmel, and Wirth, saw suburbia as the manifestation of one particular side of urbanism: the adaptive withdrawal by people from the "psychic overload" of the city. Lewis Mumford, for example, wrote that U.S. suburbs represented "a collective attempt to lead a private life."[37] Early sociological studies of suburban life endorsed this view.[38]

But subsequent investigations and, indeed, the evolution of suburbanization and the polycentric metropolis, have shown this perspective to be incomplete. Once again it was Herbert Gans who took the lead, demonstrating in his study of Levittown that suburban neighborhoods did in fact exhibit localized social networks with a considerable degree of cohesion.[39] Several factors can be identified that contribute to this cohesion, including the social and demographic homogeneity of the people in suburban neighborhoods, the physical isolation of outlying neighborhoods, and a "pioneer eagerness" to make friends in newly developed tracts. The cohesiveness of suburban communities is further strengthened by people's social networks related to local voluntary associations of various kinds (such as parent-teacher associations or charitable organizations) and to the interaction between parents before, during, and after organized activities for children (team sports, dance classes, cheerleading practice, band and orchestra recitals). This purposive kind of cohesiveness has been described as producing *communities of limited liability*. According to this perspective, suburban neighborhoods represent just one version of a series of different kinds of communities of limited liability, others being based around workplace, political, lifestyle, or other interests. So instead of suburbanization leading to the breakup of communities, it can be thought of as breaking down communality into an increasing number of semi-independent groups of people, only some of which are locality-based.

This interpretation allows us to accommodate the increasing importance of lifestyle as a basis for residential segregation. French sociologist Pierre Bourdieu has pointed to the close relationships between class and lifestyle thorough his concept of **habitus**. A social group has a habitus if it has a distinctive set of values, ideas, and practices: a collective perceptual and evaluative representation *that derives from its members' everyday experience* and operates at a subconscious level, through commonplace daily practices, dress codes, use of language, and patterns of consumption (some of this work fits within the humanistic approach). The result is a distinctive pattern in which "each dimension of lifestyle symbolizes with the others."[40] This representation, suggests Bourdieu, evolves in response to certain objective conditions (including class and lifestyle) but once established serves to reproduce and sustain those conditions. It is the subtlety of the habitus of different groups of people that provides important clues to city dwellers about social distance and group membership; and it is their richness and variety that lend vitality to city life and give color and texture to the residential kaleidoscope.

However, one of the most significant trends in terms of lifestyles was the influence of superficial, transitory, and eclectic cultural tendencies during the 1970s and 1980s, in particular. In this kind of cultural context, it is argued, people had a greater capacity to explore a much wider spectrum of sensations and experiences. Through settings such as shopping malls, museums, galleries, movie theaters, and theme parks and, above all, through media such as glossy "lifestyle" magazines, TV, and now the Internet, an aesthetic sensibility is created in which masses of people come together in temporary emotional communities.[41] These can be rooted in the materialism and hedonism of the "society of the spectacle" and centered on intermittent periods of empathy and shared self-imagery that are attached to the symbolic properties of particular urban settings, collective experiences, or material possessions. In this context the sense of belonging to a lifestyle community was often exploited (and sustained)—in the 1980s especially—by commercial advertising: the "Pepsi Generation," the "United Colors of Benetton," and the "New Generation of Olds," for example. The result was a blur of ephemeral, affective communities of people brought together around the icons of materialistic lifestyles; the inconspicuous consumption and more casual attitudes toward fashion since the 1980s notwithstanding.

Implicit in the notions of habitus, lifestyle communities, and communities of limited liability is the idea that the phenomena they are describing somehow lack the mutuality, the permanent but intangible community spirit that is characteristic of the urban village. It should not be assumed, however, that community spirit cannot develop from such settings. When territorial exclusivity, amenities, or property values are threatened, the low-key, purposive communality of suburbia can flare into an intense mutuality and a tangible sense of community spirit. This reaction has been described as *status panic* or *crisis communality*. Rarely, however, does it survive the resolution of the perceived threat. This phenomenon emphasizes yet again the complexity of relationships between society and space: It is not at all easy to say which situations reflect the existence of "community," let alone which of those also coincide with a particular territory.

Community and Territory

As we have seen, the sociological theories of Tönnies, Durkheim, Simmel, and Wirth maintain that communities should not exist at all in contemporary cities; or, at most, that they exist only in a weakened form. This theoretical perspective can be summed up as *community lost*. The view that urbanization actually promotes a variety of cultures, leading to the development of urban villages, can be summed up as *community saved*. The view that urbanization creates the possibility for people to participate in one or more of a number of communities of limited liability, only some of which are place-based, can be summed up as *community transformed*. The view that contemporary urbanization promotes ephemeral, emotional communities based on the icons of materialistic lifestyles can be summed up in terms of *community commodified*.

Apart from anything else, these different views make it necessary for us to be very careful when using the terms community and neighborhood. They are probably best used simply as general terms that cover a range of situations relating to place, space, and social interaction among people. It is useful to think in terms of a loose hierarchy in which the base unit is the *neighborhood*—a territory that contains people of broadly similar demographic, economic, and social characteristics, but that is not necessarily very significant as a basis for social interaction. We can perhaps think of some market segments, such as "Upper Crust" and "Movers and Shakers," as building blocks in which the sociospatial dialectic may or may not foster the kind of social interaction that is associated with *community*, the next level in the hierarchy. Communities (in this sense) exist where a degree of social coherence develops on the basis of interdependence among people, which in turn produces a degree of uniformity of custom, taste, and forms of thought and speech. As a result, they are taken-for-granted worlds defined by reference groups that *may* be territorially based but may equally be school-based, work-based, or lifestyle-based.

The third tier of the hierarchy is represented by *communality*, a form of community *at the level of consciousness*. This is where a degree of community spirit is apparent, where a sense of "us" and "them" is prevalent. Like community, however, communality does not necessarily have to be based on territorial cohesiveness. So, as the U.S. National Commission on Neighborhoods pointed out, each neighborhood (or community) "is what its inhabitants think it is."[42] This assertion means, in turn, that definitions and classifications of neighborhoods and communities must depend on people's conditions and perceptions of social and physical space and on the social and geographical scales of reference they use—to which we now turn.

Cognition, Perception, and Mental Maps of the City

People use the filters of their own personal experience, knowledge, and values to deal with the stream of stimuli that they encounter in urban settings. The result is that they modify their real-world, objective experiences, creating in their minds a series of partial, simplified (often distorted), and flexible inner representations, or "mental maps." These inner representations form the basis not only for various aspects of people's behavior, but also for the experiential sense of place and sense of being that are central to community and communality.

Because people's mental maps are the product of personal knowledge, experience, and values, the same social and physical stimuli may evoke different responses from different individuals, each person effectively living in his or her "own world." Most people orient their mental maps around their own neighborhood or around the **central business district (CBD)**, and most tend to tidy up and simplify reality.

One of the key elements in shaping people's mental maps is the **cognitive distance** between the objects that they include in their maps. Cognitive distance is the basis for how people arrange the spatial information in their mental maps, while **social distance** is the basis for how they arrange much of the rest of the information. Cognitive distance is generated from the brain's perception of the distance between visible objects like a courthouse or a subway station; a perception that is influenced by land use patterns, the distinctiveness of these objects, and the impression made by symbolic representations of the environment such as GPS maps or road signs.

In general, cognitive distance seems to be a function of the number, variety, and familiarity of *cues* or stimuli that are encountered along commonly used pathways through the city. These pathways constitute the framework of a person's *action space*. A simple illustration of the effect of familiarity on spatial cognition is the way that the return leg of a new trip usually seems shorter. Elasticity is also an important attribute of cognitive distance, something that is illustrated by the fact that many people judge the same physical distance in different ways, depending on orientation and context. So, for example, the cognitive distance from a suburb to a downtown shopping area is often less than the cognitive distance from the downtown to the same suburb. Similarly, cognitive distance to attractive features such as parks and lakes tends to underestimate the actual physical distance, while cognitive distance to unattractive features such as parking lots and landfills tends to overestimate actual distance.

Although cognitive distance is the basis for the arrangement of the objects that people include in their maps, the objects themselves are what give mental maps their distinctiveness. People do not have a single image or mental map that can be consulted or recalled at will. Rather, each person appears to be able to draw upon a series of latent images that are unconsciously operationalized in response to specific spatial tasks: navigating across town, walking around a shopping district, or searching listings for new homes, for example. These latent images, like the cognitive distances between the objects they contain, are largely a function of people's action space, but they are also derived in part from a broader *information space* or *awareness space* that is built up from a variety of sources and experiences: TV coverage, the Internet, newspaper stories, e-books, conversations, and so on. As these sources of information change, so the latent images change. Meanwhile, as people's tasks change, so different images are called upon. In this context, a useful distinction can be made between the *designative images* that relate to the cognitive organization of space and the *appraisive images* that represent various feelings and judgments about particular components of the city.

Following the work of urban planner Kevin Lynch, we can recognize five principal elements in people's designative mental maps of cities: paths, edges, districts, nodes, and landmarks (Figure 7).[43]

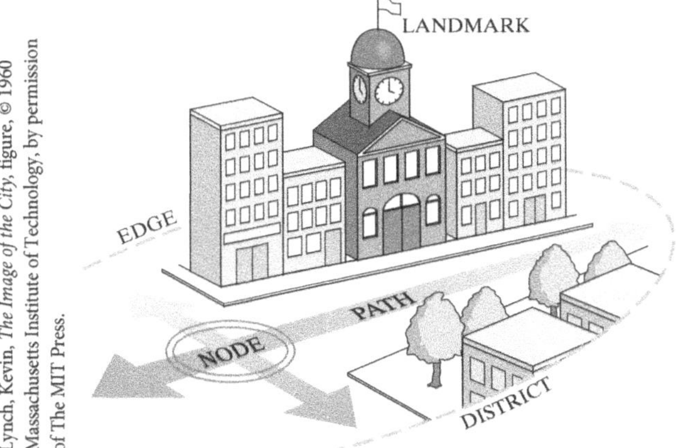

FIGURE 7 The five principal elements in people's designative mental maps of cities: paths, edges, districts, nodes, and landmarks based on the work of Kevin Lynch.

1. Paths are the channels along which people customarily, occasionally, or potentially move. They may be streets, walkways, transit lines, railroads. For many people, these are the predominant elements in their image. People observe the city while moving through it, and it is along these paths the other elements are arranged and related.
2. Edges are the linear elements that people do not use or consider as paths. They are boundaries, linear breaks in continuity: lake shores, railroad cuts, edges of development, walls. They may be barriers that are more or less penetrable, which close one district off from another; or they may be seams, lines along which two districts are related and joined together.
3. Districts are the medium-to-large sections of the city, which people actually or mentally enter, and which are recognizable as having some common, identifying character.
4. Nodes are the strategic spots in a city into which people can enter, and each is an intensive focus to and from which they are traveling. Nodes may be primarily junctions, places of a break in transportation, a crossing or convergence of paths, or they may simply be concentrations which gain their importance from being the focus of some use or physical character, such as a street-corner hangout or an enclosed square. Some nodes are the focus and embodiment of a district, over which their influence radiates and for which they stand as a symbol (Figure 8).
5. Landmarks are another type of point-reference, but in this case people do not enter them. They are usually a rather simply defined physical object: a sign, a clocktower, or a tall or otherwise distinctive building.

Lynch derived these five elements from a series of interviews and an examination of people's sketches of their mental maps. He then pioneered the construction of maps of *aggregate* perceptions of urban space, using lighter or darker symbols to indicate the proportion of people who had mentioned or sketched a particular element in recalling their own individual mental map.

Lynch used his work to focus on the *legibility*, or imageability, of cities, believing that people would prefer to live in more legible environments. These, he argued, are those settings where built form is clear and straightforward, with continuity and rhythm in paths and edges, where districts are clearly separated, and where there are dominant nodes and landmarks. So, for example, Lynch suggested that a city like Los Angeles, with few of these qualities, would be less legible (and therefore less livable) than a city like Boston, with its highly structured elements.

Subsequent research has provided broad support for such an interpretation. It has also suggested, however, that it is unwise to generalize too much about mental maps. Urban environments can be legible in different ways to different people. Some cities are legible merely because of the dominance of a clearly connected set of paths, others because of the clarity of a series of landmarks. Some people prefer *illegible*

imagebroker.net/SuperStock

FIGURE 8 Pioneer Courthouse Square, a "node" that has been planned into the structure of downtown Portland, Oregon.

environments, possibly because of abstract qualities such as "quaintness" or "intimacy." Most important of all from the point of view of relating people's mental maps to their behavior (and so being able to shed more light on the processes involved in the sociospatial dialectic) is that different groups—by age, gender, socioeconomic background, and neighborhood—tend to have very different mental maps of the same environment. This difference is the result, of course, of differences in their respective action and awareness spaces.

Appraisive Images In many circumstances it is not the sophistication of these designative images that is important so much as the feelings that people attach to specific elements in their mental maps. A specific node or district, for example, may be regarded as attractive or unattractive, exciting or dull; or, more likely, it may evoke a combination of feelings. These reflect the *appraisive images* that people draw upon in order to steer their way through the city, that they use in making decisions about where to live, shop, or play, and that they refer to in making subjective judgments about neighborhoods and their residents. One very vivid and specific aspect of appraisive imagery concerns people's perceptions of threats to personal safety. In larger cities certain districts are generally perceived to be dangerous. At a more detailed scale, residents familiar with these neighborhoods develop "fear maps" (Figures 9 and 10) that highlight danger points near gang hangouts, abandoned buildings, crack houses, drug markets, and so on.

In overall terms, appraisive imagery can be thought of as the product of three main evaluative dimensions:[44]

1. The *interpersonal environment* that is based on the social attributes of different parts of the city and that is particularly influenced by perceived social distance, by perceptions of the friendliness, helpfulness, snobbishness, pride, power, contentment, and so on of different groups of people, and by perceptions of the tranquility and safety of different settings.
2. The *impersonal environment* of streets and buildings that is based on the aesthetics of particular settings, their tidiness, level of maintenance, amenity value, and traffic noise.

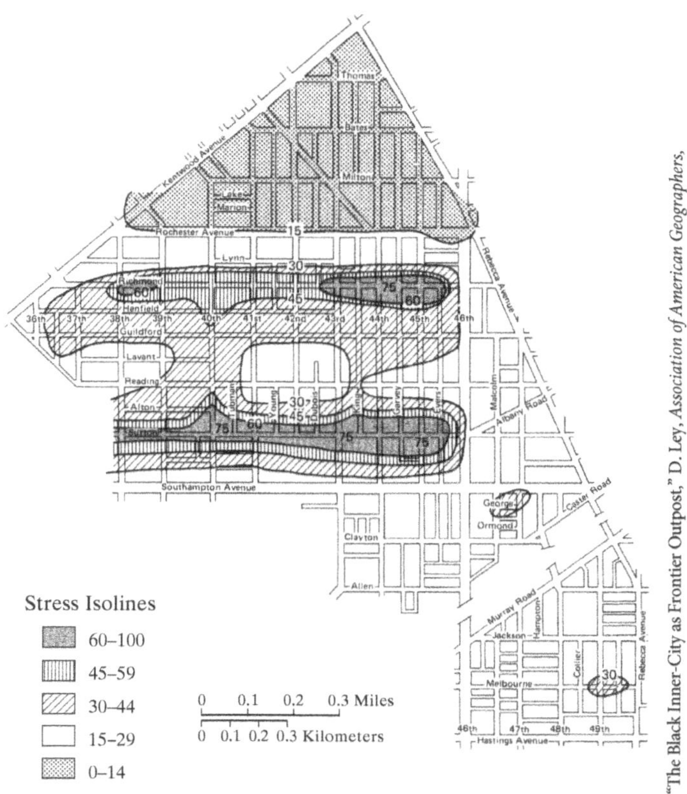

"The Black Inner-City as Frontier Outpost," D. Ley, *Association of American Geographers, Monograph Series no. 7*, Reprinted with permission.

FIGURE 9 Perceived stress in the Monroe district of Philadelphia (the darker the shading, the higher the stress).

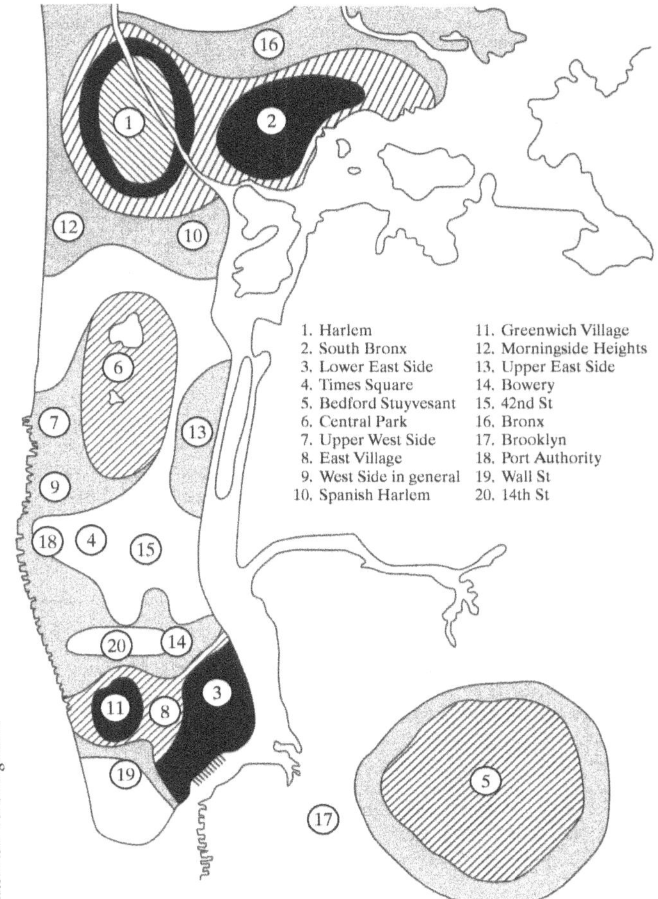

FIGURE 10 The geography of fear in New York. The darker the shading, the more widespread the fear among New Yorkers of that particular part of the city. The numbered locations list, in rank order, the 20 "most fearsome" neighborhoods.

3. The *locational* attributes of different parts of the city—accessibility to freeways and transportation systems and proximity to amenities, attractive settings, or potential nuisances.

Perhaps the best-developed aspects of appraisive imagery concern people's images of their *home area*. They are also the most important aspects of appraisive imagery in the present context because they relate directly to the issues of community and territory. As we might expect from a reading of Gans and Fischer, plenty of evidence points to the existence of some kind of neighborhood attachment among people who have lived in a particular area for any length of time. Research has shown that around 80 percent of people are able to draw upon a mental image of the area in which their daily lives are played out and that they identify with as their home area. Although these images are highly personal and idiosyncratic (with neighbors varying a great deal in the boundaries they place on their particular home area), they exhibit some interesting regularities that provide important clues about the relationships between community and territoriality. First, the area covered by people's mental images is broadly consistent at about 100 acres. Second, this territorial extent is not influenced by variations in population density: the home areas of suburban residents cover the same amount of "turf" as central city residents. Social psychologist Terence Lee has developed a hierarchical topology, based on analyses of people's responses to questions about their home area:[45]

- *The social acquaintance neighborhood*—a small area within which, apart from a few close ties of family or friends or both, people keep to themselves.
- *The homogeneous neighborhood*—a larger area that extends to a greater awareness of the physical aspects of the setting as well as the social aspects of neighbors. The homogeneous neighborhood comprises not only "people like us" but also "people who live in homes like ours."
- *The unit neighborhood*—an even larger territory, a "named" district: "our neighborhood as others see it." Unit neighborhoods usually contain a variety of shops and amenities and some variability in socioeconomic background and household types. They approximate conceptions of urban neighborhoods by planners and politicians.

This topology not only reflects the importance of *spatial scale* in relationships between community and territory but also points to the multidimensionality of people's sense of place. We now turn, therefore, to a closer examination of the affective bonds between people and places, looking in particular at the importance of localized lifeworlds and the significance of the symbolism that people associate with particular sociospatial settings.

Lifeworlds and the "Structuration" of Social Life

People's images of their home area provide an important framework for many aspects of daily life in cities. This dimension of urbanization, however, can be understood only through insights that are derived from the insiders' points of view. It requires an appreciation of how meanings are assigned to people and places, how people obtain and process information about people and places, how they "learn" sociospatial settings, and how they react to changes in them. Put another way, it requires "the contextual interpretation of subjectively meaningful social action" by a person, a "form of empathetic understanding" that is gained from adopting that person's own perspective.[46] Using this approach, it becomes apparent that every place, every locality within the city is socially constructed. That is, they are given meanings that are contingent on the people who use them and live in them. Cultural geographer Yi-Fu Tuan put it this way: "Place is a center of meaning constructed by experience."[47] So it is possible for localities to have "multiple realities," as different groups of people construct different meanings around them. As social geographer David Ley has emphasized, "Place is relational to a purpose and an intent. As such, any place has a *multiple reality* for the plurality of social groups that encounter it."[48]

A good starting point in understanding these insiders' views is the concept of the *lifeworld*, the setting (or settings)

Rudi Von Briel/PhotoEdit

FIGURE 11 The taken-for-granted lifeworld. This photograph captures something of the distinctive lifeworld of part of Harlem in New York City.

for everyday life, lived reflexively, with natural, unreflective attitudes and behaviors constituting a taken-for-granted social world (Figure 11) and a meaningful personal context in which every object has its purpose and every place its significance.[49] It follows that lifeworlds form as a result of established routines—in the workplace, around the home, in a bar, at a club. But for individual routine to become a collective lifeworld, a certain amount of *intersubjectivity* is required. People need to have access to one another's feelings and meanings for the experience of everyday life to be a shared experience. This intersubjectivity is constructed and maintained in specialized systems of communication and symbolization. Conformity to a given code of communication and symbolization—through styles of speech, specialized vocabulary, dress codes, use of personal space and body language, use of humor, unspoken rules of interaction, etiquette, and rituals—defines an in-group (members of the lifeworld) and an out-group (the rest). Such conformity requires the conscious suppression of individualistic traits, and it brings about the *un*conscious suppression of individualistic and idiosyncratic traits and the development of cognitive and attitudinal consensus.

A useful concept in this context is the notion of a *structure of feeling* developed by Raymond Williams.[50] This concept incorporates both the sense of place and the views of the world that stem from the intersubjectivity of the lifeworld, but it extends to the conscious attitudes and feelings of people about the future as well as the past and present. These feelings add up to a "structure" in the sense that they comprise an interlocking set of cognitive elements. As cultural geographer Peter Jackson observes, the idea of a structure of feeling has more than a passing similarity with Pierre Bourdieu's concept of habitus.[51] In practice, however, both depend on intersubjectivity that is maintained by people through shared codes of symbolism and communication, and these in turn depend heavily on the existence of people's established routines in time and space.

URBAN VIEW 4
Disability and the City[52]

In most studies of cities the able-bodied character of people is taken for granted.[53] Increasingly, however, geographical and other social scientific research is addressing issues of disability.[54] A key issue is what is meant by disability. Rather than being a physical defect affecting particular individuals, some argue that disability is above all a **social construction**. So the social construction of "disability" is bound up with the attitudes and structures of oppression in an able-bodied society, rather than the failing of a particular individual.

In societies that view the role of medicine as primarily one of making sick people well, disability is often seen as something unhealthy. But if sufficient facilities were provided, being disabled could be regarded as something akin to being shortsighted—wearing glasses or contact lenses not generally being regarded as a disability. This is a good example of how social scientific research can challenge taken-for-granted assumptions. What seems like common sense can often be just a highly mistaken function of the way we have been taught to think about the world.

The socially constructed character of disability was highlighted in a classic study of attitudes toward blindness. In the United States blindness was regarded as an experience of loss requiring counseling; in the United Kingdom as a technical issue requiring aids and equipment; and in Italy as a need to seek consolation and salvation through the church and religion.[55] Different societies, therefore, produce differing definitions of disability. In the United States, for example, following the politicization of disability rights, a multimillion-dollar disability industry has emerged. Disability has, therefore, become a major source of

URBAN VIEW 4
Disability and the City (*continued*)

income for doctors, lawyers, rehabilitation professionals, and disability activists.

Cities frequently contain numerous barriers to mobility and access for disabled people. Typical problems include high curbs, steep steps, no ramps for wheelchairs, narrow doors, and the absence of information in Braille. In addition, elevators for those who are disabled are often badly signposted, with inaccessible buttons, and in unattractive locations (such as when they are the service elevators next to kitchens). Public transportation systems can also pose problems for people with disabilities. Often the problem boils down to one of cost, with inadequate funds being made available for adequate conversion of premises for disabled access.

The problem is also one of dominant attitudes in society.[56] All too often, architects, planners, and the public at large have assumed that disability leads to immobility. Consequently, the needs of disabled people are ignored. The barriers are therefore social and psychological as much as physical. Cost constraints on the provision of disabled access therefore reflect wider sets of social values toward disability. These barriers prevent disabled people from fully participating in social life and in paid employment. A study in Ontario, Canada, found that 80 percent of disabled people lived in relative poverty because of their exclusion from the job market and limited support programs from both the public and private sectors.[57] In the United States, Census figures indicate that 13.3 million, or 7 percent of people of working age (16 to 64), reported difficulties finding or keeping a job due to a health-related condition. Overall, nearly 23 million people (about 12 percent of workers) reported that they were limited in the kind or amount of work they could do because of a physical, mental, or other health condition.[58]

In many cities the issue of disability is being taken more seriously. In the United States, the Americans with Disabilities Act of 1992 requires businesses to provide wheelchair access. In the United Kingdom, many local governments now have Disability Officers whose task it is to improve disabled access in city centers. These improvements are in part a response to the increased activities of various disability rights movements. These groups have begun to campaign on issues of income, employment, civil rights, and community living rather than the traditional issues of institutional care.

As with all **urban social movements**, there have been setbacks as well as advances. For example, in the United States corporate interests lobbied before the judiciary that the Americans with Disabilities Act is an unnecessary restriction on private property rights and therefore an infringement of the Fifth Amendment of the Constitution.[59] In the late 1990s in the United Kingdom, the Labor government set about reforming the benefits system to limit entitlements for disabled people in an effort to encourage disabled people to take up paid employment, given that an estimated 75 percent of disabled adults relied on some form of government support.[60]

Time-Space Routines

The importance of people's established routines in facilitating the development of codes of symbolism and communication, intersubjectivity, and conformity has been emphasized by geographer David Seamon, who has used the concept of *time-space routines* to describe the existential importance of repetitive situations for individuals and the concepts of *place ballets* to capture how multiple time-space routines unfold within a single setting.[61] The most useful approach to time-space routines, however, comes from the work of Torsten Hägerstrand and the Lund School of Geography in Sweden. In a simple but effective way, Hägerstrand captured the constraints of both space and time on individuals in their everyday lives. His model (Figure 12) details how people trace out "paths" in time and space, moving from one place (or "station," in Hägerstrand's terminology) to another in order to fulfill particular purposes ("projects").

According to Hägerstrand, the combined effects of three kinds of constraints define any individual's capacity to move from one station or project to another within the broader environment:

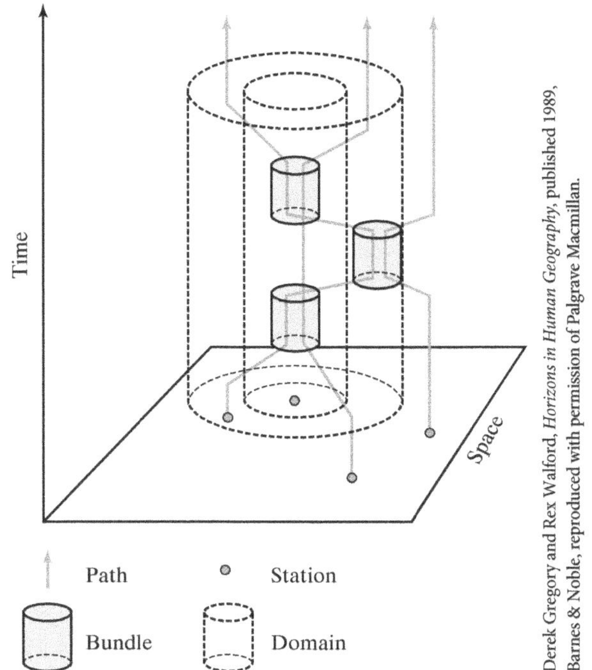

FIGURE 12 The notation of Hägerstrand's time geography.

- *Capability constraints*—which include the need for people to get a minimum amount of sleep (which limits the time available for traveling) as well as the kinds of transportation available (so determining the radius of the territory that can be included in the daily life path).
- *Authority constraints*—which include laws, rules, and customs that limit people's accessibility to certain places and curtail their participation in certain activities.

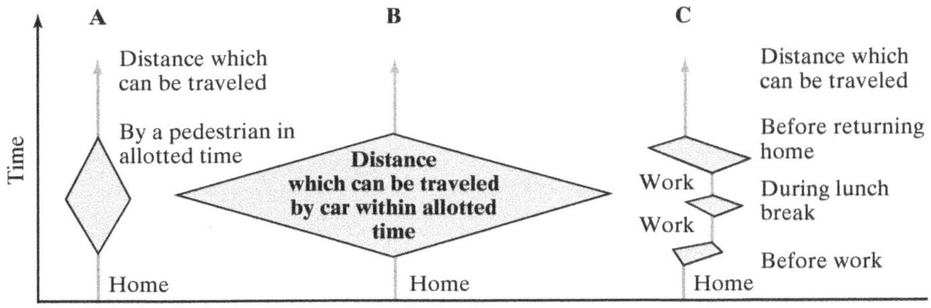

Time

A
Distance which can be traveled
By a pedestrian in allotted time
Home

B
Distance which can be traveled by car within allotted time
Home

C
Distance which can be traveled
Before returning home
Work
During lunch break
Work
Before work
Home

FIGURE 13 Time budgets and daily prisms.

An Invitation to Geography, Lanegran et al., McGraw-Hill, 1978.

Examples include highway speed limits and laws governing alcohol consumption.

- *Coupling constraints*—which stem from the limited periods during which particular projects are accessible to people. These are determined by the opening hours of businesses, the scheduling of public transport, and the timing and duration of rush hours.[62]

The *time budget* of any individual within his or her environment can be traced graphically, as in Figure 13; the effective range of that individual being described by a prism (or series of prisms) whose shape depends on capability constraints. Every stop, for whatever project, causes the range of the prism to shrink in proportion to the length of stay. In addition, both people's time budgets and their activity patterns

URBAN VIEW 5
Structuration: Time and Space in People's Everyday Life

"Bundles" of activity in time and space are central to the important concept of the "*structuration*" of social life advanced by social theorist Anthony Giddens (Figure 12).[63] Giddens emphasizes the fundamental importance of the "binding" of time and space on the very fabric of social organization and the conduct of social life. The concept of structuration is based on the interdependence between (1) social systems and structures and (2) purposeful individual behavior. It is often couched in terms of (social) *structure* and (human) *agency* respectively. According to Giddens, the continuity of people's day-to-day life and the existence of bundles of routinized activity encourage *social integration* as individuals respond reflexively and purposively to their shared experiences in time and place. In this way, people influence and are influenced by the structural properties of social systems—including the frameworks of class, gender roles, educational stratification, and legal codes.

To the extent that people's routinized social practices are consistent over particular spans of time and space, these practices flow from and channel back into structural relations that reach beyond localized bundles to influence interactions with and between others who are *absent* in time or space. This influence helps to sustain *system integration*, the extension of people's social relations across both time and space, which Giddens calls *time-space distanciation* (Figure 14). At an intermediate scale, and connecting the *time-space routinization* of the lifeworld with the time-space distanciation of society, is a hierarchy of locales, similar to Hägerstrand's domains (Figure 12). This hierarchy represents a "regionalization" that reflects multiple sociospatial networks of people based on the differential distribution of resources (economic power) and authority (political power).

This idea gives us a very useful framework for conceptualizing urban life and the way that human landscapes are created and sustained. Geographers Michael Dear and Jennifer Wolch use a similar approach in showing, as they put it, "how territory shapes social life":

> Human landscapes are created by knowledgeable actors (or agents) operating within a specific social context (or structure). The structure-agency relationship is mediated

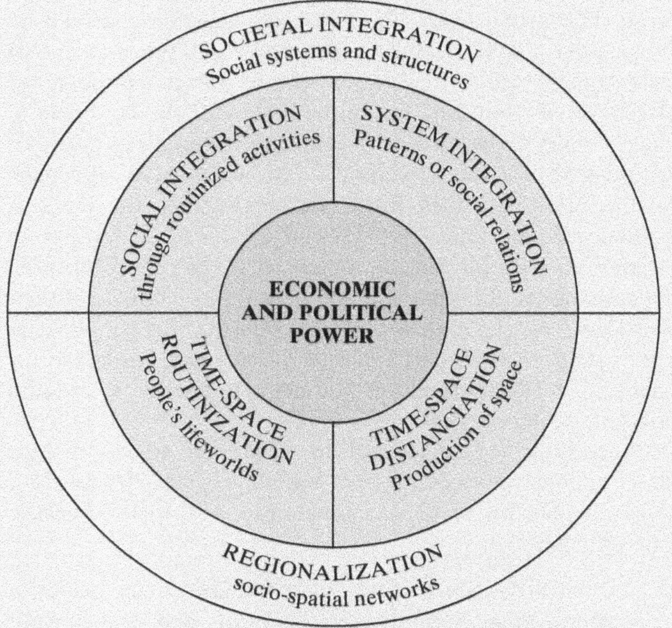

FIGURE 14 Elements in Giddens's concept of "structuration."

Derek Gregory and Rex Walford, *Horizons in Human Geography*, published 1989, Barnes & Noble, reproduced with permission of Palgrave Macmillan.

URBAN VIEW 5
Structuration: Time and Space in People's Everyday Life (*continued*)

by a series of institutional arrangements which both enable and constrain action. Hence three "levels of analysis" can be identified: structures, institutions and agents. Structures include the long-term, deep-seated social practices which govern daily life, such as law and the family. Institutions represent the phenomenal forms of structures, including, for example, the state apparatus. And agents are those influential individual human actors who determine the precise, observable outcomes of any social interaction.[64]

The concept of structuration, together with the idea of a hierarchy of locales each containing various lifeworlds, provides a strong but flexible framework for dealing with the social construction of reality. All physical space is open to the social construction of meaningful reality for people. Some spaces are even the object of multiple realities, being socially constructed in different ways by different groups of people. In this way, space is invested with an ideological content that reflects the values, attitudes, and cognitive structures of the groups of people involved. This ideological content has important implications for the overall trajectory of urbanization because of the way that it serves to reproduce and sustain the structure and dynamics of the relations between social groups.

are recognized as being shaped by—and contributing to (the sociospatial dialectic again)—wider structural features of the social systems in which they operate. These features include the various institutions that control the authority constraints and coupling constraints relevant to particular constellations of stations ("domains")(Figure 12). Shared domains and routines create coherent time geographies, the intersections of individual paths forming "bundles" of social activity in time and space.

GENDERED SPACES

One important dimension of the sociospatial dialectic that tends to be concealed by an emphasis on particular social groups or particular localities is that of gender. When the existence of women is acknowledged, neighborhoods and lifeworlds look more complex. "Middle-income" neighborhoods, for example, may well be fairly similar in terms of the occupation and socioeconomic background of the often male heads of households that society and many social scientists use in categorizing neighborhoods, but these neighborhoods also include, among others, women whose labor-market status as unskilled, semiskilled, and low-paid workers would categorize them as low-income. Conversely, "low-income" neighborhoods are often home to white-collar female workers.

Geraldine Pratt and Susan Hanson have used this point in arguing that social status is reproduced in different ways for men and women. Men often experience class consistently, with their family-class position being similar to their sense of class developed through their work. Some married women experience class less consistently. Their family-class position, if defined by their husband's occupation, may not coincide with their own class position as experienced through their own paid employment.[65] In addition, traditionally many women's experience of social status has varied significantly through time as a result of childrearing roles that have taken them out of the paid labor force for long periods, often forcing them to return in inferior positions or in totally different occupations. For the same reason, some women experience locality differently, "homemakers," for example, spending a much greater proportion of their time in localized lifeworlds centered on the home and the social acquaintance neighborhood.

The Creation of Women's Spaces

We have to go back to the early industrial city in order to understand the evolution of women's roles and women's spaces. For households caught up in the unprecedented vortex of industrialization and urbanization, the demands of daily routine required a rationalization of family life. The new rhythm of life dictated by office clocks and factory whistles made it difficult for the family to fulfill some of its traditional roles, while increased residential mobility tended to isolate the families from their extended family network. With fewer relatives available nearby to perform informal service functions, families found themselves having to give up not only their economic function (as a unit of production) but also having to shed many traditional responsibilities: educating children, supporting the elderly, and caring for the sick, for example. The corollary of these changes was the emergence of a whole range of new service professions and establishments such as tailors, barbers, doctors, restaurants, and laundries. In many ways the major distinction was between workplaces and home; but this distinction had very different consequences for men and women.

Although women were drawn increasingly into the formal economy, their participation was often transitory, depending on their stage in the **household life cycle** and on the cyclical fluctuations of the economy. For less-skilled and less-well-educated women in working-class households, opportunities for paid work outside the home were further constrained by perceptions that cast women as both physically weaker and emotionally less stable than men. As a result, many women could obtain money to supplement family income only in domestic settings (theirs or other people's), undertaking piecework such as sewing or providing domestic services such as casual cleaning, childcare, and taking in washing, or providing room and board to paying lodgers. The domestic setting also became the basis for the lifeworld of middle-income women, though for rather different reasons.

For middle-income women, it was the *suburban* domestic setting that framed the routine of everyday life. In this context it is important to appreciate the suburbs themselves as socially constructed. According to historian Margaret Marsh, this construction originally had two components. Although the domesticity now associated with suburban settings was initially a female ideal, centered around the cultural institution of family, it was seen at the time as something that could best flourish amid the markets, bakeries, and neighborly support provided by cities, not suburbs. Women domestic reformers did not urge their readers to leave the city but to develop domestic ideals within it. The initial suburban ideal, Marsh points out, was a male construct, derived from Jeffersonian political philosophy and Emersonian social philosophy, serving the Romantic ideals of Nature and the Frontier and recognizing the imperatives of property ownership.[66] It was the success of that ideal that brought the middle-class female ideal of domesticity to the suburbs, where, by the late nineteenth century, the two ideals fused to form the basis for the **social construction** of suburban life. So:

> The clear separation of work from home, the insistence on social distancing, the treatment of the home as a feminine domain, the importance attached to domestic privacy and the exclusion of the vulgar prying multitude can all be seen as parts of a code of individual responsibility, male economic dominance, female domestic subordination, and family-nurtured morality *which served to give the bourgeoisie a social identity and marked them off from the upper class and lower orders.*[67]

None of this could happen simply through the force of ideals, however. As well as being socially constructed, suburban middle-class life had to be physically designed and built, and models of appropriate behavior had to be made known to everyone. An important influence in this context was Andrew Jackson Downing's book *The Architecture of Country Houses*,[68] which went through nine printings between its publication in 1850 and the end of the Civil War. Downing took the Jeffersonian and Emersonian equation between Nature/Rurality and Morality/Democracy to the level of individual dwellings. The home, he argued, must express in size, style, layout, and ornamentation the morality of its inhabitants. He presented a series of illustrative designs to illustrate this ideal. The market was soon flooded with pattern books "whose authors considered themselves a critical component of the creation of a democratic republic."[69]

The female ideal of domesticity was similarly advanced through print media. Here, articles and advertisements in popular magazines played the most crucial roles. One of the most important was the *Ladies Home Journal*, which, under the editorship of Edward Bok between 1889 and 1912, spelled out for its readers the canons of Good Taste in everything from domestic architecture and interior design to domestic manners and social comportment. Bok's evangelical mission was:

> to reform and simplify the American home and to keep women in it. He enlisted architects, suffrage leaders, reformers, novelists, statesmen, and even presidents to assure his readers that the right kind of home environment could preserve the family, strengthen the nation, and thereby give women more than enough meaningful work to do. For him, women belonged in these homes, and each one was to be separate from the others.[70]

Meanwhile, advertisements in popular journals further advanced the twin ideal of suburban domesticity in selling consumer goods. A detailed analysis of advertisements in the *Saturday Evening Post*, for example, shows that women were given responsibility through the earliest advertising campaigns for steering the household's adaptation to its new suburban surroundings. This program not only helped to reaffirm the ideal of suburban domesticity but also helped to crystallize gender roles and gender relations.[71]

Referring for a moment to the concept of **social reproduction**, we can see how these changes created a sociospatial setting that ensured not only the social reproduction of the middle class but also the social reproduction of gender roles (Figure 15). Men set off into the world of work while women, left to make what they could of day-long isolation, busied themselves with creating a comforting and uplifting home environment.

H. Armstrong Roberts/Corbis Images

FIGURE 15 The "ideal of suburban domesticity."

Over time, as more and more people in the United States aspired to replicate the lifestyle of suburban affluence and as levels of living steadily rose, the social construction of suburban domesticity, together with the associated gender roles, came to incorporate a broad band of socioeconomic groups. The process was consolidated in various ways by the federal government and private corporations, who "made clear that the nuclear family, with a male breadwinner and a female homemaker, ought to be considered the "normal" type of family. All three **greenbelt cities**, for example, excluded all families with wives employed outside the home; banks, schools and insurance companies all across the country began to bar the employment of married women, and the women's magazines urged married women with jobs to quit so that men could take their places."[72] By the 1960s feminist Betty Friedan could justifiably claim that there had emerged a "mystique" of feminine fulfillment—the idea that the suburban housewife could find true fulfillment through her husband, children, and home—that had become "the cherished and self-perpetuating core of contemporary American culture."[73]

Changing Roles, Changing Spaces

Yet economic and social change was already overtaking the ideal of suburban family life. Some women had been drawn into the labor force during World War I to fill civilian jobs vacated by enlisted men; some were drawn into "women's work" as industrialization involved more and more product lines that could take advantage of mass production (Figure 16). World War II had been associated with an increase in the participation of married women in the labor force to an unprecedented 25 percent (Figure 17); divorce rates, having increased slowly throughout the first half of the twentieth century (with the exception of the sudden peak around the time of World War II), suddenly accelerated in the late 1960s (Figure 18) in response to changing social attitudes. Today only about 20 percent of all households in the United States consist of a married couple with their own children (down from 40 percent in 1970), while about 25 percent are headed by women with children.[74] Meanwhile, the labor force participation rate of women has risen to almost 60 percent.[75] The context of these demographic and social changes is described in relation to shifts in the dynamics of the **urban system.** But the same changes were implicated in restructuring urban life and urban space. The increase in couples' average age at marriage, the increases in the number of single and childless women, and the dramatic rise in the number of female-headed households all contributed to a huge pool of women whose work and residential choices were being exercised without reference to male preferences and without financial assistance from (or dependence on) men.

Social geographers Robin Law and Jennifer Wolch have noted three principal ways in which these trends have affected urban space:

1. The transformation of suburbia as the participation of women in the paid labor force undermined the communality of suburban neighborhoods by dramatically reducing the number of full-time housewives available to perform the unpaid volunteer community services and sustain the spatially bound friendships and acquaintanceships necessary for a sense of community.
2. The creation of distinctive new urban spaces through the residential behavior of two-worker households without children, reflecting their generally higher disposable income, a different set of consumption patterns (such as priority on restaurants rather than schools), and the need to balance the commuting trips of the two workers.
3. The creation of distinctive new urban spaces through the residential constraints experienced by female-headed families, who tend to become localized in central city neighborhoods because of low incomes and the need for

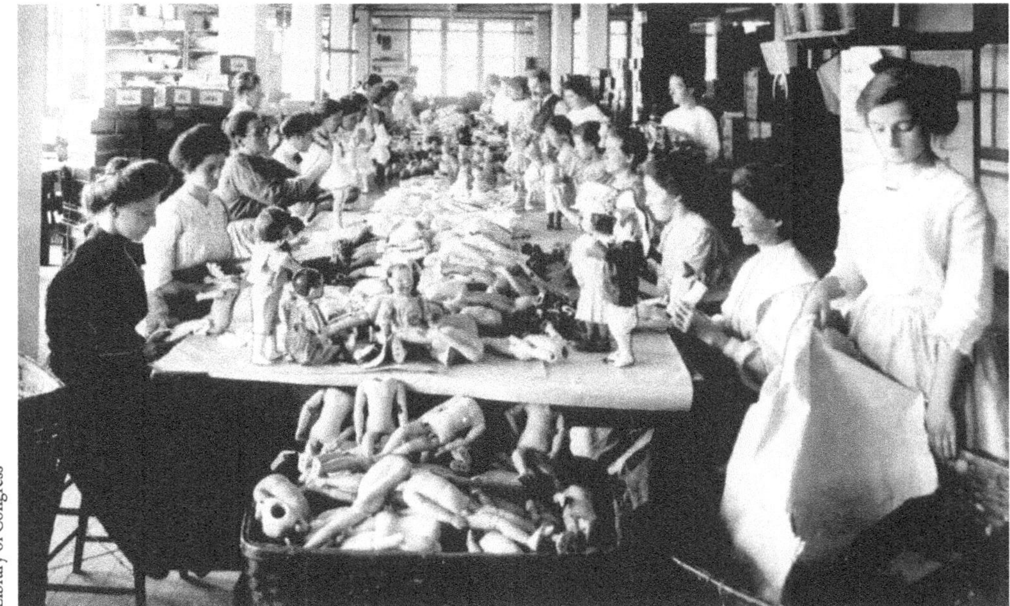

Library of Congress

FIGURE 16 Women at work in 1912. Assembling dolls at the Shrenhat Toy Company, Philadelphia.

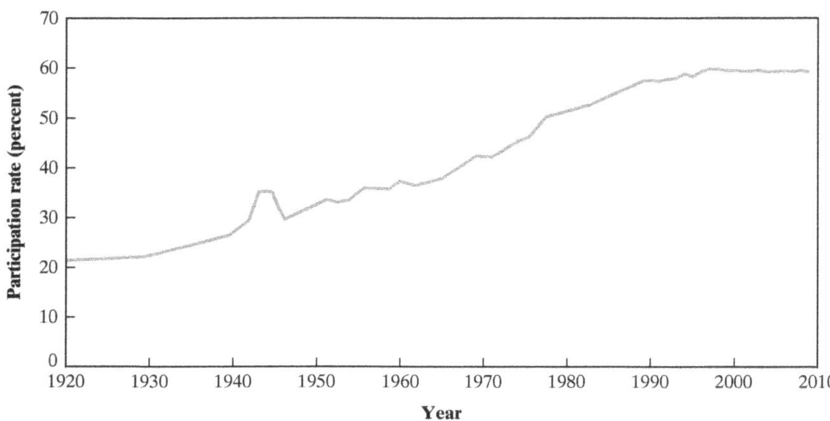

FIGURE 17 Labor force participation rate of females aged 16 and over in the United States.

> urban services such as public transport and accessibility to services such as child care.[76]

Law and Wolch also pointed out several ways in which these changes have affected urban life, time-space routines, and (therefore) lifeworlds. The increased labor market participation by women, for example, is clearly bound up in increased materialism and consumerism. But access to consumer goods and services is closely circumscribed by people's income, so differences have emerged in the *form* in which materialism is inscribed onto urban life. There is an increasing difference between the mass-produced goods and standardized services consumed by lower-income groups and the customized, personalized consumption of upper-income groups. One effect of this difference, as Law and Wolch point out,

> is that the spaces and times of personal fulfillment, entertainment and leisure for the affluent are also spaces of work (exploitation) for those in labor-intensive service jobs. Evenings, weekends, public holidays are times of work for more and more people, as are the public sites of entertainment, restaurants, shopping malls and resorts. And more and more, the requirement of that labor is that the act of labor (work-for-wage) be presented as an act of care (work-for-love). The requirement is not only that workers perform the task but that they accompany it with emotion: the sincere smile, the warm greeting . . . Thus personal life and "home" (true affection, sincerity, personal commitment) become colonized and blurred with "work."

The overall increase in female participation in the labor market has also, they point out, accelerated the time-space routine of both middle-income and low-income households. Two-worker families with children, together with single-parent families, have had increasingly to rush themselves and their children through the day, until (perhaps) an extended bedtime allows for a few moments of "quality time." Because women traditionally tended to remain burdened with an unequal domestic division of labor as a result of the legacy of the ideal of suburban domesticity and the "mystique" of feminine fulfillment, they tended to suffer most from pressures on the use of time. For them the "prism" of opportunities in time and space was too often replaced by a "prison" of coupling constraints. For this reason, many women now perceive central city environments as being most supportive and appropriate to their needs, just as they did before the advent of the male suburban ideal. In these settings, it is suggested, higher densities make it easier for women to develop social support networks and to organize collectively to assist with domestic responsibilities such as childcare. So the domestic ideal, fused for so long with the suburban ideal, promises to break free, bringing a further round of change in the sociospatial dialectic.

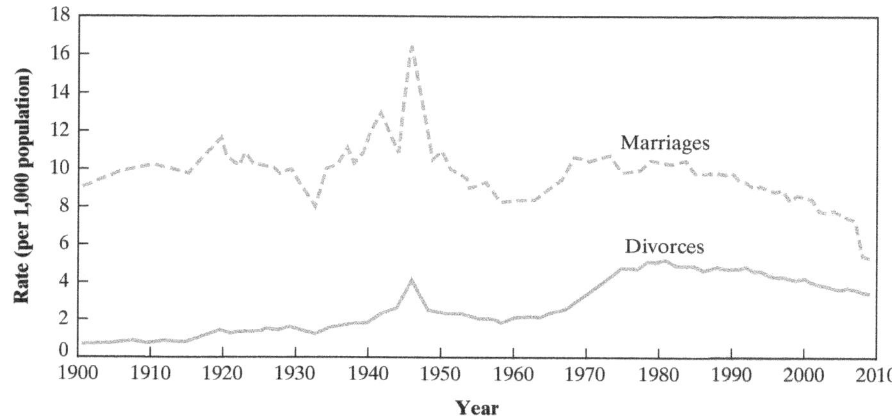

FIGURE 18 Marriage and divorce rates in the United States.

Discrimination by Design: Domestic Architecture and Gender Differences[77]

One well-worn theme in architectural theory has been the manifestation of "masculine" and "feminine" elements of design. For the most part this has involved crude anatomical referencing. Skyscrapers, as phallic towers for example, can be seen to embody the masculine character of capital investment. As some feminist interpretations of architectural history have shown, however, the silences of architecture can be more revealing than crude anatomical metaphors. Elizabeth Wilson highlighted the way that Modernist architecture, although being self-consciously progressive, had nothing to say about the relations between the sexes.[78] It changed the shape of housing units without challenging the functions of the domestic unit. Indeed, the **Bauhaus School**, vanguard of the Modern Movement, helped reinforce the gender division of labor within households through Breuer's functional Modern kitchen.

The internal structure of buildings embodies the taken-for-granted rules that govern the relations of individuals to each other and to society just as much as the overall plan and internal structure of cities. The floor plans, decor, and use of domestic architecture have in fact represented some of the most important encodings of patriarchal values (involving the broad system of social arrangements and institutional structures that enable men to dominate women). As architects themselves have so often emphasized, houses cannot be regarded simply as utilitarian structures but as "designs for living." The strong gender coding built into domestic architecture has been demonstrated in analyses ranging from Victorian country houses to bungalows and tenements.[79]

Today, the conventional interpretation of suburban domestic architecture recognizes the way that the ideals of domesticity and the wholesomeness of nuclear family living are embodied in the feminine coding given to the "nurturing" environments afforded by single-family homes that center on functional kitchens and a series of gendered domestic spaces: "her" utility room, bathroom, bedroom; "his" garage, workshop, study. The importance of these codings rests in the way that they present gender differences as "natural" and so universalize and legitimize a particular form of gender differentiation and domestic division of labor.[80]

We must recognize, however, that even as buildings and domestic spaces are designed to symbolize and codify gender roles, their meanings become contested and unstable, particularly in relation to the complex conflicts and compromises between class and gender interests that characterize the sociospatial dialectic. Meanwhile, changes in the dynamics of household formation, in employment patterns, and in the design professions themselves all conspire to continuously modify the received meaning of domestic design.[81]

FOLLOW UP

Key Terms

anomie
appraisive imagery
cognitive distance
greenbelt cities
habitus
lifeworld
mechanical solidarity
neighborhood effects
organic solidarity
place ballets
psychic overload
social disorganization
sociospatial dialectic
structuration
time-space distanciation
urbanism

Review Activities

1. At the beginning of the chapter it is argued that the idea of the city as an arena of conflict and as an incubator of loneliness, isolation, and deviance is one that is reflected in movies, TV, and music. Compile a list of examples that illustrate this point.
2. Construct a time-space path of your typical day and annotate it to show when, where, and how your path coincides with those of others in ways that contribute to promoting intersubjectivity and distinctive lifeworlds.
3. Watch a movie: *Paid in Full* (Roc-A-Fella Films, 2002; Director Charles Stone III) provides a vivid portrayal of neighborhood life in Harlem. Critically review this portrayal in the context of concepts such as lifeworld, community, neighborhood, structuration, and the sociospatial dialectic.
4. Add to your *portfolio*. One thing you can do to personalize this section is to draw your own mental map of the city or town where you live. Now analyze your mental map to see what paths, edges, districts, nodes, and landmarks you included.

Log in to **www.mygeoscienceplace.com** for self-study quizzes, *MapMaster* layered thematic and place name interactive maps, *Urban View* Google Earth™ tours, key resources and suggested readings, related websites, "In the News" RSS feeds, and additional references and resources to enhance your study of urbanization, urban life, and urban spaces.

NOTES

1. The Salvation Army South Central Los Angeles Center website: *http://www1.usw.salvationarmy.org/usw/www_usw_southla.nsf/.*
2. E. Brenner, "Combating the Spread of Graffiti," *New York Times*, May 30, 1993.
3. R. Faturechi, "Painted in Neutral Colors," *Los Angeles Times*, December 8, 2009.
4. Ibid.
5. See, for example, Y-F. Tuan, "The City: Its Distance from Nature," *Geographical Review* 68 (1978): 1–12; M. C. Jaye and A. C. Watts, eds., *Literature and the Urban Experience: Essays on the City and Literature* (New Brunswick, N.J.: Rutgers University Press, 1981).
6. E. Soja, "The Socio-Spatial Dialectic," *Annals of the Association of American Geographers* 70 (1980): 207–25.
7. See D. Ley, "Social Geography and the Taken-for-Granted World," *Transactions of the Institute of British Geographers* 2 (1977): 498–512; and A. Buttimer, "Grasping the Dynamism of the Lifeworld," *Annals of the Association of American Geographers* 66 (1976): 277–92.
8. Tönnies's work was first published in 1887 and was translated into English by C. Loomis in 1957 as *Community and Society* (East Lansing, Mich.: Michigan State University Press).
9. Durkheim's work was first published in 1893 and translated by G. Simpson in 1933 as *The Division of Labor in Society*. It was published by the Free Press, New York, in 1964.
10. This discussion is based on Knox and Pinch, *Urban Social Geography*, 314–15.
11. J. Perkin, *Victorian Women* (London: John Murray, 1993).
12. S. P. Ryder and Johnno, *Casebook: Jack the Ripper* (*http://www.casebook.org/*).
13. L. Bondi, "Sexing the City," in *Cities of Difference*, eds. R. Fincher and J. M. Jacobs (New York: Guildford Press, 1998), 177–200.
14. N. Duncan, "Renegotiating Gender and Sexuality in Public and Private Spaces," in *Bodyspace*, ed. N. Duncan (London: Routledge, 1996), 127–145.
15. E. Durkheim, *Suicide* (New York: Free Press, 1897).
16. G. Simmel, "The Metropolis and Mental Life," in *Classic Essays on the Culture of Cities*, ed. P. Sennett (New York: Appleton-Century-Crofts, 1961), 47–60.
17. See, for example, C. Beaudelaire, *The Painter of Modern Life and Other Essays* (Oxford, UK: Phaidon Press, 1964); W. Benjamin, *Das Passagen-Werk*, 2 vols., ed. R. Tiedermann (Frankfurt: Suhrkamp, 1982).
18. M. Foucault, "What is Enlightenment?" in *The Foucault Reader*, ed. P. Rabinow (Harmondsworth, UK: Penguin, 1984), 41–42.
19. L. Wirth, "Urbanism as a Way of Life," *American Journal of Sociology* 44 (1938): 1–24.
20. A. Toffler, *Future Shock* (New York: Bantam, 1970).
21. S. Milgram, "The Experience of Living in Cities," *Science* 167 (1970): 1461–68.
22. For a general review of this work see P. Knox and S. Pinch, *Urban Social Geography*, 4th ed. (Harlow, UK: Pearson Education, 2000), Chapter 10; for a more-detailed review, see C. S. Fischer, "The Public and Private Worlds of City Life," *American Sociological Review* 46 (1981): 306–16; C. S. Fischer, *The Urban Experience* (New York: Harcourt Brace Jovanovich, 1976).
23. L. Lofland, *A World of Strangers* (New York: Basic Books, 1973), 178. Emphases added.
24. Fischer, "Public and Private Worlds," 306.
25. This discussion is based on Knox and Pinch, *Urban Social Geography*, 316–21.
26. B. Weightman, "Commentary: Towards a Geography of the Gay Community," *Journal of Cultural Geography* 1 (1981): 106–12.
27. L. Knopp, "Some Theoretical Implications of Gay Involvement in an Urban Land Market," *Political Geography Quarterly* 9 (1990): 337–52; "Exploiting the Rent-Gap: The Theoretical Significance of Using Illegal Appraisal Schemes to Encourage Gentrification in New Orleans," *Urban Geography* 11 (1990): 48–64.
28. J. Egerton, "Out but Not Down: Lesbians' Experience of Housing," *Feminist Review* 36 (1990): 75–88; H. P. M. Winchester and P. White, "The Location of Marginalized Groups in the Inner City," *Environment and Planning D: Society and Space* 6 (1988): 37–54.
29. G. Valentine, "Out and About: Geographies of Lesbian Landscapes," *International Journal of Urban and Regional Research* 19 (1995): 96–111.
30. S. Kirby and I. Hay, "(Hetero)sexing Space: Gay Men and 'Straight' Space in Adelaide, South Australia," *Professional Geographer* 49 (1997): 295–305.
31. L. Johnston and G. Valentine, "Wherever I Lay my Girlfriend, That's My Home: The Performance and Surveillance of Lesbian Identities in Domestic Environments," in *Mapping Desire: Geographies of Sexuality*, eds. D. Bell and G. Valentine (London: Routledge, 1995), 99–113; L. McDowell "Body Work: Heterosexual Gender Performances in City Workplaces," in *Mapping Desire: Geographies of Sexuality*, eds. D. Bell and G. Valentine (London: Routledge, 1995), 75–95; L. McDowell, "Spatializing Feminism: Geographic Perspectives," in *Bodyspace: Destabilizing Geographies of Gender and Sexuality*, ed. N. Duncan (London: Routledge, 1996), 28–44.
32. H. J. Gans, *The Urban Villagers* (New York: Free Press, 1962).
33. In controlled laboratory experiments crowding has been shown to cause rat populations to exhibit aggression, listlessness, promiscuity, homosexuality, and the rodent equivalent of juvenile delinquency. Projecting these ideas directly to human behavior leads to the notion of crowded urbanites as "killer apes." But people are not rats and in any case never approach the degree of crowding experienced by laboratory animals. Evidence from studies of environmental psychology is mixed, and the whole debate continues to attract controversy in all of the social and environmental disciplines. For a brief summary, see Knox and Pinch, *Urban Social Geography*, 281–83.
34. Gans, *The Urban Villagers*, 249.
35. C. S. Fischer, "Toward a Subcultural Theory of Urbanism," *American Journal of Sociology* 80 (1975): 1319–41.
36. See, for example, C. R. Shaw and H. D. McKay, *Juvenile Delinquency and Urban Areas* (Chicago, Ill.: University of Chicago Press, 1942).
37. L. Mumford, *The Culture of Cities* (London: Secker and Warburg, 1940), 215.
38. R. S. Lynd and H. M. Lynd, *Middletown* (New York: Harcourt, Brace and World, 1956); W. L. Warner and P. S. Lunt, *The Social Life of a Modern Community* (New Haven, Conn.: Yale University Press, 1941).
39. H. J. Gans, *The Levittowners* (London: Allen Lane, 1967).
40. P. Bourdieu, *Distinction: A Social Critique of the Judgement of Taste* (London: Routledge & Kegan Paul, 1984).
41. M. Maffesoli, "Affectual Postmodernism and the Metropolis," *Threshold* 4 (1988): 1; M. Maffesoli, "Jeux de Masques: Postmoderne Tribalisme," *Design Issues* 4 (1988): 1–2.

42. U.S. National Commission on Neighborhoods, *People, Building Neighborhoods.* Final Report to the President and Congress of the United States (Washington, D.C.: U.S. Government Printing Office, 1979), 7.
43. K. Lynch, *The Image of the City* (Cambridge, Mass.: MIT Press, 1960), 47–48. Emphases added. See also K. Lynch, "Reconsidering 'The Image of the City,'" in *Cities of the Mind*, eds. R. M. Hollister and L. Rodwin (New York: Plenum, 1984), 151–62.
44. R. J. Johnston, "Spatial Patterns in Suburban Evaluations," *Environment and Planning A* 5 (1973): 385–95; F. M. Carp et al., "Dimensions of Urban Environmental Quality," *Environment and Behavior* 8 (1976): 239–64.
45. T. Lee, "Urban Neighborhood as a Socio-Spatial Schema," *Human Relations* 21 (1968): 241–68.
46. P. Jackson and S. J. Smith, *Exploring Social Geography* (Boston: Allen and Unwin, 1984), 9.
47. Y-F. Tuan, "Place: An Experiential Perspective," *Geographical Review* 65 (1975): 152.
48. G. Pratt and S. Hanson, "On Theoretical Subtlety, Gender, Class and Space," *Society and Space* 9 (1991): 241.
49. M. Marsh. *Suburban Lives* (New Brunswick, N.J.: Rutgers University Press, 1990).
50. F. M. L. Thompson, "The Rise of Suburbia," in *The Rise of Suburbia*, ed. F. M. L. Thompson (Leicester, UK: Leicester University Press, 1982), 13. Emphasis added.
51. A. J. Downing, *The Architecture of Country Houses.* First published in 1850; reprinted in 1969 by (Dover Publications, New York).
52. This discussion is based on Knox and Pinch, *Urban Social Geography*, 322–26.
53. D. C. Park, J. P. Radford, and M. H. Vickers, "Disability Studies in Human Geography," *Progress in Human Geography* 22 (1998): 208–33.
54. See R. Butler and S. Bowlby, "Bodies and Spaces: An Exploration of Disabled People's Experience of Public Space," *Environment and Planning D: Society and Space* 15 (1997): 411–33; M. Dear, R. Wilton, S. L. Gaber, and L. Takahashi, "Seeing People Differently: The Sociospatial Construction of Disability," *Environment and Planning D: Society and Space* 15 (1997): 455–80.
55. M. Oliver, "Theories of Disability in Health Practice and Research," *British Medical Journal* 7170 (1998): 1446–49.
56. R. Imrie, *Disability and the City: International Perspectives* (London: Paul Chapman, 1996).
57. V. Chouinard and A. Grant, "On Being Not Even Anywhere Near 'The Project:' Ways of Putting Ourselves in the Picture," in *Bodyspace*, ed. N. Duncan (London: Routledge, 1996), 170–93. See also K. England, "Disabilities, Gender and Employment: Social Exclusion, Employment Equity and Canadian Banking," *The Canadian Geographer* 47 (2003): 429–50.
58. M. Brault, *Americans with Disabilities: 2005*, Current Population Report P70-117 (Washington, D.C.: U.S. Department of Commerce, December 2008).
59. B. Gleeson, "Justice and the Disabling City," in *Cities of Difference*, eds. R. Fincher and J. M. Jacobs (New York: Guildford Press, 1998), 89–119.
60. R. Imrie and P. E. Wells, "Disablism, Planning and the Built Environment," *Environment and Planning C: Government and Policy* 11 (1993): 213–31.
61. G. Wright, *Building the Dream: A Social History of Housing in America* (New York: Pantheon, 1981).
62. C. Rock, S. Torre, and G. Wright, "The Appropriation of the House: Changes in House Design and Concepts of Domesticity," in *New Space for Women*, ed. G. R. Wekerle (Boulder, Colo.: Westview Press, 1980), 84.
63. A. Giddens, *The Constitution of Society: Outline of the Theory of Structuration* (Cambridge, UK: Polity Press, 1984).
64. M. Dear and J. Wolch, "How Territory Shapes Social Life," in *The Power of Geography: How Territory Shapes Social Life*, eds. M. Dear and J. Wolch (Boston: Unwin Hyman, 1989), 6.
65. D. Ley, *A Social Geography of the City* (New York: Harper & Row, 1983), 135.
66. A. Schutz, "The Social World and the Theory of Social Action," *Social Research* 27 (1960): 205–21.
67. R. Williams, *Marxism and Literature* (Oxford, UK: Oxford University Press, 1977).
68. P. Jackson, *Maps of Meaning* (Boston: Unwin Hyman, 1989), 39.
69. D. Seamon, *A Geography of the Lifeworld* (London: Croom Helm, 1979).
70. For an introduction to time-geography, see T. Carlstein, *Time, Resources, Society and Ecology* (Boston: Allen and Unwin, 1982); and T. Carlstein, D. Parkes and N. Thrift, eds., *Human Activity and Time Geography* (London: Arnold, 1978).
71. R. Miller, "Selling Mrs. Consumer: Advertising and the Creation of Suburban Socio-Spatial Relations. 1910–1930," *Antipode* 23 (1991): 263–301.
72. Marsh, *Suburban Lives*, 184–85.
73. B. Friedan, *The Feminine Mystique* (New York: W.W. Norton, 1963).
74. U.S. Census Bureau's American Community Survey 2009 data release for metropolitan statistical areas (*http://www.census.gov/acs/www/*) Table S1101; U.S. Congress Joint Economic Committee, *Women and the Economy.*
75. U.S. Congress Joint Economic Committee, *Women and the Economy 2010: 25 Years of Progress but Challenges Remain* (Washington, D.C.: U.S. Congress, 2010).
76. R. M. Law and J. R. Wolch, "Social Reproduction in the City: Restructuring in Time and Space," in *The Restless Urban Landscape*, ed. P. L. Knox (Englewood Cliffs, N.J.: Prentice Hall, 1993), 165–206.
77. Ibid., 194.
78. This discussion is based on Knox and Pinch, *Urban Social Geography*, 187–88.
79. E. Wilson, *The Sphinx in the City: Urban Life, the Control of Disorder, and Women* (Berkeley, Calif.: University of California Press, 1991), 94.
80. See D. Spain, *Gendered Spaces* (Chapel Hill, N.C.: University of North Carolina Press, 1992); L. K. Weisman, *Discrimination by Design* (Urbana, Ill.: University of Illinois Press, 1992).
81. L. Bondi, "Gender, Symbols and Urban Landscapes," *Progress in Human Geography* 16 (1992): 157–70; N. Gregson and M. Lowe, "Home-Making: On the Spatiality of Daily Social Reproduction in Contemporary Middle-Class Britain," *Transactions of the Institute of British Geographers* 20 (1995): 224–35.
82. R. Madigan, M. Munro, and S. Smith, "Gender and the Meaning of Home," *International Journal of Urban and Regional Research* 16 (1992): 625–27.

The City as Text: Architecture and Urban Design

Patrick Batchelder/Alamy

The City as Text: Architecture and Urban Design

For many people the design of the built environment is what gives expression, meaning, and identity to the broad sweep of forces involved in urbanization. It provides cues for all kinds of behavior. It is symbolic of all sorts of political, social, and cultural forces that affect urban residents. As a result, each group of structures of a given period and type tends to be a carrier of the zeitgeist, or "spirit of its time." Any city, therefore, can be "read" as a multilayered "text," a narrative of signs and symbols. If we think in this way of the city as a text, the built environment becomes a biography of urbanization and urban life. The text metaphor can be developed in several ways. It is possible, for example, to analyze particular settings in terms of their "grammar"—the way that buildings and spaces take on meaning in relation to one another. Similarly, it is possible to search for meanings that are conveyed by means of the graphic equivalent of allegory. Perhaps the most illuminating way to develop the metaphor, however, is to look for the equivalent of key pages or passages in the narrative: settings that are emblematic of the overall political economy, or fragments of the built environment that mark turning points in the story, or significant events or relationships. Yet in reading any text carefully we must also take note of the subtext, "reading between the lines" in order to understand the full story.

LEARNING OUTCOMES

After reading this chapter, you should be able to:

- Understand how the built environment can be "read" as a "text" that embodies certain meanings.
- Highlight the essential elements of Arcadian Classicism, Beaux Arts and the City Beautiful, and the early skyscraper tradition of the U.S. city.
- Describe the fundamental principles of Modernist architecture.
- Compare and contrast the ideas and design elements of Le Corbusier and Frank Lloyd Wright.
- Articulate the recent relationship between socio-economic change and architectural styles.

CHAPTER PREVIEW

The purpose of this chapter is to show how architecture and urban design are linked into the dynamics of urban social, cultural, and economic change. Most of the chapter is a review of successive phases of design that have left a distinctive legacy on contemporary cityscapes. First, however, we will establish the importance of architecture and design for people in more general terms—as they relate to the political economy of urbanization. We also note that architectural styles are not simply imposed, layer on layer, onto metropolitan form; they are diffused differentially through time and space, creating distinctive morphological areas.

We will see that with the advent of the industrial city, architecture and urban design developed some especially significant relationships with the urban political economy. The interdependencies between society

---

The design of the built environment is for many people what gives expression, meaning, and identity to the broad sweep of forces involved in urbanization. In fact, a city like New York can be "read" as a multilayered "text," a narrative of signs and symbols. The built environment, such as the neighborhood around Hearst Tower, then becomes a biography of urbanization and urban life.

URBAN VIEW 1

Disney's Celebration: Designing the Happiest Place on Earth?

> Celebration is the first comprehensively planned new town of any size since Columbia, Maryland, and Reston, Virginia, which were built [more than] 40 years ago. However, Robert A.M. Stern and Jaquelin Robertson, who planned Celebration, didn't look to these projects for inspiration. Instead, they took their cue from the classic garden suburbs of the early 1900s. This wasn't nostalgia. The garden suburb represents the acme of suburban planning: compact, townlike, both individualistic and comfortably civic. It is one of our great design achievements, right up there with the 19th-century urban park and the skyscraper. The white clapboard, verandas, and garden fences along the street are as unmistakably American as the rocker on the porch.[1]

Begun in 1994 on land that the Walt Disney Company owned just south of Orlando and their Disney World theme park complex, Celebration was to be the realization of Walt Disney's personal dream of creating a utopian community that would be the happiest place on earth (Figure 1).[2] In 1989, Peter Rummell, president of the Disney Development Corporation wrote to Disney C.E.O. Michael Eisner that Celebration would be a "wonderful town east of [highway] I-4 that has a human scale with sidewalks and bicycles and parks and the kind of architecture that is sophisticated and timeless."[3]

The original residents' rule book for this upscale master-planned community was several inches thick because it specified so much: the acceptable types of garden plants and even the kind of compost were spelled out; the curtains or blinds had to be white on the outside; no two adjoining homes could be the same choice from the permitted house designs (Colonial, Victorian, or Arts and Crafts styles). Some observers critical of New Urbanism have pointed out that the architectural and other requirements in Celebration were an unconscious or perhaps even conscious reflection of the social values that Disney was trying to recreate.[4]

In fact, this seemingly perfect small-town America community has been unfavorably compared to movies such as *The Truman Show* in which the unwitting lead character lives in a fake town on a reality TV show, or the creepy Connecticut community in *The Stepford Wives* in which all the docile wives are robots created by their husbands. Although not a movie set, in Celebration during the Christmas season, the ice in the rink that the children skate on is a sheet of white plastic; the snow that they play in is a soapy foam, nicknamed "snoap," that falls every hour, on the hour; and the Christmas carols come from mini-speakers hidden under the palm trees.[5]

> But locals just smiled at the jibes, a small price to pay, they said, for living in a place where you could leave your front door unlocked, where children could play freely in the streets, and where the most serious crime was a stolen bicycle . . . Locals like to joke that they live in "The Bubble."[6]

But the bubble burst in December 2010 when Celebration recorded its first murder and then its first suicide. A homeless man strangled and took an ax and bludgeoned to death a former teacher who had been counseling him in his home, just a block away from the ice rink. Less than a week later, another resident shot himself in his foreclosed home after a 14-hour police standoff; the late 2000s economic downturn had hit Celebration too (the median single-family home price had fallen to less than $400,000 from its 2006 peak of nearly $700,000); the man's business had failed and this wife had divorced him. Residents were shocked. "I guess we are not immune to what goes on in other communities" was a common reaction. As if architecture and urban design could somehow keep Celebration's residents safe!

Linda McCarthy

FIGURE 1 Celebration, Florida, an upscale master-planned community developed by the Walt Disney Company that was to be the realization of Walt Disney's personal dream of creating a utopian community that could be the happiest place on earth.

URBAN VIEW 1

Disney's Celebration: Designing the Happiest Place on Earth? (*continued*)

By that time, Disney, as planned, had already sold its stake in Celebration in 2004 to a property company that had promised to stay true to the company's original vision. Then in June 2010, Disney unveiled plans for its Golden Oak Residential Resort community located this time completely within the Disney World theme park complex and offering custom homes chosen from approved architectural types priced between $1.5 and $8 million. Some observers have noted that Golden Oak may represent Disney's now starkly changed vision of the good life. Where Celebration reflected an idealistic belief that small towns are the best model, Golden Oak reflects acceptance of the idea of separate gated, segregated communities for extremely wealthy people.[7]

and design shifted rapidly to accommodate people's new ideas and new economic, technological, social, cultural, and political forces. We give particular attention here to three successive styles: the Arcadian Classicism of the early industrial era, the Beaux Arts style adopted by the City Beautiful movement, and the early skyscraper style.

By the end of the industrial era the most influential style of all—Modernism—had begun to emerge. We will review Modern architecture and urban design in some detail, since it has left a strong imprint on contemporary urbanization. As we will see, however, many people's reactions to Modernist designs, together with new political-economic circumstances, prompted a good deal of recent city building and rebuilding. Some of this has been undertaken in the form of postmodern designs and new urbanism, through a strategy of historic preservation, or even as design for dystopia in response to intensifying social polarization associated with neoliberalism, that incorporates a combination of electronic surveillance and carefully designed architectural policing of social boundaries, as in gated communities.

ARCHITECTURE AND THE DYNAMICS OF URBAN CHANGE

As Lewis Mumford put it in 1938, "in the state of building at any period one may discover, in legible script, the complicated process and changes that are taking place within civilization itself."[8] Thirty years later, sociologist Ruth Glass was to characterize the city as "a mirror . . . of history, class structure and culture."[9] Both comments echo the structuralist approach to the built environment: that it is part of the structure of dominant forces of production by people in cities, while also reflecting underlying relationships, tensions, and contradictions in society and its institutions.

So we can see downtown landscapes as emblematic not only of the rise of great cities but also of corporate power, whose fortunes have been so closely interwoven with the process of urbanization. The landscapes of civic architecture are another common element in the urban text. We can see that the lavish city halls of the nineteenth century were built to show that a city had "arrived"; that the Beaux Arts libraries and museums of the early twentieth century were built to establish a moral order; that the anonymous Modernist office blocks of the mid-twentieth century were built to house the growing civic bureaucracy in appropriately functional settings; and that the more recent festival marketplace settings of upscale specialty stores in refurbished historic buildings have been created to reassert the image of the city, to attract economic development and, some would argue, to mask the inequality and decay of the central city.

Architecture and Exchange Value

But architecture not only reflects underlying economic and institutional structures and contexts, it also serves as one of the means through which they are sustained, protected from opposing forces, and legitimized. In this context, one of the most obvious roles for architecture is in helping to stimulate consumption through *product differentiation* aimed at tapping particular market segments of consumers.

In this perspective, if the capitalist economy is to be successfully maintained, it is vital for urbanization to facilitate the manipulation of consumer demand not only to increase consumption but also to ensure the constant circulation of capital investment in the economy. Architects, by virtue of the prestige and mystique socially accorded to creativity, add exchange value to a building through their decisions about design, "so that the label 'architect designed' confers a presumption of quality even though, like the emperor's clothes, this quality may not be apparent to the observer."[10]

Architecture and the Circulation of Capital

The professional ideology and career structure that reward innovation and the ability to feel the pulse of fashion also serve to promote the circulation of capital investment in the economy. Without a steady supply of new fashions in domestic architecture, the **filtering** mechanism on which the whole homeowner market is based would slow down to a level unacceptable not only to developers but also to financial institutions.

As Homer Hoyt recognized, one way in which the dynamics of urban growth are sustained is through the cachet of fashionable design and state-of-the-art technology that induces affluent households to move from comfortable and prestigious homes into newer ones; hence the rapid succession of themes that can be revived and "rereleased," much like the contrived revivals of some *haute couture* fashion styles. More importantly, a parallel interpretation of the role of architecture can be applied over a longer time horizon. Here, the emphasis is on the way that major shifts in design—such as the shift from Modernism to postmodernism—help to resolve problems of underconsumption by consumers during **overaccumulation crisis** periods.

Architecture and Legitimation

Another role for architecture is that of legitimation. A major theme in the literature on critical architectural history is the way that architecture has repeatedly veiled and obscured the realities of economic and social relations among groups of people. The physical arrangement and appearance of the built environment can help to suggest stability amid change (or vice versa), to create order amid uncertainty, and to make the social order appear natural and permanent to society at large.

Part of this effect is achieved through what political scientist Harold Lasswell calls the "signature of power."[11] It is demonstrated in two ways: (1) through majestic displays of power inherent in urban design and (2) through a "strategy of admiration," aimed at diverting the attention of the audience with spectacular and histrionic design effects. It must be recognized, however, that it may not always be desirable to display power. Legitimation may, therefore, involve modest or low-profile architectural motifs. Conversely, it is by no means only "high" architecture that sustains the social order. The everyday settings of people's workplaces and neighborhoods also help to structure and reproduce class relations—part of the "sociospatial dialectic."

Meaning and Symbolism Any consideration of these connotative roles of architecture, however, soon runs into complex issues of meaning and symbolism. When we move down from high-level generalizations about socioeconomic processes and urban design, we find that people often endow buildings with meanings in ways that can be highly individualistic and often independent of their class or power. If architecture communicates different things to different people, or groups of people, we have to look more closely at questions of communication *by whom*, to *what audience*, for *what purpose*, and with *what results*.

The first distinction to make here is that between the *intended* meaning of architecture (on the part of architects and their clients) and its *perceived* meaning as interpreted by other people. So a building or space, for example, can become a social symbol when it is intended or perceived as a representation of someone or of some social group, when social meaning exerts an influential role relative to its other functions. Consequently there are two sides to social symbolism, on the one hand when a building or space is intended by the architect and client to convey social meanings, that is, as a symbolic action, and on the other hand when the building or space is perceived as a social symbol for people, whether this was intended or not by the architect and client.[12]

Sometimes, however, both sides of social symbolism can be seen at once. Lasswell's "signatures of power," for example, may serve to reassure the rich, strong, and self-confident while reinforcing feelings of deference among the poor and the weak; but the same symbolism may provoke and radicalize *some* people among the poor and the weak. *The point is that much of the social meaning of architecture depends on the audience.* Meanwhile, of course, designers' and developers' preconceptions of the audience(s) will help to determine the kinds of messages that are sent in the first place.

Architecture versus "Mere Building"

People count on the built environment in several significant ways. Depending on our tastes, we look for the built environment to be functional and attractive—not only to have a sense of identity and continuity but also to contain variety and to provide a sense of security. Yet for the most part the structures that constitute the built environment were not designed with these considerations. A great deal of the built environment is not "designed" at all, in any sense that architects would recognize. The builders of many homes and smaller retail, commercial, and industrial structures simply work from pattern books, copy or hybridize other buildings, or use prefabricated technology; hence the snooty distinction made by British architectural critic Sir Nicholas Pevsner between architecture and "mere building."

For the people responsible for mere building, the prime considerations are cost and marketability. Because their markets are generally mass markets of one kind or another, aesthetics and design tend to finish a poor second in their calculations. For the creators of architecture, in contrast, their product aspires to be Art (or at the very least a contribution to the research and development (R&D) division of the culture industry). As architects' professional turf came increasingly under threat from engineers and other building specialists after World War I, it was art, aesthetics, and design that secured the profession's status and legitimacy. As a result, architecture has often been uncoupled—both in practice and analysis—from the larger built environment. The profession tends to rate its members on their artistic achievements; influential glossy trade magazines have always stressed the aesthetic over the practical; and schools of architecture have consistently instilled in their students an ethic of avant-gardism.

Nevertheless, we cannot afford to dismiss High Architecture as the product of, or for, a self-referential elite. It is High Architecture that not only punctuates the built environment with major public and commercial structures but also sets the style for others to follow: journeyman architects and, eventually, the perpetrators of "mere building." Besides, these linkages have become increasingly influential as some societies have become more affluent and better educated, which has increased the number of "cultivated" consumers. Our task, therefore, is to establish some guideposts in terms of the evolution of major architectural styles and design themes that have emerged in each of the major phases of urbanization.

THE STYLE OF PRODUCTION/ THE PRODUCTION OF STYLE

The early industrial city created unprecedented challenges for architects. Like other artists, they found themselves having to come to terms intellectually with industrialization, modernization, and, indeed, urbanization itself. Meanwhile, the tremendous increase in the scale and pace of change in cities required their new buildings to fulfill all kinds of new economic, social, and cultural imperatives. In the turmoil, it was some time before distinctive new styles emerged and even longer before any consensus could be reached on what was "good" architecture for the industrial city.

Meanwhile, the architects commissioned to build the new public buildings, factories, and mansions of the industrial period resorted to a variety of borrowed themes and revivals. One of the most pronounced underlying themes was a reaction against industrialization, modernization, and urbanization that found expression in *Romanticism* and Classicism. As a result, state capitols and city halls ended up looking like Roman forums, railway stations like medieval cathedrals, offices like Renaissance palazzi, banks and university buildings like Greek temples, and hotels like Jacobean mansions (Figure 2).

Arcadian Classicism and the "Middle Landscape"

Although this reactionary Romanticism and Classicism heavily influenced the general Western intellectual climate of the mid-nineteenth century, U.S. design professionals were especially influenced by the literary culture of the so-called **American Renaissance**. This was rooted in the works of Ralph Waldo Emerson and Henry David Thoreau and propagated by Walt Whitman and Herman Melville, whose objective was to confront the future rather than turn away from it; to define ideals appropriate to the time. Emerson had drawn, in the 1830s, on the European Romantics' notion of the pastoral ideal in arguing for settlements that incorporated the benefits of both city and country.[13] Thoreau, a disciple of Emerson's, popularized the idea of Nature as a spiritual wellspring for city dwellers in his book *Walden* (1854).[14]

By the time of the industrial period, people in the United States had come to think of their relationship with Nature and the Great Outdoors as something distinctively "American." The historian Frederick Jackson Turner advanced the influential idea that the "frontier experience" was the single most significant factor in determining U.S. character,[15] and it became broadly understood that "access to undefiled, bountiful and sublime Nature is what accounts for the virtue and special good fortune of Americans."[16]

This response came from a perspective that was also influenced by a disdain for the values and ethics of the businesspeople in commerce and industry that were transforming urban society. The principles of industrial economic development were associated in the minds of many people with corruption, exploitation, and moral decline. At the same time, there was an abiding fear of the social and physical consequences of this exploitation and degeneration: the alienated "mob" and its squalid and unhealthy neighborhoods that were seen as a threat to physical well-being and social order in the industrial city.

Against this backdrop, the American Renaissance took Nature as a fundamental spiritual wellspring, defining the ideal as a setting in which people and Nature had achieved a state of balance—what landscape architect Leo Marx has described as a "middle landscape" of pastoral and picturesque settings.[17] At the same time, the intellectuals of the period emphasized the moral superiority of domesticity and the virtues of republicanism and sanitary reform. This attitude led to a vision of ideal urban landscapes that combined the morality attributed to Nature with the enriching and refining influences of cultural, political, and social institutions. Progressive intellectuals like Andrew Jackson Downing advocated a program of "popular refinement" involving the creation of a whole series of institutions and settings such as public libraries, galleries, museums, and parks, in order to bring out the best in "ordinary" people.

Nikreates/Alamy

FIGURE 2 Neoclassical designs, such as the one used for the First Bank of the United States in Philadelphia, reflected a reaction during the early years of the industrial city against industrialization, modernization, and urbanization.

Public Parks The first manifestations of these ideals came in the arcadian landscapes of "rural" cemeteries like Mount Auburn in Cambridge, Laurel Hill in Philadelphia, and Green-Wood in Brooklyn. They were not only a response to the crowded and distasteful conditions of the unplanned and unregulated burial grounds in cities but also an attempt to express religious and social ideals. Religious ideals were expressed through monumental portals and sculptures, while social ideals were expressed in spaciousness and naturalistic scenery.

These cemeteries were influential because they inspired the creation of public parks as a means of carrying these ideals into the heart of the city, where, it was hoped, they might provide a civilizing, spiritually uplifting, and socially instructive setting for people. Very quickly, parks became one of the major issues in the struggle for political and institutional reform. For the present, however, we must focus on the design aspects of the park movement, and here the most influential figure was Frederick Law Olmsted, Sr.

Olmsted saw his work as serving the psychological and social needs of city residents to have access to a naturalistic landscape, a secluded escape from the dirt and noise of the city, a place for leisure and recreation, and an environment that would foster restraint and decorum. New York City's Central Park was Olmsted's first opportunity to achieve those goals on any scale. In 1858 he and Calvert Vaux won the competition for the design of the park, which was completed in 1862. Their design included a succession of specific areas for sport, recreation, and culture, all embedded in a picturesque landscape.

The park was integrated with the city by means of four avenues laid out with an elaborate system of independent traffic lanes, bridges, and underpasses that were designed not to interrupt the continuity of the landscape. The result was widely acclaimed, and Olmsted went on to design park projects in other cities (including Boston, Brooklyn, Buffalo, Chicago, Detroit, Milwaukee, Newark, Philadelphia, and San Francisco) and campuses for the University of California at Berkeley and Columbia University in New York, as well as collaborating again with Vaux on the Romantic-styled railway suburb of Riverside near Chicago.

Beaux Arts and the City Beautiful

The Beaux Arts movement took its name from L'Ecole des Beaux Arts in Paris, where, from the mid-nineteenth century, architects were trained to draw on Classical, Renaissance, and Baroque styles, synthesizing them in new buildings that might blend artfully with the significant buildings from earlier centuries that dominated European city centers. Though U.S. cities had no buildings of comparable age that needed to be complemented in this way, the Beaux Arts style provided a convenient packaging of High Culture that promised to resolve confusion and uncertainty about how to build the industrial city. The idiom was showcased by Daniel Burnham's neoclassical architecture for the World's Columbian Exposition in Chicago in 1893 (Figure 3). The temporary structures of the exposition showed what might be done: high culture could be married with American Arcadian Classicism.

A few years later, in 1899, this marriage produced a short-lived but widely influential movement of its own: the City Beautiful movement.[18] The thrust of the movement was decisively toward the role of the built environment as an uplifting and civilizing influence on people. The preferred architectural style was neoclassical, accompanied by matching statuary, monuments, and triumphal arches—all, if possible, laid out like Burnham's **White City** at the Chicago Exposition, with uniform building heights and imposing avenues with dramatic perspectives.

In 1901, Burnham collaborated with several others (including Frederick Law Olmsted, Jr.) on the McMillan Plan for Washington, D.C. (named after Senator James McMillan,

Corbis Images

FIGURE 3 The Beaux Arts style. The World's Columbian Exposition in Chicago, 1893, designed by Daniel Burnham.

Chairman of the Senate Committee on the District of Columbia). The purpose of the McMillan Plan was to rescue the Mall area from the neglected and unfinished framework derived from Pierre Charles L'Enfant's original plan of 1791. The centerpiece of the new plan was the redeveloped Mall and Federal Triangle, with neoclassical buildings along the Mall, a terminal memorial (the Lincoln Memorial (Figure 4)), a pantheon (the Jefferson Monument), the Memorial Bridge, and a water basin.

Although the scheme was not completed until 1922, the plans and sketches provided enough publicity to ensure the immediate future of the City Beautiful movement. Burnham went on to draw up plans for Cleveland (1902), San Francisco (1905), and Chicago (1909) before his death in 1913. These plans were also very influential, though the entire plans were not actually realized in any of these three cities. Meanwhile, John Olmsted (F. L. Olmsted, Sr.'s stepson) devised a plan for Seattle; others worked on schemes for Kansas City, Denver, and Harrisburg, Pennsylvania. In cities already too densely built up to accommodate a City Beautiful plan, the movement's themes found expression in a variety of inspirational statues, monuments, columns, and arches.

Linda McCarthy

FIGURE 4 The City Beautiful. This view of part of the Mall in Washington, D.C., shows one of the neoclassical buildings, the terminal memorial (the Lincoln Memorial) (with the more recent World War II Memorial in the lower foreground), aligned with the Washington Monument (from the top of which this photo was taken) and the Capitol.

The success of the movement helped to foster two key developments in architecture and urban design. The first was professionalization. In 1897, for example, the American Park and Outdoor Art Association was founded, followed by the American Society of Landscape Architects in 1899 and the National Playground Association in 1906; courses in landscape architecture were introduced at Harvard University in 1900. The second development was an intensification of discourse and critical analysis on the aesthetics of the built environment. The City Beautiful movement had brought architecture and urban design to the forefront of public debate, provoking an intense critique of bourgeois art, architecture, and urban design that, in turn, helped energize the emerging design professions.

It is clear, looking back, that the City Beautiful movement was an explicit and rather authoritarian attempt to create moral and social order in the face of urbanization processes that seemed to threaten to bring about disorder and instability. It was successful because it allowed private enterprise to function more efficiently while symbolizing a noble idealism that was endorsed by the pedigree of Beaux Arts neoclassicism. "Over and beyond the good intentions of its leading figures and technicians, the City Beautiful movement perfectly fulfilled its true function of matching maximum planning with maximum speculation."[19]

Its success did not last long, however, because the dynamics of urbanization were changing even as Burnham, Olmsted, and their followers were prescribing remedies to old ailments. Streetcars and electric power had begun to turn cities inside out, the severe recession of 1893–1894 had shaken up relations between classes and introduced profound changes to the social and political organization of the city, and by the time of Burnham's death in 1912, cars were beginning to make their mark. In this new context monumentality was seen as impractical, while the movement's total lack of concern for housing was seen as elitist. The immediate cause of the movement's demise, however, was that it was completely unable to accommodate the raw energy and exuberance of architects and clients who wanted to build skyscrapers.

The American Way: Skyscrapers

Even as Burnham was drawing up his plans for the Chicago Exposition, the skyscraper had established its presence in the city. Indeed, Burnham himself had worked on the 16-story Monadnock Building in Chicago (completed in 1891) and derived most of his income from designing high-rise structures like the famous Flatiron Building in New York (1902) (Figure 5). Chicago had in fact become the seedbed for skyscraper development because the technical preconditions for its development had coincided with the need to rebuild the center of the city after the fire of 1871. The first of these preconditions was fulfilled as early as the 1850s, when Elisha Graves Otis perfected the passenger elevator. In the same decade architect James Bogardus used wrought iron beams in designing a six-story building for Harper's, the publisher, in New York. But it was not until the late 1860s, with the construction of the Equitable

Alexander Alland/Corbis Images

FIGURE 5 The Flatiron Building in New York shortly after completion. Designed by Daniel Burnham and built in 1902, it is a fine example of a Beaux Arts skyscraper.

Michael Sayles/Alamy

FIGURE 6 American Gothic. Early skyscrapers in Chicago, showing the highly decorated form that was intended to be awesomely spectacular.

Insurance Company Building (also in New York), that an iron-case framework rather than masonry was successfully used to carry the dead weight of a building.

Freed from the crushing weight of masonry that had to be spread in thick walls and across a wide base, buildings were suddenly allowed to rise to unprecedented heights on relatively small lots. Given the enabling technology of elevators and iron-cage construction, skyscrapers found their fundamental economic rationale in the inflated land values and increasingly specialized patterns of land use in the **central business district (CBD)** of the late nineteenth century. Even so, it took another technological innovation—the telephone—to make skyscrapers practicable settings for businesses. The telephone allowed businesses to dispense with human messengers, who would have clogged the elevators of tall buildings.

It was the awesome spectacle of skyscrapers, however, that led to them first appearing in urban landscapes. Businesses—newspapers and insurance companies, in particular—saw the immediate benefits of advertising their presence, their success, and their dependability through highly visible, symbolic structures. From the start, therefore, the skyscraper form was less of an architectural style than it was an expression of economics and advertising (Figure 6).[20]

As skyscrapers proliferated and grew taller, they began to present problems. The skyscrapers caused the streets of larger CBDs to become like canyons, in which sunlight was cut off and air circulated poorly. As a result, some cities enacted **land use zoning** ordinances that required **setbacks**: stepping back the upper stories of a building, beyond a certain height, in order to allow light and air to reach the street (Figure 7). This restriction resulted in the characteristic wedding-cake profile of interwar skyscrapers. By the 1920s skyscrapers had come to dominate the skyline of most large U.S. cities, while even smaller cities had a centerpiece skyscraper of 20–30 stories. By this time, however, cityscapes were beginning to reflect the influence of Modernism.

Modernism: Architecture as Social Redemption

The roots of the *Modern Movement* are deep and tangled. They can be traced to several independent reactions of people to the conspicuous consumption of bourgeois Victorian society and to the richness of conventional tastes in the decorative arts. These reactions, however, collided with one another, provoked further reactions, and fused into a Modern Movement with an avant-garde culture riven by dozens of twists, turns, contradictions, and controversies. Running through these tangled roots was a fundamental shift in people's aesthetic values that had a profound effect on the appearance and physical arrangement of urban landscapes. Stated simply, this shift rested on the conviction that the best design is pure, simple, and timeless;

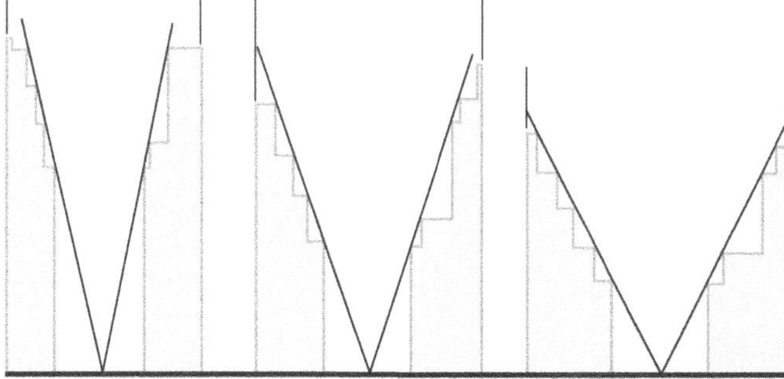

FIGURE 7 Zoning diagrams for New York height districts, 1916. These diagrams described the maximum permitted building height, in relation to the width of the street. Above a certain level, the mass of the building had to step back one foot for each additional three feet in height, in order to allow air and sunlight to street level. These laws produced skyscrapers that were stepped in form—a characteristic element in today's cityscape in midtown Manhattan.

Corbis Images

FIGURE 8 The U.S.'s most celebrated architect, Frank Lloyd Wright. Although his visions for urban design were never fulfilled, his influence on domestic architectural style has been inestimable.

that good design should be functional; and that through good design it is possible to create the matrix for new, progressive social conditions.

Arts and Crafts and Art Nouveau The prologue to the Modern Movement was written in the 1890s and early 1900s by the **Arts and Crafts movement** in England, by the **Art Nouveau** movement (called *Jugendstil* in central Europe) in France, and by Frank Lloyd Wright (Figure 8) in the United States.

The Arts and Crafts movement, led by William Morris, was a reaction by people to the extravagance of Victorian taste. Morris wanted to bring artists and craftspeople together to synthesize honest, simple, and popular forms. In architecture this meant drawing heavily on vernacular themes, something that proved convenient and attractive to the planners of the first garden cities. The Art Nouveau movement emphasized sensuous shapes and organic themes as an alternative to Victorian extravagance. Art Nouveau style became widely used in graphic art, but in architecture it was limited to the embellishment of buildings rather than being reflected in their form. The terra-cotta transoms on Louis Sullivan's Guaranty Building in Buffalo, New York, for example, were decorated in an intricate Art Nouveau style.

The Early Modernists Meanwhile, Frank Lloyd Wright, who had worked for a spell in Sullivan's Chicago office, had introduced a radically different style to residential architecture. Wright seems to have been influenced by the Japanese pavilion at the Chicago Exposition. His houses were long and low, with a geometry that owed nothing to Romanticism or neoclassicism. They were low-slung buildings with a massive central chimney, a hip roof with deep overhangs (Figure 9), and no attics or basements, which Wright declared "unwholesome." This style is generally known as Wright's Prairie Style, after the pattern-book design he published in 1901 in the *Ladies' Home Journal* that he called "A Home in a Prairie Town."

Wright's departures from earlier styles were mild, however, compared to the revolutionary ideas that emerged as artists struggled not only to free themselves from the canons of bourgeois taste but also to come to grips with the age of machinery, technology, and unprecedented speed. Modernism, in pursuit of progress, strove to break with history.

Many critics have set the turning point at 1907, the year Picasso painted *Les Demoiselles d'Avignon*, which initiated the abstractionist challenge to representational art. Art critic Robert Hughes has suggested that the Cubism of Picasso and Braque was at least in part an attempt to capture the experience of constantly altering landscapes as seen from a moving train or car.[21] Architects took their cue from this abstractionism, seeking to design buildings appropriate to the needs and experiences of the machine age, and translating the angularity of Cubism into built form.

Several future-oriented schools of thought had already emerged in response to the challenge of the machine age. One was the **Secessionists**, headed by Austrian architect Adolf

FIGURE 9 The Robie House in Chicago, designed by Frank Lloyd Wright. For this house, one of the famous private residences in the world, Wright's distinctive horizontal emphasis, known as the Prairie Style, was adapted to a narrow city lot. The influence of the Prairie Style has echoed throughout suburbia ever since.

Everett Collection Inc/Alamy

Loos, whose principal motivation was the elimination of all "useless" ornamentation from architecture. Another was the **Deutscher Werkbund**, led by Peter Behrens (architect and chief designer for the German electrical manufacturer AEG). The Werkbund was a loose coalition of artists and craft firms founded in 1906 to reform the relationship between artists and industry, their guiding principle being that quantity and quality should complement each other. Behrens saw industrialization as the manifest destiny of the German people, and his factory designs were for muscular temples to technological power. Other members of the Werkbund (including architects Bruno Taut and Walter Gropius), meanwhile, emphasized the need to overcome the alienating aspects of traditional society by leaving behind all the architectural refinements and symbolism of the old order and replacing them with a no-nonsense style.

The Bauhaus and the Modern Movement Such sentiments were to become central to the Modern Movement in architecture as it developed in the 1920s. In general, the Modernists developed the idea of the importance of redesigning entire cities in the name of technological progress and social democracy. One subset, the **Futurists**, wanted nothing at all to do with the past. Led by Filippo Marinetti and represented in the field of architecture and urban design by Antonio Sant'Elia, they sought to provoke social and institutional change through cities that were to be stages for permanent revolt, with huge and spectacular edifices that were at once monuments to the masses and to technology. Sant'Elia's drawings of *La Citta Nuova* (1914) showed massive concrete buildings with soaring high-rises and huge parapets, enormous generating stations, and huge factories and airfields, the like of which were not to appear in actual urban landscapes for another 50 years.

It was from these roots that there emerged in the 1920s a Modernism that was founded in the idea of architecture and design as agents of social redemption. Through industrialized production, modern materials and functional design, architecture (as opposed to mere building) could be produced inexpensively, become available to all city residents, and thus improve the physical, social, moral, and aesthetic condition of cities.

It was the **Bauhaus School** that did more than any other single movement to promote this ideal. Founded by Walter Gropius in Weimar in 1919 and later (1926) relocated to Dessau, the Bauhaus became the "refinery of . . . European avant-gardes,"[22] setting the standards for modern design in everything from teapots to workers' housing. The unifying themes were simplicity of line, plain surfaces, and suitability for mass production. The socialist ideology that was clearly embodied in the school's approach did not sit well, however, with Nazi ideology, and in 1933 the school closed when Adolf Hitler came to power.

Many of the leading practitioners of the Bauhaus moved to the United States, and it was at this point that Modern architecture and urban design began to appear in U.S. urban landscapes. The decisive event was an exhibition at the Museum of Modern Art in New York City, organized in 1932 by two American architects, Philip Johnson and Henry-Russell Hitchcock. The exhibition included examples of work by Ludwig Mies van der Rohe (the last director of the Bauhaus and author of the famous dictum "Less is more") and many other European Modernists. More in hope than as a reflection of reality, the exhibition was called *The International Style*. It did not take long, however, for the Modernism of the International Style to become truly international as the principal language of design:

> Both Walter Gropius and Mies van der Rohe ended up in America, where, after World War II, their design influence would be almost unquestioningly accepted by the executives of great business corporations and institutions anxious to install themselves in buildings which symbolized progress and prosperity.[23]

Le Corbusier The international impact of Modernism on urban design was sharpened by the energy and self-promotion of another key figure: the French-based Swiss architect Charles-Édouard Jeanneret (Figure 10), who is better known by his pseudonym, *Le Corbusier* ("crowlike"). In an artistic and intellectual climate that had become highly politicized and dominated by utopian and futuristic manifestos, Le Corbusier got himself noticed not just for his undoubted design talent but also as a result of his deliberately shocking slogans and outrageous claims for his prescriptions. Le Corbusier would not tolerate any thought of moderation in his approach. Standardization, anonymity, and purity of form were essential, he asserted, in creating a "mass production spirit" that would properly shape the daily routines, desires, and leisure activities of the industrial proletariat in modern cities. (Meanwhile, though, he was not above taking commissions for luxury homes from wealthy individuals.)

Paul Almasy/Corbis Images

FIGURE 10 Le Corbusier, an ardent advocate of the need for architecture to embrace the message of the machine age.

By 1919 Le Corbusier had conceived his basic formula for the relationship between architecture and cities: both should be a "machine for living." In 1922 he published his ideas on the principles of urban design in a book entitled *La Ville Contemporaine*. The key, he argued, was to reduce the congestion of city centers by increasing their density—by building upwards, in other words. High-density, high-rise city cores, he pointed out, would leave plenty of space for wide avenues to carry automobile traffic and for green space for recreation (Figure 11).

La Ville Contemporaine was conceived as a class-segregated city, with the best-located, most spacious and best-appointed high-rises reserved for elite groups of industrialists, scientists, and artists. Blue-collar workers were to have smaller garden apartments located in satellite units at some distance from the central cultural and entertainment complex. His plan for Paris, the Plan Voisin, reflected this strategy without any concessions to the existing city or to its inhabitants. The Plan Voisin called for eighteen 700-foot high-rises that would have required the demolition of most of historic Paris north of the Seine (Figure 12). Inside the high-rises the apartments ("cells," as he called them) were to be uniform, with the same standard furniture.

Not surprisingly, such a radical, totalitarian plan attracted a lot of attention and a great deal of opposition. Le Corbusier claimed not to be able to understand people's opposition, choosing instead to regard his own ideas as beyond the grasp of ordinary citizens. After the power elite failed to support his ideas and the Great Depression took away the ability of industrialists to back him, he revised his ideas on urban design. In *La Ville Radieuse* (1933) he adopted a rather different stance, arguing that everyone should live in giant collective apartment blocks called *unités*, with a minimum of interior space.

Although Le Corbusier's reformulated prescriptions came no nearer to being realized at the time than his earlier ones, they became widely influential in urban design circles. They were incorporated into the discourse of CIAM (the Congrés Internationaux d'Architecture Moderne), an international association of avant-garde architects formed in 1928 to promote Modernist architecture, and they found expression in the *Athens Charter*, a document published by CIAM in 1943 that set out the ideological basis of Modernist urban design. But the reasons for Le Corbusier's influence have as much to do with professional dynamics and avant-gardism as with aesthetics or functionality. The heroic scale of his ideas and his sheer irrepressibility drew admiration from architects and urban designers who wanted leadership and recognition, while his willingness to confront the automobile era drew admiration from technocrats.

From this admiration grew a conventional wisdom that was centered on the need to modernize cities through ruthless redevelopment, tearing out their centers and replacing them with high-rise housing linked by intrusive freeways. This impulse, indeed, was what began to materialize at an alarming rate in U.S. cities in the 1960s, prompting urban geographer Sir Peter Hall to observe that "The evil that Le

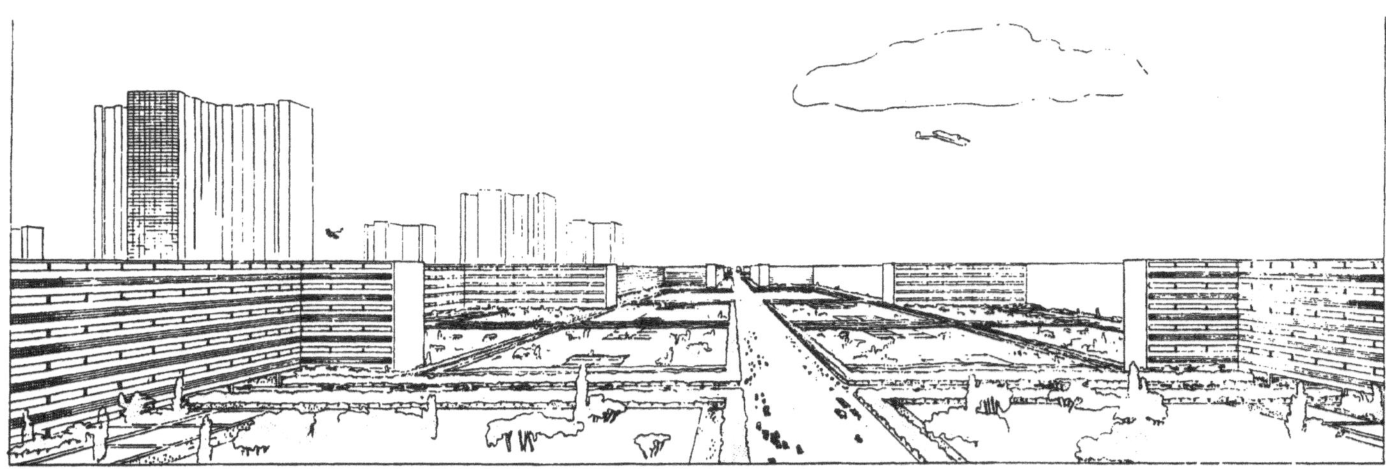

FIGURE 11 Illustration for *La Ville Contemporaine*, by Le Corbusier.

Corbusier did lives after him; the good perhaps interred with his books, which are seldom read for the simple reason that most are unreadable."[24] Hall's criticism of Le Corbusier and his followers is not directed at his designs as much as at the mindless bureaucratic arrogance and political naïveté that led the designs to be inserted into an urban process to which they were not actually suited.

But Le Corbusier's work and influence did not end with *La Ville Radieuse.* After World War II he gave up the heroic scale of urban design in favor of individual structures. His work ranged from expressionist designs like his famous chapel at Ronchamp in France (built 1950–1955; interpreted variously as resembling a nun's three-cornered hat, hands held in prayer, or the prow of a ship) to Cubist-inspired angular concrete buildings. The most celebrated example of his Cubist-inspired design is an apartment block in Marseilles—*L'Unité d'Habitation* (Figure 13)—that carried through his formula of a "machine for living" with its integrated community services, daycare facilities, and shops.

It was the style of L'Unité, however, that proved to be Le Corbusier's second legacy to mid-twentieth century urbanization. The poured concrete sections and panels, textured and sculpted with recessed windows and balconies, not only struck the Modernist chord but also proved to be relatively inexpensive to construct and amenable to prefabrication.

It was not long before cities everywhere were being recast in the image of L'Unité. Urban geographer Edward Relph, describing the evolution of contemporary urban landscapes, wrote:

> The dominant lines can be made vertical or horizontal or into a grid; the colours of the glass and metal can be changed; the proportions of the windows can be varied; the shape of the building can be altered to fit the site . . . [B]ut . . . [t]he result is that most Modernist buildings are almost indistinguishable one from another. They merge both in perception and in memory into a confusion of pale tones, sharp angles and cereal box forms, standing on end downtown and lying flat in the suburbs.[25]

This pervasive anonymity eventually contributed, as we will see, to a strong reaction by corporate owners, developers, the general public, and, indeed, many professional architects. We should note, however, that it was not so much Corbusian design per se that was at the root of the problem. Rather, it was a

FIGURE 12 Illustration for the Plan Voisin, by Le Corbusier.

Thomas A. Heinz/Corbis Images

FIGURE 13 L'Unité d'Habitation, Le Corbusier's influential apartment block, a "machine for living" that became the model for countless Modern buildings throughout the world. The Society for the General Esthetics of France called it "an eyesore and a public nuisance."

question of the cheapness of the execution of Corbusian forms, coupled with their imposition in thoughtless (and sometimes ruthless) ways. We should also note that within architectural and design circles, Le Corbusier has been elevated alongside Wright, Gropius, and van der Rohe as the embodiment of artistic genius.

An American Response While Le Corbusier was developing his ideas, Frank Lloyd Wright was pursuing the possibility of a U.S. response to the challenge of machine age, automobile-based urbanization. His vision was for a "Usonian" (a word-play on U.S. own) future. It was a vision that was heavily influenced by the individualism and naturalism of Jefferson, Thoreau, and Emerson. Wright, like other U.S. intellectuals of the automobile era, saw the popularity of the car as a threat to the special good fortune of the United States.

Wright hoped that, through planned metropolitan decentralization, cities might be spared from the sprawl and congestion threatened by cars. From this position it was only a short step to the promise of the car as the means of gaining access—for some, at least—to the blessings of Nature. In this way, two powerful U.S. traits—a pioneer affinity with the outdoors and a love of the car—might be reconciled.

Frank Lloyd Wright became the U.S.'s homegrown architectural hero by pursuing this very formula. **Usonia** was to be the physical framework for a new U.S. way of life where workers would be emancipated from the trap of congested but expensive cities, living instead in "Prairie Style" homes in a semirural setting. Where Le Corbusier wanted to increase densities and build tall, Wright wanted to decrease densities and spread out. Where Le Corbusier wanted people to live in identical apartments, Wright wanted them to live in differentiated and individualized homes that were designed to be in harmony with their natural surroundings. His ideal city, conceived in the mid-1920s but not written up until some 20 years later,[26] was **Broadacre City**. It was to be built on the basis of two new technologies: the automobile and mass-production building technology using high-pressure concrete, plywood, and plastic.

The inaccessibility of the large lots and low densities of Broadacre City was to be conquered by a network of landscaped parkways and freeways, with the focal point of semirural neighborhoods being provided by huge public service stations, architectural centerpieces that would provide a broad range of regularly purchased goods and services. Much to Wright's frustration and annoyance, Broadacre City drew not only widespread attention but also criticism. "For his pains, he was attacked by almost everyone: for naïveté, for architectural determinism, for encouraging suburbanization, for wasteful use of resources, for lack of urbanity, above all for being insufficiently collective in his philosophy."[27]

Meanwhile, however, some of Wright's designs for individual buildings—such as administration buildings—for example, the administration building for the S.C. Johnson Wax Company in Racine, Wisconsin and the Kaufmann House ("Fallingwater": Figure 14) in Bear Run, Pennsylvania—had established him as an architect of first rank. And despite being unsuccessful in seeing it widely adopted, Wright never gave up his Usonian vision—indeed, he came to identify himself completely with his own myth. But after World War II he concentrated more on the design of individual buildings in a new phase, marked by the use of curvilinear forms, arches, and spirals (the Solomon R. Guggenheim Museum in New York (Figure 15) is the best-known example), that itself was very influential in introducing a Modernist **Expressionism** into U.S. urban landscapes.

Linda McCarthy

FIGURE 14 Fallingwater, Frank Lloyd Wright's most celebrated attempt to harmonize architecture with nature. The stream actually runs through the house.

The Critique of Modernism According to some commentators, the symbolic death of Modern architecture took place at 3:32 P.M., on July 15, 1972, with the dynamiting of the Pruitt-Igoe project, a group of 33 public housing apartment blocks in St. Louis (Figure 16). Seventeen years before, its architect, Minoru Yamasaki, had won an award from the American Institute of Architects. Good as Pruitt-Igoe looked on the drawing board, it turned out to be unlivable. It was the tenants themselves who suggested that their homes be dynamited.

Although Modern architecture has never actually died, the demolition of Pruitt-Igoe did draw attention to a crisis in architecture and urban design that had first been articulated in 1961 by Jane Jacobs in her famous book *The Death and Life of Great American Cities*.[28] Jacobs reasoned that the design professions had taken away the life and vitality of cities, tearing out their sclerotic hearts only to replace them with a "great blight of Dullness" in the form of Corbusian high-rises.

A decade later, as Pruitt-Igoe was being prepared for demolition, Oscar Newman published an equally influential attack on Modernist housing. He argued that Modern architecture had been too preoccupied with form, with architecture as sculpture, and insufficiently attentive to people's need for functional, defensible spaces.[29] Specifically, he suggested that much of the petty crime, vandalism, muggings, and burglaries in modern public housing projects was related to

Russell Kord/Alamy

FIGURE 15 The Guggenheim Museum in New York City. Designed by Frank Lloyd Wright and built between 1943 and 1957, this is an example of Modernist Expressionism.

s68/ZUMA Press/Newscom

FIGURE 16 The demolition of prize-winning apartment blocks in the Pruitt-Igoe project in St. Louis has come to symbolize the public's loss of faith in Modernist urban design.

a weakening of community life and a withdrawal of local social order caused by the inability of residents to identify with, or exert any control over, the space beyond their own front doors. Sociability among residents, in short, had been "designed out" by Modernists, along with color, variety, and ornamentation.

In fact, there have been many worthy heirs to the early Modernists, architects whose contributions to the built environment in cities around the world have been both positive and influential. But although their designs were successful refinements of the clarity of form intrinsic to early Modernist design, they all more or less abandoned the idea of any social purpose to Modernism. This was highlighted by the abstract theorizing of iconoclastic "art compound" architects like Peter Eisenman and Bernard Tschumi, the fantasy architecture of celebrity architects like Frank Gehry and Zaha Hadid (Figure 17), and the downright elitism of high-profile commercial architects like Robert Stern (Figure 18), Charles Gwathmey, and Charles Moore. As a result, Modernist design in general came increasingly to be regarded as elitist and dysfunctional. It was in this climate that postmodern architecture began to prosper when the post-1973 recession burst the bubble of postwar growth. When economic prosperity returned a few

Peter Jordan_EU/Alamy

FIGURE 17 This station on the Nordpark Cable Railway, Innsbruck, Austria, is an example of the fantasy architecture of celebrity architect Zaha Hadid.

Arcaid Images/Alamy

FIGURE 18 The Garey House in Kent, Connecticut is an example of the downright elitism of high-profile commercial architects such as Charles Gwathmey.

years later, it was with a very different cultural sensibility: consumer materialism rather than radical progressivism. As a result, the idea of a futuristic, Modernist Utopia became increasingly anachronistic.

The Postmodern Interlude

> The explicit beginnings of postmodern architecture stem from the work of architect Robert Venturi whose response to the Modern architecture aphorism, "Less is more," was "Less is a *bore*." In arguing that architects should "Learn from Las Vegas,"[30] Venturi was really arguing for an architecture that was appropriate to a new phase of urbanization: In the Venturi cosmology, the people could no longer be thought of in terms of the industrial proletariat, the workers with raised fists, engorged brachial arteries, and necks wider than their heads . . . The people were now the "middle-middle" class, as Venturi called them . . . They were the "sprawling" masses, as opposed to the huddled ones.[31]

What the middle classes wanted most, it transpired, was to live in comfortable and affluent-looking houses in neighborhoods with human scale and decorative detail and to work and shop in grand and spectacular settings. This vision was not lost on developers with upscale businesses and consumers in mind: They soon realized that although it would cost more to build a "rich" building, it would sell or rent more quickly—and often at a premium—because it can project the appropriate "look." The shift in attitudes was soon reflected in the pages of trade magazines.

In contrast to the abstract formalism of Modern architecture, postmodern buildings were to be scenographic, decorative, and full of signs and symbols. Postmodern architecture was by definition wide-ranging and eclectic. One of its principal characteristics was "double coding," combining Modernist styling or materials with something else—usually historical or vernacular motifs. This mixing allowed the use of the symbolism of everything from historicism and revivalism to metaphysical and metaphorical references to eclectic pastiche. Postmodern architectural style was the style of styles (Figure 19). It was this self-conscious stylishness that made it attractive to the middle classes and developers.

Beginning in the 1980s commercial and residential townscapes throughout metropolitan areas were suffused with the hallmarks of postmodern design. New single-family homes were built in revival styles or historicist modes, while shoebox stores and offices were peppered with arches, atria, columns, keystones, semicircular windows, and cornices; and office villages and shopping centers came to resemble period stage settings—"lite" architecture, the built environment's equivalent of easy-listening music, lite beer, and lo-cal snacks.

Postmodern design was sustained by a new alliance of taste and capital. Instead of an alliance between a liberal elite and public capital or between a cultural avant-garde and corporate capital, it developed through an alliance between representatives of status-oriented and consumption-oriented society and managers of flexible capital investment.

In terms of the broader context of urbanization, postmodern architecture, along with postmodern culture and philosophy, can be interpreted as having emerged in tandem with the development of globalized, more flexible forms of capitalist enterprise. David Harvey argues that, although this transition and the critique of Modernism had been under way for some time, it was not until the international economic crisis of 1973 that the relationship between art and society was sufficiently shaken to allow postmodernism to became both accepted and institutionalized as the "cultural clothing" of neoliberalism.[32] The real estate development industry was quick to adopt postmodern design in order to ensure product differentiation and maximize exchange values. At the same time, the new conservatism and

David R. Frazier Photolibrary, Inc./Alamy

(a)

Martin Thomas Photography/Alamy

(b)

FIGURE 19 Postmodern style. (a) The Denver Public Library in Denver Colo., designed by Michael Graves, is an example of radical eclecticism in Postmodern architecture. (b) The Chiat/Day office building, also known as the "Binoculars Building," Venice, California, designed by Frank Gehry. Metaphorical and metaphysical references are some of the many wide-ranging characteristics of Postmodern architecture.

materialism of Western society that emerged in the 1980s and lasted until the economic downturn of the late 2000s produced an eager marketplace of consumers.

New Urbanism New Urbanism is a school of thought within urban design circles that advocates one particular kind of privately planned residential community: "neotraditional communities" that seek to reproduce the simple, communitarian, white-picket fence ambience, appearance, and serenity of small-town, walkable neighborhoods (Figure 20). It is based on a prescriptive system of design that seeks to replicate an older way of making places through the adoption of land use zoning regulations that require traditional street grids, mixed-use zoning, careful regulation of building materials and massing, and an emphasis on creating pedestrian-friendly environments. It is conventionally championed as a response to urban sprawl, as an appropriate residential form for edge cities, as a sustainable setting that minimizes car use, and as a new urban form appropriate for new lifestyles.

But New Urbanism has been widely criticized as catering to pretentious materialism, social exclusion and fear, and as being based on a selective recall of urban history. Critics of New Urbanism argue that there is nothing either new or urban about it, and that the outcome at best is very bland rear-guard buildings; at worst, kitschy, Disneyesque architecture that is more about imageability than liveability. In contrast to the hype surrounding New Urbanism, it has succeeded only in producing a plethora of middle-class subdivisions that are isolated from their host communities by private management and retro architecture that is too naively artificial to bear its own stylistic weight. New Urbanism's claims about car use and sustainability are belied by actual commuting patterns and levels of SUV ownership.

Linda McCarthy

FIGURE 20 New Urbanism. Celebration, Florida, an upscale master-planned community designed by Robert A.M. Stern and Jaquelin Robertson for the Walt Disney Company to embody "neo-traditional" urban design.

New Urbanism relies on webs of "servitude regimes" (covenants, controls, and restrictions that regulate both the physical environment and social comportment) that serve to frame a distinctive physical and social environment that appeals both to producers and to consumers. For producers, a key mechanism is the residential community association, which is used to establish by-laws, a constitution, and an overarching servitude regime in order to ensure stability until the subdivision is completed and sold off. As such, developers become benevolent dictators, establishing a tyranny of taste that defines the "legal landscapes" of U.S. suburbia. For consumers, these servitude regimes offer a means of narrowing uncertainty, protecting equity values, and, above all, establishing the proscenia—the backdrop and "stage"—for their lifestyles.

Historic Preservation Historic preservation has been a significant architectural movement during the past several decades in the United States.[33] As John C. Sawhill, former president of the Nature Conservancy, put it: "A society is defined not only by what it creates but by what it refuses to destroy."[34] Historic preservation has saved, refurbished, and provided a new lease on life to downtown department stores, office blocks, hotels, government buildings, railroad stations, and warehouses for future generations. It is no accident that many people's support for historic preservation coincided with the emergence of postmodern design: historic preservation must be interpreted as drawing heavily on some of the same economic and sociocultural forces. Historic buildings and districts lend both distinctiveness and identity; they also form an obvious linkage for people with that aspect of postmodern culture that emphasizes the past, the vernacular, and the decorative.

In the United States the National Historic Preservation Act of 1966 established the National Register of Historic Places as the official list of national historic resources worthy of preservation.[35] Since then, more than 80,000 properties have been listed in the National Register. During that time the number of towns and cities that have established historic district commissions or adopted historic preservation ordinances and design guidelines for protecting and rehabilitating historic structures increased to well over 2,000.[36] The availability of tax credits and accelerated depreciation benefits for investments in historic property have bolstered historic preservation in U.S. cities.

But in most cities historic preservation was initiated by citizen groups that had been produced by the "counterculture" of the late 1960s. Confronting developers and public agencies, they gained widespread support by tapping the anti-Modernist attitude that had developed among the public. One of the landmark cases was the victory by Don't Tear It Down, a Washington, D.C., preservation group that succeeded in stopping the demolition of the Old Post Office, a Romanesque Revival federal building on Pennsylvania Avenue that developers subsequently adapted to offices as well as shops, entertainment space, and a food court in its expansive interior atrium (Figure 21).

It did not take long before historic preservation became institutionalized, with city councils passing legislation and city planners making plans that reflected the shift in values among the electorate. Developers were also quick to respond. Instead of offering resistance, developers have taken the initiative, hiring architectural historians and conservation experts and seeking out older buildings ripe for reconditioning. Preservation groups, meanwhile, having established a secure power base, began

Linda McCarthy

FIGURE 21 The Old Post Office, a Romanesque Revival federal building on Pennsylvania Avenue in Washington, D.C., survived as a landmark case representing a victory by Don't Tear It Down, a preservation group that succeeded in stopping its demolition. Now refurbished as the Old Post Office Pavilion, it houses offices as well as shops, entertainment space, and a food court in its expansive interior atrium, and includes tours of its Clock Tower that offer great views of the surrounding area.

to invite the newly enlightened developers to serve on their boards, to accept donations from the development industry, and to compromise with, rather than confront, redevelopment projects. The result has been that parts of some downtowns, particularly on the U.S. East Coast, now amount to tableaux of preservation projects. In a graphic reflection of the compromises between developers and preservation groups, many of the elements in these tableaux are buildings whose facades have been saved but whose interior spaces have been entirely remodeled to accommodate contemporary office layouts, air conditioning, and wiring for new telecommunications systems—a practice referred to irreverently as "facodomy" among Modernist designers. This practice is also prevalent in cities in Europe and other parts of the world where historic preservation regulations tend to be much more stringent than in the United States (Figure 22).

Design for Dystopia

Within cities that contain intensifying concentrations of poor, minority households, growing numbers of homeless people, and increasing levels of crime, violence, and vandalism, security has become a "positional good" (a measure of social status). The casualties of economic and social change—the people who are poor, socially marginalized, and disadvantaged—are perceived as an unnecessary and undesirable part of the contemporary landscape of planned communities, gallerias, malls, office plazas, festival marketplaces, and so on.

Linda McCarthy

FIGURE 22 An example of facodomy from Macao.

As a result, architecture and urban design have increasingly incorporated a combination of built-in electronic surveillance and carefully designed architectural policing of social boundaries (Figure 23). This is a direct reflection of the intensifying social polarization associated with neoliberalism. Gated subdivisions are a form of privately planned residential community. More than 9 million households in the United States live in developments behind walls and fences—about 5 million of them in communities where access is controlled by gates, entry codes, key cards, or security guards. Gated developments are more prevalent in Sunbelt metropolitan areas such as Dallas, Houston, Phoenix, and Los Angeles, but they are rapidly becoming popular in older metropolitan areas like New Orleans, Long Island, N.Y., Chicago, Atlanta, and Washington, D.C. A significant number of gated communities in the United States are also inhabited by minority households—especially Hispanics—with moderate incomes.

Fortress L.A. In "Fortress L.A.," described by urban design critic Mike Davis,[37] affluent neighborhoods are themselves miniature fortresses, with surrounding walls, restricted entry points, privatized roadways, overlapping public and private police services, homes that bristle with alarm and surveillance systems, and lawns staked with signs that advise would-be trespassers of an armed response. In less-affluent neighborhoods, gates are replaced by random police checkpoints, people have wrought-iron grilles installed over their home's windows and surround their properties with chain-link fences, and thousands of people have had their rooftops painted with identifying street numbers that can be seen from above by the police in helicopters.

In downtowns, and in the commercial areas of edge cities, "public" spaces are monitored by cameras and security personnel. Street furniture and landscaping features convex "bum-proof" benches and sprinkler systems programmed to drench unsuspecting sleeping homeless people at random times during the night. Office builders maintain their distance from unwanted and inappropriate visitors by removing logos and names from street-level access ways and retreating behind mirror-glass facades. Retailers find safety in gallerias or in developments that can be reached only by elevated pedestrian ways that are easily monitored.

All this security and surveillance paraphernalia has been translated into High Design through the architecture of Frank Gehry. Beginning in the 1960s, Gehry found ways of designing for upscale uses amid decaying neighborhoods. His "stealth houses" (Mike Davis's term) hid their luxurious interiors behind blank or proletarian facades. The frontage for his Danziger Studio in Hollywood, for example (Figure 24), is simply a massive gray wall, "treated with a rough finish to ensure that it would collect dust from passing traffic and weather into a simulacrum of nearby porn studios and garages."[38]

Later, encouraged by the postmodern penchant for irony, Gehry began to recycle elements of decaying and polarized urban landscape into his buildings—roughly cast concrete,

FIGURE 23 Security and surveillance systems have become an important component of urban design as more and more city spaces become exclusive and exclusionary, as in the case of Black Diamond Ranch, a gated community in Citrus County, Florida.

Mark J. Barrett/Alamy

for example, and chain-link fencing—that made his 1980s designs the equivalent of post-Punk, radical-chic clothing. Gehry transformed the security theme from a secondary dimension of design, where it had been low-profile and high-tech, to an explicit fortress style, high-profile and low-tech. So his Frances Howard Goldwyn Hollywood Regional Branch Library looks as if it is built to withstand a siege, with 15-foot security walls of stucco-covered concrete block, antigraffiti barricades covered in ceramic tile, a sunken entrance, and stylized sentry boxes (Figure 25). We can read this architecture either as a mordantly witty parody or as a prologue to the text of twenty-first century urbanization.

"Starchitects," "Starchitecture," and World Cities

Globalization amid civic entrepreneurialism and the politics of image have made star architects like Gehry important figures in shaping the cityscapes of world cities. The ability of

FIGURE 24 The Danziger House and Studio, Los Angeles. Architect Frank Gehry.

Michael Moran

Michael Moran

FIGURE 25 Frances Howard Goldwyn Hollywood Branch Library. Architect Frank Gehry.

an iconic building to put a city on the global map was first demonstrated by the Sydney Opera House (Figure 26), designed by Danish architect Jørn Utzon in the late 1950s and completed in 1973. Utzon was not a star architect—indeed, he was relatively unknown and the Opera House was his first project outside Denmark and Sweden. But the structure, initially very controversial, is now recognized by UNESCO as a World Heritage Site and has become iconic of both Sydney and, indeed, of Australia as a whole. In spite of massive cost overruns in the construction of the building, the return on investment for Australia's government has been extraordinary.

This lesson was not lost on the leadership of Bilbao, Spain, a decaying industrial city that embarked on an ambitious revitalization process featuring signature structures by celebrity architects as symbols of modernity and economic revitalization. The strategy—and in particular Gehry's Guggenheim Museum building (Figure 27)—was highly successful in re-branding Bilbao and elevating its perceived status within the global economy. Its success has prompted many other city officials to engage "**starchitects**" in attempts to replicate what has come to be referred to as the **"Bilbao Effect."** In addition to Frank Gehry, today's star architects include Mario Botta, Santiago Calatrava, Pierre de Meuron, Peter Eisenman, Norman Foster, Massimiliano Fuksas, Michael Graves, Zaha Hadid, Jacques Herzog, Rem Koolhaas, Daniel Libeskind, Richard Meier, Jean Nouvel, César Pelli, Renzo Piano, Christian de Portzamparc, Richard Rogers, Robert A. M. Stern, and Peter Zumthor (Figure 28). Their celebrity depends in part on having a portfolio of work that extends to a variety of world cities. Based in (or as frequent visitors to) world cities, starchitects can also become public intellectuals, involved in discourse on a wide range of topics, contributing to influential cultural ideas and trends, and fueling and reinforcing economic and cultural globalization. This, in turn, allows celebrity architects like Michael Graves and Philippe Starck to engage in "brand extension," lending their names to "designer" consumer product lines from kitchenware and tableware to fountain pens and desk accessories.

Meanwhile, just as starchitects derive some of their standing from the visible presence of their built work in

Hemis/Alamy

FIGURE 26 The Sydney Opera House, designed by Danish architect Jørn Utzon, was initially very controversial, but is now recognized by UNESCO as a World Heritage Site and has become iconic of both Sydney and, indeed, Australia as a whole.

Marka/SuperStock

FIGURE 27 The Guggenheim Museum, Bilbao. Architect Frank Gehry.

major cities, so the potency of the symbolic capital of world cities is derived in part from their association with starchitects, starchitecture, and the associated nexus of fashion, design, and luxury consumption. Stardom and city branding have become mutually self-reinforcing as real estate developers realize that celebrity architects can add value to their projects, world-city leaders compete to retain the services of the top names to design signature buildings that will keep their city on the map, and the signature buildings of star architects provide the backdrop for fashion shoots, movie scenes, TV commercials, music videos, and satellite news broadcasts.

Peter Barritt/Alamy

FIGURE 28 The Bilbao Effect, as evidenced by the Imperial War Museum North, Salford Quays, Manchester, United Kingdom, designed by starchitect, Daniel Libeskind.

FOLLOW UP

Key Terms

- Art Nouveau movement
- Arts and Crafts movement
- Beaux Arts movement
- "Bilbao Effect"
- design for dystopia
- Expressionism
- Futurists
- historic preservation
- International Style
- Modern Movement
- New Urbanism
- Postmodernism
- Romanticism
- Secessionists
- starchitects

Review Activities

1. With what characteristics of architecture and urban design do you associate the following architects? Go online to Google Images, for example, and look at some of their buildings. Which do you personally prefer and why?

   Frank Lloyd Wright
   Le Corbusier
   Robert Venturi
   Frank Gehry
   Andres Duany and Elizabeth Plater-Zyberk

2. Select a large town, city, or metropolitan area you know well and attempt to "read" its biography through the "text" of its built environment. What districts and neighborhoods are distinctive in terms of their built environment, and why? Are there distinctive architectural landmarks or "signature" structures? What do they symbolize, and to whom? Are there town- or cityscapes that are typical of U.S. towns or cities in general? Of the region? Of particular social groups? Are there settings that are unique to the place you have selected and, if so, what is it about them that makes them unique?
3. If you have time, read a paperback with a story that has as its backdrop some aspect of architecture and urban design covered in this chapter. There are many to choose from, including *The Celebration Chronicles: Life, Liberty, and the Pursuit of Property Value in Disney's New Town*, by Andrew Ross (New York: Balantine, 1999).
4. Work on your *portfolio*. The material in this chapter lends itself especially well to your own sketches as well as photos and videos recorded on your camera or phone. Be selective in the way you put these together—choose specific themes, or particular kinds of townscapes. Provide captions, annotations, or commentaries that explain or interpret them in relation to the overall "text" of the city.

   In addition, you could select a landmark building in your nearest town or city, find out as much as you can about who commissioned it and who designed it; and then write a short essay or put together a video, Microsoft PowerPoint, or web-based presentation that explains the building in relation to its geographical and historical context. You could also investigate buildings and districts that are currently designated for historic preservation: How many are there? Where are they? What sorts of places are involved, in what parts of the city? If you live in the United States, you could start with the *National Register of Historic Places* (see the National Park Service website [*http://www.cr.nps.gov/nr/about.htm*] or try the historic preservation office of your state or city).

Log in to **www.mygeoscienceplace.com** for self-study quizzes, *MapMaster* layered thematic and place name interactive maps, *Urban View* Google Earth™ tours, key resources and suggested readings, related websites, "In the News" RSS feeds, and additional references and resources to enhance your study of architecture and urban design and your reading of the city as text.

NOTES

1. Slate Magazine, *Disney's Celebration* (*http://www.slate.com/id/2113107/slideshow/2113258/*).
2. T. Leonard, "The Dark Heart of Disney's Dream Town: Celebration has Wife-Swapping, Suicide, Vandals . . . and Now Even a Brutal Murder," *Daily Mail*, December 9, 2010 (*http://www.dailymail.co.uk/*).
3. "Murder in Disney Town: Idyllic Community Built by Disney Rocked by Its First-Ever Murder," *Daily Mail*, December 2, 2010 (*http://www.dailymail.co.uk/*).
4. Leonard, *Daily Mail.*
5. Ibid.
6. Ibid.
7. B. Watson, "Golden Oak: Why Disney's Latest Real Estate Gamble Isn't Such a Goofy Move" *DailyFinance*, June 23, 2010 (*http://www.dailyfinance.com/*); J. Chung, "Disney World's New Thrill Ride: Selling Luxury Vacation Homes," *Wall Street Journal*, June 23, 2010 (*http://online.wsj.com/*).
8. L. Mumford, *The Culture of Cities* (New York: Harcourt, Brace and World, 1938), 403.
9. R. Glass, "Urban Sociology in Great Britain," in *Readings in Urban Sociology*, ed. R. E. Pahl (Oxford, UK: Pergamon, 1968), 21–46.
10. R. Darke and J. Darke, "Towards a Sociology of the Built Environment," *Architectural Psychology Newsletter* 11 (1981): 13.
11. H. Lasswell, *The Signature of Power* (New York: Transaction Books, 1979).
12. D. Appleyard, "The Environment as a Social Symbol," *Ekistics* 278 (1979): 272.
13. The argument is made in Emerson's *Nature* (Boston: James Munroe & Co., 1836). See also J. Woods, "Build, Therefore, Your Own World": The New England Village as Settlement Ideal," *Annals of the Association of American Geographers* 81 (1991): 32–50.
14. H. D. Thoreau, *Walden* (New York: Holt, Rinehart and Winston, 1961). First published in 1854.
15. F. J. Turner, *The Frontier in American History* (New York: Henry Holt, 1920).
16. L. Marx, *The Machine in the Garden* (New York: Oxford University Press, 1964), 13.
17. Ibid.
18. For a detailed review of the movement and its legacy, see W. H. Wilson, *The City Beautiful Movement* (Baltimore: Johns Hopkins University Press, 1989).
19. M. Tafuri and F. D. Co, *Modern Architecture: 1* (New York: Rizzoli, 1986), 40.
20. M. Domosh, "The Symbolism of the Skyscraper," *Journal of Urban History* 14 (1988): 321–45.
21. R. Hughes, *The Shock of the New* (New York: Knopf, 1980).
22. Tafuri and Co, *Modern Architecture*, 116.
23. E. Relph, *The Modern Urban Landscape* (London: Croom Helm, 1987), 118.
24. P. Hall, *Cities of Tomorrow: An Intellectual History of Urban Planning and Design in the Twentieth Century*, 3rd ed. (New York: Basil Blackwell, 2002), 219.
25. Relph, *The Modern Urban Landscape*, 198.
26. F. L. Wright, *When Democracy Builds* (Chicago: University of Chicago Press, 1945).
27. Hall, *Cities of Tomorrow*, 314–15.
28. J. Jacobs, *The Death and Life of Great American Cities* (New York: Vintage Books, 1961).
29. O. Newman, *Defensible Space* (New York: Macmillan, 1972).
30. R. Venturi, *Complexity and Contradiction in Architecture* (New York: Museum of Modern Art, 1966).
31. T. Wolfe, *From Bauhaus to Our House* (New York: Farrar, Straus, Giroux, 1981), 109–10.
32. D. Harvey, *The Condition of Postmodernity* (Oxford, UK: Blackwell, 1989).
33. V. Scully, *American Architecture and Urbanism* (New York: Henry Holt, 1988), 287–88. Historic preservation involves the identification, evaluation, and protection of historic and archaeological resources.
34. J. C. Sawhill, "The Nature Conservancy," *Journal of Soil and Water Conservation*, September 1, 2006.
35. Aside from recognition of significance, the benefits of National Register listing include eligibility for federal investment tax credits for renovating income-producing properties and consideration in projects involving federal funding. Although National Register listing does not protect against alteration or demolition by private property owners using their own funds, listing by local historic district commissions does.
36. Chestertown, Maryland Town Council website (*http://www.chestertown.com/gov/Historic_district/index.php*); National Park Service, *National Register of Historic Places* (*http://www.cr.nps.gov/nr/about.htm*).
37. M. Davis, *City of Quartz: Excavating the Future of Los Angeles* (New York: Verso, 1990), 226.
38. Ibid., 238.

Problems of Urbanization

blickwinkel/Alamy

Problems of Urbanization

Cities have always been seen as problem-ridden: so much so that in the United States, cities have been viewed very negatively by many people. Cities have been widely regarded as necessary evils, essential for economic development but somehow unnatural, dangerous, and corrupt. It should not be surprising, of course, that cities are so often where problems are quite visible. With so much economic, demographic, social, political, and cultural change taking place in cities, it is inevitable that there are unanticipated and unintended consequences, and that tensions, contradictions, and conflicts occur between different groups of people. Some of these problems stem from society as a whole; they are particularly stark in cities simply because metropolitan areas contain the bulk of the population and are crucibles of change. Some of the problems, however, are caused or at least intensified by urbanization and the nature of urban settings.

LEARNING OUTCOMES

After reading this chapter, you should be able to:

- Recognize that as U.S. cities have evolved through time the perceived and real problems of those cities changed also.
- Understand the relationship between urbanization and the cycle of poverty.
- Explain the linkages between the urban context and violent crime.
- Describe the complex nature and root causes of homelessness.
- Recognize why the continued viability of the urban environment is dependent upon efficient governance and planning and continuous capital investment.

CHAPTER PREVIEW

The purpose of this chapter is not to provide an exhaustive review and inventory of urban problems. Rather, the objective is to show how the processes involved in urbanization are inherently problematic, with space and place often playing key roles. We focus on four major sets of problems—poverty, criminal violence, homelessness, and infrastructure problems. First, however, we discuss how society came to perceive and define certain issues as "problems" and how certain perceptions by people have changed, persisted, or been modified as urbanization has moved from the early industrial city to today's metropolis.

One of the most persistent themes to emerge from this review is that of poverty, slum neighborhoods, and people's behavior that is associated with slum settings. These issues are examined in detail, providing an opportunity to consider theories and concepts related to the causes and effects of poverty, slums, and localized social problems. Discussion of these issues is followed by an examination of the geography of criminal violence and a review of the effects of crime on urbanization and urban life. Once again, the objective here is to demonstrate the two-way causality inherent to urbanization: Certain settings give rise to high levels of criminal violence, which, in turn, affect the whole dynamic of spatial organization. Homelessness is likewise presented in the context of its wider causes and effects, with particular attention being given to the ways in which recent changes in the political economy of urbanization have led to new forms of homelessness.

Cities have always been seen as problem-ridden and have been regarded by many people as necessary evils, essential for economic development but somehow unnatural, dangerous, and corrupt. It should not be surprising, of course, that cities are so often where urban problems are quite visible, as in this photograph from Brooklyn, New York City.

URBAN VIEW 1

Homeless Students Struggle to Keep Up

In a corner, ten-year-old Daniel Crowell quietly reads *The Swiss Family Robinson*, a story about a shipwrecked family on a tropical island. Like the family in his book, Daniel's family is marooned but his story is not fiction. He lives with his mother and two brothers in a garage with no heat or running water in a modest neighborhood in Los Angeles.[1] His mother gets food stamps and some cash aid, some of which she uses to pay her cousin $150 a month in rent for the garage. The family became homeless after Daniel's parents divorced, his mother lost their home to foreclosure, and then could not keep up with the rent on their apartment after she was laid off from her job at a car parts factory. His father is also unemployed and does not pay child support. His mother has applied for jobs in retail stores and fast-food restaurants and is still looking for work.

> "He's depressed a lot," she said of Daniel, whom she says has been the most affected of her sons by the loss of their home. "He does his work for class, but very slowly, like he's thinking. He worries a lot about living like this."[2]

The stigma of homelessness prevents some families from acknowledging their dire situation. Other parents are afraid of losing custody of their children because they can be charged with neglect if their children live in an emergency shelter, car, tent, or garage. When a parent cannot get their child to school, they may call in to say that the family is going to Disneyland or tell their children to lie about where they live. Following the downturn in the economy during the late 2000s recession and the foreclosure crisis, school districts saw more students from middle-income, working-class, and working-poor families becoming homeless.

The U.S. Department of Housing and Urban Development estimated that there were about 643,067 homeless people in the United States in 2009. About 37 percent of these people were families with children. These figures are based on an annual one-night survey by service agencies and volunteers who count the number of people living on the street, in cars, tents, or homeless shelters and report the number to HUD. This survey has been severely criticized for undercounting the homeless because it excludes families who are doubled up with other families or living in trailers or garages.

Other estimates of the homeless population are much higher. In 2007–08, the last school year for which data are available, the country's 14,000 public school districts reported more than 780,000 homeless students, representing a 15 percent increase from the previous year. A late 2008 survey by two nonprofits, the National Association for the Education of Homeless Children and Youth and First Focus, estimated the number of homeless students to be close to 1 million, higher than the numbers following Hurricanes Katrina and Rita. The National Law Center on Homeless and Poverty's 2009 estimate was one out of 50 or nearly 1.5 million children in the United States are homeless at some point each year.

Back in the garage, Daniel's routine remains the same. After school, his mother takes them to his grandmother's apartment where his mother's sister and brother also live. It is somewhere for the boys to eat and study, but it is difficult to concentrate on their homework. They cannot stay overnight because his grandmother's landlord would evict them if he found out.

> Unaware of the tough odds he faces, Daniel says he plans to finish high school. Like other boys his age, he still has big dreams—of becoming a basketball star and working at something important someday. But first and foremost, he dreams of "a very beautiful house . . . with a room of my own." The walls would be decorated, he said, "with posters, and pictures that I have drawn, and tests that I did in school."[3]

In the final section of the chapter the focus shifts away from social problems toward the physical problems associated with environmental degradation and decaying infrastructure. As we will see, these problems are intimately linked to the economic, political, and fiscal issues that dominate politics, policymaking, and planning.

PROBLEM? WHAT PROBLEM?

It is important to make a distinction at the beginning of this chapter between problems *in* cities and problems *of* cities.[4] Problems *of* cities are those that are brought about or amplified in some way by the attributes (environmental, demographic, social, economic, political, cultural, and so on) of urban settings. The anomie, social disorganization, and **deviant behavior** of people in certain neighborhoods, as interpreted by Durkheim and the Chicago School, provide classic examples. Another example stems from the **neighborhood effects** that we have already encountered: local value systems that socialize residents into distinctive (and sometimes problematic) attitudes and behaviors. Problems *in* cities are problems associated with the broader sweep of socioeconomic change. They are "urban" problems only to the extent that people and jobs are concentrated in urban settings. The settings themselves are not seen as causal factors that create the problems. Rather, the roots of the problems are found (1) in the tensions and contradictions for people that are associated with the economic and social relations that underpin particular forms of economic and social organization and (2) in the conflicts and frictions among people arising from ongoing economic and social restructuring.

In practice it is not always easy to make this distinction between problems in and problems of the city. A good example is provided by one of the biggest problems of all: poverty and deprivation. Poverty is a problem *in* cities in that it is fundamentally the product of an income distribution that reflects the workings of a competitive social and economic system based on unequal distributions of resources, skills, and power among people. It is also a problem *of* cities in that it is compounded and intensified by the settings in which the majority of poor people find themselves: physical and social environments that inhibit an individual's life chances and contribute to cumulative, negative spirals of community well-being.

It is also helpful if we recognize at the outset that some issues are widely regarded as problems while some are tolerated (or overlooked) as part of life, something that "comes with the territory." In short, situations are problematic only if they are, first, perceived to be a problem and, second, generally regarded as being very important. The first question, therefore, is: "Problem? What problem?" Environmental pollution provides a good example of the kind of situation that can represent a serious threat long before it is acknowledged as a problem. Some problems, in contrast, become recognized only as a result of their **social construction**. So, for example, Chinese people in towns and cities along the U.S. West Coast came to be seen as a problem in the mid-nineteenth century as a result of a mixture of beliefs, misconceptions, and prejudices that characterized Chinese customs and behavior as a threat to public health. "All great cities have their slums and localities where filth, disease, crime and misery abound," conceded a Special Committee of San Francisco's Board of Supervisors in 1885, "but in the very best aspect which Chinatown can be made to present, it must stand apart, conspicuous and beyond them all in the extreme degree of all these horrible attributes, the rankest outgrowth of human degradation that can be found upon this continent."[5] In fact, subsequent scientific research showed that Chinese neighborhoods were no worse than other slums: The Chinese had been made scapegoats for a more general breakdown of social order and the failure of public health systems.[6]

The second key question is: "Whose problem?" Homelessness provides a good example of a long-standing issue that for many years was considered not to be a serious social problem. Even in the early 1980s, when U.S. cities were estimated to contain between 250,000 and 300,000 homeless people, the "problem" was invisible to many people and denied by the country's leadership. Only after intensive publicity campaigns has homelessness come to be widely recognized as a problem. This example also points to another source of ambiguity. Because problems are very much a function of the eye of the beholder, they are often *relative*. How much homelessness must there be before it constitutes a problem? The answer often depends on how much of a threat that homelessness poses: Do other city residents believe that homeless people threaten their lifestyle or their property values? Do they offend people's sense of equity or intrude on their sense of aesthetics? Such differences in perception are the stuff of politics. Perceptions also change over time, however, as economic and social restructuring alters social relations among people.

In this chapter we begin by reviewing very briefly some of the changes that have occurred in the major problems associated with urbanization (i.e., problems *of* cities) in more affluent countries like the United States. This is an important exercise, because it gives us some perspective on contemporary problems. Against this background, four of the major problems of contemporary cities—localized poverty, criminal violence, homelessness, and environmental and infrastructure problems—are examined in closer detail.

FROM HAUNTS OF VICE TO GANG WASTELANDS—AND BACK

The central paradox of urbanization has always been that although cities have been essential as crucibles of economic development and production, they have brought with them a variety of negative and unprecedented side effects for the people in them.

Problems of the Early Industrial City

The industrial logic that underpinned urbanization in the early industrial cities of Europe and the United States was based on a division of labor and on **external economies**, **agglomeration economies**, and **localization economies** that required vast numbers of workers to be in close proximity to staff the complex and interdependent new systems of production, storage, and distribution. But this same concentration of workers also resulted in acute problems of overcrowding, pressure on the infrastructure of roads, bridges, and waste disposal systems, and greatly increased risk from fire and infectious diseases.

Within this context, the **division of labor** was starkly reflected in economic inequality among different groups of people. Social interaction was also affected: The heterogeneity of city populations, as well as their sheer size and density, had the potential to lead to anomie, social disorganization, and deviant behavior. Meanwhile, social conflict and political volatility increased partly in response to all these pressures and partly as a result of increased political consciousness and sensitivity (facilitated by the widespread emergence of pamphlets and popular newspapers but underpinned by the increased literacy and numeracy of workers required within the new division of labor).

It was the physical dimensions of these side effects that were most clearly perceived as problems. They were also broadly seen as *urban* problems rather than economic, social, or political problems: problems *of* the city rather than problems *in* the city. Adna Weber, summarizing in 1899 the literature on nineteenth-century urbanization, wrote in terms of a "theory of city degeneracy," pointing to what he saw as incontrovertible evidence that cities were the site, and city life the cause, of the deteriorated behavior of people.[7] Among the evidence available to Weber was the conclusion of British statistician G. B. Longstaff:

> That the town life is not as healthy as the country is a proposition that cannot be contradicted. . . . The narrow chest, the pale face, the weak eyes, the bad teeth, of the town-bred child are but too often apparent. It is easy to take an exaggerated view either way, but the broad facts are evident enough; long life in towns is accompanied by . . . degeneration.[8]

Another of Weber's sources went further, emphasizing mental "degeneration":

> The inhabitant of a large town, even the richest who is surrounded with the greatest luxury, is continually exposed to the unfavorable influences which diminish his

vital powers. . . . He breathes an atmosphere charged with organic detritus; he eats stale, contaminated, adulterated food; he feels himself in a state of constant nervous excitement, and one can compare him without exaggeration to the inhabitant of a marshy district. The children of large towns who are not carried off at an early age . . . develop more or less normally until they are 14 or 15 years of age, are up to that time alert, sometimes brilliantly endowed, and give highest promise. Then suddenly there is a standstill. The mind loses its facility of comprehension and the boy, who only yesterday was a model scholar, becomes an obtuse, clumsy dunce, who can only be steered with the greatest difficulty through his examinations. With these mental changes, bodily modifications go hand in hand.[9]

Although cities in general were seen as contributing to physical and intellectual degeneration, the slums that were discovered in every large city between 1840 and 1875 were seen as "haunts of vice" that harbored *moral* degeneration, which was in turn seen as a fundamental cause of both poverty and disease. In the words of a leading New York City housing reformer:

Vice, crime, drunkenness, lust, disease and death here hold sway, in spite of the most powerful moral and religious influences. . . . Their intellects are so blunted and their perceptions so perverted by the noxious atmosphere which they breathe, and the all-pervading filth in which they live, move, and have their being, that they are not susceptible to moral or religious influences. . . . [This is] a depraved physical condition which explains the moral deterioration of these people and which can never be overcome until we surround them with the conditions of sound health.[10]

It was this focus on the degeneracy inherent to urban life (Figure 1) that fueled the antiurbanism of the Romantics, encouraged the pastoral ideals of the **American Renaissance**, and added momentum to the European Garden City movement, and to, in the United States, the first planned suburbs, Arcadian Classicism, and the City Beautiful movement. It also gave direction to the first crucial efforts at urban reform and planning in Europe and the United States.

Problems of the Industrial City

By the late nineteenth century in the United States, many people's perceptions about the nature of urban problems had changed, along with changes in the overall dynamics of urbanization. The poverty and disease of the slums were still perceived as the main problems, but they came to be associated more with concentrations of particular ethnic groups. Instead of city life in general and the slums in particular being perceived

Jacob August Riss/Corbis Images

FIGURE 1 Haunt of vice. The Short Tail Gang in its hideaway beneath a pier at Jackson Street in New York, photographed by Jacob Riis in 1888.

Jacob August Riss/Corbis Images

FIGURE 2 Immigrants, formerly portrayed as the cornerstone of a new country, came to be portrayed in the late nineteenth century as a problem and a menace.

as "breeding" the physical and moral degeneration that seemed for some people to lead to depravity, drunkenness, poverty, prostitution, and abandoning their religious faith, it was now concentrations of immigrants that represented the root of these perceived urban problems.

The slum problem became the ghetto problem. Immigrants (Figure 2), previously seen as the backbone of the country, were recast in the role of public menace. As we have seen, Chinese people (Figure 3) bore the brunt of this prejudice in U.S. West Coast towns and cities. Elsewhere, it was the newcomers from southern and eastern Europe (who by then were responsible for much of the demographic growth and increased labor power of cities) who found themselves held accountable for many of the shortcomings of urban life. Joseph Kirkland, describing life among the poor in Chicago in 1895, argued that:

> For depth of shadow in Chicago low life one must look to the foreign elements . . . Among them may be found a certain degree of isolation, and therefore of clannish crowding; also of contented squalor, jealous of inspection and interference. It is in the quarters inhabited by these that there are to be found the worst parts of Chicago, the most unsavory spots in their moral and material aspects.[11]

Arnold Genthe/Library of Congress

FIGURE 3 Chinatown, San Francisco, about 1900. An ethnic community that was unjustly blamed, like its counterparts in other cities, for citywide public health problems.

Within just a couple of decades, the movement of many of the immigrants and their children upward socioeconomically and outward geographically had undermined this perspective on urban problems. Immigrant ghettos consequently came to be seen as short-term havens that, over the longer term, became "ramps" for economic advancement. It was organized crime, rather than endemic criminality, that instead appeared to many people to be the main obstacle to civic order. Meanwhile, African-American ghettos, newly formed by the migration of rural African Americans from the South after 1910, replaced immigrant ghettos (in the social construction of urban problems) as the main places of social and economic problems.

By that time cities had also changed in other ways, bringing new issues and new perceptions about the nature of urban problems. Cities were not only much bigger, but they were also more "modern," faster-paced, and with much more specialization and compartmentalization in economic affairs and social life. Alienation, along with mental and physical stress, came to be more widely perceived as major problems. As the ideas of Louis Wirth and the Chicago School took hold, these problems came to be seen increasingly as a product of the breakdown of "normal" patterns of social interaction and social organization among people. Once again, though, it was the slum that came to be perceived as the location of the problem: the place where stresses were intensified to the point where deviant behavior of all kinds was the result.

AP Images

FIGURE 4 Urban decay in the South Bronx.

Problems of the Post-War City (1945–73)

After World War II, a combination of

1. **Keynesian** economic management;
2. improved public health due to modern medicine, improved diets, and environmental regulation;
3. an unusually long period of economic growth and prosperity; and
4. an unprecedented array of welfare programs

made impressive inroads into the previously intractable problem of slums and poverty. In this period of optimism and can-do attitudes, one of the persistent problems was seen as social and institutional discrimination against black people, not least because it was trapping black households in central city neighborhoods.

The central cities themselves also came to be seen as problematic. Their aging neighborhoods, suddenly reaching structural and technological obsolescence just as cities were decentralizing, now contained blighted tracts (Figure 4) that became emblematic of urban problems. Nevertheless, the overall affluence of the period led to a feeling of optimism. Urban problems, it was felt by many people, were a residual legacy that would be tidied up quickly: Slums and poverty would soon be a thing of the past. The maturing Baby Boom generation, taking prosperity for granted, soon began to look beyond economic and social inequality, uncovering environmental issues as a much greater cause for concern.

Problems of the Neoliberal City

In the mid-1960s many people in the United States were rudely awakened from their feelings of affluence and optimism that had been growing since World War II. Widespread civil unrest, including riots, looting, and arson, served notice that racial tension and economic inequality were still major problems. With attention focused once more on central city slums, poverty was rediscovered in the midst of affluence. But before the magnitude of the problem could be properly grasped by most people, the overall dynamics of urbanization were jarred by the economic recession that was triggered by the 1973 oil price increases. **Deindustrialization** and its problems of unemployment and **underemployment** now came into focus, and price inflation became problematic, bringing with it the issue of housing affordability. City budgets were suddenly unable to sustain the services and welfare programs for residents that had been built up during more affluent times, resulting in acute fiscal crises that were soon followed by drastic changes in the role and orientation of city governments.

As the economic boom of the 1980s followed these crises, the nature of urban problems came to be redefined yet again. Prosperity brought renewed concern, by people in many cities, for the costs of urban growth: traffic congestion, air pollution, escalating house prices, and backlogs in public service provision. For some observers, though, the sociocultural changes that began in the 1980s gave new cause for concern. New kinds of stresses and strains emerged from the combination of intensified materialism and labor market changes that drew more and more women into the labor force and contributed to the restructuring of family and community life. In this context it was not only the slums but also the suburbs that seemed to be the location of urban problems. Teenybopper mall rats and "Valley Girls" came to be emblematic of the shallow materialism endemic to the outer fringes of the polycentric metropolis. Much worse, however, was the darker side of social and economic change in the suburbs: a landscape of latchkey kids and teenagers with socialization problems; a place where economic, social, and family pressures contributed to unprecedented rates of teen suicide and behavioral problems.[12]

Yet the more things change, the more they can seem to stay the same. Poverty and slums were rediscovered once again beginning in the 1980s, with the relationships between them reformulated this time to reflect the latest impulses in the dynamics of urban change. Economic restructuring and the retrenchment of welfare programs had spread and intensified poverty and contributed to homelessness on an unprecedented scale. Changing social attitudes, meanwhile, had contributed to an increase in divorce and single-parent households. One result was the **feminization of poverty** and homelessness. This in turn was compounded by spatial restructuring, which resulted not only in the decentralization of many better-paid jobs but also in the departure of many role models from low-income neighborhoods.

The result was that the problem came to be reformulated in terms of a geographically, socially, and economically isolated group of truly disadvantaged poor people who were seen at once as victims and as incubators. Here were residual populations of vulnerable and disadvantaged households, casualties of structural economic change. But here also were the inheritors of urban degeneracy, within a whole spectrum of new problems that included senseless and unprovoked violence, premeditated, predatory violence, domestic violence, the organized but mostly latent violence of street gangs, and high levels of HIV/AIDS—all closely associated with drug taking and drug dealing. With this interpretation, then, we have returned to the theme of cities as haunts of vice.

What can we learn from this brief sketch of the history of urban problems? First, we should recognize that problems have to be seen not only in the context of the broader sweep of urban change but also through the eyes of beholders whose own position in this same history colors their perceptions. Second, we should recognize the complexity of the cause-and-effect web inherent to all urban problems: Each strand of the web is part of the sociospatial dialectic, not simply a series of irritations and unwanted by-products of urbanization. The remainder of this chapter is designed to illustrate this complexity in relation to some of the most critical problems of contemporary urban change in more affluent countries like the United States. Third, given this perspective, we can appreciate the changing imperatives of urban governance, policy making, and planning.

POVERTY

Poverty is arguably the most compelling problem both *in* and *of* cities. In reality, if someone is living in poverty it means not only low income, few material assets, and a low quality of life, but also *persistent* low income and little prospect of acquiring material assets or an improved quality of life. More important, poverty comes as a package for a person: low income is at the core, but it is inextricably linked to poor diet, poor environment, poor health, the psychological stress of continuously having to make ends meet, and the economic, social, and political disadvantages of being often unfairly stigmatized as a "loser" in an increasingly competitive society.

In general terms, the core of poverty—a person's low income—can be understood as a product of:

1. the income distribution inherent to an economic system based on relatively free competition and private profit;
2. uneven distributions of resources and opportunities between places and between socioeconomic and ethnic groups; and/or
3. institutional shortcomings, specifically involving:
   - inadequate participation and representation within the political process, and/or
   - inappropriate structure or malfunctioning of the welfare system.

To the extent that these factors may well be more intense in some cities or neighborhoods, the resultant poverty can be considered to be a problem *of* urbanization. Cities and their residents who are affected by deindustrialization, for example, are likely to experience high levels of unemployment and depressed wage rates within their local labor markets.

But poverty also needs to be understood in terms of the full package of disadvantages that come with people's low incomes, and it is in this context that poverty can be seen more clearly as a problem of urbanization. The logic of urban land use dictates that low-income households are left to live at high densities in the cheapest housing. Such households live in crowded conditions in run-down housing in older neighborhoods: classically in the **zone in transition** around the **central business district (CBD)** or adjacent to older industrial districts and corridors. Filled with low-income households and nearing the end of their physical lifecycle, these neighborhoods are subject to a **spiral of decay** that can result in their becoming *slums*, where maintenance and repairs are, at best, makeshift and where physical deterioration is accompanied by graffiti and garbage piling up.

The spiral of decay begins with substandard housing (a product of physical deterioration and structural and technological obsolescence occupied by low-income households who can afford to rent only a minimal amount of space. The resultant overcrowding not only causes greater wear and tear on the housing itself but also puts intense pressure on the neighborhood infrastructure of streets, parks, schools, and so on. The need for maintenance and repairs increases very quickly, but is not met. Individual households cannot afford to pay for dwelling maintenance or repair on any significant scale. Landlords have no incentive to do so, because they have a "captive" market. The public officials in government agencies are often indifferent to the needs of such neighborhoods because of their relative lack of political power, or they simply feel overwhelmed by the costs of rehabilitating and maintaining the neighborhood infrastructure. The same spiral of decay hurts local businesses such as shops, restaurants, and hairdressers. With a low-income clientele, profit margins must be kept low, leaving little to spare for upkeep or improvement. Many small businesses fail or relocate to more profitable settings, leaving commercial property vacant for long periods. In extreme cases, this property becomes abandoned, the owners unable to find either renters or buyers.

Some residential buildings may also be abandoned by their owners, though abandonment of residential rental property usually results from the loss of a significant proportion of a landlord's captive market of low-income households. In the 1960s residences began to be abandoned by their owners in some of the older industrial urban centers as a result of the decentralization of manufacturing jobs. Faced with escalating maintenance costs and rising property taxes and unable to increase revenue because of rent controls and the depressed state of the central city housing market, some landlords simply wrote off their property by abandoning it to long-term vacancy. In desperation, some property owners (both commercial and residential) resorted to arson in an attempt to at least salvage insurance monies.

Abandonment in any case increases the risk of accidental fire caused by homeless people or others such as drug addicts. Fire-blackened ruins or just boarded-up, abandoned property can be an important element in the spiral of neighborhood decay because of the **contagion effect** that they introduce (Figure 5). A detailed study of abandoned buildings in Philadelphia by geographer Michael Dear suggests that abandonment can be viewed as a contagious process that affects other property owners in two broad stages.[13] Initially there emerges a broad scatter of loosely defined clusters of abandoned structures, each consisting of several groups of two or three abandoned units. Over several years, these patterns are intensified, the broad scatter being maintained but the number of abandoned units increasing.

The reasons for this localized contagion effect are closely related to the depressing effects of long-term vacancy on the value and desirability of adjacent property. The fact that abandoned buildings make good sites for vandalism and crime (especially the sale and use of illegal drugs) provides a further impetus for owners of nearby property to abandon them. The downward spiral is further reinforced by what Dear calls the "psychological abandonment" of the wider area by realtors, financiers, and landlords (who become more systematic in their disinvestments) and by public agencies (that begin to cut back on maintenance and service delivery).

The Cycle of Poverty

There is, meanwhile, a dismal cycle that intersects with these localized spirals of decay. The **cycle of poverty** for a person or household (Figure 6) also begins with low incomes, poor housing, and overcrowded conditions. These conditions are unhealthy, poor housing and overcrowding cause people to be more prone to poor health, which is compounded by poor diets that are also a result of low incomes. Poor health causes absenteeism from work, which results in less income. Absenteeism from school through illness may also contribute to the cycle of poverty by hindering educational achievement, limiting occupational skills and so leading to low wages. More significantly, overcrowding also produces psychological stress that contributes not only to people's poor health (and therefore reinforces the cycle of poverty) but also to behavioral responses.

This cycle, of course, is related to the ecological ideas first put forward by Louis Wirth and the Chicago School sociologists: High-density slums represent distinctive "natural areas" in which "normal" adaptive behavior by people is often not enough to cope with the dreadful pressures not only of overcrowding but also of a heterogeneous population living under chronic economic stress (Figure 7). As a result, these areas tend

Joseph Sohm/Corbis Images

FIGURE 5 Abandoned property such as this represents a risk to public health and safety, provides a niche for illegal activity, and has a negative effect on the value of neighboring property because of the contagion effect.

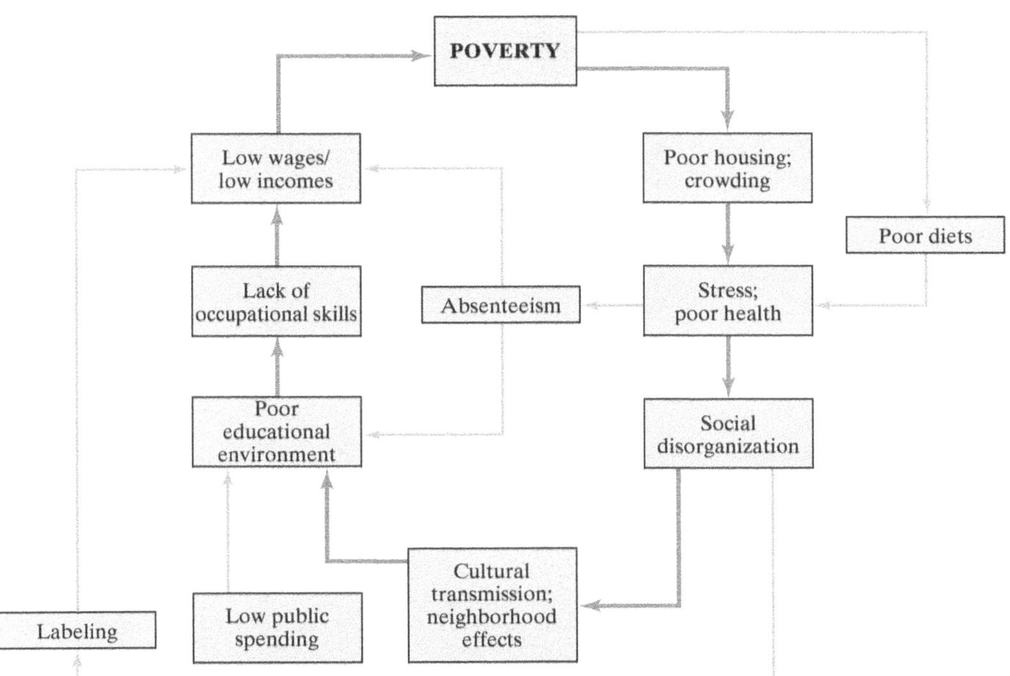

FIGURE 6 The cycle of poverty.

to be associated with **anomie**, **social disorganization**, and a variety of pathological behaviors, including crime and vandalism. Such conditions not only affect individuals' educational achievement and employment opportunities but can also lead to **labeling**, through which all residents may find that their neighborhood's poor image affects their employment opportunities. The lifeworld of slum neighborhoods, meanwhile, is often seen as a critical element in the cycle of poverty, a combination of **cultural transmission** and neighborhood effects leading to the transmission of modes of speech, codes of behavior, and attitudes to schooling, work, and society that operate as handicaps in wider educational and labor markets, and so contributing to the "trap" of low incomes and slum settings.

Probably the most important element in this cycle of poverty, however, is the educational setting of slum neighborhoods. Schools tend to be not only obsolete and physically deteriorated, like their surroundings, but also unattractive to teachers—partly because of the physical environment and partly because of the social and disciplinary environment. Because of the traditional indifference of public agencies to the political and social significance of their surrounding communities, often the schools are also resource-poor with relatively small budgets for staff, equipment, and materials. Over the long term, poor educational resources translate into poor education, however positive the values of the students and their parents. Poor education limits occupational choice and, ultimately, results in lower incomes. Faced with the evidence all around them of unemployment and low-wage jobs at the end of school careers, some students and parents inevitably find it difficult to sustain positive values about education. The result then becomes a self-fulfilling prophecy.

Some observers have gone so far as to give particular emphasis to the role of people's values and attitudes in perpetuating

Joseph Sohm/Corbis Images

FIGURE 7 The disadvantages are not insurmountable, but growing up in this kind of setting brings a high probability of staying trapped in a localized cycle of poverty.

the cycle of poverty, arguing that there eventually emerges a **culture of poverty**.[14] In essence, this view points to a vicious cycle that links lack of opportunity and a resultant lack of aspiration. Poor people, according to this perspective, see the improbability of achieving any kind of material or social success and so adapt their expectations and behavior to the situation, becoming so accustomed to deprivation, so defeated, and so withdrawn that they become both unable and unwilling to seize opportunities for educational or occupational advancement when they do occur. This view has been particularly popular among observers whose political inclination is to "blame the victim." There has been, however, considerable debate about the validity of this notion of a culture of poverty. And in fact, there is no empirical evidence to suggest that the urban poor have values and aspirations—let alone a culture—significantly different from the rest of society.

URBAN VIEW 2
Poverty, Stress, and Civil Disorder in U.S. Cities

Slum settings inevitably create severe stress and tensions for the people living in them, and so it is not surprising to find that rioting and civil disorder break out from time to time. There has been a long but intermittent history of such civil disorder in the United States. Many of the early incidents involved minority migrants and immigrants being attacked by mobs of established residents who had been incited to anger at the perceived threat to their health, their wages, and their jobs. Significant incidents occurred in East St. Louis in 1917, in Chicago in 1919, and in Detroit in 1943. Rioting *within* slums, more a form of mass protest and an expression of anger and desperation, first occurred in Harlem, New York City, in 1935 and 1943, the riots in both cases consisting of African Americans massing in the streets, breaking store windows, looting, and starting fires. The incidents were confined entirely to Harlem; the only whites victimized being those who were encountered on local streets.

Riots like these occurred in every major U.S. city during the mid- to late 1960s. Although their immediate effects were quite localized, their impact on U.S. politics and urban affairs was profound. Poverty was rediscovered, racism was confronted, and policymakers developed specifically targeted urban policies.[15]

As the "long hot summers" of the late 1960s progressed, a clear pattern emerged. The riots were typically triggered by a police incident. Many African Americans saw the prejudice and brutality of the white police officers who typically patrolled their neighborhoods as symbolic of the broader discrimination they experienced. Specific incidents served as catalysts for chain reactions of frustration and resentment. Once under way, the typical riot lasted for several days and nights, with massive crowds in the streets looting and setting fire to stores and preventing police and firefighters from entering the area. Large numbers of local people were involved—30 to 50 percent of residents was common—but few attempts were made to spread riot activities into other areas of the city.

The scale and intensity of the riots can be conveyed by a few examples. In two of the earliest incidents, riots in Harlem and Bedford-Stuyvesant (in Brooklyn) in 1964 lasted for six nights, resulting in one death, 118 injuries, and 465 arrests. The following year saw many more incidents, including the notorious Watts riots in Los Angeles, in which 30,000–35,000 people participated, burning, looting, or damaging almost 1,000 buildings, creating property damage estimated at $40 million, and resulting in 34 deaths, over 1,000 injuries, and nearly 4,000 arrests. The peak year of rioting was 1967, with major incidents in Cleveland, Newark, and New Haven. The most serious incident, however, was in Detroit, where more than 75,000 people were on the streets for three days and nights. This time 43 people were left dead, over 2,000 were injured, 5,642 were arrested, and the bill for property damage was $200 million. In that fourth year of rioting, President Lyndon Johnson appointed the Kerner Commission to investigate the causes of what had become a major national issue. *The commission's report pointed to the underlying racism of U.S. society, acknowledged bad relations between black communities and the police and public agencies as the immediate cause of most incidents, and stressed the economic and social distress of African Americans in central city neighborhoods.*

A generation later, the country's worst-ever incident provided a depressing echo of the 1960s (Figure 8). After the acquittal in May 1992 of police officers who had beaten Rodney King, an African-American motorist, rioting in Los Angeles (including in the Watts neighborhood) resulted in more than 1,000 fires, left more than 5,000 buildings damaged, and produced a death toll of 52. More than 8,000 people were arrested, the bill for property damage was more than $5 billion, and thousands of workers were temporarily unemployed because of riot damage.

In the early twenty-first century the underlying causes of these incidents have not yet been adequately addressed, as evidenced by the riots that continue to be sparked in black central city neighborhoods. In June 2003 rioting broke out in Benton Harbor, Mich., after the death of Terrance Shurn, a black Benton Harbor resident, whose speeding motorcycle crashed into a building as police were chasing him into the city. Benton Harbor is a city of 12,000 people about 100 miles northeast of Chicago that is 92 percent black. Across the river, St. Joseph, a city of 8,800, is 90 percent white. Hundreds of people rioted in the streets near where the crash occurred, setting fires, and attacking passers-by. At least 11 people were hurt. Rioters fired shots at emergency personnel, and eight police and fire vehicles and two private vehicles were damaged. Property damage was estimated at $500,000—the fires completely burned 21 homes, including the one that Shurn struck, while seven more were damaged.

URBAN VIEW 2
Poverty, Stress, and Civil Disorder in U.S. Cities (*continued*)

Ted Soqui/Corbis Images

FIGURE 8 A neighborhood in South Central Los Angeles following the 1992 riots that swept the city for days after the acquittal of police officers who had beaten Rodney King, an African-American motorist.

Poverty in U.S. Metropolitan Areas We can put conceptual frameworks such as the cycle of poverty into better perspective by reviewing the actual pattern and extent of poverty in U.S. metropolitan areas. In the United States poverty is officially defined in terms of a set of money income thresholds that vary by family size and composition, indexed against the Consumer Price Index. In 2011 the federal poverty level for a family of four was $22,350.[16] By this measure, the number of people living in poverty in the United States had risen to nearly 44 million by 2010, with almost 82 percent of these people living in metropolitan areas.[17] As Figure 9 shows, the poverty rate has fluctuated at about the same level (between 12 and 15 percent of the population) since falling dramatically in the early 1960s because of economic prosperity and the introduction of a welfare safety net.

For metropolitan areas, the period since 1970 is especially illuminating, because the poverty data cover a period of particularly intense restructuring as a result of the impacts of neoliberalism. Since 1970, the number of people living below the poverty rate in metropolitan areas increased more than two-and-a-half fold (from 13.4 million or 10.2 percent of the population in 2007 to 35.7 million or 13.9 percent of the population in 2009). During the same time, the percentage of people living below the poverty rate in central cities more than doubled (from 8.2 million to 18.3 million) as the rate rose from 14.3 to 18.7 percent; the suburbanization of poverty as a trend also began, with the percentage of people below the poverty rate in the suburbs rising from 7.1 to 11.0 percent, as the number of poor people more than doubled, from 5.2 million to 11.0 million (Figure 10).[18] During the economic expansion in the 1990s,

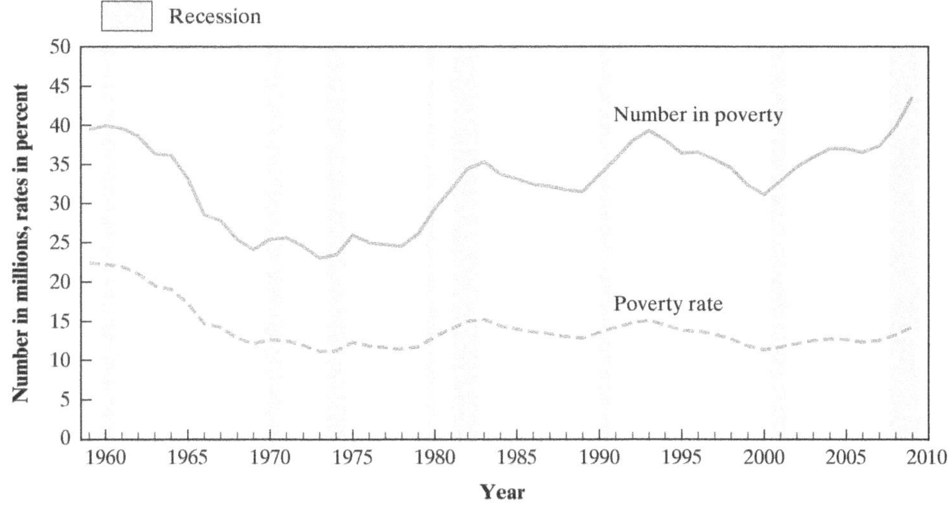

FIGURE 9 Poverty in the United States.

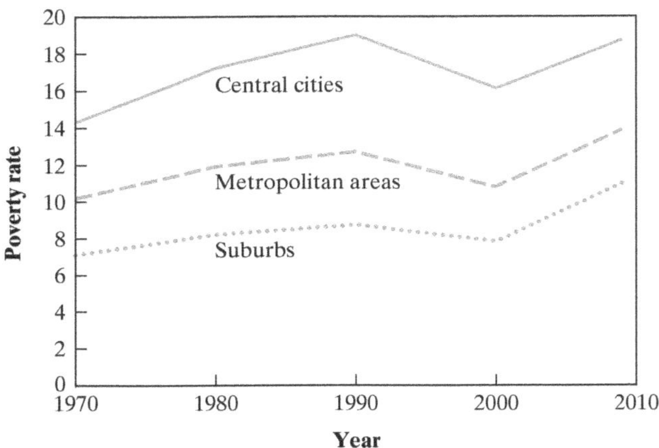

FIGURE 10 Percentage of people below the poverty rate in U.S. metropolitan areas.

poverty rates in metropolitan areas fell before rising again in the 2000s with an upward spike occurring as the United States entered the recession.

The late 2000s recession exacerbated the poverty trends in the central cities and suburbs of metropolitan areas. Between 2000 and 2009 alone, the number of people living below the poverty rate in metropolitan areas increased one-and-a-half fold (from 24.3 million or 10.8 percent of the population in 2000 to 35.7 million or 13.9 percent of the population in 2009). During the same time, the percentage of people living below the poverty rate in central cities rose from 16.1 to 18.7 (increasing by nearly 12 percent, from 16.1 million to 18.3 million), while the percentage of people below the poverty rate in the suburbs rose from 7.8 to 11.0 percent, an increase of nearly 54 percent (from 11.3 million to 17.4 million) (Figure 10).

The composition and share of the poor living in metropolitan areas has also changed a good deal. By 2008, 25 percent of the poor population in central cities was white, while this figure was 46 percent for the suburbs. In contrast, 32 percent of poor central city residents were black, versus 17 percent in the suburbs. Although some of the differences in the makeup of the central city and suburban poor can be explained by differences in the racial and ethnic composition of their total populations, African Americans and Hispanics still comprise a disproportionate share of poor people in both central cities and suburbs. Only in the outer suburbs and in the exurbs do white people comprise the majority of the poor people, and, even there, minorities make up a disproportionate share of the poor. In both central cities and low-density exurban communities, African Americans account for a disproportionate share of the poor, whereas in older and denser suburbs, the poor people are disproportionately Hispanic (Figure 11).[19]

Figure 12 shows the growth of poverty in Detroit. Although poverty grew between 1970 and 1990, this trend reversed during the economic expansion in the 1990s.[20] By 2010, Detroit was one of the cities that had been hardest hit by the recession. The pattern of residential demolition orders in 2010, for example, captures the deepening poverty in the already poor neighborhoods of the city (Figure 13).[21] In 2010, the City of Detroit began tearing down 3,000 vacant houses as part of a four-year plan to demolish 10,000 of the houses that are in the most dangerous condition of the more than 33,000 vacant houses that are located in some of the poorest neighborhoods in the city. Of the 50 largest U.S. metropolitan areas, Detroit was ranked the most stressful place to live and work in 2010 based on having the third worst unemployment rate, being in the top ten for murders, having the most robberies, heart attacks, and families in poverty, and the fewest sunny days.[22]

Dual Cities? Some evidence supports the idea that very poor neighborhoods are the expression of localized cycles of poverty *in which disadvantages are accumulated and deprivation is intensified*. Poor people in these areas experience significantly higher rates of unemployment than the poor who lived in areas of less-concentrated poverty, and they are more

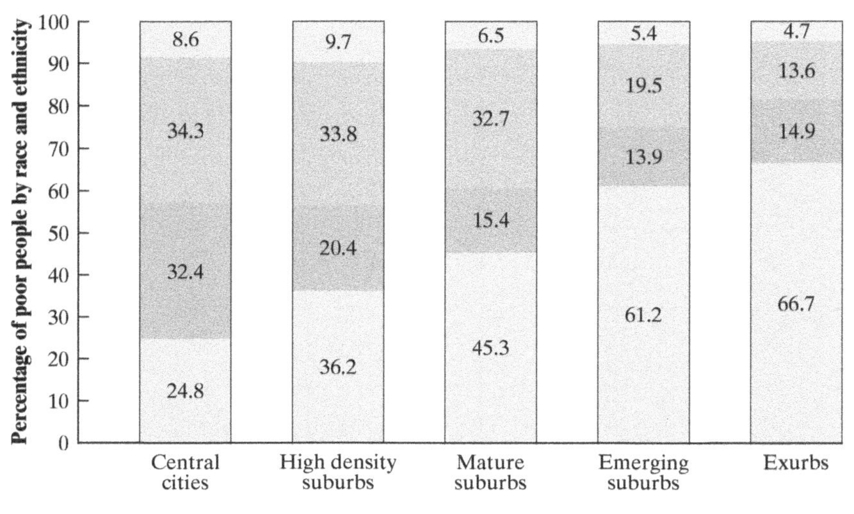

State of Metropolitan America: On the Front Lines of Demographic Transformation. Brookings Institution. www.brookings.edu/metro. Reprinted with permission.

FIGURE 11 Race and ethnicity of the poor in U.S. metropolitan areas, 2008.

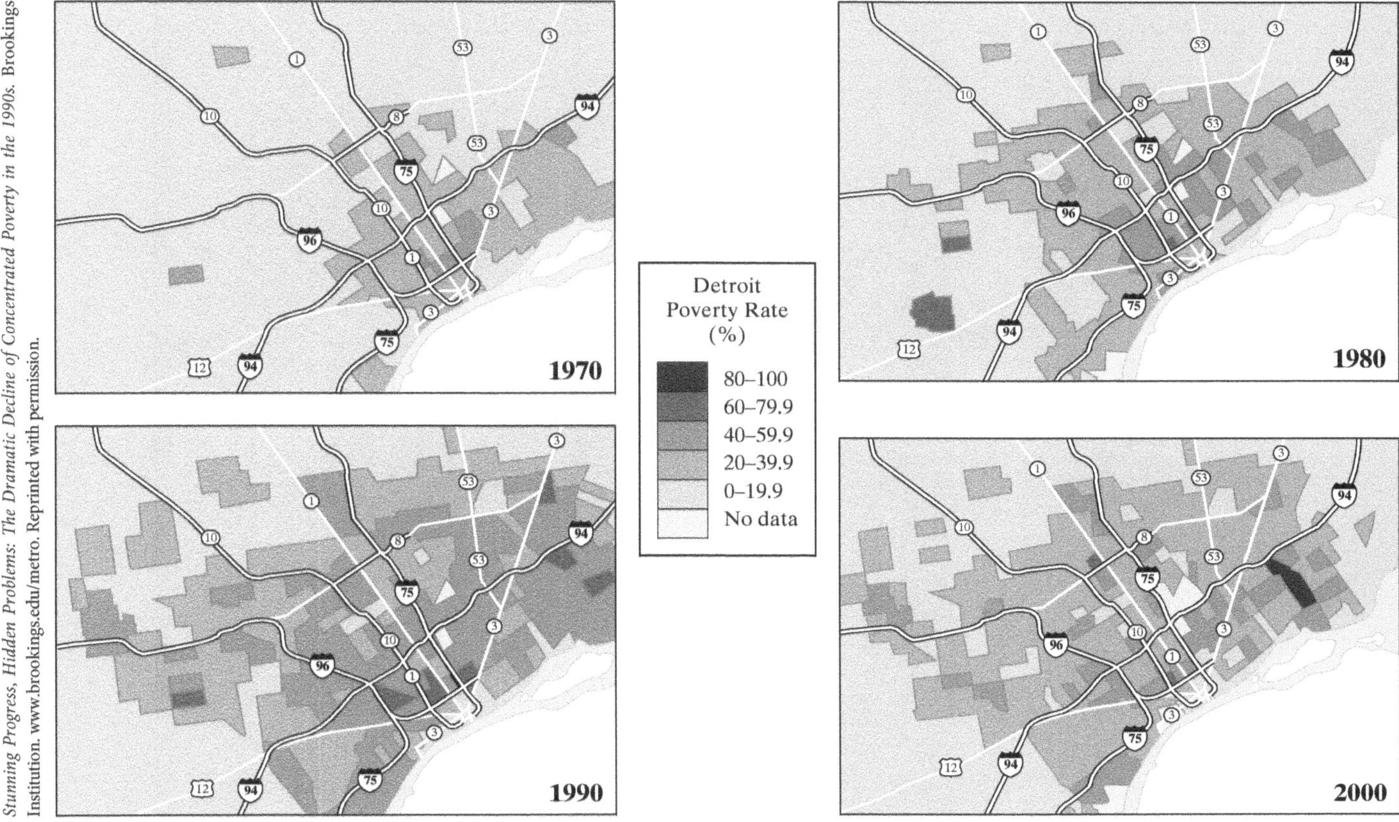

Stunning Progress, Hidden Problems: The Dramatic Decline of Concentrated Poverty in the 1990s. Brookings Institution. www.brookings.edu/metro. Reprinted with permission.

FIGURE 12 Poverty in Detroit.

dependent on welfare (Figure 14), more likely to live in single-parent households, and more likely to have dropped out of high school. They are also *much* more likely to be African American.

Sociologist William Julius Wilson defined the truly disadvantaged as:

> . . . that *heterogeneous grouping of families and individuals who are outside the mainstream of the American occupational system*. Included . . . are individuals who lack training and skills and either experience long-term unemployment or are not members of the labor force, individuals who are engaged in street crime and other forms of aberrant behavior, and families that experience long-term spells of poverty and/or welfare dependency.[23]

A more concise conceptual definition has been suggested by Martha Van Reitsma:

> . . . those who are weakly connected to the formal labor force and whose social context tends to maintain or further weaken this attachment.[24]

Van Reitsma, like Wilson, pointed to macroeconomic restructuring as the fundamental reason for this situation. The existence of large numbers of people with only weak connections to the formal labor force is attributed to the **spatial mismatch** between people and jobs that intensified as many of the low-skill manufacturing and services jobs traditionally found in central city areas have been relocated to the suburbs and beyond, to be replaced often by jobs requiring higher skills.[25] Figure 15 shows the extent of the spatial mismatch between the more central city location of minorities and poor people and the more suburban location of jobs in the 50 largest U.S. metropolitan areas. Looking at the top third of the bar chart, we can see that, although more than 70 percent of all residents, and more than 75 percent of white residents, live in the suburbs, the percentages of blacks and Hispanics in the suburbs are only about 50 and just over 60 percent respectively. The middle third of the bar chart shows how the percentages of people who are poor and live in the suburbs are even lower. The bottom third of the bar chart shows how job sprawl in these largest metropolitan areas has put many low-skill jobs out of the reach of poor minorities especially. More than 70 percent of total employment is located more than five miles from the CBD in these metropolitan areas, and of that, about 75 percent of low-skill services jobs and nearly 80 percent of manufacturing/blue collar jobs are now located more than five miles from the CBD.

Table 1 shows the top five large metropolitan areas with the most decentralized (suburban) employment. In terms the spatial mismatch, Detroit and Chicago are among the most segregated U.S. metropolitan areas for African Americans and are also among the metropolitan areas with the most decentralized employment.

Together, Figure 15 and Table 1 capture the fact that a large proportion of minorities and poor people remain

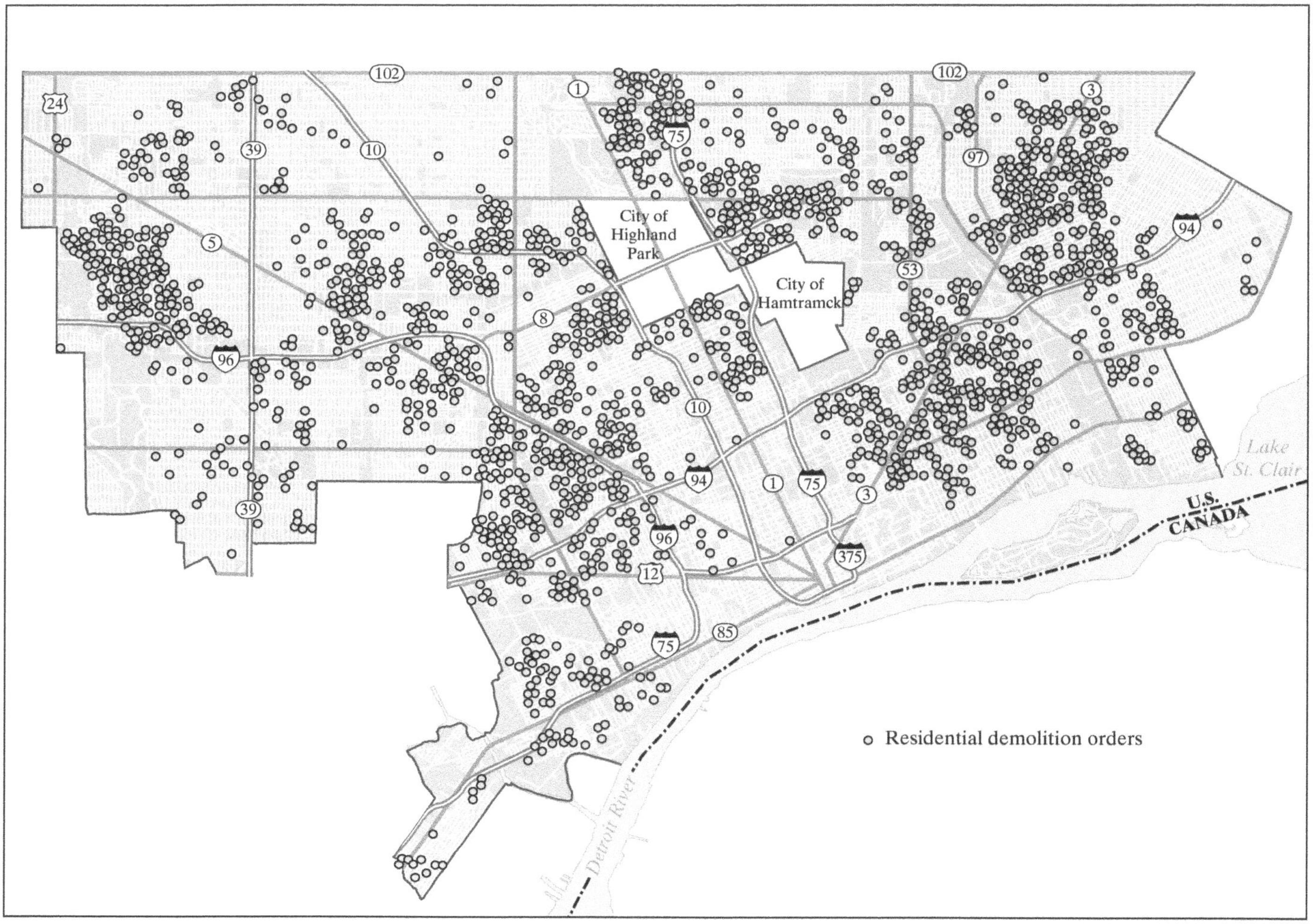

o Residential demolition orders

Residential Demolition Orders, Detroit Free Press, April 4, 2010.

FIGURE 13 Residential demolition orders in Detroit, 2010.

"trapped" in central cities, unable to afford housing or transportation that would make suburban jobs accessible to them, and dependent on welfare or various "informal" sources of income, including illegal activities such as prostitution and drug dealing.

The "social context" referred to by Van Rietsma is partly a reference to the role of local attitudes and behaviors in relation to the idea of a cycle of poverty—the tendency to drop out of school, for example. But it also refers to the outcomes of some of the social, cultural, political, and demographic trends that

ZUMA Press/NewsCom

FIGURE 14 People waiting to apply for benefits, including Medicaid and Food Stamps, at the Gwinnett County Department of Family and Childrens Services to the northeast of Atlanta in Georgia.

Job Sprawl and the Suburbanization of Poverty. Brookings Institution. www.brookings.edu/metro. Reprinted with permission.

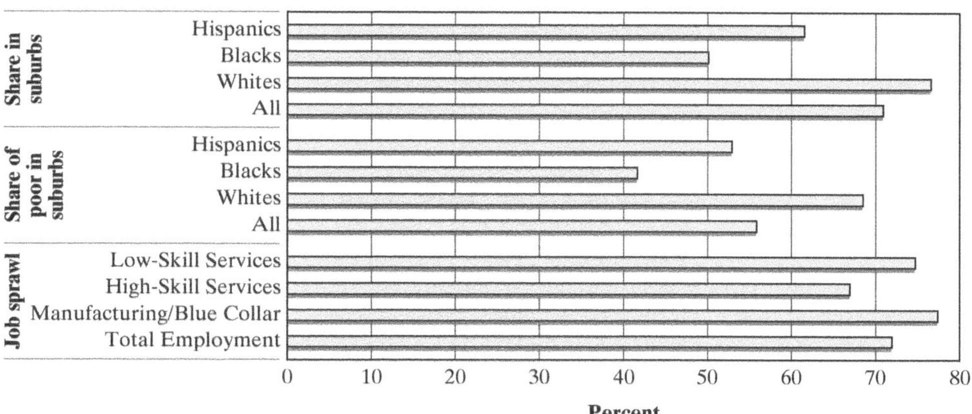

FIGURE 15 The spatial mismatch: suburbanization, poverty, and job sprawl in the United States, 2007.

have been associated with the economic and metropolitan restructuring since the 1960s. One of the most important of these trends is the increase in the number of black teenage unwed mothers that has led to dramatic increases in black single-parent households and to the feminization of poverty. Another is the very selective migration of younger and higher-skilled blacks from central cities to the suburbs. This migration has left a residual population of unskilled and elderly people along with female-headed households in central city neighborhoods. It has left them, additionally, with significantly fewer role models. As a result, there has been an erosion of mainstream social norms and an increase in social disorganization and deviant behavior.

But these linkages between labor markets, poverty, migration, household structure, race, gender, attitudes, and behavior have proven very difficult to measure scientifically. Even so, the economic, social, and metropolitan restructuring since the 1960s has resulted in both qualitative and quantitative changes in the poorest U.S. neighborhoods. In this context it is important to see these changes as part of a broader shift in the dynamics of urbanization—one that has introduced new patterns of vulnerability, fragmentation, and disadvantage for some people while intensifying the occupational and spatial polarization of the city. Sociologist Manuel Castells has argued that the overall result is a **Dual City** in which downsized workers, the poor, and the homeless share excluded spaces, fragmented along racial and ethnic lines. Just as the slums of nineteenth-century U.S. cities were "haunts of vice," so these "excluded" spaces are today the places that are most often associated with the most intense concentrations of social problems. As Castells observes, "Downgraded areas of the city serve as refuges for the criminal element of the informal economy, as well as reservations for displaced labor, barely maintained on welfare."[26] In the next section we examine one particular aspect of pathological behavior—criminal violence—that is widely identified as one of the salient problems of contemporary urbanization.

TABLE 1 Most Decentralized Large U.S. Metropolitan Areas by Employment Share, 2006

	Share of Jobs (%)		
Highest share more than 10 miles away	**More than 10 miles**	**3–10 miles**	**Within 3 miles**
Detroit	77.4	15.7	7.0
Chicago	68.7	13.4	17.9
Dallas	66.9	22.5	10.6
Los Angeles	65.6	26.2	8.2
Philadelphia	63.7	20.8	15.5

Source: E. Kneebone. *Job Sprawl Revisited: The Changing Geography of Metropolitan Employment*. Washington, D.C.: Metropolitan Policy Program at Brookings, 2009, Table 2.

CRIMINAL VIOLENCE

The United States is "the most violent and self-destructive nation on earth" was the conclusion of a 1991 report of the Senate Judiciary Committee on violent crime (homicide, rape, robbery, assault). Since the early 1990s, however, violent crime rates have declined to the lowest levels recorded recently in the United States (from about 48 incidents per thousand people in 1973 to about 17 per thousand by 2010) (Figure 16).[27] Like poverty, social problems such as criminal violence are the product of the interplay of several processes; like poverty, they are at once problems *in*, as well as problems *of*, the city. In addition, poverty and social pathology have for a long time been seen as causally linked, poverty and its associated environments seeming to foster deviant attitudes and to create stress and resentment that trigger deviant behavior by some people.

We can recognize four main explanatory linkages between urban context and violent crime.[28]

1. *Economic Effects.* Among the reasons for hypothesizing direct causal linkages between poverty and violent crime are ideas that:
   - the poor as a group develop a culture that embraces violence as an expression of toughness or a measure of status and identity;
   - people's feelings of relative deprivation lead to anger and resentment that are channeled into violence;

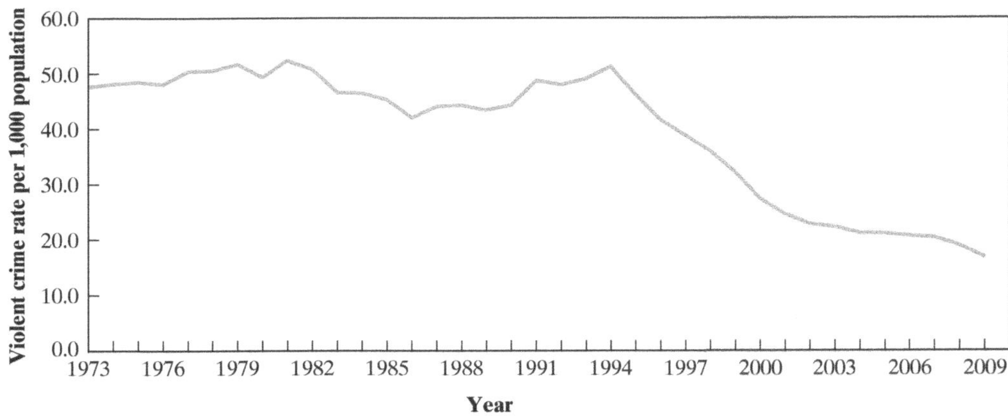

FIGURE 16 Violent crime rate (assault, rape, robbery, and homicide) per 1,000 people in the United States.

- psychological reactions to feelings of relative deprivation, together with psychological stresses resulting from poverty, serve to limit interaction among people, which undermines both formal and informal mechanisms of social control and so removes many of the restraints that would otherwise suppress violence.

Research has confirmed clear links between poverty, inequality, and violent crime—particularly homicide (Figure 17). But researchers have found it difficult to isolate these effects from the mediating and interactive effects of other factors, such as racial composition, racial inequality, family composition, and social disorganization.

2. *Social Disorganization.* Even without the presence of significant levels of poverty or inequality, high rates of population turnover and high levels of diversity in the social, demographic, cultural, and ethnic attributes of community populations can sometimes inhibit the formation of local ties by people, weaken the effectiveness of local social institutions, and leave communities ill-equipped to maintain social order. Research has shown clear links between criminal violence (particularly assaults) and rapid social and demographic change, even when the change is positive, involving neighborhood upgrading.
3. *Demographic Effects.* Since the age at which many people commit offenses shows a clear peak in the late teens and early twenties, an argument is that sociodemographic changes in neighborhood composition are related to changes in urban crime rates. Research has provided some evidence of such a linkage, though it was underpinned by differential effects of economic restructuring on specific sociodemographic groups.
4. *Effects of Lifestyle and Routine Activities.* The linkage between crime and people's lifestyle or routine patterns of activity is based on the observation that for a crime to occur there are three preconditions: (a) the presence of a motivated offender, (b) the absence of capable guardians or protectors, and (c) the presence of a suitable victim.

ZUMA Press/Newscom

FIGURE 17 Violent crime has become endemic to many poor central city neighborhoods in the United States.

URBAN VIEW 3
High School Student Drug Abuse

Based on a study of death certificates, the Center for Disease Control determined that accidental drug overdose is now the leading cause of accidental death in the United States, ahead of alcohol-induced or firearm-related deaths. The Center attributes the rise partly to an increase in prescriptions for drugs such as Vicodin and OxyContin. In fact, the newest and most widespread development in substance abuse is the intentional abuse of prescription drugs. It is more common than heroin, cocaine, and ecstasy use, and is believed to have contributed to a 113 percent rise in teenage fatal drug overdoses.[29]

Since 1975, the Monitoring the Future (MTF) study of the U.S. National Institute on Drug Abuse has surveyed adolescent students about their drug, alcohol, and cigarette use and attitudes. Some of the negative consequences of substance abuse in adolescence include low academic outcomes, delinquency, risky sexual behavior, and even death.[30]

Daily use of marijuana has been increasing among students in their final year of high school (12th grade) and is now at its highest point since the early 1980s. This upward trend is expected to continue because the perceived risk of regular marijuana use by students has also declined. In 2010, 21.4 percent of high school seniors reported using marijuana during the past 30 days, while 19.2 percent said they had smoked cigarettes during the past month.

After marijuana, prescription and over the counter medications are the most commonly abused drugs by 12th grade students (Figure 18). Eight percent of these students reported the non-medical use of Vicodin, while 5.1 percent used OxyContin. Non-medical use of Adderall and over-the-counter cough and cold medicines remains high at 6.5 and 6.6 percent respectively.

Alcohol use has declined among high school seniors, falling to 41.2 percent of students reporting taking a drink in the past month in 2010 (with the use of flavored alcoholic beverages also declining, to 47.9 percent). Binge drinking (defined as five or more drinks in a row during the past two weeks) declined to 23.2 percent.

Among the most dramatic differences among students are among whites, African Americans, and Hispanics. African-American students have substantially lower rates of drug, alcohol, and cigarette use than white students. In fact, African-American students' cigarette smoking is dramatically lower than that of white students. Hispanic students have rates of use that fall between the other two groups, but are usually closer to the higher rates for white students compared to African-American students.

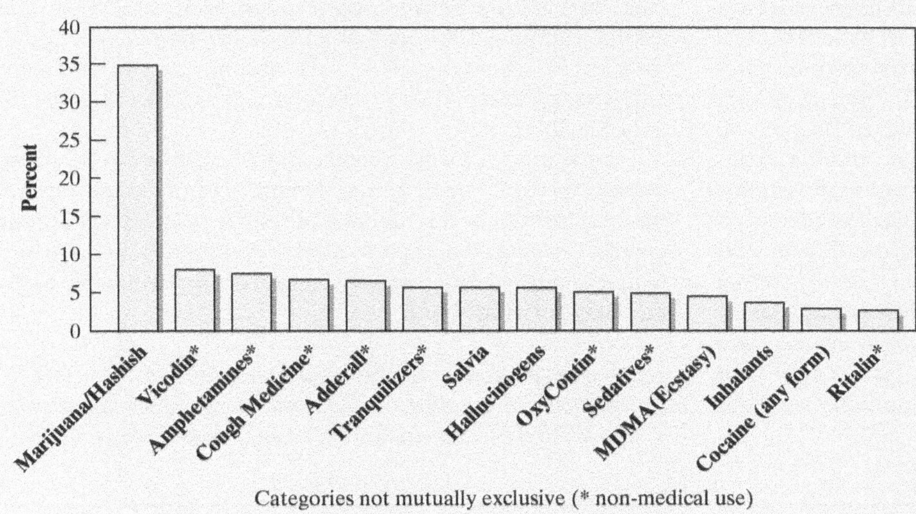

FIGURE 18 Drug abuse during the past year by U.S. high school students (12th grade), reported in 2010.

It follows that victimization rates vary from one community to another because the time-space routines of different socio-economic, demographic, ethnic, and lifestyle groups result in different degrees of proximity, *in space and time*, between potential offenders and potential victims. Research on such linkages, however, has shown that while they help to explain patterns of property-related crimes they do not explain patterns of criminal violence.

Spatial Patterns

Given this discussion, the spatial patterns associated with violent crime hold few surprises, whether the focus is on the geography of offenders' residences or the geography of crime locations. The classic pattern of *offenders' residences* was established long ago in the ecological studies of the Chicago School of the 1930s: a steady gradient, with low rates in the suburbs and a peak in the zone in transition that contains the worst slums. This basic pattern varies according to the nature of the offense and has been modified by the changes that have occurred in metropolitan structure since the 1930s. Nevertheless, the conclusions of a landmark study in 1960 of the geography of crime in Seattle hold true: High crime areas "are generally characterized by all or most of the following factors: low social cohesion, weak family life, low socio-economic status, physical deterioration, high rates of population mobility and personal disorganization."[31]

In broad terms, classic *occurrence patterns* are very similar. A large literature on the ecology of violent crime has not only established a strong correlation between low-income neighborhoods and crimes of violence but also points to the tendency for victims to be drawn overwhelmingly from the family, friends, neighbors, and local associates of criminals.[32] This literature also points to the importance of the microscale physical environment because of its influence on *opportunities* for crimes of violence. A report for the U.S. National Commission on the Causes and Prevention of Violence concluded that accessibility, visibility, control of property, residential density, and state of physical repair were the most significant aspects of the micro-environment of violent crime.[33]

It was factors such as these that led architect Oscar Newman to stress the importance of **defensible space**. Newman was the author of a damaging attack on Modernist housing that argued that much of the vandalism, burglary, and criminal violence of central city neighborhoods was related to the "designing out" of territorial definition and delineation in the Modernist apartment blocks that had come to dominate many central city neighborhoods. Once the space immediately outside the dwelling becomes public, Newman suggested, nobody feels inclined to supervise it or defend it against intruders.[34]

Although the overall trend is down, in recent years the patterns and intensity of offenses have seemed to many observers to have changed, generating increased concern about criminal violence as a key element in an increasingly dystopian scenario of central city life. Watching the news on TV and computer screens, it seemed to many people that criminal violence was on the increase because it had not only spilled out onto the streets but also had become a daytime as well as a nighttime activity. At the same time, the menace of violence had extended beyond the slums. In some cases, delinquency had escalated from vandalism and larceny to senseless violence: beatings and shootings triggered because an innocent passer-by has the wrong look or is wearing the wrong colors or because a driver in the next lane at traffic lights is playing the wrong music. Fists, knives, and handguns have been augmented by semiautomatic weapons and sniper rifles. The city's most fearsome threat is now that of becoming embroiled in "random" violence: muggings, drive-by shootings, gang rapes. As a result, the city's slums and poor neighborhoods are not just excluded from the mainstream life of the city, they are carefully avoided by all outsiders except those whose work takes them there. This despite the fact that the murder, aggravated assault, forcible rape, and robbery rates in U.S. cities have all decreased since the early 1990s (Figures 16 and 19).

As we would expect, violent crime differs considerably among individual cities (Figure 19). In 2009, for example, the incidence of aggravated assaults was about five times as high in Detroit and St. Louis as it was in San Diego and Honolulu; the incidence of rapes, however, was more than twice as high in Fort Worth and Jacksonville as in Washington, D.C. or Baltimore. The murder rate in Dallas in 2009 was more than twice as high as that in San Francisco. In 2009, New Orleans was the country's Murder Capital yet again. With 174 homicides in 2009 (down from 363 in 1995), the city had a homicide rate of 51.7 per 100,000 residents, compared to 40.3 and 40.2 per 100,000, respectively, in St. Louis and Detroit.

Within cities, the overall spatial pattern of criminal violence shows little change, despite the recently declining rates of violence. In Washington, D.C., for example, there is a very close association between patterns of homicide, rape, and assault and neighborhoods of poor blacks and of individuals with low levels of education.[35] The vast majority of the homicide victims in the District of Columbia are, like their assailants, black.[36]

In part, the criminal violence described here can be attributed to the same shifts in the dynamics of urbanization that have intensified occupational and spatial polarization among people, introduced new patterns of vulnerability and disadvantage, and created the "**dual city.**" In part, it is attributable to the escalation of drug abuse.

Drug abuse and addiction are a major burden to society. Estimates of the total overall costs of substance abuse in the United States—including health- and crime-related

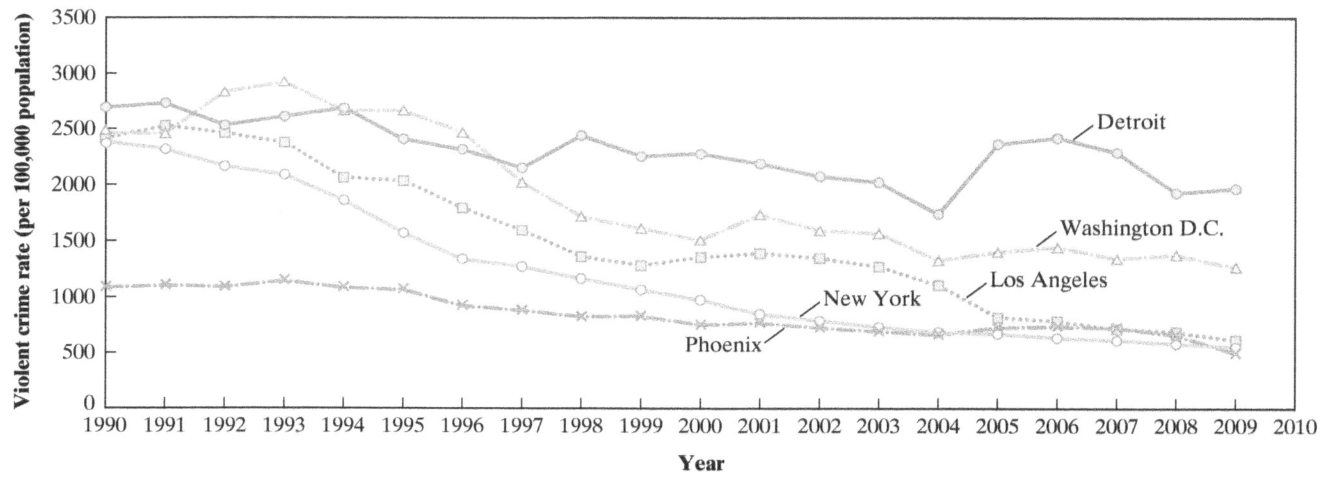

FIGURE 19 Violent crime rate (assault, rape, robbery, and homicide) per 100,000 people, selected U.S. cities.

costs as well as losses in productivity—exceed half a trillion dollars annually. This includes approximately $181 billion for illicit drugs, $168 billion for tobacco, and $185 billion for alcohol. Staggering as these numbers are, however, they do not fully describe the breadth of deleterious public health—and safety—implications, which include family disintegration, loss of employment, failure in school, domestic violence, child abuse, and other crimes.[37]

By the 1980s heroin addiction had become an endemic problem in the central parts of U.S. cities. When crack cocaine appeared in the mid-1980s, the problems multiplied and intensified. The link between homicide and drugs, in particular, became direct. Two-thirds of the homicide victims in the District of Columbia in recent years have in fact been found to have drugs in their systems, and a similar proportion of defendants in the District's Superior Court have been found to be drug users. Some of the homicides are the result of execution-style killings, casualties of the turf wars that have proliferated in the absence of organized crime and the presence of an oversupply of aggressive people seeking to escape poverty or pay for their own habit by dealing drugs.[38]

The Effects of Crime on Urbanization and Urban Life

Just as criminal violence can be seen partly to be produced and substantially mediated by processes of urbanization, it can also be seen to influence these processes. As such, it is a good example of the **sociospatial dialectic**, the continuous relationship in which urban space and urban society shape and reshape each other.

URBAN VIEW 4

Russian Mafia Crime and Corruption—Not Just in Russian Cities Anymore[39]

As the Soviet Union broke up, crime and corruption appeared in cities on an unprecedented scale. There was a proliferation of ethnic *mafiyas*—Russian, Chechens, Azeris, Georgians—and a rapid spread of corruption that has since become ingrained in the businessworld and the political system. Organized crime and widespread corruption have subverted democratic governance, affecting the nature of economic and social development in many cities.

In the institutional vacuum that followed the breakup of the Soviet Union, there was no accepted code of business behavior, no civil code, no effective banking or accounting systems, and no procedures for declaring bankruptcy. Security agencies were disorganized and bureaucratic lines of command were blurred. Before long almost all small private businesses were paying protection money to criminal groups. Canadian Doug Steele (of Moosehead fame), who used to own a Moscow nightclub called The Hungry Duck, reckons that he paid out more than US$1 million in payoffs to police, officials, and the mafia. Having survived one kidnapping attempt, he said: "You have to grease the palm or you won't be in business."[40]

Organized crime did not consist just of the traditional activities of prostitution, drugs, auto theft, and protection rackets; it also included the more lucrative business of illegal traffic in weapons, nuclear materials, oil, rare metals, other natural resources, and currency. Levels of criminal violence escalated in parallel with rising levels of crime and corruption, so that by as early as the mid-1990s contract killings had become a way of life in the business world of Russian cities.

One result of this has been that crime and corruption have taken away a significant proportion of resources that would otherwise have gone into restructuring the urban economies in Russia. This loss of investment for cities, combined with the uncertainties created by crime and corruption, has restricted the role of cities as engines of economic growth. Crime and corruption have also stifled the emergence of a civil society with a democratic base in cities. The scale of crime and corruption in Russia has also become an issue of geopolitical importance. Russian *mafiyas* have extensive connections to international organized crime; corrupt politicians and businesses have laundered so much money through the international financial system that the ramifications are truly global in scale. Global organized crime, particularly the Sicilian Mafia and the Colombian drug cartels, seized the chance to link up with Russian *mafiyas* to launder huge sums of money, to circulate counterfeit dollars by the millions, and to establish smuggling networks. The Russian *mafiyas,* in turn, seized the chance to extend their operations to the rest of the world. Today the scope of Russian *mafiya* organizations ranges from prostitution rings in major tourist hubs outside Bangkok, Thailand, to heroin trafficking between Afghanistan and major cities in Europe.

In the United States, Brooklyn's Brighton Beach neighborhood now has the nickname "Little Odessa" because it reminds its thousands of Russian immigrant residents of the Black Sea resort. But although it is home to many law-abiding Russian immigrants who came to the United States for a better life, Little Odessa is also home to the Russian Mafia. When most people imagine organized crime, they think of movies like *The Godfather* or *The Sopranos* that portray the Italian Mafia. But since the breakup of the Soviet Union, Russian organized crime has grown to the point where it is viewed as a greater threat by law enforcement officials than that posed by La Cosa Nostra. As Robert I. Friedman, author of *Red Mafiya: How the Russian Mob has Invaded America*, who had a $100,000 price put on his head by the Russian Mafia, put it "Italian organized crime in America is a pimple on a horse's ass compared with Russian organized crime in America—and globally."[41] Their smuggling operations are run out of a port in Newark, N.J. One of their clever scams to avoid paying taxes involved tanker loads of wood-grain alcohol shipped back to Russia. It was dyed blue and labeled windshield washer fluid in the U.S., and un-dyed and sold as vodka in Russia.[42]

Bob Daemmrich/The Image Works

FIGURE 20 Crime and fear of crime can have important effects on urban life. The large number of Crime Watch signs in U.S. residential neighborhoods of all kinds attests to the extent of the problem.

One of the best-documented effects of crime on urbanization concerns house values and rents.[43] A relatively small change in the incidence of crime—a shift of 4 to 5 percent—can affect aggregate house values in a large city by hundreds of millions of dollars (and affect property tax revenues by tens of millions of dollars, which impacts the city's ability to deliver services and maintain infrastructure). In fact, of course, these changes are highly localized, so that one of the main effects, in terms of the sociospatial dialectic of urbanization processes, is on the dynamics of neighborhood change.

Closely related to changes in house values are changes in people's housing and neighborhood satisfaction. Yet although high crime rates have been shown to produce significant increases in local residents' desire to move out, research has shown that actual residential mobility is not affected. Quite simply, the residents of high-crime neighborhoods cannot afford to move out—everywhere else is more expensive. Being trapped amid the prospect of crime and criminal violence reduces community quality of life and induces strong feelings of fear and powerlessness among people that become contagious. Sensationalist reporting and repressive styles of policing often amplify these feelings. Fear of crime consequently becomes a significant element in the politics and behavior of other neighborhoods (Figure 20), even of the whole city. Fear also leads to changes in the built environment, as targets are "hardened" (Figure 21), and in methods of policing and social control, as electronic surveillance becomes the norm.

Meanwhile, back in the high-crime areas, there often emerges a social cleavage between those involved with crime or drugs, or both and everyone else, usually a majority. This majority of people have to adapt their lifeworld to the presence of the people involved in crime and/or drugs. In practice, they

Steve McDonough/Corbis Images

(a)

Kim Karpeles/Alamy

(b)

FIGURE 21 Urban spaces have become "hardened" in response to increasing crime rates.

have to watch cars carefully to avoid being caught in a drive-by shooting, steer clear of the neighborhood whenever possible, lock themselves into their apartments when they are home, draw shades and blinds and watch TV well away from windows, and stuff towels around doors to keep out crack fumes from the stairwells.[44]

An additional form of social cleavage can emerge within the group of people who are involved in crime and drugs, if there are sufficient numbers. In addition to the hierarchies within and between gangs and the divisions between offenders whose activities are accorded different degrees of status, the type of drug preferred by users can provide the basis for individual patterns of social interaction. "[Heroin] junkies despise the 'pipeheads' whose entire daily existence is often spent getting high or looking for rocks. Crack smokers, meanwhile, consider themselves superior to heroin junkies, who litter the complex with needles"—and will shoot up anywhere, cooking their dope with rainwater from the ground and using the mirrors of parked cars to locate veins in necks or groins.[45]

At the scale of central cities, the effect of high crime rates is to inhibit in-migration by the upwardly mobile young professionals who might otherwise be likely to act as a catalyst for neighborhood upgrading. Similarly, investment in retail and service businesses is deterred and disinvestment accelerated, while local government services are cut back or eliminated. In short, high crime rates can accelerate and intensify the spiral of decay and reinforce the cycle of poverty.

URBAN VIEW 5

Terrorism and Cities[46]

Not only has terrorism taken on a global reach, it is also apparent that cities have become a preferred location for large-scale terrorist attacks. Even before the terrorist attacks on the World Trade Center and the Pentagon in September 2001, cities had become the central venues of terrorist attacks.

There are several reasons for this. First, cities—especially **world cities**—have considerable symbolic value. They are not only dense agglomerations of people and buildings but also symbols of national prestige and military, political, and financial power. A bomb in London's Underground or a poison gas release in a Tokyo metro arouses international alarm. This kind of event will be communicated instantly to a world audience. Second, the assets of cities—densely packed and with a large mix of industrial and commercial infrastructure—make them rich targets for terrorists. Third, cities have become nodes in vast international networks of communications. This is a reflection not only of their power, but also of their vulnerability. A well-placed explosion can produce enormous reverberations, paralyze a city, and spread fear and economic dislocation. Finally, word gets around more quickly and socialization proceeds more rapidly in high-density localities. These kinds of environments can be an abundant source of recruits for terrorist organizations.

The impacts of the 2001 terrorist attacks on New York City have been significant—and not limited to the death and destruction that was targeted on lower Manhattan on September 11. The cost of doing business in New York has gone up, even for workers and companies quite distant from the attack site. Some of these costs include[47]

- Greater security and counterterrorism spending.
- Higher insurance premiums.
- Corporate decisions dictated by concerns about terrorism rather than business efficiency.
- The emotional toll on workers caused by concerns about future attacks.

Central London has sought to reduce the real and perceived threat of terrorist attacks that were until recently almost exclusively associated with the Irish Republican cause. Physical and increasingly technological approaches to security have been adopted at increasingly expanded scales. In 1989 the prime minister, Margaret Thatcher, ordered iron security gates to be installed at the entrance to Downing Street as a means of controlling public access (Figure 22).

In 1993 the 30 entrances to the central financial zone of the City of London (the Square Mile) were reduced to 7, with road-checks manned by armed police. Restricted access was based primarily on funneling traffic through rows of plastic traffic cones. This was viewed as a more symbolic and technologically advanced approach to security that avoided the previous "barrier mentality" in cities like Baghdad or Belfast. Over time the spatial scale of the security cordon was increased to cover 75 percent of the Square Mile (Figure 23).[48]

AP Images

FIGURE 22 The iron security gates at the entrance to Downing Street in London.

(continued)

URBAN VIEW 5
Terrorism and Cities (*continued*)

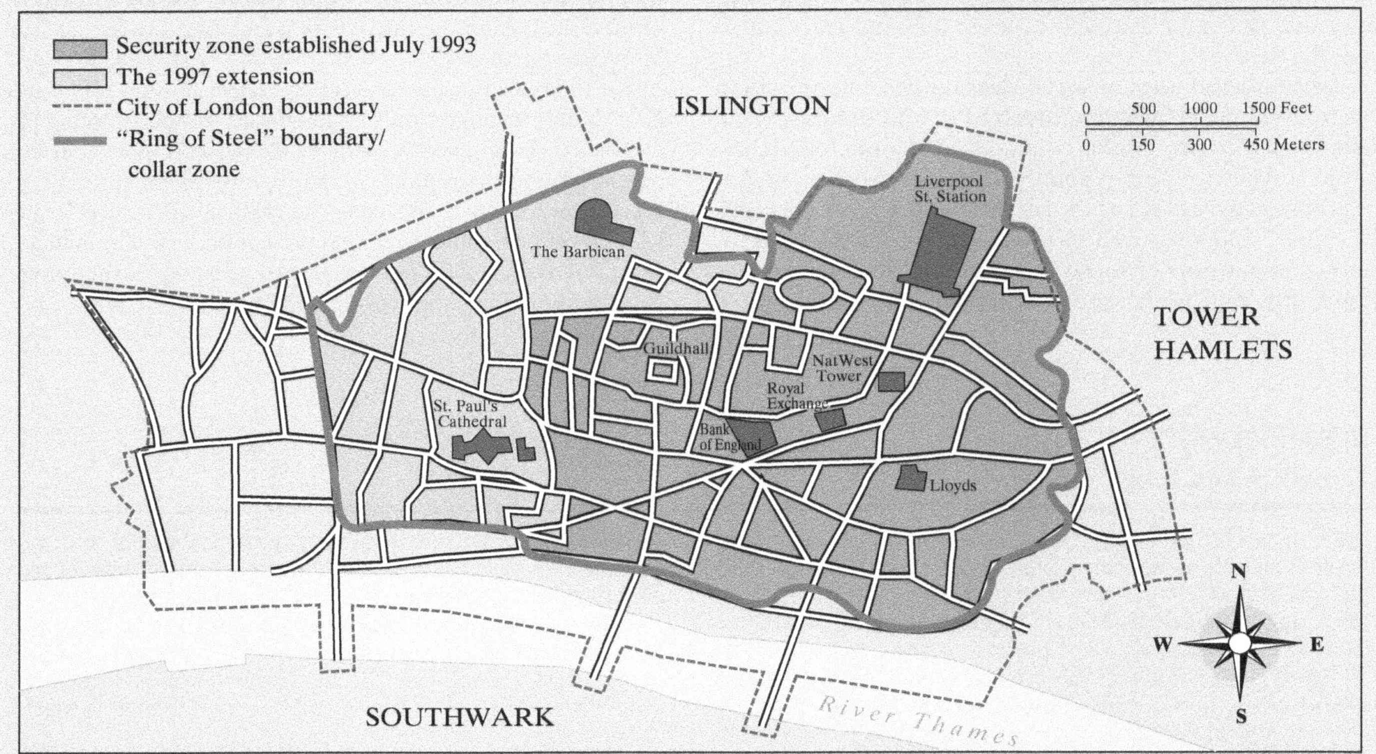

FIGURE 23 The security cordon covering the City of London.

"Rings of Steel, Rings of Concrete and Rings of Confidence: Designing out Terrorism in Central London pre and post September 11th," J Coaffee, *International Journal of Urban and Regional Research* 28(2004). Reprinted with permission.

The security cordon, as a territorial approach to security, was augmented by retrofitting the closed circuit TV (CCTV) system. The police, through its "CameraWatch" partnership effort, encouraged private companies to install CCTV. At the seven entrances to the security cordon, 24-hour Automated Number Plate Recording (ANPR) cameras, linked to police databases, were installed. Within a decade the City of London had been transformed into the most surveilled space in the United Kingdom, and perhaps in the world; it has more than 600 surveillance cameras in operation, many linked to the ANPR system.[49] With a population of only about 9,000, this represents nearly 70 cameras per 1,000 people.

HOMELESSNESS

"Homelessness" wrote sociologist Peter Rossi, "is most properly viewed as the most aggravated state of a more prevalent problem, *extreme poverty*."[50] There is a very fine line between someone being precariously housed and being literally homeless; and between short-term (episodic) homelessness and long-term (chronic) homelessness. These distinctions make for some slippery definitional problems. The most widely used definitions of homelessness involve chronic homelessness—not having customary and regular access to a conventional dwelling. They therefore include people sleeping in homeless shelters as well as those sleeping in doorways, bus stations, cars, tents, temporary shacks, and cardboard boxes and on park benches and steam grates. They do not include people living in **single-room occupancy (SRO) hotel** rooms, in lodgings, rooming houses, or the homes of relatives or friends. Nor do they include people living in backyard structures and taped-up garages without kitchen or toilet facilities.[51] Homelessness, then, is not simply the lack of a regular place to sleep and receive mail but also a condition of extreme poverty and a state of *disaffiliation*, without lasting or supporting ties to family, friends, and neighbors.

Like poverty, homelessness has always been a feature of U.S. cities. It was first perceived to be a serious problem during the Great Depression of the 1930s, when economic dislocation led to an accumulation of an estimated 200,000 hobos and tramps. The problem receded quickly during World War II and virtually disappeared in the affluence of the postwar era. But the number of homeless people began to rise sharply again with the economic and social polarization and dislocation resulting from the restructuring of the 1970s. By the mid-1980s, homelessness was once more beginning to be perceived as a serious problem.

Estimates of the number of homeless people vary a great deal. The National Law Center on Homelessness and Poverty

AP Images

FIGURE 24 Homeless men in Washington, D.C.

has estimated that about 3.5 million people, nearly 1.5 million of them children, are likely to experience a spell of homelessness at least once during a year.[52] The U.S. Department of Housing and Urban Development's annual estimate put the total number of homeless at 643,067 in 2009 (63 percent were in places like homeless shelters and 37 percent were on the street or in places not meant for human habitation). The HUD annual estimate of the number of homeless people has been criticized for undercounting the homeless population because of its methodology (it excludes people and families who are doubled up with other families or living in trailers or garages). What is not disputed is that homelessness increased significantly from the mid-1970s onward.

What has made the homelessness since the mid-1970s such a striking problem is not just the scale but also the nature of this homelessness. Whereas homelessness had previously involved mostly older white adult males, with relatively few of them actually sleeping outdoors, the "new" homeless people are of mixed race, include significant numbers of women and children, and are less likely to be able to find shelter indoors (Figures 24 and 25).

There are several characteristics of the "new" homeless:[53]

1. They are worse off than the homeless in the past in terms of employment and income. Compared with the homeless in a benchmark study of 1958, the "new" homeless

Rafael Ben-Ari/NewsCom

FIGURE 25 The new homelessness. One of many homeless women in New York City, with the Statue of Liberty in the distance.

are less likely to have a steady job (20 percent now compared with 28 percent in 1958); in addition, the real (inflation-adjusted) value of the minimum wage is now far less than it was when it was introduced in the 1970s.

2. They are worse off in terms of shelter. Whereas, for example, only about 100 people slept out on the streets of Chicago in 1958, of the 5,124 homeless people in 2009, 884 slept on the streets, while 1,691 were served in Chicago homeless shelters.[54]
3. More of them are families. Families with children comprise about 40 percent of the homeless population and are the fastest growing segment. It has been estimated that almost 1.5 million children experience homelessness at least once during a year.
4. More of them are women. Using the 1958 benchmark again, the proportion of women among the homeless population has gone up from 2 or 3 percent to more than 36 percent. Most homeless families are now headed by single women.[55]
5. They are younger, the median age being the early years of adulthood rather than middle age. In addition, there are proportionally fewer elderly among the homeless now (although their absolute number has grown during the past few decades).
6. More of them are African American. Although the figures vary from city to city, African Americans now account for almost 40 percent of the homeless, compared with less than 25 percent in the 1950s.

Another important change is that homelessness has spilled out from its traditional "Skid Row" settings. The geography of homelessness is in fact complex, reflecting several aspects of change in metropolitan structure and dynamics. One of the most significant of these was the emergence of **service-dependent neighborhoods** in the central city as welfare dependency came to be increasingly prevalent as slum and very poor neighborhoods increasingly became the location for the services offered by an expanding array of public and voluntary welfare agencies—institutions that in turn attracted the service-dependent populations of the poor, the disadvantaged, and the homeless.[56] At the same time, new, flexible **land use zoning** schemes made it possible for shelters and other services for homeless and impoverished people to be created in a wider range of central city locations.

Meanwhile, urban renewal projects were systematically targeting traditional skid row settings. Another important factor was the decriminalization of public drunkenness and vagrancy. Swamped by increasing numbers of cases and with more pressing crimes to deal with, police no longer warned the homeless away from public places or picked them up for sleeping on the streets. As a result, homelessness became more widely suffused through urban space and, consequently, much more visible. This visibility, in turn, led to more repressive reactions: the use of urban design to discourage the homeless from entering and (where such subtlety might not be sufficient) the installation of convex "bum-proof" benches and of sprinkler systems programmed to switch on at random times through the night, and the deployment of private security forces. Meanwhile, the dynamics of central city revitalization began to encroach on the central city spaces that had previously been the preserve of welfare-dependent populations and welfare agencies. More of the homeless were pushed out of the central city altogether into shelters that were opened in converted properties in the first-tier and second-tier suburbs and into the parks and public spaces of edge cities.

The Causes of Homelessness

An immediate cause of homelessness is the excess demand for the very cheapest accommodations. People who do not have enough money to compete effectively at the bottom end of **housing submarkets**, and who do not have friends or relatives with enough room to offer them permanent accommodation, become homeless. There are, however, many reasons why people become homeless, as Figure 26 shows. In broad terms, the dramatic increase in homelessness beginning in the 1970s was the product of interaction between four sets of trends: economic restructuring, sociodemographic change, metropolitan restructuring, and policy changes. The nature of these trends is already familiar: They are central to the evolutionary adaptation, restlessness, and cyclical changes that have been recurring themes of this book. Some of the major factors that have contributed to the increase in homelessness are surveyed here (Figure 26).

The occupational shifts and new **class factions** generated by economic restructuring and technological changes have contributed to a *sociospatial fragmentation* that has resulted in, among other things, **gentrification**. Gentrification, in turn, has removed substantial amounts of moderately low-cost housing from the market, displacing many households into lower-cost segments of the market, ultimately putting pressure on the lowest-cost housing.

The *income polarization* associated with metropolitan restructuring and changing labor markets, as we have seen in this chapter, has contributed to increases in poverty and economic marginalization, leading directly to increased competition by people for cheap accommodation. In 1975, the least affluent 20 percent of U.S. households accounted for only 4.3 percent of total (*pre*-tax) household income; but by 2009, the inflation-adjusted percentage had fallen to 3.4. The average income of the bottom 20 percent of households in 2009 was only $11,552, up but not by a great deal from the 1975 average of $10,545 in constant (2009) dollars[57]. The slow income growth trend of lower-income households was compounded by the effects of policy changes that were themselves a product of a political climate associated with the same economic shifts. The **new conservatism** of the Reagan and first Bush presidencies featured two policy shifts that were particularly important in this respect:

1. *Tax Changes.* Designed to cater to middle- and upper-income voters and to stimulate economic growth through increased materialism, tax changes resulted in a highly regressive shift in the burden of taxation away

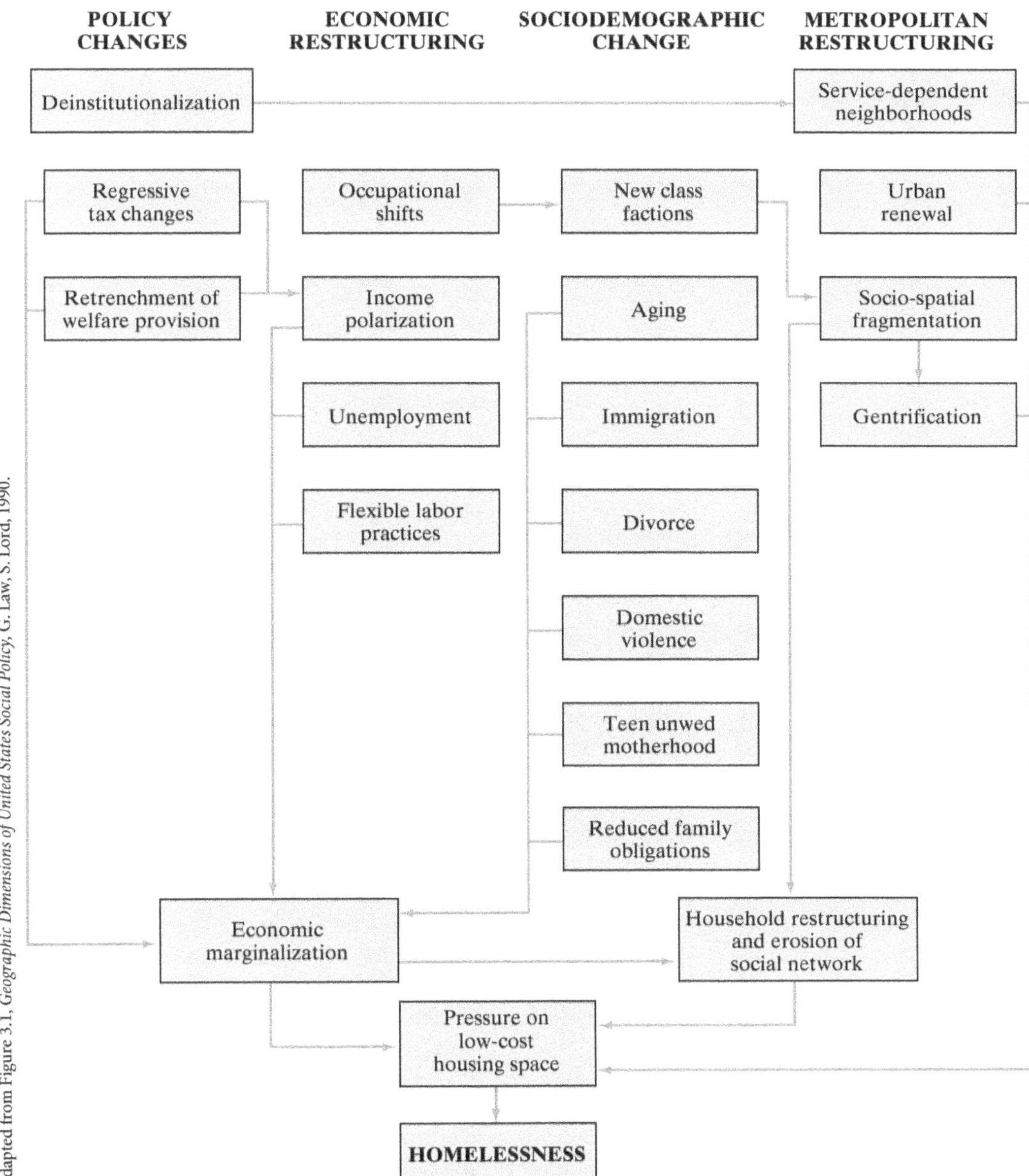

Adapted from Figure 3.1, *Geographic Dimensions of United States Social Policy*, G. Law, S. Lord, 1990.

FIGURE 26 The causes of homelessness.

from wealthier people and households. Between 1979 and 2006 the average *after*-tax income of the poorest 20 percent of households increased by only 10.7 percent (compared to 86.5 percent for the richest 20 percent). During that time, the share of after-tax income of the poorest 20 percent of households fell from about 6.8 to 4.7 percent (while that of the richest 20 percent rose from about 42.4 to 52.1 percent).

2. *Cutbacks on Public Spending for Welfare Services.* When this retrenchment began in the late 1970s, discretionary programs (including most housing programs) bore the brunt of the cutbacks. In 1982 $1 billion was cut from child nutrition programs and $1.7 billion was removed from the food stamp program. Public housing programs came to a virtual standstill. More than 150,000 people (many of them with psychiatric disabilities) were removed from Social Security Disability Insurance rolls as a result of more stringent interpretations of eligibility. As retrenchment gathered momentum during the 1980s, the erosion of income support and housing programs for the poor became so pronounced that the period came to be known in policy circles as the War on Welfare. The average monthly value of Aid to Families with Dependent Children (AFDC) and Temporary Assistance for Needy Families (TANF) benefits declined by more than a third, in constant dollars, between 1977 and 2006 (Figure 27).

Structural unemployment, which rose significantly in the 1970s as a result of economic recession and restructuring, had

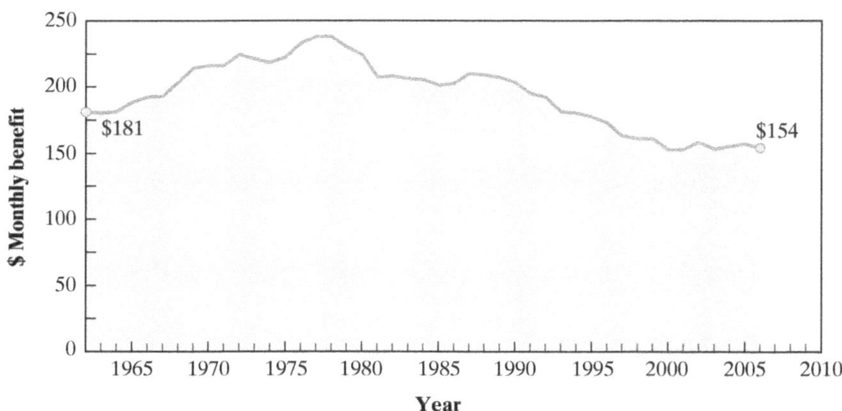

FIGURE 27 Average monthly Aid to Families with Dependent Children (AFDC) and Temporary Assistance for Needy Families (TANF) benefits per recipient in constant U.S. dollars.

a differential impact, affecting young males more than others. Whereas the unemployment rate in 1968 for men under 35 years of age was less than 5 percent, it rose to a high of 15 percent in 1980, declining to 13 percent in 1984 and fluctuating between 12 and 14 percent to end at a high of almost 15 percent in 2010. For younger black males, unemployment reached disastrous proportions: 40 percent among black males aged under 25 in 1985, declining to almost 34 percent in 2010. These figures, of course, do not reflect potential workers who had given up looking for work and had become disengaged from the labor market.[58]

Flexible labor strategies (involving temporary, part-time, and short-term contract jobs) that emerged with the shift away from **Fordist** approaches segmented the labor force in new ways, creating a large "secondary" labor market in which many workers had low wages, little or no job security, and no healthcare or retirement benefits. This secondary labor market added to the number of economically marginal households, creating still further pressure on the limited amount of low-cost housing space.

A major policy shift in the care of mentally disabled and impoverished people resulted in the **deinstitutionalization** of many individuals. The traditional settings of care, such as psychiatric hospitals and orphanages, had come into question by the 1950s because of the shortcomings of crowded and antiquated institutions, an increasing conviction that prolonged institutionalization often did more harm than good, and an increasing sensitivity to the civil rights of impoverished and mentally disabled people. With the expansion of spending on welfare in the 1960s, federal funds were made available to encourage the transition from institutionalized care to grassroots care that typically involved a combination of independent living and local service delivery points. More recently, some people with HIV/AIDS have lost their jobs due to discrimination, fatigue, or periodic hospitalization caused by HIV-related illnesses; they have also found their incomes drained by healthcare costs. They and others can be denied services or be prematurely discharged from healthcare facilities due to managed care arrangements. As a result, large numbers of vulnerable, disabled, and low-income individuals gravitated to the service-dependent neighborhoods that had developed in central city areas, adding more pressure to the localized submarkets of low-cost housing.

Urban renewal has eroded the stock of older, low-cost housing, partly in response to public policies and partly in response to changing patterns of land values and **ground rents**. Almost all the cubicle hotels that had characterized the skid rows of the 1950s and 1960s had been demolished by 1980. Many single-room occupancy (SRO) hotels also disappeared—about half the total stock of SRO units was destroyed in the 1970s. Between 1970 and the mid-1980s, an estimated 1 million SRO units were demolished.[59] Of New York City's 129,000 $200-per-month or less SRO units in 1960, 113,000 units—87 percent—were gone by 1983 (and since then SROs have continued to be refurbished and converted to higher income accommodation); in Chicago, more than 19,000 SRO units—70 percent of the stock—have been lost since 1973;[60] and in Los Angeles, more than half of the downtown SRO units disappeared between 1970 and 1985.[61] As a direct result, the price of the remaining SRO housing has risen significantly.

Sociodemographic changes contributed to pressure on this diminishing pool of low-cost housing through the increased numbers of vulnerable households seeking shelter as a result of several interconnected trends. Dramatic increases in divorce rates and the number of teenage unwed mothers led to the feminization of poverty. Meanwhile, the combination of economic hardship and an increasing gap in lifestyles and norms of behavior between different generations led to the evolution of new limits on family obligations. In short, fewer families were both able and willing to shelter and care for children and grandchildren who had fallen on hard times. The resulting social dislocation and economic marginalization was compounded by the *sociospatial fragmentation* that, in the process of restructuring urban space, inevitably eroded some of the long-established social networks of traditional low-income neighborhoods. And despite a recent decline, domestic violence has also been identified as an important factor in homelessness among women.[62]

Finally, two other sociodemographic trends added to the number of low-income households and, therefore, intensified the pressure on low-cost housing. One is the *aging* of the

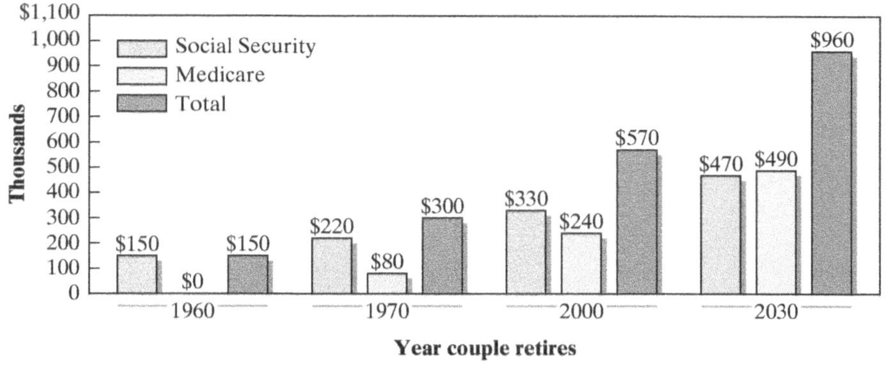

"Lifetime Social Security and Medicare Benefits, Straight Talk on Social Security and Retirement Policy," E. Steuerle, A. Carasso, *The Urban Institute* 36:2003. Reprinted with permission.

FIGURE 28 Growth in U.S. Social Security and Medicare lifetime benefits for an average-wage/low-wage two-earner couple (discounted to present value at age 65).

population, which has also contributed to an increase in the number of single-person households. The effects of this trend have been mitigated, however, by the fact that welfare retrenchment did not as adversely affect the elderly (largely because of their weight of influence at the ballot box). In fact, the constant-dollar value of Social Security and Medicare benefits increased by 280 percent between 1960 and 2000, thanks to increases in benefit levels and the indexing of benefits to the Consumer Price Index; the current projection is for an increase of another 68 percent by 2030 (Figure 28). In contrast, increased *immigration* brought large numbers of households, many of them unaware of their welfare rights, who, as a group, often faced discrimination in both the welfare system and the labor market.

INFRASTRUCTURE AND ENVIRONMENTAL PROBLEMS

For many people the stereotypical image of urbanization in the older industrialized national economies includes congested traffic, smoke-belching factories, and a striking lack of vegetation and wildlife. Implicit in these images are an overloaded infrastructure and an abused and degraded environment, all part of the "necessary evil" that comes with cities' primary roles as economic production and distribution points. Thanks to improved public health, urban planning, more rigorous environmental regulation, and sustained investment in public works, significant advances have been made in managing urban environmental quality. Nevertheless, cities are more than ever like monstrous organisms that consume vast quantities of oxygen, water and organic matter, fossil fuels, and foodstuffs and that release huge quantities of waste products into the air, into rivers, lakes, and oceans, and onto the land.

In 2009 in the United States alone, more than 3 billion pounds of toxic chemicals were released into the air, water, and land.[63] The spatial pattern of these incidents reflects the division of labor within the urban system, with some of the heaviest releases occurring in counties in the older, well-established industrial centers of the eastern United States and with concentrations in the "chemical corridor" of the Northeast and in the petrochemical complexes surrounding the Gulf Coast (Figure 29).

Although urban environments are now highly regulated and although sophisticated transportation and waste disposal systems have been developed to allow them to continue to function, they are, as physical geographer Ian Douglas notes, *dependent systems*. They are dependent on (1) continuous expenditure and capital investment; (2) efficient governance; and (3) planning, management, and monitoring. Ultimately, therefore, they are dependent on finance, both public and private, in order to sustain their life-support functions.[64] In this context many of the infrastructural and environmental problems of cities can be seen as the result of investment and operating expenditures that:

- are persistently insufficient.
- significantly lag behind urban growth.
- decrease faster than city population.

It is also important to recognize the functional interdependence between the economic and environmental dynamics of cities and their relationships to the need for services and infrastructure by the people living in them:

> The material demands of the urban population are the cause of profound changes in the physical character of the city and its surroundings. Modification of the natural flows of energy, water and materials goes hand in hand with the creation and maintenance of artificial flows of the same commodities. The character of the land surface changes, affecting the radiation balance, the rainfall-runoff relationship, sources and supplies of sediments and solutes, infiltration and groundwater levels, soil chemistry and plant and animal habitats. Such changes are both continuous and episodic. The impacts become apparent irregularly, often in extreme or unusual climatic conditions such as floods caused by exceptional rains whose height and duration is affected by the large proportion of impermeable roofed and paved surfaces in the urban catchment area. Similarly, temperature inversions can trap the gaseous pollutants and particulates emitted by urban chimneys and exhausts, causing smog. Both floods and smog lead to accidents, damage, injury and even loss of life, which in turn place demands on urban services.[65]

It is not possible to do full justice here to this kind of complexity. It is possible, however, to illustrate the functional

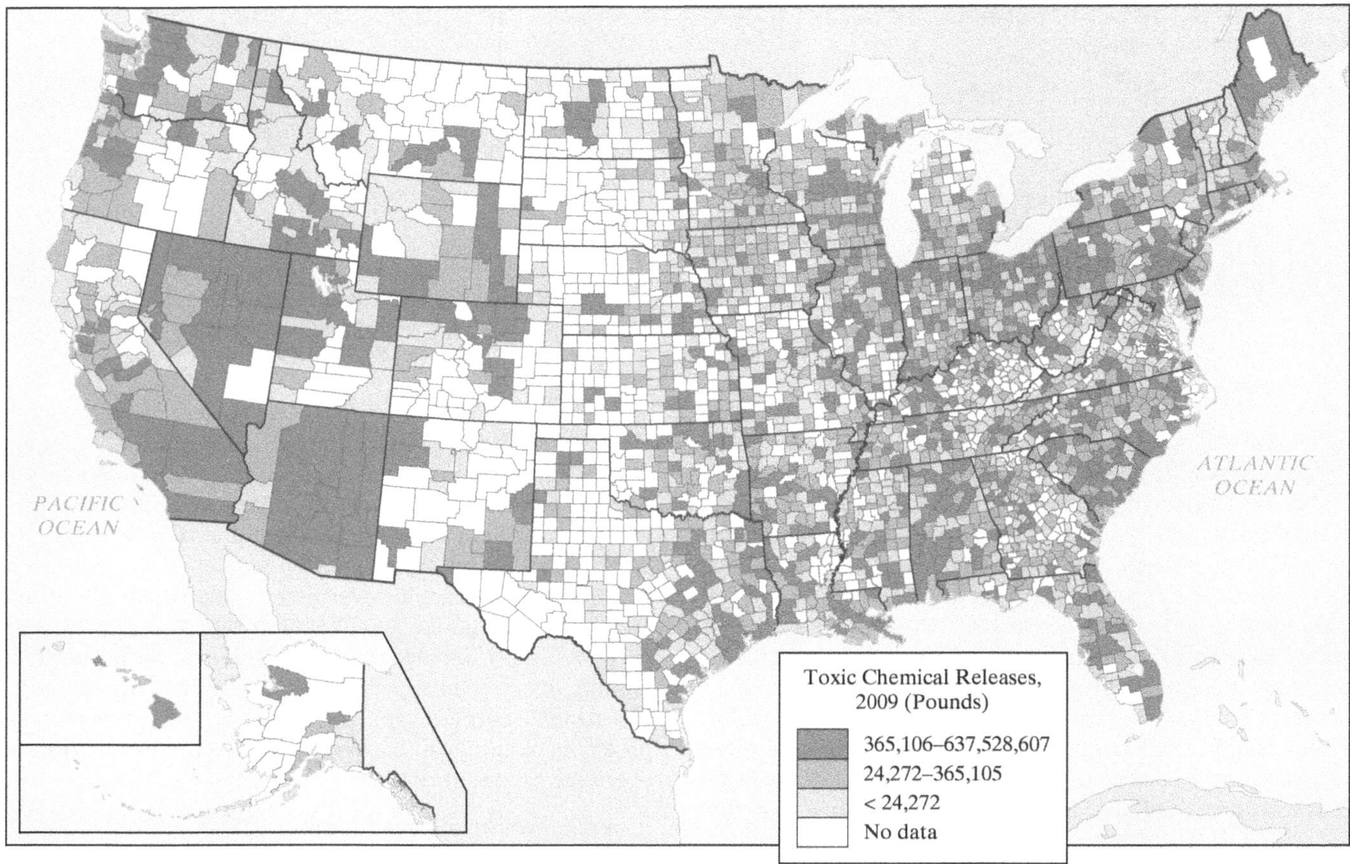

FIGURE 29 The U.S. hazardscape, as measured by on- and off-site toxic chemical releases (in pounds) into the air, water, and land, by facilities in all industries in 2009. "High" = the 33 percent of counties with releases between 365,106 and 637,528,607 pounds; "medium" = the 34 percent of counties with releases between 24,272 and 365,105 pounds; "low" = the 33 percent of counties with releases less than 24,272 pounds.

interdependence between economic and environmental change and the relationship of both to infrastructure and service needs. The following sections attempt to do this by pointing to three important issues for cities and the people who live in them: problems of water supply, problems of air pollution, and the threatening crisis in urban infrastructure.

Water Supply Problems

There are two main dimensions to the problem of urban water supply for people: quantity and quality. The people in cities consume vast amounts of water. More important, urban water consumption has increased dramatically. Public-supply withdrawals in the United States (that are supplied by public and private freshwater suppliers) were more than 44 billion gallons in 2005, compared to only 14 billion gallons in 1950. Public-supply water is supplied for domestic, commercial, industrial, or thermoelectric power uses. About 258 million people depend on water from public suppliers (with about another 43 million people self-supplied—usually from a well).[66]

Not only do metropolitan areas in the United States accommodate more people than ever, but people's consumption of water has risen in parallel with increased consumerism. The increased prevalence of washing machines, dishwashers, showers and whirlpool tubs, car washes, and suburban lawns has helped to push consumption to unprecedented levels. A typical family of four uses about 400 gallons of water *each* day. About 27 percent of indoor water consumption is used to flush toilets, 17 percent is used in showers and bathtubs, 22 percent is used by washing machines, 16 percent for faucet use (such as for cooking and drinking), and 14 percent is lost to leaks (and the rest is used for other purposes). Despite the Energy Policy Act of 1992 requirement that all new home-use toilets be "low-flow" (using 1.6 gallons per flush instead of the 3.5 to 5 gallons in traditional toilets), the United States still uses about three times more fresh water per person than countries such as Germany, France, China, and India.[67]

Consumption on this scale has become a problem in many U.S. cities. Inland cities that depend heavily on groundwater have suffered from serious *subsidence*. The withdrawal of large quantities of groundwater lowers the water table and reduces the water held in voids between mineral particles. The sheer weight of urban development causes these particles to become more closely compacted, leading to subsidence. Coastal cities that depend heavily on groundwater have suffered from a different problem: *seawater intrusion* that ruins the quality of groundwater supplies. Seawater intrusion has been a problem for many cities along the East and Gulf coasts and a severe problem for some cities in Florida.

In some other regions, because groundwater supplies are insufficient to support urbanization, transfers of water from other water basins have had to be arranged. Los Angeles, for example, had to tap the Owens River watershed, east of the Sierra Nevada Mountains, as long ago as 1913. Less than 20 years later this source had to be supplemented by an aqueduct that tapped into the Colorado River (Figure 30). Federal subsidies for such transfers have been reduced, however, while many politicians, businesses, and residents in states such as Nevada and Arizona, aware that their own economic growth depends on urbanization that requires water supplies, are reluctant to sell their state's water rights. The growth of municipal and industrial demands for water has led to conflicts over the distribution of water rights. City planners now see water resources as a major constraint to growth and increased economic development, especially in the arid regions of the West and Southwest.[68] Finally, these fundamental problems of supply are compounded by infrastructure problems; leaky water mains, for example, afflict most cities. The U.S. Geological Survey estimates the water lost from water distribution systems at 1.7 trillion gallons each year, at a national cost of $2.6 billion annually.[69] Detroit, Pittsburgh, and Philadelphia, for example, lose an estimated 25 to 30 percent of the water they treat each year, costing millions of dollars for water that never reaches homes and businesses.[70] The U.S. Environmental Protection Agency has estimated that the investment in the drinking water systems in the United States would need to be $334.8 billion over the next two decades, with most of the money needed for improvements to the transmission and distribution systems in order to maintain desired levels of service for customers, meet standards for water quality, and maintain and replace their assets cost-effectively.[71] Concerns over water *quality* are important not only because of potential public health hazards, potential employment losses in industries that require sound water, and potential hazards to wildlife and vegetation, but also because of potential reductions in the quantity of available supplies. Beginning in the 1970s, every U.S. state has had reported cases of groundwater contamination. According to the National Research Council, initial remediation of the 300,000 contaminated groundwater sites in the United States will cost as much as $1 trillion over the next few decades.[72]

Jim West/ZUMA Press/Newscom

FIGURE 30 The California Aqueduct, a federally subsidized lifeline for Los Angeles.

The 28 waterborne disease outbreaks associated with drinking water that the Centers for Disease Control and Prevention tracked in 2005–06 (excluding the 78 outbreaks in recreational situations such as water parks and public swimming pools) caused illness among more than 600 people and were linked to four deaths. These outbreaks were associated with bacteria, viruses, or parasites and caused acute respiratory illness, acute gastrointestinal illness, or hepatitis. The worst outbreak in the United States was in Milwaukee back in April 1993, when 403,000 people, about half the city's population, became ill from an outbreak of the intestinal parasite *Cryptosporidium*. In healthy people this parasite—which is excreted by cattle or other animals and washed into lakes and reservoirs, easily finding its way through obsolescent water purification systems—causes stomach upsets and diarrhea. It can be dangerous for people with weak immune systems: 104 died in Milwaukee's outbreak.[73]

In addition to coliform bacteria from septic tanks and sewage, chlorides, oil, phosphates, and nondegradable toxic chemicals from industrial wastewater, dissolved salts and chemicals from highway deicing, and nitrates and ammonia from fertilizers and sewage are all regular threats to city water supplies. One reason for the increasing scale of the problem is that many older local water treatment systems were designed to cope with bacteria rather than chemicals and toxic substances. Another is that in hundreds of cities like Milwaukee the sewer system was engineered in combination with the storm drainage system. Treatment plants attached to such systems are easily overloaded after heavy rain, with the result that plant managers open floodgates as many as 75,000 times each year, releasing a yearly three to 10 billion gallons of storm water overflows from combined sewer systems that carry untreated sewage into rivers and lakes.[74]

A third reason is the breakdown of treatment plants that has become increasingly common as urban growth has outstripped investment in maintenance and construction. In 2008, the U.S. Environmental Protection Agency estimated the total investment needed in publicly owned treatment plants at $298.1 billion.[75]

URBAN VIEW 6

Brownfield Redevelopment[76]

On a once-contaminated 28-acre property in Little Rock, Ark., sits the headquarters of Heifer International, a nonprofit organization that works with communities around the world to help end hunger and poverty and to care for the environment by providing livestock, seeds, and training. Heifer completed the largest brownfield cleanup in Arkansas, which involved removing more than 75,000 cubic yards of contaminated soil. For over 100 years, the site had been used by Union Pacific Railroad as a rail yard. Chemicals and residues from the railroad ties and tracks contaminated the site. It was also contaminated when used by warehouse, light industry, and trucking companies. For more than 50 years, Superior Trucking contaminated the site in its vehicle maintenance and other industrial activities. Today, many "green" building features have been incorporated into the new headquarters building. An environmentally friendly parking plaza drains rainwater through bio-swale filters into a holding pond for use in landscaping and irrigation. A new wetland area helps control, store, and reuse surface water, filter storm water runoff, and provide new habitat for local animals and plants. Other green aspects included reusing crushed masonry from the disused buildings on the site, installing lighting fixtures that dim automatically as natural light increases, using recyclable fast growing bamboo for the building's flooring, and incorporating a four-story water tower to capture rainwater from the roof.[77]

Brownfield sites are abandoned or underutilized industrial and commercial facilities where real or perceived contamination complicates expansion or redevelopment. This definition, used by the U.S. Environmental Protection Agency (EPA) and Department of Housing and Urban Development (HUD), characterizes a tremendous number of properties as brownfields—because the severity of contamination is not specified and the environmental problems can be merely *suspected* as well as actually documented.

Many brownfields are found in central cities and industrial suburbs with a history of traditional manufacturing. These sites can be abandoned industrial and railroad facilities or manufacturing plants that are operating but show signs of pollution. Brownfields can also be small commercial or even residential lots with only suspected contamination. The most contaminated brownfields in the United States are the 1,300 or so Superfund sites on the National Priorities List. These sites contain waste with amounts of toxic chemicals, like lead or mercury, considered hazardous by the EPA. Most brownfields, though, have only low to medium levels of known or suspected contamination from ordinary waste like nonhazardous garbage.

Brownfield cleanup and redevelopment involve a wide range of citymakers. For landowners, speculators, developers, builders, real estate agents, and financiers, brownfields can offer opportunities for profit from a large underexploited source of land within established communities. The benefits for government agencies and households have focused traditionally on employment and tax base generation.

Brownfield redevelopment, however, presents a dual land-use policy challenge: reducing the barriers to private-sector redevelopment while connecting reuse to broader community goals. The first part of this challenge involves addressing the uncertainties and risks—involving cost and time—created for the private sector by four main issues: legal liability for contamination; lack of information about the level of contamination and uncertainties about cleanup standards; availability of funding for site investigation, cleanup, and redevelopment; and complicated regulatory requirements. The second part of this challenge involves connecting brownfield redevelopment and the reuse of these sites to wider community efforts—involving sustainable development and environmental justice—to achieve environmental and health protection, improved public safety, targeted jobs and training, central city revitalization, and reduced metropolitan sprawl.

During the last few decades, federal, state, and local governments in the United States have attempted to reduce the barriers to brownfield reuse for landowners, speculators, developers, and builders in a variety of ways, including legislative changes addressing landowner and lender legal liability, increased financial incentives, and efforts to improve inter-agency coordination at all levels of government. In comparison to the United States, the national governments in Europe typically are more involved in brownfield redevelopment because of their stronger role in urban policy and planning. In the United Kingdom, for example, the central government in 1998 set a target of 60 percent of all new housing to be built on previously developed land by 2008. The target was exceeded. Both brownfield and **greyfield sites** (abandoned or underutilized sites such as shopping malls with huge parking and paved surfaces but without the environmental issues of brownfields) were redeveloped. The government offered favorable tax treatment to encourage brownfield redevelopment while introducing strict planning policies to protect **greenfield sites** at the periphery of cities from uncontrolled development. In addition, despite the different national regulatory frameworks across Europe, the involvement of the European Union (EU) has facilitated a more coordinated approach to environmental policymaking in general and to brownfield redevelopment in particular as part of government efforts to address the problem of contaminated sites in the cities across Europe.

Air Pollution

Cities not only generate a great deal of air pollution but also create microclimates and change atmospheric conditions in ways that intensify the problem. The built environment, overall, has a lower *albedo* than natural earth surfaces. That is, it is less able to reflect incoming solar radiation and so absorbs more heat. In addition, the atmosphere in cities is warmed by the release of heat from fossil fuels and, in summer, from air conditioning units. The aggregate effect is for cities to become **heat islands**, with average temperatures one or two degrees Fahrenheit above those of the surrounding countryside. During heat waves, this heat island effect is particularly pronounced and can lead to average temperatures that are 10 to 15 degrees Fahrenheit above those of the surrounding countryside. Several successive days of such intensified heat can increase both physical and mental stress for urban residents. On such days mortality increases among sick and elderly people, tempers fray, and tensions rise.

The heat island effect also leads to a distinctive pattern of local air circulation on days when there is no regional air movement. Light surface winds are drawn toward the city center, then rise above the city to descend slowly to the ground at the edge of the built-up area. This pattern is associated with the development of an **urban dust dome** in which fumes, soot, and chemicals are trapped in the air over the city. This kind of situation becomes particularly pronounced in circumstances where a temperature inversion develops and a lid of relatively warm air effectively flattens the dust dome, keeping pollution low over the city. When exposed to strong sunshine, such concentrations of pollution can be transformed through photochemical reactions into an orange-brown *smog* (Figure 31).

Improved air quality for much of the United States has resulted from the introduction of cleaner cars, industries, and consumer products. Since 1990, air quality has improved for six common air pollutants (ground-level ozone, particle pollution, lead, nitrogen dioxide, carbon monoxide, and sulfur dioxide). In fact, pollution released into the air in the United States has fallen to about 107 million tons each year.[78] Of this, about half is generated by transportation, just under a fifth is generated by stationary fuel combustion (in power plants, etc.), and just under one-tenth is generated by industrial processes. Of the common air pollutants, the main ones are

- *Sulfur oxides* that are chiefly the product of burning fossil fuels (mainly coal and oil). In addition to aggravating respiratory conditions, they combine with atmospheric water to form dilute sulfuric acid that is a major component of **acid rain**, which can kill trees and erode the surfaces of buildings, monuments, and statues.
- *Nitrogen oxides* that are the product of burning fossil fuels at high temperature (as in power plants and cars). Like sulfur oxides, these emissions aggravate respiratory tract problems and combine with atmospheric water to produce acid rain. In addition, nitrogen oxide reacts with oxygen to form nitrogen dioxide, a reddish-brown gas that reduces visibility.
- *Carbon monoxide.* Chiefly derived from automobile emissions (when carbon in fuel is not burned completely), this colorless and odorless gas is poisonous in high concentrations and even at lower concentrations can impair psychomotor functions and the cardiovascular system. Motor vehicle exhaust contributes almost 60 percent of all U.S. carbon monoxide emissions.[79]
- *Carbon dioxide.* Another by-product of burning fossil fuels, it has no directly harmful effects on plant or animal life but is now recognized as a serious threat because of the human role in enhancing the **greenhouse effect**. The possibility of global warming poses the specific problem for coastal cities of rising sea levels, which would increase the probability of flooding; it might also cause an intensification of the hydrological cycle, which would in turn exacerbate problems of water supply.
- *Photochemical oxidants*, such as ground-level ozone (the primary constituent of smog), that are produced by the effects of heat and sunlight on nitrogen oxides and volatile organic compounds (VOCs). The resultant smog is an irritant to eyes and the respiratory system and can kill some trees and ornamental plants. The 1990 U.S. Clean Air Act amendments require that reformulated gasoline containing oxygen to reduce harmful emissions of ozone be used in cities with the worst smog pollution. Reformulated gasoline is gasoline that is blended to burn cleaner and reduce smog-forming and toxic pollutants in the air.
- *Particulate matter,* an airborne mixture of solid particles and liquid droplets, that is formed by sulfur dioxide, nitrogen oxides, ammonia, VOCs, and direct particle

Mathew Richardson/Alamy

FIGURE 31 Urban air pollution: downtown Los Angeles skyline shrouded in smog.

emissions. The resultant smog (containing dust from roads or soot [black carbon] from combustion sources) can reduce visibility and be an irritant to the respiratory system. Airborne particles can also damage vegetation and cause damage to paints and building materials.

- *Fluorides*, mainly the product of industrial processes, cause eye irritation, respiratory inflammation, nosebleeds, and damage to trees and flowering plants.
- *Lead*, a heavy metal pollutant produced mainly by industrial processes, primarily metals processing, with the highest air concentrations usually found in the vicinity of smelters and battery manufacturers. It is a cumulative poison that can be inhaled, absorbed through the skin, or ingested from vegetables that have absorbed lead via air pollution. It has adverse effects on metabolism, blood, and kidney functions. Even at low doses, lead exposure is associated with damage to the nervous systems of fetuses and young children, resulting in learning deficits and lowered IQ. Thanks to the phaseout of leaded gasoline by the U.S. Environmental Protection Agency since the 1970s, culminating in the banning of lead in motor vehicle gasoline after 1995, lead pollution has decreased significantly. The 2009 average air quality concentration for lead was 93 percent lower than in 1980.

Finally, it is worth noting that air pollution is associated with definite spatial distributions, with significant gradients between one part of a city and another. In general, the highest pollution occurs near the largest concentrations of traffic and industry. What is less obvious but of much greater significance is that the residential settings with the highest levels of air pollution tend to be low-income neighborhoods and, in particular, poor neighborhoods and slums. Children are at greater risk because they are generally more active outdoors and their lungs are still developing. The elderly and people with heart or lung diseases are also more sensitive to some types of air pollution.[80]

Infrastructure Crisis

The urban infrastructure of roads, bridges, parking spaces, transit systems, communications systems, power lines, gas supplies, street lighting, water mains, sewers, and storm drains is crucial not only to economic efficiency and productivity but also to public health, safety, and the quality of life. In the early 1980s a series of highly publicized infrastructure failures (including the collapse in 1983 of the Mianus River Bridge on the Connecticut Turnpike that killed three motorists and the broken water main in the same year that interrupted power to businesses in New York's garment district for a week) underlined the importance of urban infrastructure and drew attention to some important shortcomings. In 1983, Pat Choate and Susan Walter published a systematic survey whose title—*America in Ruins*—conveyed the sense of crisis that had developed.[82]

The federal government, meanwhile, had created a National Infrastructure Advisory Committee, which submitted its final report[83] to the Joint Economic Committee of Congress in 1984. The report suggested that $1.trillion would have to be spent by the year 2000 to repair, maintain, and develop basic infrastructure (transportation, water, and sewer systems) simply to avoid major quality of life and productivity problems. The committee estimated that $714 billion would be available from federal and state revenues. This sum left a shortfall of $436 billion that seemed unlikely to be met because of the inherent nature of infrastructure provision: unglamorous and easily postponed because it rarely enjoys priority support from any politically significant group of people. Congress immediately established the National Council on Public Works Improvement to study the problems in more detail; its 1988 report, *Fragile Foundations: A Report on America's Public Works*, which concluded that the quality of the U.S.'s infrastructure was barely adequate to meet current requirements and inadequate to meet the demands of future economic development, stressed the importance of creating an agenda upon which Congress could act. By 2001, however, the American Society of Civil Engineers (ASCE) estimated

URBAN VIEW 7
High-Speed Rail in Europe

The increasing integration of Europe, following fundamental geographic principles, relies heavily on policies that increase accessibility and spatial interaction among cities. The idea is to link more remote cities like Lisbon, Seville, and Rome with cities like Paris and Brussels that are located at the heart of the European Union (EU) territory. The goal is to boost economic efficiency, reinforce the social and political cohesiveness of the EU, and in the process create up to 1 million new jobs.

The EU has a far-reaching plan for a series of trans-European networks that would link together Europe's patchwork of separate national road, rail, air, and water transportation systems into a single multi-nodal network. The €900-billion (US$1.3 trillion) budget includes €269 billion (US$376 billion) for investment in new and upgraded high-speed track (Figure 32).[81]

The relatively short distances between major cities in Europe makes it ideally suited for rail travel; Europe is less suited to air travel because of the high population densities and traffic congestion around airports. Even before the terrorist attacks of September 11, 2001, the delays caused by check-in times and accessibility to airport terminals meant that it was quicker to travel between many major European cities by rail than by air. Europe's high-speed rail network—in direct competition with the airlines—already connects a number of cities at the core of Europe. The PBKAL network connects Paris, Brussels, Köln (Cologne), Amsterdam, and London (via the so called Chunnel—Channel Tunnel).

Improved locomotive technologies and specially engineered tracks and rolling stock make it possible to achieve speeds of 250 kilometers (155 miles) per hour. New tilt-technology railway

URBAN VIEW 7
High-Speed Rail in Europe (*continued*)

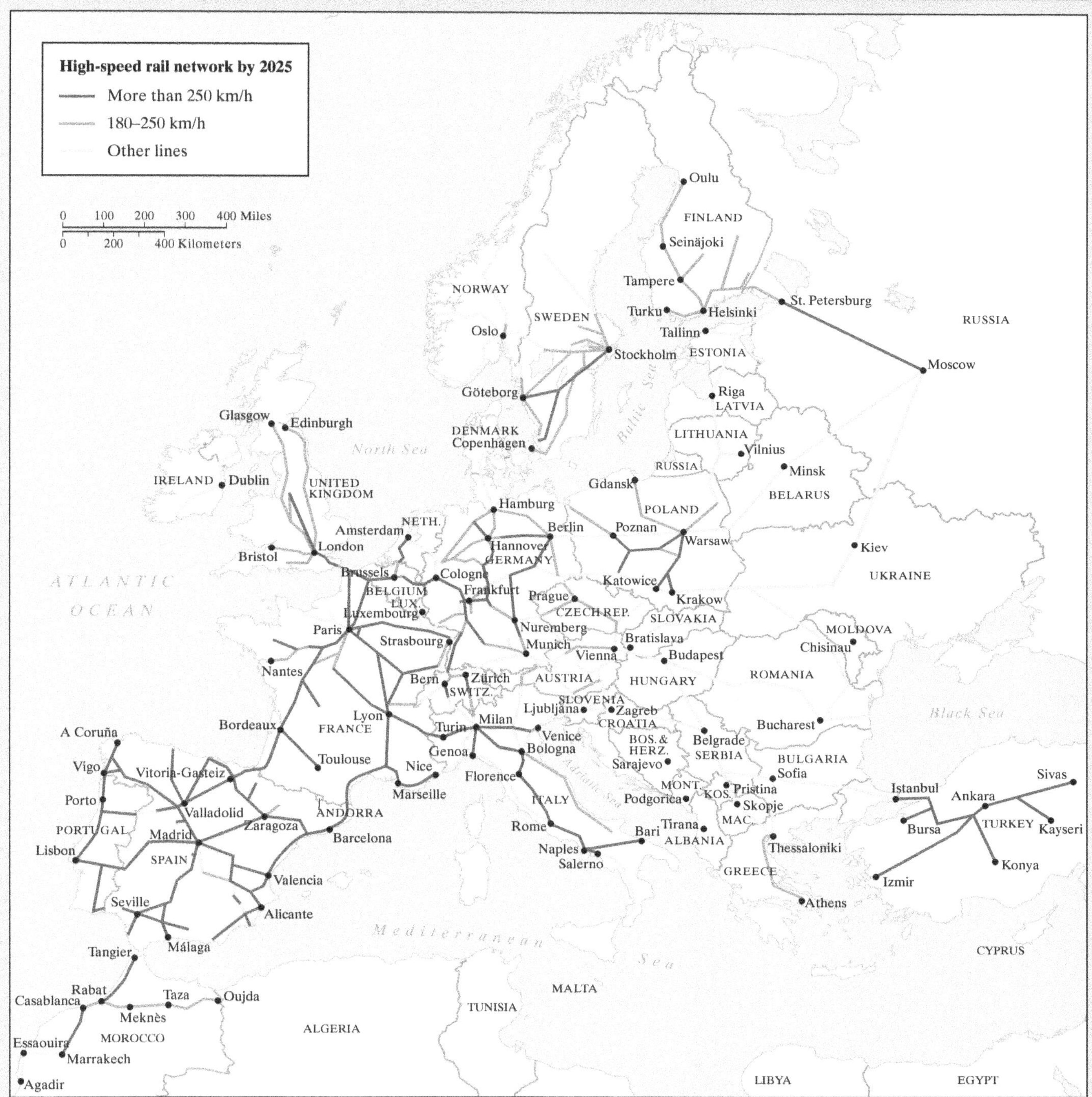

FIGURE 32 High-speed rail network by 2025 in Europe.

cars—designed to negotiate tight curves by tilting the train body into turns to counteract the effects of centrifugal force—are being introduced in many parts of Europe to maximize speeds on conventional rail tracks.

Europe's high-speed rail network is already leading to a restructuring of the urban hierarchy. Because the high-speed rail routes have only a few scheduled stops—to avoid the time penalties incurred by decelerating, stopping, and accelerating—the system has created winners and losers across the urban system. In relative terms, the cities with scheduled stops are now more accessible and so more attractive for economic development than those cities without scheduled stops.

FIGURE 33 Infrastructure collapse. In April 1992 the collapse of part of a wall in disused underground freight tunnels beneath Chicago's business district caused over 250 million gallons of river water to flood a large section of the downtown, causing $500 million in damage—all for want of a $30,000 repair.

that it would cost $1.3 trillion over five years to bring the country's infrastructure up to acceptable standards; its 2009 report found little improvement and upped this estimate to $2.2 trillion.[84] Deadly incidents, such as the Minneapolis bridge collapse in August 2007, add the human toll to the findings of these studies.

It is now very clear that the root of the infrastructure crisis is a combination of aging central cities and declining capital investment in the face of public budgetary crises, voter opposition to infrastructure projects, and—given the threat of possible terrorist attacks on critical infrastructure—the diversion of available public funding away from infrastructure maintenance and toward infrastructure security measures. The key infrastructural elements of most central cities were put in place more than a century ago and have long outlasted their designed life span (Figure 33). Most investment in new infrastructure, meanwhile, has taken place in new suburbs, boomburbs, and edge cities. Federally funded infrastructure investment increased significantly in the 1930s as a result of New Deal make-work programs, and federal, state, and local outlays on infrastructure increased in the affluent postwar period. During the period between the late 1950s and 2007, however, transportation and water infrastructure was neglected compared to other public expenditures and declined as a percentage of gross domestic product (GDP) (Figure 34). The impact of this trend on infrastructure provision has been compounded by inflation in the costs of building construction, which has consistently been higher than general economic inflation.

As a result, many cities, if not immediately facing the irretrievable decay and ruin of their infrastructure, face serious problems. For example, past neglect of the City of New York's public infrastructure created an identified need for nearly $5.6 billion in repairs for existing facilities in order to bring them up to a "state of good repair" as defined by engineers. The City's

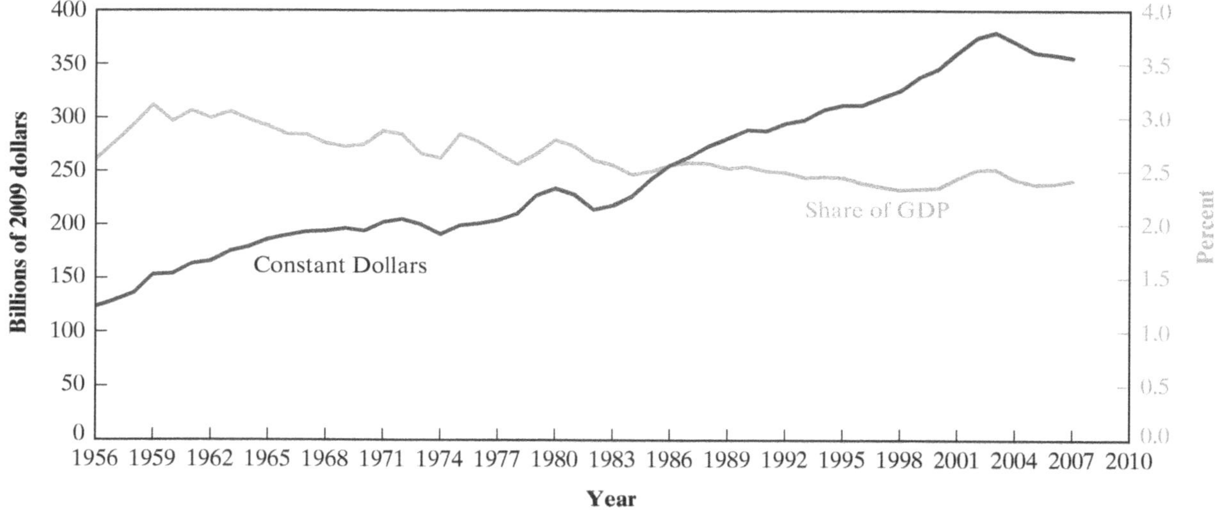

FIGURE 34 Total U.S. (federal, state, and local) public spending for transportation and water infrastructure (encompassing the facilities and systems that support transportation (e.g., highways and airports), provide water resources, supply drinking water, and treat wastewater).

Gabriel Boys/NewsCom

FIGURE 35 Gridlocked traffic on Highway 10 in Los Angeles, California.

capital budget for 2011 allocated just over half (53 percent) of this amount, with the shortfall especially large for streets (40 percent). Only 70 percent of the needed amount for ongoing maintenance was funded. The city is responsible for more than 6,000 miles of water mains, 600 miles of subway tracks, 5,700 linear miles of streets, 78 police precincts, 223 firehouses, 6 community colleges, 23 health facilities, and more than 1,000 school buildings (half of which are more than 50 years old).[85]

In New York City alone, up to 3,000 miles of water mains and 1,000 miles of sewers are at or near the end of their expected life spans. In the United States, there are 240,000 water main breaks each year, equivalent to about 650 every day or one every two minutes.[86] Across the United States, chronic leaks in pipes already consume between 10 and 20 percent of the water carried by many aging city supply systems, and more water mains crack or fail each year. By 2016 the U.S. Environmental Protection Agency predicts that more than 50 percent of the country's sewer pipes will be in poor or very poor condition or broken, up from only 8 percent of the estimated 600,000 miles of sewer lines that were in poor shape or worse in 2000.[87] In many cities the most visible and acute infrastructure problems are those associated with the shortcomings of roads and transit systems. Suburban gridlock became increasingly common during the 1970s as new car and truck registrations overtook highway capacity and outstripped highway construction (Figure 35). Freeway rush-hour traffic that was officially "congested"[88] was virtually unheard of during the 1960s. By 1975 it had reached 41 percent of freeway rush-hour traffic, and by 1985 it had reached 56 percent. By the beginning of the 1990s, rush-hour traffic on the likes of Interstate 25 southeast of Denver, the Dan Ryan Expressway in Chicago, the Ventura Freeway in Los Angeles, Route 101 south of San Francisco, and the Beltway (I-495) around Washington, D.C., was routinely stop-and-go, averaging about 12 mph. Between rush hours, traffic was merely "congested."

Based on analysis of the performance of the transportation systems in the U.S.'s metropolitan areas, traffic congestion remains a problem (despite reduced driving as a result of the Great Recession and slow economic recovery). The average annual delay per person in 2009 was 34 hours (up from 14 hours in 1982). There were 15 very large metropolitan areas with delays per person in excess of 50 hours, reflecting the very large delays in metropolitan areas with populations larger than 3 million (Figure 36). The most congested metropolitan areas were Chicago, Washington, and Los Angeles (with 70, 70, and 63 hours of delay each year respectively). In 2009 the cost of the extra time and fuel caused by congestion was $115 billion, based on travel delays of 4.8 billion hours and 3.9 billion gallons of wasted fuel (compared with $24 billion (in 2009 dollars) based on 1 billion hours of travel delays and 0.7 billion gallons of wasted fuel in 1982); meanwhile, the average cost per person was $808.[89]

Part of the reason for this congestion was of course the continual increase in car and truck registrations. This factor

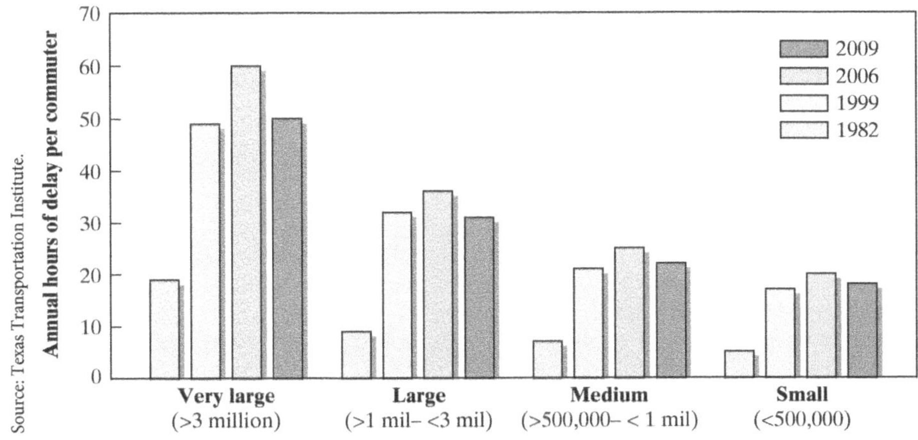

Source: Texas Transportation Institute.

FIGURE 36 Trends in traffic delays in the U.S.'s metropolitan areas.

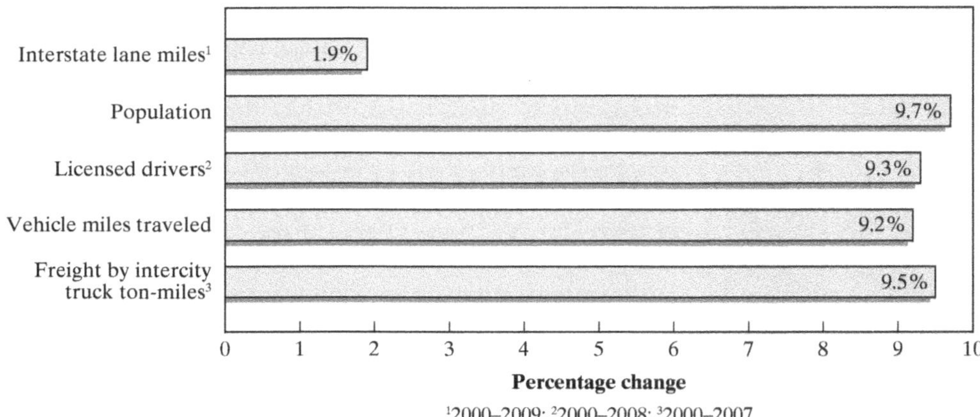

FIGURE 37 Variables contributing to U.S. traffic congestion, percent change, 2000–2010.

was compounded by the traffic that is crosstown or between suburban areas that has resulted from the decentralization of both jobs and homes since the 1950s. Since 1960 the share of work trips that began and ended in suburban locations has increased from less than 30 percent to more than 60 percent. It was also compounded by a slowdown in capital investment in highway infrastructure and by delays caused by maintenance and repairs to an increasingly degraded infrastructure. By the early 1980s more than 70 percent of the 125,000 miles of interstate highways within the boundaries of metropolitan areas were classified as being in only "fair" or "poor" condition. By 1990 almost all these highways were approaching the end of their design lives, with the resulting rehabilitation and improvements causing additional traffic delays and escalating expenditures.

Meanwhile, after the General Accounting Office criticized the state of the interstate highway system in 1991, Congress focused its funding efforts on this aging infrastructure. Capital outlays—expenditures of federal, state, and other funds—for projects on interstate highways have generally increased since 1992. State spending of federal, state, and local funds for interstates grew from $13 billion in 1992 to $16.5 billion in 2006. Besides capital outlays, states also had routine interstate maintenance outlays that averaged well over $1 billion each year. These outlays, which generally consist of nonfederal funds, are for routine tasks such as sealing cracks and patching potholes that help keep the road surfaces in good condition but do not cover capital improvements like resurfacing, rehabilitation, or reconstruction.[90] As a result, the interstate highway system's physical condition (road surface and bridges) has improved (with 8.7 percent of urban interstate road surface in poor condition in 1998 compared to 5.4 percent in 2008,[91] while the number of structurally deficient urban bridges declined by 8 percent between 1998 and 2007).[92] Meanwhile, traffic congestion on interstate highways has grown because the pressures from the many factors that contribute to congestion—population growth, the number of licensed drivers, vehicle miles traveled, and freight movement by truck—have continued to outpace the increase in capacity in terms of new lane miles (Figure 37).

Although the transit systems supported by federal policies helped to prevent highway congestion from becoming even worse—with transit ridership increasing by 9.3 percent between 2000 and 2010 (to 2.5 million passenger trips in the third quarter of 2010)[93]—they could not compete with the U.S. population's love for—and dependency on—cars. Meanwhile, many people cannot take advantage of public transportation. About 140 million people live more than a quarter of a mile from a public transit stop (planners in the United States generally assume that most transit riders will not walk more than a quarter of a mile to a bus stop). In small urban and rural communities, 41 percent do not have access to transit at all.[94]

URBAN VIEW 8
London's Traffic Congestion Charge

The number of passenger cars worldwide more than doubled between 1980 and 2010 and continues to rise.[95] Highways in many cities are heading toward gridlock. Because it took so long to clear, most people think of the 2010 60-mile long backup that stretched from Beijing to the northern province of Inner Mongolia as the world's longest traffic jam. But the record is still held by the road between Paris and Lyon when a combination of bad weather and people returning from skiing vacations caused a 109 mile traffic jam in 1980. The United Kingdom is Europe's most congested country (Table 2). Governments are struggling to address the problem.

Building more roads as part of the solution to traffic congestion is incredibly expensive, cannot keep up with increasing demand, and often encounters voter objection. In the central parts of cities like London and Paris, there is just no room to build new roads. Some cities have tried new regulations to reduce congestion. Many European cities have created extensive pedestrian shopping areas along the narrowest streets in the oldest parts of the urban core. Rationing has been tried—with varying degrees of success. Traffic congestion charges are another option. Congestion charging to deter drivers from driving

URBAN VIEW 8
London's Traffic Congestion Charge (*continued*)

TABLE 2 Traffic Congestion in Europe

Countries and most congested cities	Travel-time tax*	Hours lost annually on average
United Kingdom	*22.5*	*49.6*
London	25.0	54.9
Manchester	32.8	72.2
Belfast	28.2	62.0
Newcastle	28.2	62.1
*Benelux***	*21.1*	*41.5*
Brussels	29.8	65.5
Amsterdam	29.7	65.3
Antwerp	29.5	64.9
Luxembourg	21.6	47.4
Germany	*19.7*	*43.3*
Ruhr region***	23.3	51.4
Hamburg	20.3	44.6
Berlin	16.0	35.2
Frankfurt	21.6	47.6
France	*14.3*	*31.4*
Paris	32.0	70.4
Lyon	15.7	34.6
Lille	22.9	50.4
Limoge	15.6	34.4

*Additional percentage time taken during peak times.
**Belgium, Netherlands, Luxembourg.
***Includes Dortmund, Duisburg, Essen.
Source: Inrix, *National Traffic Scorecard 2010 Edition*, Kirkland, WA: Inrix, 2010.

Russell Boyce/Reuters Pictures

FIGURE 38 Motorists who drive into the eight square mile congestion charging zone in London—the most central and heavily congested part of the city—must pay a fee to enter.

into the city center already operates in a number of urban areas such as Singapore and Stockholm.

In London, with average traffic speed having fallen to less than 10 miles per hour during the day—the lowest average speed since cars were invented—and air pollution and traffic noise a constant problem, city officials in 2003 introduced a £5 per day congestion charge for motorists driving into an eight-square-mile zone within the Inner Ring Road—the most central and heavily congested part of the city (Figure 38). Motorists pay in advance or on the day of travel using a variety of options (online, by telephone or text message, in person at selected retail outlets, post offices, gasoline stations, at self-service machines in some car parks, or by mail). Between 7:00 a.m. and 6:30 p.m., cameras at 400 locations scan the license plates of all the cars that enter the zone during the work week. This information is matched each night using an automatic number plate recognition (ANPR) system against a database of drivers who have paid the charge, and anyone who has failed to pay by midnight is liable for a hefty fine. Persistent nonpayers can find their vehicles "clamped" or towed. Exceptions include taxis, emergency services, moped riders, vehicles powered by alternative fuels, and disabled people. Residents of the zone are given a 90 percent discount. In 2007, the congestion charge was raised to £8 and the hours reduced by 30 minutes to 6:00 p.m. to encourage more people to visit the theaters, restaurants, and bars in the city. In 2011, the charge was again raised, to £11.

While intended to reduce traffic levels and ease severely congested roads, the congestion charge, unlike regulation, also raises revenue to reinvest in the city's transportation system. The estimated £148 million annual revenues are earmarked for improving transport in London. But although there has been a 6 percent increase in bus passengers during charging hours, many motorists who drive into central London prefer not to take the bus. In fact, although congestion fell after the introduction of the congestion charge, it has risen back to pre-charging levels.

PERSISTENT FUTURE PROBLEMS

Cities of mile-high skyscrapers, spaceship-style pods, or bio-ecological harmony are for the e-books on i-Pads of dreamers. We now know that the triumphs of progress and the conquests of technology are uneven in their imprint on urbanization. The notion of a futuristic metropolis (of whatever sort: benign or malevolent, ecological or technological) is a chimera, an unrealizable dream/nightmare. Rather, only bits and pieces of the metropolis and of urban life in the future will bear the imprint of radical change. The basic fabric of the metropolis and urban life, however, will be a continuation and evolution of what is currently in place and recognizable to most people.

The future is already here, embedded in the physical, institutional, and social structures of the twenty-first century city. It is a future that will grow from the various dimensions of urbanization that have been described. Although we cannot map the future in detail, we can be fairly confident about the general shape of our urban future, what it will and will not involve.

This last point should remind us that the future will bring no escape from the problems of urbanization. As with the overall shape of the urban future, we can identify the main sources of these problems from the current trajectory of urban change. So, for example, it is almost inevitable that problems of traffic congestion will persist and probably intensify because polycentric development and central city redevelopment will make the spatial relationships between people's jobs and homes more complex, leaving little prospect of a reduction in commuting distances. It is even more certain that problems of inadequate and decayed *infrastructure* will intensify: a large proportion of roads, bridges, water lines, and sewer lines are currently at or near the end of their expected life span in some more affluent countries such as the United States; in many less developed countries there are limited municipal funds for providing adequate infrastructure to meet the demands of rapid urban growth.

Problems of *pollution and environmental degradation* are a little less certain, although these problems are escalating rapidly in those cities in the less developed countries that are unable to devote many resources to environmental issues due to pressing problems of poverty, poor housing, and inadequate service and transportation infrastructures. Political change could conceivably encompass **sustainable urban development** and a **green urbanism** movement that might significantly moderate some of the air and water quality problems through tough legislation in richer countries, for example, in Europe. The aging Baby Boomers, whose political track record already includes the radical counterculture movements of the 1960s, represent a potent source for change in this respect. Now established with homes, families, and areers, and some already in retirement, they may feel able to afford—using disposable income or free time—to rekindle their radicalism around the cause of environmental quality. Figure 39 shows some of the changes that will need to be made to ensure the future sustainability of metropolitan areas. It has to be acknowledged, though, that there is no real evidence to suggest that a very dramatic shift in attitudes among many people in countries where car ownership is entrenched is actually about to happen. A more sober prognosis, therefore, would be that problems of pollution and environmental degradation will spread and intensify along with the spread of urbanization in many parts of the world.

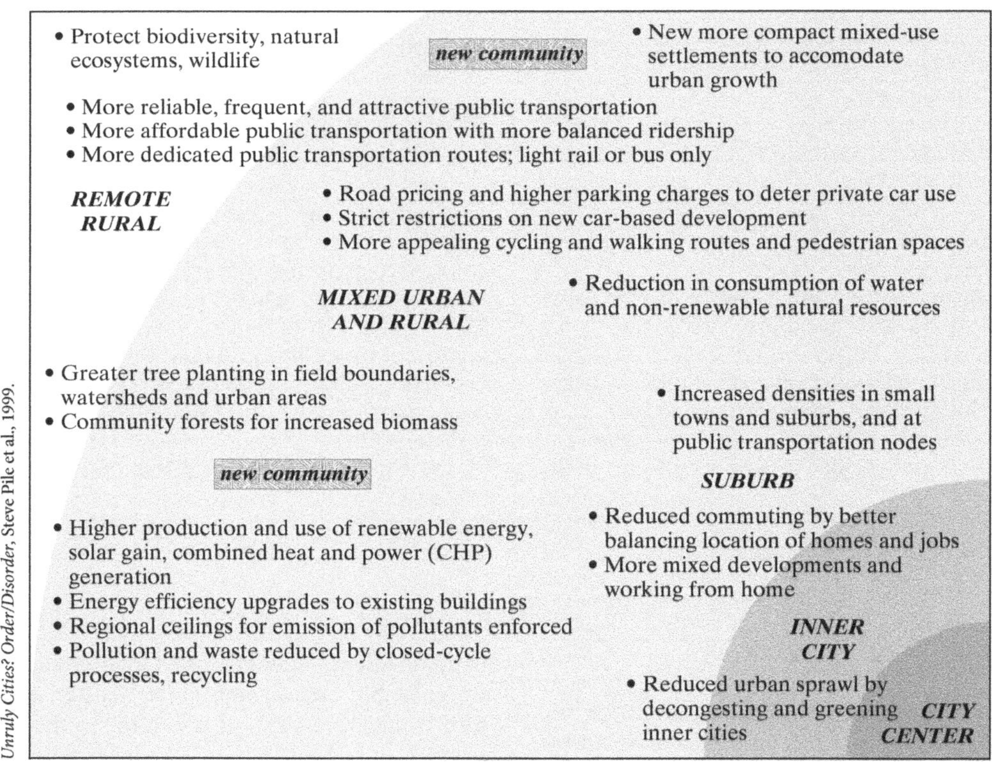

FIGURE 39 Some of the changes that will need to be made to ensure the sustainability of metropolitan areas.

Unless current patterns and structures of urban governance are radically reformed, the future will also see a continuation of problems of *fiscal stress* and, consequently, of problems associated with inadequate *municipal service provision* in cities around the world. In the United States, for example, central city governments will face increasing difficulty in raising the necessary resources to fund schools, police, and other important public services so long as their populations continue to be dominated by people of low-income and so long as they continue to lose semiskilled, mid-level jobs. Both are trends that appear set to continue through the near future, at least until central city areas begin to be redeveloped on a large scale, at which point low-income people will be pushed into older suburban settings. This trend will not put an end to problems of fiscal stress—it will merely displace the problem spatially, so that in the future it is likely that the people in some older suburban neighborhoods will experience the kind of fiscal stress that so far has been associated only with central cities.

Finally, we can be confident that problems of *poverty* will persist. Among them will be problems of **slums,** although, like fiscal stress, they may move around metropolitan areas in response to the broad changes that will occur in metropolitan form. Along with these fundamental and seemingly intractable problems of economic and social inequality, we can also expect certain other problems to persist: in more affluent countries like the United States and Germany, the particular sociocultural problems experienced by the most disadvantaged members of some ethnic minority groups, the problems of containing criminal violence, and the problems of dealing with drug abuse; in the less developed countries, the growing number of people in cities subsisting below the World Bank's international poverty line of US$1.25 a day, the large proportion of the population dependent on the informal sector of the economy, and the problems of dealing with urban health issues. Given the probable dynamics of economic change in the immediate future, it is reasonable to expect that poverty-related problems will not only persist but intensify. This prognosis holds the possibility that another crisis (or perhaps merely a continued sense of imminent crisis in wake of September 11, 2001) will trigger a new round of activity in terms of policy and planning. The critical question, of course, is: Will we be able to find more effective ways to plan and manage urban change and urban problems for the benefit of the people in cities?

FOLLOW UP

Key Terms

brownfield
contagion effect
culture of poverty
cycle of poverty
defensible space
dual city
feminization of poverty
greenfield
greyfield
labeling
social construction of problems
spatial mismatch
spiral of decay
slums
urban dust domes

Review Activities

1. If you have access to a library with DVDs or online videos, you should arrange to watch an educational video about central city problems and then consider what you think about the arguments and explanations put forward. A classic is *Solving Black Inner-City Poverty*, in which Bill Moyers interviews Dr. William Julius Wilson about central city black poverty (Princeton, N.J.: Films for the Humanities and Sciences, 2004).
2. Poverty is arguably the most compelling problem both *in* and *of* cities. Find out how many people in your town or city are below the official poverty line (you will need to look online for census materials and publications). What can you find out about the characteristics of these people—their age, gender, race, and place of residence? To what extent do they seem to be localized in particular neighborhoods?
3. What do you think are some of the main opportunities and challenges for achieving sustainable development in metropolitan areas? What are some urban policies that you would like to see implemented that would make cities more sustainable?

4. Add the results of your research to your *portfolio*. You may also find that a systematic online review of your city newspaper is revealing; look for newspaper stories that deal with urban problems and classify them according to the type of problem (crime, environment, infrastructure). Review your findings in relation to the themes discussed in this chapter, paying special attention (1) to the issues of problems of and in cities and of the social construction of problems and (2) to any consistent spatial patterns that show up in news stories about particular kinds of problems. How do these findings relate to what you were able to find out about poverty in the city?

Log in to **www.mygeoscienceplace.com** for self-study quizzes, *MapMaster* layered thematic and place name interactive maps, *Urban View* Google Earth™ tours, key resources and suggested readings, related websites, "In the News" RSS feeds, and additional references and resources to enhance your study of problems of urbanization.

NOTES

1. Urban View 15.1 is based on K. Huus, " 'Tidal Wave' of Homeless Students hits Schools," *msnbc.com*, March 2, 2009.
2. Ibid.
3. Ibid.
4. This distinction was first made by D. T. Herbert and R. J. Johnston in *Social Areas in Cities: Spatial Processes and Form* (London: Wiley, 1976).
5. *Report of the Special Committee of the Board of Supervisors of San Francisco on the Conditions of the Chinese Quarter and the Chinese in San Francisco* (San Francisco, Calif.: San Francisco Board of Supervisors, July 1885).
6. J. B. Trauner, "The Chinese as Medical Scapegoats in San Francisco, 1870–1905," *California History* 57 (1978): 70–87.
7. A. F. Weber, *The Growth of Cities in the Nineteenth Century: A Study in Statistics* (Ithaca, N.Y.: Cornell University Press, 1963), 368 (first published in 1899 by the Macmillan Company, New York).
8. Ibid., 369.
9. Ibid., 368. The quote is from M. S. Nordau, *Degeneration* (New York: D. Appleton, 1895), 35.
10. S. Smith, *The City That Was* (New York: Allaben, 1911). Quoted in D. Ward, *Poverty, Ethnicity and the American City, 1840–1920* (New York: Cambridge University Press, 1989).
11. J. Kirkland, "Among the Poor of Chicago," in *The Poor in Great Cities*, eds. R. A. Woods et al. (New York: Arno Press, 1971), 198. Originally published in New York by Scribner's in 1895.
12. D. Gaines, *Teenage Wasteland: Suburbia's Dead End Kids* (New York: Pantheon, 1991).
13. M. J. Dear, "Abandoned Housing," in *Urban Policymaking and Metropolitan Dynamics*, ed. J. S. Adams (Cambridge, Mass.: Ballinger, 1976), 59–99.
14. Sociologist Oscar Lewis first outlined the concept in relation to village life in Mexico, but it has since been deployed in relation to the slums of American cities. See O. Lewis, "The Culture of Poverty," *Scientific American* 215 (1966): 19–25.
15. For brief summaries, see K. Fox, Chapter 5 in *Metropolitan America: Urban Life and Urban Policy in the United States, 1940–1980* (New Brunswick, N.J.: Rutgers University Press, 1985); and J. C. Teaford, Chapter 6 in *The Twentieth Century American City* (Baltimore, Md.: Johns Hopkins University Press, 1986). For a more-detailed review, see J. R. Feagin and H. Hahn, *Ghetto Revolts* (New York: Macmillan, 1973).
16. Despite criticisms, the federal poverty standard continues to be used both as an administrative measure to determine government program eligibility and in academic research. See M. Orshansky, "Counting the Poor: Another Look at the Poverty Profile,"*Social Security Bulletin* 28 (1965): 3–29; G. M. Fisher, "The Development and History of the Poverty Thresholds," *Social Security Bulletin* 55 (1992): 3–14; and the U.S. Department of Health and Human Services website (*http://aspe.hhs.gov/poverty/*).
17. C. DeNavas-Walt, B. D. Proctor, and J. C. Smith, *Income, Poverty, and Health Insurance Coverage in the United States: 2009* (Washington, D.C.: U.S. Department of Commerce, 2010).
18. 1970–2009 data from U.S. Census Bureau's *Consumer Income Reports (P60)* (*http://www.census.gov/prod/www/abs/p60.html*).
19. W. Frey et al. *The State of Metropolitan America* (Washington, D.C.: Brookings Institution, 2010).
20. P. A. Jargowsky, *Stunning Progress, Hidden Problems: The Dramatic Decline of Concentrated Poverty in the 1990s* (Washington, D.C.: The Brookings Institution) (*http://www.brookings.edu/es/urban/publications/ jargowskypoverty.pdf*).
21. *www.urbanophile.com/category/cities/detroit.*
22. *http://www.portfolio.com/resources/Portfolio-2010-Metro-Area-Stress-Rank.pdf.*
23. W. J. Wilson, *The Truly Disadvantaged: The Inner City, The Underclass, and Public Policy* (Chicago, Ill.: University of Chicago Press, 1990), 8.
24. M. Van Reitsma, "A Conceptual Definition of the Underclass," *Focus* 12 (1989): 28.
25. J. Kasarda, "Jobs, Migration, and Emerging Urban Mismatches," in *Urban Change and Poverty*, eds. M. G. H. McGeary and L. E. Lynn, Jr. (Washington, D.C.: National Academy Press, 1988), 148–98; see also C. Jencks and S. E. Mayer, "Residential Segregation, Job Proximity, and Black Job Opportunities," in *Inner-City Poverty in the United States*, eds. L. E. Lynn, Jr. and M. G. H. McGeary (Washington, D.C.: National Academy Press, 1990), 187–222.
26. M. Castells, *The Informational City* (Cambridge, Mass.: Blackwell, 1989), 227.
27. See U.S. Bureau of Justice Statistics website (*http://www.ojp.usdoj.gov/bjs/*).
28. R. B. Taylor, "Urban Communities and Crime," in *Urban Life in Transition*, eds. M. Gottdiener and C. G. Pickvance (Newbury Park, Calif.: Sage, 1991), 106–34.
29. L. J. Paulozzi, *Drug-Induced Deaths: United States, 2003–2007* (Atlanta, GA: Centers for Disease Control and Prevention, 2011).
30. Based on National Institute on Drug Abuse (NIDA), *InfoFacts: High School and Youth Trends* (Washington, D.C.: National Institutes of Health, 2011); National Center for Education Evaluation and Regional Assistance, *The Effectiveness of*

Mandatory-Random Student Drug Testing (Washington, D.C.: U.S. Department of Education, 2010).

31. C. F. Schmid, "Urban Crime Areas," *American Sociological Review* 25 (1960): 678.
32. S. Smith, *Crime, Space and Society* (Cambridge, UK: Cambridge University Press, 1986).
33. D. L. Lange, R. K. Baker, and S. J. Ball, *Mass Media and Violence: A Report to the National Commission on the Causes and Prevention of Violence* (Washington, D.C.: U.S. Government Printing Office, 1969).
34. O. Newman, *Defensible Space* (New York: Macmillan, 1972).
35. D. Davis, Jr., *Violent Crimes in the Capital City.* Paper presented at the annual meeting of the American Association for the Advancement of Science (Washington, D.C., February 14–19, 1991).
36. P. L. Knox, "The Postmodern Urban Matrix," in *The Restless Urban Landscape*, ed. P. L. Knox (Englewood Cliffs, N.J.: Prentice Hall, 1993).
37. National Institute on Drug Abuse (NIDA), *Understanding Drug Abuse and Addiction* (Washington, D.C.: National Institutes of Health, 2008).
38. Knox, "The Postmodern Urban Matrix."
39. This discussion is based on S. A. Marston, P. L. Knox, and D. M. Liverman, *World Regions in Global Context: Peoples, Places, and Environments* (Upper Saddle River, N.J.: Prentice Hall, 2002), 181.
40. BBC News, "So Who Are The Russian Mafia?" April 1,1998.
41. CBS News, "Russian Mafia's Worldwide Grip" July 21, 2000.
42. Ibid.
43. See Taylor, "Urban Communities and Crime," 127–28.
44. Knox, "The Postmodern Urban Matrix."
45. L. Duke and D. M. Price, "A Microcosm of Despair in Washington, D.C.," *Washington Post* (April 2, 1989): A7.
46. This discussion is based on H. V. Savitch and G. Ardashev, "Does Terror Have an Urban Future?" *Urban Studies* 38 (2001): 2515–33; P. L. Knox and S. A. Marston, *Places and Regions in Global Context: Human Geography*, 2nd ed. (Upper Saddle River, N.J.: Pearson Education, Inc., 2004), 406.
47. J. Harrigan and P. Martin, "Terrorism and the Resilience of Cities," *Economic Policy Review* 8 (2002): 97–116.
48. Coaffee, "Rings of Steel."
49. Ibid.
50. P. Rossi, *Down and Out in America: The Origins of Homelessness* (Chicago, Ill.: University of Chicago Press, 1989).
51. The Bring LA Home, The Partnership to End Homelessness, Economic Roundtable estimated in 2004 that on a typical night in Los Angeles County, 10 percent of homeless residents were doubled up with friends and relatives; 11 percent were in rehabilitation facilities, jail, or hospital; 24 percent were in emergency shelters and transitional housing; and the sleeping arrangements of the remaining 55 percent were sidewalks, cars, public transit, empty buildings, roadways, or parks (see *www.bringlahome.org/*).
52. National Law Center on Homelessness and Poverty, 2009 Annual Report (Washington, D.C.: NLCHP, 2009).
53. Rossi, *Down and Out in America*, Chapter 2; Burt and Aron, *Homelessness.*
54. U.S. Conference of Mayors, *Hunger and Homelessness Survey: A Status Report on Hunger and Homelessness in America's Cities: A 27-City Survey* (Washington, D.C.: U.S. Conference of Mayors, 2009) (*http://www.usmayors.org/pressreleases/uploads/USCMHungercompleteWEB2009.pdf*).
55. Ibid.
56. See M. J. Dear and J. R. Wolch, *Landscapes of Despair* (Princeton, N.J.: Princeton University Press, 1987).
57. C. DeNavas-Walt, B. D. Proctor, and J. C. Smith, *Income, Poverty, and Health Insurance Coverage in the United States: 2009* (Washington, D.C.: U.S. Census Bureau U.S. Department of Commerce, 2010).
58. Calculated from Bureau of Labor Statistics unemployment data (*http://www.bls.gov/cps/cpsaat3.pdf*).
59. C. Dolbeare, "Housing Policy: A General Consideration," in *Homelessness in America*, National Coalition for the Homeless (Washington, D.C.: Oryx Press, 1996).
60. C. Lawrence, "Chicago's Homeless Lose Out While Housing Agencies Feud Over Rehab Funds," *The Chicago Reporter,* August 13, 2007.
61. U.S. Conference of Mayors, *Hunger and Homelessness Survey*; P. Koegel et al., "The Causes of Homelessness," in *Homelessness in America*, National Coalition for the Homeless (Washington, D.C.: Oryx Press, 1996); G. Laws and S. Lord, "The Politics of Homelessness," in *Geographic Dimensions of United States Social Policy*, eds. J. Kodras and J. P. Jones III (London: Edward Arnold, 1990), 59–85; Rossi, *Down and Out in America*, 182.
62. See U.S. Conference of Mayors, *Hunger and Homelessness Survey.*
63. U.S. Environmental Protection Agency, *TRI (Toxics Release Inventory) Explorer: Releases: Chemical Report* (*http://www.epa.gov/triexplorer/maps.htm*). The Toxics Release Inventory contains information on releases of nearly 650 chemicals and chemical categories from industries including manufacturing, metal and coal mining, electric utilities, and commercial hazardous waste treatment.
64. I. Douglas, "The Rain on the Roof: A Geography of the Urban Environment," in *Horizons in Human Geography*, eds. D. Gregory and R. Walford (Totowa, N.J.: Barnes and Noble, 1989), 217–38.
65. Ibid., 220.
66. U.S.Geological Survey, *Summary of Estimated Water Use in the United States in 2005* (Washington, D.C.: U.S. Department of the Interior, 2009) (*http://pubs.usgs.gov/fs/2009/3098/*).
67. Ibid; U.S. Environmental Protection Agency, *Indoor Water Use in the United States* (Washington, D.C,: EPA, 2008); P. H. Gleick, *The World's Water 2008-09*, Data Table 2, Freshwater Withdrawal by Country and Sector (Washington, D.C.: Island Press, 2009).
68. United Nations Environmental Programme, *GEO-2000: Global Environment Outlook* (*http://www.unep.org/Geo2000/ov-e/index.htm*).
69. U.S. Environmental Protection Agency, *Aging Water Infrastructure Research Project* (Washington, D.C.: National Risk Management Research Laboratory, 2007); ITT, Failing Infrastructure and Water Scarcity n.d. (*http://itt.com/valueofwater/media/1019_ITT_BACKGROUNDER2_NOMARKS.pdf*).
70. S. McKinon, "Money down the Drain," *Arizona Republic*, May 20, 2007 (*http://www.azcentral.com/specials/special26/articles/0520leaks-home0520.html?&wired*).
71. U.S. Environmental Protection Agency, EPA's 2007 Drinking Water Infrastructure Needs Survey and Assessment (Washington, D.C.: Office of Water, 2009) (*http://www.epa.gov/ogwdw/needssurvey/pdfs/2007/fs_needssurvey_2007.pdf*).
72. Quoted in B. Howard, "Scientists and Citizens Are Stymied by Water Crisis," *E/The Environmental Magazine* (January 8, 2004) (*http://www.ecoearth.info/shared/reader/welcome.aspx?linkid=28241&keybold=climate%20AND%20%20wildfire%20AND%20%20emergency*).
73. Centers for Disease Control and Prevention, *Morbidity and Mortality Weekly Report*, Vol. 57, 2008 (*http://www.cdc.gov/*

mmwr/pdf/ss/ss5709.pdf); M. Lavelle and J. Kurlantzick, "The Coming Water Crisis," *U.S. News and World Report* (August 12, 2002) (*http://www.purewatergazette.net/comingwatercrisis.htm*); P. L. Knox and S. A. Marston, *Places and Regions in Global Context: Human Geography*, 2nd ed. (Upper Saddle River, N.J.: Pearson Education, Inc., 2004), 431.

74. 111th U.S. Congress, *House Report 11-026: Water Quality Investment Act of 2009.*
75. U.S. Environmental Protection Agency, *Clean Water Needs Survey* (Washington, D.C.: Office of Water, 2008)(*http://water.epa.gov/scitech/datait/databases/cwns/2008reportdata.cfm*).
76. L. McCarthy, "Off the Mark? Economic Efficiency in Targeting the Most Marketable Sites rather than Spatial and Social Equity in Public Support for Brownfield Redevelopment," *Economic Development Quarterly*, 23 (2009): 211–28; L. McCarthy, "The Brownfield Dual Land-Use Policy Challenge: Reducing Barriers to Private Development while Connecting Reuse to Broader Community Goals," *Land Use Policy* 19 (2002): 287–96.
77. U.S. Environmental Protection Agency website (*http://www.epa.gov/region6/6sf/pdffiles/heifersuccess2007.pdf*); Heifer International website (*www.heifer.org/*).
78. Data for 2009 from U.S. Environmental Protection Agency, *Air Quality Trends* (*http://www.epa.gov/airtrends/aqtrends.html*).
79. 2005 data from U.S. Environmental Protection Agency, *Air Emissions Sources: Carbon Monoxide* (*http://www.epa.gov/air/emissions/co.htm#conat*).
80. R. J. Earickson and I. H. Billick, "The Areal Association of Urban Air Pollutants and Residential Characteristics: Louisville and Detroit," *Applied Geography* 8 (1988): 5–23; U.S. Environmental Protection Agency, *Latest Findings on National Air Quality.*
81. European Commission, *High Speed Europe* (Brussels: Directorate-General for Mobility and Transport, 2010) (*http://ec.europa.eu/transport/infrastructure/studies/doc/2010_high_speed_rail_en.pdf*).
82. P. Choate and S. Walter, *American in Ruins: The Decaying Infrastructure* (Durham, N.C.: Duke University Press, 1983).
83. M. Kaplan et al., *Hard Choices: A Report on the Increasing Gap Between America's Infrastructure Needs and Our Ability to Pay for Them* (Washington, D.C.: US Government Printing Office, 1984).
84. American Society of Civil Engineers, *Renewing America's Infrastructure, 2001*; *2009 Report Card for America's Infrastructure* (*http://www.asce.org/reportcard/*).
85. Citizens Budget Commission, *Mind the Gap: Funding Repair and Maintenance of New York City's Infrastructure*, July 27, 2010 [based on the OMB's *Asset Information Management System (AIMS) Reports*] (*http://www.cbcny.org/cbc-blogs/blogs/mind-gap-funding-repair-and-maintenance-new-york-city-infrastructure*); A. G. Hevesi, *Dilemma in the Millennium: Capital Needs of the World's Capital City* (New York: Office of the Comptroller, City of New York, August 1998) (*http://www.comptroller.nyc.gov/bureaus/eng/complete.pdf*); J. Obser, "The Road Ahead for New York City," *Future.Newsday.Com* (May 16, 1999).
86. U.S. Environmental Protection Agency, *Aging Water Infrastructure Research Project* (Washington, D.C.: National Risk Management Research Laboratory, 2007); ITT, Failing Infrastructure and Water Scarcity n.d. (*http://itt.com/valueofwater/media/1019_ITT_BACKGROUNDER2_NOMARKS.pdf*).
87. A. C. Revkin, "Federal Study Calls Spending on Water Systems Perilously Inadequate," *New York Times* (April 10, 2002).
88. With an average speed of less than 35 mph.
89. Texas Transportation Institute, *Urban Mobility Report* (College Station, Tex.: Texas A&M University System) (*http://mobility.tamu.edu/ums/*).
90. United States Department of Transportation, *2008 Status of the Nation's Highways, Bridges, and Transit: Condition and Performance*, Report to Congress (Washington, D.C.: Federal Highway Administration, 2008) (*http://www.fhwa.dot.gov/policy/2008cpr/pdfs/cp2008.pdf*).
91. D. T. Hartgen, R. K. Karanam, M. G. Fields, and T. A. Kerscher, *19th Annual Report on the Performance of State Highway Systems* (1984–2008) (Los Angeles, CA: Reason Foundation, 2010)(*http://reason.org/news/show/19th-annual-highway-report*).
92. American Society of Civil Engineers, *Renewing America's Infrastructure, 2001; 2009 Report Card for America's Infrastructure* (*http://www.asce.org/reportcard/*)(*http://www.infrastructurereportcard.org/fact-sheet/bridges*).
93. American Public Transportation Association, Transit Ridership Reports (*http://www.apta.com/resources/statistics/Documents/Forms/AllItems.aspx?RootFolder=http%3a%2f%2fwww%2eapta%2ecom%2fresources%2fstatistics%2fdocuments%2fridership&FolderCTID=0x0120006BE9AB89F00BAE4583BADAD635F81BA6*).
94. Sierra Club, Transit Fact Sheet: America Needs More Transit (San Francisco, CA: Sierra Club, 2003)(*http://www.sierraclub.org/sprawl/reports/transit_factsheet.pdf*).
95. International Road Federation (Geneva, Switzerland) data.

Map: World Urbanization

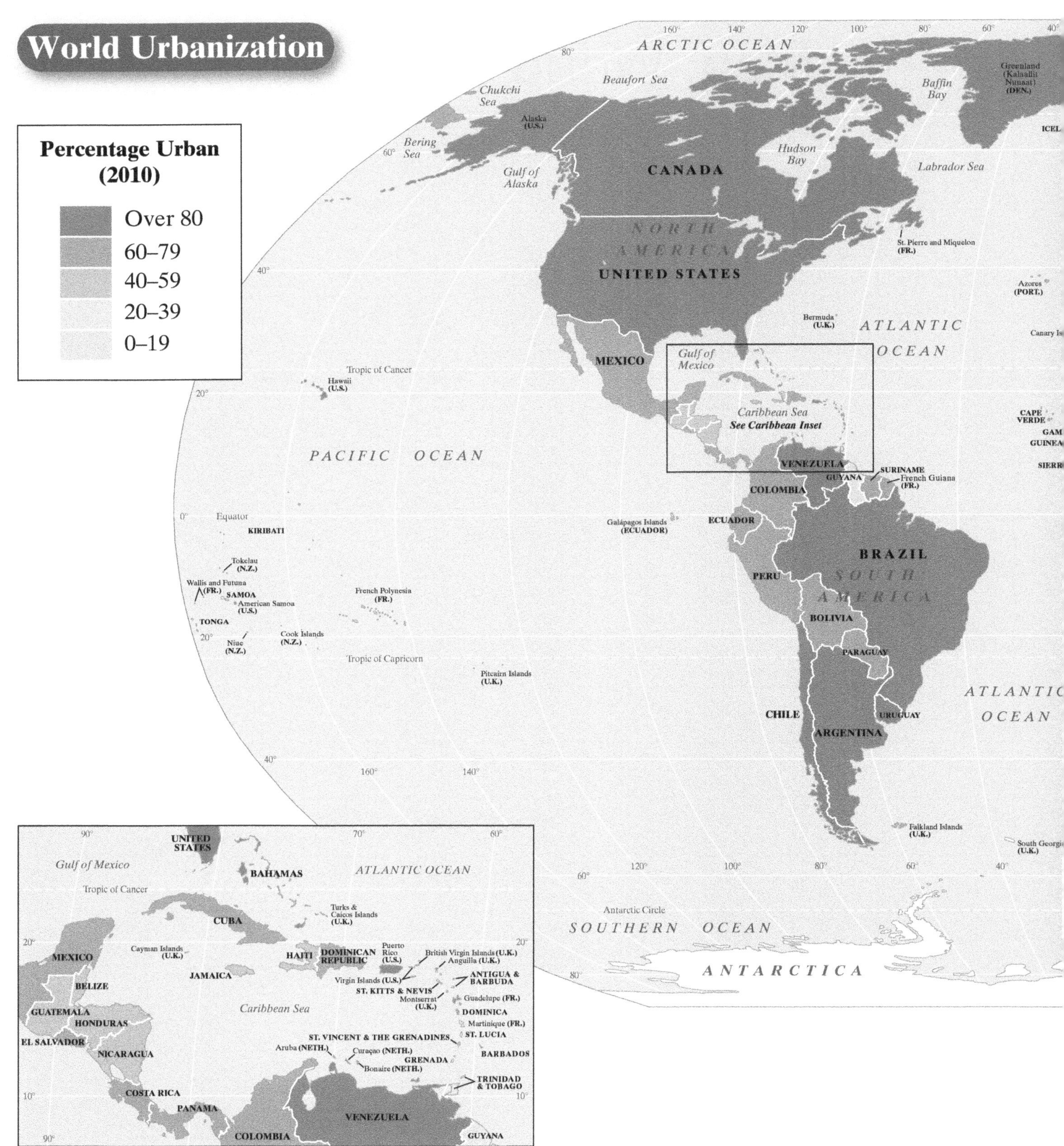
World Urbanization
Percentage Urban (2010)
Over 80
60–79
40–59
20–39
0–19
ARCTIC OCEAN
Beaufort Sea
Chukchi Sea
Bering Sea
Gulf of Alaska
Alaska (U.S.)
CANADA
Hudson Bay
Baffin Bay
Labrador Sea
Greenland (Kalaallit Nunaat) (DEN.)
ICEL
NORTH AMERICA
UNITED STATES
St. Pierre and Miquelon (FR.)
Azores (PORT.)
Bermuda (U.K.)
ATLANTIC OCEAN
Canary Is
MEXICO
Gulf of Mexico
Caribbean Sea
See Caribbean Inset
Tropic of Cancer
Hawaii (U.S.)
CAPE VERDE
GAM
GUINEA
SIERR
PACIFIC OCEAN
VENEZUELA
GUYANA
SURINAME
French Guiana (FR.)
COLOMBIA
Equator
KIRIBATI
Galápagos Islands (ECUADOR)
ECUADOR
BRAZIL
SOUTH AMERICA
Tokelau (N.Z.)
Wallis and Futuna (FR.)
SAMOA
American Samoa (U.S.)
TONGA
Niue (N.Z.)
Cook Islands (N.Z.)
French Polynesia (FR.)
PERU
BOLIVIA
PARAGUAY
Tropic of Capricorn
Pitcairn Islands (U.K.)
ATLANTIC OCEAN
CHILE
URUGUAY
ARGENTINA
Falkland Islands (U.K.)
South Georgi (U.K.)
Antarctic Circle
SOUTHERN OCEAN
ANTARCTICA
UNITED STATES
Gulf of Mexico
Tropic of Cancer
BAHAMAS
ATLANTIC OCEAN
Turks & Caicos Islands (U.K.)
CUBA
MEXICO
Cayman Islands (U.K.)
HAITI
DOMINICAN REPUBLIC
Puerto Rico (U.S.)
British Virgin Islands (U.K.)
Anguilla (U.K.)
BELIZE
JAMAICA
Virgin Islands (U.S.)
ANTIGUA & BARBUDA
ST. KITTS & NEVIS
Montserrat (U.K.)
Guadelupe (FR.)
GUATEMALA
Caribbean Sea
DOMINICA
HONDURAS
Martinique (FR.)
EL SALVADOR
ST. VINCENT & THE GRENADINES
ST. LUCIA
NICARAGUA
Aruba (NETH.)
Curaçao (NETH.)
BARBADOS
GRENADA
Bonaire (NETH.)
TRINIDAD & TOBAGO
COSTA RICA
PANAMA
VENEZUELA
COLOMBIA
GUYANA

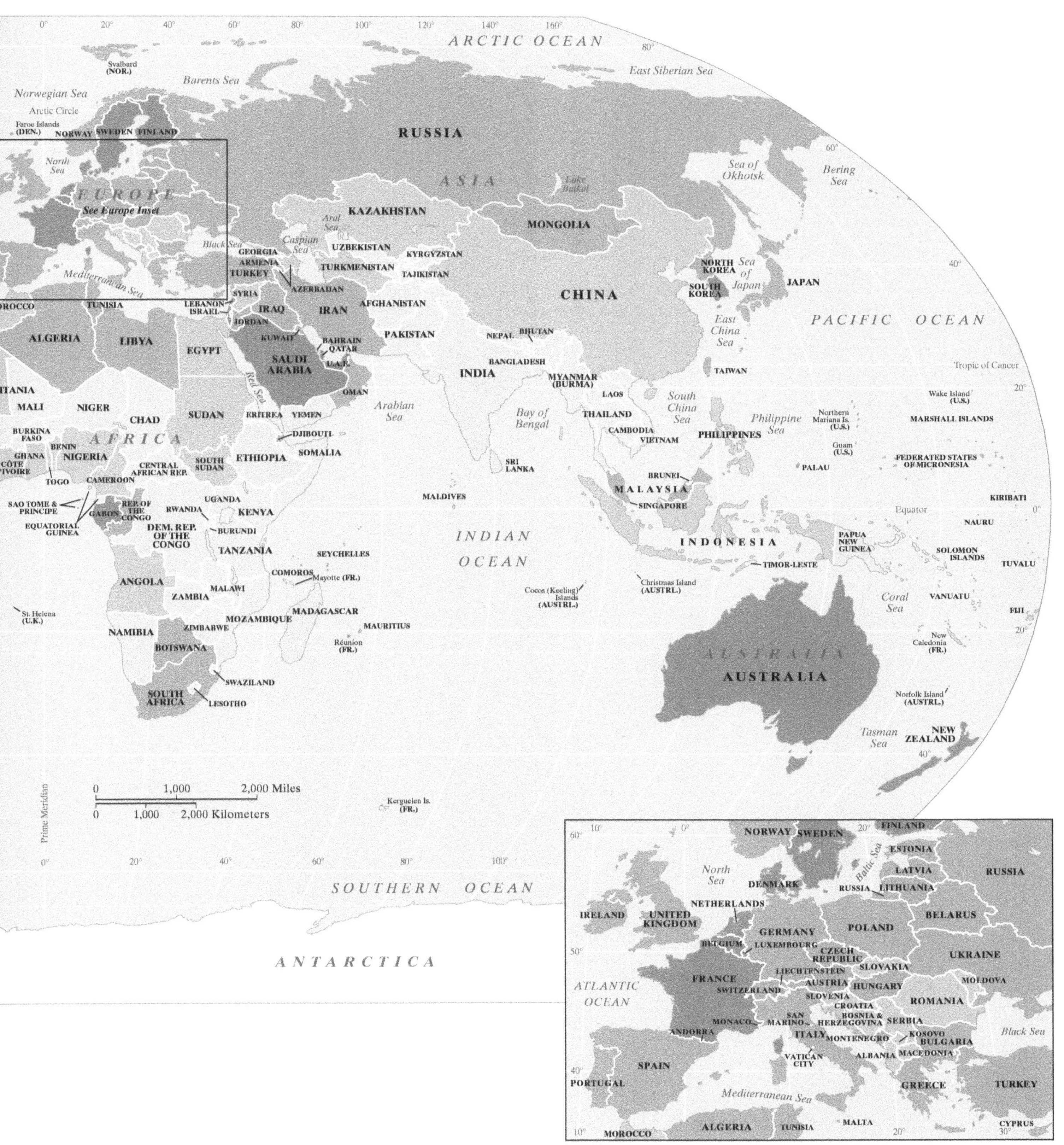

ARCTIC OCEAN
East Siberian Sea
Svalbard (NOR.)
Barents Sea
Norwegian Sea
Arctic Circle
Faroe Islands (DEN.)
NORWAY
SWEDEN
FINLAND
RUSSIA
North Sea
EUROPE
See Europe Inset
ASIA
Lake Baikal
Sea of Okhotsk
Bering Sea
KAZAKHSTAN
Aral Sea
MONGOLIA
Caspian Sea
Black Sea
GEORGIA
ARMENIA
TURKEY
AZERBAIJAN
UZBEKISTAN
KYRGYZSTAN
TURKMENISTAN
TAJIKISTAN
NORTH KOREA
Sea of Japan
JAPAN
SOUTH KOREA
Mediterranean Sea
SYRIA
CHINA
TUNISIA
LEBANON
ISRAEL
IRAQ
IRAN
AFGHANISTAN
JORDAN
East China Sea
PACIFIC OCEAN
ALGERIA
LIBYA
EGYPT
KUWAIT
BAHRAIN
QATAR
PAKISTAN
NEPAL
BHUTAN
SAUDI ARABIA
U.A.E.
BANGLADESH
INDIA
TAIWAN
Tropic of Cancer
MYANMAR (BURMA)
Red Sea
OMAN
MALI
NIGER
LAOS
South China Sea
Wake Island (U.S.)
CHAD
SUDAN
ERITREA
YEMEN
Arabian Sea
Bay of Bengal
THAILAND
Philippine Sea
Northern Mariana Is. (U.S.)
MARSHALL ISLANDS
BURKINA FASO
AFRICA
DJIBOUTI
CAMBODIA
VIETNAM
PHILIPPINES
BENIN
GHANA
NIGERIA
SOMALIA
Guam (U.S.)
CÔTE D'IVOIRE
CENTRAL AFRICAN REP.
SOUTH SUDAN
ETHIOPIA
SRI LANKA
PALAU
FEDERATED STATES OF MICRONESIA
TOGO
CAMEROON
BRUNEI
MALAYSIA
MALDIVES
UGANDA
KIRIBATI
SAO TOME & PRINCIPE
GABON
REP. OF THE CONGO
RWANDA
KENYA
SINGAPORE
Equator
EQUATORIAL GUINEA
DEM. REP. OF THE CONGO
BURUNDI
NAURU
INDIAN OCEAN
PAPUA NEW GUINEA
TANZANIA
SEYCHELLES
INDONESIA
SOLOMON ISLANDS
TUVALU
COMOROS
Mayotte (FR.)
TIMOR-LESTE
ANGOLA
MALAWI
Christmas Island (AUSTRL.)
ZAMBIA
Cocos (Keeling) Islands (AUSTRL.)
Coral Sea
VANUATU
St. Helena (U.K.)
MADAGASCAR
FIJI
MOZAMBIQUE
MAURITIUS
NAMIBIA
ZIMBABWE
Réunion (FR.)
New Caledonia (FR.)
BOTSWANA
AUSTRALIA
SWAZILAND
SOUTH AFRICA
LESOTHO
Norfolk Island (AUSTRL.)
Tasman Sea
NEW ZEALAND
Prime Meridian
0 1,000 2,000 Miles
0 1,000 2,000 Kilometers
Kerguelen Is. (FR.)
SOUTHERN OCEAN
ANTARCTICA
NORWAY
SWEDEN
FINLAND
ESTONIA
North Sea
Baltic Sea
LATVIA
RUSSIA
DENMARK
RUSSIA
LITHUANIA
NETHERLANDS
IRELAND
UNITED KINGDOM
BELARUS
POLAND
GERMANY
BELGIUM
LUXEMBOURG
CZECH REPUBLIC
UKRAINE
SLOVAKIA
LIECHTENSTEIN
FRANCE
AUSTRIA
MOLDOVA
ATLANTIC OCEAN
SWITZERLAND
HUNGARY
SLOVENIA
ROMANIA
CROATIA
BOSNIA & HERZEGOVINA
MONACO
SAN MARINO
SERBIA
ANDORRA
ITALY
MONTENEGRO
KOSOVO
BULGARIA
Black Sea
VATICAN CITY
ALBANIA
MACEDONIA
SPAIN
PORTUGAL
GREECE
TURKEY
Mediterranean Sea
ALGERIA
TUNISIA
MALTA
CYPRUS
MOROCCO

Map: Major Urban Agglomerations

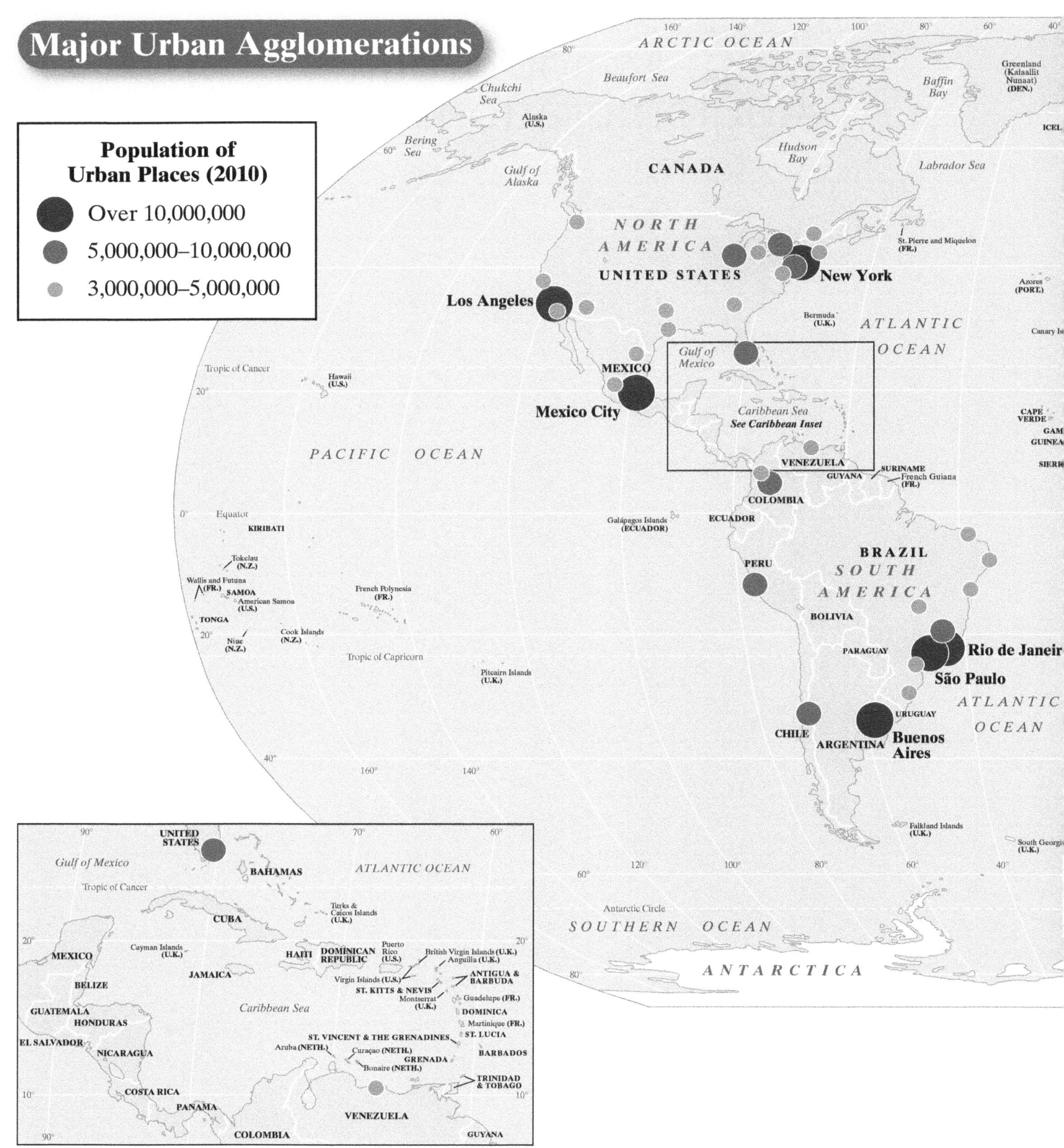

Major Urban Agglomerations
Population of Urban Places (2010)
Over 10,000,000
5,000,000–10,000,000
3,000,000–5,000,000
ARCTIC OCEAN
Beaufort Sea
Chukchi Sea
Alaska (U.S.)
Bering Sea
Gulf of Alaska
CANADA
Hudson Bay
Baffin Bay
Labrador Sea
Greenland (Kalaallit Nunaat) (DEN.)
ICEL
NORTH AMERICA
UNITED STATES
St. Pierre and Miquelon (FR.)
New York
Los Angeles
Bermuda (U.K.)
ATLANTIC OCEAN
Azores (PORT.)
Canary Is
MEXICO
Mexico City
Gulf of Mexico
Caribbean Sea
See Caribbean Inset
Tropic of Cancer
Hawaii (U.S.)
PACIFIC OCEAN
CAPE VERDE
GAM
GUINEA
SIERR
VENEZUELA
GUYANA
SURINAME
French Guiana (FR.)
COLOMBIA
ECUADOR
Galápagos Islands (ECUADOR)
Equator
KIRIBATI
Tokelau (N.Z.)
Wallis and Futuna (FR.)
SAMOA
American Samoa (U.S.)
TONGA
Niue (N.Z.)
Cook Islands (N.Z.)
French Polynesia (FR.)
Tropic of Capricorn
Pitcairn Islands (U.K.)
BRAZIL
SOUTH AMERICA
PERU
BOLIVIA
PARAGUAY
Rio de Janeir
São Paulo
URUGUAY
ATLANTIC OCEAN
CHILE
ARGENTINA
Buenos Aires
Falkland Islands (U.K.)
South Georgi (U.K.)
Antarctic Circle
SOUTHERN OCEAN
ANTARCTICA
UNITED STATES
Gulf of Mexico
BAHAMAS
ATLANTIC OCEAN
Tropic of Cancer
Turks & Caicos Islands (U.K.)
CUBA
MEXICO
Cayman Islands (U.K.)
HAITI
DOMINICAN REPUBLIC
Puerto Rico (U.S.)
British Virgin Islands (U.K.)
Anguilla (U.K.)
Virgin Islands (U.S.)
ANTIGUA & BARBUDA
ST. KITTS & NEVIS
Montserrat (U.K.)
Guadeloupe (FR.)
JAMAICA
BELIZE
GUATEMALA
HONDURAS
EL SALVADOR
NICARAGUA
Caribbean Sea
DOMINICA
Martinique (FR.)
ST. LUCIA
ST. VINCENT & THE GRENADINES
BARBADOS
Aruba (NETH.)
Curaçao (NETH.)
GRENADA
Bonaire (NETH.)
TRINIDAD & TOBAGO
COSTA RICA
PANAMA
VENEZUELA
COLOMBIA
GUYANA

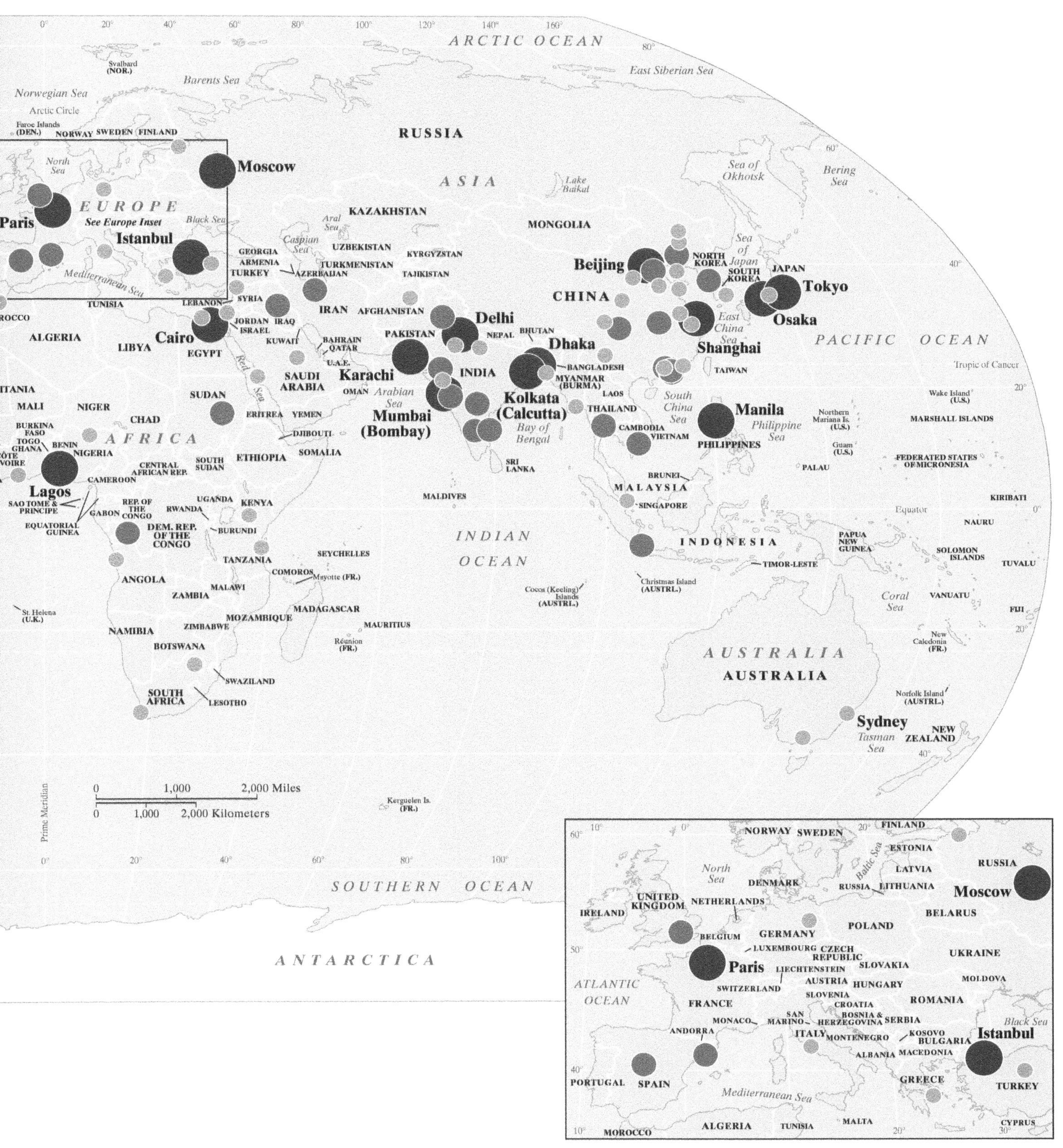
ARCTIC OCEAN
East Siberian Sea
Barents Sea
Norwegian Sea
Arctic Circle
Svalbard (NOR.)
Faroe Islands (DEN.)
NORWAY
SWEDEN
FINLAND
RUSSIA
ASIA
Moscow
North Sea
EUROPE
See Europe Inset
Paris
Istanbul
Black Sea
Mediterranean Sea
Lake Baikal
Sea of Okhotsk
Bering Sea
KAZAKHSTAN
Aral Sea
Caspian Sea
UZBEKISTAN
KYRGYZSTAN
TURKMENISTAN
TAJIKISTAN
MONGOLIA
GEORGIA
ARMENIA
AZERBAIJAN
TURKEY
Beijing
NORTH KOREA
SOUTH KOREA
Sea of Japan
JAPAN
Tokyo
Osaka
CHINA
TUNISIA
LEBANON
SYRIA
IRAN
AFGHANISTAN
ROCCO
JORDAN
IRAQ
ISRAEL
KUWAIT
BAHRAIN
QATAR
U.A.E.
PAKISTAN
Delhi
NEPAL
BHUTAN
East China Sea
Shanghai
PACIFIC OCEAN
ALGERIA
LIBYA
Cairo
EGYPT
Red Sea
SAUDI ARABIA
Karachi
OMAN
Arabian Sea
INDIA
Dhaka
BANGLADESH
MYANMAR (BURMA)
TAIWAN
Tropic of Cancer
LAOS
Kolkata (Calcutta)
Mumbai (Bombay)
THAILAND
South China Sea
Manila
Philippine Sea
Northern Mariana Is. (U.S.)
Wake Island (U.S.)
MARSHALL ISLANDS
MALI
NIGER
SUDAN
ERITREA
YEMEN
CHAD
BURKINA FASO
TOGO
GHANA
BENIN
CÔTE D'IVOIRE
NIGERIA
AFRICA
DJIBOUTI
SOMALIA
ETHIOPIA
Bay of Bengal
CAMBODIA
VIETNAM
PHILIPPINES
Guam (U.S.)
PALAU
FEDERATED STATES OF MICRONESIA
CENTRAL AFRICAN REP.
SOUTH SUDAN
CAMEROON
SRI LANKA
BRUNEI
MALAYSIA
SINGAPORE
Lagos
SAO TOME & PRINCIPE
EQUATORIAL GUINEA
GABON
REP. OF THE CONGO
UGANDA
KENYA
RWANDA
BURUNDI
DEM. REP. OF THE CONGO
MALDIVES
KIRIBATI
Equator
NAURU
INDIAN OCEAN
INDONESIA
PAPUA NEW GUINEA
SOLOMON ISLANDS
TUVALU
SEYCHELLES
TANZANIA
ANGOLA
COMOROS
Mayotte (FR.)
TIMOR-LESTE
ZAMBIA
MALAWI
Christmas Island (AUSTRL.)
Cocos (Keeling) Islands (AUSTRL.)
Coral Sea
VANUATU
FIJI
St. Helena (U.K.)
MADAGASCAR
MOZAMBIQUE
MAURITIUS
NAMIBIA
ZIMBABWE
Réunion (FR.)
New Caledonia (FR.)
BOTSWANA
AUSTRALIA
SWAZILAND
SOUTH AFRICA
LESOTHO
Norfolk Island (AUSTRL.)
Sydney
Tasman Sea
NEW ZEALAND
0 1,000 2,000 Miles
0 1,000 2,000 Kilometers
Prime Meridian
Kerguelen Is. (FR.)
SOUTHERN OCEAN
ANTARCTICA
NORWAY
SWEDEN
FINLAND
ESTONIA
RUSSIA
North Sea
Baltic Sea
LATVIA
DENMARK
RUSSIA
LITHUANIA
Moscow
UNITED KINGDOM
NETHERLANDS
IRELAND
BELARUS
POLAND
BELGIUM
GERMANY
LUXEMBOURG
CZECH REPUBLIC
UKRAINE
Paris
LIECHTENSTEIN
SLOVAKIA
AUSTRIA
MOLDOVA
ATLANTIC OCEAN
SWITZERLAND
HUNGARY
SLOVENIA
FRANCE
CROATIA
ROMANIA
SAN MARINO
BOSNIA & HERZEGOVINA
SERBIA
MONACO
ANDORRA
ITALY
MONTENEGRO
KOSOVO
BULGARIA
Black Sea
Istanbul
ALBANIA
MACEDONIA
PORTUGAL
SPAIN
GREECE
TURKEY
Mediterranean Sea
MOROCCO
ALGERIA
TUNISIA
MALTA
CYPRUS

Index

Page references followed by "f" indicate illustrated figures or photographs; followed by "t" indicates a table.

E

F

G

Q

R

S